GW00600630

·THE·
HUMAN BODY

A comprehensive guide to the structure and functions of the human body

·THE· HUMAN BODY

A comprehensive guide to the structure and functions of the human body

GALLEY PRESS

Published in this edition 1989 by Galley Press
an imprint of W H Smith Limited

Registered No. 237811 England

Trading as WHS Distributors, St John's House,
East Street, Leicester LE1 6NE

ISBN 0 86136 191 1

2 3 4 5 93 92 91 90

Printed in Spain

Consultant Editor	John O. E. Clark
Managing Editor	Ruth Binney
Art Editor	Eddie Poulton
Picture Editor	Zilda Tandy
Production	Barry Baker Janice Storr
Contributors	Sally Burningham Hugh Johnstone Ann Kramer Pip Morgan Stephen Parker Clint Twist
Editors	Louise Bostock Stephen Luck Sam Merrell
Proof reader	Fred Gill
Indexer	Kathie Gill

Typeset by MS Filmsetting Limited, Frome, Somerset

Originated by Gilchrist Brothers Limited, Leeds and Regent
Publishing Services Limited, Hong Kong

Printed and bound in Spain by Printer Industria Gráfica S.A.,
Barcelona

Contents

Introduction

The human body has been described as a wonder, a marvel, a miracle – yet no one word seems adequate to describe it fully. Perhaps that is because of its amazing versatility. As human beings we live, breathe, move and think. And only through a better understanding of these and other life processes can we truly appreciate the wonder.

The purpose of *The Human Body* is to explain these principles of structure and function, illustrating them with some of the finest medical artwork, which captures the workings of the body in all their glory. At the same time the text, skillfully written by experts who know how to explain complex phenomena to nonexperts, identifies the major organs and tissues, and reveals how they work alone and in combination.

The Human Body begins by describing the growth and development of a newborn baby through infancy, puberty and adulthood. This is followed by an account of the basic structure of the body: the bones of the skeleton, the sturdy and reliable framework for the body's flesh, and the muscles that work together with the joints to provide a whole range of brilliantly synchronized movements. To protect ourselves from, and harmonize with, the external environment we are covered in the thin, elastic skin.

Like all body tissues, skin is made up of cells, the smallest units of living matter in which basic life processes take place. These processes include energy-generating reactions which require oxygen, taken to the cells by blood oxygenated in the lungs during respiration. The blood and blood vessels are the transportation system, which is powered by the tireless pumping action of the heart.

The other ingredients for generating energy and rebuilding worn-out body tissues come from the food we eat, through the breakdown processes of digestion. Undigested food is excreted, and potentially poisonous waste products from the body's biochemical activity are removed by the kidneys, the non-stop filtering mechanism that also regulates the pressure, volume and acidity of the blood.

Two body systems have the job of ensuring that all the organs and tissues carry out the tasks for which they are designed. The endocrine system, through the action of its chemical messengers the hormones, regulates functions as diverse as digestion and reproduction. The nervous system, reporting to and under the coordinating command of the brain, carries nerve impulses throughout the body to control muscles and glands. Knowledge about our surroundings is provided by the sensory organs, most of which pass their information directly to the brain.

The story of the human body is brought full circle by an explanation of yet another miracle – the fascinating process of reproduction, the means by which men and women continue the human species. For human beings are what we are; *who* we are – how we look and behave – is determined by genetics and heredity, and by the environment in which we are raised.

CHAPTER 1

Growth and Development

Human beings are arguably the most complex organisms on this planet. They are all built to the same basic plan, yet all are different – everybody is an individual. In many countries, people have an average lifespan of 70 plus. During the years between birth and death they undergo a remarkable and continual process of physical, emotional and intellectual development that begins in the womb and continues until life's end. The way in which human beings grow and develop has been seen as a miracle which still fascinates philosophers, psychologists and educationalists alike. Its study is also a major branch of medical science.

Even today, despite the theories of Sigmund Freud, the Gessels, Jean Piaget and others, there are still many unanswered questions. Scientists can chart human growth and development quite accurately, but they still do not know exactly why certain changes take place, or what are the major influences that lead to the creation of an individual person with unique characteristics.

In particular some controversy still surrounds the fundamental question of how far any person is the product of the environment and how much simply the result of the mingling of two sets of genes. What is certain, however, is that with the act of conception a train of developmental events is set in motion which continues uninterrupted until death.

Growth before birth

Human development in the nine months before birth is faster than at any other time after. Human life begins with conception when a male sperm fuses with a female ovum to produce a single cell. Over the next 40 weeks this single fertilized cell divides again and again, millions of times, in a process that transforms it from a minuscule speck no larger than a pinhead into a fully-formed miniature human individual completely equipped for independent life outside the womb.

The mechanism by which this occurs is called mitotic cell division, a copying and splitting by which cells constantly divide and reproduce themselves. This process, which begins in the womb, continues throughout life. It is controlled by genes, minute particles within the cells. Each gene carries within it a "blueprint" of information. Together, the millions of genes determine the unique physical and mental characteristics inherited by the new human being from its parents.

At conception, which usually takes place in one of the woman's Fallopian tubes, the newly fertilized cell contains a full complement of gene-carrying chromosomes. Half of these come from the male sperm and half from the female ovum. Cell division begins almost immediately. The fertilized ovum divides, first into two, then four, then eight, and so on. The first division takes about 24 hours; subsequent divisions occur more rapidly with individual cells, called blastomeres, becoming increasingly smaller. By the fourth day, the original egg, now consisting of 16 cells and resembling a mulberry, leaves the Fallopian tube and enters the uterus, or womb. There it implants in the womb lining,

Old meets young as an adult gently reaches down to take hold of a baby's hand. These tiny fingers will gradually grow to adult size, while the child's maturing nervous system and developing muscles will provide the dexterity and coordination unique to the human hand.

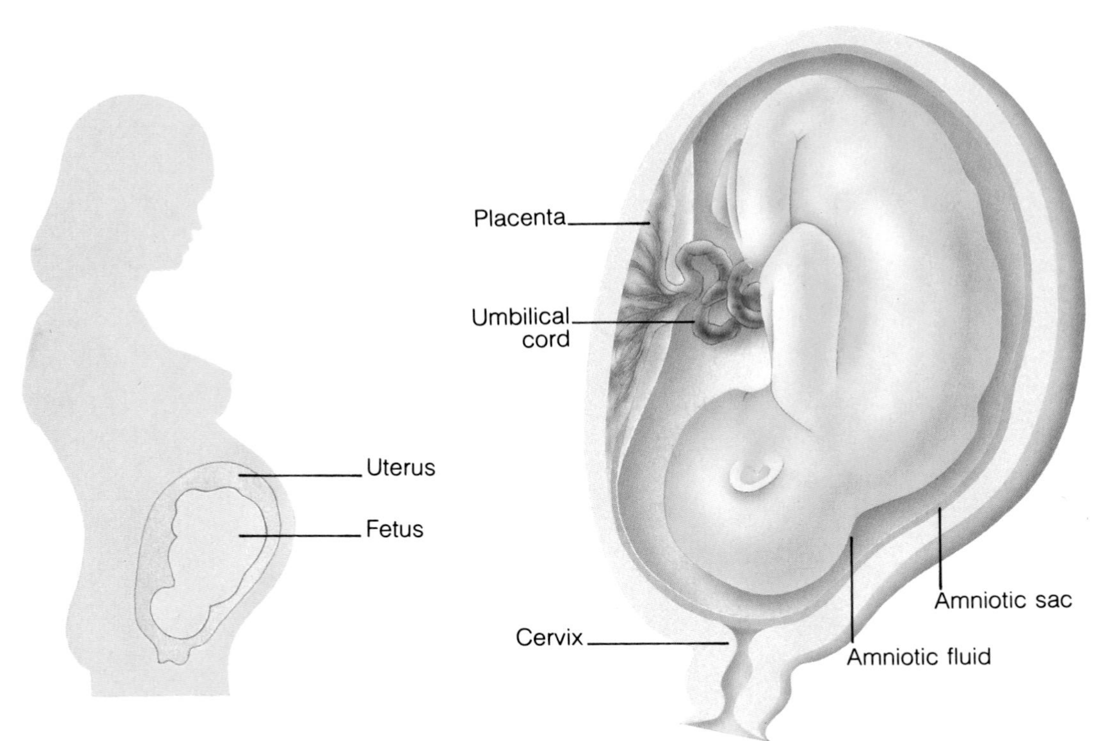

Within the sheltered domain of its mother's womb, a fetus undergoes nine months of miraculous development. Safe inside the amniotic sac and bathed in amniotic fluid, the fetus receives oxygen and nourishment extracted from the mother's blood by the placenta. They travel along blood vessels in the umbilical cord, which also contains vessels that carry waste products back to the placenta.

or endometrium. As the cells divide they also change from being all the same and differentiate into a range of specialized tissues – from muscles to brain cells – each designed to do a specific job.

The remarkable gene

Cell division and differentiation is a highly complex process, much of which is still not completely understood. Among the most puzzling questions is how genetic instructions within the fertilized egg are translated so that the new human being contains all the different tissues required to sustain life, and how the different pieces of the body are put together correctly. The answers to these enigmas seem to lie in the remarkable capability and versatility of the genes themselves.

Essentially the genes function by passing out instructions to the cells in which they are contained. The cell follows these instructions and manufactures proteins, which are used both to build cell structures and to determine and control chemical reactions that take place within the cell. Different genes seem to have different functions.

As the fertilized cell divides and grows, different genes are activated at different times. If, for instance, muscle cells are required, only those genes responsible for manufacturing proteins specific to muscle cells are activated. In the same way, if the genes for making blood proteins are "turned on," the cells become blood cells. For this selection process to take place, it seems as though some genes act as controllers, determining which genes should be activated and which suppressed. Quite different genes also seem to be programmed to give instructions about cell building and division. At the same time, chemical messengers within each cell – also controlled by the genes – influence how each one relates to its neighbors.

It is now believed that the way in which body tissues in the developing embryo make contact with each other is crucial to cell behavior and relationships. If the cellular pattern is wrong, the resulting body may end up incorrectly pieced together, endowed with developmental disorders such as spina bifida, in which the membranes surrounding the spinal cord are imperfectly formed.

Cell differentiation is a vital stage in human growth and development. Any mistake can result in subsequent disability. As in the case of spina bifida, mistakes in genetic instructions can manifest themselves as congenital defects. Alternatively, external influences such as drugs or alcohol may adversely affect the way structures are put together during this early stage in the womb.

Development in the womb

During the third week after conception, the bundle of cells, now firmly planted in the uterus, develops three cellular layers – the ectoderm, mesoderm and endoderm. From these develop all the various tissues and organs of the body. The notochord – the structure around which the vertebral column forms – also emerges at this time. From now until the eighth week after conception this developing being is known as an embryo. By the end of the fourth week the minuscule hunched creature with a tiny, curved tail measures one-sixth of an inch in length.

The embryonic period is a critical stage in human development because it is during these few weeks that all the organ systems of the body are laid down. Any external influences now, such as drugs or viral infections, can cause great damage.

The embryo grows rapidly. From the fourth to the eighth week it more than quadruples in length, and by the eighth week is recognizably human. Although no more than one inch in length the initial formation of the organs is complete, and the embryo has arms, legs, nose, eyes, mouth and even eyelids.

From the eighth week until birth, the developing child is called a fetus. During these seven months all the tissues and organs of the body – which are already formed but are still simple in structure – grow to maturity. The fetus itself also increases in size and weight at an astonishing rate. By the time the baby is born, weight will have increased from about one ounce at eight weeks to between six and nine pounds. Length too will have increased from about one inch at eight weeks to an average of 20 inches.

Besides general fetal growth, head and body hair and nails are growing by the twentieth week. Bone formation takes place between weeks 13 and 16, and by week 30 fat is deposited under the skin, making the fetus smoother and more rounded. By week 12 the fetus begins to produce urine, the external genitals have developed, and the sex of the fetus can be distinguished by ultrasound. Eyelids, which have been growing, close over the eyes in the ninth week to open again in the 22nd week.

By week 14 the fetal heart begins to pump blood, and by the eighteenth week the heart can be heard by placing an ear on the mother's abdomen. By this time, too, the mother can feel the fetus moving within her womb.

Muscular reflexes also develop on the eyelids, palms and feet and the swallowing reflex starts at the fourteenth week after conception. Thumbsucking can also begin and the fetus develops a whole range of abilities – hiccuping, sucking, swallowing, somersaulting, hand clasping, sleeping and even responding to external stimuli such as loud noises.

Substantial weight gain in the fetus is made from the twentieth week. Between weeks 26 and 29 the lungs mature ready for breathing air, the eyes which have closed in week nine now open again, and the fetal body begins to grow plumper. The fetus also changes position, most infants turning head down.

Out of the womb

A full-term pregnancy lasts 40 weeks, or 288 days, although birth can be expected anytime from the end of the 38th week or day 266 – the birth date is calculated by the obstetrician from the onset of the last menstrual period. Labor is thought to be triggered off by various hormones, including oxytocin, which is produced by the pituitary gland, prostaglandins and catecholamines.

The effect of birth on a baby's growth and development is still a matter of medical speculation. Some experts, such as the French obstetrician Frederick Leboyer, believe that birth can be a violent and traumatic experience for the baby, who leaves the peace of the womb to enter what Leboyer describes as an "overwhelming confusing world." Leboyer, who recommends a highly relaxed "birth without violence" approach, is particularly opposed to the technologically based practices of the modern delivery room. Among these he includes Caesarean section and Apgar assessment, the range of tests carried out on a newborn baby within the first five minutes of birth. Advocates of high-tech delivery, however, argue that such developments have either saved lives of infants or, in the case of Agpar testing, have alerted physicians to babies' needs for immediate care.

An average full-term newborn weighs between six and nine pounds and is about 20 inches long; the head accounts for about one-quarter of this length. The skin is dry, soft and wrinkled and may still be covered with a soft down or lanugo. The eyes are almost always blue, although they may change color in later months, and the infant's vision is blurred. Hearing, too, may be slightly impaired because of amniotic fluid in the ears.

Born survivor

Survival is the keynote to the first few days of life. The new infant, who has been protected in the warm, liquid environment

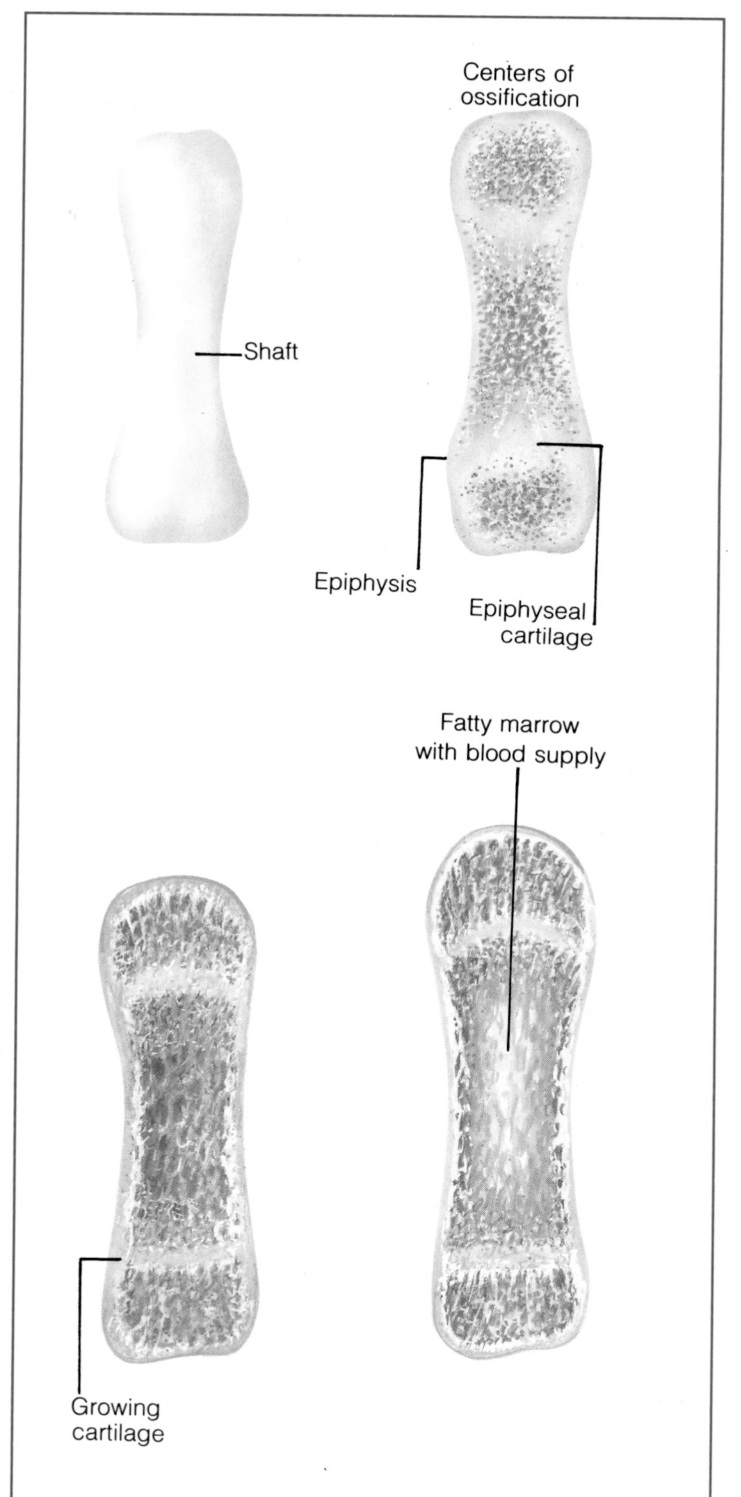

Ossification, the process of bone formation, occurs in the long bones of the fetus at the shaft and ends of cartilage "models." Cartilage continues to form in the epiphyses, gaps between the ossification centers, as the bone grows. Capillaries bring blood to bone-forming cells, or osteoblasts, and fatty marrow forms in the bone's hollow center. The marrow starts manufacturing blood cells.

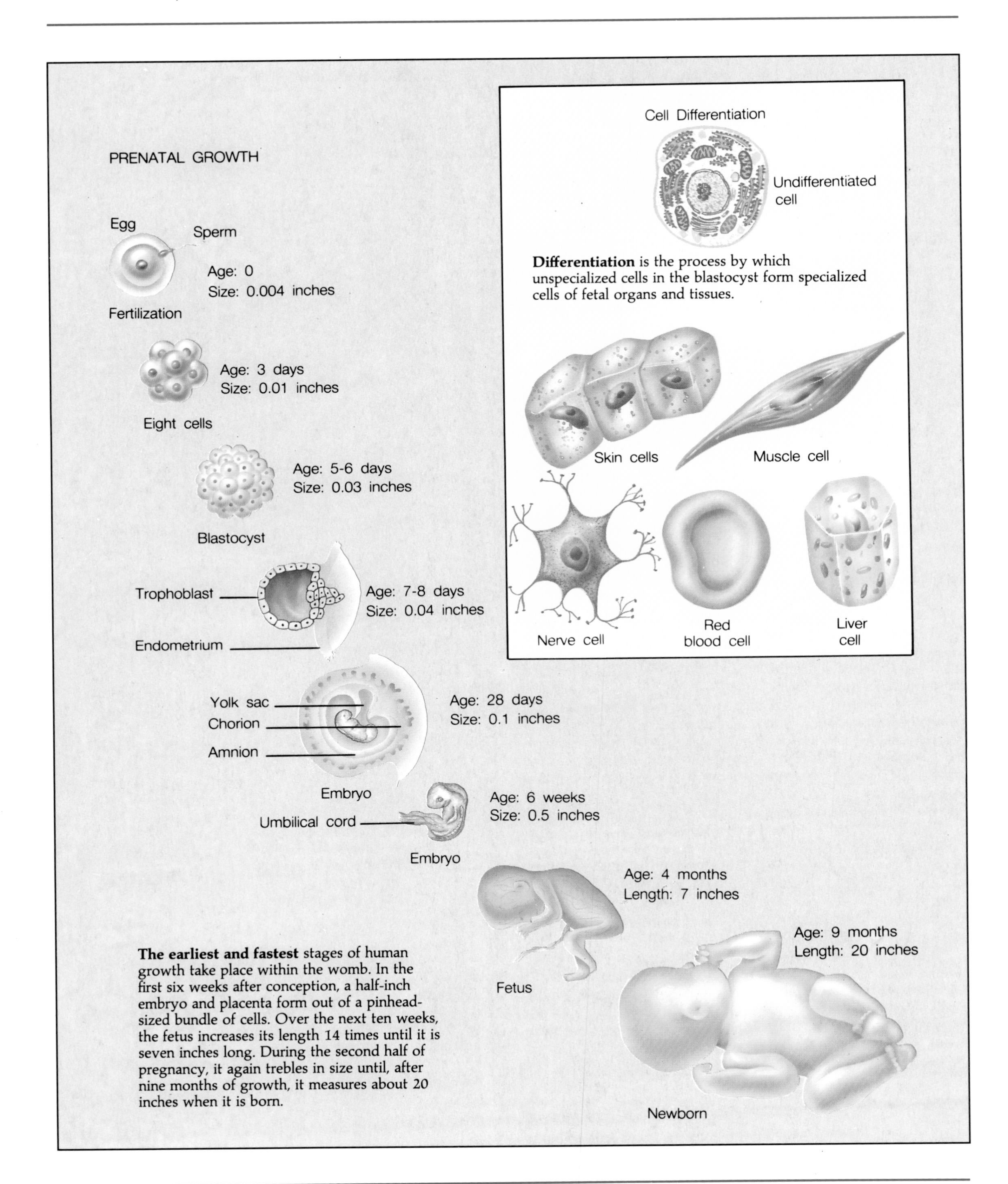

Differentiation is the process by which unspecialized cells in the blastocyst form specialized cells of fetal organs and tissues.

The earliest and fastest stages of human growth take place within the womb. In the first six weeks after conception, a half-inch embryo and placenta form out of a pinhead-sized bundle of cells. Over the next ten weeks, the fetus increases its length 14 times until it is seven inches long. During the second half of pregnancy, it again trebles in size until, after nine months of growth, it measures about 20 inches when it is born.

In this spiral of early development, the colored bands define the average age ranges of milestones of attainment. Physical skills, such as sitting, crawling and walking, have to wait for the parallel development of the nervous system and the necessary muscles – strength and coordination are both needed. Most babies acquire these skills in the order shown, although some infants reach each stage before others. The beginnings and ends of the colored bands pinpoint the ages at which 25 percent and 90 percent of infants have attained a particular stage of development. The ages at which 50 percent have mastered a skill are indicated by the positions of the babies within the colored bands.

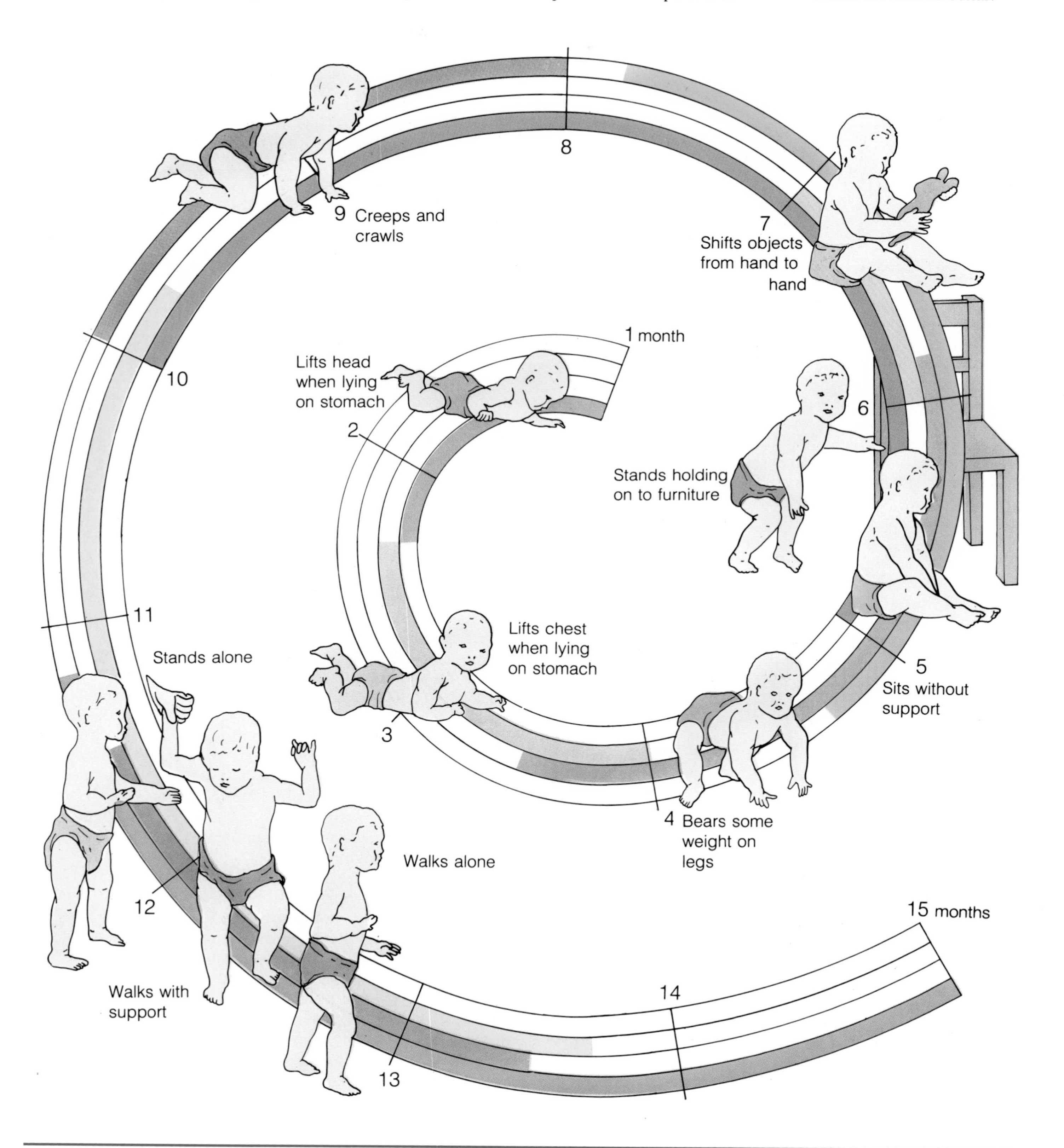

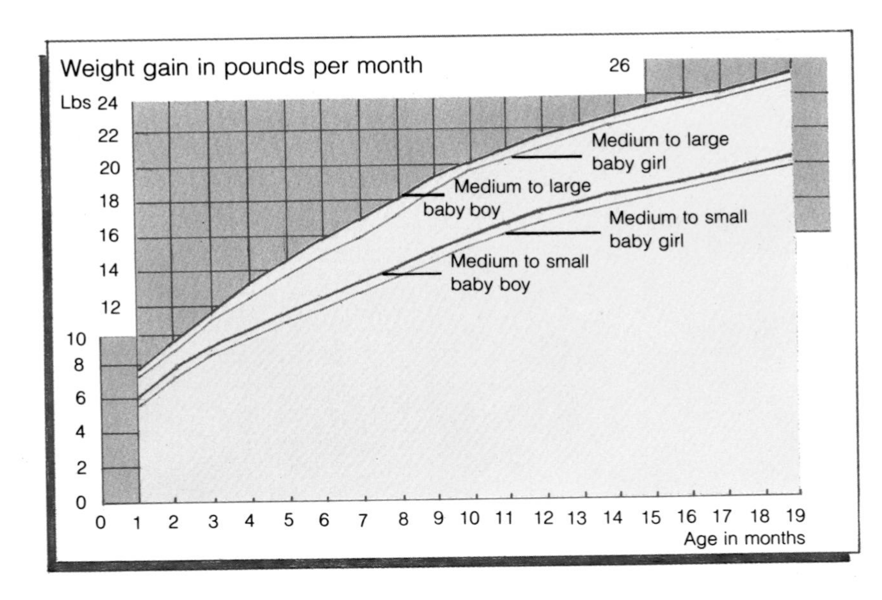

In the first weeks of life, most newborns lose a little weight – usually only a few ounces. Then for the next 13 weeks babies gain about 5 to 7 ounces per week, after which the rate of weight increase slows down. Each day, a newborn baby needs $2\frac{1}{2}$ to 3 ounces of milk for each pound of body weight – up to 21 ounces for the "average" 7-pound baby. Once the baby weighs 10 pounds or has reached the age of three months, solid foods can be added to the diet. By the age of 18 months, an "average" baby weighs 25 pounds.

of the womb for nine months relying on the life-support system of the placenta, has to make an immediate adjustment to the outside air-breathing world. Respiration is the first concern. At birth the lungs are still partly filled with amniotic fluid. During labor the fetal adrenal glands release the hormone epinephrine (adrenaline) to absorb the fluid so that it can be replaced by air. As a result the lungs should inflate as soon as the baby is born. Sometimes this fails to happen, perhaps because the umbilical cord is wrapped around the windpipe. Even so, the newborn is surprisingly resilient and can survive for short periods even with an inadequate supply of oxygen.

The circulatory system switches from total dependence on the placenta to becoming an independently functioning unit, a process that involves the closure of various pathways which, in the womb, allowed the fetus to receive oxygen without using the lungs. The kidneys are also immature, but are helped through the first critical days by hormones which maintain the correct balance of body fluids until the kidneys can take complete control of this function.

Finally, the newborn baby also has to adjust to digesting milk rather than receiving nutrients directly into the bloodstream via the placenta. Again the mechanisms for adjustment are highly efficient. In a breast-fed baby the first feed of colostrum, the yellowish fluid which precedes breast milk, encourages the release of hormones which, in turn, stimulate the gastrointestinal system, and in this way enable digestion and absorption to take place properly.

A newborn baby appears completely helpless. With a large head on a scrawny neck that cannot support it, small legs, and an uncoordinated nervous system, the baby seems unlikely to be capable of achieving an independent existence. But the survival instinct and the drive toward achievement are very strong indeed. During the first two years, growth and development proceed at a remarkable pace.

The timing of the various developmental stages varies, with different children arriving at key stages at slightly different ages. Nevertheless by their second birthday most children have traveled a long way from babyhood. By that time most can walk, talk, feed themselves, solve simple problems – in fact they are now, in many ways, mini-adults.

Survival systems

Babies need constant loving care in order to thrive and develop both physically and emotionally. The newborn baby arrives in the world with a deeply built-in survival system of reflexes and behavior, much of it directed toward obtaining the care that is needed to sustain development.

Crying is a baby's first means of communication. A cry may mean hunger, tiredness, or discomfort, but it is a sound that few parents can ignore. If food is not what is needed, the parent picks up the baby and provides comfort and love, forming an emotional bond by which the infant can thrive and to which he or she increasingly responds.

Together with crying, the newborn baby has other reflexes, some of which date back to the very earliest days of human existence. The sucking reflex is extremely strong. When the baby's cheek is touched, he or she automatically turns toward the touch and the lips pucker. In this way when the lips touch the breast or a feeding nipple, the baby immediately begins to suck, so ensuring successful feeding.

The grasp reflex causes a baby automatically to clench the fist if an object such as a finger is placed in the palm. This reflex disappears after the first three months, but is so strong that a baby can support its entire weight for a few seconds. The startle reflex consists of rapid movements. If startled or handled too briskly babies will throw back their heads and fling out their arms and legs. The movements are usually followed by crying. Most of these reflexes disappear as voluntary actions take over from this instinctive behavior.

In the first few weeks after birth, weight gain is the most usual way of checking that growth and development is on course. A baby's weight at birth may be affected in many ways. Boys, for instance, tend to be slightly heavier than girls; first babies are often slightly lighter than subsequent infants. Heredity also plays a part. Large parents usually produce babies who are larger than average, whereas small parents have smaller babies. Like many other human characteristics, such as color of eyes or hair, height, too, is genetically controlled. This control is exercised by polygenes, groups of genes which are passed down as blocks through the generations. Environmental factors such as smoking during pregnancy or maternal illness may also affect birthweight.

Most babies lose weight in the very first few days, but birthweight is usually regained within ten days. Thereafter the baby's weight should steadily increase in a consistent fashion.

On the move

Newborn babies have little or no control over physical movement. They lie in the birth position, on the stomach, knees bent underneath, buttocks in the air. To the anxiety of many first-time parents, the baby flops helplessly if lifted unsupported. By six weeks, however, the baby has gained some control of the head, and as the muscles gain in strength the head gradually becomes less heavy in relation to the body. By seven to eight weeks most babies are strong enough to keep their bodies in line with the head when lifted from a lying position, and by six months they can raise their heads when lying down.

Gradually too the baby uncurls from the initial birth position. Lying on their stomachs, babies soon learn to push up and support the weight of head and shoulders on the forearms. Physical control increases rapidly. By ten weeks most babies can roll from a side position onto their backs, and a month later can roll from front to back.

At three months the back still sags in a sitting position, but by six months the baby can sit upright supported by cushions or in the corner of an armchair. As physical ability develops, so too the baby increasingly responds to the people around, waving arms and legs, giving and receiving pleasure from interacting with other people.

The ability to balance is a major milestone in an infant's development. At first the baby balances in a sitting position by leaning forward and placing one hand on the ground. By the age of one year, however, he or she can turn and twist around, reaching outward to grasp a toy. Few babies can crawl before the age of six months, but many are quite mobile, propelling themselves along on their stomachs by pushing with the feet. By ten months most babies can pull themselves along by their elbows and a month later the majority are crawling with the stomach off the ground.

All of these early movements are preparation for the hurdle of walking. By about 11 months many babies are pulling themselves into an upright position using a helping hand or the edge of furniture. Balance is still uncertain, however, and without support the baby falls back down again. In the next stage, the baby, still using a support, gains the confidence to shift weight slightly and take a first, shuffling step to the side.

The age at which a baby takes those first solo steps varies considerably. Initially, the baby travels only short distances, from one piece of furniture to another, then increases the range until he or she can cross the room tottering unsteadily on two feet. Progress is not always continuous. Often the baby reverts to crawling, saving energy for another attempt. Illness too may cause a setback, but in general during this period the baby steadily moves toward the walking stage with all the changes that accompany this development.

Physical and mental skills

During the first year the baby gains considerable physical and mental skills. These occur because of the way in which the brain and nervous system mature during this period. At birth many of the baby's nerves are still immature. They are not coated with the insulating, fatty myelin sheath required for the complex interaction between nerves and muscles that produces voluntary

Pattern recognition is a well-studied aspect of behavior in very young babies. Researchers record the times at which infants show interest in various patterns by noting reflections in their eyes. Not surprisingly, two-month-old babies repeatedly prefer the "real" face to a scrambled face or a non-face *(upper diagram)*. Pattern is more important than color or brightness, as shown by the timings in the lower part of the diagram which demonstrate a strong preference for a face, printed letters and a bull's eye as opposed to plain colored disks.

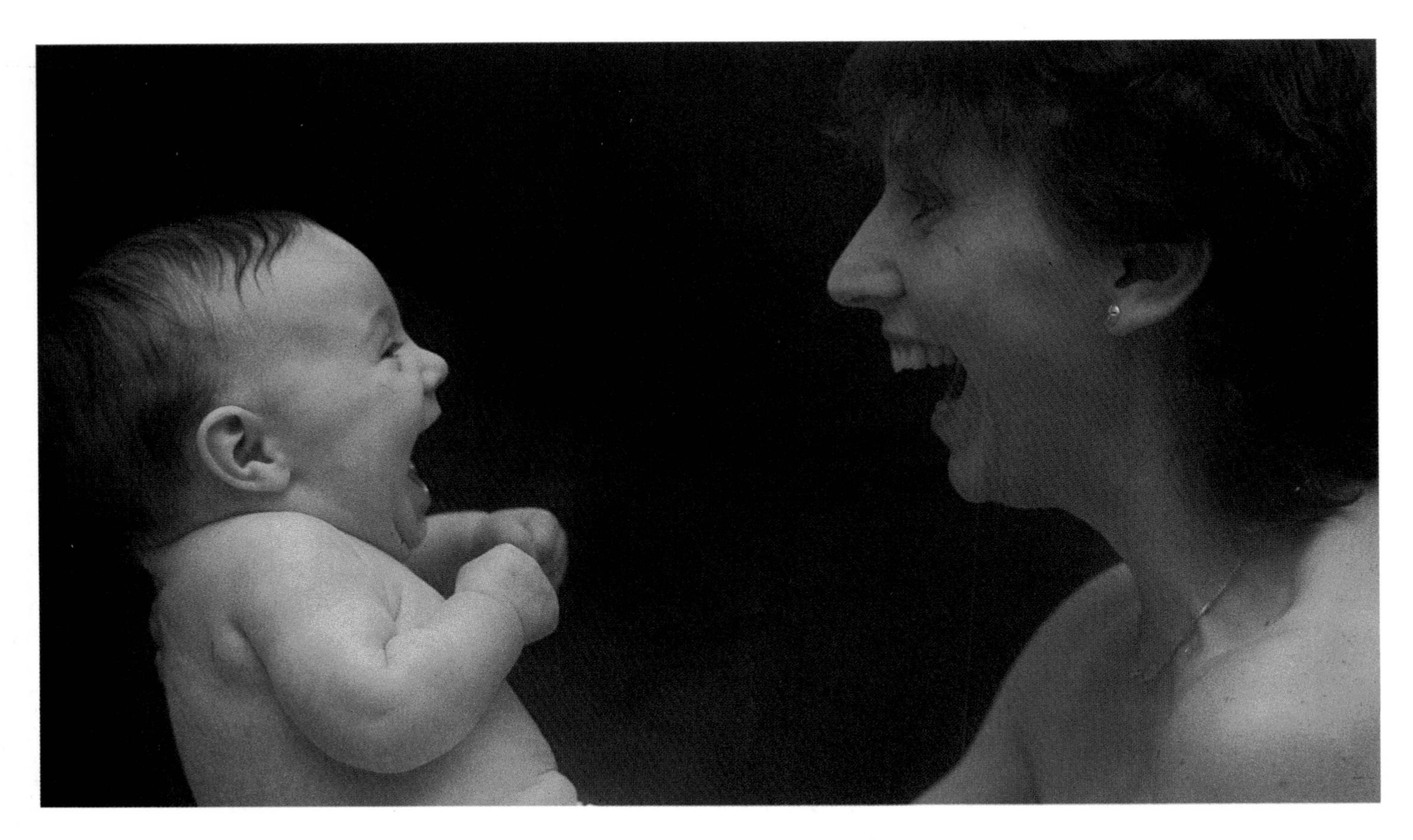

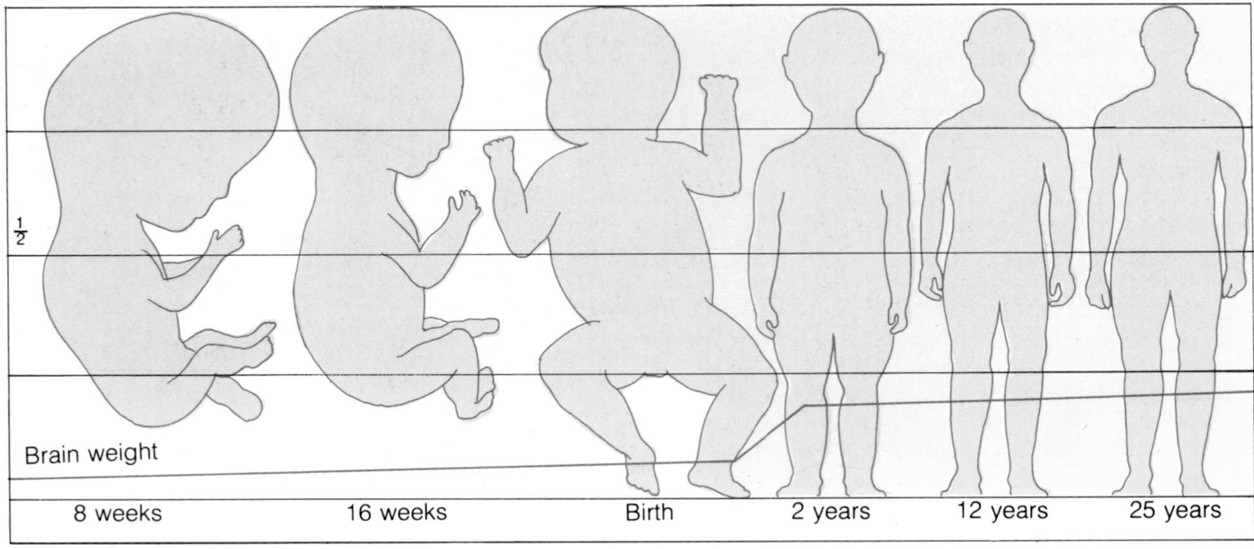

Body proportions change during the whole period of growth from conception to maturity. The head occupies nearly half the total body length of an eight-week-old embryo, whereas in an adult it accounts for about an eighth of overall height. The brain gains weight steadily during the period in the womb, grows rapidly in the first two years of life, and then settles once more to a steady rate of growth. Other body proportions also change. In a fetus, the arms and legs are shorter than the torso, whereas in an adult they are longer. Growth rate is not constant, however, and varies with age; there are two noticeable growth spurts in about the seventh year and during early puberty, which occurs a year or so earlier in girls than in boys. There are also seasonal variations; height increases among adolescents seem to take place fastest in summer, and the fastest weight gain takes place in winter.

physical achievement. By the end of the first year, however, myelination of the nerves is virtually complete, enabling the baby to walk and achieve other physical skills.

The brain also undergoes considerable development during the first 12 months. At birth an average baby's brain weighs 26 percent of its adult weight. But by the age of 12 months, it already weighs more than half its adult weight. By the second birthday the brain has virtually achieved maturity and grows only slightly thereafter. Such massive growth is allowed for by the membranous joints between bones of the skull, which are not finally fused together until the age of four.

By 15 months most babies are toddling happily. By 18 months most can pull themselves upright from a sitting position, and by their second birthday the average toddler is not only walking but running, dodging, stopping in mid-run and may be sufficiently coordinated and confident to play and kick a ball without tumbling over.

Entering the social world

A baby can see and hear from birth, and responds to sudden or startling sounds by blinking or stiffening, and to calming sounds by relaxing. At first, however, there is little or no coordination between hearing and seeing. Although the baby responds to sound, the head is not turned toward it. At this stage a baby is very nearsighted and it takes some months for the visual part of the brain to mature. The baby focuses at nine inches – almost exactly the distance between the baby's and mother's face during feeding – and studies show that the baby's first interest is in faces

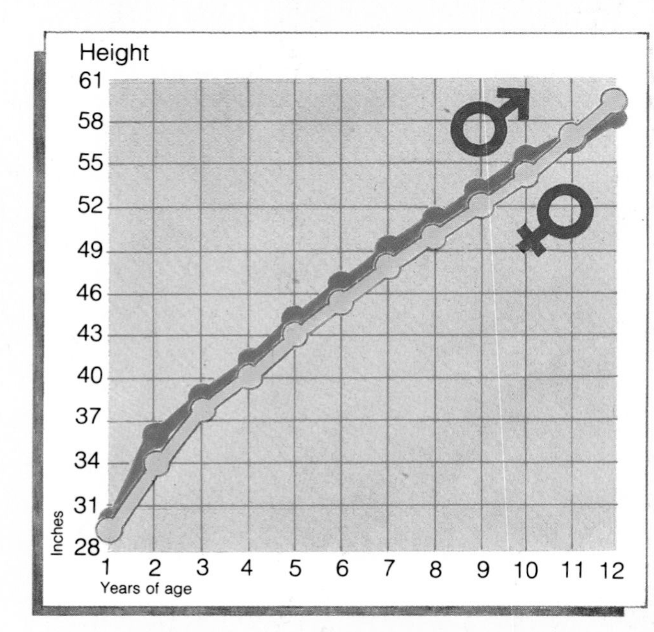

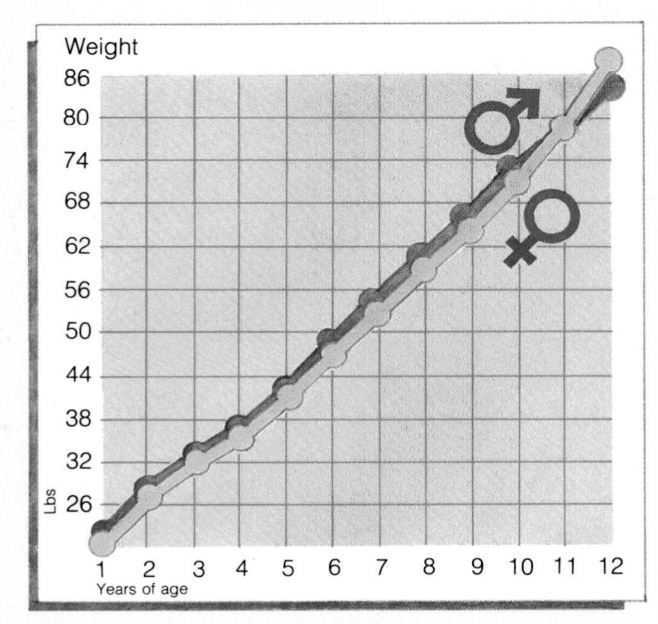

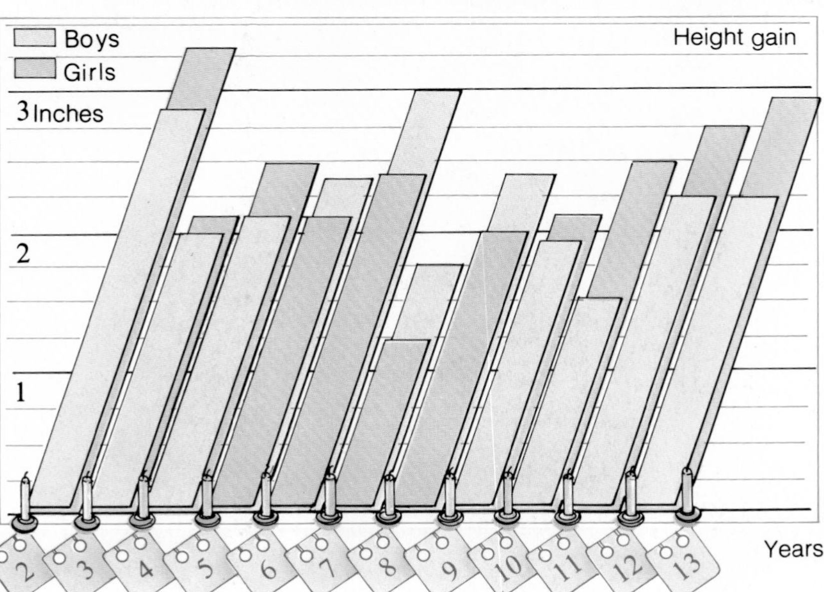

The rate of physical growth during childhood differs for boys and girls. For most of the time, boys (blue lines in the upper diagrams) are taller and heavier than girls, but at age 11 girls (yellow lines) overtake boys. The rate of growth also differs between the sexes (lower diagram), with girls gaining height more rapidly than boys except in the four years between ages five and nine. All the statistics in these charts are averages taken from large samples of children; an individual child's height and weight, or the rate at which he or she grows, may be quite different from the average at one particular age or at a particular stage of development.

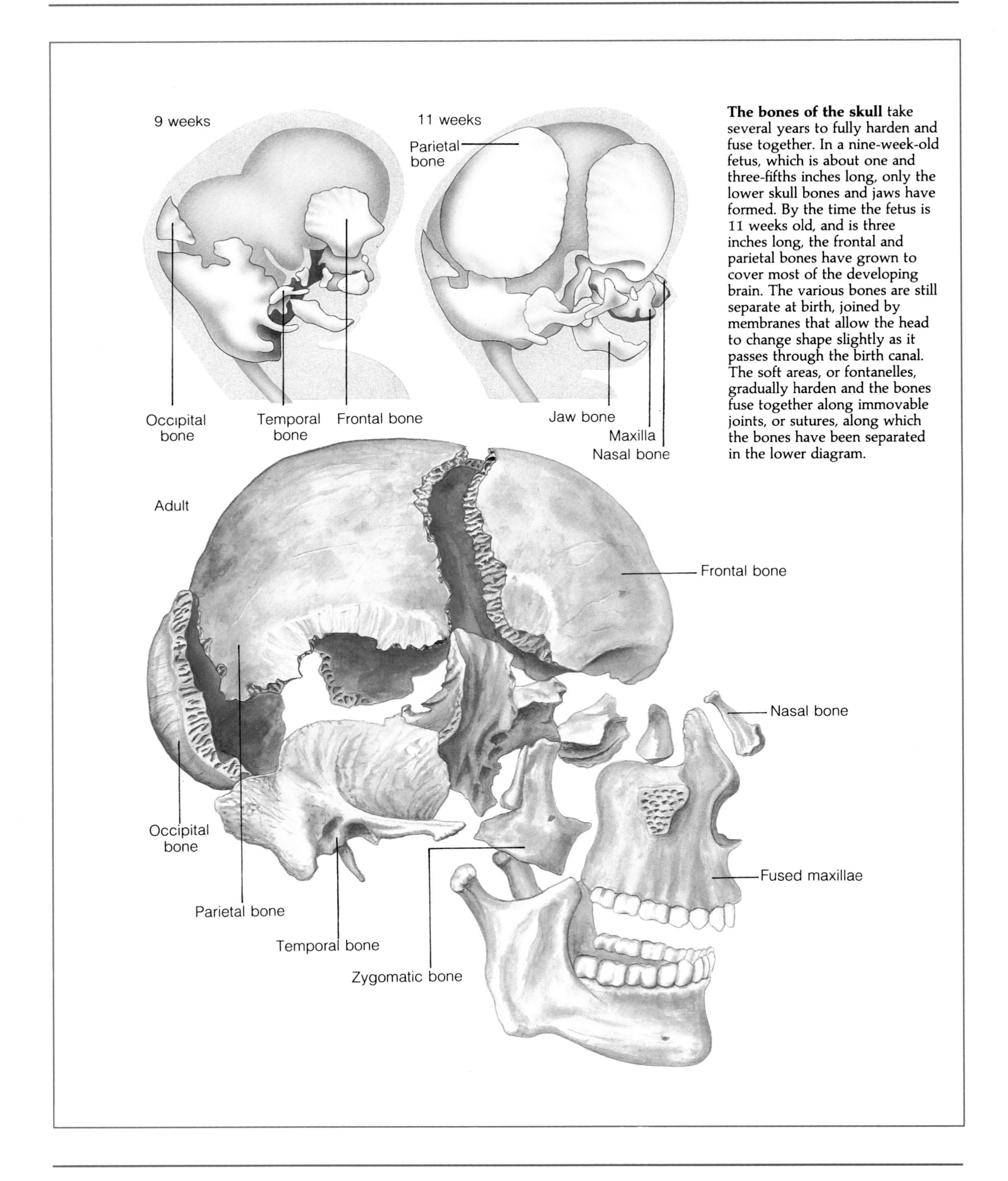

The bones of the skull take several years to fully harden and fuse together. In a nine-week-old fetus, which is about one and three-fifths inches long, only the lower skull bones and jaws have formed. By the time the fetus is 11 weeks old, and is three inches long, the frontal and parietal bones have grown to cover most of the developing brain. The various bones are still separate at birth, joined by membranes that allow the head to change shape slightly as it passes through the birth canal. The soft areas, or fontanelles, gradually harden and the bones fuse together along immovable joints, or sutures, along which the bones have been separated in the lower diagram.

There are 32 teeth in a full adult set, eight in each half of each jaw *(upper diagram)*. The ages at which the permanent teeth erupt vary slightly between boys and girls, beginning with the bottom two incisors at about age six and being complete (apart from the third molars, or wisdom teeth) at about age 12. As the permanent teeth move into place, they displace the first, or milk, teeth, which fall out. Some adults never cut their wisdom teeth. The lower diagrams show the ages at which the different teeth erupt – boys' teeth on the left-hand side of each diagram, and girls' teeth on the right-hand side.

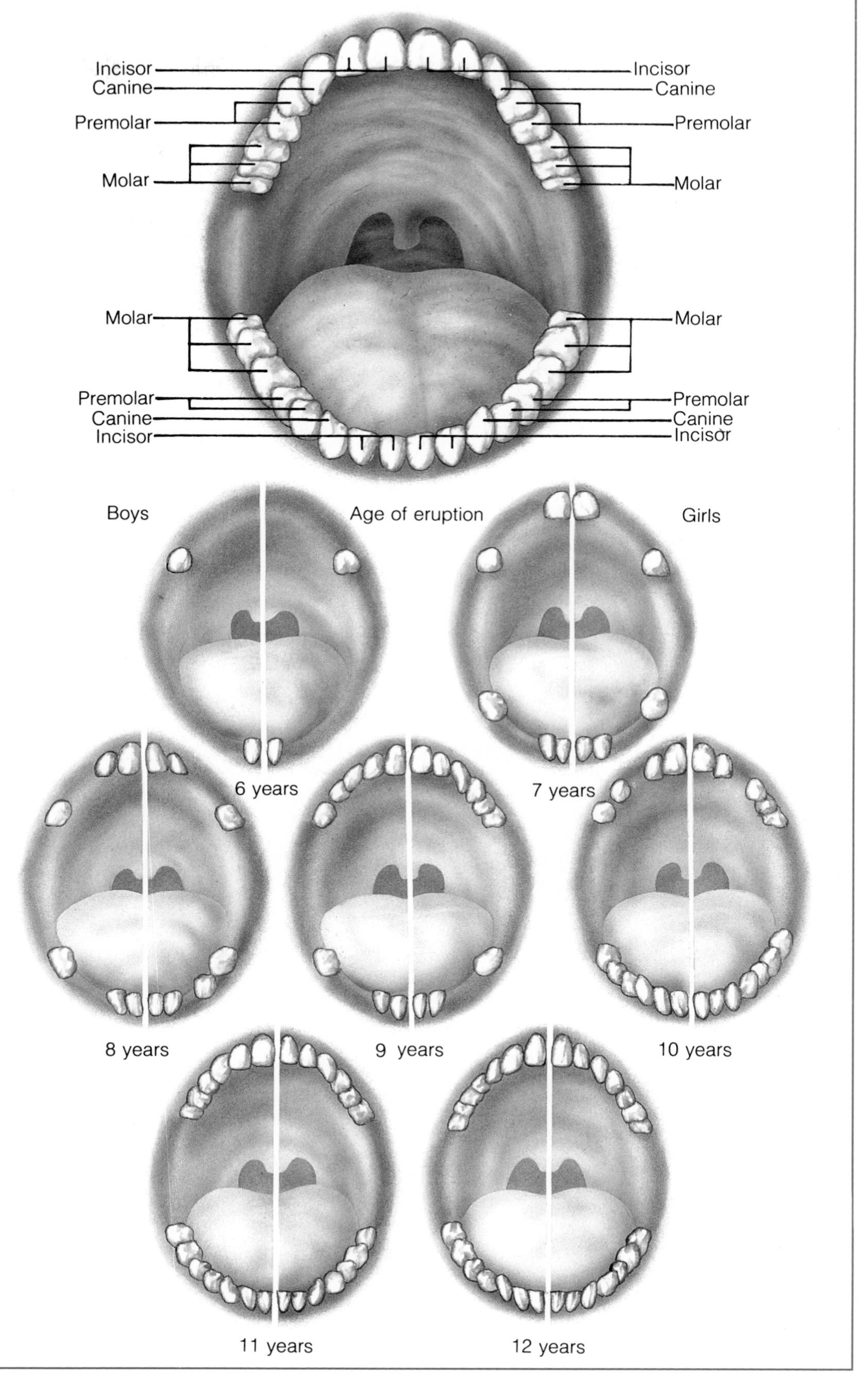

or face shapes. By six weeks, however, the baby has begun to focus more accurately and responds immediately to a familiar face with smiles and sounds.

Socializing is a crucial aspect of development, and young babies are just as social as adults. The more babies socialize the more they learn, copying the people around them in every detail. A baby reacts to speech and smiles from the very earliest days, producing his or her own sounds in imitation. Initially the sounds are based on vowels; by six months consonant sounds are added, and toward the end of the first 12 months the baby may have developed and learned a number of key words.

As babies learn about the world around them, so too do they discover more about their own bodies. At six weeks they can clasp their hands together. Every baby spends considerable time watching the hands, sucking them, and exploring them.

This is a crucial stage in the development of hand–eye coordination and one that can be encouraged by providing extra stimulation, such as a mobile or wall posters. Many authorities now believe that the nerve fibers themselves become more efficient the more the baby is stimulated and encouraged. As time passes, hand–eye coordination develops rapidly. By nine months the baby can perform the delicate operation of linking forefinger to thumb, and by the first birthday can pick up small objects.

The childhood years

By the age of two the helplessness of babyhood has been left far behind. Over the coming years the developing child becomes increasingly independent both physically and emotionally. Growing physical dexterity, plus a wealth of social experiences, enable the child to develop the abilities to think, reason, and adapt to new situations and people.

One major hurdle on the way to independence is scaled around the age of two when preschool children begin to learn how to control bladder and bowel movements – achievements that require physical development but which also demand the ability to understand what is expected.

During the childhood years growth is slower and steadier than during infancy or adolescence. The child becomes longer and slimmer, and body proportions change until they are nearer those of an adult. Girls tend to grow more quickly than boys, although by the age of five both sexes have reached half their adult height. From then until puberty the two sexes alternate, until by the age of 12 girls tend to be slightly the taller.

Growth is controlled by hormones. The hypothalamus in the brain produces growth hormone releasing factor (GRF), which acts on the pituitary gland at the base of the brain and stimulates it to release the growth hormone somatotropin. This hormone is carried in the bloodstream to the liver and kidneys, where it is modified into somatomedin. It is this substance that finally stimulates bone growth and the calcification of cartilage, a process which hardens it to bone.

Toward the end of adolescence growth stops as the production of growth hormone inhibiting factor (GIF) rises, which critically prevents the release of somatotropin. If this does not happen, a number of disorders may result, including gigantism. Dwarfism, in contrast, is caused by failure to produce enough growth hormone during childhood.

Secondary, or permanent, teeth appear from about the age of six, the first to erupt being the back molars. Thereafter teeth appear in much the same order as the milk teeth, with the new teeth displacing the old ones. Overcrowding may be a problem

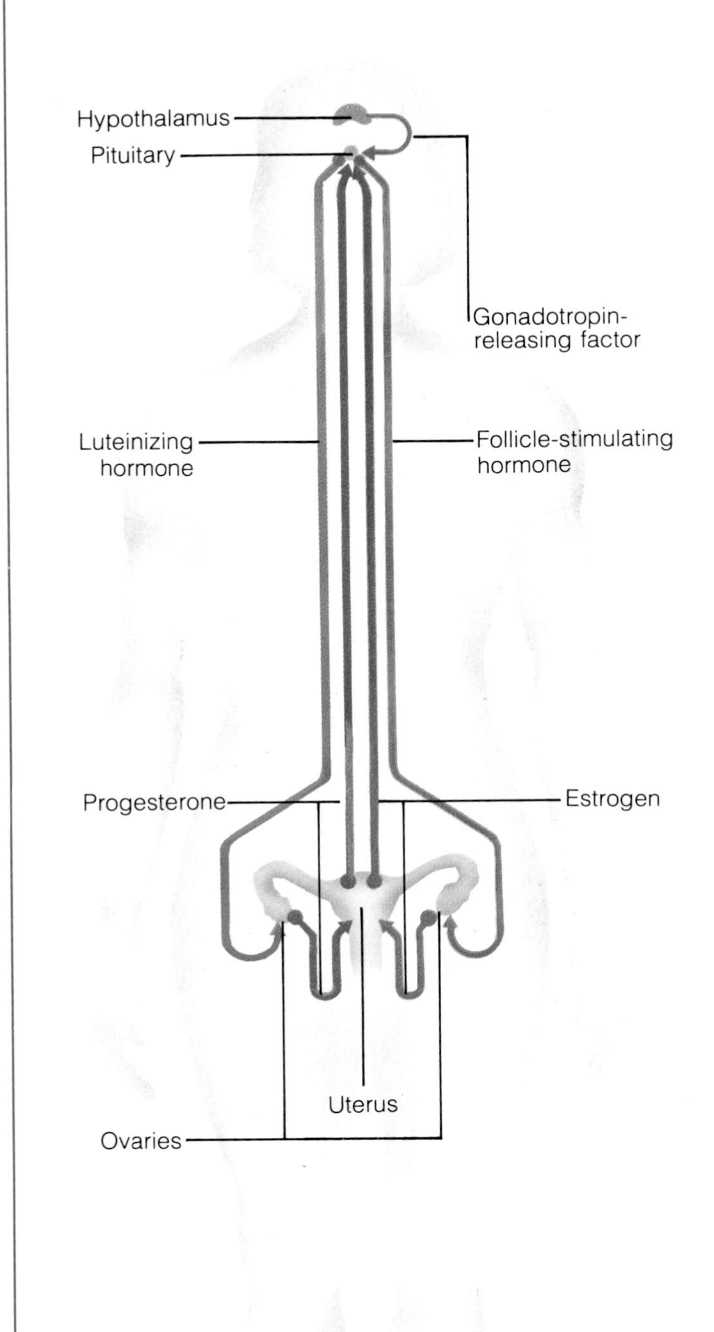

The physical and emotional changes that take place during puberty are brought about by increasing levels of sex hormones coursing through the bloodstream. They begin as the hypothalamus in the brain stirs into action. In girls *(left)* at about the age of ten or 11, it secretes gonadotropin-releasing factor, which makes the nearby pituitary release luteinizing hormone (LH) and follicle-stimulating hormone (FSH). LH and FSH travel in the blood to the ovaries, where LH induces them to produce and release progesterone. This hormone initiates development of the breasts and starts the periodic changes that accompany the menstrual cycle. FSH causes the ovaries to release estrogen, which is responsible for the development of secondary sexual characteristics such as the pattern of pubic hair growth and the accumulation of subcutaneous fat, which gives the girl a rounded, womanly figure. The levels of ovarian hormones in the bloodstream are detected and fed back to the hypothalamus to control the pituitary's output of LH and FSH.

A similar chain of events happens in boys at puberty *(right)*, but in them the pituitary hormones act on the testes to produce testosterone, the principal male sex hormone. Testosterone stimulates the growth of facial and body hair, causes the larynx to enlarge so that the boy's voice breaks, starts up the production of sperm in the testes, and initiates the development of the genitals. Once again a feedback mechanism, this time of testosterone carried in the bloodstream to the hypothalamus, controls the release of LH and FSH by the neighboring pituitary.

In both boys and girls during puberty, the ebb and flow of hormones can produce other effects, from the distressing symptoms of acne to the violent swings of mood characteristic of the turbulent teens.

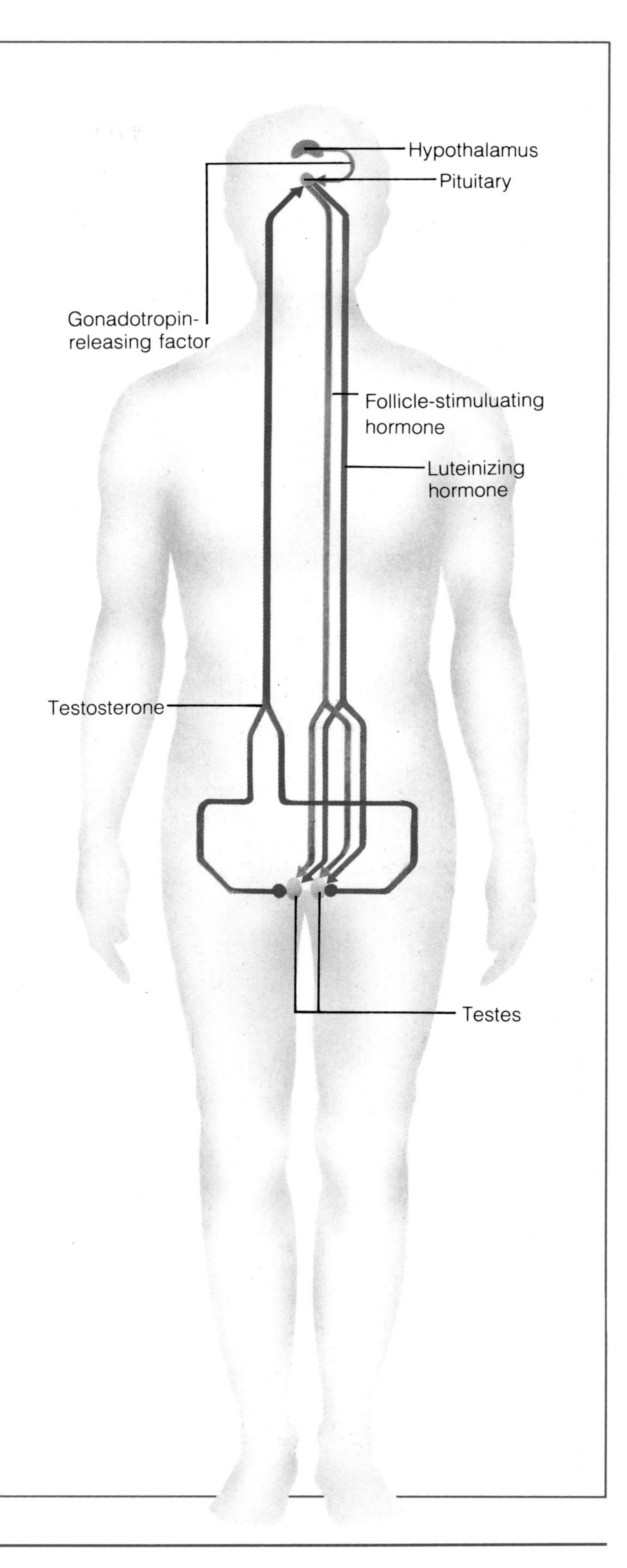

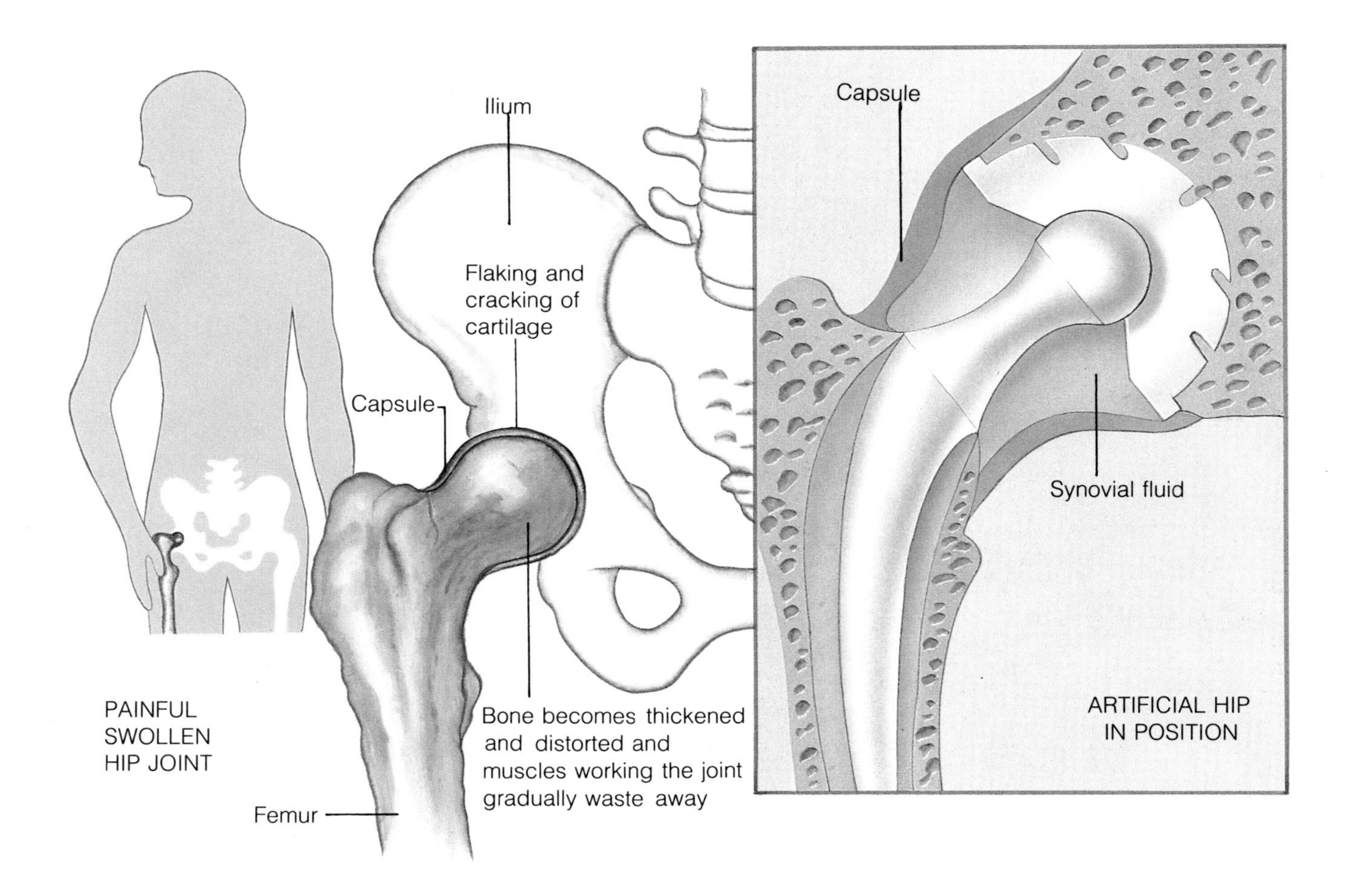

One of the penalties of aging is the inevitable wear and tear on the larger weight-bearing joints, such as the hips, knees and spine. The result may be the crippling disorder osteoarthritis, in which the smooth layer of cartilage that normally lines and lubricates the joint gradually erodes away. An excellent remedy, particularly for the hip joint, is an artificial replacement. The head of the thigh bone, or femur, is removed and replaced with one made of metal or metal and plastic. Such joint-replacement operations have done much to restore mobility and improve the quality of life of thousands of elderly people.

and require attention if it occurs, but as with other aspects of growth and development, the course of teething depends partly on genetic factors and partly on the environment – in this instance on diet in particular.

Mind and muscle

A human child has a lot to learn. Between the ages of two and 12, physical and intellectual abilities progress rapidly as the nervous and muscular systems develop. By the age of two most children can use building bricks; by the age of three many children can pedal a tricycle and dress themselves. By the age of four physical coordination is sufficiently advanced for the child to play ball games. Over the next few years such physical abilities advance even further, as with increasing confidence children experiment with tree climbing, sport, swimming and so on, pushing themselves to ever greater efforts.

A growing child matures in other areas too. As concentration increases, the child moves on to develop quite precise skills such as sewing, drawing, writing and, eventually, reading. All require good hand–eye coordination. With reading and writing they also require the ability to recognize images and translate them.

Drawing is another example. Most children begin to draw and paint quite independently of instruction somewhere between the ages of two and three. The first emphasis is on the face, but gradually the child adds extra detail such as eyes or clothing by recognizing these, then translating them onto paper. The complex processes of reading and writing are learned in much the same way, as the child first learns to recognize a name or other familiar words and word shapes, and gradually translates them onto paper by re-creating the shapes or reading them.

With increased communication children enter a new development phase. Using language, they can reason, argue, and explain themselves. Although all still need the security adults must provide, they are increasingly able to exercise personal control.

Sometimes a child's desires conflict with those of an adult, and this can lead to confrontation and confusion about whose needs should prevail. Too much control limits the child's ability to make decisions and infringes on children's rights. At the same time the child is learning to fit into society and make allowances for the needs of others. In fact, once they begin to attend school, most children actively seek acceptance from their peer group, a process often characterized by stereotypical sexual attitudes, among other things.

The turbulent teens

Adolescence or the teen-age years mark the transition from childhood to adulthood. Like other stages of development, it is marked by physical and psychological changes, both closely interlinked. These can be confusing years as the adolescent struggles with self-consciousness, the drive for independence, and a search for personal identity. Family problems, too, may impinge on an adolescent's development because by this stage parents, who are no longer responsible for young children, may be making major changes to their own lives.

Puberty dominates the early teen-age years. In both sexes its onset is stimulated by increasing levels of sex hormones in the bloodstream, and in girls occurs at about the age of 12, on average two years earlier than in boys.

In girls puberty is controlled by luteinizing hormone (LH) and follicle-stimulating hormone (FSH) secreted by the pituitary gland. Luteinizing hormone stimulates the ovaries to produce another hormone, progesterone, which is responsible for breast development and the onset of menstruation. Follicle-stimulating hormone brings about the release of estrogen, which initiates the development of female sexual characteristics.

The first external sign of puberty in girls is nipple development. It may begin from the age of eight, and is usually noticeable by the time the girl is 11. The areola, the pigmented area around the nipple, enlarges and over the next two and half years the breasts become fully formed. Pubic hair also begins to grow, vaginal secretions increase and about two years later the menarche, or first menstrual period, occurs. Regular ovulation usually begins about a year later.

Puberty in boys usually begins between the ages of 13 and 15½. As in girls, the onset of puberty is controlled by hormones released by the pituitary gland. Testosterone, the male sex hormone, is produced by the testes and is responsible for sexual development. The testes and penis increase in size, the scrotum grows, and pubic hair develops. Facial hair also develops and the boy's voice "breaks" or deepens as the larynx increases in size.

The adult years

Strange as it may seem the aging process begins with late adolescence, but it is not noticeable until much later on. For most people the twenties and thirties are a time of peak physical fitness. Growth is complete and there are very few age-specific disorders to contend with. Instead health is closely linked to lifestyle: regular exercise, a balanced diet, and abstinence from cigarette smoking and excessive alcohol consumption prevent serious and life-threatening problems such as cancer and heart disease later in life.

The turbulence of adolescence has now passed and most young adults are concerned with making choices about careers, relationships and lifestyle. The conventional path toward marriage and children is changing. Statistics today show that people in the West are choosing to marry later or to remain single for life. And surveys conducted during the mid-1980s showed that more than half of American households consisted of only one or two people. Thus the traditional two-parent family is now in a minority.

People are living longer and life expectancy continues to rise. In the United States since 1900, life expectancy at birth as risen from 47 to 69 years for males and from 51 to 75 for females. The chart shows how the death rate as a percentage (i.e. deaths per 100 people) varies with age, being highest at about 71.

Having made choices, most people find the years from 30 to 45 are ones of consolidation of career, partnerships, and social activities. Old age still seems a long way away, although the fortieth birthday is often viewed with apprehension.

The middle years

Physical and emotional changes begin again during the middle years. For many people middle age is a time of crisis, often comparable with the upheaval of adolescence. Children may be leaving home, there may be worries about sexuality, and there is an increasing anxiety about the onset of old age.

For women the middle years are marked by the menopause, the ending of their reproductive period. It usually occurs between the ages of 45 and 55, its onset coinciding with a decrease in the levels of the sex hormones. Menstruation and ovulation cease, and fertility comes to an end.

The menopause and post-menopausal period is usually accompanied by various physical symptoms, such as hot flashes, and physical changes such as vaginal atrophy and osteoporosis (weakening of the bones). Hormone replacement therapy (HRT) may ease these symptoms, but it is a controversial form of treatment because of a possible link with uterine and breast cancer.

Later middle age may also be marked by various social changes, most especially retirement. For some retirement is a welcomed rest from the rigors of work, for others it is a distressing experience involving separation from colleagues and lifelong routines. Because of the modern trend toward younger retirement, many people finishing work are still physically and emotionally capable of making much of the retirement years. Retirement is a time of new opportunities best faced with thorough and sufficient preparation.

The later years

It is difficult to say when old age begins, particularly today when life expectancy has increased by more than 25 years over the last century. Even so, the years from 55 onward are marked by the obvious signs of the aging process.

Most changes are gradual. As a person gets older the skin loses elasticity and subcutaneous fat disappears, causing wrinkling and sagging. Pigmentation also changes and hair loses its color. Muscles shrink as they are replaced with fibrous tissue, a process that begins in the thirties and, for women, accelerates after the menopause. It is thought to be caused by impaired calcium absorption and calcium loss. The bones become brittle and are more liable to fracture. Cartilage thickens and joints become stiffer.

Disease and debility, however, are not inevitable in old age. Normal aging does involve a decline in physical and mental ability, but research shows that bad habits such as smoking, a sedentary lifestyle and poor diet are major contributory factors in disease in the later years. Switching to a healthy lifestyle can do a great deal to combat the adverse effects of aging.

Aging itself is a gradual process, and decline is most usually caused by illness rather than age in itself. Today the main concern of gerontology, the study of aging processes, is to find a way of providing remedies and treatments for the common disorders of aging, and to discover a means to postpone old age or extend longevity. Much work has been done in this area, but for the moment it is a matter of luck – notably in the way you choose your parents – and healthy lifestyle.

Wrinkled skin and gray hair typify the aging process. Other noticeable changes include a loss of height due to a shortening of the spine, often exaggerated by a slight stoop. Bones become more porous and brittle, and the senses of hearing, taste and smell become less acute.

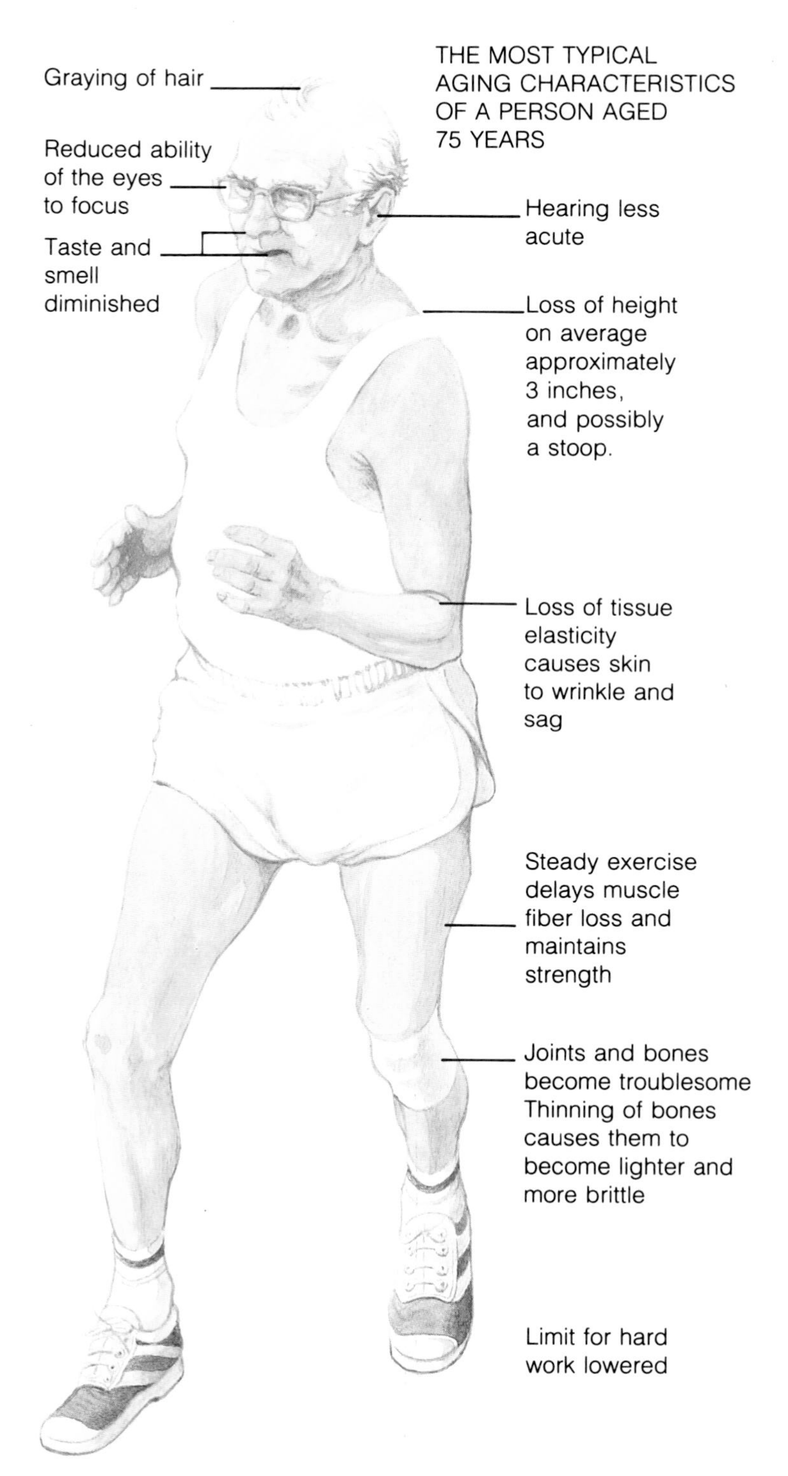

Organs and body systems have reduced capacities in the elderly. Assuming that they work at 100 percent capacity at age 20, the diagram indicates the percentages for somebody in his or her late seventies. Most affected are the waste-processing ability of the liver and kidneys, the respiratory capacity of the lungs and the pumping efficiency of the heart. Nerve conduction also slows down, and brain weight declines.

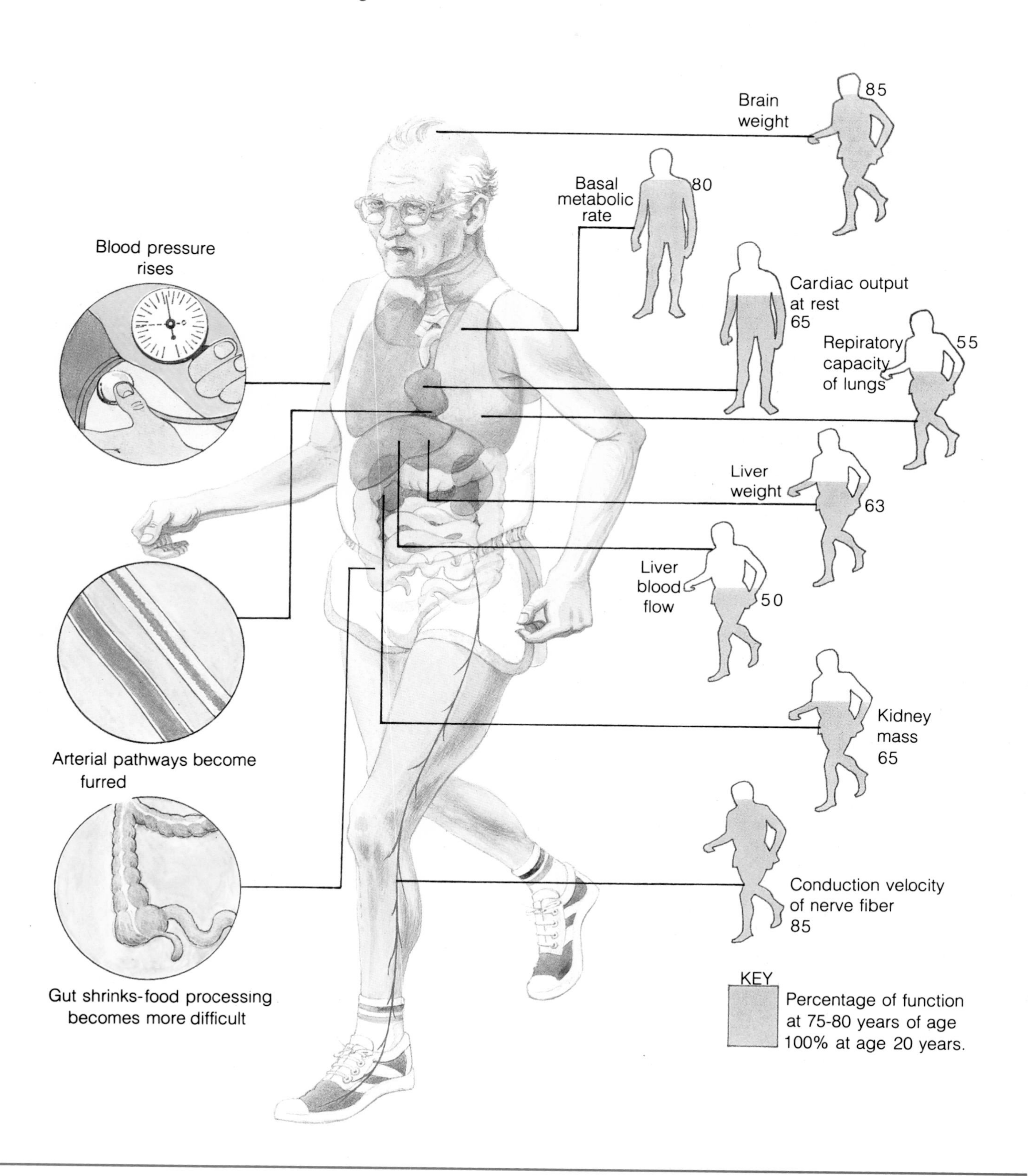

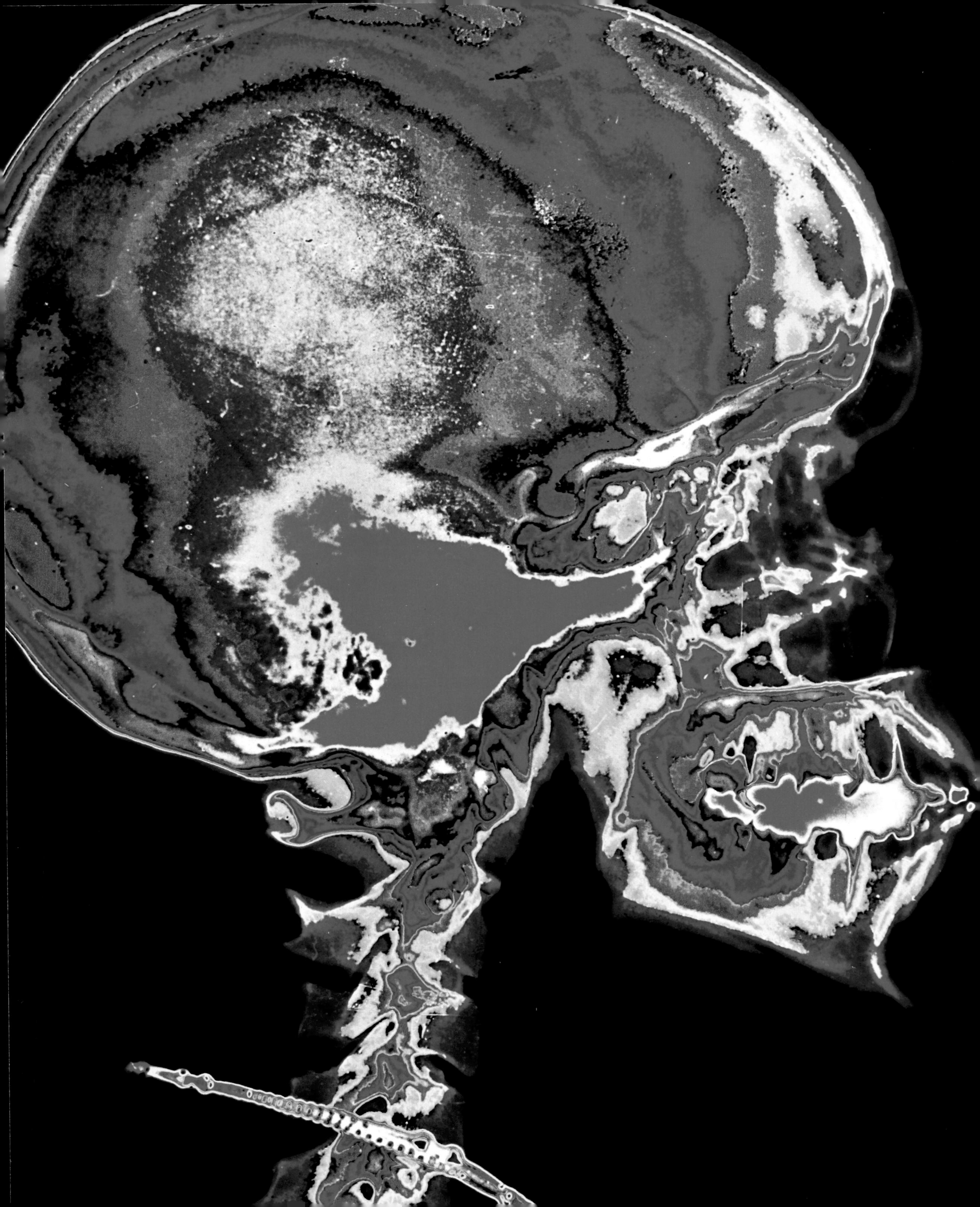

CHAPTER 2

The Skeleton

Humans are vertebrates, creatures with backbones. They are reliant on a sturdy internal frame centered on a prominent spine. Bone is their being's cornerstone. When broken, it heals itself without scarring – a characteristic that sets it apart from nearly all other tissues in the body. Light yet strong, and seemingly simple in structure, the skeleton is nonetheless a masterpiece of architecture.

The human skeleton is built with the strength of an oak, yet it can also bend with a sapling's ease. It surrounds and protects the organs, supports the body and, bound by muscles, bestows the grace of movement. Ever building and breaking down, bone is dynamic tissue that forms in proportion to the task required of it. The bones in a ballerina's feet, a sculptor's hands or a bricklayer's arm gain mass and alter in shape in response to the stresses their varied pursuits impose.

A flexible framework

Without the skeleton humans would find movement impossible. It provides a base for attaching muscles and the leverage to assist in their pulling. The skeleton protects the body's vital organs and provides life-giving substances – blood cells from red bone marrow and minerals from its bony storehouse. An ever-changing structure, it is also sensitive to the user's needs. The skeleton grows rapidly through childhood and adapts itself to our lifestyles, reinforcing areas where we, from sport, hobby or occupation, add to the forces that already burden it.

The skeleton's most important function is that of support. Like beams that hold up buildings, the bones of the skeleton, assisted by muscles, hold the body erect. But the body does not remain rock-still like a building. Where bone meets bone, joints provide the skeleton with flexibility and enable it to move.

Anatomists divide the skeleton into two broad categories – the axial and appendicular. The axial skeleton comprises the bones of the body's central core: the skull and spinal column which protect brain and spinal cord, and the rib cage, which shelters the heart and lungs. The appendicular skeleton includes all bones in the arms and legs, as well as the shoulder and pelvic girdles, which join the limbs to the axial skeleton.

Distinct skeletal parts are recognizable in an embryo only five weeks old. Even when the embryo is the size of a pea, the prominent spinal column forms a single majestic curve. At birth, babies have about 350 individual bones. The adult skeleton totals 206 bones on average. Some people are born with an extra pair of ribs, some with one pair less than normal. The number of the bones in the spine can also vary. And bones that generally fuse together during life can remain separate.

A multifaceted column

The mature spine is a series of alternating convex and concave curves that supports the body and absorbs shock. The column is built of 33 small bones, the vertebrae. Seven cervical vertebrae in

The protective bones of the skull are revealed to have different densities in this false-color X-ray of a woman wearing a necklace. A bony cage that encloses the brain, the skull is just part of the structural framework of more than 200 bones that make up the human skeleton.

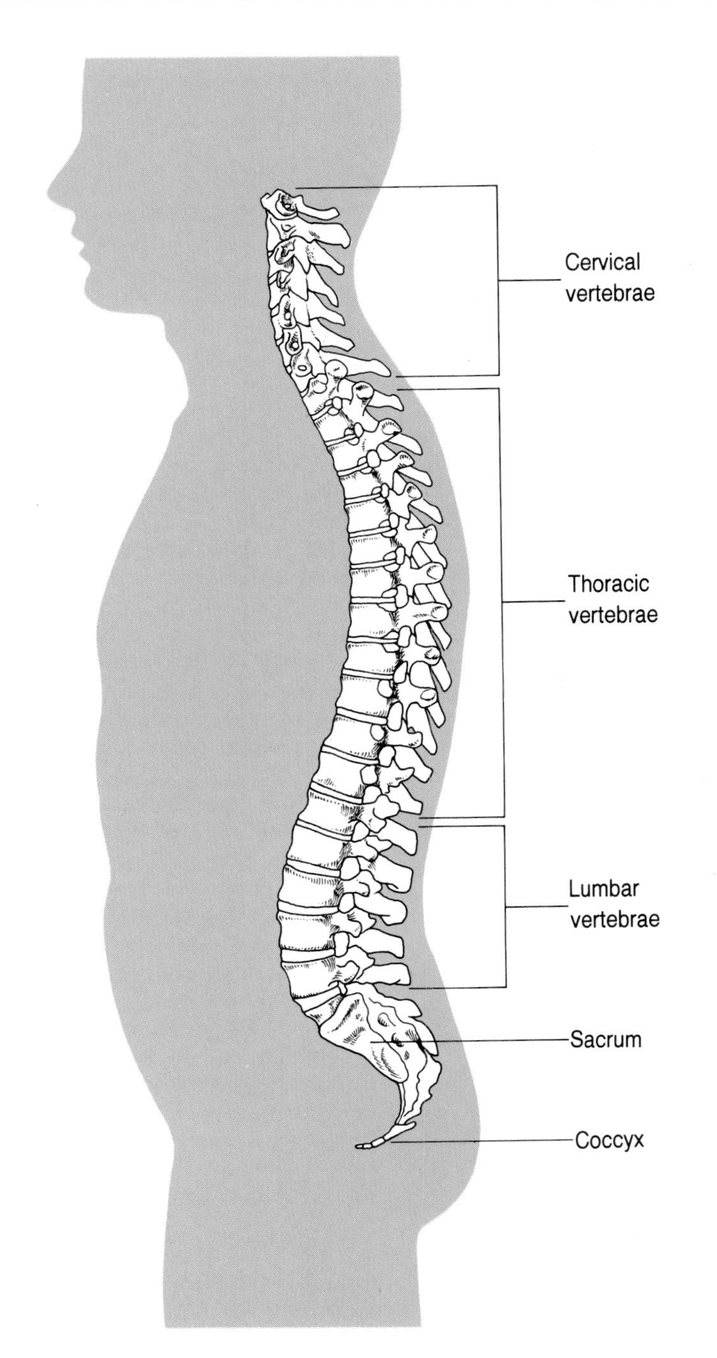

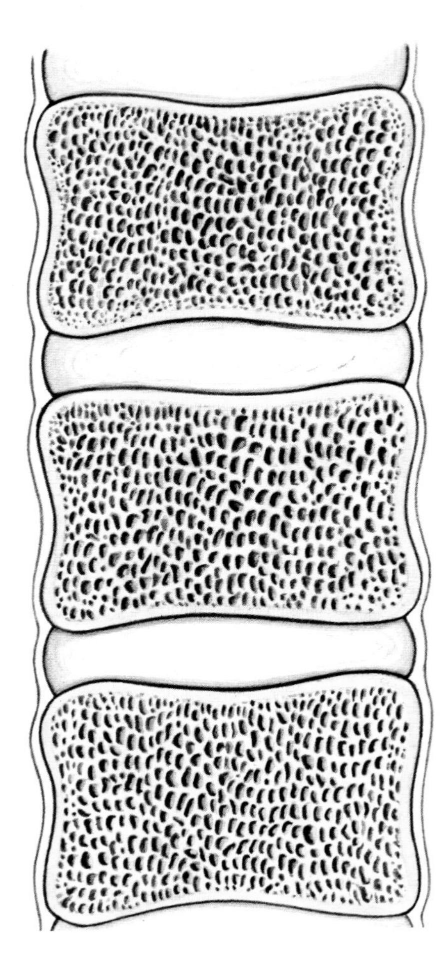

Doubly curved like an elongated S, the spinal column *(left)* has evolved to support human beings in their unique upright posture. Its 35 vertebrae provide strength and flexibility. Intervertebral disks between the vertebrae *(above)* allow a range of movements and cushion the spine against everyday jolting and jarring.

the neck are the smallest and allow the widest range of movement along the spine. Twelve heavier bones, the thoracic vertebrae, lie below them, forming the upper back. These bones also hold the 12 pairs of ribs in place. Special dish-shaped indentations called facets help anchor the ribs along each side of the vertebrae.

Five lumbar vertebrae in the small of the back are the largest bones of the spine and bear most of the body's weight. Below these, five smaller bones, individual at birth, fuse at about age 25 to form the wedge-shaped sacrum, a single bone fitted between the two hip bones and forming the back of the bowl-shaped pelvis. Tucked beneath the sacrum is the coccyx, a small, tapering bone built of four fused vertebrae. The coccyx is functionless and probably the vestige of a tail.

The building blocks of the spine

With the exception of the fused vertebrae in the sacrum and coccyx, and two highly specialized vertebrae at the top of the column, all the spine's bones are structurally similar. The main part of a vertebra, its body, is a flattened, oval block of bone. Short columns, the pedicles, arise from the back of the body. Small plates called laminae seal the opening between the two pedicles, creating a circular canal through which the spinal cord passes. Three winglike extensions crown the laminae. Spaced at 90-degree angles from each other, they anchor muscles and give the spine its knobby appearance under the skin. The articular processes, two smooth knobs on top of each vertebra and two beneath, link each vertebra with its neighbors.

The spine's topmost vertebra, the atlas, supports the skull. It does not have the solid body of bone typical of the other vertebrae. Rather, it forms a bony ring with a large central opening. Rounded projections of the skull's occipital bone – the bone making up much of the skull's rugged base – fit into two large hollows atop the atlas. Tough ligaments bind skull to spine. The second neck vertebra, the axis, sends a bony projection into the base of the atlas. This articulation permits the atlas, holding the skull, to rotate at the top of the neck.

All along the spine, strong ligaments, muscles and facets on the vertebrae bind the spine into a single column. Between adjacent vertebrae, there is a special cartilage padding called an intervertebral disk. Occupying about a quarter of an adult's spine, disks absorb shock and prevent the bones from grinding against each other. They allow motion between the vertebrae, bolstering the spine with added flexibility and strength.

The disks also play a role in shaping the spinal column, which rests at the body's vertical center of gravity. The spine crosses the gravity line in several places. The resulting curves provide much more stability and strength than a straight column could. To create curves, the disks subtly change shape, with portions narrowing so that the vertebrae do not sit directly on top of each other but rest at a slight angle. The overall effect is a series of alternating curves that run gracefully down the spine.

The bony cage of the chest

The spine is, in every sense, the backbone of the body and it directly or indirectly anchors all other bones. The ribs are joined to the thoracic vertebrae. Twelve pairs of resilient ribbons of bone spring from the sides of these vertebrae, their heads nestled in shallow facets. The upper seven pairs of ribs, the "true" ribs, arch around the body and attach to the sternum – the breastbone – via shafts of cartilage called costal cartilage.

The remaining five pairs of ribs are termed "false" because they join with the sternum indirectly. Costal cartilage links the upper three false ribs, which are connected, in turn, to the lowest pair of true ribs. The lower two pairs of false ribs "float," barely reaching around the sides of the body. They do not connect with the sternum at all, but terminate in cartilage connected to muscle of the abdominal wall.

The ribs demonstrate the essence of the skeleton's protective yet flexible framework. The thorax, the bony cage they form, hugs the respiratory organs it houses. Yet the lungs must expand to do their work. Accordingly, the ribs swing outward and upward to accommodate the taking in of air. The sternum aids this process by maintaining a flexible space between two of its three main portions. Although shaped like the blade of a sword, this plate of bone acts as a shield, sheltering the heart and lungs.

The protective skull

Like the sternum, many bones in the body contribute to the protection of vital organs. The skull is made of 22 bones united in protecting the brain and major sense organs. Large at birth, compared to the rest of the body, the infant's skull is compressible. Soft spots lie between large pieces of bone that have not yet grown together. Passage through the birth canal usually squeezes the skull and elongates it slightly, but the head reassumes its natural shape within a few days.

The skull's soft spots, the fontanelles, are fibrous membranes that eventually harden and grow together until the bones meet and mesh much like the teeth of a zipper. The largest of the six main fontanelles is a diamond-shaped region near the center of the top of the skull. It is the last one to close, a process completed when a child is about 18 months old.

The cranium is built of eight bones that interlock at immovable joints called sutures. The frontal bone curves around the skull to create the forehead and roof of the orbits, bony sockets holding the eyes. Two temporal bones, forming the sides and a portion of the skull's floor, contain canals that lead to the middle and inner ears.

In the middle ear, just behind the eardrum, lie the smallest bones in the body. Each ear has a set of three bones: the malleus, incus and stapes. The malleus picks up sound vibrations from the eardrum to which it is attached and passes them along to the incus and stapes. This relay method intensifies the vibrations; by the time they reach the stapes, their force has increased 20-fold.

The centerpiece of the cranium, the sphenoid bone, is bat-shaped with two large wings and two smaller "feet" extending from its central body. It forms much of the cranium's base and helps anchor most other cranial bones.

Fourteen bones combine to make the face. The main bones, the two maxillae, act together as a keystone for other facial

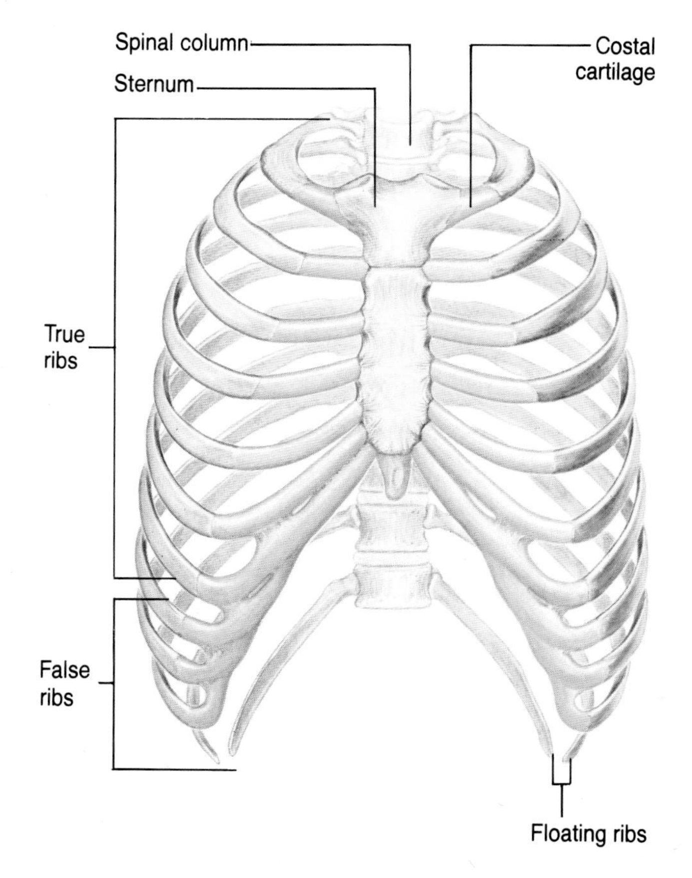

Twelve pairs of ribs encircle and protect the organs of the chest. Each pair is jointed by means of hollow facets to one of the thoracic vertebrae of the spine at the back, and all except the two lower pairs of short floating ribs curve round to join with the sternum, or breastbone, at the front. The upper seven pairs of true ribs meet the sternum directly, while the next three pairs fuse with each other and the seventh rib. The joints themselves are made of rubbery cartilage, which allows the rib cage to move slightly with the movements of breathing.

bones. The maxillae articulate, or link, with all but the lower jaw. They form the entire upper jaw, part of the orbits, the nasal cavity and the section of the roof of the mouth that contains the deep sockets for the upper teeth. At each side of the face, the ends of the maxillae swell forward and link with the small cheekbones.

The largest facial bone, and the only one in the skull with a movable joint, is the mandible, or lower jaw. It is a strong, curved bar of bone that holds the lower teeth. Arms of bone reach up its sides to articulate with the skull's temporal bones, creating the jaw's hinge joint. Tucked beneath the mandible is the U-shaped hyoid, the only bone in the body that does not articulate with any other. It anchors muscles, particularly those of the tongue.

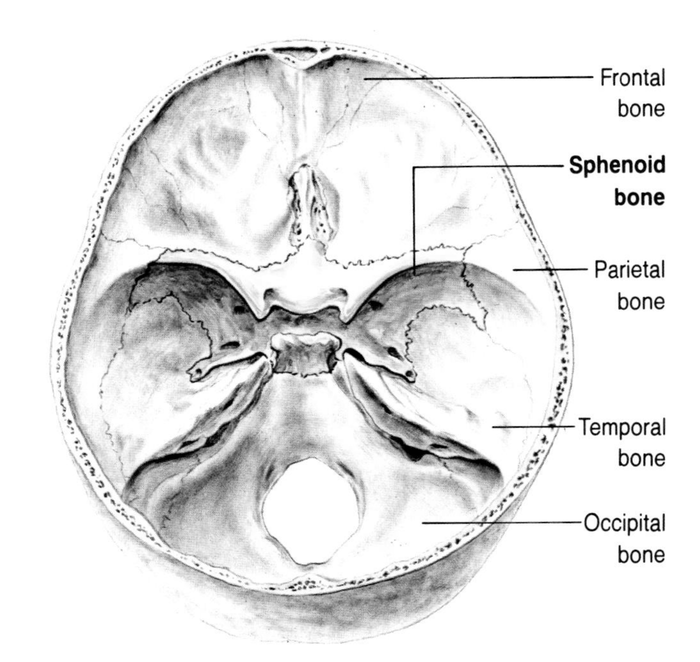

The skull is not a single bone, but is made up of several curved panels linked jigsaw fashion by immovable joints. The key piece in the jigsaw is the bat-shaped sphenoid bone, which locks the others together.

Legs and arms

The pelvic girdle is the bony construction that connects the legs to the rest of the skeleton. Charged with bearing the weight of much of the body, helping to hold it upright and move it forward, the pelvic girdle consists of two heavy, substantial hip bones joined by the sacrum at the back.

The leg bones, the body's supports, are much stronger than the bones of the arm, but are capable of much less movement. The largest bone in the body is the femur, or thigh bone. Long and elegantly proportioned, it is also the strongest and heaviest. At the top of its shaft, its smooth rounded surface snugly articulates with a socket in the outside of the hip bone.

The lower leg's main bone, the tibia, lies prominently at the front of the leg, where it is known as the shin. The lower leg's other bone, the fibula, resembles the jointed side of a safety pin and articulates with the tibia at its upper and lower ends. This slender, twisted bone acts mainly to anchor leg muscles. The two lumpy ankle bones are actually the prominent ends of each lower leg bone. The tibia surfaces at the inside of the ankle, the fibula on the outside.

The humerus is the long single bone in the upper arm. Like the femur, it has a smoothly rounded head which fits into a socket joint. The socket in the shoulder blade, or scapula, is shallow and much smaller than the head of the humerus. This arrangement, which allows the arm to move at many angles away from the body but with little stability, makes the joint the most prone to dislocation.

The opposite end of the humerus meets two bones of the forearm, the ulna and the radius, to create the elbow joint. The predominant bone in the elbow is the topmost part of the ulna. The head of the radius is buttonlike, rounded at the edges and flat on top. This flat top glides against the surface of the humerus, and the rounded edges swivel against the ulna. Such a design greatly increases the range of movement of the arms.

The hand's great flexibility begins in the wrist, where eight carpal bones are arranged in two rows of four each. They look somewhat like small pebbles, yet their placement and articulation are precise. Beyond them, five metacarpal bones fan across the hand. Although largely alike they differ in length. Each of these bones articulates with one of the five digits. In each hand, there are 14 bones called phalanges, three in each of the fingers and two in the thumb.

The thumb is the most specialized of the digits and is the most important contributor to the hand's dexterity. This is due to the thumb's great individual movement. Its metacarpal bone is placed lower in the hand than those of the other digits and has a special joint, which links it to the wrist and allows the thumb to rotate freely across the surface of the palm. This permits the important hand movement known as "opposition," in which the thumb reaches across the palm to meet the other digits, fingertip to fingertip. Without this ability, the hand would be but a claw or an ungainly pair of forceps, with its precise movement and powerful grip severely limited.

The foot is hugely resilient to the great pressures and strains it encounters and sacrifices dexterity to provide stability to the body. The ankle has only one bone less than the wrist, but the arrangement of its seven tarsal bones – different from that of the carpal bones in the wrist – increases their weight-bearing efficiency. When the body is standing, its entire weight funnels down through one of the tarsals, the talus. There, the weight divides evenly. Half travels to the calcaneus, the heel bone, while the other half channels to the five remaining tarsal bones and into the bones that make up the arch of the foot.

Slender metatarsal bones join in the central portion of the foot itself. These five bones articulate with the ankle's tarsals above and the bones of the toes below. They form the arch, the most important structural part of the foot. The foot disperses body weight through three arches – two over its length, one across its width. The arches also provide leverage for locomotion.

Toes have the same number of bones as the fingers and have the same anatomical name, phalanges. They are also distributed similarly. The big toe has two bones and the others have three. Yet the bones of the toes have little in common with those of the fingers. Their stout shape enables them to bear weight. In a singular act of force, these bones help the foot push off with each stride as you walk or run. Firmly gripping the ground, they help us balance as we stand.

Compact bone – one of the hardest tissues in the human body – is not completely solid, but is honeycombed with channels and spaces. Like the growth rings of a tree, bunches of columnar Haversian systems run the length of the bone. Each "ring" gets its strength and hardness from mineral crystals embedded in microscopic strands of the protein collagen. Cavities called lacunae contain osteocytes, which manufacture bone cells, and a central Haversian canal houses blood vessels carrying essential nutrients. The bone's outer surface is covered by the tough "skin" of the periosteum.

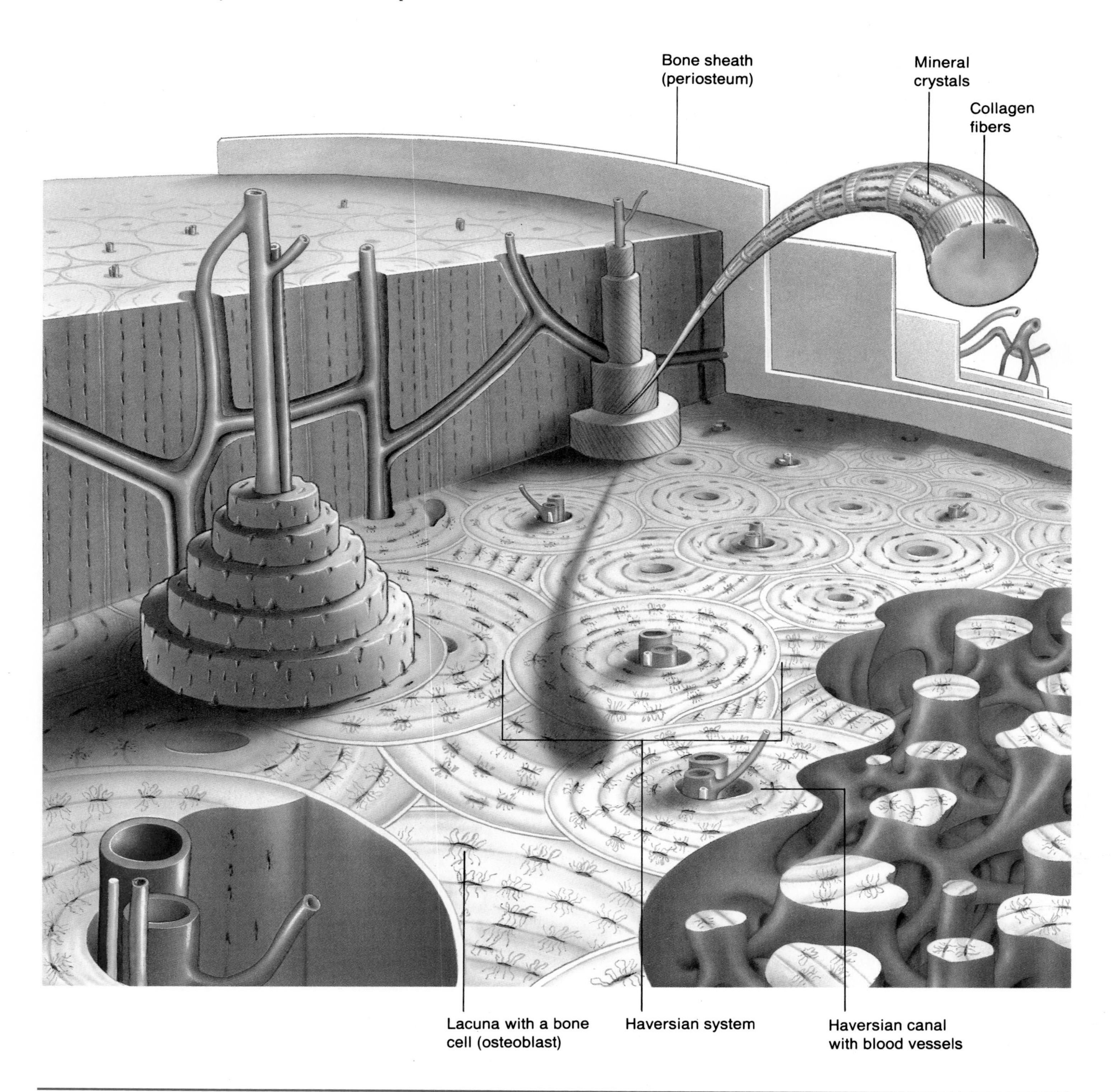

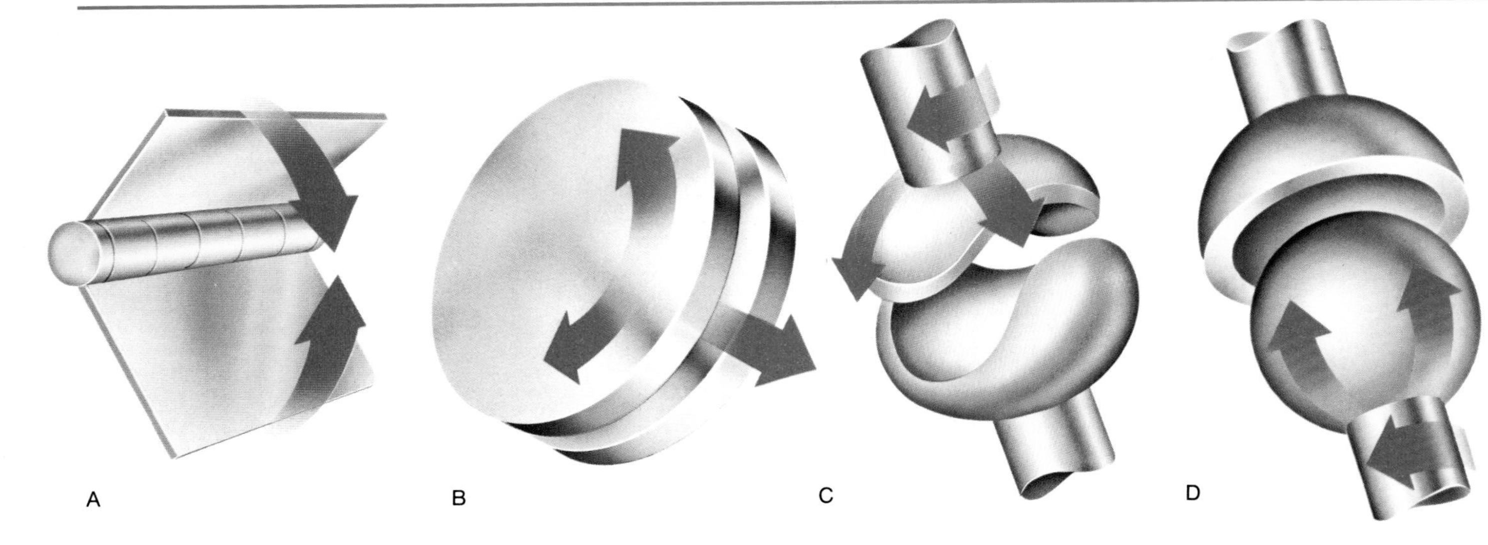

The skeleton incorporates several kinds of movable joints. The bearing surfaces are usually made smooth by a layer of slippery cartilage, which reduces friction. In larger joints lubrication is increased by synovial fluid. The bones are held in position by ligaments, which limit movements to the directions dictated by the joint's construction and the activating muscles' ability. The simplest type of joint is the hinge (A), as found in the elbows and the joints of the fingers and toes. It allows movement in only one direction. Gliding joints (B) permit a wider range of mostly sideways movement and occur in the wrists, ankles and spine. A saddle joint (C) is yet more versatile, and gives the human thumb its unique ability to "cross over" the palm. The widest range of movement is provided by a ball-and-socket joint (D), in shoulders and hips.

The strength of bone

By almost any measure, bone is among the strongest materials devised by nature. One cubic inch can withstand loads of at least 19,000 pounds – about the weight of five ordinary pickup trucks. This is roughly four times the strength of concrete. Indeed, bone's resistance to loads equals that of aluminum.

Bone seems even more remarkable considering its lightness. The skeleton accounts for only 14 percent of total body weight, or about 20 pounds. Steel bars of comparable size would weigh four or fives times as much. Bone, ounce for ounce, is actually stronger than steel and reinforced concrete.

Unlike a steel shaft, which is useless when broken, bone is living tissue that repairs itself. Bone derives its strength by weaving protein and mineral into a resilient fabric. Each component enhances the strengths of the other. Nearly two-thirds of bone consists of various salts, mainly complex compounds of calcium and phosphorus. These salts form rod-shaped crystals, which lend bone hardness and rigidity. The remainder of bone is composed of collagen, an elastic protein which is converted to glue when bones are boiled. Collagen fibers are studded with the mineral crystals and wind around themselves like fibers in a rope. Remove the minerals and bone would become so rubbery that it could be tied in a knot, like a garden hose. Without collagen, it would be as brittle as glass.

Within bone tissue, the composite of collagen and mineral crystal forms elaborate structures. Compact bone is concentrated in long bones such as the femur. A cross-section of compact bone reveals an intricate pattern of concentric circles bunched together. Each self-contained cluster of circles, called a Haversian system, resembles the cut end of a tree trunk. Corresponding to tree rings are lamellae, layers of bony tissue made of crystal-studded collagen. At the heart of large systems lie Haversian canals, pipelines containing blood vessels, lymph vessels, nerve filaments and delicate connective tissue.

Within the bone are small cavities in the lamellae called lacunae. A single cubic inch of compact bone contains more than four million lacunae. Each contains an osteocyte, or bone cell, which maintains mature bone tissue and supplies nutrients through minute channels. Radiating from each Haversian canal, the channels direct nutrients and remove waste from bone cells nestled in the lacunae.

The periosteum is a thin membrane which envelops the surface of compact bones like skin. It is laced with blood vessels, bringing nourishment to bones. If arteries that run through Haversian canals are blocked, vessels from the membrane can still supply local bone cells. Thin strands from the periosteum penetrate the bone and unite the two tissues. Other fibers entwine with tendons, securely anchoring muscle to bone.

Midway along the femur, the bone becomes hollow, yet does not lose mechanical strength. The resultant saving in weight is considerable. If the femur was a solid shaft, it would weight 25 percent more. Nature has efficiently employed such hollow bones by filling them with marrow. In the center of ribs, vertebrae, pelvic and skull bones lies red marrow, one of the body's most biologically active tissues. Each minute, red marrow delivers millions of red cells, white cells and platelets into the bloodstream.

Long bones, such as the femur, are filled with yellow marrow, consisting mainly of fat cells. Yellow marrow is a strategic store of energy reserves for occasions when the body's fat becomes depleted. Likewise, when blood becomes anemic through a lack of red blood cells, yellow marrow is quickly transformed into red marrow for the manufacture of red cells.

Where bone meets bone

The need for strength makes bones rigid. If the skeleton were cast as one solid bone, movement would be impossible. In all vertebrates, nature has solved this problem by dividing the skeleton into many bones and creating joints where bones intersect. Joints come in an array of designs, each custom-built for the limb it serves. Lashed together by fibers of collagen called ligaments, and continuously lubricated to offset friction, joints permit movement.

Some joints, such as those binding the skull's protective plates of bone, allow no movement. Others permit only limited movement: the joints between the spinal vertebrae are united by disks of cartilage and other tissues, which allow some movement in several directions.

Most joints offer far greater play of movement. Connections called synovial joints are sturdy enough to hold the skeleton together while permitting a range of movements. The structure of synovial joints helps transmit power and motion between the bones. The ends of bones at synovial joints are coated with a tough articular cartilage, which reduces friction and cushions the joint against jolts. Lying between the bones is a narrow space known as the joint cavity, which offers freedom of movement. Ligaments bind the bones, prevent dislocations and limit the joint's mobility.

Synovial joints are fulcrums, the bones they connect are levers and the muscles attached to them apply force. The joint between the skull and the atlas vertebrae of the spine is the fulcrum across which muscles lift the head. Nodding one's head would be impossible without it. When you lift a book, the elbow joint is the fulcrum across which the biceps muscle performs the work.

A sleevelike extension of the periosteum envelops every synovial joint. This joint capsule permits movement within the joint, but it also has sufficient tensile strength to prevent dislocation. Lining the joint capsule, a fine membrane secretes the lubricant synovial fluid. The fluid oils the joint while cells within it remove microorganisms and debris.

Bursae are closed sacs with a lining not unlike that of synovial joints. Bursal walls secrete a substance similar to, but less viscous than, synovial fluid. The sacs help one musculoskeletal structure to glide over another, for example skin or tendon over bone. The knee has at least 12 such sacs.

Each type of synovial joint is precisely designed for a specific movement. The most freely moving are ball-and-socket joints in which the hemispherical head of one bone lodges in the hollow cavity of another. In the shoulder joint, the humerus bone of the upper arm fits into the socket of the shoulder blade. Because the socket is shallow and the joint loose, the shoulder is the body's most mobile joint.

The ball-and-socket of the hip joint is less mobile than the shoulder, but more stable. The ball of the femur's head fits tightly into a deep socket in the hip bone. A rim of cartilage lining the socket helps grip the femur firmly; the ligament binding the two bones is among the strongest in the human body.

The saddle joint connecting thumb to hand permits movement in two directions. Hinge joints, such as elbow and finger joints, are less mobile and allow movement in only one direction, like the hinge on a door. A pivot joint near the top of the spine allows the head to swivel and bend. Another pivot joint, in the forearm, allows the wrist to twist.

The hinge joint of the knee, the body's largest joint, is unusual because it can swivel on its axis, allowing the foot to turn from side to side. Thus the knee is constantly rolling and gliding during walking. Two crescent-shaped wedges of cartilage called menisci permit the joint to glide easily because of their slippery surfaces. Menisci are also shock absorbers, cushioning the blows the joint endures in everyday use. They are the knee's weak link; torn menisci account for 90 percent of all knee surgery.

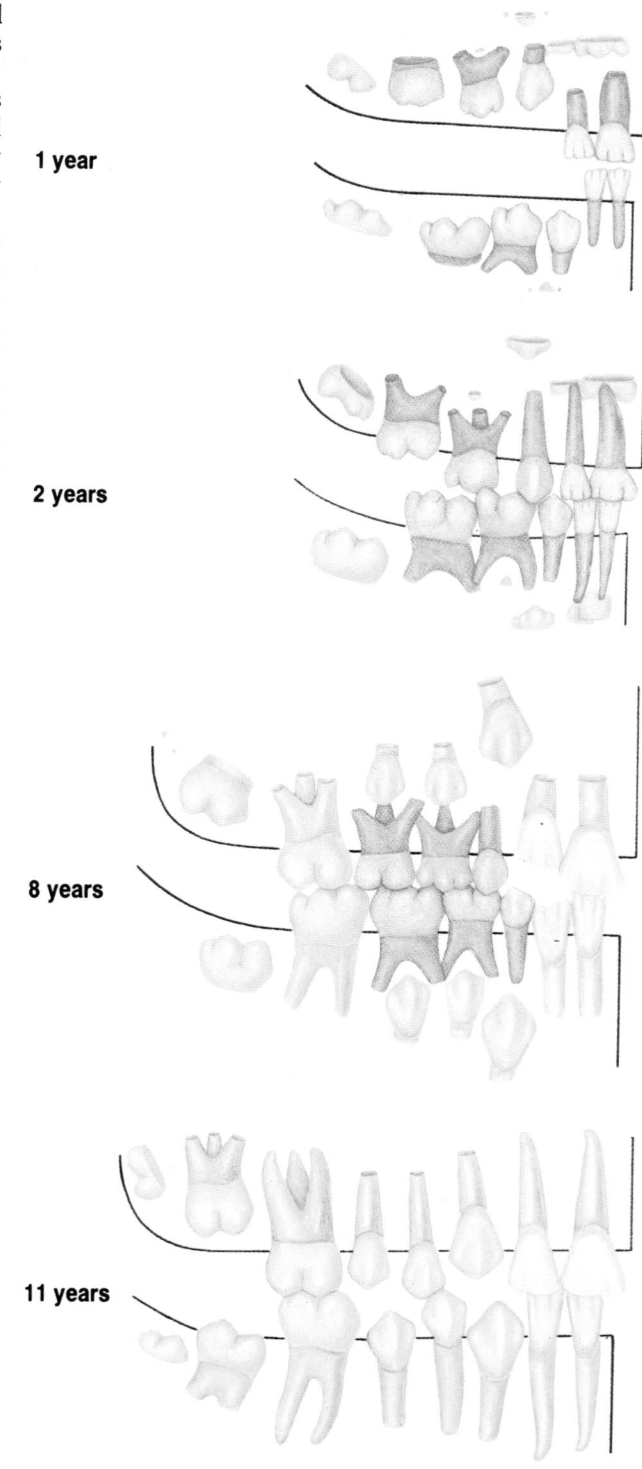

Teeth are even harder than bones, but like them are composed mainly of calcium-containing minerals. The diagram shows the order of eruption of a child's 20 milk, or first, teeth (blue), which are followed by the 32 permanent, or second, teeth (brown). The third molars, commonly called wisdom teeth, may never erupt. The illustration represents the upper and lower halves of the right side of the jaw.

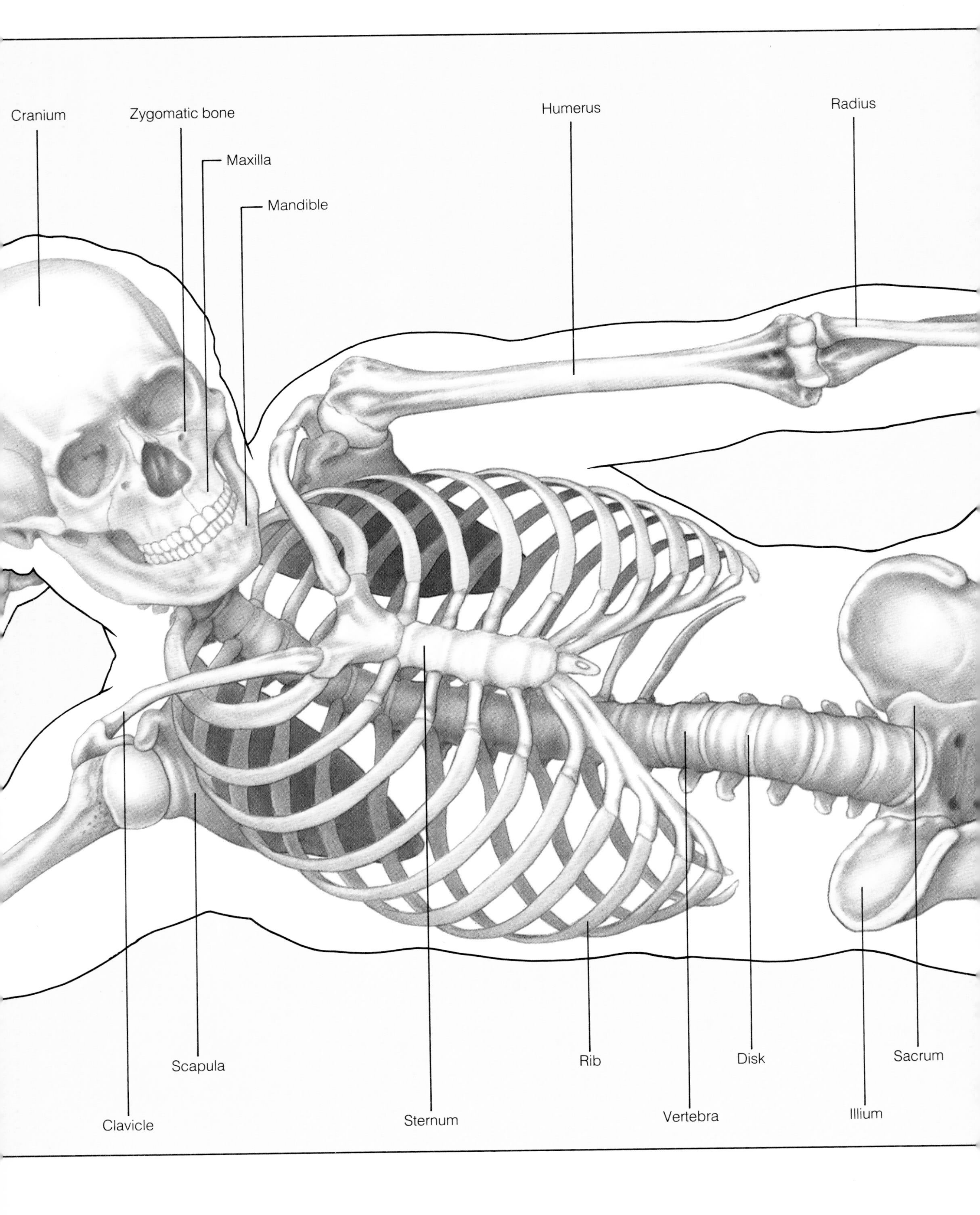

Cranium
Zygomatic bone
Maxilla
Mandible
Humerus
Radius
Scapula
Clavicle
Sternum
Rib
Vertebra
Disk
Illium
Sacrum

Back of hand

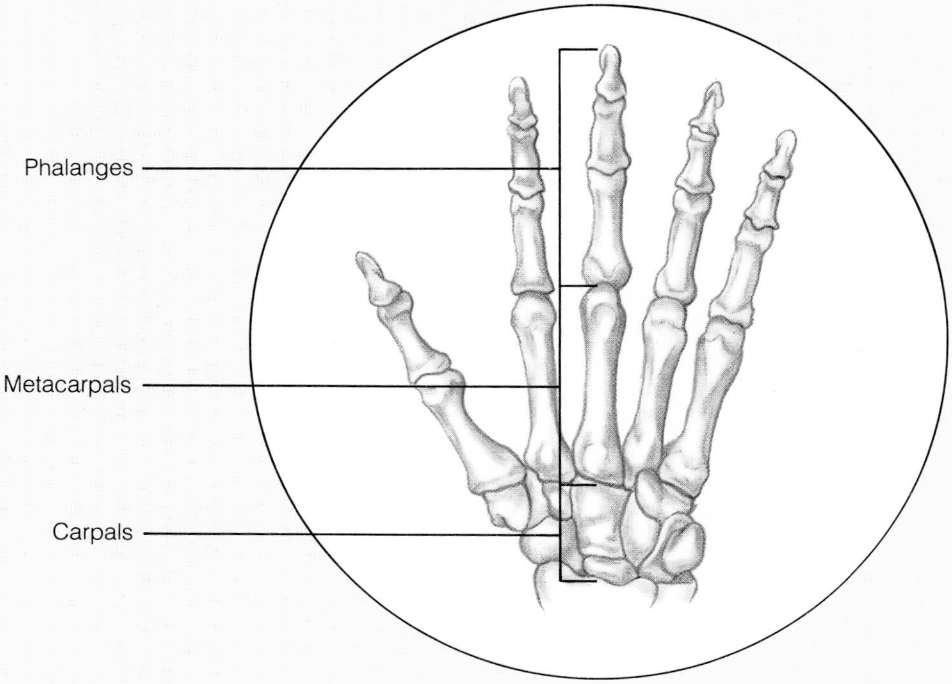

Phalanges

Metacarpals

Carpals

Front of elbow

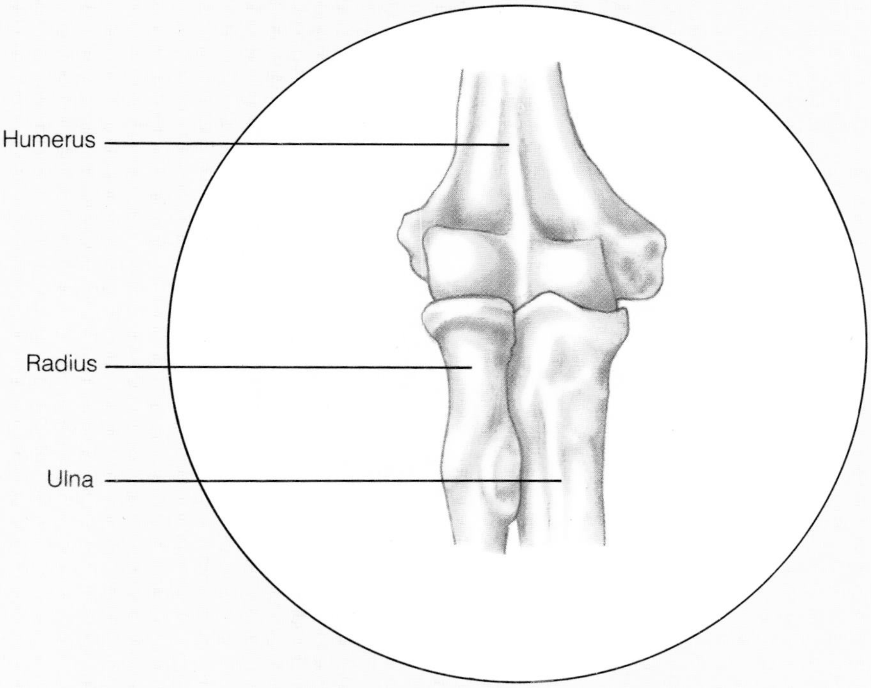

Humerus

Radius

Ulna

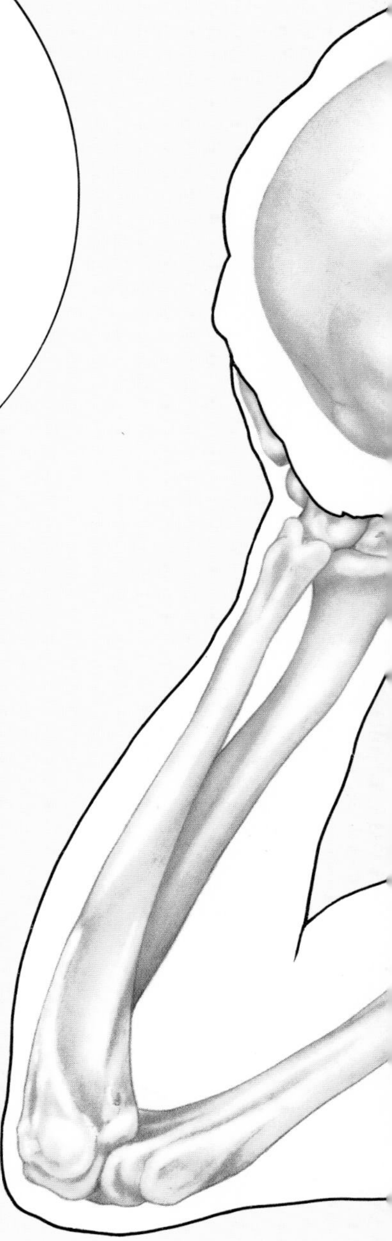

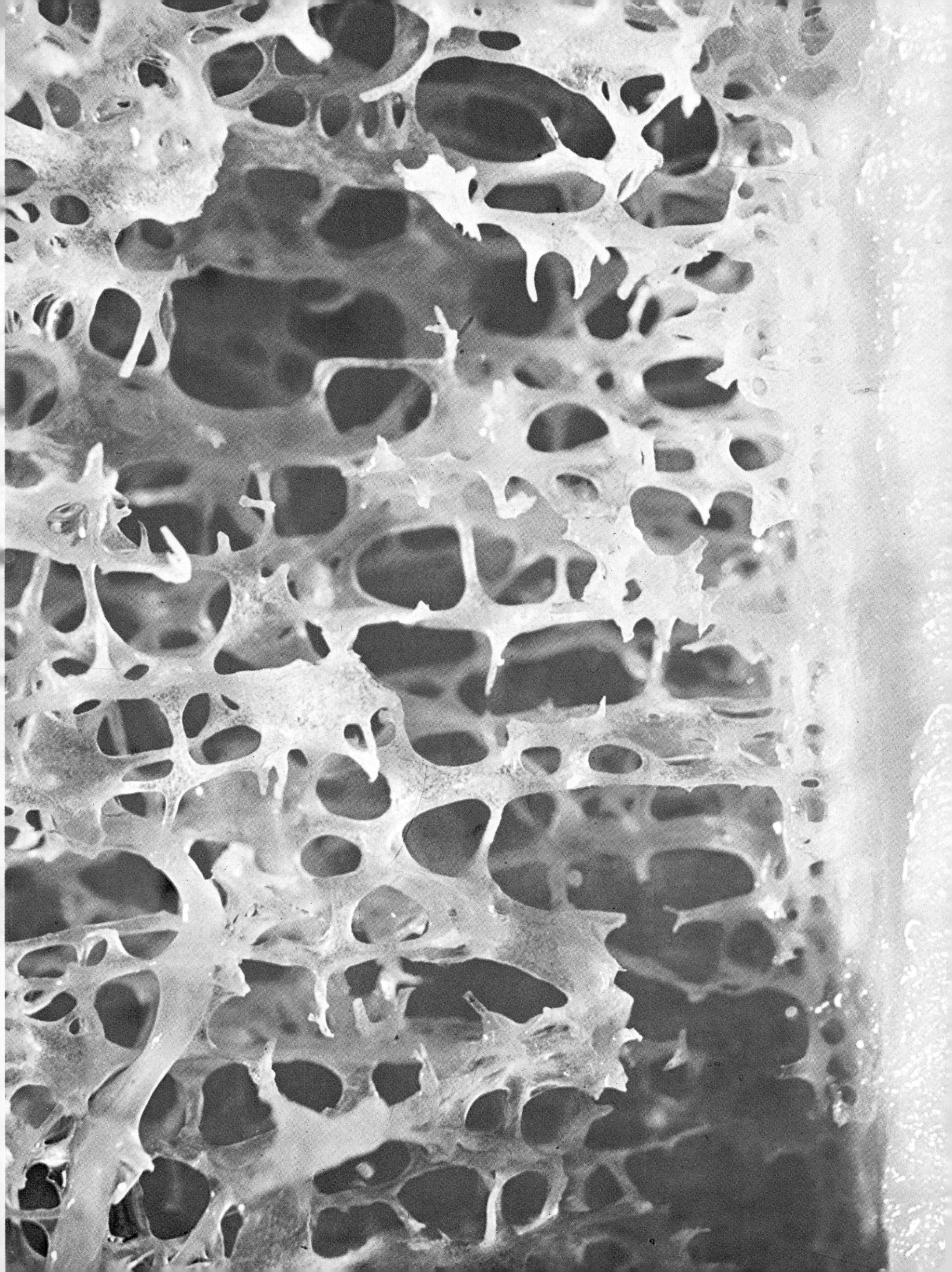

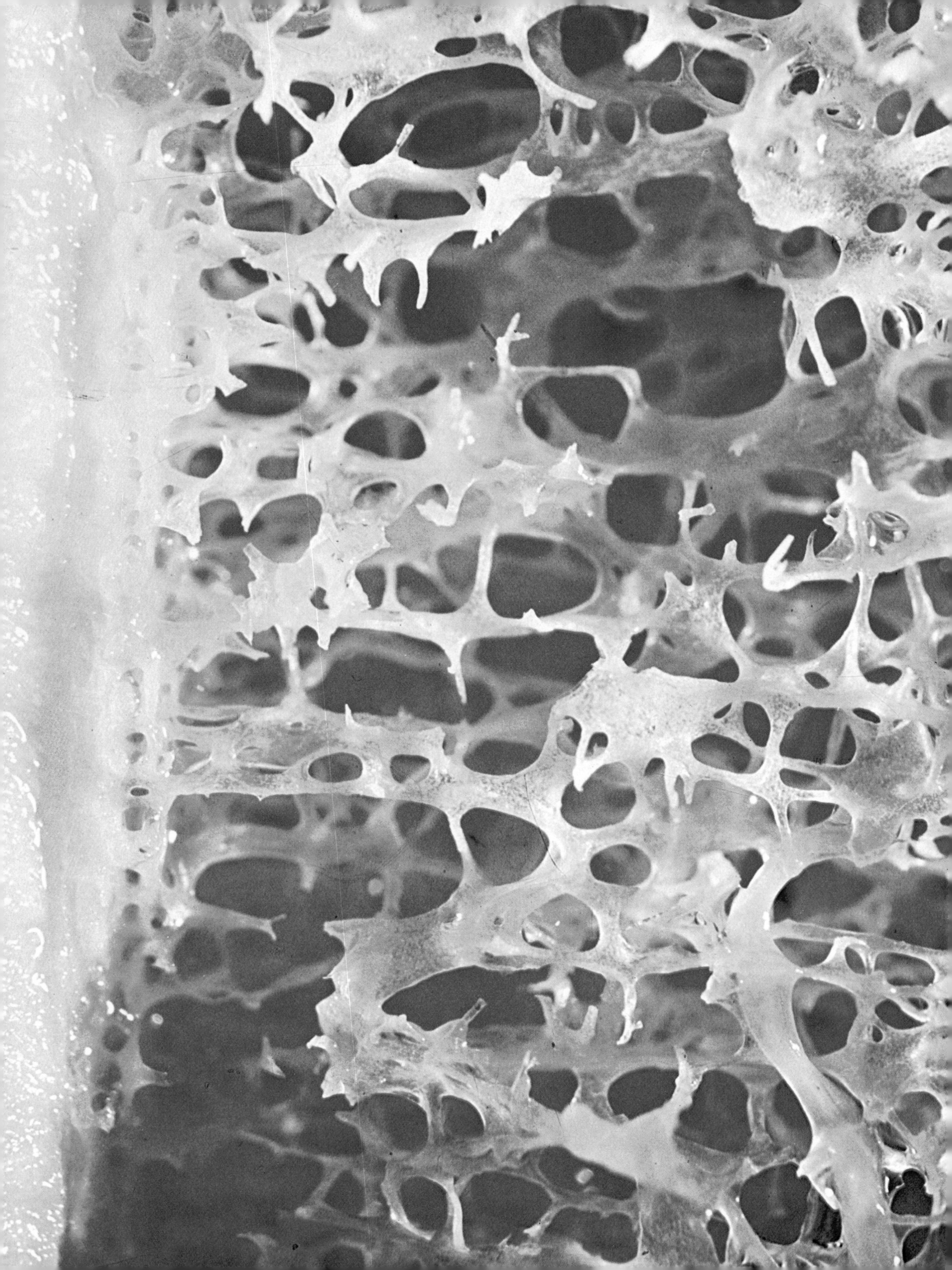

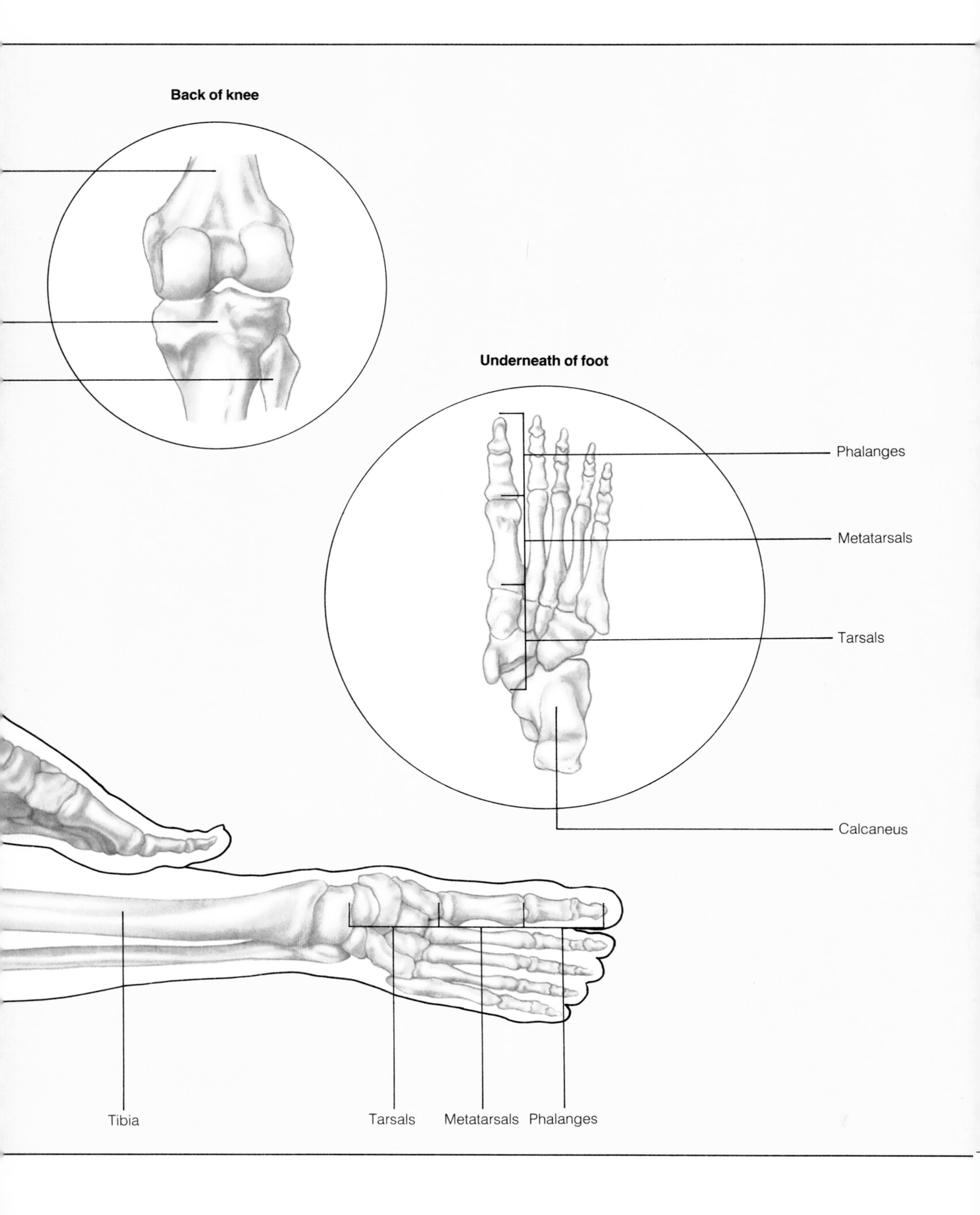

Back of knee
Underneath of foot
Phalanges
Metatarsals
Tarsals
Calcaneus
Tibia
Tarsals
Metatarsals
Phalanges

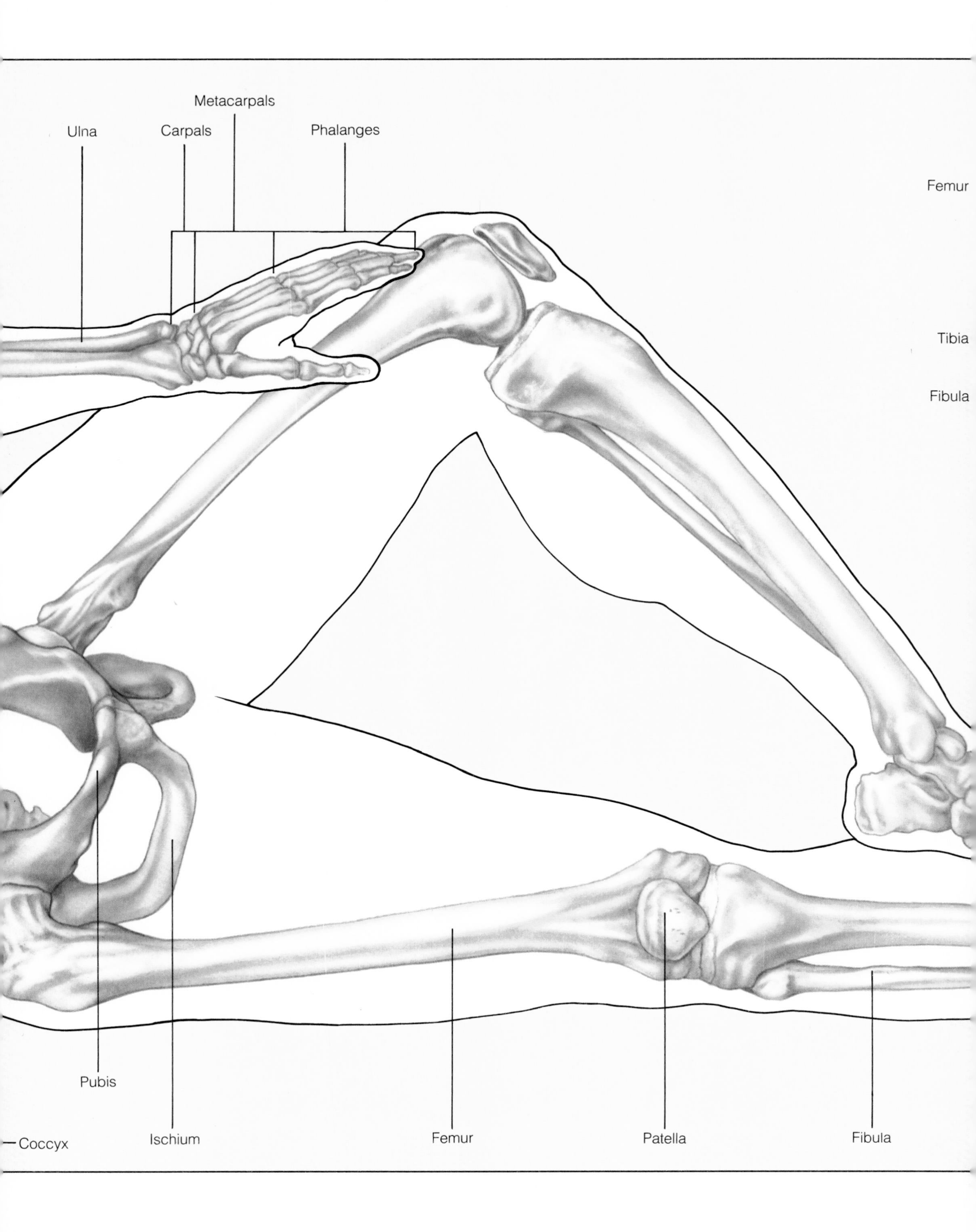
Metacarpals
Ulna
Carpals
Phalanges
Femur
Tibia
Fibula
Pubis
Coccyx
Ischium
Femur
Patella
Fibula

Hollow but strong describes the spongy structure that forms the core of many of the skeleton's long bones *(see previous page)*. An interconnected lattice of thorny spicules provides strength without the penalty of too much weight.

Weight-bearing joints, such as those of the hips, knees and spine, rely on their cartilage linings to operate smoothly. Breakdown of the cartilage, as occurs in osteoarthritis, leads to friction, erosion of the joint, and pain. The joint becomes inflamed and stiff, and in severe cases may seize up altogether and cause crippling disability. Osteoarthritis in the joints of the hand results in nodular swelling of the knuckles.

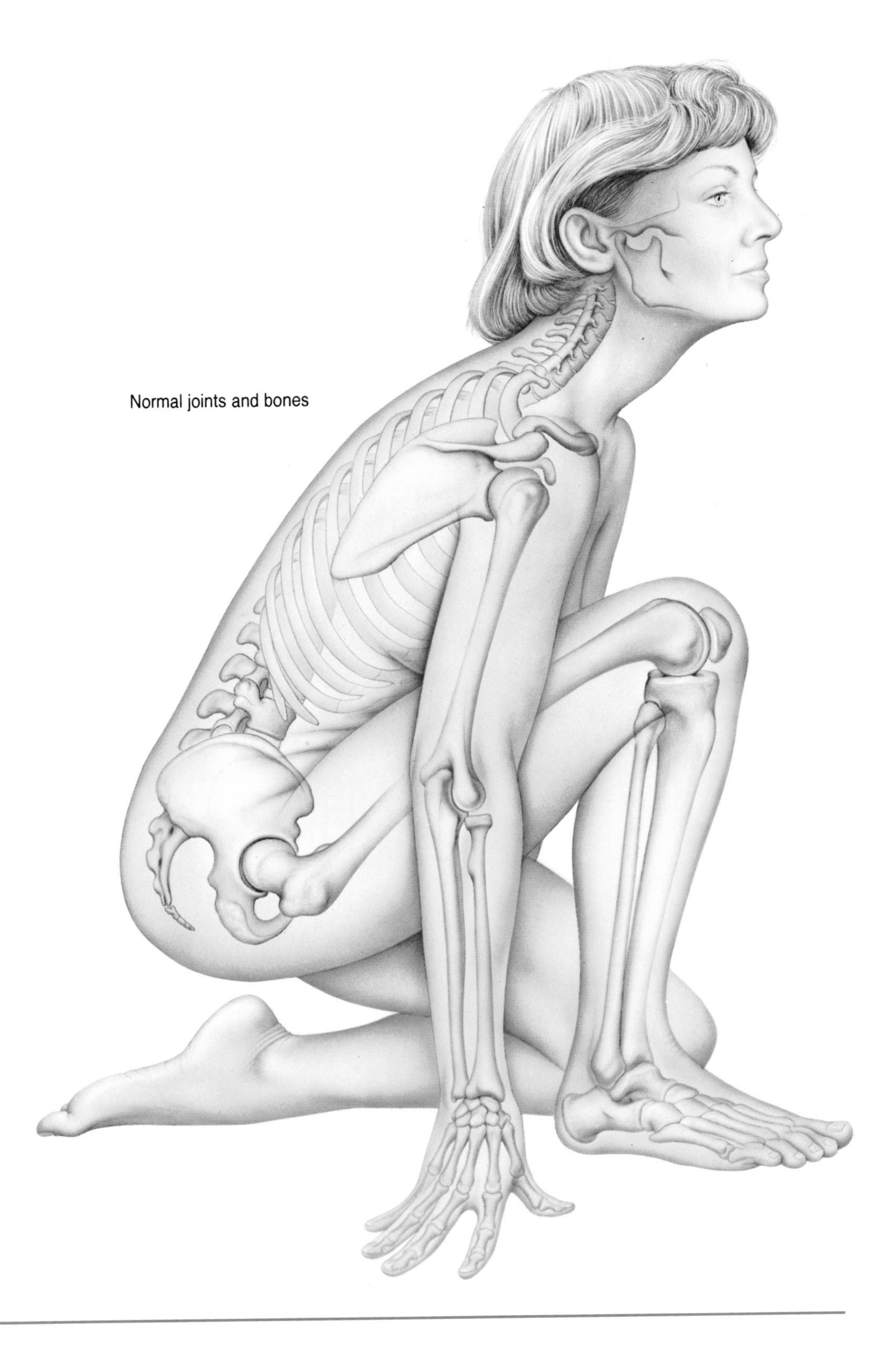

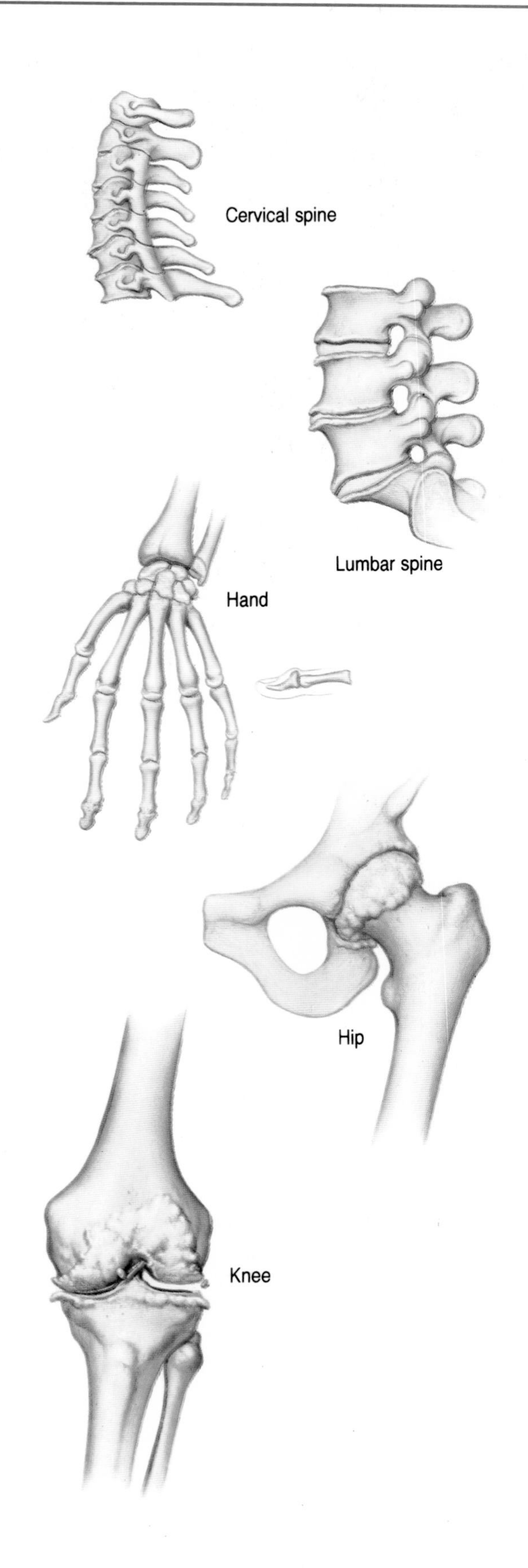

Growth and renewal of bone

Bone grows by slow accretion, adding layer upon layer until adulthood. It changes constantly, growing and repairing itself like the rest of the body. Bones develop and harden through the process of ossification, which begins around the fifth week of pregnancy. Most of the developing bones in a human fetus are first evident in a flexible skeleton of cartilage, a tough gristlelike substance the color of milky glass. For many people, ossification continues until about the age of 25, by which time bone has replaced cartilage, ending further growth.

Skeletal growth is regulated by secretions from the pituitary, thyroid, adrenal and sex glands. Proper nutrition strengthens bone, particularly when adequate amounts of vitamins A, C and D are present. Pressure on the ends of bones, whether from cartilage, muscular development or injury, can also influence growth.

The most prevalent reminder of the skeleton's ultimate tendency to fail mechanically is arthritis, a disease with a long legacy. Arthritis is broadly defined as the chronic inflammation of the joints and their subsequent degeneration. More than 100 varieties of arthritis are known, but these can be grouped into three categories: gout, osteoarthritis and rheumatoid arthritis.

Gout is the best understood and most treatable of the three. It is brought on by a rich diet high in purines, compounds that yield uric acid when metabolized by the body. This acid crystalizes if it accumulates in the blood in sufficient quantity. The crystals then migrate to joints, especially in the big toe, where they cause inflammation and pain.

Osteoarthritis is believed to be essentially mechanical in nature, because it comes about mainly from a lifetime's wear and tear. The disease turns up mainly in the load-bearing joints – hips, knees and spine – and in the hands. Curiously it afflicts women twice as often as men. It occurs when cartilage, serving as a shock absorber between bones, begins to break down. Bone starts to rub against bone, causing severe pain and reduced mobility. The regenerative chemistry of bone tissue alters, unable to prevent the soft edges of the worn cartilage hardening into bone spurs.

In rheumatoid arthritis, the synovial membrane lining a joint becomes inflamed. Malfunctioning white blood cells and joint tissue cells form a deposit called pannus on the joint surface. Enzymes released by the inflammation mysteriously turn on the joint and start to digest it. Scar tissue forms between adjacent bones and hardens into bone, which fuses the joint. Rheumatoid arthritis may be triggered by a volatile mix of a yet-to-be identified virus and a faulty immune response. In this scenario, white blood cells alerted by the infection mistake the joint for the enemy and attack it.

Bone begets bone

When a cut or a burn heals, the missing sensitive, hair-bearing skin is replaced by inflexible scar tissue. Containing no pigment and no nerve endings, the new skin looks and feels dead. But when the ends of a fractured bone grow together again, the repair is made with living bone. This ability is crucial to recovery after injury, and has been regarded as essential to the survival of the human race.

The key to bone regeneration is a substance known as bone morphogenic protein (BMP). It induces fibroblasts – scarlike bone cells in a healing fracture – to function once more as cartilage-producing chondroblasts. And once cartilage has

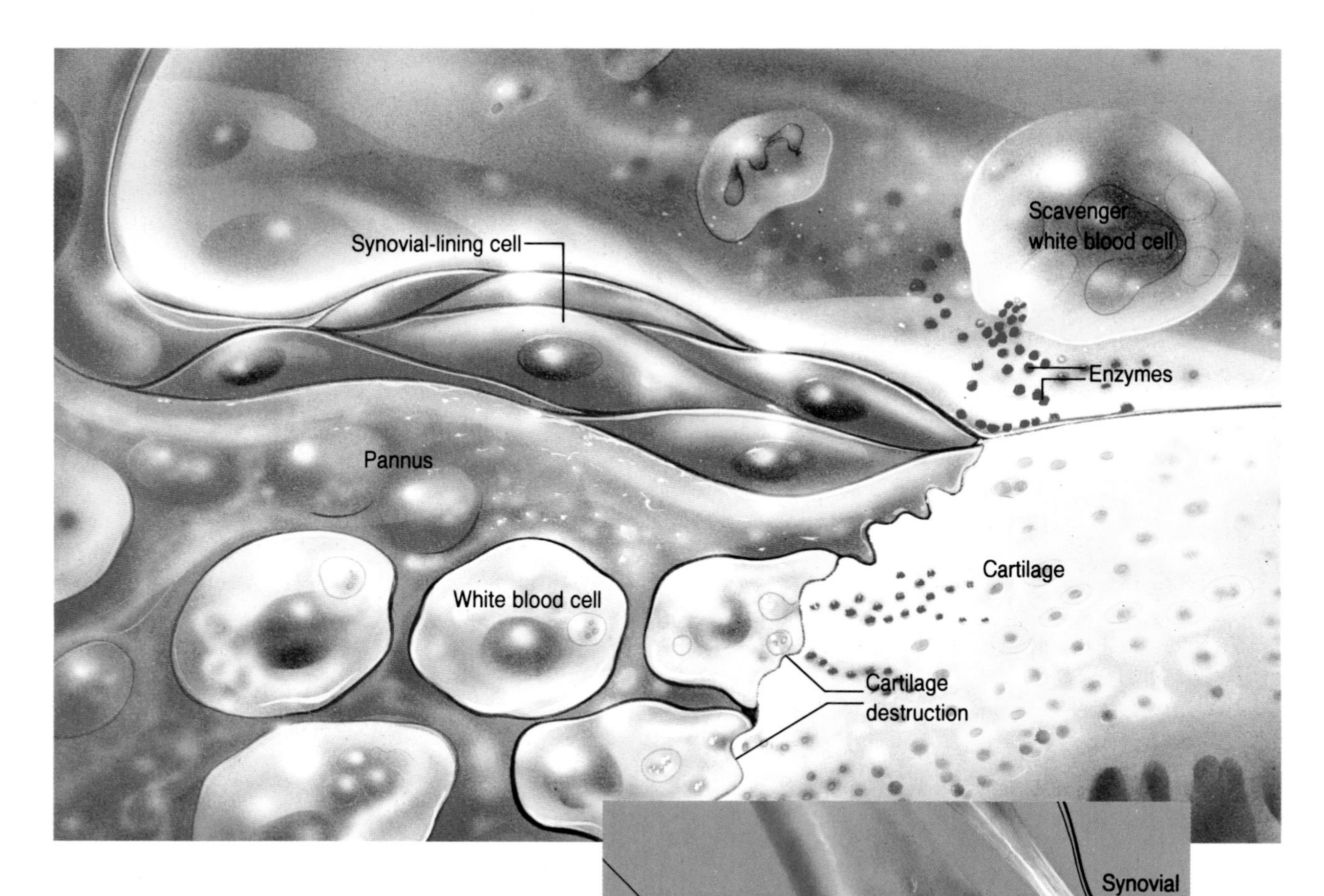

formed, it can mineralize into proper bone. Surgeons are conducting clinical trials that use BMP to encourage healing of fractures or to help "glue" transplanted bone into position. It is believed that it may be of particular value to surgeons rebuilding the bones of a patient who has traumatic or congenital malformation of the face.

Bone banks

Repairing joints, implanting artificial joint replacements and patching together the jaw of an accident victim are some of the marvels now performed by orthopedic surgeons. To do so they often need bone grafts, or transplants. Ideally such pieces of bone are autografts, taken from elsewhere in the patient's body. But autografts have limitations and are impossible in children (who cannot afford to "lose" any part of their growing bones).

The alternative is an allograft, using a piece of bone from a donor. To prevent problems with rejection or reabsorption, the graft is treated so that the recipient's body does not recognize it as "foreign." Pieces of donor bone, taken from cadavers, are freeze-dried and stored at room temperature, or deep-frozen and stored in a bone bank.

Seeing inside the knee

Every time you walk or turn or kneel, your knee has to glide, rotate or bend in response to the stresses that motion puts on it. The knee is probably the most ingenious joint in the skeleton, but it is also one of the most vulnerable.

Stability is provided by four major ligaments binding the joint and 13 muscles supporting it. With every movement, the muscles and ligaments have to contract, relax, twist and turn.

Destruction of cartilage in an arthritic joint is initiated by the build-up of a deposit known as pannus. In it lurk leukocytes (white blood cells) and cellular debris. Other white cells called phagocytes, the scavengers of the body's immune system, arrive at the scene to mop up the remains. But they also release enzymes that attack the healthy tissues of the joint lining *(upper diagram).* A commonly affected joint is the knee *(lower diagram),* in which eroded cartilage first becomes scar tissue and then rough and unyielding bone. The progress of rheumatoid arthritis is now well known, but despite intensive research the cause remains a mystery.

Two crescent-shaped pieces of cartilage, called menisci, lie between the thigh bone (femur) and shin bone (tibia). Their function is to lubricate the joint, but because they have a poor blood supply they have little capacity for repairing themselves if they are damaged. A torn cartilage in the knee – the dread of people who play sport professionally – usually has to be removed surgically. Left in place, it begins to wear and may eventually give rise to arthritis.

Examination of the knee joint is termed arthroscopy. A 9-inch tube containing lenses and carrying its own optic-fiber illumination, an arthroscope can be passed through a small quarter-inch incision into the joint. A surgeon can use the same instrument to introduce tiny scalpels to cut away the torn part of the cartilage, leaving the rest intact and so minimizing the chances of subsequent osteoarthritis. Replacement ligaments can also be grafted into position, to stabilize a weak knee joint.

Delicate disk surgery

Perhaps nowhere in the body is cartilage more integral to functioning of the skeleton than in the spine. From each of the 33 vertebrae, except for the top two, a pair of nerves branches and extends through the body, transmitting and receiving a range of sensations which they relay to the spinal cord and brain. Three membranous layers of tissue sheathe the spinal cord. It, in turn, threads through the rock-hard protective casing of the bony spinal column.

Between the vertebrae lie the disks, rubbery cartilaginous material that absorbs and disperses shock waves rolling up the spine. Like any other anatomical feature, disks vary from one person to the next. In some people, disks are more dense and better able to absorb shock.

A cross-sectional cut of a disk resembles a slice of onion. At its center lies a jellylike substance surrounded by concentric rings of fibrous tissue. Undue stress on the disk, coupled with a degenerating outer ring, can force the gel to migrate outward, cracking successive rings in its path. If the stress on the disk is severe enough, the gel eventually breaks through the outer ring and pinches the nerve root leading from the vertebra. This is a "slipped" or herniated disk, and is accompanied by pain, numbness or muscle weakness in the extremities of the limbs served by the affected nerve.

Lacking exercise, the network of muscles and ligaments supporting the edifice of the spine falls into disuse and weakens. Without support, the disks become more prone to injury when stressed. Most particularly, such injury centers in the lower back – the lumbar segment of the spine – which bears the heaviest loads. Leaning forward to lift a weight, a man increases the load on his lumbar disks nearly 100 percent. If at the same time he rotates his spine, he increases the load by 400 per cent. Because of such conditions, back pain – particularly lower back pain, or lumbago – is one of the most widespread ailments to afflict members of modern, affluent societies.

When a disk does slip, it needs urgent medical attention. Traction to stretch the spine slightly may relieve the pressure and the pain, but often surgery is necessary. In a standard laminectomy operation, pressure on the nerve is cured by removing most of the offending disk, which is then gone for ever. As a result, the affected vertebrae touch each other and may eventually fuse together. In microlumbar diskectomy, the surgeon delicately removes only the bulge of the herniated disk, leaving the rest of the cartilage intact.

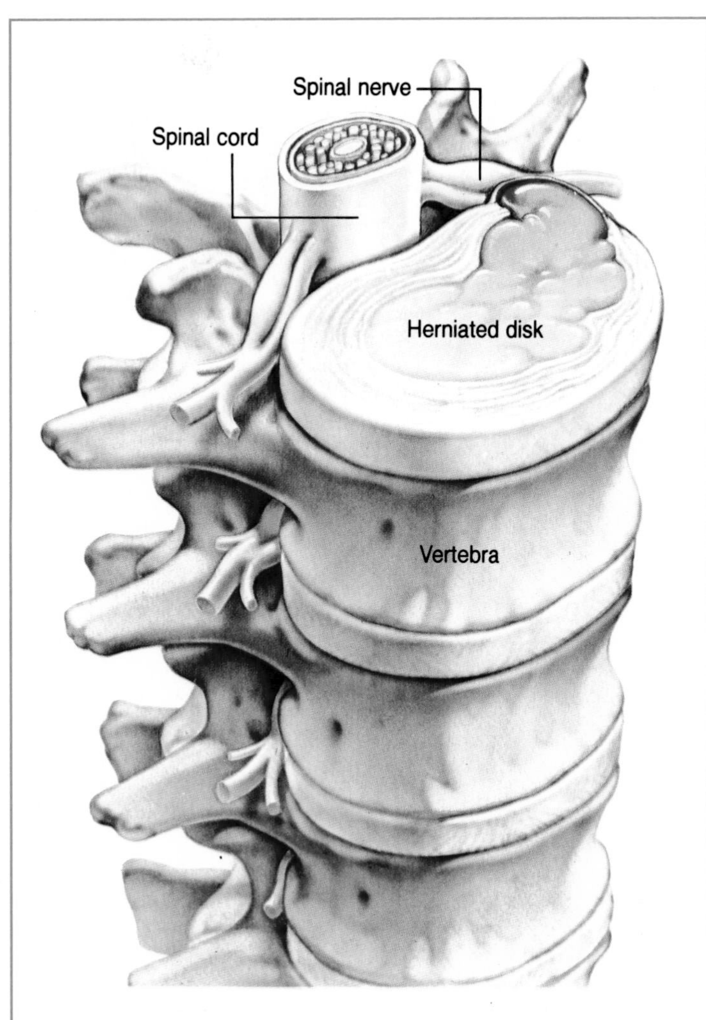

The spinal cord and the nerves emerging from it are protected by the bony canal formed by the stack of vertebrae in the spinal column. Flexibility and shock-proofing of the spine are provided by the spongy cartilage of the intervertebral disks. But if the spine is overstrained or subjected to sudden shock, the jellylike center of a disk may bulge out on one side, resulting in a herniated disk – commonly known as a slipped disk. A bulge pressing on one of the emerging nerves causes pain or numbness in the part of the body served by the nerve.

CHAPTER 3

The Muscles

All motion follows mechanical principles, and the motion of the human body is no exception. Like the machines made by humankind, the body is a set of levers whose movements copy the geometry of classical mechanics. These levers are powered by muscles, the elegant, efficient body components whose actions are as simple as their character is complex. The day-to-day operation of muscles, which we all take for granted, belies a division of labor so elaborate that scientists have only just begun to unravel it.

Each of more than 600 muscles is served by nerves. Linking muscles to the brain and spinal cord, a network of nerve circuits carries signals that direct the ebb and flow of muscular energy. Many muscles must work together to perform even the simplest tasks. And much muscular activity occurs outside the realm of the conscious mind as the body, through the neuromuscular network, manages its own motion.

Muscles move – and by their motion we move. Yet despite the variety of actions we are capable of performing, muscle itself moves only by becoming shorter – it pulls but it cannot push. And while whole muscles can be seen to contract in this way, they consist, in reality, of millions of tiny, finely tuned protein filaments working together in superb synchrony.

Bodily functions demand that muscles accomplish different chores. Accordingly, there are three distinct types. Cardiac muscle, found only in the heart, powers lifelong pumping. Smooth muscle surrounds or is part of internal organs and the blood vessels that fuel them. Both are described as involuntary, since they are not usually under our conscious control. But the word "muscle" brings something else to mind. What we will into action, what gives the body form, what aches after a ten-mile hike is skeletal muscle, muscle that carries out voluntary or conscious movements. Skeletal muscle is anchored to bones and pulls them to initiate movement. Accounting for 23 percent of body weight in women and 40 percent in men, skeletal muscle is the body's most abundant tissue.

Weightlifters, baseball pitchers and sprinters exhibit some of the many remarkable displays of force the human body is capable of performing. Yet such force is possible only through the arrangement of the muscles, bones and joints that make up the body's lever systems. Bones act as levers, while joints perform as living fulcrums, the pivots about which vital powers are exercised. Muscle, attached to bones by tendons and other connective tissue, exerts force by converting chemical energy into tension and contraction. When a muscle contracts it shortens, in many cases pulling bone like a lever across its hinge. This movement depends on many factors, such as the stabilizing influences of other muscles that contract at the same time.

Muscular action also involves the senses. Progressive movements constituting behavior, from a smile to a ballerina's arabesque, are possible only through incessant communication between senses and muscles. This constant sensory-motor

With his legs flashing like the spokes of a spinning wheel, American gymnast Kurt Thomas whips his body around the pommel horse with mechanical symmetry. A set of bony levers joined by sinewy hinges and driven by powerful muscles, the human body moves like a machine – but one under the control of a mind.

There are more than 600 voluntary muscles in the body, which together account for about 40 percent of a man's weight. The strongest are the major skeletal muscles shown here. Their contraction causes all body movements, from bending and running to lifting and gripping. Even subtle changes of facial expression are brought about by delicate adjustment of the small muscles in the face. Strong non-elastic tendons join muscles to bones, where they exert their pull. The fingers, for instance, are bent and straightened by muscles in the forearm connected to the finger bones by long tendons.

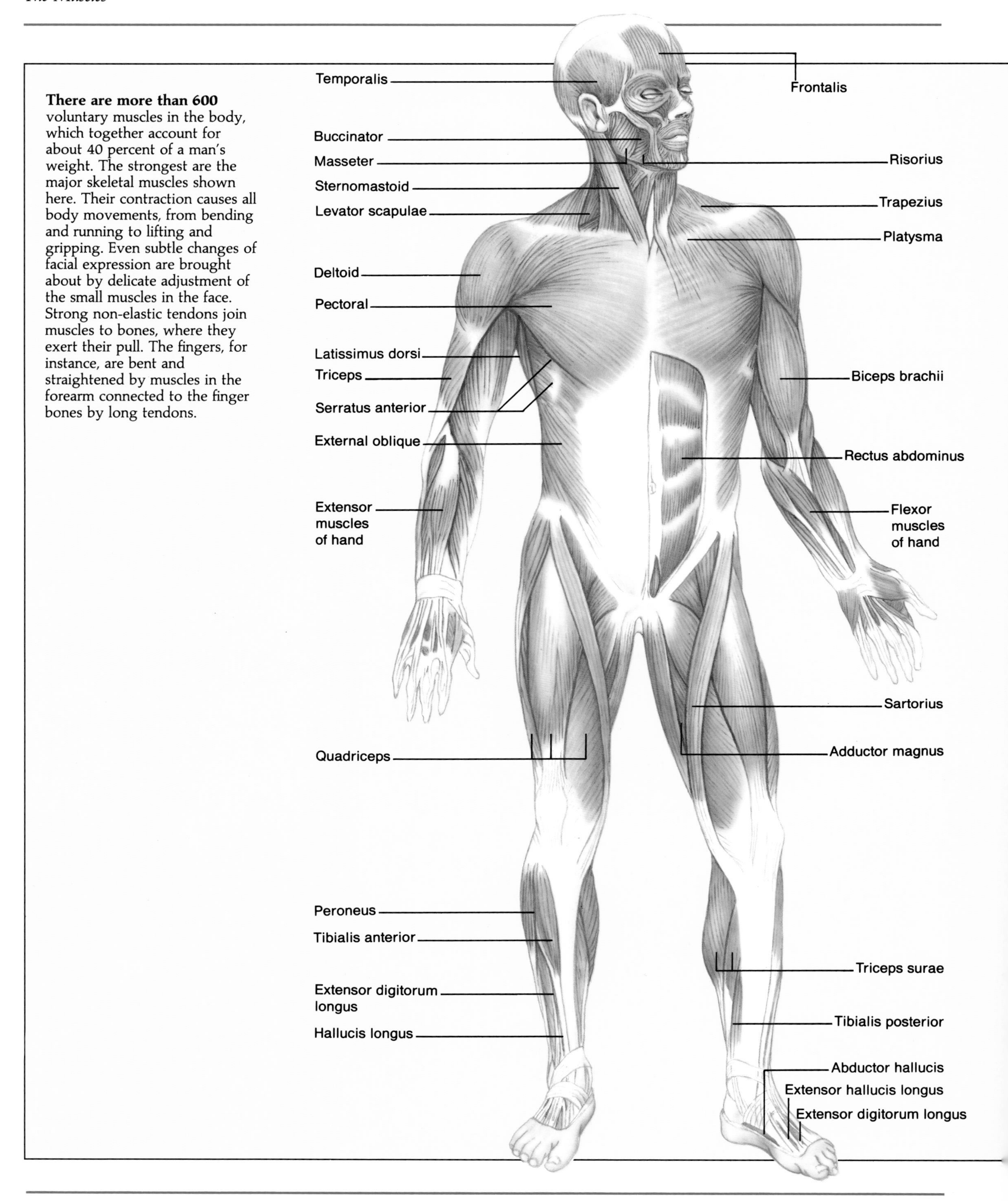

Muscles need a constant supply of oxygen in order to work. This vital gas is carried in the bloodstream along arteries, whose small branches encircle each bundle of muscle fibers. Even individual fibers have their own network of capillaries, in which gas exchange takes place: oxygen is supplied to the fiber in exchange for carbon dioxide. This waste gas is carried away in the blood along veins, which transport it to the lungs for disposal.

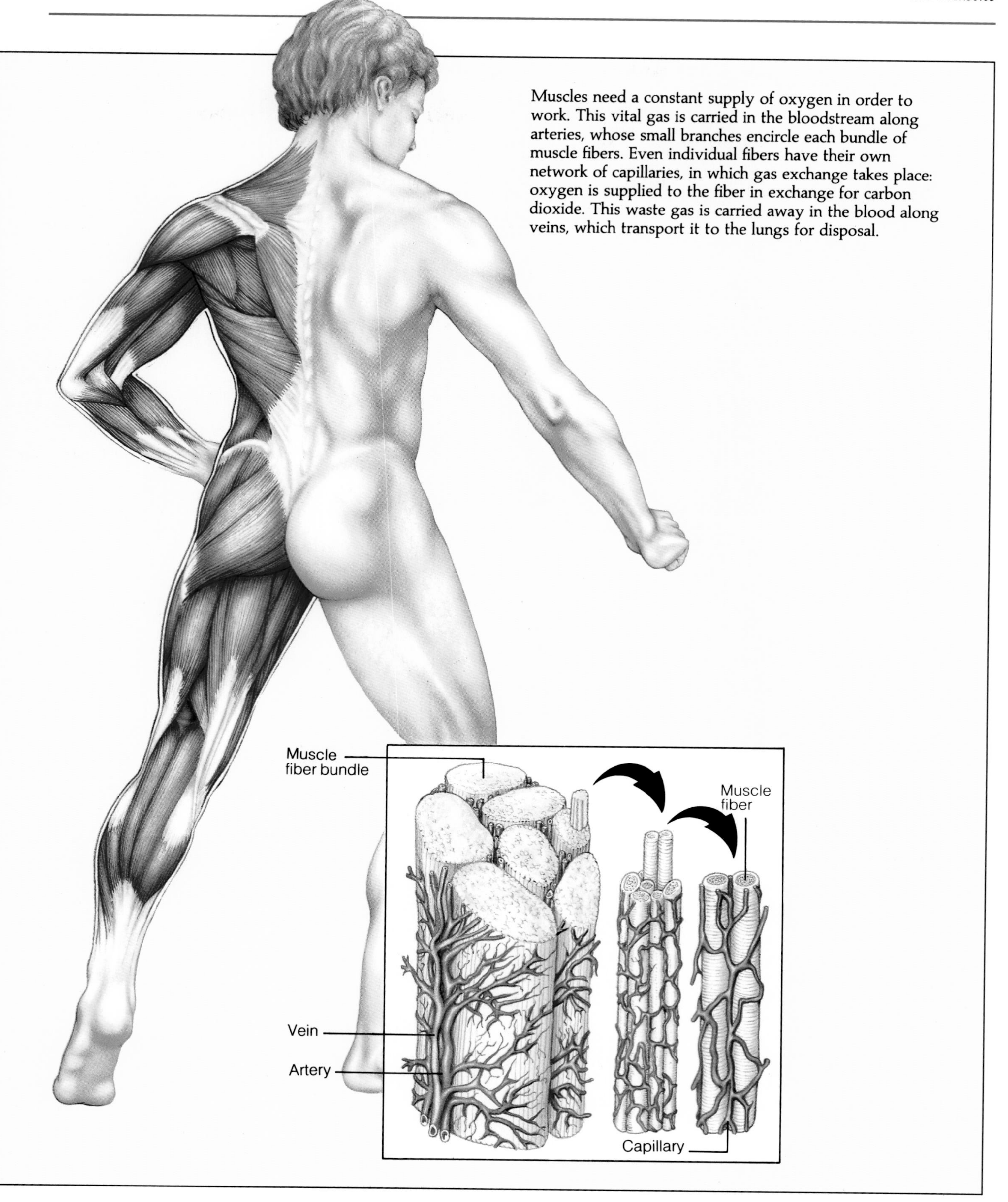

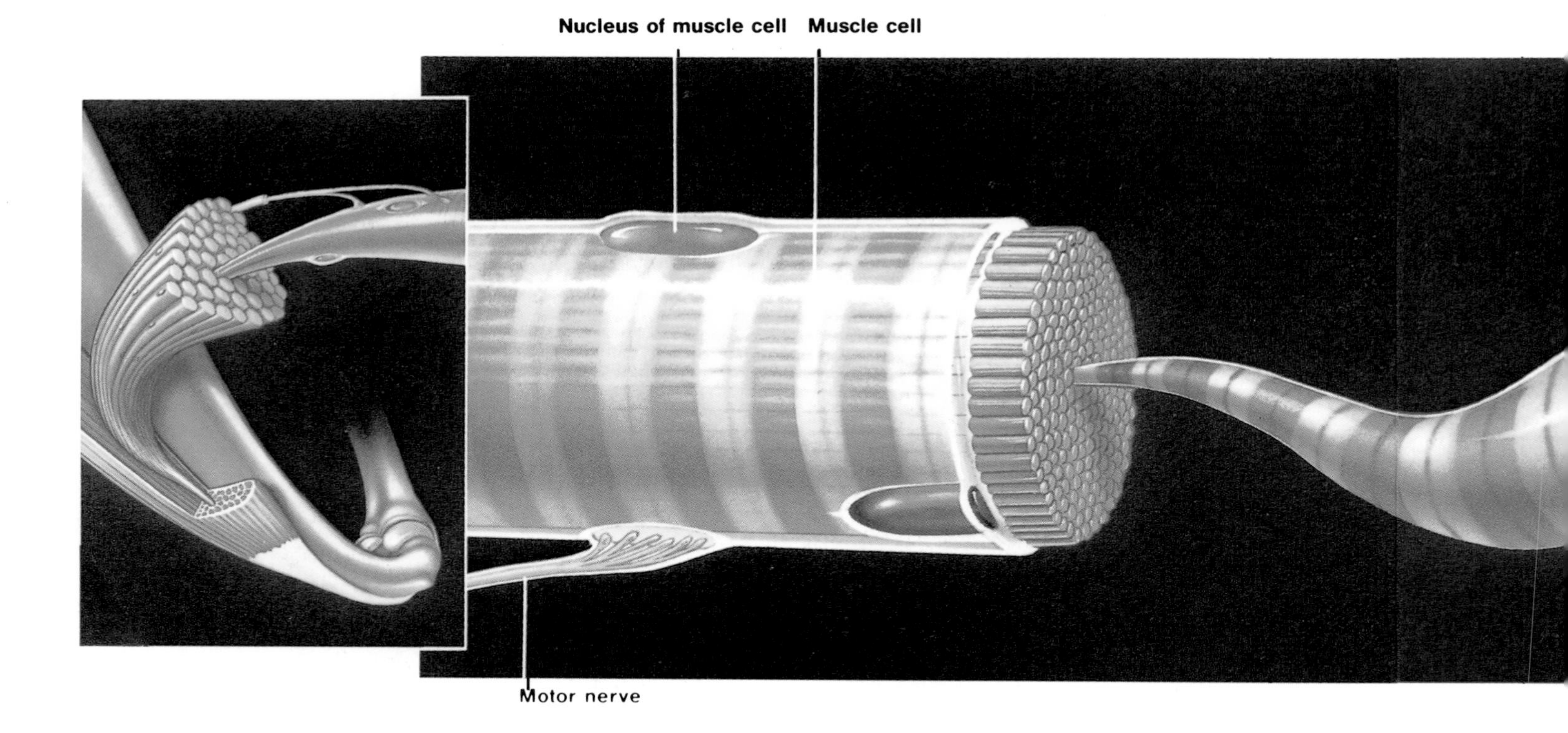

Muscles look solid but are really made up of bundles of tiny fibers *(left)*. Each fiber is a muscle cell, which has a fibrous microstructure of its own, consisting of thick and thin filaments in units called sarcomeres. Signals arriving along a motor nerve make the filaments slide over each other, causing the whole muscle to contract.

integration occurs between the spinal cord and the part of the brain that coordinates sensation and motor response.

Muscles and motion

"Motion is the cause of all life," wrote Leonardo da Vinci in his notebooks. Indeed, the movements of the human body, and many of its vital internal functions, are guided and coordinated by muscles of varying shapes and sizes. Yet before they can move the body, the muscles must first be able to maintain its characteristic upright posture against the collapsing force of gravity. Just standing still and upright depends on the contraction of a whole range of different muscles.

Control of the muscles necessary to standing and moving is not something babies are born with. But gradually, over the early years of life, they learn the coordination and control that will, barring accidents, disease and the ravages of old age, last them a lifetime. Babies learn to control their muscles from the head down: muscles of the neck, followed by those of the shoulders and arms, then the body. Only with mastery over muscles of the pelvis and legs are standing and walking possible.

Oddly, babies often run before they can walk. Typically, babies on the verge of true walking stand with the feet wide apart, ensuring a wide base of support for the body. They lean forward, thus advancing the center of gravity and, in a natural though erratic sequence of steps, can maintain an upright posture for a few seconds as they scamper to a person or piece of furniture for support.

A fluent synchrony

The apparently simple act of walking thus takes time to perfect. It also involves specific actions by many different muscles. In the human body, every muscle has a particular function, but each works in a fluent synchrony with others to achieve its role. Muscles called prime movers, such as the deltoid muscle in the shoulder, are powerful initiators of force. The triangular deltoid is a weightlifter's prime mover when he raises a heavy load above his head.

Because their contraction results in movements, such muscles are called agonists. Their action is opposed by muscles called antagonists: any muscle that extends a limb is acting antagonistically to a muscle that flexes or bends it. The triceps in the upper arm, for example, is antagonistic to the biceps when the arm is bent at the elbow.

Such paired muscles as the biceps and triceps faithfully alternate roles as agonists and antagonists in perfect collaboration, making possible the cooperation necessary to smooth and efficient muscular effort. When a prime mover contracts, the tension in its antagonist slackens and stabilizes movement. Assistant movers, muscles that contribute to a specific movement, often assist the prime mover in making muscular expression possible. The hamstring muscle of the thigh is the prime mover in bending the knee, the sartorius is its assistant and the quadriceps its antagonist.

The role of fixator or stabilizer muscle is to hold a bone or other body part steady, so providing a firm foundation upon which active muscles can pull. In weightlifting, for example, abdominal muscles contract to prevent sagging of the hips and trunk and to enable the transfer of momentum from the body to the weight as the weightlifter stands up.

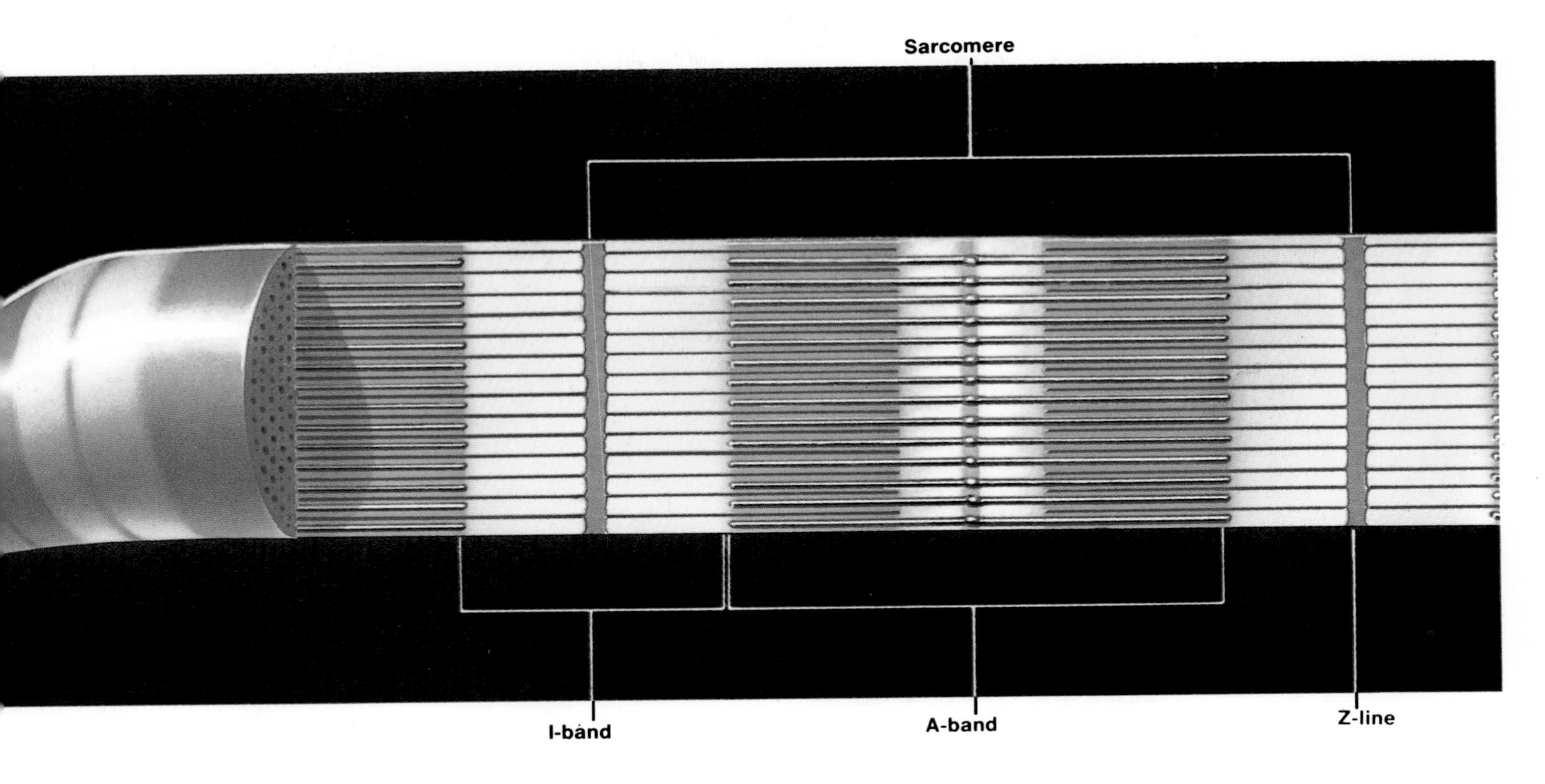

Systems of levers

To keep the body in action, muscles work in concert with bones and joints in lever systems that obey the basic physical rules of motion and energy. To return to the business of walking: placing one foot in front of the other, pace by pace, is a process in which balance is continually lost and regained as each step establishes a new base. The legs and arms act as lever systems, pivoting on fulcrums of ankles, shoulders and hips. The legs generate a rotary motion that propels the body forward and momentarily upsets the body's equilibrium. By synchronizing the movements of the arms and legs, the muscles ensure that balance is restored before the body becomes unstable and falls.

The impact of muscle groups working together is called power. For more power, the nervous system must animate more muscle fibers. Muscles exert this power upon the points where they are attached to bones (their points of insertion). Bones linked together by joints form lever systems: the arm lever bends, for example, at the elbow's hinge joint and swings or rotates at the shoulder's ball-and-socket joint.

In all, there are only three types of lever system in the body. Each lever has a load arm – the zone between the load to be managed and the fulcrum – and a power arm, which is the region between the fulcrum and the muscle. This power arm is nearly always shorter than the load arm.

The muscles that act on the power arm develop their greatest power by shortening slowly while, at the same time, developing high tension. The force is amplified in the lever system producing rapid movement of the load arm. Called first-class levers, they have a fulcrum sited between the force and the resistance. The throwing action of a baseball pitcher uses this kind of lever – short, relatively slow movements of arm muscles produce rapid movements of the hand.

In second-class levers, the resistance is between the fulcrum and the muscular force that is pitted against it. Here, power rules rather than speed. A ballet dancer, on half-point in an arabesque, balances the body weight with an upward pulling calf muscle, using the toes as a fulcrum.

In third-class levers, the force is applied between the fulcrum and the resistance. The faster the muscle contracts, the greater is the force that can be applied at the end of the moving lever. The slower the contraction, the less energy is required to produce it. The action of the biceps as it exerts an upward pulling force on the forearm when the elbow is bent is an example of such a lever.

In the human body, speed and range of motion prevail over force. The body is best at tasks involving fast or delicate movement, and the movement of light objects. When great force is demanded, as during heavy work, the body usually proves inadequate. To compensate for this, the human mind has invented machines and tools, such as the crowbar, which extend muscular leverage to gain a force advantage.

Physical work most often involves the use of more than one of the body's lever systems. Even when movement of a single lever does take place, many other parts of the body must be held in place. When force depends on speed at the lever's end, other levers team up. Working in sequence, each exerts force as the previous one reaches its peak speed, much like the windup before a pitcher releases a baseball. But when many levers engage in a heavy task, like pushing against a door with all one's weight, they function in unison.

Fibrous bundles

The structure of muscles is designed to enable them to contract and relax, and they provide the body with a lifetime of movement. All muscles are made up of fibers, but each of the three types of muscle has a different microscopic structure related to its exact role in the body machine.

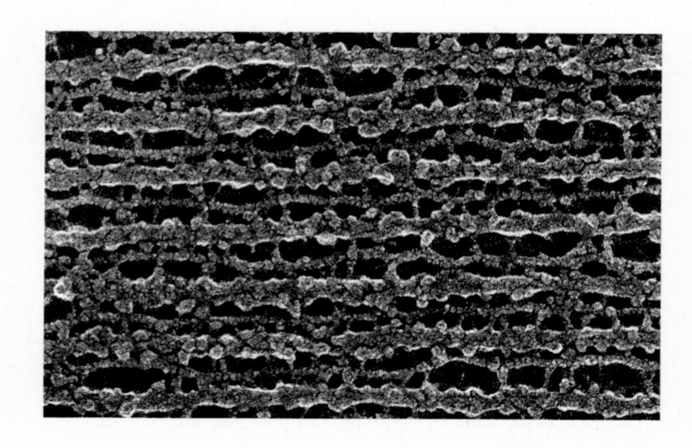

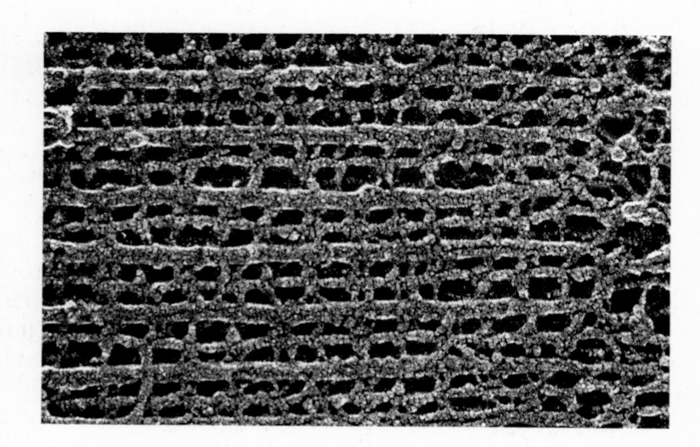

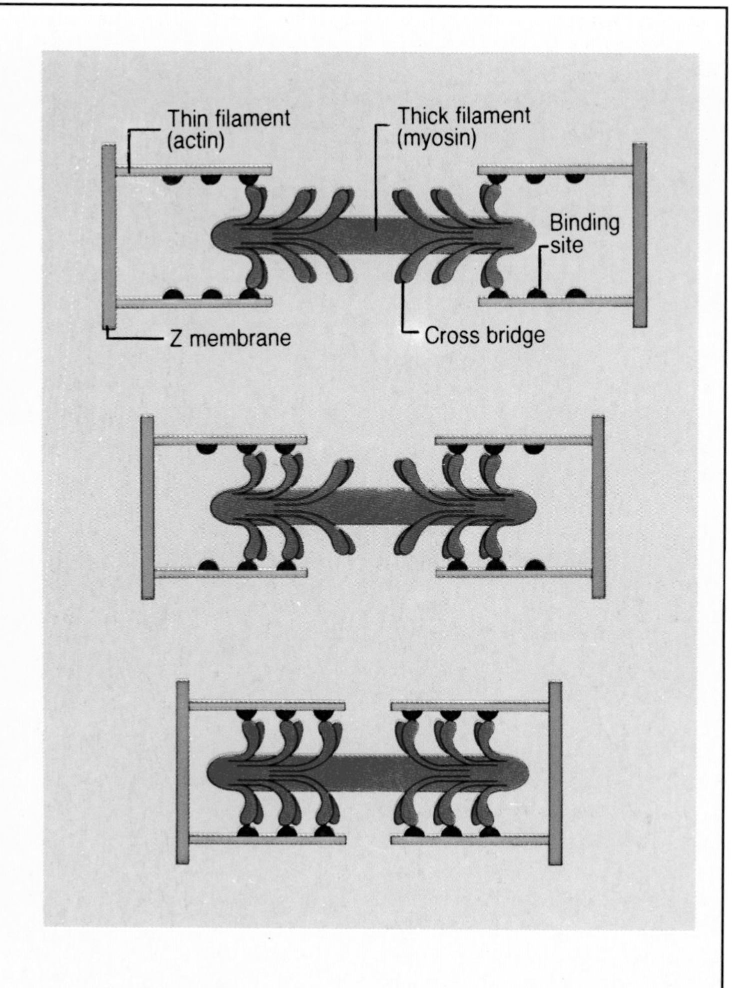

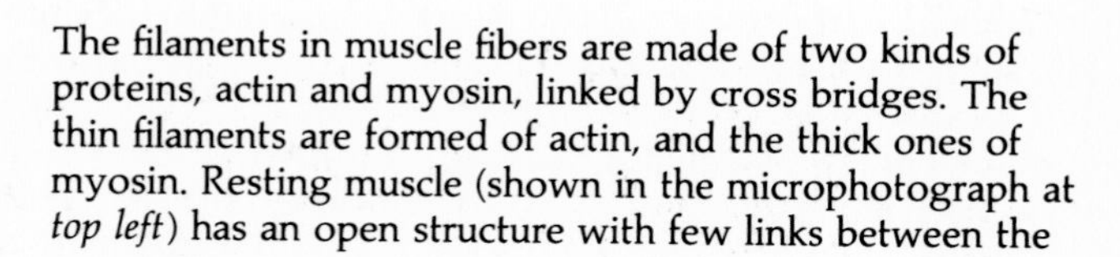

The filaments in muscle fibers are made of two kinds of proteins, actin and myosin, linked by cross bridges. The thin filaments are formed of actin, and the thick ones of myosin. Resting muscle (shown in the microphotograph at *top left*) has an open structure with few links between the two types of filaments. As muscle contracts, and the filaments slide over each other, myosin "fingers" latch on to a series of binding sites on the thin actin filaments to form more cross bridges and create a much stronger and denser structure *(bottom left).*

In skeletal muscle the fibers are elongated cylinders. Each fiber contains several nuclei (the sites where genetic material is housed). These nuclei originally belonged to myoblasts, smaller "pre-muscle" cells that merged together before birth. Because skeletal muscle fibers are much larger than cardiac or smooth muscle fibers, many are visible to the naked eye. Some, like those in the sartorius muscle of the thigh, are more than 12 inches long.

Individual skeletal muscle fibers can extend the whole length of a muscle. Usually, however, one end attaches to tendon, the tough tissue that binds muscle to bone, and the other attaches to connective tissue in the muscle itself. The firm, white tendons form a kind of core for a muscle by extending far inside it and emerging at the muscle ends to link to bone.

Fibers of muscle and tendon are completely different materials, however, and do not merge. Instead, connective tissue extending from the tendon links with the end of the muscle fiber.

Surrounding each muscle fiber is the endomysium, a thin sheath of connective tissue. Another sheath, the internal perimysium, bundles the individual fibers into fasciculi, groups of about 12 fibers. These groups are themselves bound together by another layer of connective tissue called external perimysium or epimysium. It is this final grouping of bundles that is commonly referred to as muscle.

The number of fibers in any given muscle is fixed at birth; damaged fibers can never be replaced, even by a healthy body. A weightlifter has no more fibers than the proverbial 97-pound weakling, but he does have larger muscle fibers and muscles with more connective tissue.

Muscles enlarge with use because exercise, especially lifting weights, stimulates the production of greater amounts of actin and myosin, the special proteins that muscle contains, so expanding the fibers. Primed for heavy work, such muscles become well defined under the skin. But muscle enlargement, or hypertrophy, is not as great in women as in men because it is partly regulated by the male sex hormone testosterone.

Time or lack of use causes muscle to waste or atrophy; muscular strength usually peaks at the age of 30. As people get older muscle cells degenerate and the number and size of muscle fibers dwindle. Connective or "filler" tissue replaces the lost fibers, making muscles more rigid and slowing their reactions. This decline diminishes muscle performance. The process is not as unforgiving as it may seem, however. Steady exercise is a valuable aspect of preventive medicine that may also help delay fiber loss and maintain strength. It is a fact that muscle works better the more it is used.

Fast-twitch and slow-twitch

Skeletal muscles are composed of two kinds of fiber in proportions that vary according to function. Fast-twitch fibers provide strength and power; they are the "white meat" of muscle. They contract quickly, yielding short bursts of energy and are recruited most heavily for brief, intense exercises – sprinting, weightlifting, putting the shot or swinging a golf club. But these muscles are quick to exhaust – cramp sets in as they become vulnerable to build-up of lactic acid, a by-product of their own metabolism.

Slow-twitch fibers produce a steadier tug and are the cords of endurance, tiring only when their fuel supplies are consumed. Slightly smaller than fast-twitch fibers, and containing fewer nerve endings, they draw more oxygen from the blood. They are the dark meat of muscle, their color imparted by their abundant blood supply. Exercises requiring enormous stamina – long-distance running, swimming or cycling – rely on the sturdy strength of slow-twitch fibers.

The strands of muscles

The dynamics of muscle are locked in its basic component, the fiber. Each individual fiber is surrounded by a thin plasma membrane, the sarcolemma. Some 80 percent of the fiber's volume is filled with tiny fibrils, known as myofibrils – from several hundred to several thousand, depending on the width of the muscle fiber. The remainder of the fiber is filled with a jellylike intracellular sarcoplasm, the many nuclei and other constituents of any typical body cell, such as mitochondria in which energy-producing reactions take place.

Skeletal muscle is often referred to as striated muscle for the simple reason that it appears striped or banded when viewed through a microscope. The striations arise not from the surface of the fiber but from its many myofibrils, each of which is banded. The parallel arrangement of the myofibrils within the fiber give its characteristic appearance.

Thin dark bands called Z-membranes separate the myofibrils into cylindrical compartments called sarcomeres – the basic units of a muscle's contraction. A prominent dark strip, the A-band, occupies the center of each sarcomere; the space between each A-band is taken up by the paler I-band. This exact pattern is repeated down the length of every myofibril.

Just as a single fiber contains many myofibrils, so each myofibril contains many smaller filaments arranged in a repeating pattern along the fibril's length. There are thick filaments made up of the protein myosin and thin filaments composed of the protein actin. The arrangement of the filaments gives rise to the striations. The thick filaments form the dark regions of the sarcomere; the thin filaments the light areas. Thus, the dark A-band in the center of the sarcomere consists mostly of myosin filaments anchored at the M-line, the center of the A-band.

The thin actin filaments, anchored to the Z-membranes, form the light I-bands. The darkest striations occur where myosin and actin filaments overlap. The A-band is darkest because it normally includes both types of filaments. However, the actin filaments do not quite stretch into the center of the sarcomere, which is why the H-zone appears lighter.

The powers of contraction

Muscles must counteract gravity. Although they may relax fully, they also remain alert in case they are called into play to prevent the body from falling. Scientists use the word "tone" to describe this constant state of readiness, which is determined by a healthy nervous system. Without tone, jaws would hang open and muscles could not support parts of the body.

The term "contraction" does not always refer to the shortening of a muscle. Technically, it refers only to the development of tension within a muscle. There are two major types of contraction. A contraction in which the muscle develops tension but does not shorten is isometric, and one in which the muscle shortens but retains constant tension is said to be isotonic. Both are determined by the amount of resistance the muscle meets as it contracts.

A person trying to lift too many books in a box strains against the weight. His arm muscles develop tension but do not shorten because the amount of resistance offered by the box is greater than the muscle's tension. But when he lightens the load, the working muscles shorten as they contract. This is an isotonic contraction. Because his muscles shorten in overcoming the resistance, the isotonic contraction is said to be concentric. If he wants to put the box down on a table, he must gradually extend his arms. To do so, the biceps in the upper arms lengthen, maintaining tension to counter the weight of the box. The biceps approach, but do not reach, their resting state. In this case the isotonic contraction is described as eccentric because the muscles lengthen as they act to maintain tension.

When a fiber contracts, the length of the dark A-bands remains constant, but the two pale regions (the I-band and the H-zone) shorten. The protein filaments themselves do not shorten but slide past one another, like lines of soldiers marching in opposite directions.

This sliding-filament theory of muscle contraction was announced almost simultaneously in 1954 by two of the world's leading research teams studying contraction – Hugh Huxley and Jean Hanson of M.I.T., and Sir Andrew Huxley and Rolf Niedergerke of England's Cambridge University. It is still the leading theory of muscle contraction, although much remains to be discovered.

The muscle machine swings into motion only when it receives impulses from the central nervous system which tell it to do so. The nerves terminate near the muscle fiber's delicate membrane, where they release transmitter chemicals. These neurotransmitters initiate a wave of electrical activity that spreads through the whole fiber. This causes the fiber's membrane to release calcium ions, electrically charged calcium atoms that spark the mechanical process of contraction.

The calcium ions spread throughout the fiber via a network of fine tubules and diffuse into the myofibrils, coming into contact with the fiber's contractile proteins, the thick and thin filaments. There they interact with two other proteins, troponin and tropomyosin, which, as a team, circle around the thin actin filaments like delicate embroidery. In a chemical reaction, the calcium binds with the troponin, somehow causing it to influence the activity of tropomyosin. The tropomyosin threads shift their hold on the actin filament, uncovering spots along the shaft of the actin filament that are receptive to binding with the myosin filaments. This entire chain reaction takes place in a few thousandths of a second.

Cross bridges between filaments

Emerging from the myosin filaments are pairs of rounded buds. Each pair forms the head of a single myosin molecule. These highly-ordered projections, called cross bridges, form the central part of the muscle's contractile mechanism and take their name from that role.

Cross bridges are crowned with a remarkable substance, adenosine triphosphate (ATP). ATP is an organic compound, and the main source of life's energy derived from the food that we eat. The high-energy ATP is transformed into two low-energy products: adenosine diphosphate (ADP) – which has two phosphates incorporated in it instead of ATP's three – and the inorganic phosphate that splits away. The energy lost in the split is then available for use in the body's metabolism.

ATP and the myosin molecules are said to have a high affinity for each other – so much so that in normal muscle almost every myosin head has ATP resting on it. Each of the head's two buds has a different function. One contains the enzyme adenosine triphosphatase (ATPase), which has the ability to split ATP molecules and liberate energy as a result. The other portion of the head binds to the actin filaments.

As the protein tropomyosin shifts away from the binding sites on the actin filaments, the myosin arms are able to link with actin. The myosin ATPase then splits ATP and its energy fuels rapid contraction of the muscle machinery by throwing the cross bridges into action. Without ATP the actin and myosin filaments remain locked together, and the muscle is unable to contract. This is what causes rigor mortis, the stiffening of muscles shortly after death. Dead muscle cells have no ATP.

Contractions normally shorten a sarcomere by a fifth or more, cross bridges making and breaking links with the actin filament in repeated cycles. Like a team working hand over hand to pull a rope, cross bridges repeat this action until the contraction is complete.

The more cross bridges there are the more powerful is the fiber's contraction. When a muscle is stretched too much there is little or no overlap between the thick and thin filaments, so cross bridges cannot make enough connections to create tension. On the other hand, with too much overlap, thin filaments on either side of the sarcomere begin to overlap themselves, interfering with the action of the cross bridges. And, in a greatly shortened fiber, the thick filaments are compressed between the two Z-membranes, making further shortening difficult. For this reason, a muscle has one particular length at which it contracts most efficiently. Normally, resting muscle is quite near this length. The greatest tension is produced when a muscle is stretched slightly beyond this length, and it contracts more forcefully.

Movement and the brain

From all parts of the body nerves known as sensory neurons carry impulses to the brain or spinal cord, carrying messages about the state of affairs in every body part, including the muscles. Conversely, motor neurons transmit impulses to muscles, often via intermediate connections or interneurons in the spinal cord. Motor impulses travel to the muscle fibers where they trigger the release of the neurotransmitter substance acetylcholine. This crosses the gap or synapse at the junction between nerve and muscle, setting off a chain of events that ends in contraction. Within a second, millions of impulses reach the motor neurons, some sent from various parts of the brain and spinal cord, others from special sense organs located in joints, ligaments, tendons and muscles themselves.

The seeds of movement, however, are sown by the brain, in its primary cortex, a region of the brain's wrinkled surface spanning both cerebral hemispheres. This motor cortex lies in a ridge just in front of the central sulcus, a deep furrow running vertically along each of the cerebral hemispheres. And a patch of cortex directly in front of the primary area also houses neurons involved in movement. This secondary motor area is thought to be crucial to speech and intricately coordinated movements such as those the hands perform.

Electrical impulses from many regions of the brain feed into the motor areas. Before the nervous system can orchestrate a coordinated movement, it must collect and integrate all sensory messages. In the somatic sensory cortex lying on the other side of the central sulcus the brain registers sensations. Interplay between the senses and movement is continuous and elaborate. Sight, sound, smell, pressure and pain are naturally important, but so are the messages relaying information about the angles and position of joints, the length and tension of muscles, even the speed of movements.

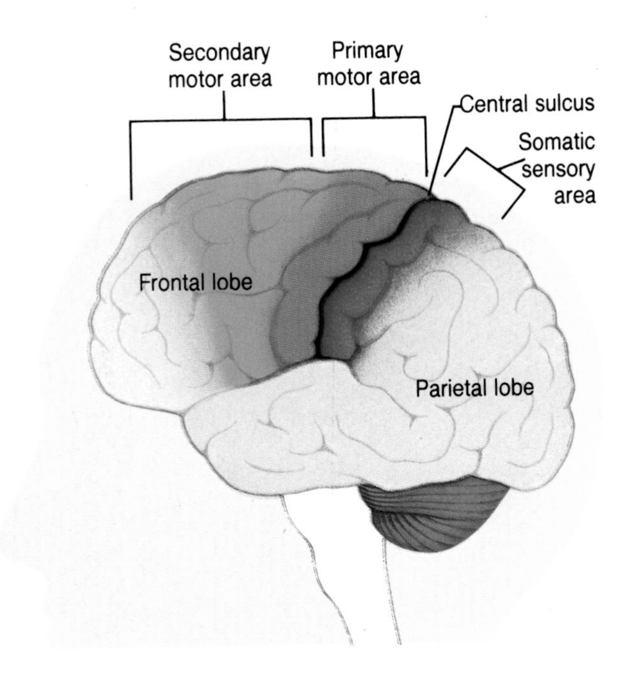

Muscular movement is coordinated by various areas of the brain, principally in its outer layer, the cerebral cortex. The primary motor area, a band across the center of the cortex, controls gross movements; but when fine, intricate movements are required the secondary motor area comes into play. Input from the various senses passes first to the somatic sensory area, which supplies feedback vital to any voluntary muscular action, before reaching the motor areas.

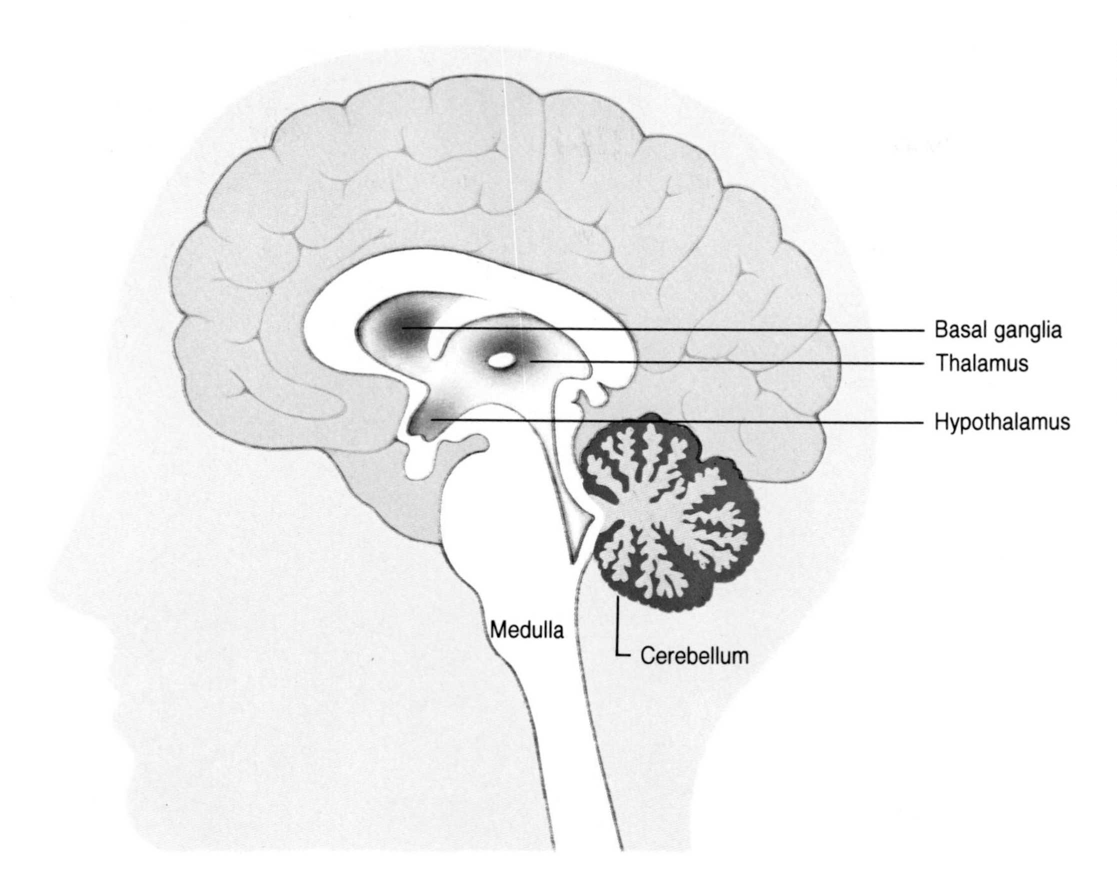

No matter how well tuned the muscles, they cannot work correctly without instructions from the brain. This is the role of the cerebellum, the major part of the hindbrain, which is connected by motor pathways to the cerebral regions above it. Together with the basal ganglia it initiates all movements and maintains muscle tone, the continuous state of slight muscle tension which, during waking hours, keeps all the body's muscles ready for action.

Motor systems

The brain's central stalk cradles the brainstem, a control center for breathing, heartbeat and blood pressure. The area called the medulla, at the brainstem's base, is actually a bulbous extension of the spinal cord, and is a crossroads for one of the two major transmission networks linking brain and skeletal muscles. Beds of nerve fibers swell into twin pyramids on the surface of the medulla, then descend to the spinal cord in nerve tracts known as the pyramidal system.

The pyramidal system is an important mediator of delicate, skilled muscle movements. Its nerve fibers have their cell bodies in the cortex. About 60 percent come from the motor areas and 40 percent from the somatic sensory region. Pyramidal nerves are among the body's longest, some stretching two feet.

Inside the medulla, an estimated 80 percent of the pyramid nerves cross over from one side of the brain to the other, so that skeletal muscles on the right side of the body are controlled primarily by neurons in the left side of the brain, and vice versa. The nerves course down through the spinal cord on interneurons, which in turn fire motor neurons that supply muscle fibers. About 10 percent of all pyramidal nerves terminate directly on motor neurons, providing a direct path for quick muscle activation.

At every step along the descent from brain to muscle, impulses can influence numerous interneurons, and so vary the precision of muscular control. And an average motor neuron may have as many as 15,000 synapses, connections which provide information from all over the body. Parts of the body such as the back, which have a limited precision of motion, are supplied by relatively few pyramidal neurons – perhaps 50,000. Hand muscles, which perform delicate movements, are driven by impulses from roughly 200,000 pyramidal neurons.

The second major transmission network is the extrapyramidal system. Because it produces contractions of groups of muscles either simultaneously or in sequence, it is responsible for larger, more automatic, body movements – such as those of running, walking or swimming.

The motor unit

The basic building block for all voluntary movement, which is under conscious control, is the motor unit: a single motor neuron (nerve cell) and all of the muscle fibers it supplies. Each muscle contains many motor units, but the type of muscle determines the size of the units. Muscles that require precise control and that act rapidly have small motor units of a few muscle fibers, enabling them to increase tension gradually. In large motor units, tension increases swiftly as the help of each additional motor unit is enlisted. On average there are about 150 muscle fibers to a unit. A single motor unit in the tiny muscles of the eye may contain only three muscle fibers, whereas those in the relatively slow-moving large muscles such as the calf muscles house a thousand or more.

Within each motor unit, muscle fibers obey the "all or none" principle, meaning that all contract or none contracts. If the muscle fibers of a motor unit are sufficiently stimulated by nerve impulses to contract at all, they contract maximally.

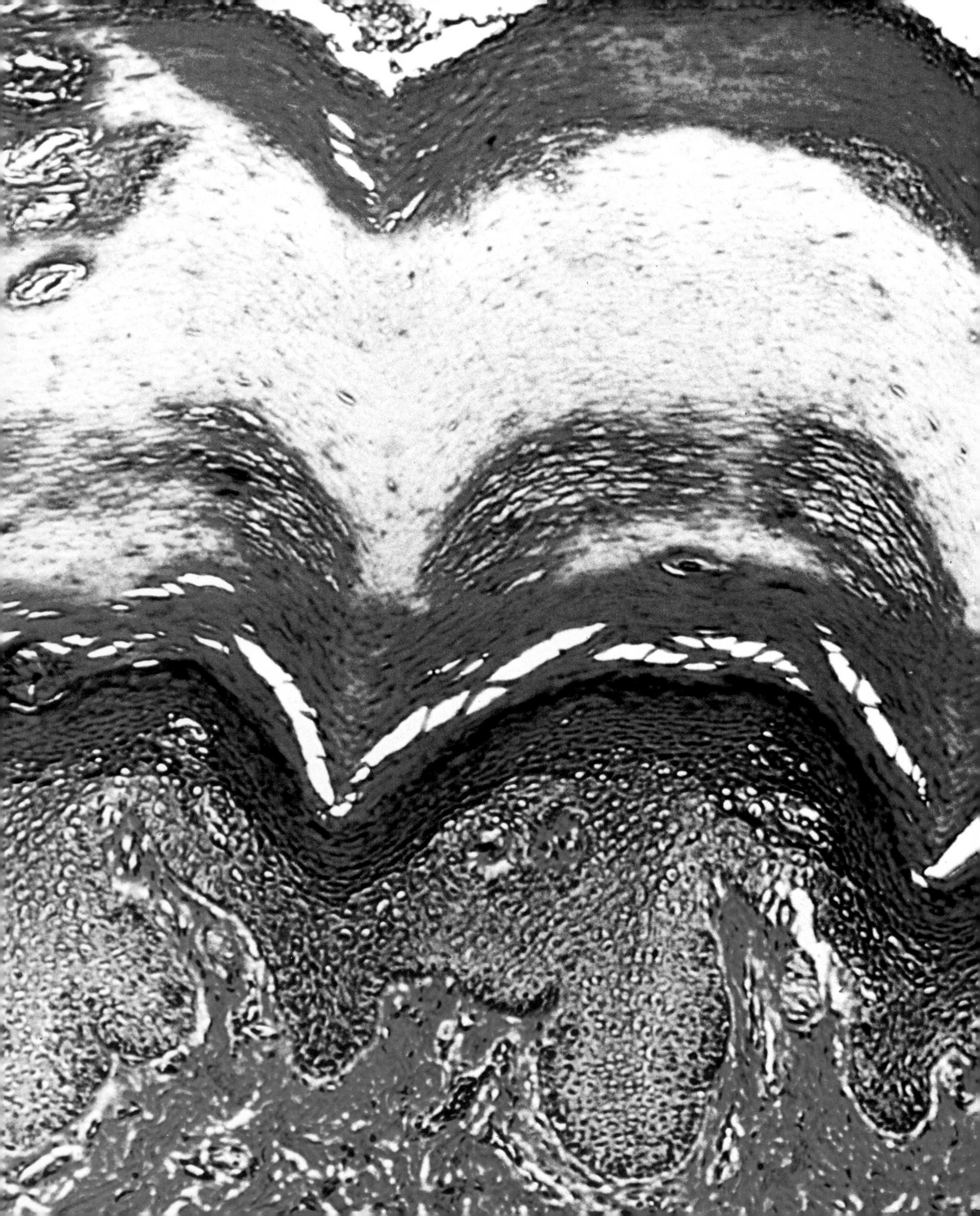

CHAPTER 4

The Skin

Silent and immobile, skin registers sensation constantly and supports a teeming, unseen population of tiny organisms. It is adapted to its various functions with remarkable versatility. Not only does it harden from use, but it molds into different shapes. And as it responds to the most delicate touch, skin becomes an organ of communication, sometimes more eloquent than words.

Knitted together with tough cells, skin is the first line of the body's defenses against disease-causing invaders. Repelling countless microorganisms it protects the soft tissue within the body. Nevertheless, its durable exterior is not an impassable barrier. By allowing water and heat to permeate its fabric, the skin helps control the body's temperature.

Skin is a living boundary that separates the inside of the body from the outside world. Constantly in contact with its surroundings, skin is tough enough to resist countless chemical and environmental assaults, yet soft and sensitive enough to respond to the gentlest touch. A versatile organ, skin regulates the movement of substances from the interior to the exterior. Our bodies are made mostly of water – as much as 75 percent at birth and somewhat less later in life. Skin protects this bodily content from a considerably drier environment.

The skin barrier is selectively penetrable. Skin secretes fluids that lubricate, barricade toxic substances and maintain a stable internal environment. It absorbs other substances, particularly those soluble in oils. This absorptive function has proved effective in administering certain medications.

A barrier to the world

The skin totals between 12 and 20 square feet in area and accounts for 12 percent of body weight. It is composed of three integrated layers – the epidermis, dermis and subcutis. The terms refer to the positions of these layers as "overskin," "skin" and "underskin." The epidermis is outermost, forming an overall protective covering for the whole body.

The epidermis is never more than one twenty-fifth of an inch thick. Like the other layers, its thickness varies over different parts of the body. It is thickest on the palms of the hands and soles of the feet, where friction is needed for gripping and walking. It is thinnest on the eyelids, which must be light and flexible. The most protean of the layers, the epidermis, also grows into fingernails, toenails and hair.

As in all tissue, the skin's basic unit is the cell. The epidermis is woven of three kinds of cell – keratinocytes, melanocytes and Langerhans cells. The first synthesize the protein keratin, an essential ingredient of outer skin, hair and nails. Melanocytes inject granules of pigment into the neighboring cells. Melanin, the dark pigment produced, protects the skin by absorbing the sun's ultraviolet radiation. This radiation can damage the genetic make-up of skin cells, causing skin cancer. Melanin darkens skin exposed to sunlight. The pigment also accounts for different colors of skin, hair and eyes, Langerhans cells aid the body's

Dipping and curving like a mountain range, a single fingerprint ridge in microscopic cross section reveals the layered structure of human skin. The skin is sensitive to touch, heat and pain, and sheathes the body as a protective barrier against disease-causing germs. And if it is penetrated in an injury, it rapidly heals and reestablishes itself as the body's first line of defense.

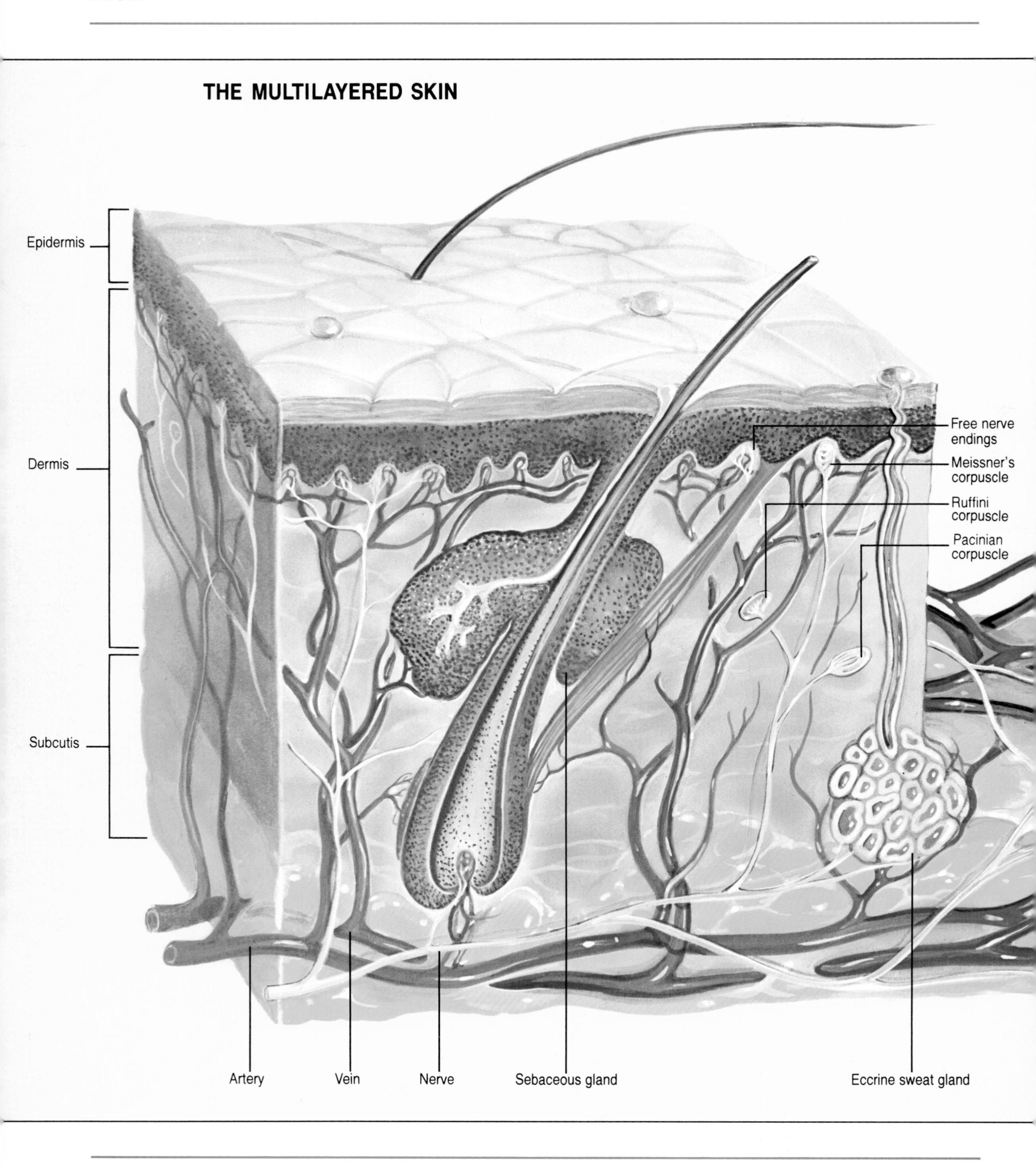
THE MULTILAYERED SKIN
Epidermis
Dermis
Subcutis
Free nerve endings
Meissner's corpuscle
Ruffini corpuscle
Pacinian corpuscle
Artery
Vein
Nerve
Sebaceous gland
Eccrine sweat gland

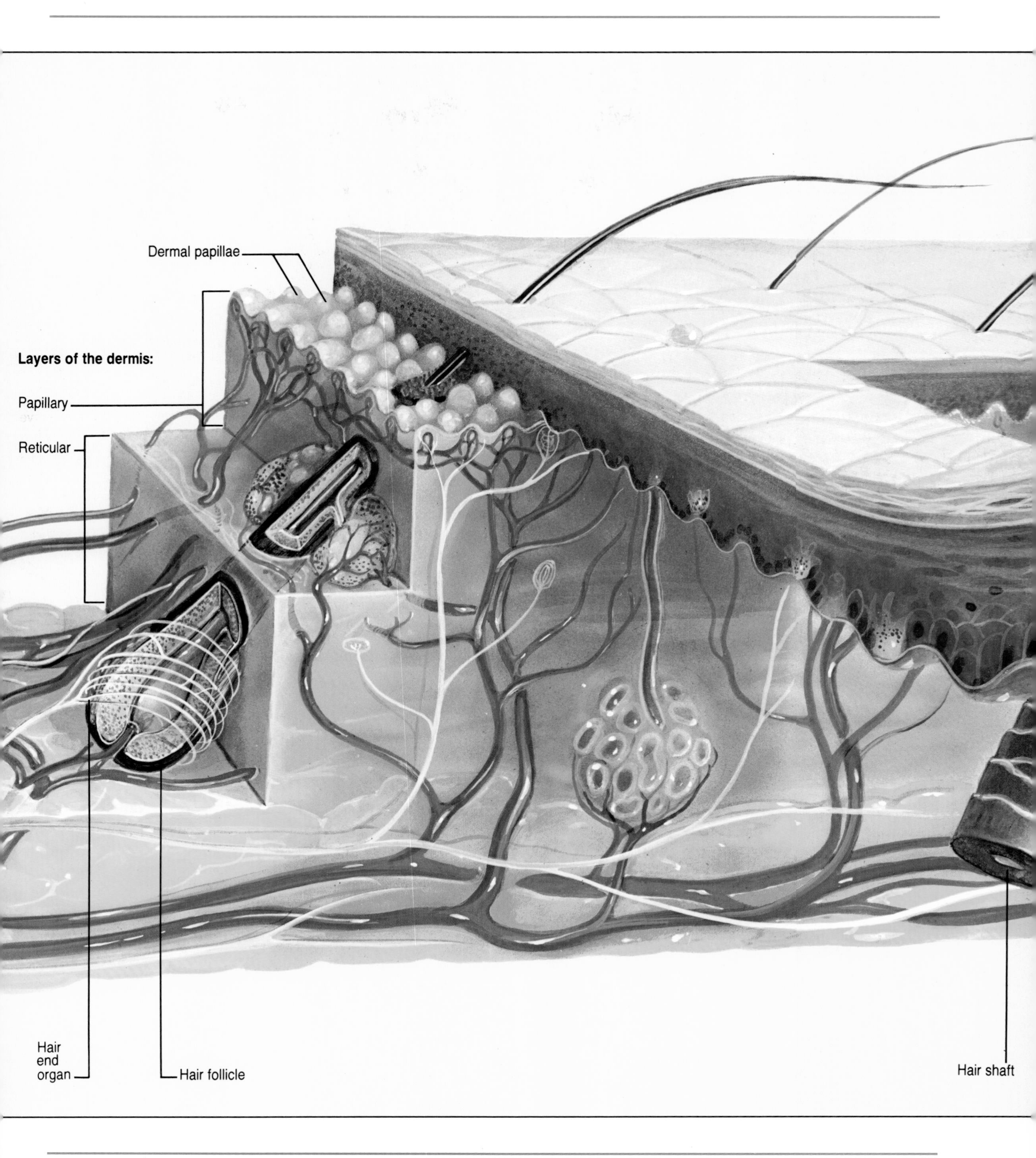
Dermal papillae
Layers of the dermis:
Papillary
Reticular
Hair end organ
Hair follicle
Hair shaft

One of the commonest skin blemishes – and for teenagers one of the most distressing – is the comedo, or blackhead. Large numbers of blackheads occurring together result in acne. The chief culprit is sebum, the mixture of waxes and fatty acids produced by the skin's sebaceous glands. During puberty, increased hormonal activity encourages these glands to produce more sebum. Dead skin cells accumulate in the minute pit round the base of a hair, mix with sebum and form a plug that stops further sebum from escaping. Enzymes from bacteria in neighboring sebaceous glands then get to work, forming irritant acid substances and pus. With nowhere else to go, the pus bursts the hair follicle and seeps into the dermal tissue. The skin becomes inflamed, and a white-headed pimple forms. Deeper skin lesions may form pus-filled cysts.

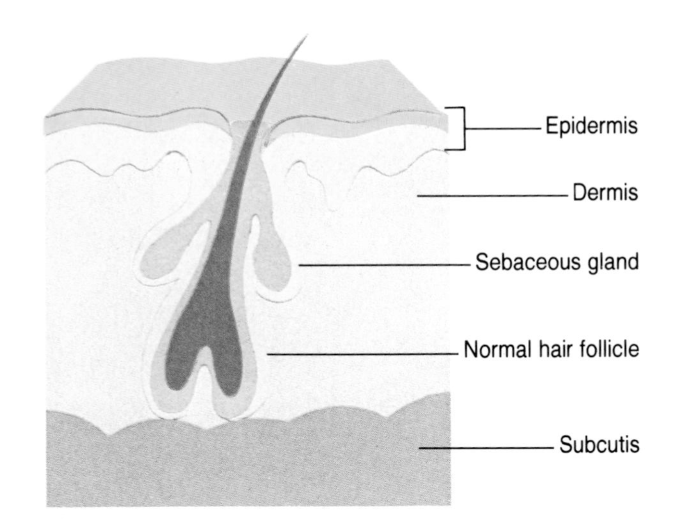

immune system by intercepting alien material in the skin.

At the bottom of the epidermis lies the basal layer, where cell division generates new cells every day. Because the rate of multiplication depends on the body's available energy, the process usually occurs during the four hours after midnight when the body's metabolism has slowed down. In a cycle lasting about 27 days, the new cells move upward through the epidermis, gradually changing from the soft, columnar cells of the basal layer from which they are eventually shed. The cells are attached to each other by plaques called desmosomes and ascend toward the surface of the skin in a continuous impermeable layer.

The horny corneal layer, composed of dead epidermal cells averaging 20 cells deep, creates a durable, protective barrier for the layers beneath it. A thin structure called the Reins barrier is resistant to salt and water, further protecting living cells from the damaging effects of excessive dehydration.

A fibrous and fatty fabric

The dermis, or true skin, is thick, sturdy and subtle, Rich in nerves and blood vessels, as well as sebaceous and sweat glands, it shields and repairs injured tissue. Also known as the corium, the dermis, like the epidermis, is thickest on the palms and soles. This layer consists primarily of collagen, which originates from cells called fibroblasts, and is one of the strongest proteins found in nature. It gives skin durability and resilience.

Certain fibroblasts produce another protein, elastin, a relative of collagen but more easily stretched. Elastin fibers, while especially abundant in the scalp and face, are found everywhere. Collagen and elastin permit the skin to stretch easily, yet quickly regain its shape. They enable the skin to bend and fold with the body's unceasing motions. Collagen also builds a scar tissue to heal skin damaged by cuts and abrasions.

The dermal papillae carry nerve endings and capillaries to the living layers of the epidermis. When capillaries rupture, blood leaks into surrounding tissue, giving rise to the black-and-blue discoloration of a bruise. Papillae also equip the epidermis with lymph capillaries which carry away cellular wastes and help dispatch toxins and dangerous organisms. Visible evidence of the dermal papillae also shows up in skin lines, most notably in the ridges of fingerprints.

Deeper in the dermis lies the reticular layer, a dense but elastic fabric of collagen fibers. Here, junctions of blood vessels control blood circulation through the skin and so help regulate body temperature and blood pressure. Also packed into this layer are sweat glands, hair follicles and oil-producing sebaceous glands.

The subcutis, joined to the bottom of the dermis, is the deepest layer of the skin. Just as keratinocytes synthesize keratin for the epidermis and fibroblasts furnish collagen for the dermis, so lipocytes make lipids for the subcutaneous tissue. This fatty layer cushions muscles, bones and internal organs against shocks, and acts as an insulator and source of energy during lean times.

The all-purpose oil

Like the hard cells of the corneal layer, hair grows from the living tissue of the epidermis. The hair root, however, reaches deep into the subcutaneous tissue and rises from a special structure, a follicle, which is a thin sac of epidermal tissue with a bulb at its base.

Adjacent to the hair follicle, and connected to it by a short duct, are two sebaceous glands that provide oil for the hair and outer skin. These glands produce sebum, a mixture of waxes, fatty acids, cholesterol and debris of dead cells. Unique to mammals, sebum coats both hair and fur with a waterproof shield that helps insulate the body from the rain and cold. Since humans have lost all but a few patches of hair, and have clothed and sheltered themselves against the elements, they no longer need much sebum for insulation. But their glands still produce it in abundance.

Aside from enhancing the insulating properties of hair, sebum serves several other functions. By coating the dead keratin cells of the corneal layer and the hair, sebum retains moisture, keeping hair glossy and skin pliable. Sebum also contains a chemical, which when catalyzed by the ultraviolet rays of the sun, becomes vitamin D. In addition sebum can kill certain forms of harmful bacteria.

Earwax, dandruff and the crusty substance that collects around

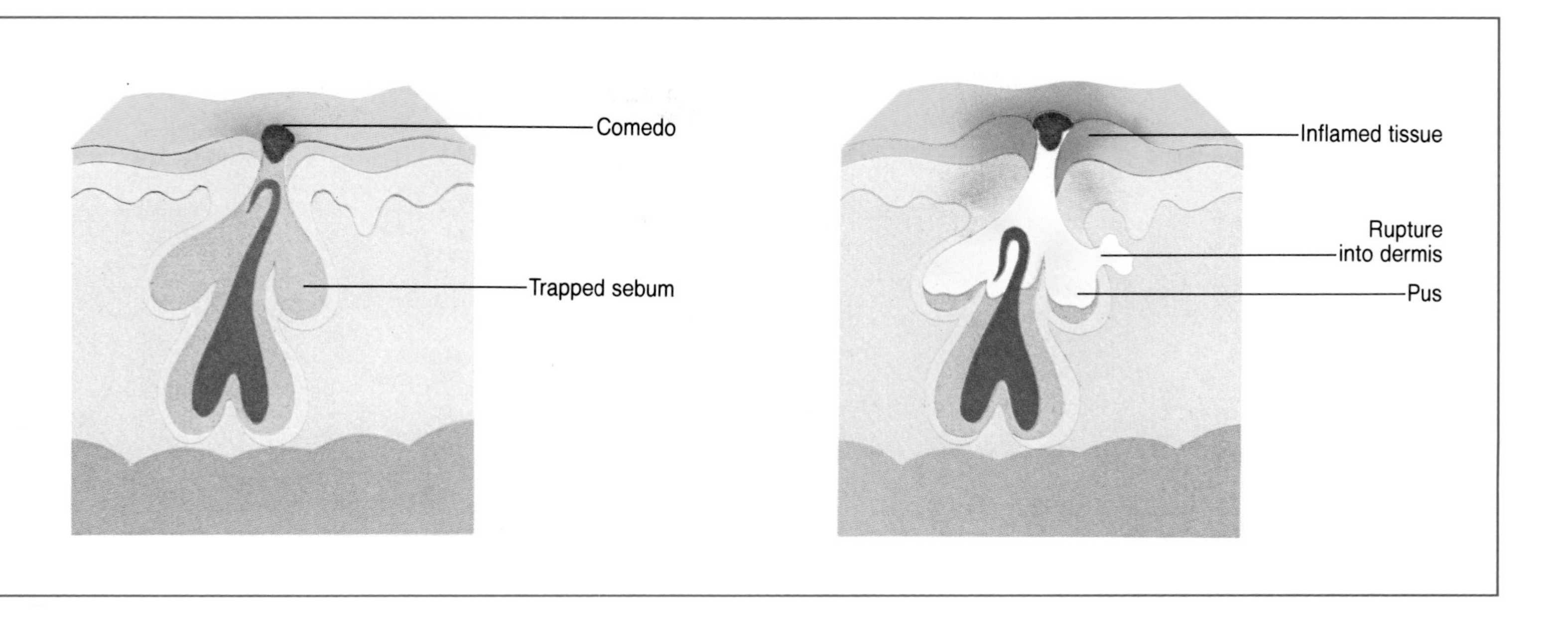

the eyes during sleep are all forms of sebum. Indeed, the face and scalp, where there are proportionately large numbers of sebaceous glands, are particularly rich in it. Sebaceous glands coat the skin continuously, more copiously in men than in women because of the effect of the male sex hormone testosterone.

Occasionally, the glands supply too much sebum. This condition, called seborrhea, gives the hair and skin a greasy sheen. Even with a normal rate of sebum production and regular washing, the sebaceous glands and hair follicles of the face, neck and upper back create conditions favorable to acne, perhaps the most common disorder of the skin.

Two reasons to sweat

The body produces different kinds of sweat from two types of gland, the apocrine and eccrine. Apocrine glands become active during adolescence and are therefore considered a sexual characteristic. They respond not to heat but to the excitement of fear, anger, sexual arousal and other strong feelings. The apocrine glands respond immediately to these emotions, secreting a small amount of cloudy fluid, squeezed out by a muscular sheath. Like the sebaceous glands, they open into the hair follicle to gain access to the skin.

Apocrine glands are found only in the armpits, ear canals, and around the nipples and genitals. They may be the vestigial remains of a once prominent scent system that also played a role in social behaviour, perhaps producing a sexual stimulant, or a territorial marker.

The sweat that dampens the brow comes from eccrine glands. Spread over the body they number between two and three million and secrete a greater profusion of sweat than apocrine glands. The output of eccrine glands can total as much as three gallons a day in hot weather.

As a thermoregulator, the eccrine system reacts to stimulation from the hypothalamus in the brain by moistening the skin. The sweat, almost totally composed of water, evaporates and draws excess heat out of the body, thereby maintaining a constant internal temperature. Certain eccrine glands respond to other stimuli. Glands situated in the forehead, underarms, palms and soles, for instance, work at times of psychological stress, independent of heat or muscular exertion.

The skin's residents

Human skin appears lifeless to the naked eye. But the number of living organisms on a person's skin roughly equals the number of people on the planet. The fauna and flora of the skin are permanent residents. Harmful organisms attract the most attention, but the overwhelming majority of the skin's residents are harmless.

The skin varies from one region of the body to another. Different populations of bacteria and yeasts have adapted to specific environments. The dry expanse of the forearm, the dense tangle of the scalp and the oily surface of the nose all harbor particular species.

One of the largest residents of the skin lives in the hair follicles of the eyelashes, nose, chin and scalp of most adults. A narrow, wormlike mite, *Demodex folliculorum*, lives most of its life in the hair follicle and lays its eggs in sebaceous glands. The young mites molt twice in the follicle, then journey across the skin at night in search of another follicle to colonize.

The bacterial frontline

The largest community of the skin's residents are bacteria, which are acquired at birth. In natural deliveries, babies pick up bacteria when passing through the mother's vagina. In the days following birth, the bacteria population expands rapidly. Within a day, bacteria in the armpit will number about 36,000 per square inch. In another four days, their number reaches 144,000 and by the ninth day the population levels off at around 490,000. Bacteria spread from person to person both by contact and by constant shedding of skin cells.

In addition to the skirmishing lines of resident bacteria, the skin relies on several other defensive strategies to ward off harmful microorganisms. Daily skin loss prevents many would-be colonists from gaining a foothold on the skin. The skin also presents an acidic mantle that deters certain types of bacteria. When the skin's bacteria break down sebum, fatty acids that

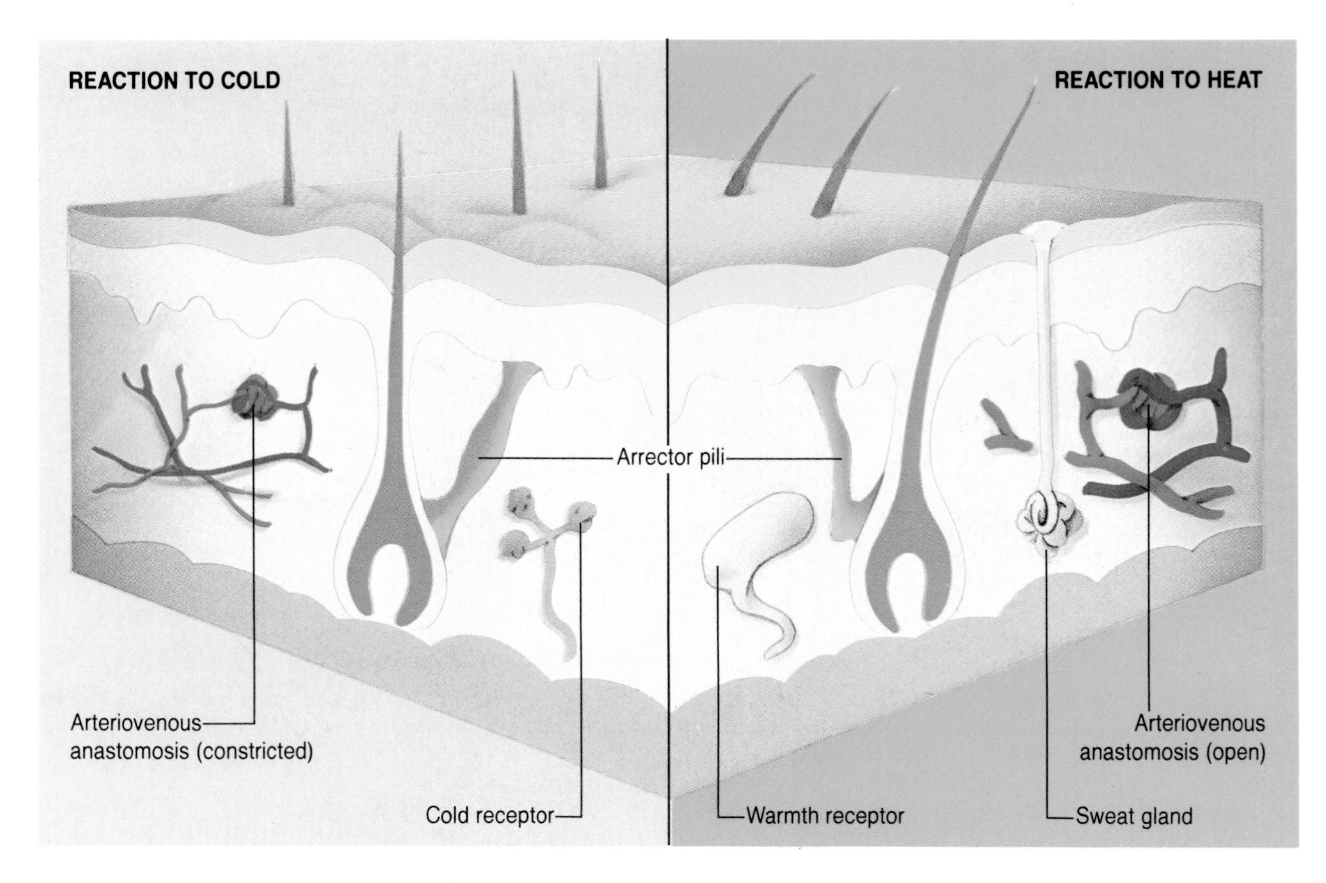

In winter's cold or summer's heat, the skin reacts to the body's needs. In the cold *(left)*, tiny arrector muscles pull hairs erect to trap an insulating layer of air next to the skin. The anastomoses between capillary veins and arteries become narrowed to reduce blood flow to the cool skin surface. Heat or vigorous exercise causes opposite effects *(right)*. The arrector muscles relax, and sweat glands pour moisture onto the skin, where its evaporation produces cooling. Anastomoses widen, so that blood reaches the skin surface to shed heat.

increase the skin's acidity are produced. Bacteria thrive in moist areas, so the dryness of the skin probably accounts for most of its resistance to bacterial infections. When the skin becomes too moist, its resistance decreases and harmful bacteria break through normal defenses.

Keeping the temperature normal

The skin plays a prominent role in maintaining the body's temperature. The job is a constant one, requiring small adjustments to ensure that the body's core temperature does not stray from the narrow range at which the organs function efficiently. This temperature – which differs slightly from person to person – averages 98.6°F. Called the set point, it is analogous to the setting of a thermostat. The system is so delicately balanced that if the core temperature varies by 1.5 degrees, the body's metabolism is altered by about 20 percent.

When body heat shifts slightly from the set point, the heating or cooling mechanisms quickly restore the proper temperature. The skin achieves this regulation largely by controlling the amount of heat lost. To do so, it works in concert with the hypothalamus, a cluster of nerve cells at the center of the brain. Specialized regions of the hypothalamus contain heat-sensitive and cold-sensitive cells which respond to changes in blood temperature by increasing the number of nerve impulses they transmit. On receiving these commands, the skin hastens to make the appropriate adjustments in its domain.

Even in stable surroundings, this temperature-regulating system continues to function all the time, for although the body constantly produces heat, it also constantly loses heat. At a moderate room temperature, the body loses most heat through what scientists call radiation – rays of heat that emanate in all directions. Other objects in the room constantly radiate heat back to the body, but a body warmer than its surroundings always loses more heat than it gains.

Responding to the cold and heat

Most cold detection occurs on the body's periphery, which contains abundant specialized nerve endings called thermoreceptors. The skin has far more receptors to detect coldness than receptors to detect warmth. The number of cold receptors also varies from one region of the body to the next. The skin on the lips contains about 20 times more cold receptors than the skin on the chest or legs.

Cold receptors relay their signals via the spinal cord to the hypothalamus, which sends impulses to structures in the skin called arteriovenous anastomoses. These are specialized connections between veins and arteries that intermittently substitute for capillaries throughout the circulatory system when blood flow needs to be rerouted. They are particularly abundant in the hands, feet, eyelids, nose and lips. The middle portion of an anastomosis is a thick, muscular wall that contracts, restricting blood flow to the extremities and thus reducing heat loss. This is why our lips and fingernails turn slightly blue when we are cold.

With restricted blood flow, tiny lumps, resulting from erection of the hair shafts, cause goose bumps to appear on the skin. This action, or piloerection, is largely useless in humans. In animals, it makes fur stand on end and helps create an extra layer of warmth by trapping warm air next to the body.

The body reacts to excess heat by taking measures to encourage heat loss. Sweating begins and blood vessels in the skin dilate to allow more blood to reach the surface. The blood dissipates more body heat, accounting for the flushed, rosy glow of the skin during vigorous exercise. This same mechanism causes blushing. The layer of sweat that builds up on the skin's surface helps cool it by evaporating.

The pigmentation of the skin

The wide variety of human skin color is a direct measure of the diverse climates to which man has adapted. All skin color stems from the same substance, the pigment melanin. Two forms of melanin color the skin, hair and eyes of humans. The major pigment, eumelanin, produces shades of brown and black; the other, phaeomelanin, is the pigment of red hair.

Melanocytes are spider-shaped, with long irregular arms that reach out from the cell body. The arms of each melanocyte link it with about ten surrounding cells. Melanocytes inject pigment granules, melanosomes, into these neighboring cells, thus spreading pigment across the skin. Freckles, a result of the pigment's gathering in clusters, can occur in all skin colors but are most prominent in light skin.

Regardless of skin color, all human beings have approximately the same number of melanocytes in the skin – around one percent of all skin cells. Differences in skin color are due to the amount of melanin the cells produce, which in turn depends on how much ultraviolet radiation they must absorb. Melanin absorbs ultraviolet rays and converts them into harmless infrared rays. The more melanin there is in the body the more efficient this absorption becomes, and the better the protection afforded.

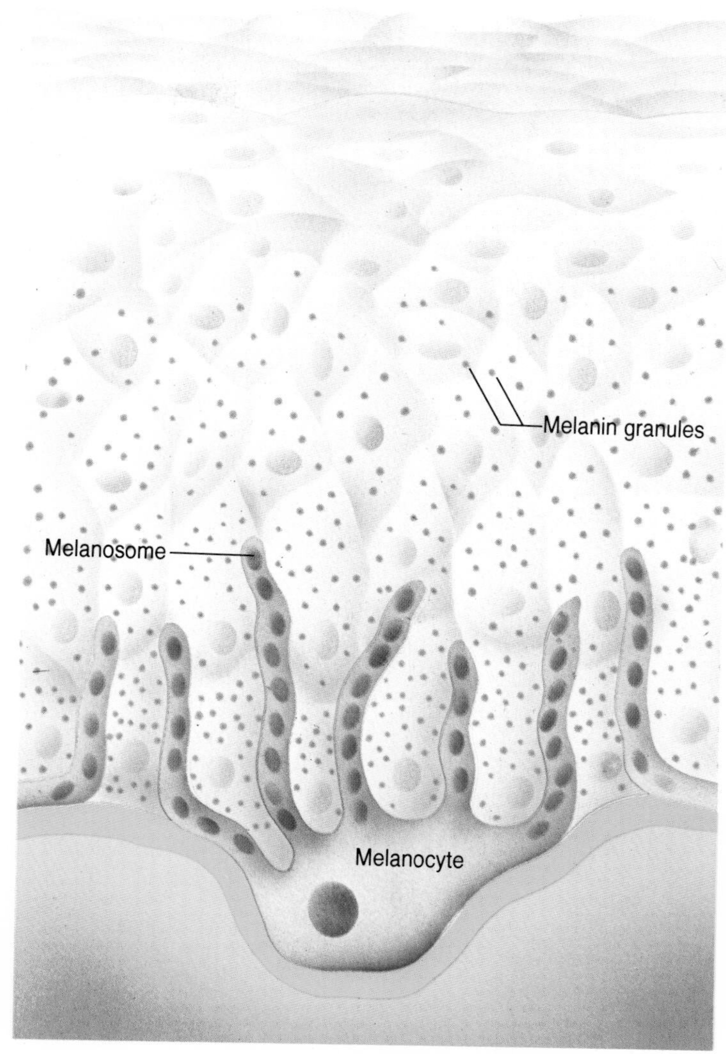

A golden suntan results from the activities of pigment cells in the skin. Called melanocytes, they are triggered into action by ultraviolet radiation from the sun or a sunlamp. They increase their output of the dark pigment melanin and inject it into nearby epidermal cells. The cells undergo their normal migration toward the skin's surface, so that in four or five days they arrive there, tanning the skin and forming a filter to block some of the sun's rays and preventing possible damage.

The hairy human

Like all mammals, human beings are covered in hair. But unlike the thick fur coats of four-footed members of the group, our "fur" consists, for the most part, of fine hairs that grow few and far between. The only thick growths of hair are in the pubic region, under the arms, on the head and on the faces of men – although children have no pubic, axillary or facial hair and many men lose much of their head hair after middle age.

Every square inch of the body – except the lips, palms of the hands and soles of the feet – has hair growing on it. Some is so fine as to be almost invisible, although it reveals its presence as the site of every goosebump when the body is cold. Each hair grows like a stem of grass from a bulblike follicle, which tunnels through the surface of the skin. The root forms inside the follicle and the shaft grows out of the skin. A combination of black, brown and yellow pigments give the hair its color. Cells in the outer layer, or cuticle, of the shaft overlap like a fish's scales. A fine nerve fiber encircles the follicle, which also has a cluster of sebaceous glands draining into it. They coat the hair with oily sebum, which lubricates and protects it.

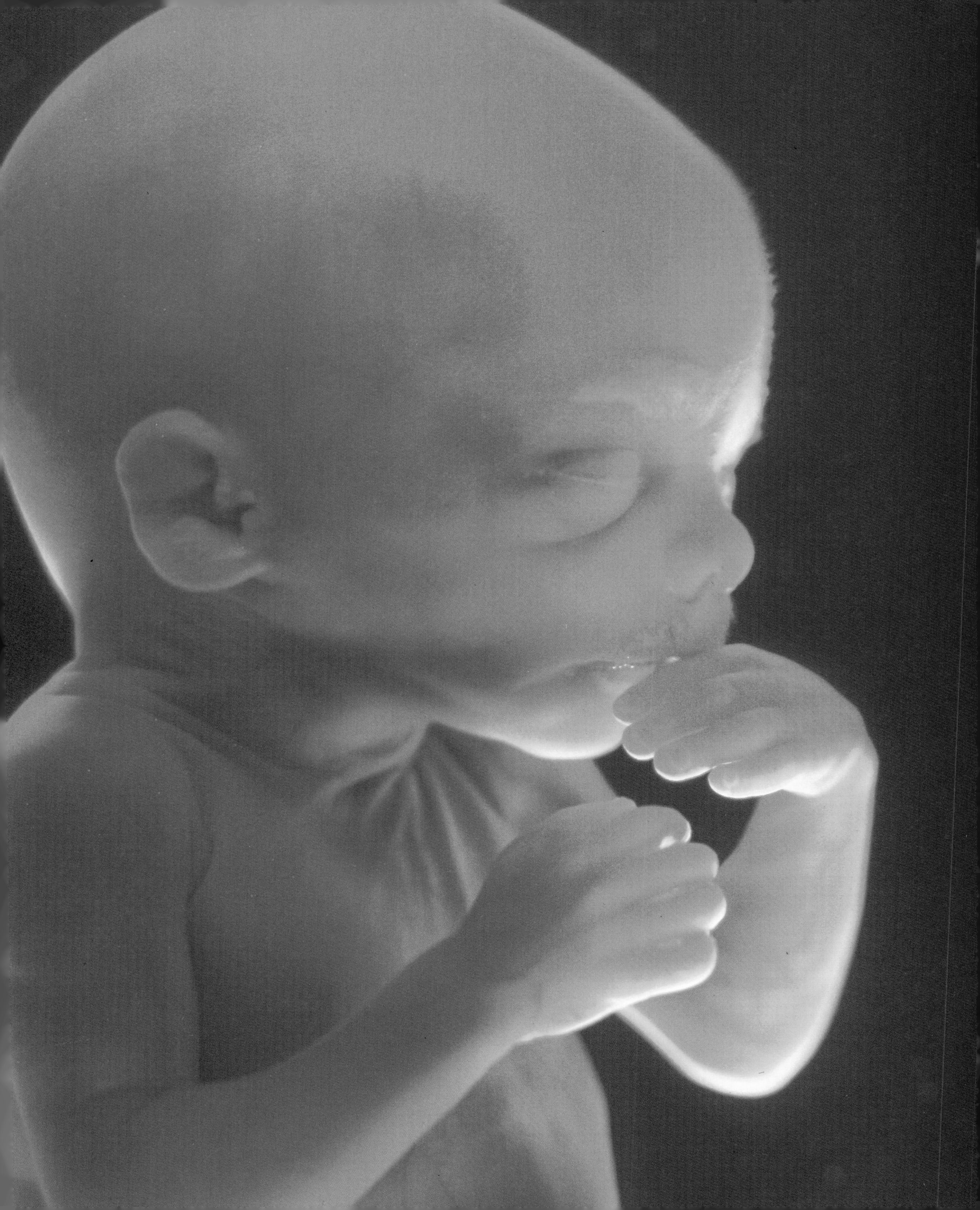

CHAPTER 5

The Cell

The cell is the smallest unit of living matter. Every type of tissue in the body, from bones and brain to muscles and skin, is made of cells. Blood is based on watery plasma fluid; but it is thicker than water because it contains millions of blood cells. And the sperm and the egg – those microscopic scraps that come together to form a new human being – are each only a single cell.

Cells group together to form tissues, and tissues in turn form organs and body systems. Each type of tissue has its own type of cell, and as a result there are hundreds of different kinds of cells in the body. Superficially a typical cell resembles a blob of jelly in a membranous bag. But it can replicate and sustain itself within its environment and it can grow. It takes in what it needs to function, and produces other substances that the body needs to stay alive.

A cell therefore consists of the essential components needed for the life process, contained within its enveloping membrane. These essentials are the chemicals and enzymes which react to provide the cell with energy and building materials, together with genetic material which bears the information for producing cell components and for cell replication. The cell is like a factory which is supplied with raw materials and processes them to make products. The products themselves are used either to make components for the factory's production lines or are transported out of the cell to be used elsewhere.

Two kinds of cells

Essentially, there are two types of cells – procaryotic cells and eucaryotic cells. The two differ fundamentally in the way that they are organized within. In procaryotic cells, the arrangement is simple. The chemicals and enzymes needed to produce energy, and those needed for cell growth and division, are contained in the cytoplasm – a complex jellylike mixture – which is packaged in a plasma membrane that forms the cell boundary. These cells do not have a nucleus, and their genetic material – deoxyribonucleic acid (DNA) – is attached to the plasma membrane. Procaryotic cells were probably the first sort of cells to appear on Earth. Most procaryotes today are single-celled organisms, notably bacteria and blue-green algae, and none is found in the human body – or in any other living animal.

Eucaryotes, on the other hand, are much larger than procaryotes, and have a far more highly evolved and complicated internal arrangement. They are typical of the cells in the human body. In them, the genetic material is contained within a nucleus, which is surrounded by nucleoplasm and bounded within its own membrane. Surrounding this nuclear membrane is the cytoplasm.

Generally, eucaryotes are highly specialized cells able to perform specific functions by the action of internal structures called organelles. These intracellular compartments, usually within the cytoplasm, coordinate cell chemistry. They also make essential cell chemicals and export the products manufactured by the cell. The organelles, by concentrating cell components

Explosive growth and multiplication is the fate of the single cell – a fusion of sperm and egg – that is the origin of a new human being. In only 40 weeks, the microscopic fertilized egg duplicates and reduplicates until it is a newborn baby 20 inches long. The first few cells that go to form the embryo are identical. But soon they begin to differentiate, becoming different kinds of cells to form different kinds of tissues. The instructions about how to do this, and the secret of the newborn's physical and mental characteristics, are contained in genes located on chromosomes in the nucleus of that first single cell.

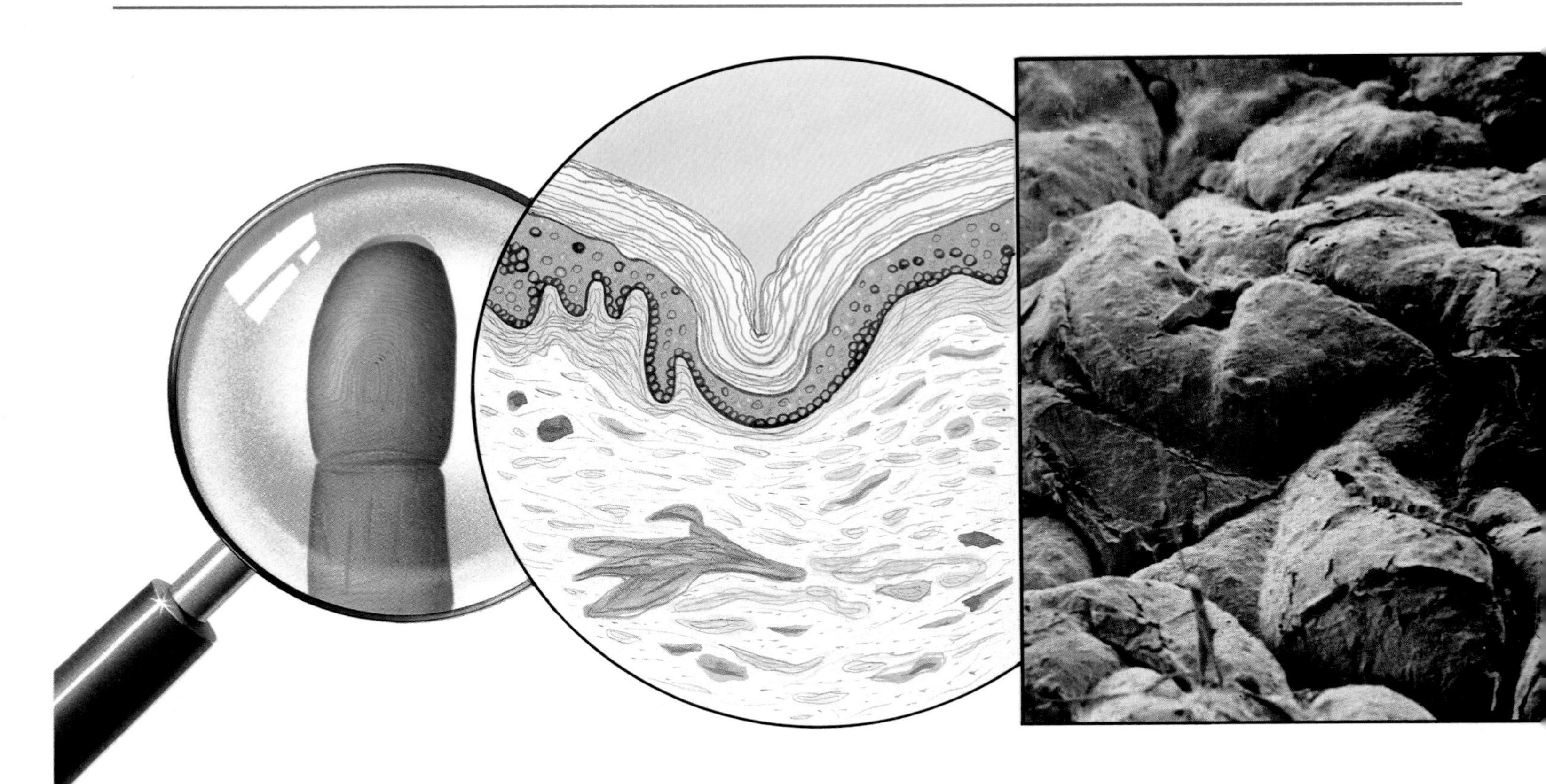

together, can make the cell's internal biochemical reactions more efficient, and can "package" potentially dangerous chemicals, effectively isolating them and preventing them from poisoning or destroying the cell from within.

Deep in the valley between two fingerprint ridges, individual skin cells can be seen flaking *(right)*, like bolders of mudstone cracking in the sun. The skin of the fingers has a unique pattern of whorled ridges and grooves, one of which is shown in colored section *(middle)*. Cells in the subcutaneous tissue just below the skin's surface die and are sloughed off as they reach the surface. The electron micrograph shows this process in remarkable close-up.

A plan for life

The central nucleus is the largest and most prominent of the many organelles in a human body cell. It contains all the information needed by the cell to carry out its functions – the manufacture of components and self-replication. It houses the blueprint or program for the life process.

Within the nucleus is the nucleolus – a structure involved in protein manufacture via ribonucleic acid (RNA), which works in collaboration with DNA – and the nucleoplasm, which is the cell's "reference library." This file of genetic information appears in the form of DNA packaged into a number of chromosomes.

These chromosomes are built of proteins, called histones, combined with tightly packed coils of DNA; the combination is called chromatin. The long DNA molecule is spooled twice around sets of eight histones to form a nucleosome, and numerous nucleosomes wound with the same DNA strand form a chromosome. When cells multiply, or replicate themselves, genetic information is passed on in exactly the same form. Just before mitosis (cell division), the chromosomes are duplicated. This means that the two daughter cells created when the cell divides receive an identical copy of each of the cell's chromosomes.

The nuclear membrane – the double-walled structure that envelops the nucleus – is perforated with submicroscopic holes. This arrangement allows a two-way flow of molecules between the nucleoplasm in the nucleus and the cytoplasm immediately outside it. In this way, proteins needed to build the nucleus and keep it working enter the nucleus from the cytoplasm, where

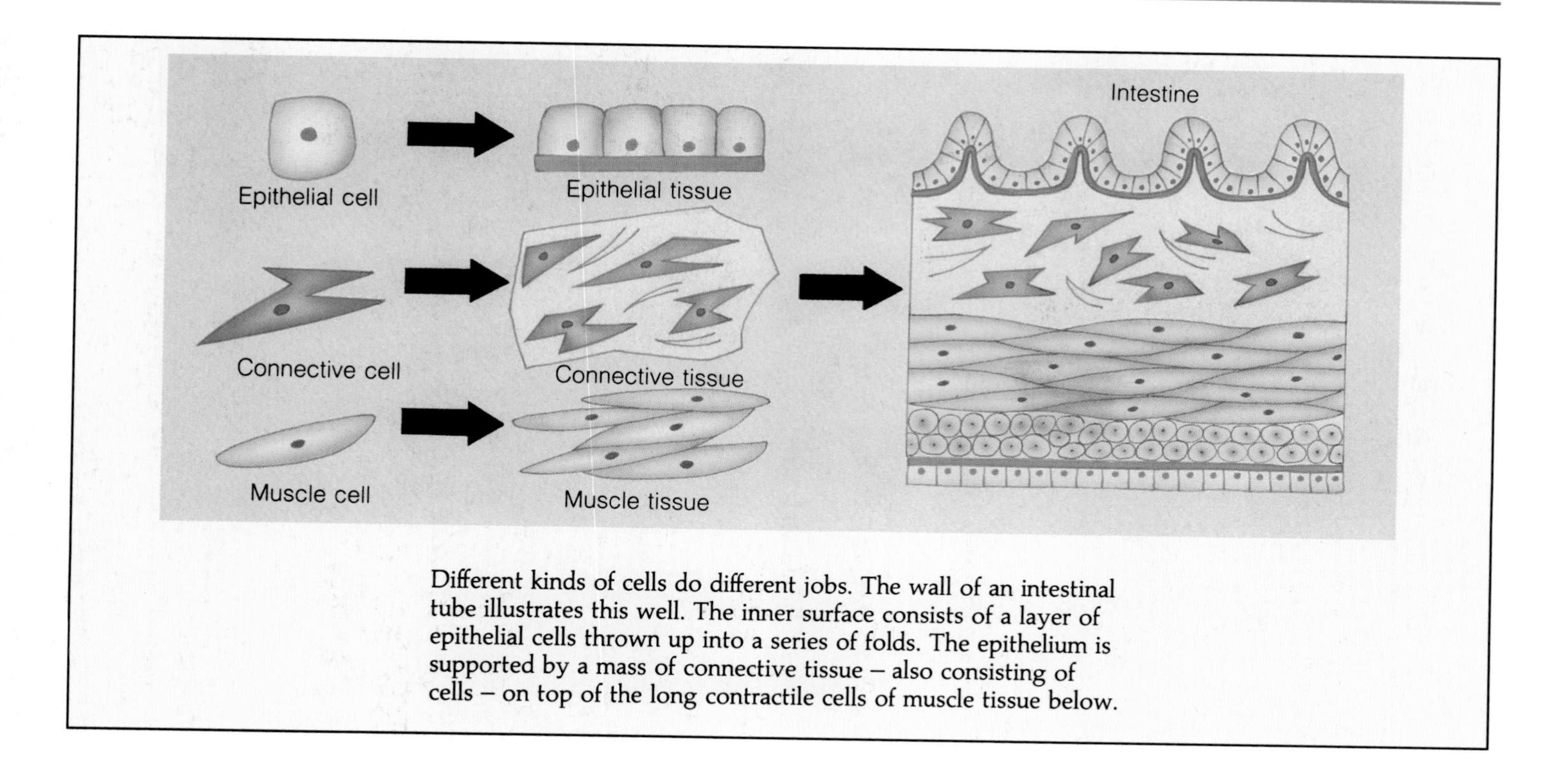

Different kinds of cells do different jobs. The wall of an intestinal tube illustrates this well. The inner surface consists of a layer of epithelial cells thrown up into a series of folds. The epithelium is supported by a mass of connective tissue – also consisting of cells – on top of the long contractile cells of muscle tissue below.

they are processed. In return, the RNA essential for protein synthesis passes from the nucleoplasm out into the cytoplasm.

The driving force

In human body cells one type of organelle has the specific job of energy production. This is the mitochondrion, the power house of the cell. Mitochondria have a double-membrane structure. The inner membrane is convoluted into folds like hanging drapes to give a large surface area. The folds trap enzymes, proteins that act as catalysts, helping to break down sugars and fatty acids to release energy. In some cells, fat droplets in the cytoplasm provide the fatty acids that are a major source of energy for the mitochondria. In cells that have a very high energy requirement, such as the liver, the mitochondria may account for as much as 20 percent of the cell.

Mitochondria behave like tiny cells within a cell. And unlike all other organelles they can reproduce themselves. Each is controlled by its own small, circular DNA molecules, not by the genetic material contained in the nucleus. In this way they are inherited differently from all other cell components. Sometimes all the mitochondria in an organism are derived from its mother, unlike nuclear DNA, which is derived equally from both parents. This feature suggests that mitochondria evolved from bacteria that once coexisted with primitive eucaryotic cells and have evolved to such a level that they now live within all living creatures and direct their own metabolism – making them the most successful life forms ever.

The bricks of life

Other types of organelles – ribosomes – are essential for protein synthesis. Ribosomes, which look like minute spheres within the cell, line up on the endoplasmic reticulum, a membrane system enfolded through the cytoplasm. The ribosomes are composed of RNA and protein, which in turn is made up of amino acids linked in chains.

Proteins are vital to life. They perform most of the cell's biochemical reactions and are the structural material from which the cell itself is made. So they function both as the toolkit and framework of the cell. What distinguishes one protein from another is the sequence in which its amino acids are linked. This, in turn, is determined by the coding of the DNA chromosome, which directs their manufacture.

In protein synthesis, first the DNA of a gene is copied into RNA. This passes out of the nucleus into the cytoplasm. There ribosomes, enzymes and other RNAs come together to translate the gene's RNA structure into the amino acid sequence of a specific protein. In this complex process, three kinds of RNA molecules interact during synthesis. Messenger RNA (mRNA) "read" from the DNA provides the coding for the amino acid sequence, transfer RNA (tRNA) carries the amino acids that are being assembled, and ribosomal RNA is bound into particles that link the amino acids to one another. The ribosomes have three slots on their surface: one holds the mRNA and the other two hold the tRNAs while enzymes bound to the ribosomes link the amino acids in the coded sequence to form the protein.

Synthesizer and sorter

The endoplasmic reticulum, on which the ribosomes are clustered, has two important jobs to do in the cell. It helps in the synthesis of large molecules such as proteins, lipids and complex carbohydrates that make up some of the other organelles in the cell. And it also separates newly synthesized molecules needed by the cytoplasm from those intended for transport to other sites. It consists of a membrane folded in the form of a crumpled egg-shaped sac that surrounds the nuclear envelope. This enclosure is the cisternal space, or lumen.

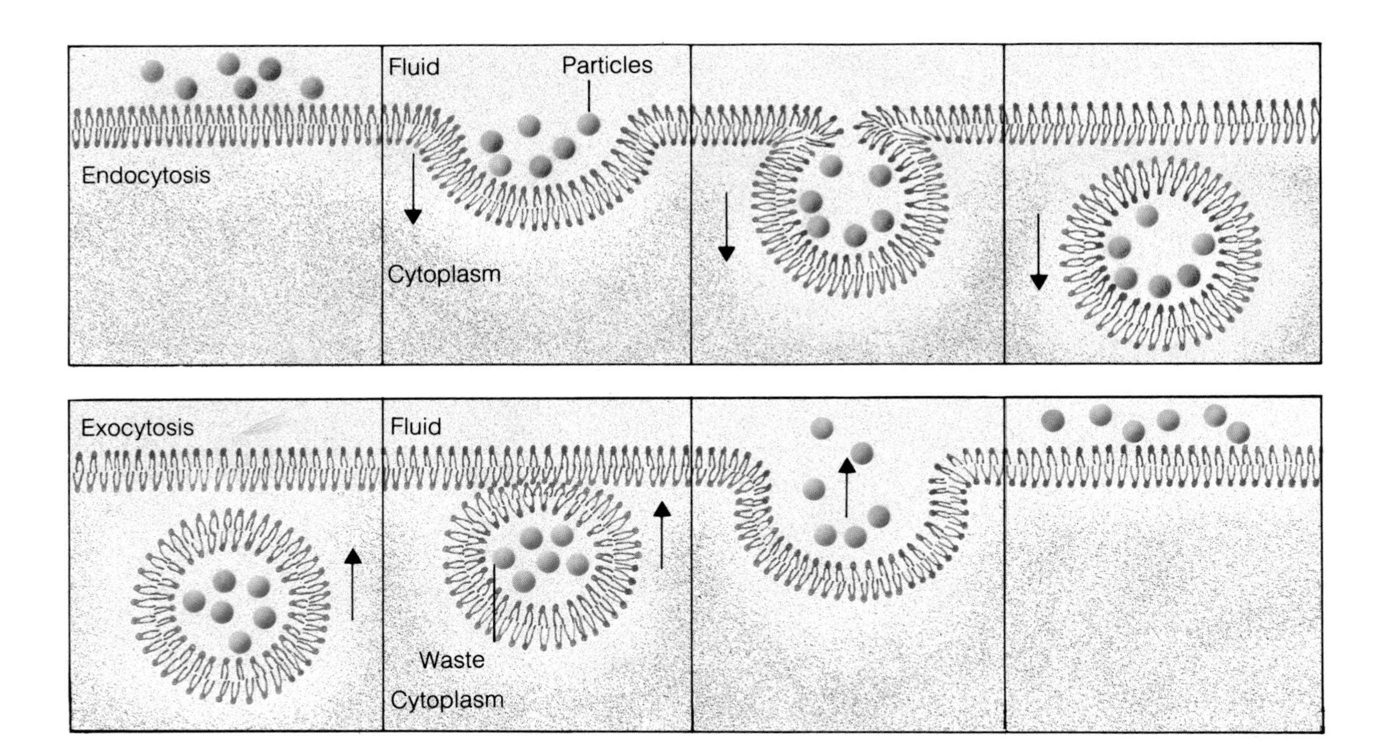

The folds of the membrane farthest from the nuclear envelope form a boundary in the cytoplasm. Proteins are assembled on the cytoplasmic side of this boundary and must pass through it into the cisternal space if they are to be exported. The endoplasmic reticulum also separates potentially dangerous proteins, such as digestive enzymes, from the rest of the cell as they are made.

Studies of the cell using an electron microscope reveal two distinct forms of endoplasmic reticulum. One is rough or granular, with ribosomes attached to its outer surface, and is a site of lipid synthesis. Smooth endoplasmic reticulum is composed mostly of fine hollow tubes, unlike the rough form, which has a flattened sac-shaped interior.

The amounts of rough and smooth endoplasmic reticulum in any cell depend on what the cell has to do in the body. Cells that make large quantities of proteins for export, such as B-lymphocytes, in which antibodies are formed, have a large proportion of the rough form, whereas cells that specialize in processing lipids – for instance in the production of steroid hormones – have much more of the smooth type.

Proteins and membranes made by the endoplasmic reticulum are transported to their destination by an ingenious mechanism in the cell – in "buds" or vesicles that sprout from the endoplasmic membrane. If the cell surface is the intended destination, then the buds fuse with it and release the newly formed substance. If a cell organelle is the destination, then the bud fuses with it and the substance is released inside it.

Packaging proteins

Many of the vesicles containing substances synthesized by the endoplasmic reticulum are destined for the Golgi apparatus – a stack of interconnecting membranes and spaces near the nucleus. This miniature packaging factory within the cell is probably important in the final processing of the newly synthesized proteins that reach it. After processing, the protein is trans-

Particles passing in and out of cells are packaged for the purpose. Infolding of the double-layered cell membrane during endocytosis allows food particles to enter the cell's cytoplasm from the fluid exterior. The membranous fold gradually encircles the particles and forms a vesicle in which they are transported within the cell. Particles of waste products are first packaged in a vesicle and transported to the cell wall, where the reverse process of exocytosis enables them to pass through the membrane and out of the cell.

The dominant structure inside a cell is the nucleus. Surrounding it, within the jellylike cytoplasm, are numerous other organelles, smaller structures which carry out various cell processes. Ribosomes (red dots) cling to the folds of the rough endoplasmic reticulum (pale blue), which is the site of protein synthesis. The sausage-shaped mitochondria (red) are the power houses, within whose folded inner membranes enzymes break down sugars to generate energy.

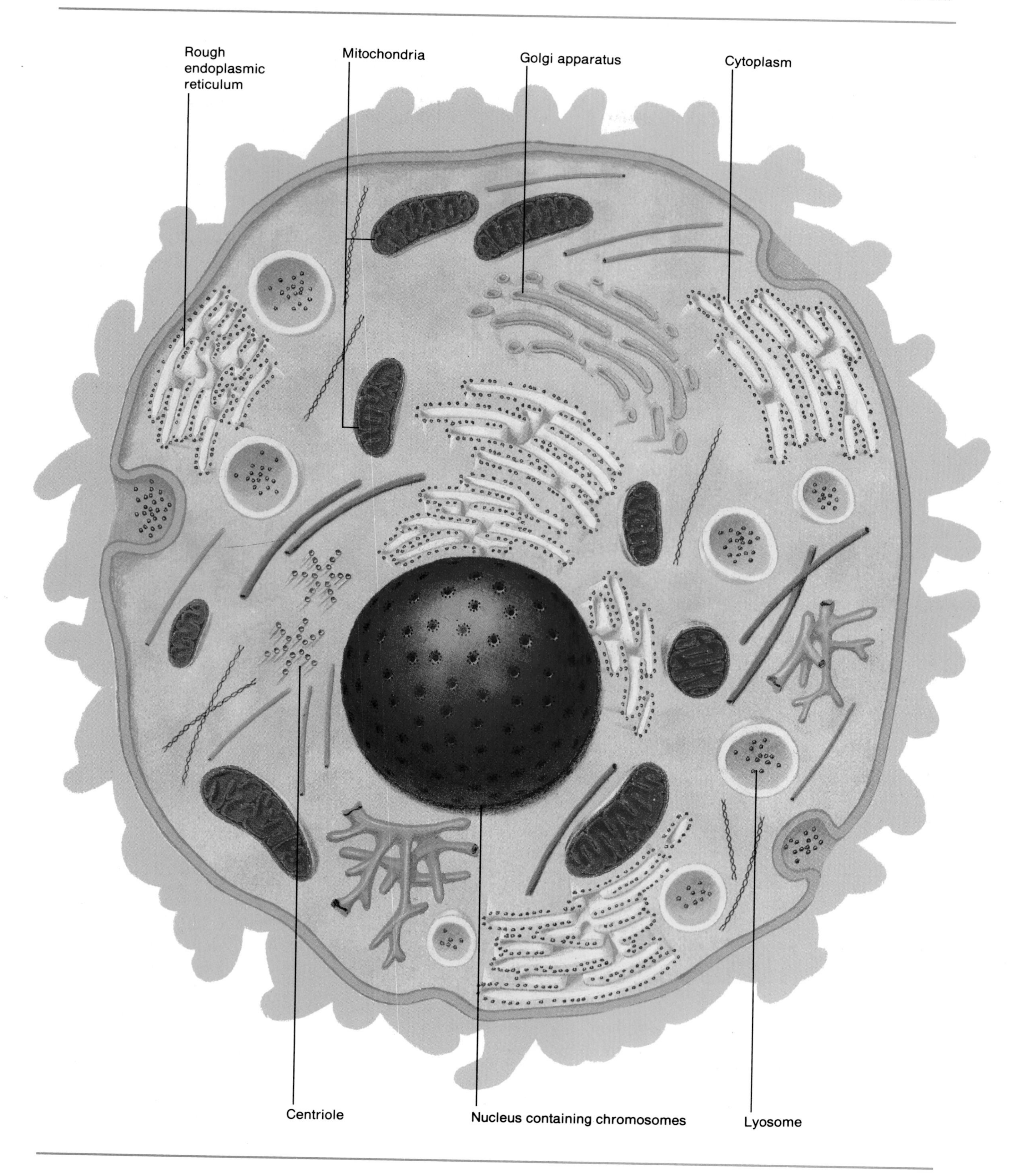
Rough endoplasmic reticulum
Mitochondria
Golgi apparatus
Cytoplasm
Centriole
Nucleus containing chromosomes
Lyosome

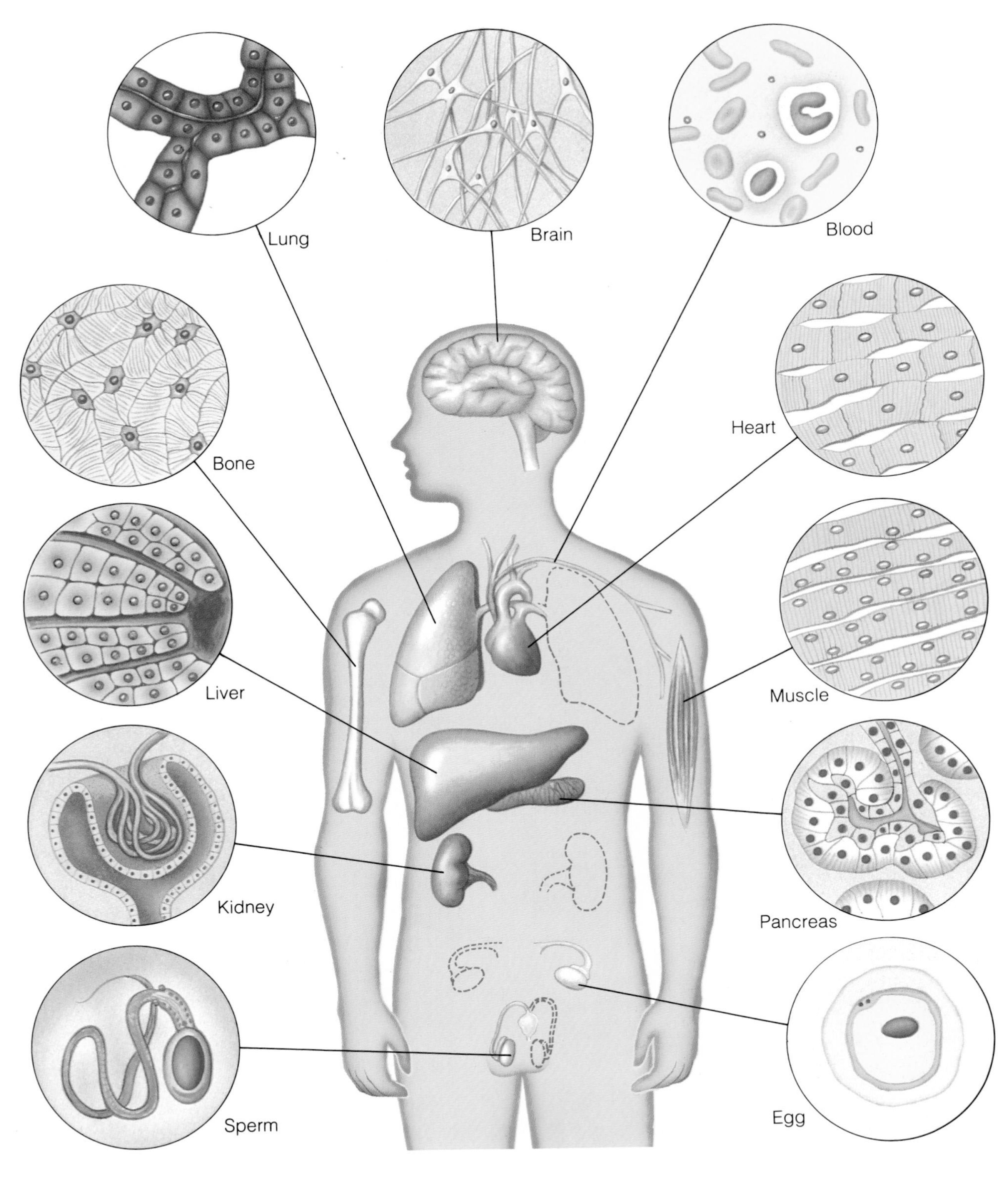

Dozens of different kinds of cells form the various body tissues. All except red blood cells share the common feature of a central nucleus (shown green), but their shapes can be very dissimilar. The neurons, or nerve cells, in the brain have many tendrillike branches to make contact with neighboring nerve cells, whereas bone cells amass mineral particles locked into a rock-hard matrix.

Like a tumbling stack of dimpled cushions, mature red blood cells pass along an artery. Instead of a nucleus and other organelles, they contain hemoglobin to carry life-giving oxygen to serve the needs of other cells.

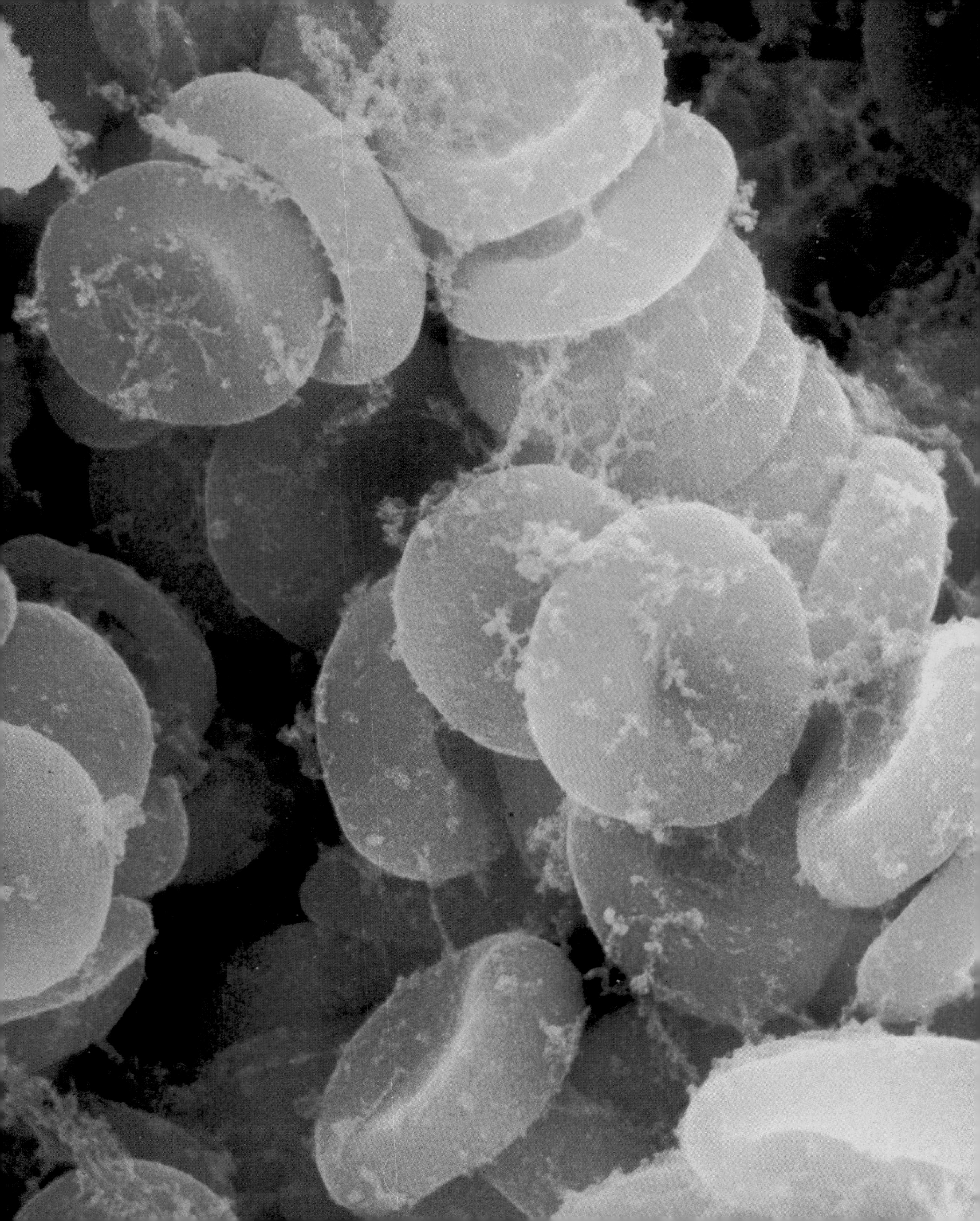

ported, also in vesicles, to its destination.

The Golgi apparatus and the endoplasmic reticulum manufacture two organelles – lysosomes, responsible for digestion inside the cell, and peroxisomes, the cell's detoxification plants, which break down alcohol and other potential poisons to prevent them from killing the cell. The packaging of substances into "cartons," or vesicles, is important to cell safety. Both digestion and detoxification involve powerful chemicals, such as enzymes and oxygen, which would damage the rest of the cell should they be free to come into contact with it.

The dividing line

All human body cells are bounded by an outer plasma membrane. It contains the cell and isolates the contents from the environment. But, equally important, the membrane is a wall with doors in it. By opening and closing its doors, it regulates the passage of substances into and out of the cell. The plasma membrane is only about one four-thousandth of an inch thick. And its structure is essentially the same in all living organisms. It is composed of proteins sandwiched between two layers of fatty phospholipids.

The phospholipids can be pictured as tadpole-shaped molecules with hydrophobic (water-hating) heads and hydrophilic (water-loving) tails. Thousands of the "tadpoles" line up side by side and tail to tail in two layers to form the plasma membrane. Scattered among the ranks of phospholipids are "islands" of protein material which act as "doors" in the membrane wall. All interactions between the environment and the cytoplasm inside the cell must involve this membrane. Molecules outside the cell may react with membrane proteins. This opens the "doors" in the wall and makes them suitable to cross the membrane. The proteins may also bind the cell to its neighbors, so that they can work together in unison.

Some small molecules, such as oxygen and carbon dioxide, can cross the plasma membrane without opening doors. They do so simply by diffusing from a region of high concentration to one of low concentration. Others, such as glucose or sodium and potassium ions, need an active transport process – the doors open only if there is energy available. The sodium-potassium pump, by which molecules of ATP (adenosine triphosphate) are broken down to release energy, which in turn controls the amount of salt inside a cell, is just such a process.

Molecules may also open the doors in the plasma membrane by means of a special cell vehicle. One such vehicle is created as a dimple, which forms in the membrane and engulfs the incoming substance – perhaps a bacterium, which is "eaten" by the cell and destroyed. This is phagocytosis. Another similar mechanism is pinocytosis, the "eating" of fluid by the membrane.

The great divide

Growth, renewal and repair are fundamental features of all life. All are made possible by mitosis – the process by which cells reproduce themselves and pass on their characteristics to the next generation. The genetic information is carried in DNA (deoxyribonucleic acid), the molecule of life which, together with protein, makes up the bulk of the nucleus. The genes are strung onto the chromosomes like beads on a necklace. But just how the genes might be handed down remained a mystery until the mid-20th century, when researchers James Watson, Francis Crick, Maurice Wilkins and Rosalind Franklin, working at Cambridge University in England, identified the DNA molecule and ex-

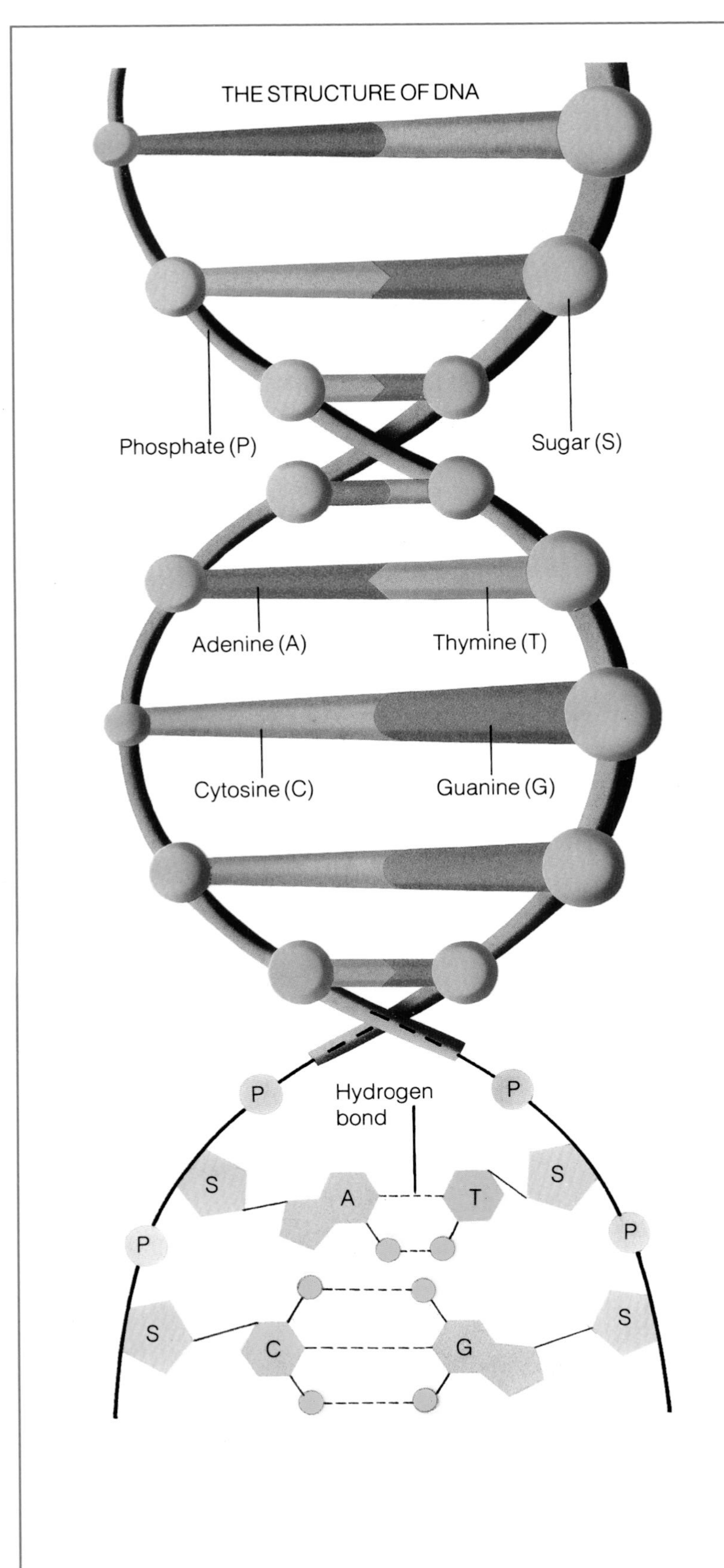

Three key components combine to form the long double helix molecule of DNA, full name deoxyribonucleic acid *(left)*. The sides of the strands, resembling a twisted ladder, are formed of alternate sugar (pale green) and phosphate (purple) groups joined in long chains. Each "rung" of the ladder consists of two nitrogen-containing bases, either adenine (dark green) and thymine (orange) or cytosine (red) and guanine (blue). The bases are held together at mid-rung by comparatively weak hydrogen bonds. DNA makes up the genetic material of every cell, located on chromosomes inside the nucleus.

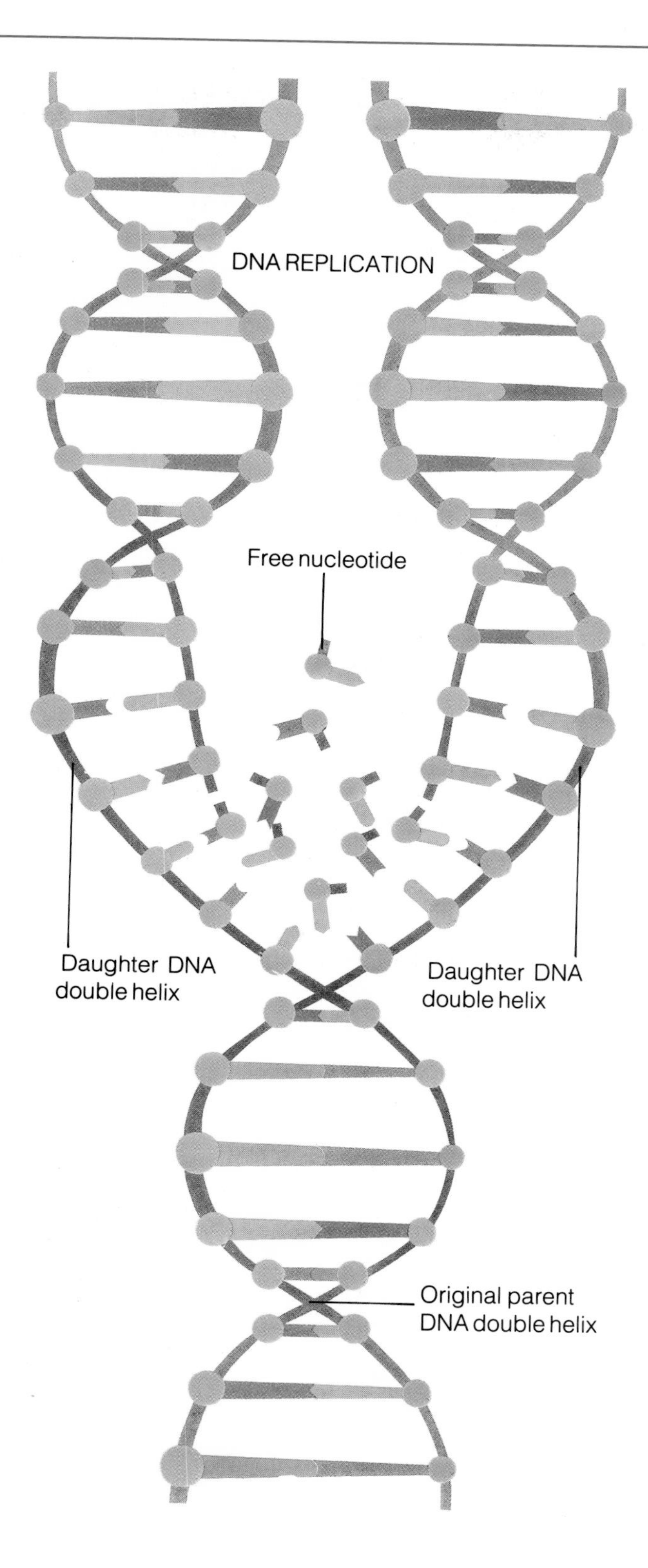

DNA unzips down the middle when it replicates during cell division. The "rungs" of the parent helical ladder (lower part of diagram, *left*) break at their weak hydrogen bonds, forming two "half ladders" (*center* of diagram). Free nucleotides – a base attached to a sugar and a phosphate – come along and attach to their usual partner, re-forming the rungs and creating two new and identical daughter DNA strands *(top)*.

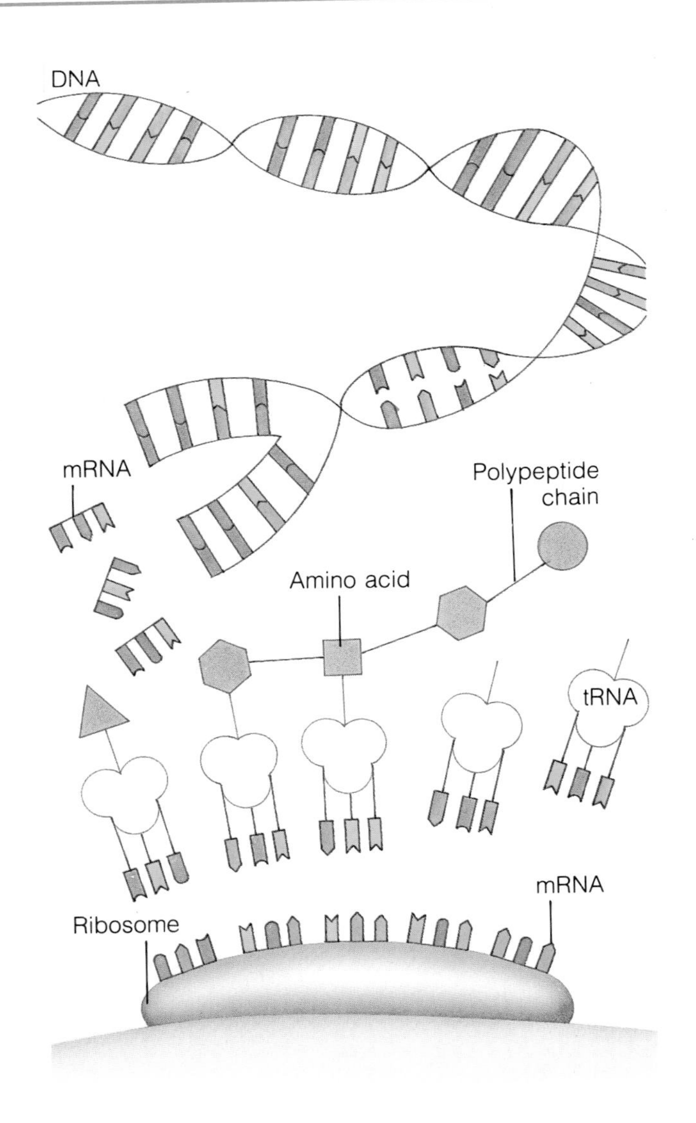

Production of proteins within a cell also involves the unzipping of DNA *(left)*. The sequence of bases on the unzipped strands acts as a code and is reproduced as messenger RNA (mRNA), which attaches itself to a ribosome. Molecules of transfer RNA (tRNA) with the complementary sequence of bases arrive, each locked onto one of 20 or so different amino acids. The acids separate from the RNA and link to each other to form a polypeptide chain, the structural unit of a protein.

Cell duplication, the basis of growth and repair, is called mitosis. Its key stages are illustrated on the opposite page. At the beginning of the protophase *(top right)* the duplicated chromosomes become visible in the nucleus and locate themselves on spindle fibers. At metaphase they align across the center of the cell, and in early anaphase they each split in two. Late anaphase sees the chromosomes move apart toward the centioles, as the cell prepares for telophase by constricting across the middle. Finally it splits completely, so that each daughter cell is an exact replica of its parent cell. During interphase the DNA in the two new nuclei replicate, ready for another division.

plained its structure and function.

DNA is a giant molecule in the form of a double helix resembling a twisted rope ladder. Each side of the ladder is a single strand made up of alternate sugar and phosphate molecules. The sugars are the points where the "rungs" meet the strand and the phosphates are the rope between the rungs. The rungs of the ladder are formed by pairs of nitrogen-containing bases linked to the sugars and joined by hydrogen bonds.

Each unit of the helix, comprising a sugar, a phosphate and a base – equivalent to a knot, the rope between knots and the half rung – is a nucleotide. There are four nitrogen bases – adenine and thymine, and cytosine and guanine – which always pair to form the rungs. The number of nitrogen bases on the helixes number many thousands. It is their exact order and sequence that form the genetic code, the vital information which controls the synthesis of various different proteins in the cell.

Before the cell divides, the cell contents go through a preparatory stage, during which the DNA replicates or forms identical copies of itself. In this remarkable natural replication mechanism, the twisted rope ladder arrangement separates lengthways as the hydrogen bonds linking the half-rungs break. On each strand, nucleotides within the nucleus attach to the bases to rebuild the structure, and so two molecules of DNA are formed. They are identical because the order of the bases is retained on each newly separated strand, and the bases always pair with the same partner.

After DNA replication, the nuclear membrane breaks down and the DNA can be seen within the cell as distinct strands wrapped in protein. These are the chromosomes, each of which contains two strands of DNA. The strands divide into two and each moves to opposite sides of the cell. As the cell divides, each daughter cell receives an identical complement of DNA with identical coding. Then the strands "unravel" in the nucleus of the new cells. They continue their essential work within the cell, but are much more difficult to discern as individual entities.

From strength to strength

Just as cells may take in food particles by first enveloping them in the cell wall or membrane, so, by a reverse process, waste products packaged in similar compartments are shipped to the membrane. The door-opening mechanism out of the cell works like this: the walls of the vesicle fuse with the membrane to form an opening to the outside of the cell, and the waste is ejected like garbage out of a rubbish chute.

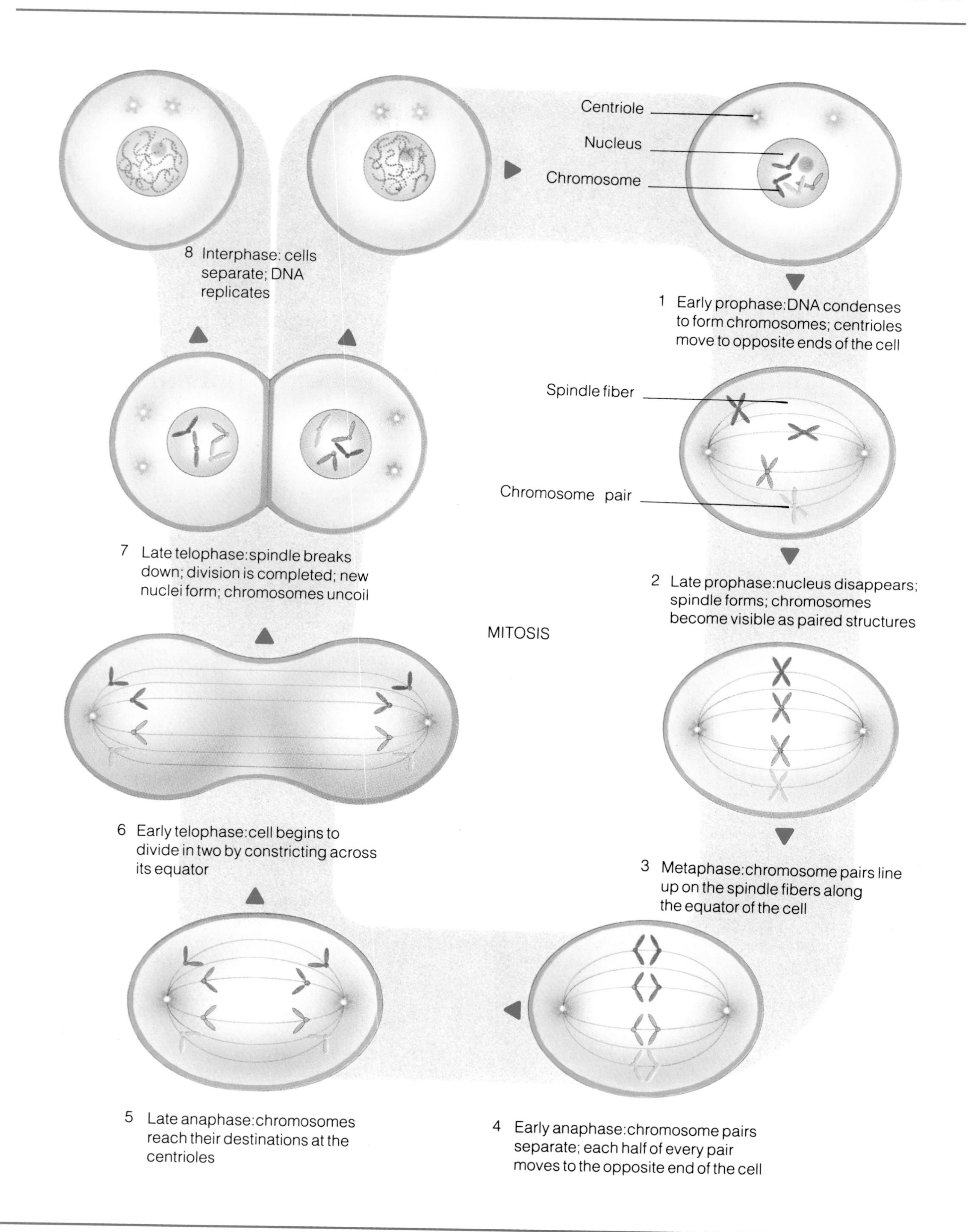
Centriole
Nucleus
Chromosome
8 Interphase: cells separate; DNA replicates
1 Early prophase: DNA condenses to form chromosomes; centrioles move to opposite ends of the cell
Spindle fiber
Chromosome pair
7 Late telophase: spindle breaks down; division is completed; new nuclei form; chromosomes uncoil
2 Late prophase: nucleus disappears; spindle forms; chromosomes become visible as paired structures
MITOSIS
6 Early telophase: cell begins to divide in two by constricting across its equator
3 Metaphase: chromosome pairs line up on the spindle fibers along the equator of the cell
5 Late anaphase: chromosomes reach their destinations at the centrioles
4 Early anaphase: chromosome pairs separate; each half of every pair moves to the opposite end of the cell

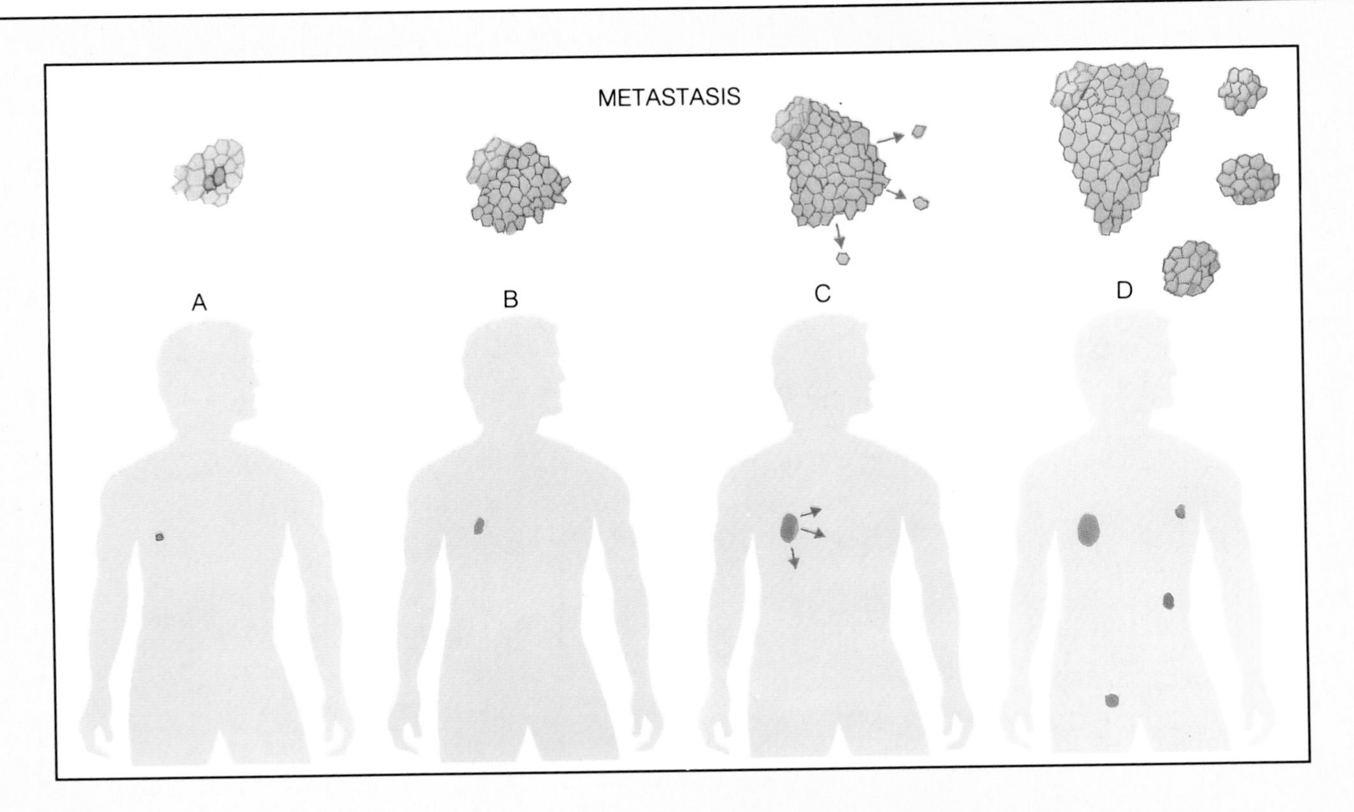

Cancer occurs when tumor cells multiply out of control and, if unchecked, spread through the body by the process of metastasis. For some as yet unknown reason, cells replicating faster than natural wastage requires (A) gives rise to a primary tumor (B), which after it exceeds a certain size begins to shed cancer cells (C). These travel in the bloodstream or through the lymphatic system, where they initiate the growth of secondary tumors at other sites in the body (D). A primary tumor in the lung, for example, may be tolerable for a while and operable if discovered. But secondary tumors in critical organs such as the liver and brain are frequently the cause of death from cancer.

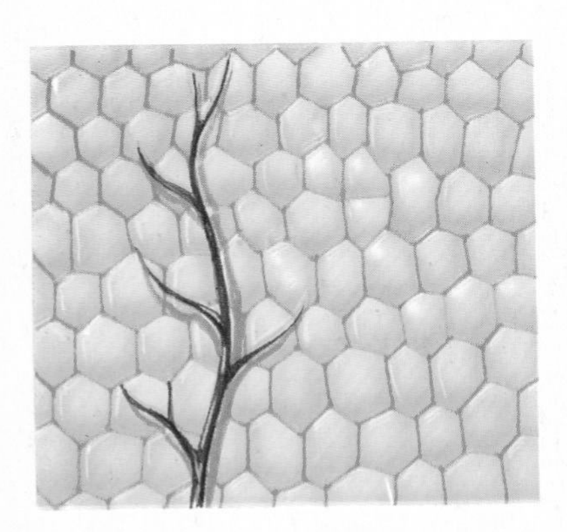

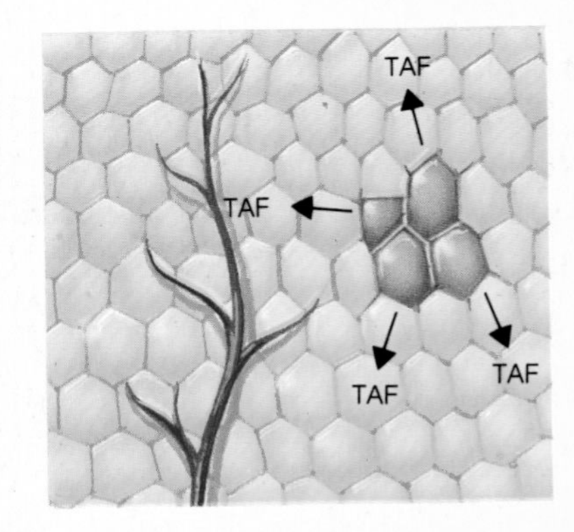

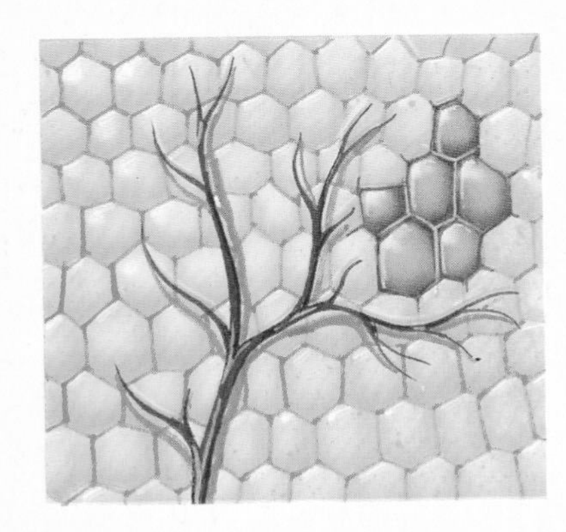

All cells need a supply of blood *(left)*, carrying nutrients and oxygen. But fast-growing tumor cells (shown in red) need more blood than do normal cells. To obtain it, they release a substance called tumor angiogenesis factor (TAF, *center*), which stimulates the growth of secondary blood vessels at the site of the tumor *(right)*. The increased nutrient supply allows the cancer cells to grow even more rapidly, setting up a vicious circle of demand and supply.

Once inside the cell, food is put to good use. It is broken down in the cell's mitochondria to provide energy for protein synthesis, waste disposal, cell replication, the movement of cell components and other cell activities. The mitochondria contain reaction-speeding enzymes – proteins made in the cell which are crucial to the metabolic process. Without these the biochemical reactions that break down the cell's food would be too slow to be effective. And enzymes are specific – they act on only one type of chemical change. The breakdown of a simple sugar involves more than 50 reactions, each with its own specific enzyme.

These energy-generating reactions within the cell are known collectively as respiration (sometimes called internal respiration to distinguish it from the external respiration that takes place in the lungs). The respiratory reactions occur in two stages. In the first stage, food is only partly broken down into intermediate substances, such as alcohol and acids. This stage does not require oxygen and is called anaerobic respiration. The second stage, aerobic respiration, can only happen if oxygen is available. In it, the intermediate substances are broken down completely, into the waste products carbon dioxide and water, and energy essential to all life is released.

Storage units

The energy generated by respiration is not immediately available for powering the cell and its processes. It is placed in temporary storage – as the energy in chemical bonds – and released only as it is needed. The storage medium is the chemical adenosine triphosphate (ATP). This crucial chemical is made during respiration from another essential chemical called adenosine diphosphate (ADP).

As their chemical names suggest, ADP has two phosphate groups (linked to the substance adenosine), whereas ATP has three. The energy from respiration goes to force a phosphate group to bond with an ADP molecule, forming an ATP molecule, which can be stored. This bonding is energy-intensive, but the bonds are easily broken to release the stored energy that went into their formation.

As ATP changes to ADP and the high-energy bond is broken, energy is released virtually instantaneously, so the cell can react rapidly to energy demands without having to go through the 50 or so reactions involved in respiration. Because the breaking of each phosphate bond releases a specific amount of energy, ATP also provides a precise metering of the amount of energy released to the cell. So there is no wasted energy which might cause heat and burn the cell, and the metabolic processes are not slowed down by being starved of energy. Another advantage of the system is that the phosphate energy bonds are transferable. They can be moved from the ATP account to other substances to increase their energy reserves without any leakage of "funds" from the cell.

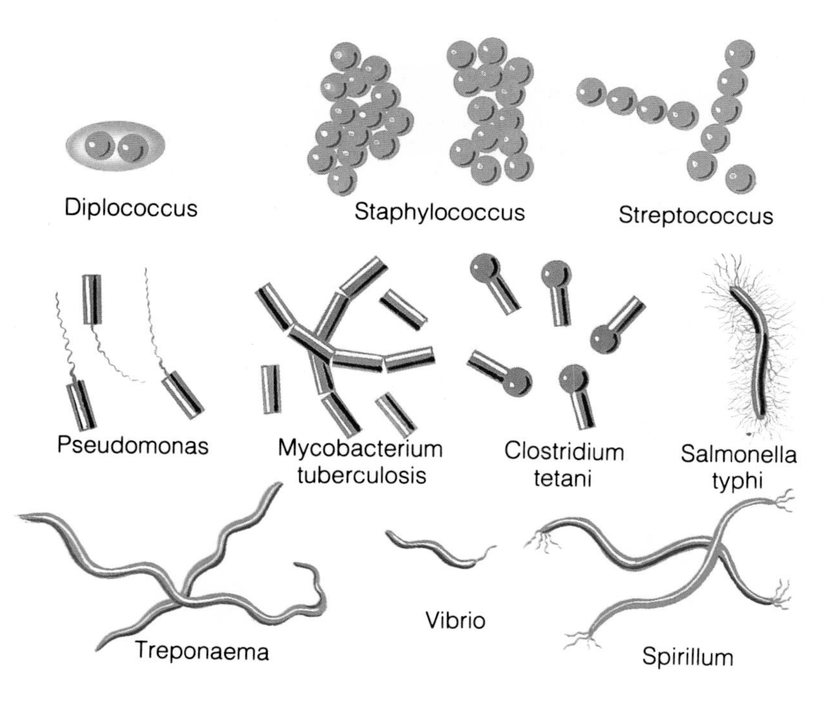

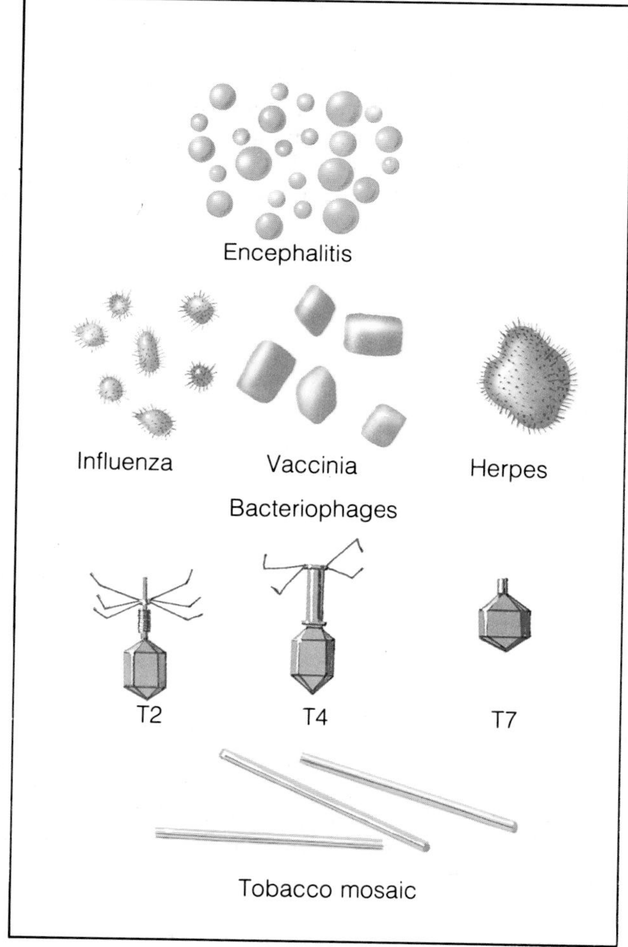

Single cells can make up complete living organisms, such as bacteria *(above)*. The spherical type are known as cocci, and exist as pairs (diplococcus), bunches (staphylococcus) or chains (streptococcus). Rod-shaped bacteria, with or without flagella as "tails," include those that cause tuberculosis and tetanus. Wormlike forms, some with cilia "whiskers," include salmonella and the organism that causes syphilis. Viruses *(right)* can live only as parasites within a host cell. Some, the phages identified by the letter "T," attack and kill bacteria.

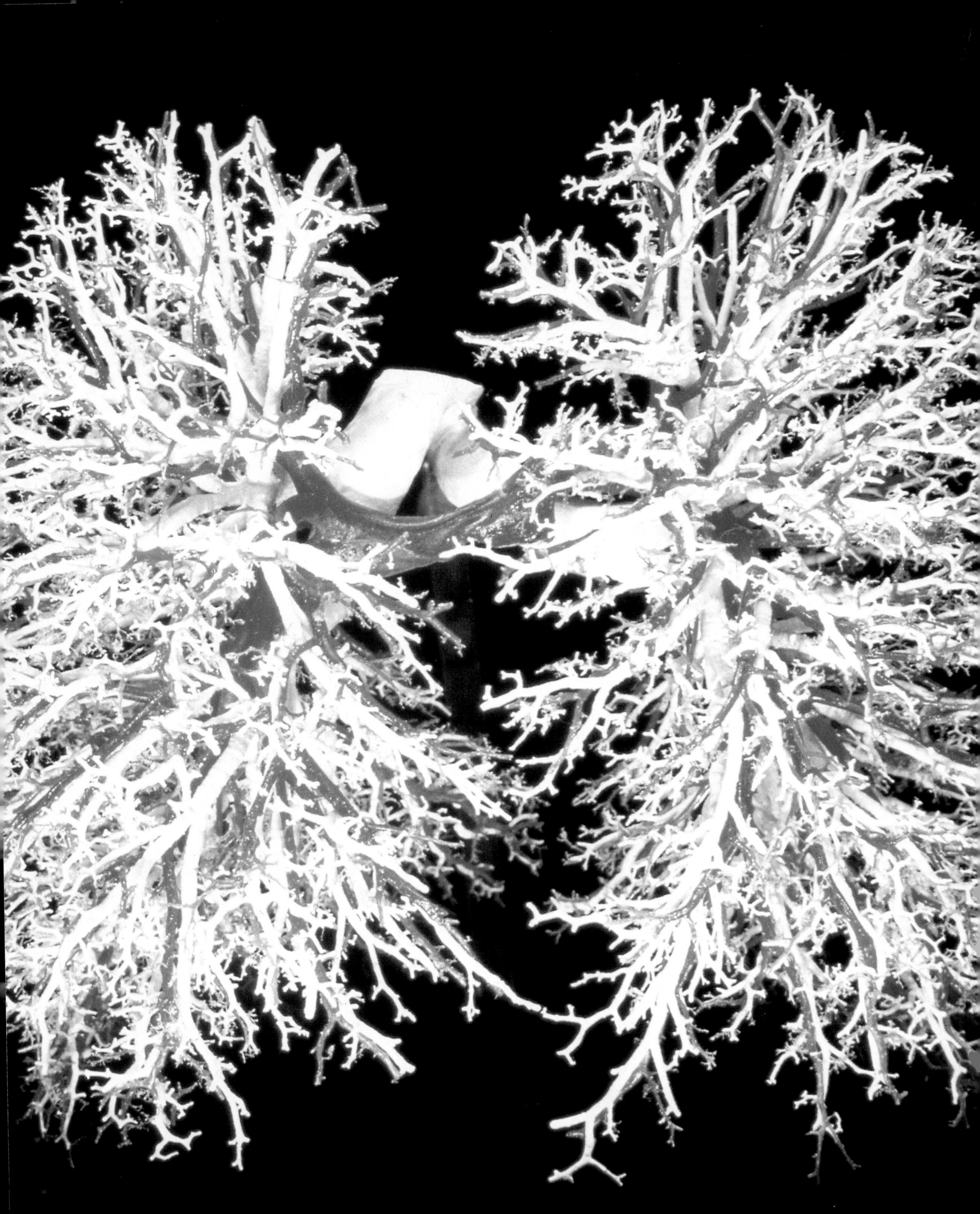

CHAPTER 6

Respiration

Breathing is the most obvious and important of the many tasks accomplished by the respiratory system. But it is also involved in yawning, sneezing, coughing, sniffing – and hiccups. Speech, from a whisper to a shout, and the subtle sense of smell, also depend on the movement of air in the respiratory passages.

Respiration, with digestion, is one of the body's two great "input" processes. The respiratory system's main job is to absorb oxygen from the atmosphere into the body, and to expel the waste product carbon dioxide from the body into the atmosphere. Its various parts take up much of the space in the face and the neck, and most of the chest. The system occupies a lot of physical space and expends hard-won energy in the muscle-powered movements of breathing.

Over thousands of years of evolution, our respiratory system has become adapted to perform various other functions. Breathing carries odor molecules on the incoming airstream, which are detected by the olfactory organs in the nose and perceived by the brain. Perhaps the odor warns us of danger, such as smoke from a fire. It might warn us off a potential meal, signaling that food is bad and not fit to eat, or advertises the presence of a mating partner. The respiratory airflow has also been "hijacked" by the larynx (voice-box), which exploits it to create the vast range of sounds that makes humans so communicatively vocal.

Vital oxygen

The respiratory system's role is fundamental to life. The oxygen it obtains is supplied via the circulatory system – the heart and blood vessels – to every cell in the body. Without oxygen, cells die. Within minutes of oxygen deprivation, tissues become irreparably damaged. This is because oxygen is a vital ingredient of metabolism, the biochemical processes that keep the body running. Digested nutrients, from food, are mixed with oxygen, enzymes and other chemicals. After several intermediate stages of chemical reaction, the result is a helping of available energy. The cells "burn" this energy, using it to drive its many other biochemical processes. It is, literally, "energy for life."

The biochemical process of burning energy, using oxygen in this case, is often called aerobic or cellular respiration. The physical process of taking in air in order to absorb the oxygen in the first place is also referred to as respiration. They are the end and beginning, respectively, of the same overall process.

The heart is often regarded as being "up front," the foremost organ. We all know that if the heart stops, life ceases. Yet the pumping action of the heart simply distributes to all corners of the body the oxygen already obtained by the respiratory system. The pulsing blood collects the carbon dioxide that the respiratory system ultimately expunges from the body. No matter how hard and fast the heart pumps, if there is no fresh oxygen, there is soon no life. Even the very muscle of the heart itself relies on this continuing supply of oxygen. Small wonder we talk of "the breath of life."

Like a pair of trees, the airways (white) and blood vessels (red) divide and branch within the structures of the lungs. This resin cast shows how the trachea, or windpipe, splits into two main bronchi, one leading to each lung. The bronchi divide and subdivide to form the bronchioles – the narrowest airways that end in tiny alveoli. There gas exchange takes place with blood in the capillaries surrounding them.

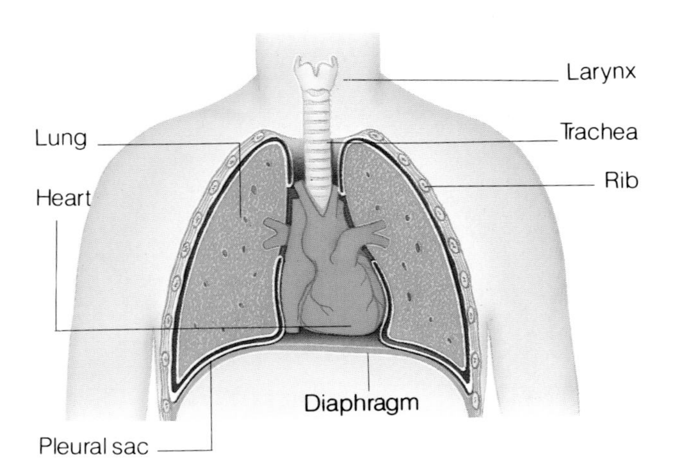

In this cutaway view of the chest cavity, the upper airways can be seen leading down to the lungs behind the heart. The lungs are surrounded by the double membrane of the pleural sac, and supported by the muscular dome of the diaphragm.

A cross-section of the chest reveals how the lungs are protected within the bony cage formed by the ribs at each side, the spine at the back and the breastbone at the front. The left lung is slightly smaller to allow room for the heart.

The treasured chest

While you have been reading, you have certainly been doing at least one other thing: breathing. Air has passed in and out, probably at the rate of about 14 breaths a minute, unless you have just completed a run around the block or been watching a torrid love scene on the television. Physical effort and emotional passion both bring on heaving chests, for within the chest reside the two lungs, the twin centers of the respiratory system.

The system can be divided into three main sections. The first contains the "pumping machinery," muscles and elastic fibers which pull and push air in and out of the lungs. The second is the system of conducting airways through which air is transported to and from the third region, the lung alveoli. These microscopic "air bubbles" within the lung tissue facilitate absorption of the vital oxygen into the bloodstream.

The lungs are large, cone-shaped, pink-gray, spongy organs. The two lungs are not quite mirror images of each other. The right one is larger and is sectioned into three main compartments, or lobes. The left lung has two lobes and a hollow scooped in its lower front, in which nestles the ever-pounding heart. Indeed, the lungs themselves are never still. In quiet, restful breathing, they draw in (and push out!) about three-quarters of a pint of air with each breath. That's around 15 pints each minute, or 790 gallons overnight – about one-third of the air in a smallish bedroom. In the gasping aftermath of a 1,000-meter race, the lungs' throughput is 20 times their resting turnover.

Air is conveyed down into the lungs by a branching system of tubes that have been likened to an upside-down, hollowed-out tree. The base of the bronchial tree consists of the nose and mouth, twin inlets for atmospheric gases. The main "trunk" is the throat and the trachea, or windpipe, a 5-inch-wide tube which links the throat to the chest. The trachea divides into two smaller tubes, the principal bronchi, one to each lung. These tubes further subdivide into smaller and smaller bronchi (the "branches" of the tree). Ultimately the smallest air passages, the bronchioles (the "twigs"), end in the minute air sacs called alveoli – more than 350 million of them in each lung. Surrounding and supporting the airways and alveoli are elastic fibers and connective tissue, the interstitial tissue or ground substance of the lungs. The whole gives the lungs a somewhat spongy texture.

The piston of the respiratory pump

The stretchy, spongy structure of the lungs permits them to continually enlarge and contract, to suck in air and blow it back out; that is, to breathe. The pumping system that powers the lungs involves the bony cage around the chest created by the spine, ribs and breastbone (sternum), and two sets of muscles.

The first set is the diaphragm, a muscular sheet that consists of three overlapping groups of crisscrossed muscle fibers. The diaphragm forms the floor of the chest, or thorax, and separates it from the abdomen below. Its edges connect to the spine at the back, to the lower ribs around the sides, and to the bottom of the breastbone at the front.

A relaxed diaphragm is domed, almost bell-shaped. Its center bulges under the pressure from the abdominal organs below and projects up into the chest, to a level only an inch or so below the nipples. A tensed, contracted diaphragm is much flatter and its net result is to enlarge the chest, pulling down on the lungs and increasing their volume. It works rather like a piston moving down inside a cylinder.

The lungs inflate because each is wrapped in a thin, slippery membrane, the pleura, which folds back on itself to line the inside of the chest. The part covering the lung is known as the visceral pleura; that lining the inside of the chest wall is the parietal pleura. The two layers of the pleura are in intimate contact, lubricated by an incredibly thin smear of oil-like pleural fluid between them. The entire set-up is an airtight fit in the chest cavity. The only way in or out is along the trachea.

Now, imagine a probably more prosaic situation: a glass, stuck to a coaster beneath by a thin film of liquid, which creates a seal. Lift the glass, and the coaster comes up with it. For a similar reason, the diaphragm is "sealed" by the pleurae and their fluid to the bases of the lungs. As the diaphragm flattens, it pulls the lungs down with it. As the lung volume goes up, the pressure inside goes down, creating a partial vacuum. Nature abhors a vacuum, and to equalize the pressure air forces its way in from the atmosphere outside – and so you breathe in.

The flexible cage

The second set of breathing muscles consists of those in the back, neck and, especially, the intercostal muscles between the ribs. As they contract in coordinated unison, they pull the ribs upward. Joints, between the ribs and spine at the rear and the ribs and breastbone at the front, allow the cage to swing up and also out, due to the slightly down-slanting position of the ribs at rest. Once again, the net result is to enlarge the chest, pulling the lungs forward and increasing their volume.

Breathing in – inhalation, or inspiration – is therefore an active process. It requires muscle power from the diaphragm, the intercostals and other chest-wall muscles. The lungs themselves play a purely passive role in the inspiratory movement: the changes that they undergo are brought about solely through changes in the capacity of the chest. Almost two-thirds of the volume of air inhaled in an average breath is brought about by the pistonlike action of the diaphragm; the rest comes from the motion of the rib cage.

Breathing out – exhalation, or expiration – is a simpler process. The diaphragm and chest muscles relax. The chest cavity elastically recoils to its previous volume. It is aided by the elasticity of the lungs, which expels air from them, and abetted slightly by gravity pulling down on the ribs. As the lungs contract they blow air back up through the trachea.

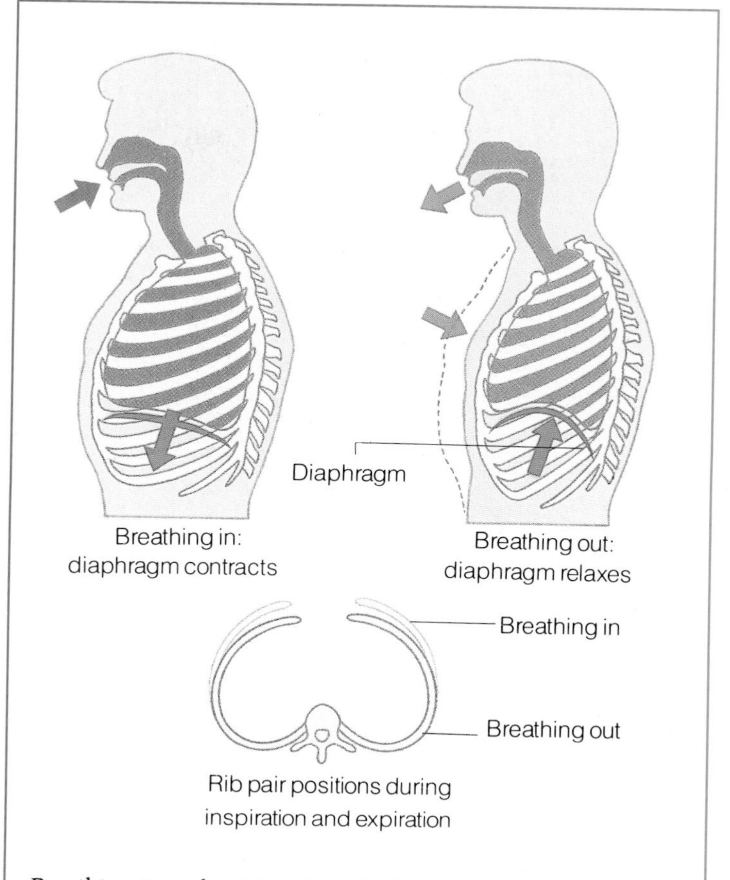

Breathing in and out involves slight movements of the ribs and larger movement of the diaphragm. As the diaphragm muscle contracts and moves downward, the lungs enlarge and air rushes into them through the respiratory tract. Upward movement of the diaphragm forces air out again.

Remembering to breathe

The movements of breathing are automatic. We can impose our will on them and change them when we want to – for example, during speech or when coughing. But for most of the time, breathing is controlled in a reflex manner by the respiratory center in the brain.

The respiratory center is situated in the brain stem, or medulla oblongata, a "primitive" region of the brain concerned with vital processes such as heartbeat, breathing and blood pressure. The respiratory center ensures, no matter how occupied the brain is elsewhere – whether concentrating on a symphony, full of rage at injustice or simply asleep – that we always remember to breathe. It happens with hardly a moment's thought, every few seconds, every day, throughout life.

Actually, this is not quite true. The first nine months of life, in the womb, are an aquatic existence. The fetus floats in a pool of fluid, which filters into the lungs. There is no air to breathe. But this creates no problem: oxygen, nutrients and other essentials are passed to the developing fetus via the placenta, which is both its lungs and its intestines. This situation requires a specific blood flow around the body, which must change after birth when the lungs become functional.

Breathing via the placenta

The heart is effectively two pumps side by side. The right pump sends blood to the lungs for oxygen; this "refreshed" blood returns to the heart's left pump, from where it is sent around the body to deliver its oxygen to the tissues. The blood then returns to the right pump again, so completing the double circuit.

Fetal blood circulation is different. During the first four or five months of development, the fetus has a "hole in the heart," between the upper chambers (atria) of the left and right pumps. This allows most blood returning to the right pump, from the body, to flow directly through into the left pump and so bypass the lungs. The lungs receive only a small "maintenance" supply to fullfil their basic requirements, much like any other organ.

As the fetus grows, the hole gradually closes up, and normally it has gone by birth. This forces more and more blood into the pulmonary artery, the main vessel that conveys it from the heart's right pump to the lungs. However at this stage there is another bypass, the ductus arteriosus. This small valvelike tube links the pulmonary artery to the aorta, the main vessel carrying

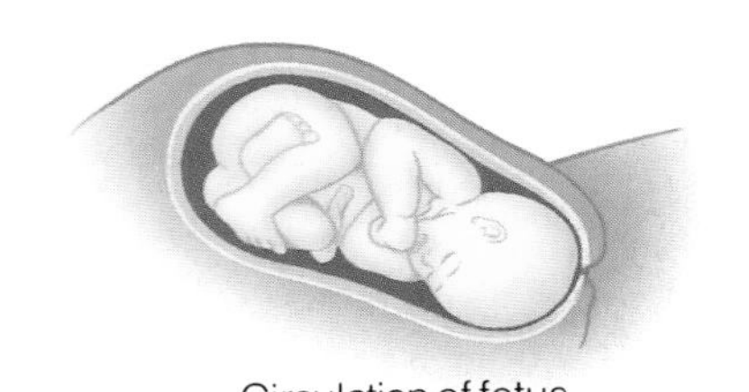

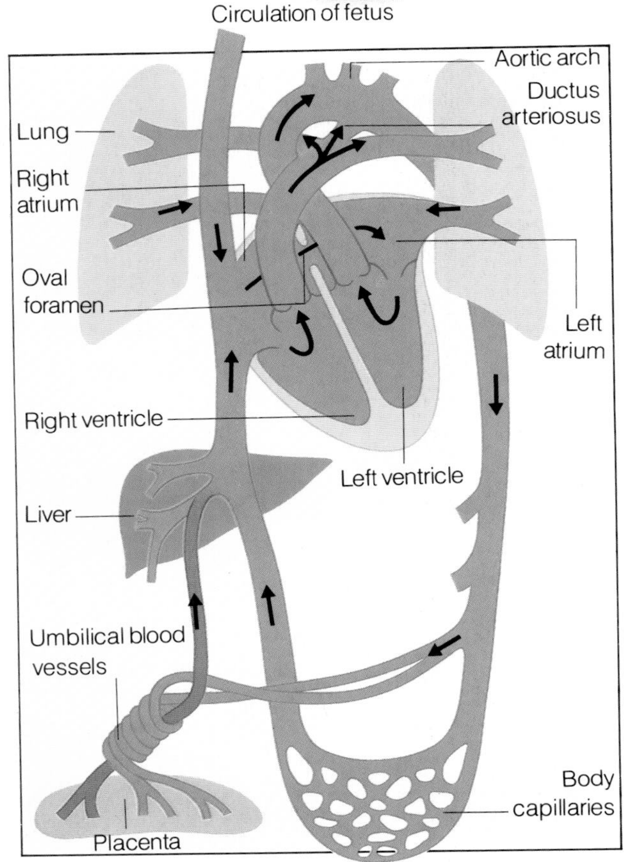

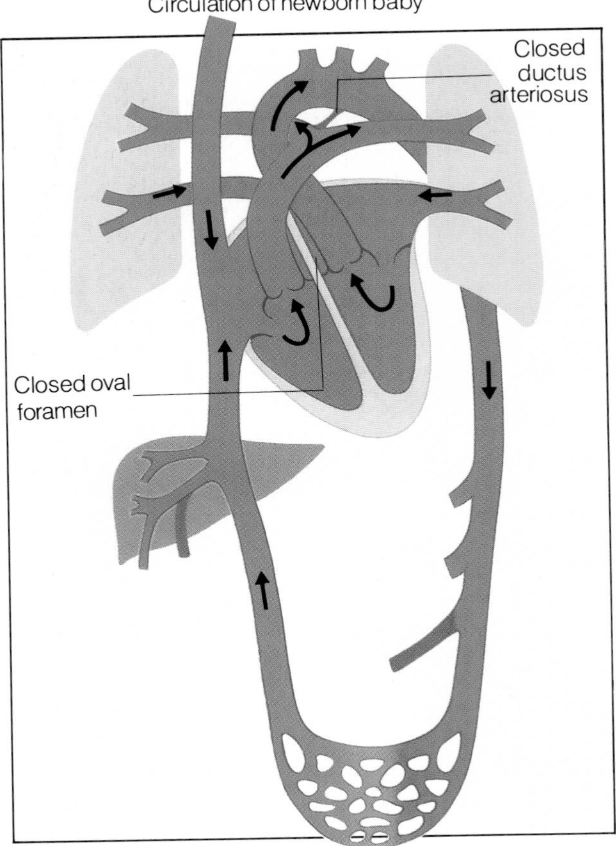

A vital change takes place in a baby's blood circulation at the moment of birth. While it is still a fetus in the womb, blood from the placenta flows first to the right side of the heart and then, via the oval foramen, to the left side and on around the body. At birth the foramen closes, and the baby's first breath forces blood from the right side of the heart into the lung circulation for oxygenation, before it returns to the heart.

A fetus within the womb *(right)* has its lungs full of amniotic fluid. The placenta provides the growing baby with oxygen in blood that travels along vessels inside the umbilical cord.

blood from the left pump to the body. Thus most blood leaving the heart's right pump, apparently destined for the lungs, is shunted into the aorta and off around the body again: it is another kind of pulmonary bypass.

The first breath

Before birth, the fetal lungs are not in a collapsed state. They are "inflated" to about two-fifths of their total capacity by a special fluid produced by alveolar cells. The fluid contains a surfactant, a form of natural detergent. It is thought that surfactant helps the baby to overcome the tremendous forces of surface tension in the alveoli, where air must replace liquid in the first breaths. The fetus is known to make small breathing movements in the few weeks before birth, and one effect of these practice breaths is to squeeze excess fluid out of the lungs.

At birth, the baby's diaphragm is jolted into action and the first breath is drawn – usually accompanied by loud and lusty crying. It involves an enormous effort on the part of the new infant. As air is pulled down into the lungs, surfactant helps to keep the alveoli open, and excess lung fluid is absorbed into the bloodstream. Surfactant continues to play this life-giving role through life.

Within a few minutes, the ductus arteriosus begins to contract and should close completely between the fourth and tenth day after birth. It is not known for certain what causes the ductus to close, but it probably involves the contraction of smooth muscle tissue in its walls, and also particular hormones called prostaglandins. If all goes well, the ductus gradually shrinks and becomes a fibrous band joining the pulmonary artery to the aorta, which persists throughout life.

Birth also sees the shrinking of the umbilical arteries and vein, which formerly transported blood to and from the placenta (afterbirth) along the umbilical cord. Modern obstetric procedure may hasten the process by tying and cutting the cord as soon as the baby's breathing becomes established.

Birth, then, brings about this amazing series of respiratory changes. The baby is no longer an appendage of the mother, but has started out on the road to independence. The first breaths are hard work, and it takes several days before uniform lung ventilation is achieved. What follows are some 500 million or more breaths, until the last one signals the end of life.

The postnatal pattern of blood flow is slightly more complex. Blood from the right ventricle, the main chamber of the heart's right pump, surges out through the pulmonary artery to the lungs. There it takes on board the vital oxygen. Then back it flows via the pulmonary veins to the left atrium, the smaller reception chamber of the left pump. This part of the circuit is termed the pulmonary circulation.

The blood then passes from the left atrium through a valve to the left ventricle, which drives it powerfully along the aorta and its branching system of arteries to the rest of the body. In the thousands of miles of capillaries, the smallest vessels that thread through all tissues, the rewards of respiration are finally seen: oxygen diffuses from blood to cells, and carbon dioxide passes in the opposite direction. The capillaries unite to make veins, which return the blood along the vena cavae to the right atrium. This is the systemic circulation.

The circulatory system permits gas exchange in the capillary network: oxygen is traded for carbon dioxide. The workings of the respiratory system involve its own capillary network, in the lungs, doing the opposite. First, however, air must permeate deep into the lungs, to reach these capillaries.

Inspired air usually enters the upper respiratory tract (nose, throat and trachea, or windpipe) through the nostrils rather than the mouth. The nose has advantages as the inlet, for a number of reasons. The nose filters: hairs in the nostrils trap dust and other particles that float in even the cleanest air. In addition, the sticky mucus lining the nasal cavity behind the nose also entraps airborne matter. And the nose warms: the copious blood supply to nostrils and nasal cavity (witnessed by the seeming "flood" of a nosebleed) passes heat to the incoming air. And the nose humidifies: moisture from the mucous lining evaporates and dampens the airstream. These built-in nasal features deliver air more suited to the lung environment.

The site of speech

After it leaves the nasal cavity, the air passes backward and downward through the larynx, or voice-box. This complex three-dimensional structure of cartilage plates, thin bone, muscles and ligaments is the site of speech. In its center, projecting from its inner walls, are two pearly, stiffish folds of tissue. Each of these vocal cords, or folds, is about two-thirds of an inch long. Every sound we make, from the softest whisper to an unbridled scream of pain or anger, emanates from the vocal cords.

There is a gap between the cords which allows air to pass silently to and fro during quiet breathing. As we prepare to speak, muscles pivot cartilages that swing the cords together. Air flowing through the slit between them now sets up vibrations along their length. Sound waves are created by the vibrations, in the same way that the reeds of a mouth organ tremble noisily as we blow or suck wind past them.

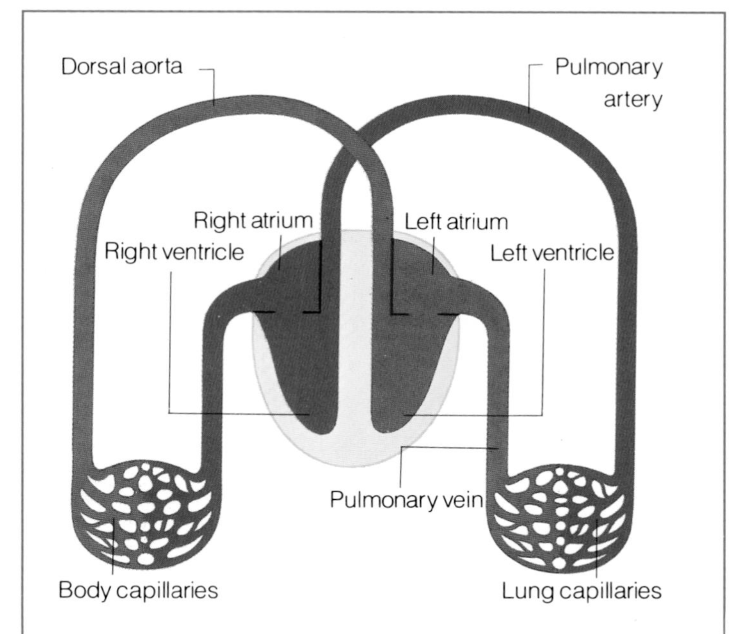

The human body has two blood circulations, called the pulmonary circulation (shown on the right of the diagram) and the systemic circulation (on the left). The right side of the heart pumps blood around the pulmonary system to be oxygenated, and the left side of the heart pumps the oxygenated blood to all parts of the body round the systemic system. Blood low in oxygen is shown blue, and oxygenated blood is red.

A tree of bronchi

Immediately beneath the larynx is the trachea. This gateway to the lungs, the beginning of the lower respiratory tract and the "trunk" of the bronchial tree, is similar to the air hose from a portable hair drier. It is flexible, to cope with bends and twists as the head and neck move. It has strengtheners in its wall, to maintain an open airway even under pressure (in the body, pressure is exerted by muscles in the chest and by the lungs themselves). Unlike the wire helix of a hair drier hose, the tracheal stiffeners are 16 to 20 C-shaped hoops of cartilage.

At its lower end, the trachea forks into two short, stubby tubes, the first of the bronchi. More forks occur as the bronchi, the "branches" of the tree, become narrower. Cartilage stiffeners hold open the larger bronchi, but they gradually become more sparse until, some 20 divisions after the trachea, they disappear altogether. These slenderest "twigs" on the tree, less than one twenty-fifth of an inch (a millimeter) across, are the bronchioles. Their structural integrity is derived from muscle fibers that run around their diameter like so many rubber bands.

Some germs, dust and dirt particles evade the nasal filtering mechanisms. But the pipes and passageways of the lower respiratory tract are fully prepared. All of these airways are lined with mucous membranes. Special cells in the membranes, known as mucous or goblet cells, make endless supplies of sticky mucus, which catches and holds settling particles. The membranes also bear countless millions of tiny hairlike cilia. Under the micro-

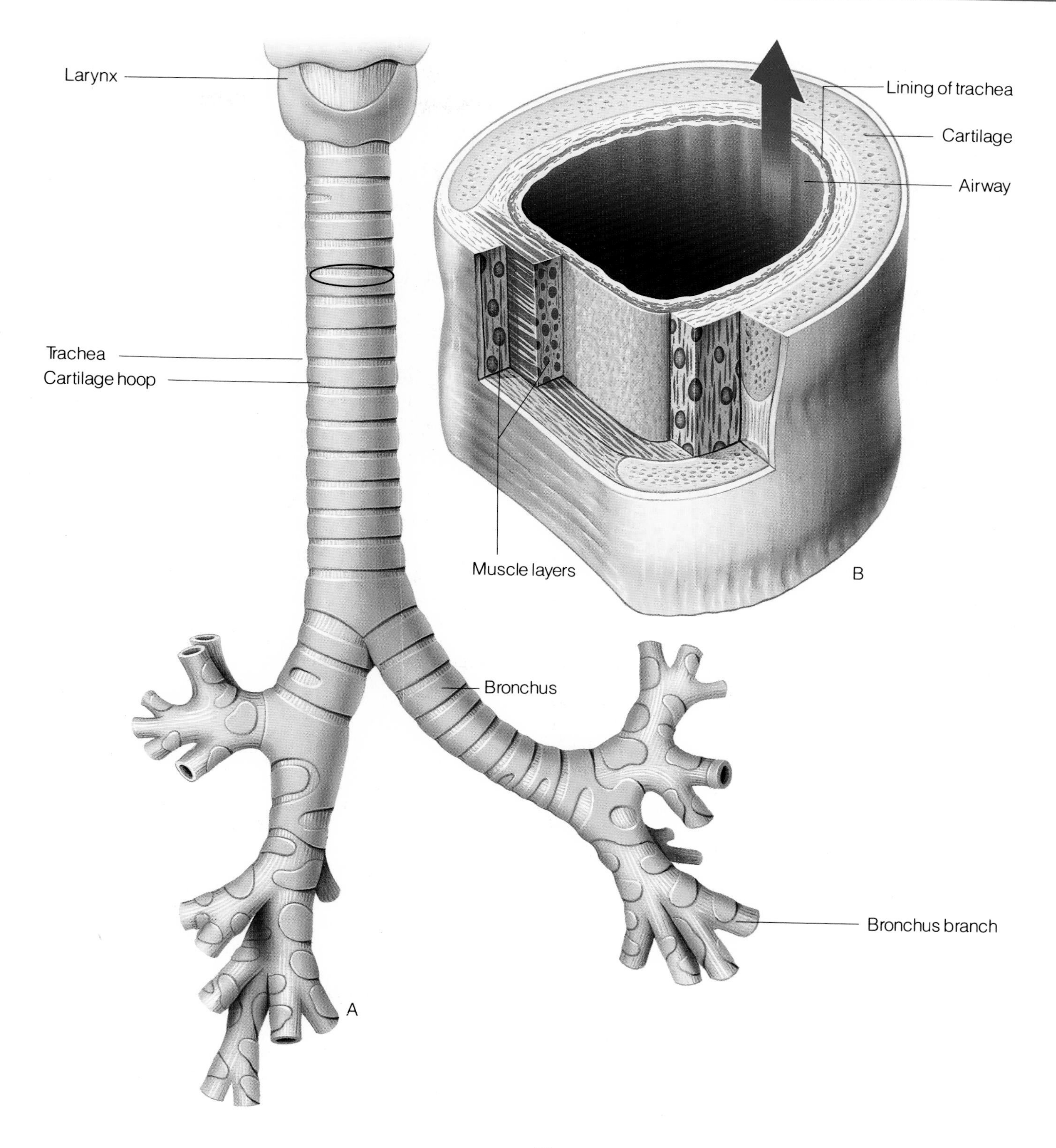

The major airways of the upper respiratory tract (A) are the trachea, or windpipe, and the bronchi. The trachea extends from the larynx to the fork of the main bronchi, which subdivide as they extend into the lungs. Hoops of cartilage stiffen the trachea, as shown in the cross-section (B), so that the airway remains open even with the wide variations in air pressure inside it. The lining bears millions of hairlike cilia, which move mucus and debris away from the lungs.

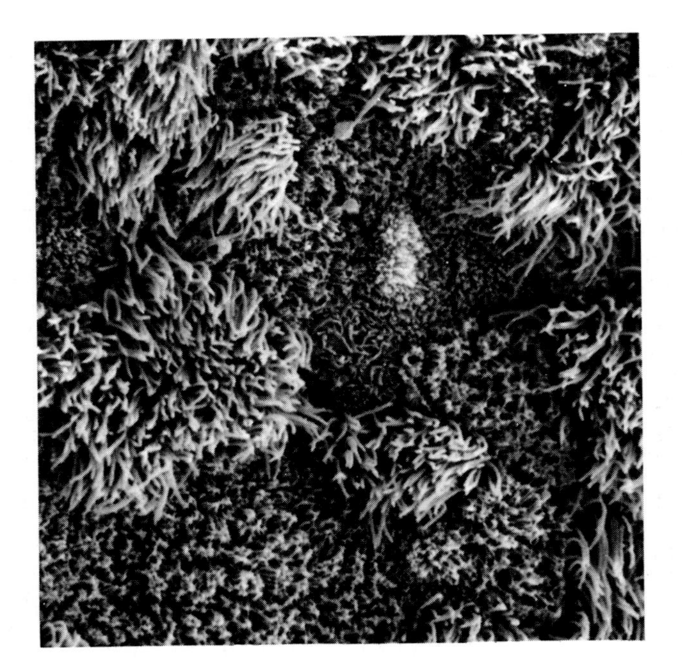

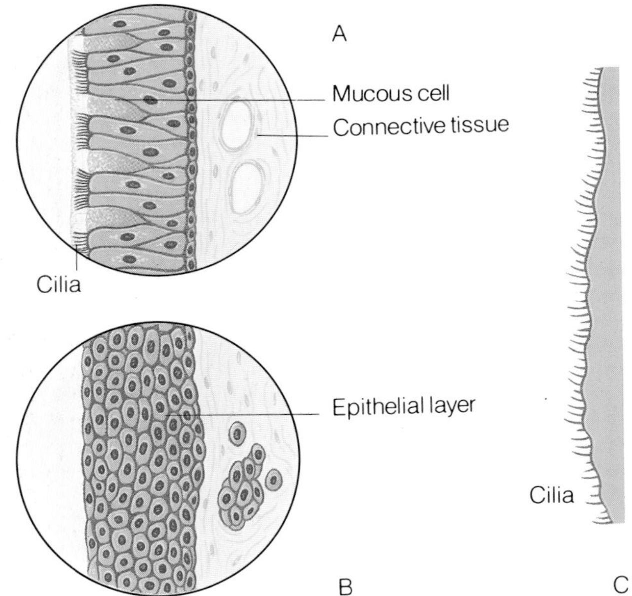

Like strands of waving seaweed *(above left)*, thousands of cilia beat rhythmically to waft away mucus and any trapped particles of dust and germs. The cilia are borne on mucous cells (A) lining the airways, and their wavelike motion (C) keeps the mucus moving. Prolonged cigarette smoking paralyzes the cilia and eventually destroys them (B), so that the smoker has regularly to cough to shift the accumulating mucus.

A computerized X-ray scan of the lungs *(right)*, false-color coded according to density, reveals dramatic detail of its living structure. Such scans are an invaluable diagnostic aid in detecting disease.

scope, the cilia can be seen beating upward like teams of rowers wielding miniature oars, driving the tide of mucus and its flotsam of trapped refuse relentlessly up and out of the lungs, to the throat. Every so often we swallow the mucus, perhaps with a small cough to "clear the throat" – or rather, to clear the lungs.

Grapes on the vine

If the bronchial tree can be likened to a vine, then the terminal bronchioles are the final twigs, and the alveoli the grapes. Each bronchiole ends in a cluster of these minute air sacs. The name of the game is now surface area, because the greater the area the lungs have for gas exchange, the higher is their oxygen-absorbing efficiency. In fact, compared to their physical size, the lungs have an immense area through which to gain oxygen and lose carbon dioxide. The 700 million or more alveoli, flattened out, would cover an area of between 50 and 100 square yards. This is nearly the size of a tennis court, neatly folded and compactly bundled into the chest cavity.

The pulmonary artery, bringing deoxygenated (low-oxygen) blood from the heart's right pump, divides into two branches, one for each lung. In yet another version of the branching tree, the artery divides many times until the vessels become capillaries. A network of these pulmonary capillaries surrounds each alveolus. Blood flows steadily through, a fluid conveyor belt bringing carbon dioxide for disposal and collecting oxygen for dispersal. At any instant, around one-fifth to one-tenth of the body's total blood volume, ten and a half pints, is in the pulmonary circulation.

The game has now switched to minimum resistance. Oxygen must be allowed to pass down a diffusion gradient, from its relatively high concentration in the air of the alveoli to its lower levels in the deoxygenated, "used" blood. The smaller the barriers placed in its path, the easier is its journey. A second high-efficiency feature of the lungs is therefore the thinness of the interface for gas exchange. Each alveolus has a wall only one cell thick – and a very thin cell at that. Likewise the capillary wall is one cell thin. The distance between air and blood is about one-thousandth of a millimeter.

The blood's oxygen vehicles are red blood cells. These donut-shaped structures are tiny even for cells, but even so only just manage to squeeze single file through the pulmonary capillaries. Most of the absorbed oxygen is transported by the red cells, bound chemically to the substance hemoglobin. Red cells are packed with hemoglobin, the blood's red pigment, which acts as a "magnet" for oxygen.

As oxygen molecules latch on to hemoglobin they convert it to oxyhemoglobin, which has a much brighter, redder color. In this way the blood is transformed from its dark, red-blue deoxygenated state to the vivid scarlet of its oxygenated form. A small amount of the oxygen dissolves in the plasma (the noncellular, liquid part of blood) and is swept away in solution.

It is said that fair exchange is no robbery, and so it is in the lungs. Carbon dioxide completes the barter as it leaves the blood, diffuses through the capillary and alveolar walls, and comes out of solution, taking its place in the air inside the alveolus.

The pulmonary capillaries reconnect to become larger and larger vessels, the pulmonary veins, which eventually return the blood to the heart. The lungs can be regarded as an intricate, intertwined network of three trees in one: airways and arteries becoming smaller, and veins getting bigger.

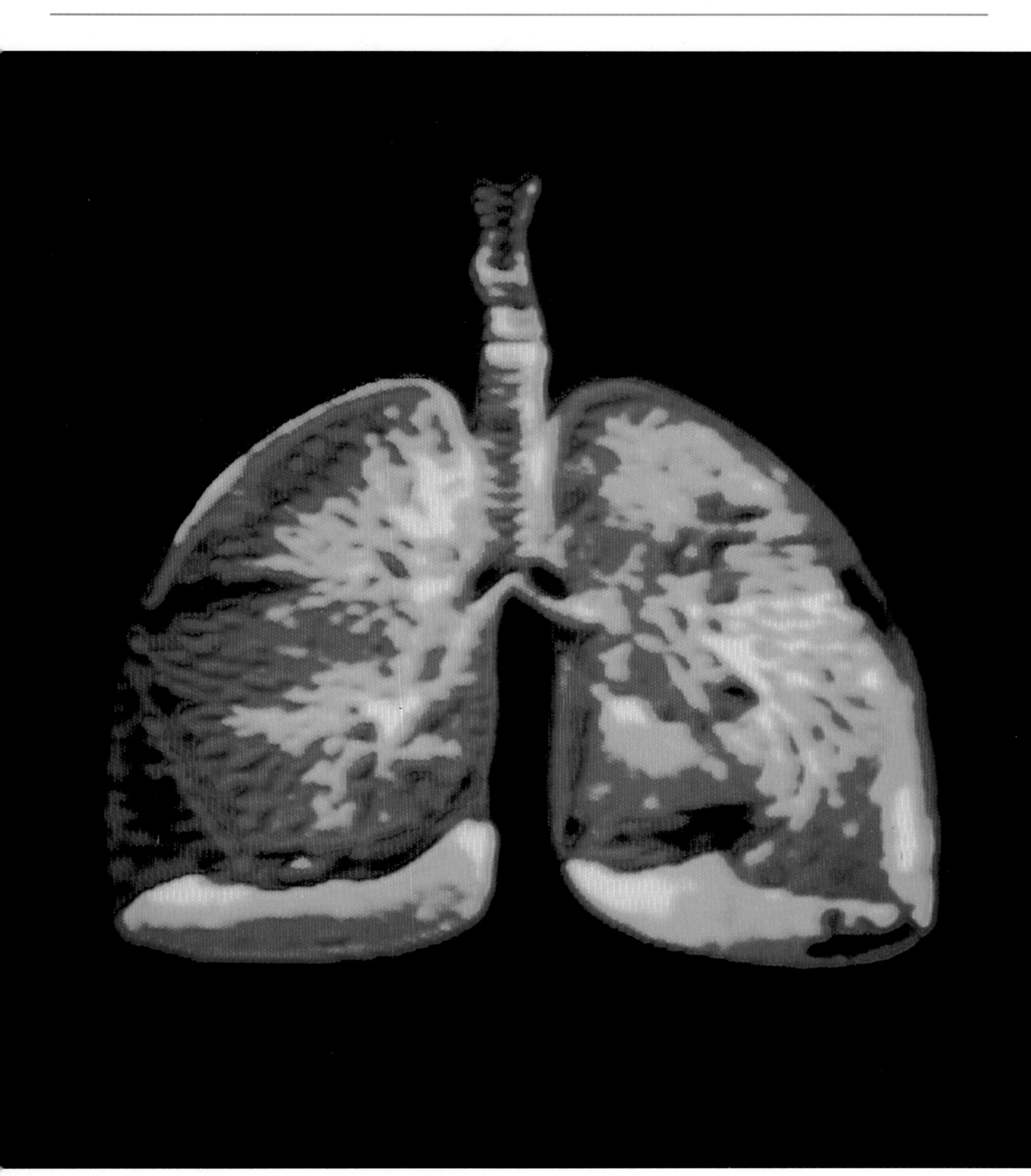

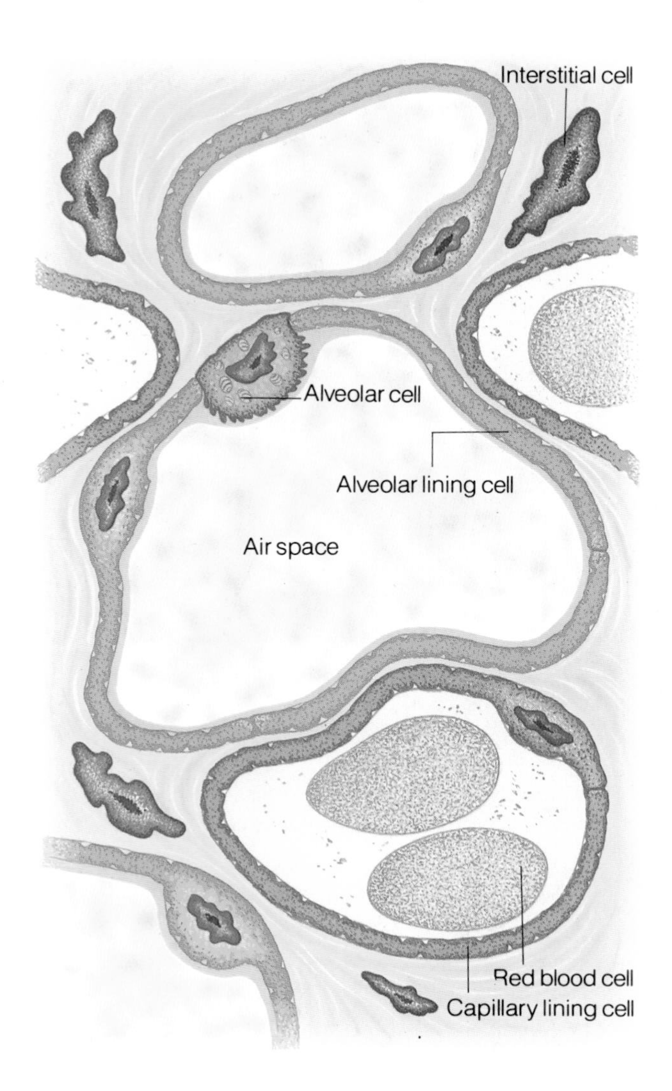

For efficient gas exchange to take place in the lungs, the alveolar air spaces and the circulating red blood cells must be very close together, as this diagrammatic section through alveoli and blood capillaries shows. Hemoglobin in the red cells takes up oxygen, which passes through the alveolar wall and cell membrane. Carbon dioxide, carried in the blood as a waste product, passes the other way into the alveoli to be breathed out.

The role of surfactant

Surfactant, the "natural detergent" in the alveoli, is a chemical vital to a baby's successful first breaths. A complicated mixture of protein and fat, with a substance called dipalmitoyl lecithin as its main component, it is present in the lungs through life. During expiration, the alveoli shrink and surfactant molecules are pushed closer together, so reducing the surface tension and preventing the alveoli from collapsing completely. When the alveoli swell upon inspiration, the surfactant molecules are farther apart, but the force tending to collapse the alveoli is much reduced.

Recall what happens when you inflate a party balloon. At first it requires considerable effort, but once the balloon is partly expanded, it becomes easier to blow up. In the same way, the forces that tend to collapse the alveoli are greatest when the alveoli are smallest, at the end of exhalation – which is when surfactant molecules, at their closest, are most effective at lowering surface tension.

Some babies, especially those born prematurely or with inadequately developed respiratory systems, may have great trouble in breathing. The general term for such conditions is respiratory distress syndrome, or RDS. One contributing factor is lack of surfactant, which the lungs of the premature baby have not yet manufactured. Insufficient surfactant means greater surface tension in the alveoli, and the baby may not be able to overcome this. Consequently, lung function is poor and the airway linings are rapidly damaged.

RDS affects some 50,000 babies each year in the United States. Developments in treatments, including intensive care, oxygen incubators and mechanically-assisted breathing, have dramatically reduced the mortality rate in the past quarter-century. In the late 1980s British researchers developed an artificial surfactant which promises to save even more new lives.

How much, and how fast?

The amount of air we can breathe, and how fast we can breathe it under various conditions of rest and activity, reveals much about the inner workings and health of the respiratory system. It therefore interests many people, from chest physicians and asthma patients to super-fit athletes looking for that extra ounce of stamina. A spirometer is an apparatus which, in its most elementary form, uses the air breathed out to displace water and make a pen trace a line across a moving chart. It measures the volume of air per unit time, which provides a measure of flow rate. More complex spirometers are linked to gas analyzers that determine the makeup of inspired and expired air.

Inspired air is usually our atmosphere, which has a relatively constant composition: nitrogen 78.08 percent, oxygen 20.95 percent, argon 0.93 percent, carbon dioxide 0.03 percent, and a miscellany of other trace gases. Expired air from quiet, restful breathing differs in two respects. The proportion of oxygen has fallen to nearer 17 percent, while that of carbon dioxide has risen to around 4 percent. The respiratory system has absorbed some oxygen and "blown off" carbon dioxide. At an average of 14 breaths per minute, at a pint each, a total of $15\frac{1}{2}$ pints of air are moved each minute. This enables some $3\frac{3}{4}$ pints of oxygen to enter the body and $3\frac{1}{2}$ pints of carbon dioxide to be removed from it every minute.

In restful breathing, the lungs work at far below their maximum capacity. The vital statistics are as follows, for an average person. The total amount of air which the lungs could hold after the largest inhalation, that is, the total lung capacity, is some 13 pints. This volume is composed of four components.

First is the residual volume, the quantity of air that remains in the lungs even when you breathe out as hard as you can; it is air that can never be expired. It measures around $3\frac{1}{3}$ pints. It is itself composed of two subvolumes. One is air hanging about in the contracted alveoli; the other is air remaining in the passages of the trachea, bronchi and bronchioles. This second subvolume is the "dead space" – air that fills the airways during each respiration cycle. It is not available for the real business of gas exchange in the alveoli, hence its name.

Second is the expiratory reserve volume. This is "extra" air you can manage to force out of your lungs after already having exhaled a normal breath. Its volume averages $2\frac{1}{5}$ pints.

Third is the tidal volume – the air shifted in and out during normal, restful breathing. As mentioned previously, it is around 1 pint, but increases rapidly with exercise.

Fourth is the inspiratory reserve volume. You inhale as normal for an average breath, but then go on inhaling as far as you can. The amount of air drawn in, over and above the tidal volume, makes up this fourth component. It averages $6\frac{1}{2}$ pints. Add the four together: the answer, and total lung capacity, is 13 pints.

There is not a lot that can be done about the lungs' residual volume. But expiratory reserve plus tidal plus inspiratory reserve volumes – together constituting the lungs' vital capacity – often

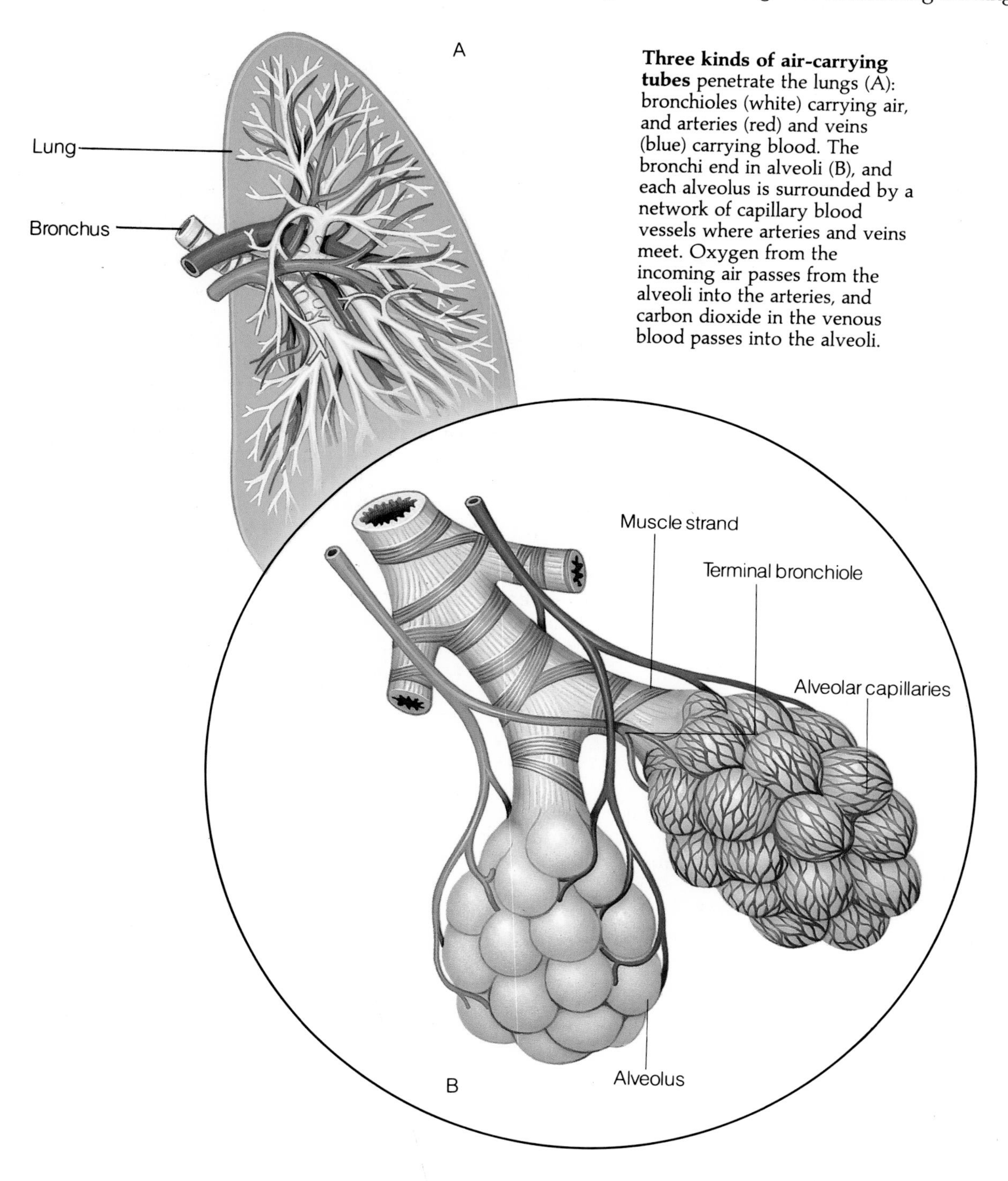

Three kinds of air-carrying tubes penetrate the lungs (A): bronchioles (white) carrying air, and arteries (red) and veins (blue) carrying blood. The bronchi end in alveoli (B), and each alveolus is surrounded by a network of capillary blood vessels where arteries and veins meet. Oxygen from the incoming air passes from the alveoli into the arteries, and carbon dioxide in the venous blood passes into the alveoli.

The efficiency of the lungs can be measured by lung function tests, which use an analyzer linked to a person on an exercise machine. The key to the equipment is a spirometer, which measures the capacity of the lungs and the volume of air breathed in and out.

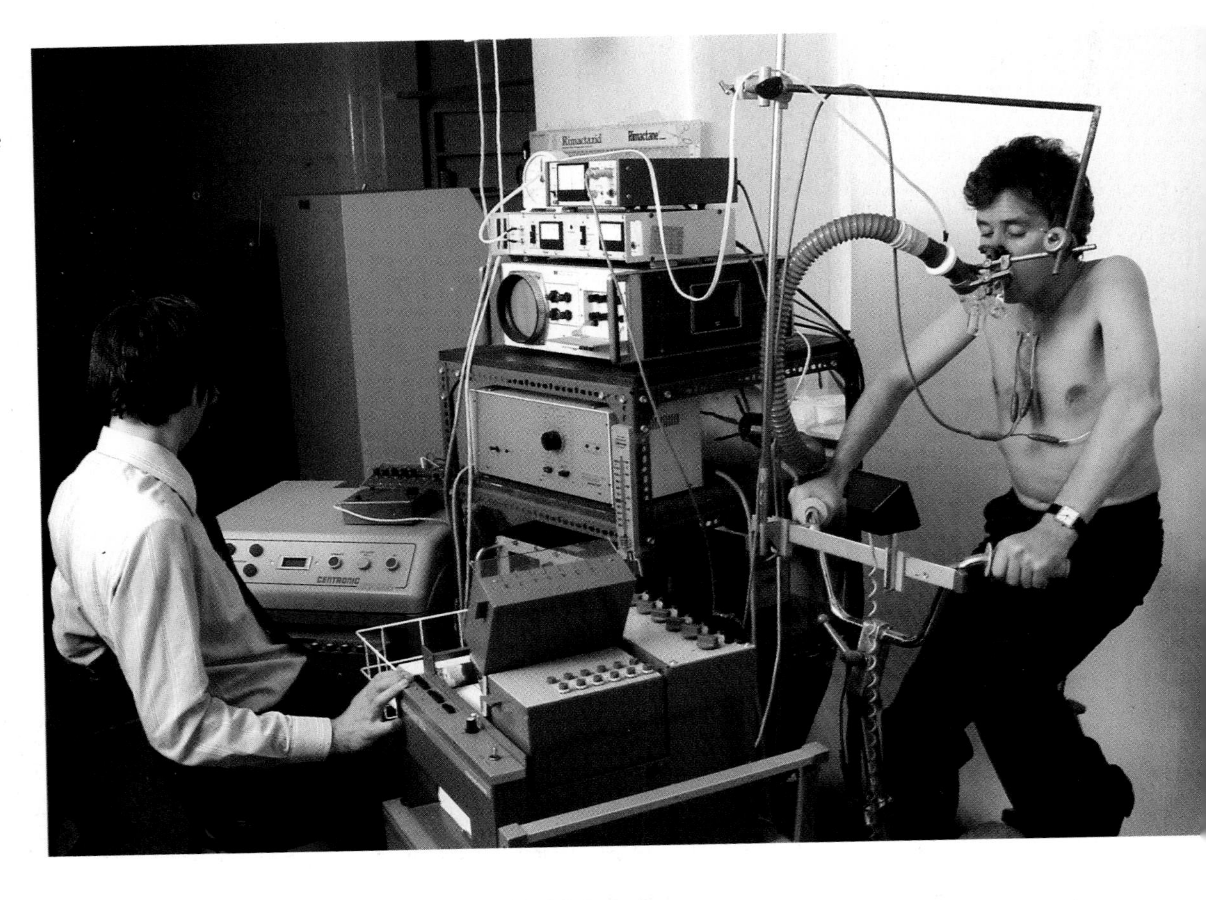

provides a useful pointer to general health. And in some respiratory illnesses, termed restrictive lung diseases, the vital capacity is reduced.

Another gadget, much simpler and cheaper than the spirometer, is a peak flow meter. It gauges the fastest (peak) rate at which air can be blown out of the mouth, and hence up from the lungs. The meter is a useful ready reckoner in respiratory illnesses, for instance, in assessing the severity of an asthma attack and the efficacy of treatment.

The stimulus to breathe

Have you ever thought what it is really like to be suffocating? People who have lived through the experience describe how the need for air is totally overwhelming. An irresistible urge to breathe comes from deep in the "primitive" brain, and overrides all higher thoughts. The victim would do anything for a lungful of fresh air. Yet the body is not responding to the need for air as such – or even the need for oxygen. The gasping comes chiefly from the desperate necessity to get rid of carbon dioxide.

The respiratory center in the brain regulates breathing, and it does so primarily in response to carbon dioxide levels in the body. In the brain stem special chemically-sensitive cells, or chemoreceptors, monitor carbon dioxide concentrations. They are not actually in contact with blood, but with cerebrospinal fluid, a clear fluid bathing the brain and spinal cord. Carbon dioxide diffuses into this fluid from the blood, however, and its concentrations in the two fluids are comparable.

When the brain stem's chemoreceptors detect an increase in carbon dioxide, nerve messages are sent down the spinal cord to the diaphragm. With such positive feedback, the noxious carbon dioxide is expelled through deeper breathing, so lowering its levels in the body, and thereby restoring the feedback loop to normal. Other chemoreceptors coordinate with those in the brain stem to influence breathing. Groups are scattered in the bloodstream, in the arch of the aorta and in the carotid artery in the neck. They also are sensitive to carbon dioxide.

Oxygen plays a role, though a more minor one. Yet other chemoreceptors, in vessels such as the carotid artery, attend to its concentration in the tissues. As the level of oxygen falls, so the chemoreceptors stimulate the breathing center to increase respiratory activity – a negative feedback loop in this case.

Faulty plumbing

The body may apparently suffer from poor respiratory function, with labored breathing and reduced oxygen supply to the tissues, when in reality the blood circulation is at fault. There might be a wrong connection between blood vessels, or there may be a vessel where there should not be one, which creates a "shunt." This is a pattern of flow that allows blood to reach the arteries system without passing through the ventilated part of the lungs for oxygenation.

In babies born with certain congenital heart defects, a significant amount of blood can be shunted through the heart or main arteries away from the lungs. In the heart, a ventricular septal defect (VSD) allows deoxygenated blood from the right ventricle to leak through into the left ventricle and so journey to the tissues once more. A pulmonary atriovenous fistula (PAF) permits some blood to short-circuit directly from the pulmonary artery to the pulmonary vein, so bypassing the oxygenating tissues of the lung.

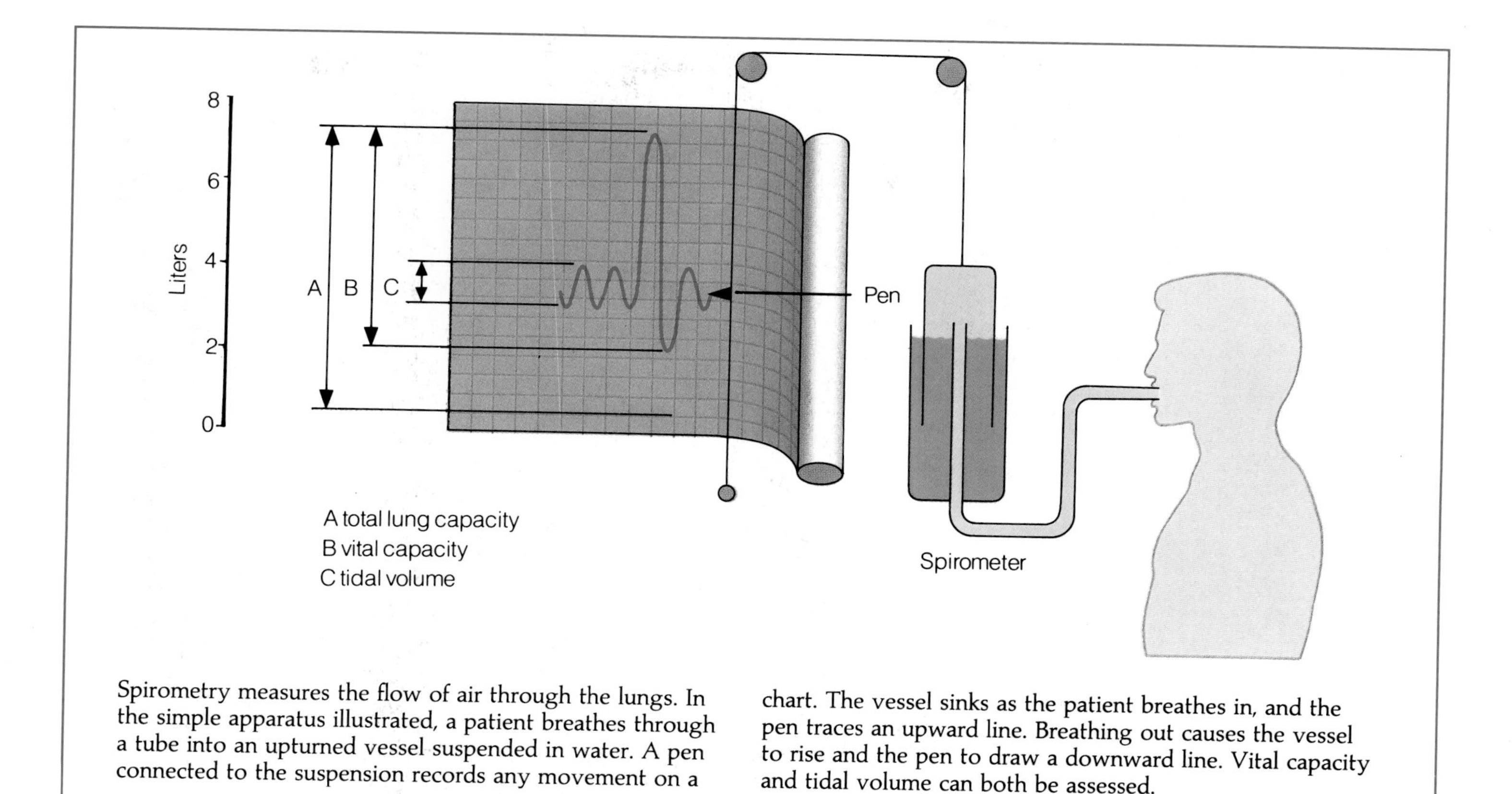

Spirometry measures the flow of air through the lungs. In the simple apparatus illustrated, a patient breathes through a tube into an upturned vessel suspended in water. A pen connected to the suspension records any movement on a chart. The vessel sinks as the patient breathes in, and the pen traces an upward line. Breathing out causes the vessel to rise and the pen to draw a downward line. Vital capacity and tidal volume can both be assessed.

These babies are often called blue babies, referring to their lack of the usual healthy pink glow, or even tinges of real blueness. Oxygen-poor, red-blue blood is circulating in their skin (and other organs) in place of the usual oxygenated, bright red form. Increasingly, modern-day surgery can correct such faulty plumbing and enable these babies to lead normal lives.

Respiratory infections

The respiratory system, with its membrane-lined airways and delicate-walled alveoli, is a prime candidate for invasion by airborne germs. The viral infections of colds and influenza are experienced by almost everybody at some time. Bacteria or viruses are also responsible for many cases of pharyngitis, laryngitis, tonsillitis and the various other "-itises" and "sore throats" that afflict us.

More serious is pneumonia, which is broadly an inflammation of the lungs. It makes the alveoli fill with mucus and fluid, so impairing gas exchange and, as a result, the functioning of the whole body. Pneumonia was – and indeed still is in some cases – a fatal complication for the frail, weakened body of an elderly person already racked by some other illness. There are three main causes: bacteria, viruses and mycoplasmas.

Bacteria are usually inhaled and, once they infest the alveoli, cause inflammation and the accumulation of fluid and microbe-fighting white cells. The infected part of the lung becomes waterlogged, or consolidated, and shows up as a shadow on a chest X-ray. Patients usually suffer from fever, a sputum-producing cough, shaking, chills and chest pain. As more lung tissue is affected, gas exchange is impaired and in severe cases this may lead to respiratory failure and even death.

Legionnaire's disease is a form of bacterial pneumonia, the culprit being the organism *Legionella pneumophila.* It was characterized following the outbreak during an American Legion conference in Philadelphia in 1976. Like most forms of bacterial pneumonia, it can be treated using antibiotic drugs.

As many as half of all pneumonia cases are caused by viruses. Symptoms resemble those of influenza, with fever, aching joints and muscles, headache, watering eyes, nasal congestion, cough and sore throat. There is no specific drug therapy, but plenty of rest and fluids, "GNC" (good nursing care) and fever-fighting drugs such as aspirin seem to help the body's own defenses to counteract the infection.

Mycoplasmas, the causative organisms in the third type of pneumonia, are best described as hybrids, halfway between bacteria and viruses. They bring on a generally mild illness in young adults, the most predominant symptom being a dry, hacking cough. A chest X-ray usually shows patchy shadows on the lung, rather than one solid, consolidated area.

A second serious lung infection, once a leading cause of death worldwide, is tuberculosis, or TB. Its cause is the microorganism *Mycobacterium tuberculosis,* and it often begins with a short pneumonialike reaction. But the bacilli become walled into pockets in the lung, scarring its tissue, and they spread to other regions of the body. At some future date, perhaps when the person's resistance is weakened by another illness, tuberculosis may reactivate and further scar, clog, disrupt and generally damage the lung tissue. Nowadays this feared infection can be treated by antituberculosis antibiotics, but it is still prevalent in areas with inadequate health care and where poverty, poor hygiene and lack of health education encourages its spread.

Lung disorders

Most lung disorders can be classified according to the pattern of functional abnormality: the effect on the lungs, rather than the cause. In obstructive lung conditions such as asthma and emphysema, the major problem is that the air flow through the conducting airways is obstructed or limited in some way. Sufferers undergoing pulmonary function tests inhale or exhale air at a rate much lower than a comparable healthy person. The obstructed airflow may result in audible wheezing, as in asthma.

The airway obstruction may be a partial plugging from build-up of mucus in the air passages, as in patients with chronic bronchitis. It may be disruption and expansion of the fibrous supporting tissue around the airway, which compresses the bore of the passage, as in emphysema. It may be a thickening of the airway wall, as in some types of "dust disease" (pneumoconiosis, caused by long-term inhalation of damaging dust particles). Or it may be contraction of the smooth muscle bands encircling the bronchial walls; asthma affects the bronchioles in this way.

In restrictive lung disorders, the problem is a lowered total lung capacity. In lung function tests, when patients with restrictive disease are asked to breathe in and fill their lungs completely, they are unable to take in an adequate volume of air compared to healthy people of the same weight and height. There is no obstruction to airflow, unless obstructive disease is also present (as happens in emphysema).

Restriction of lung capacity can be caused by any of three

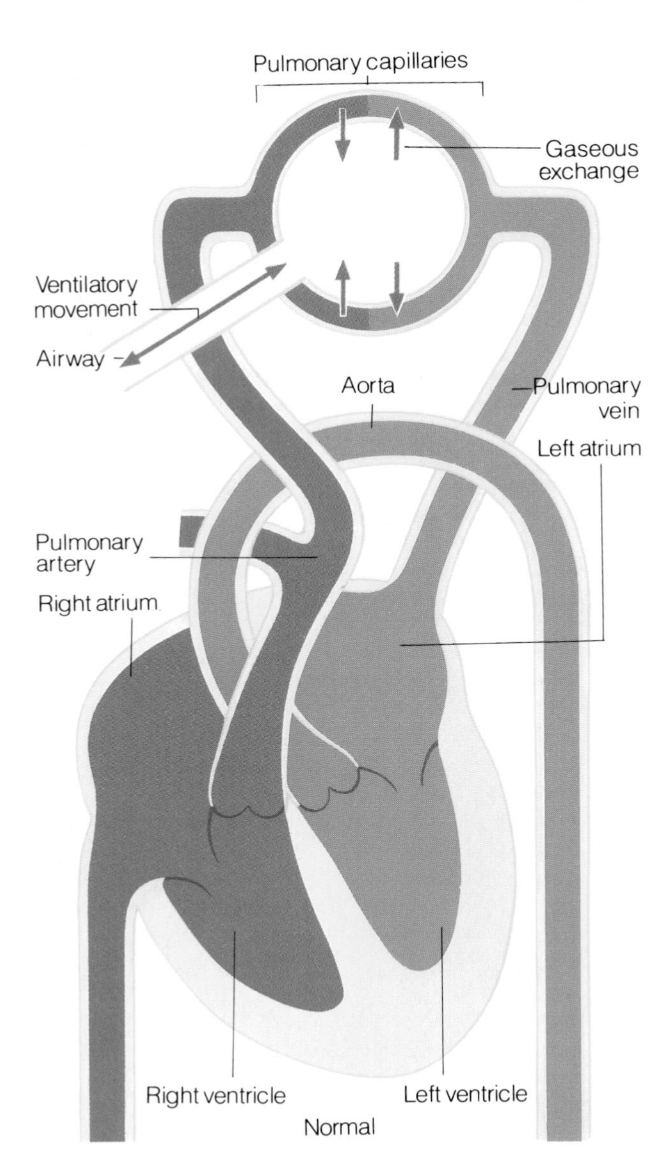

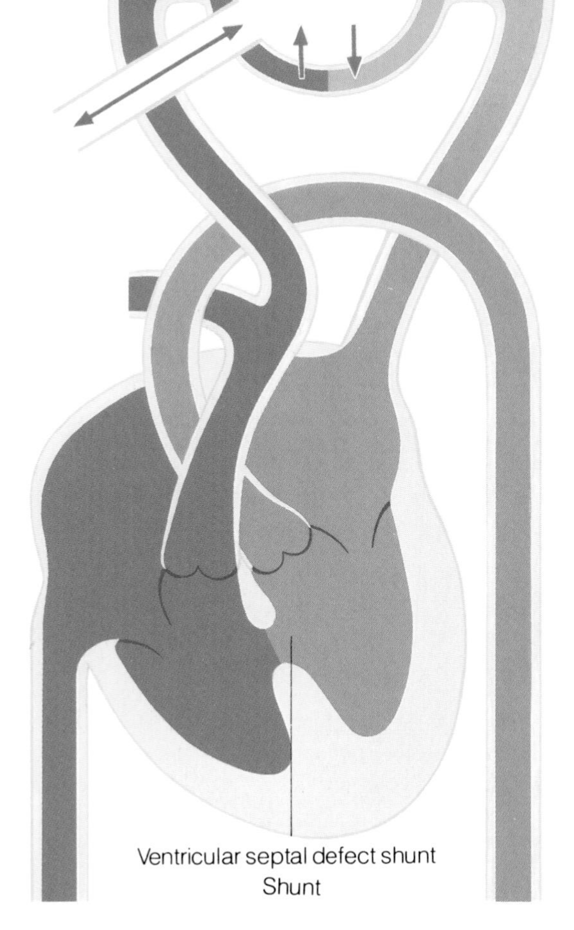

Shunts – abnormal connections in the pulmonary circulation – prevent proper oxygenation of the blood. In the normal circulation *(left)* oxygenated blood (red) and deoxygenated blood (blue) are kept completely separate. A shunt caused by a fistula between the pulmonary vein and artery or a septal defect in the heart *(right)* allows the two types of blood to mix, so that the oxygen content and therefore usefulness of the arterial blood is reduced.

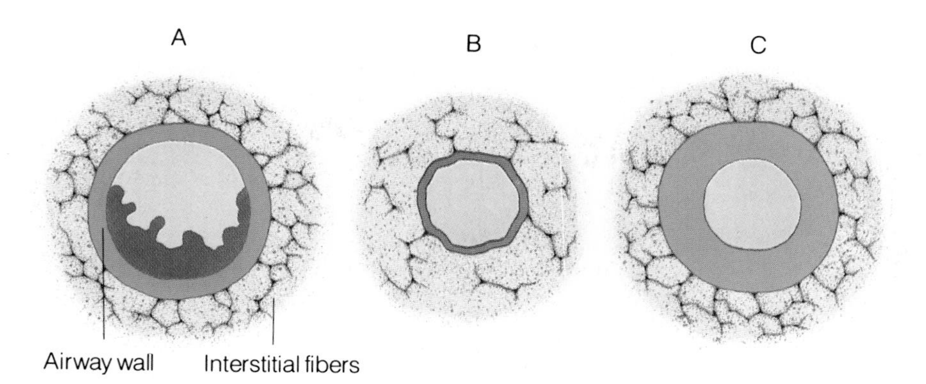

Airways become blocked when obstructive lung disease strikes. This may happen in one of three ways. Excessive mucus may block the airway (A); loss of traction in the surrounding interstitial fibers may cause narrowing (B); or the airway wall may thicken and restrict airflow (C). Whatever the cause, the result is a reduced flow of air and impaired efficiency of respiration.

factors: disease in the lung itself, problems with the pleurae covering the lungs, and malfunction of the chest wall or muscles of respiration.

In the lung itself, diseases such as sarcoidosis and asbestosis affect the interstitial tissue – that is, the "ground matter" of elastic fibers and supporting connective tissue, around and between the airways and alveoli. The lungs lose their elasticity, or compliance, and become stiff and difficult to inflate to the required degree. Paradoxically, the airways may not be involved in the disease process itself but they may become dilated, or wider than usual. These dilated small airways, surrounded by scarred, thickened connective tissue, have been likened to a honeycomb in appearance, hence the term "honeycomb lung" to describe severe interstitial restrictive lung disease.

Pleural problems can also lead to restriction of lung volume. The most striking example is pneumothorax, when air from outside enters the space between the two layers of pleural membrane around each lung. The air destroys the "seal" by which the breathing muscles and rib cage stretch the lungs upon inhalation. The lung, no longer under tension, collapses. Spontaneous pneumothorax is usually caused by the rupture of a small air space in the top of the lung. It is five times more common in men than women, and particularly affects young, tall, thin men, and more frequently the right rather than the left lung. Provided the leak is short-lived and soon sealed, the air in the pleural space is gradually absorbed over ensuing weeks and the lung expands back to normal.

If the chest wall suffers trauma and is pierced, perhaps by a knife or a bullet, or as the result of an automobile accident, the situation may be more serious. The lung collapses as air floods the pleural space. With each breath, air is sucked in and out through the wall rather than up and down the trachea; the lungs stay shrunken. Since each lung has a separate pleura, damage to one does not necessarily affect the other. But air entering through such a "sucking chest wound" may build up in the chest, compressing and threatening the other lung and the heart. As the victim struggles harder for breath, the condition worsens. Emergency treatment is vital, firstly to prevent inflow and accumulation of air, and secondly to remove the indrawn air from the pleural space and allow the lung to reinflate.

Smoking: respiratory overkill

Four or five decades ago, smoking cigarettes was variously perceived – or at least presented – as sleek, sophisticated, glamorous, mature and "cool." Today it is seen by most non-smokers (and by some smokers, too, who cannot kick the habit) as dirty, dangerous, a form of slow suicide for the smoker, a source of pollution of the atmosphere and a threat to the health of others.

Smoking affects the stomach, heart, bladder and several other body organs. But its effects on the respiratory system are perhaps more simply pictured, and they are devastating. The toxic components in tobacco smoke include carbon monoxide, nicotine and tar, inhaled as vapors. They kill millions of the tiny hairlike cilia lining the airways, which leads to accumulation and stagnation of mucus, germs and inhaled debris, which the cilia would otherwise sweep away. The smoker's cough is an attempt to bring up the mucus and clear clogged airways.

Prolonged exposure to cigarette smoke (not necessarily one's own) tends to result in increased mucus production, greater susceptibility to infection, obstructive and destructive changes in the bronchial tubes and severe gas exchange problems. Smoking is the single most important reason for chronic bronchitis and emphysema, which annually kill tens of thousands of people in the United States.

In the 1980s, more than 130,000 people died from lung cancer each year in the United States. The link between smoking and lung cancer is beyond doubt. About one in ten heavy smokers (20 or more cigarettes daily) eventually suffers from this disease. Treatments are advancing, but the survival rate remains low.

Despite the long list of devastating health risks associated with smoking, recent research has shown that it is always worthwhile giving up. Rapidly-felt benefits include easier breathing, re-sharpened senses of smell and taste, and improvement in general respiratory function. Over the longer term, the risks of developing serious smoking-linked illness, such as cardiovascular disease or lung cancer, gradually fade away.

Asthma and the airways

In the United States and in Britain about one person in 30 has asthma. In this condition, the airways (especially the bronchioles) are narrowed by the contraction of the smooth muscle strands in their walls, and further obstructed by swelling of the wall tissue and an increase in mucus secretion. Asthma is typically episodic. During an attack, the characteristic symptom is strained, wheezy breathing, sometimes accompanied by spells of breathlessness and coughing. Around three-quarters of children with asthma improve in this respect during adolescence.

Most cases of asthma in children and adolescents are associated with an allergic reaction of a particularly sensitive bronchial tree. The allergen which triggers the reaction may be pollen, certain foods or additives, house dust or mites, fragments

Effects on airways of restrictive lung disease

Interstitial fibers

Airway wall

Emphysema

Normal airway

Fibrosis

Restrictive lung disease interferes with airflow to the lungs by altering the diameter of the airways. In emphysema surrounding fibers are disrupted, reducing radial traction and causing collapse of the airway. The opposite occurs in fibrosis, in which fiber traction increases and widens the airways. A normal airway is shown *(middle)* for comparison.

of animal fur, skin or feathers, and similar organic products. Asthma can also be triggered by infections, exercise, inhalation of cold air, specific medicines (such as aspirin), and psychological or emotional factors.

An important part of asthma therapy is to identify, and subsequently avoid, the allergen or other trigger factors. Bronchodilator drugs, which relax and clear the airways, are the mainstay of chemotherapy. These are generally inhaled as a fine mist, so that they rapidly reach their site of action in the lungs and act directly on the narrowed airways.

Technology takes over

Medical technology has long been utilizing the respiratory system. English chemist and inventor Humphry Davy realized the analgesic (painkilling) properties of inhaled nitrous oxide, or "laughing gas," in 1799. This gas became the first of the inhaled anesthetics in 1844, when Connecticut dentist Horace Wells used it to relieve the pain of a tooth extraction. In 1846, Boston surgeon John Warren employed ether as an anesthetic while removing a tumor from the jaw of a patient at Massachusetts General Hospital. He was accompanied by dentist William Morton, and his former lecturer Charles Jackson, who had suggested using ether. The technique soon spread, and chloroform was also utilized. Combined with Joseph Lister's antiseptic techniques, developed around 1865–75, the era of modern surgery had arrived.

In 1837, the Royal Humane Society in London recommended that people who had almost drowned, or suffocated in some other way, could be resuscitated by "artificial respiration" carried out by external compression of the chest wall. In the late 1950s, it was shown that mouth-to-mouth resuscitation (blowing into the victim's lungs) was more effective. Both techniques have saved countless lives.

More recently, mechanical devices for artificial ventilation have been developed. Ventilators of various kinds, providing intermittent positive pressure ventilation (IPPV), and also continuous respiratory function monitors, are indispensable in today's intensive care units.

Surgical techniques are now able to bypass heart and lungs completely, for example when carrying out a heart-valve implant or some other open-heart surgery. The cardiopulmonary bypass equipment, or "heart-lung" machine, was first used successfully in 1951, but there have been continual improvements since that time. Blood is led from the vena cavae, the main veins of the

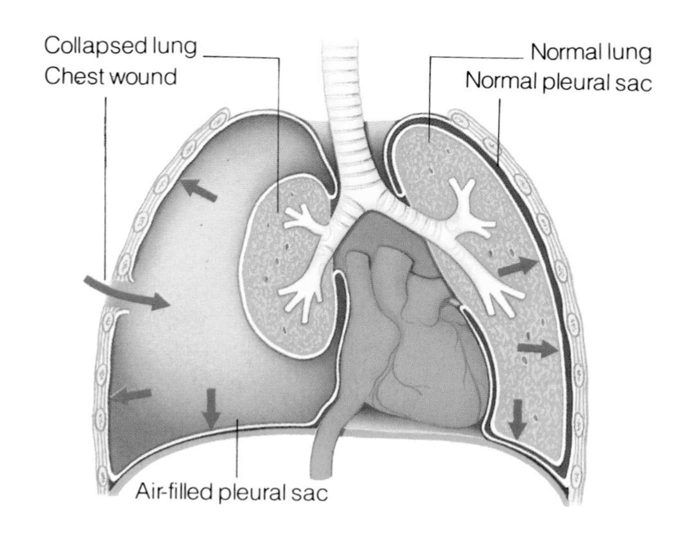

A hole in the chest wall, usually resulting from a wound, is the usual cause of pneumothorax. If the pleural sac is penetrated, each intake of breath causes air to enter the sac, which expands and rapidly collapses the lung. On breathing out, some air leaves through the hole, but even more enters. Emergency first aid treatment consists of covering the wound and making it airtight until the victim can be treated in the hospital and the damaged lung reinflated.

Secondhand smoke from somebody else's cigarette can affect you and pollute the atmosphere. This antismoking poster from the American Cancer Society draws attention to the potential dangers.

SMOKING
POLLUTES
YOU AND
EVERYTHING
ELSE
American Cancer Society

A healthy bronchus has an open airway lined with mucous cells and cilia that keep it moving. Disease can cause the surrounding ring of muscle to contract and tighten on the airway, constricting it and reducing the flow of air through it. Other disorders can cause mucus to accumulate until there is too much for the cilia to clear away, again leading to obstruction of the airway.

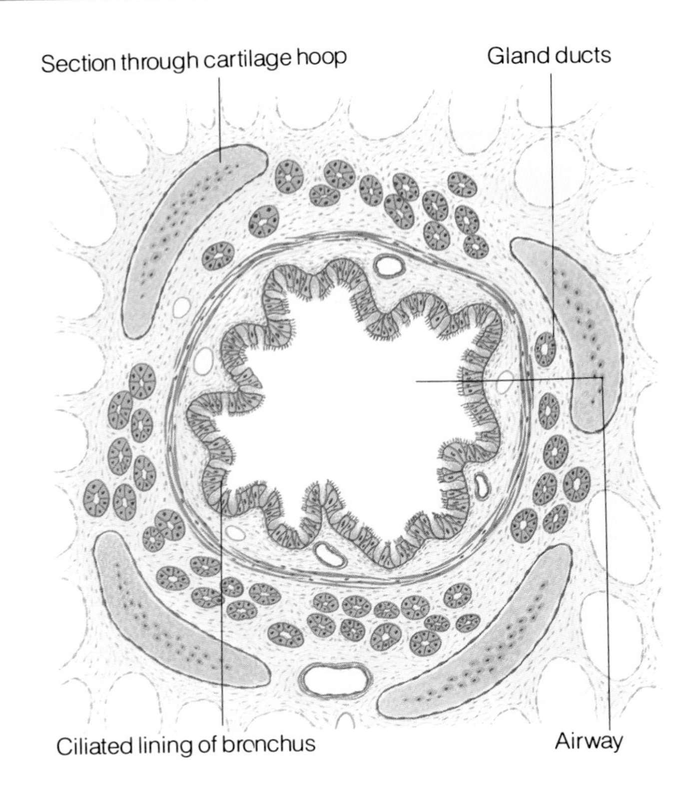

systemic circulation, along tubes to an oxygenator (which takes the place of the lungs), then to a roller or peristaltic pump (fulfilling the part of the heart), and back into the main aorta for another systemic circuit.

Two main types of oxygenators are used in cardiopulmonary bypass devices. The bubble oxygenator sends small gas bubbles, one-twelfth to one-third of an inch in diameter, through the blood reservoir. The blood must be meticulously debubbled and defoamed before being sent back into the body. In the more modern membrane oxygenator, blood flows past a semi-permeable membrane, on the other side of which is gas rich in oxygen and low in carbon dioxide. Gas, but not liquid, can pass through the membrane, which therefore functions in much the same way as the thin walls of the alveoli in the lungs.

The effects of exercise

Energy for exercise is normally obtained from aerobic metabolism, also referred to as aerobic respiration. In this process, one molecule of energy-containing glucose in a cell is converted by a series of biochemical stages to provide 36 molecules of ATP (adenosine triphosphate). ATP is the standard energy molecule of metabolism, the "energy currency" that can be fed into other biochemical reactions, to drive them along.

Aerobic metabolism, as its name implies, requires plenty of oxygen. Since we need more energy during physical activity, the respiratory system must provide more oxygen, by breathing harder. When the exercise finishes, breathing soon returns to normal.

Anaerobic ("without oxygen") metabolism is called into play during short, high-energy bursts – sprinting, for example – when there is not enough time for increased oxygen needs to be met, or simply when sufficient oxygen is temporarily unavailable. In a modified series of biochemical steps, only two molecules of ATP are generated per molecule of glucose. Also, an end-product of the reaction is lactic acid, an awkward chemical which, as it builds up, inhibits metabolism and makes the process self-limiting. At the end of exercise, breathing stays at an increased level to provide oxygen, which enters a side-arm of the metabolic pathway and converts the lactic acid to more harmless substances. As a result, during anaerobic exercise, an "oxygen debt" is incurred, which must be repaid by continued respiratory activity even after the physical activity is over.

Herein lies one lesson of good health. In general, aerobic forms of exercise such as sustained running and swimming are more beneficial than "short-burst" anaerobic types. The lungs must breathe harder, and the heart must beat faster, to supply the muscles with extra oxygen. Provided this is done regularly, and not overdone, it increases the general level of fitness.

The human respiratory system evolved to cope with occasional periods of heavy demand. It did not expect to work within a body whose only exercise is a short, slow walk to the car, or to have toxic tobacco smoke poured into it. Treated with understanding and respect, it should give good service for life.

Down the wrong way

Sometimes, perhaps because somebody is talking while eating or drinking, food misses the esophagus, the tube leading to the stomach, and passes into the windpipe. Instead of being eaten, the food is inhaled and blocks the airway to the lungs.

The usual reaction to choking is a reflex one, resulting in paroxysms of coughing. Often this is sufficient to dislodge the food, and breathing soon returns to normal. Prolonged choking is much more serious, and the victim rapidly turns blue in the face (because of asphyxiation through lack of oxygen) and may lose consciousness. If the victim is a baby, he or she can be held upside-down by the legs and given a sharp slap between the shoulder-blades. An older child should be held face-down over a seated person's knees, and again slapped sharply on the back.

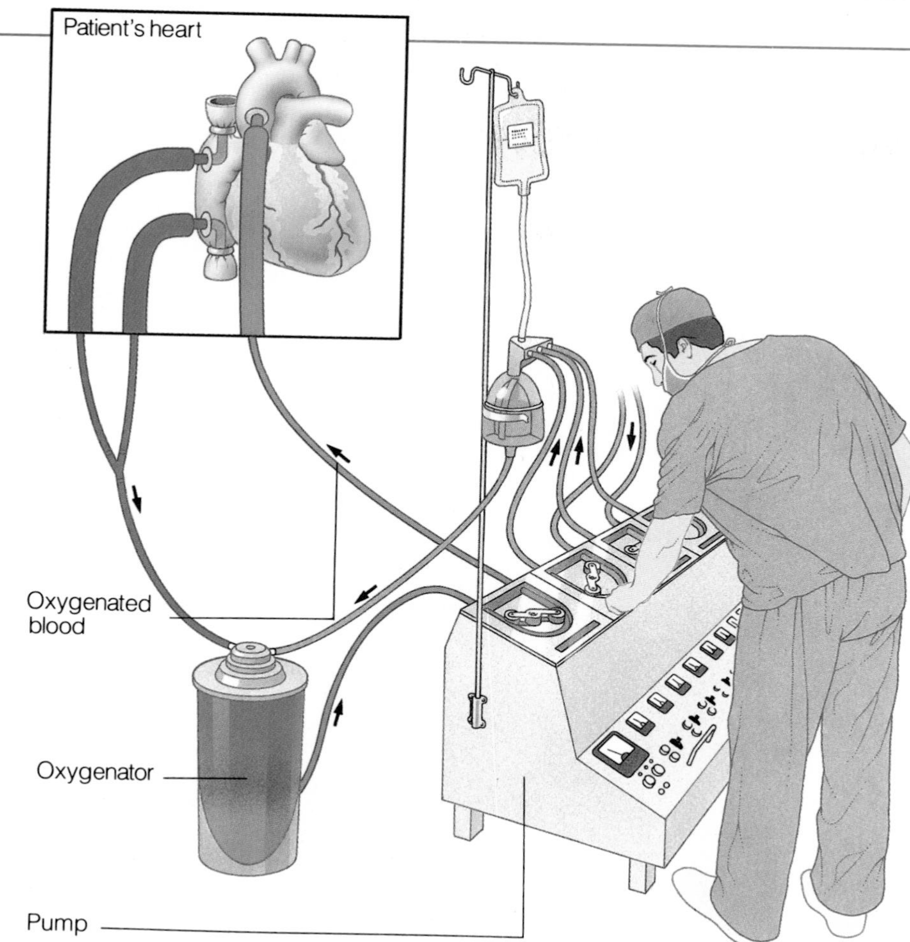

A heart-lung machine takes over both the pumping action of the heart and the respiratory function of the lungs. Deoxygenated blood from the vena cavae, the two major vessels that carry it to the heart, is diverted to an oxygenator. There oxygen is added and the oxygenated blood pumped back into the aorta, the major vessel leaving the heart, and so into the body's circulation. Secondary pumps allow plasma or other fluids to be added to the circulating blood. Open-heart surgery or other forms of cardiopulmonary bypass operation *(below)* were not possible before the development of a reasonably small and reliable heart-lung machine.

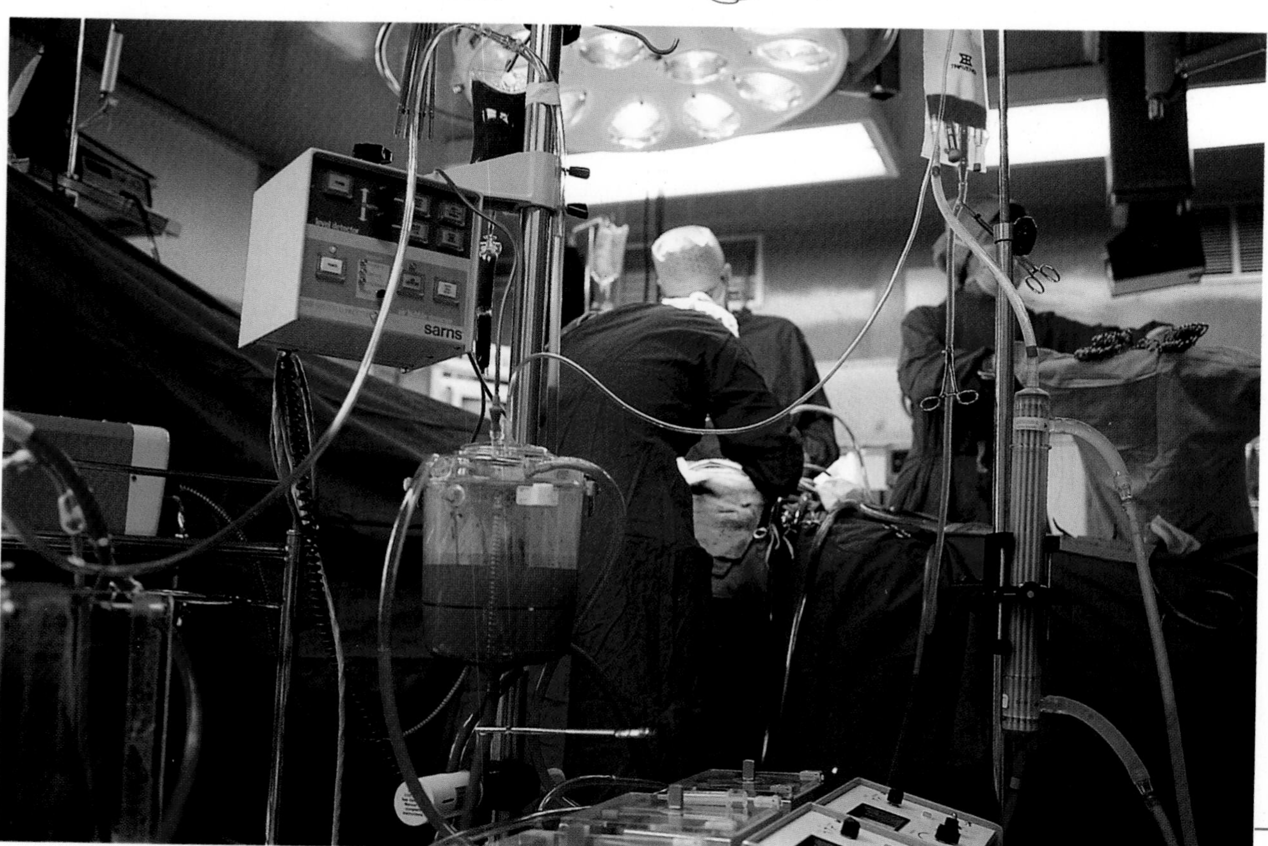

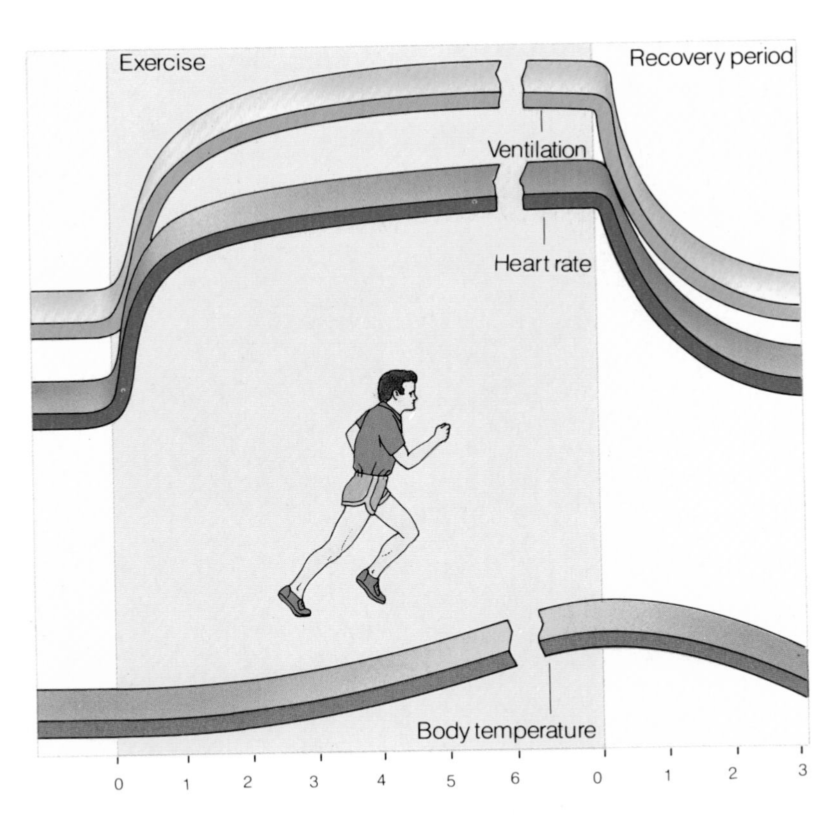

During vigorous exercise active muscles demand more oxygen, and this has to be supplied from the lungs via the blood. At the beginning of the activity there is a sudden increase in breathing rate (ventilation) with a corresponding increase in the heart rate. Both eventually level out as the slower-rising body temperature reaches a maximum. At the end of the exercise ventilation and heart rate fall rapidly, but body temperature makes a more gradual decline.

In a short burst of a very energetic activity, such as a 100-meter dash *(right)*, there is not enough time for the lungs to respond to the need for extra oxygen. Energy is produced anaerobically – without oxygen – and the sprinter may not even breathe at all during the race. But the breathing rate must catch up later to repay the oxygen debt.

Neither of these methods is possible with a chocking adult, for whom the best emergency action is the Heimlich maneuver. To perform it, stand behind the victim and put your arms around him just above his navel. Make one hand into a fist, clasp it with the other hand and thrust both hands upward and inward toward the victim's chest.

Several quick thrusts may be necessary to dislodge the obstruction. Each thrust rapidly and vigorously reproduces the action of breathing out, lifting the diaphragm and sending a forceful expulsion of air out of the lungs to the larynx. The victim will then probably cough. The Heimlich maneuver technique should be used only in a genuine emergency and then be carried out with great care, and is safest in the hands of a trained paramedic or experienced first-aider.

Deep breathing

The human lungs are designed to breathe air at atmospheric pressure – the pressure at the surface of the Earth. At high altitudes atmospheric pressure is lower, and until visitors are acclimatized to high-altitude regions they often find themselves out of breath. At very high altitudes, additional air or oxygen must be provided.

The opposite problem is encountered by divers, because of the higher pressure under water, which increases by one atmosphere for each 33 feet of depth. Even only two feet down, the water pressure on a diver's chest is too great for him or her to be able to breathe out air at atmospheric (surface) pressure – which is why 18 inches below the surface is the maximum depth for snorkeling.

To go deeper, a diver needs special breathing arrangements. These include a diving bell or caisson, which is a large inverted container filled with compressed air (which serves also to keep the water out); a diving suit, which has an air-tight helmet connected to a supply of compressed air; or, for maximum mobility, SCUBA gear. Standing for Self-Contained Underwater Breathing Apparatus, SCUBA gear consists of a breathing tube in the mouth, carrying air or oxygen from gas tanks on the diver's back. The diver's nostrils are kept closed by a spring clip, and a special valve delivers gas at the same pressure as the pressure of water on the diver's chest, and adjusts the pressure with varying depth.

Beating the bends

Although we breathe air for its oxygen content, in fact some four-fifths of air is composed of the comparatively inert gas nitrogen. Normally nitrogen just passes in and out of the lungs, acting only to dilute the oxygen. But when a diver breathes compressed air at high pressure, which is necessary at depths of 100 feet or more, nitrogen penetrates the alveolar capillaries in the lungs and dissolves in the blood and other tissues.

If the diver then ascends rapidly to the surface and normal atmospheric pressure, the nitrogen "fizzes" out of the blood like the gas in carbonated water and can form bubbles in the joints – giving the painful set of symptoms of decompression sickness, commonly called the bends. Severe cases involve giddiness, respiratory distress and even paralysis.

The prevention of the bends is simple if time-consuming. The diver has to return to atmospheric pressure very gradually, either by making an extremely slow ascent or by spending a long time (up to several days) in a decompression chamber in which the pressure is gradually reduced. An alternative is to breathe artificial air consisting of oxygen mixed with helium instead of nitrogen. However, because this has a different density from ordinary air it affects the vocal cords and hence the diver's voice.

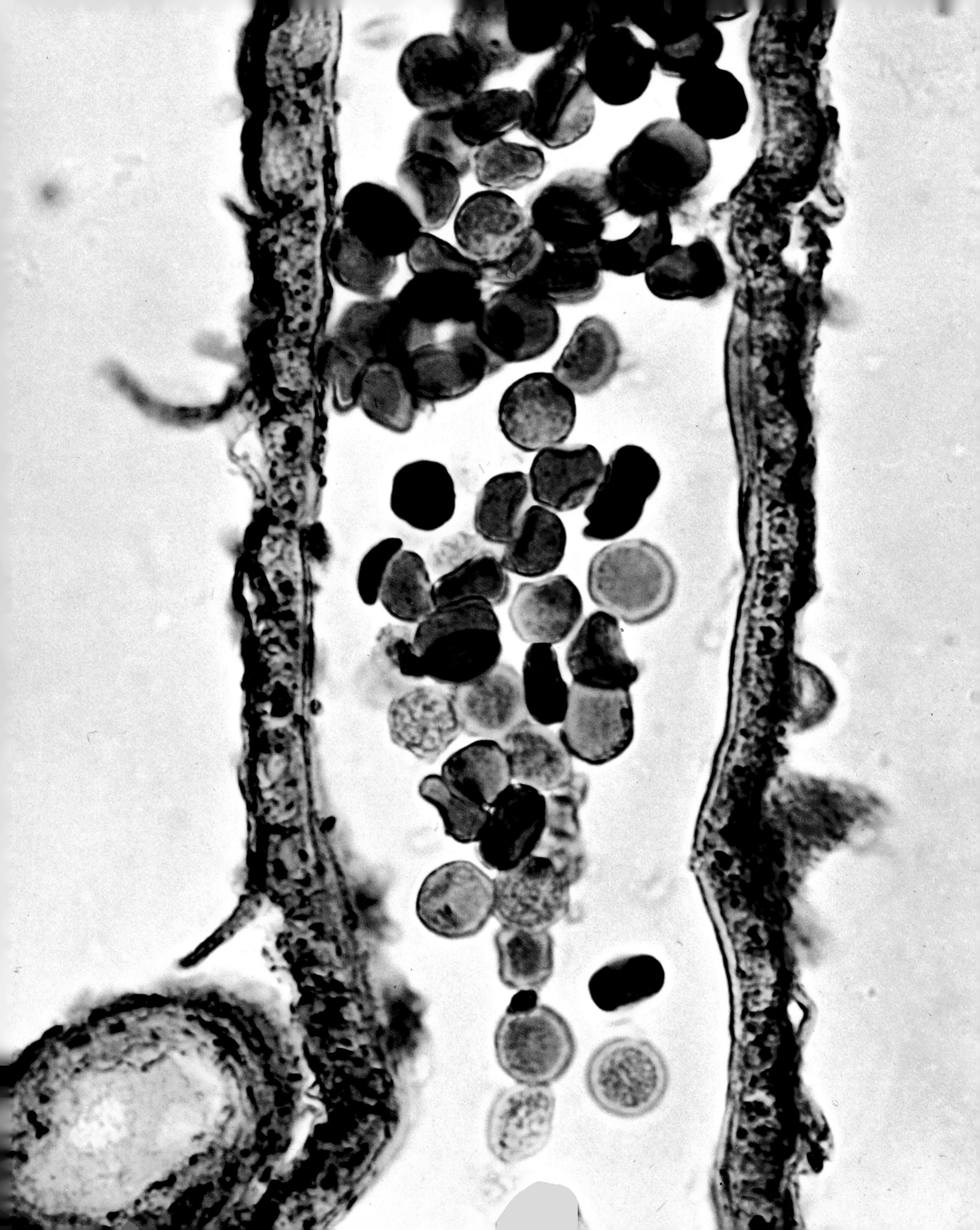

CHAPTER 7

Blood

Blood is the stuff of life. Like most liquids, it takes on the form of its container – in our case, the human body. It dwells nowhere, has no "home base" – although it belongs within a fraction of an inch of every cell in the body. Blood is red – or, at least, reddish, though at times it looks pink, scarlet, ruddy, or leaden reddish-blue. It has a characteristic chemical composition – but this varies from minute to minute, and from organ to organ, depending on what its owner is doing.

The appearance of blood signifies life itself – yet the sight of only a few drops outside its container can make grown people faint on the spot. Blood is never still, but always flowing. Nevertheless it goes nowhere, simply round and round in an endless loop. When it ceases to flow, or leaks from the body, so life itself ebbs away.

A river in reverse

The heart pumps blood, and the blood vessels channel and deliver it. Arteries convey blood away from the heart. A surgeon has to be as familiar with the route of each one, and its scientific name, as he or she is with the route to the hospital each morning. Although blood is often likened to a river, carrying its cargo of nutrients to all parts of the body, the arteries are more like a river in reverse. For whereas river tributaries coalesce, and become larger, arteries divide, and become smaller.

Arteries are thick-walled tubes. A circular sandwich of yellow elastic fibers contains a filling of muscle. The elastic design helps to absorb the tremendous pressure wave of each heartbeat, so that by the time blood reaches the tiny, fragile capillaries it is oozing rather than spurting. The pressure wave is evident after its journey along your arm to your wrist – it is your pulse.

The body's nervous system keeps control of the arterial muscles. By instructing the muscle to contract, the artery's bore is narrowed, and so less blood can flow through. Here is the basis of a flow control system. Arteries around the body are continually adjusted, becoming narrower (constricted) or wider (dilated), as the body undertakes various tasks. Run for a bus and the leg muscles need extra blood – so the arteries leading to them become wider. Eat a meal and the intestinal blood supply has to be increased, to absorb digested food – so open up its arteries. Cold? Shut down the vessels near the skin, so that less blood flows there, and less heat is lost. You turn pale, too.

Thin and slack

Eventually arteries split into arterioles, and then into capillaries, the smallest of the blood vessels. One arteriole may serve a hundred capillaries. In them, in every part of every tissue of every organ, blood's great work is done as it gives up what the cells want, and takes away what they do not want. There is little hindrance: the capillary walls are only one cell wide, and very thin cells at that, looking like curved shingles.

Now the river analogy really does apply. Capillaries join to

Blood is the body's vital fluid, carrying life-giving oxygen to all of its tissues. The close-up photograph *(left)* shows red blood cells tumbling over each other as they pass along a narrow capillary blood vessel serving the brain, which soon ceases to function and dies if starved of oxygen.

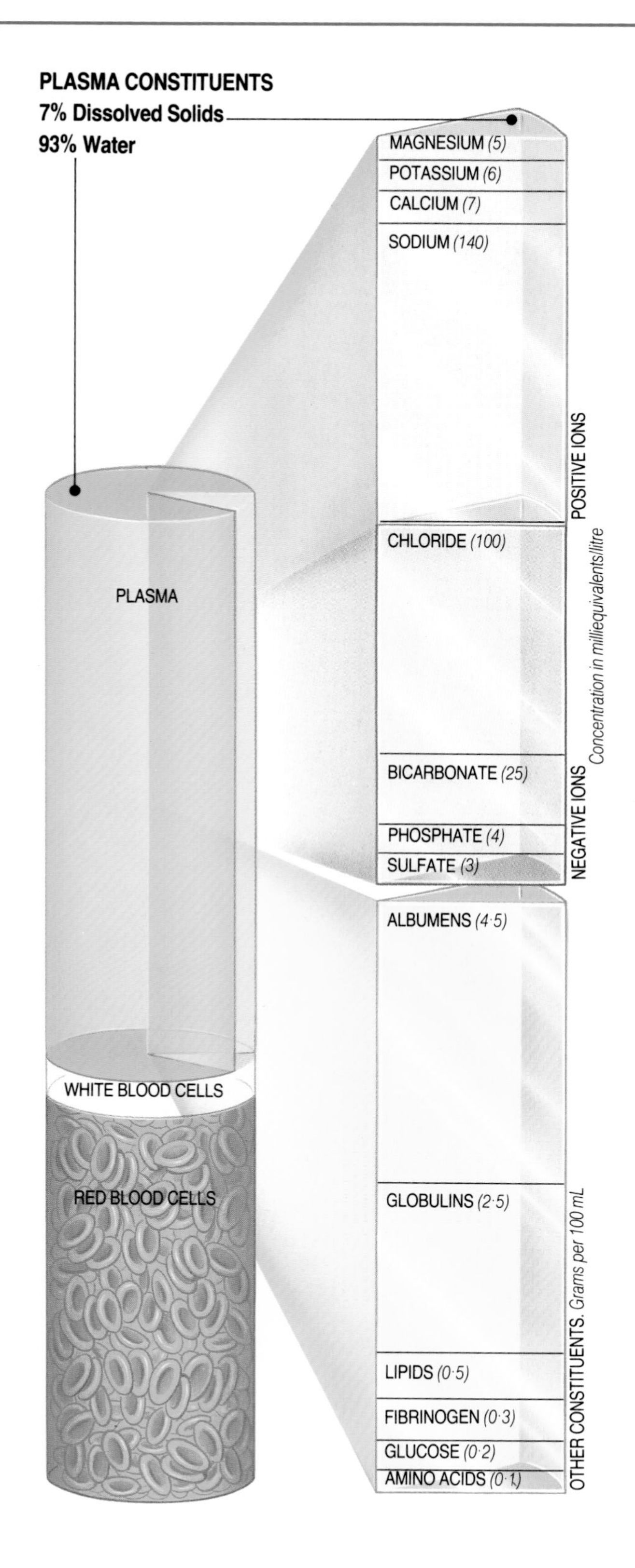

When treated with salt, blood settles into three distinct layers *(left)*, consisting of watery plasma, white blood cells and red blood cells. The plasma is mostly water containing about 7 percent of dissolved solids in the form of inorganic salts and organic substances. The salts are present as positive and negative ions, while the organics include proteins, fats and carbohydrates.

form venules, to form small veins, to form the main veins (venae cavae), and so back to the heart.

Veins are not at all like arteries. Their walls are thin and slack, because by the time it reaches them blood has almost lost the great pressure which forced it out of the heart. Dark reddy-blue, the blood oozes slowly on its way. At any one time the veins contain about 75 percent of the body's blood. Some 20 percent is in the arteries; only 5 percent is in the capillaries.

The heart is at the center of the system, filling with blood and then expelling it about once each second, throughout life. Perhaps surprisingly, although heart muscle surrounds so much blood, it has to have its own separate supply. The blood within its chambers is under too great a pressure, and anyway, on the right side of the heart it carries little oxygen, being on its way to the lungs for refreshment. So there are special vessels, the coronary arteries, that run across the heart's surface and then branch and dive into its thick, muscular wall. Coronary veins return the favor, by returning this blood to the chambers of the heart. This circuit, quickly around its own pump, is blood's shortest in the body's circulatory system.

Back at the heart, blood now begins its second journey. From the right side of the heart it flows along pulmonary arteries to the lungs, to collect vital oxygen. Then back to the heart's left side, and on around the body once more. The two-part circuit is complete, a sort of figure-eight with the heart at the cross-over.

Absorbing the breath of life

Life relies on a supply of energy. In our type of cellular chemistry, liberating this energy from the food we eat is an aerobic process – that is, it requires oxygen. Growth and repair of body tissues depend on nutrients for raw materials, also supplied in our food. The respiratory and digestive systems take care of obtaining oxygen, energy and raw materials from the outside world. But what about the inside world? Straightaway, blood is on the scene, doing the jobs it does most – fetching and carrying.

In every one of the lungs' 300 million microscopic sacs, or alveoli, blood is only four hundred-thousandths of an inch away from air (give or take a little). It courses through millions of capillaries, that clasp each alveolus like thin fingers. This blood is hungry for oxygen. It has been around, and the body's needs have almost drained it of the vital substance. In the alveoli, oxygen is plentiful, breathed in from the surrounding air and passing down a concentration gradient into every nook and cranny of the lungs.

The efficiency of oxygen exchange is staggering. Spread out flat, the double-layer of alveoli and capillaries would carpet a room 30 feet by 25 feet. But it would be a thin carpet – the capillaries hold only three ounces of blood at any time. This blood flows sufficiently near to the air for oxygen exchange to take place for only one quarter of a second.

During this time, oxygen obeys a simple physical law. It moves from a region of higher concentration (the oxygen-rich air) to one of lower concentration (the oxygen-starved blood). It dissolves in the fluid layer lining each alveolus, diffuses through the thin cells making up the alveolar wall and capillary wall, and reaches the blood. Refreshed, and its color turned from leaden reddish-blue to crimson, the blood flows on its way, via the heart, back to the body tissues.

But this is only half the story. As oxygen is required, so carbon dioxide is not. This waste product of bodily chemical reactions has been mopped up from the tissues, in dissolved form, by the

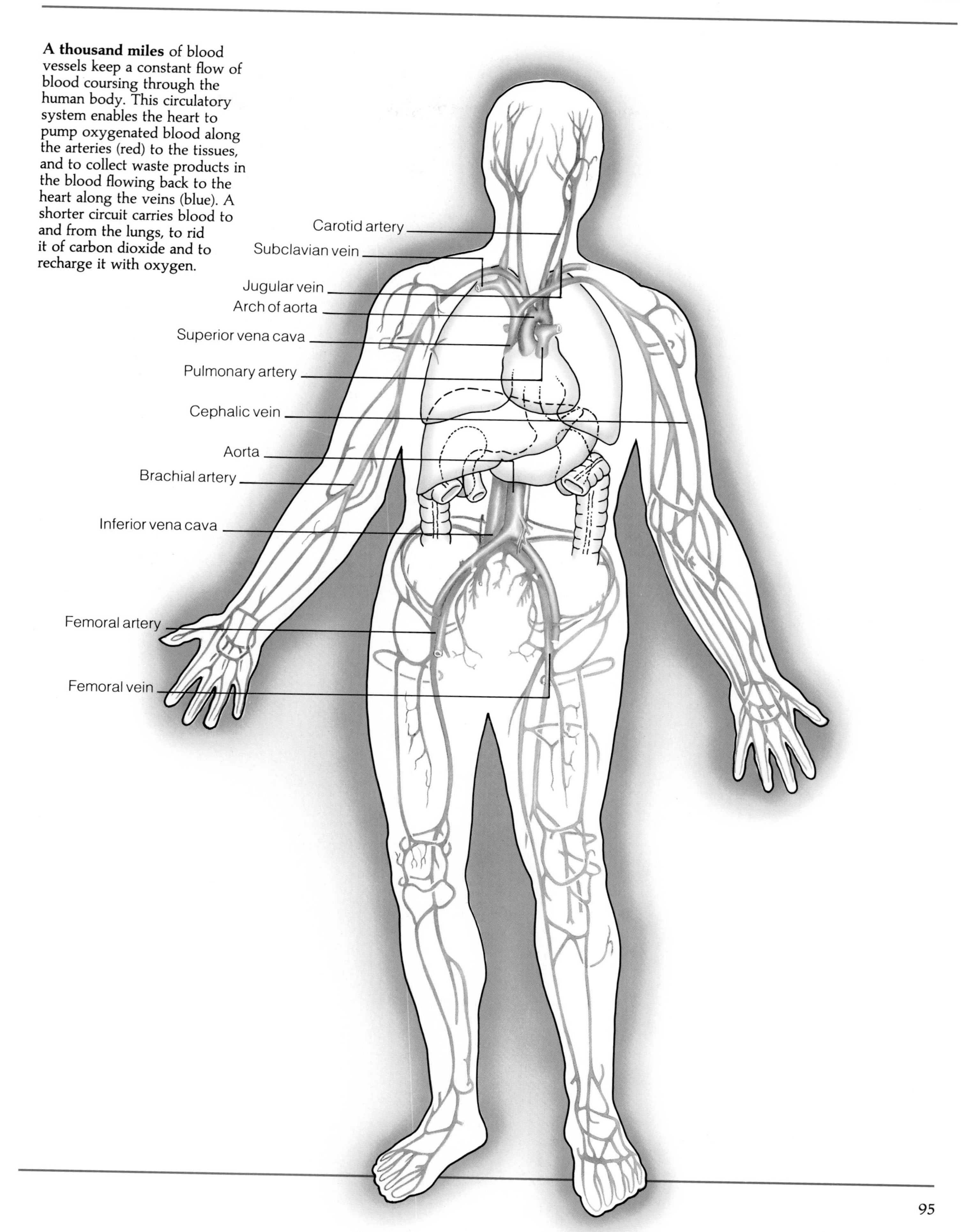

A thousand miles of blood vessels keep a constant flow of blood coursing through the human body. This circulatory system enables the heart to pump oxygenated blood along the arteries (red) to the tissues, and to collect waste products in the blood flowing back to the heart along the veins (blue). A shorter circuit carries blood to and from the lungs, to rid it of carbon dioxide and to recharge it with oxygen.

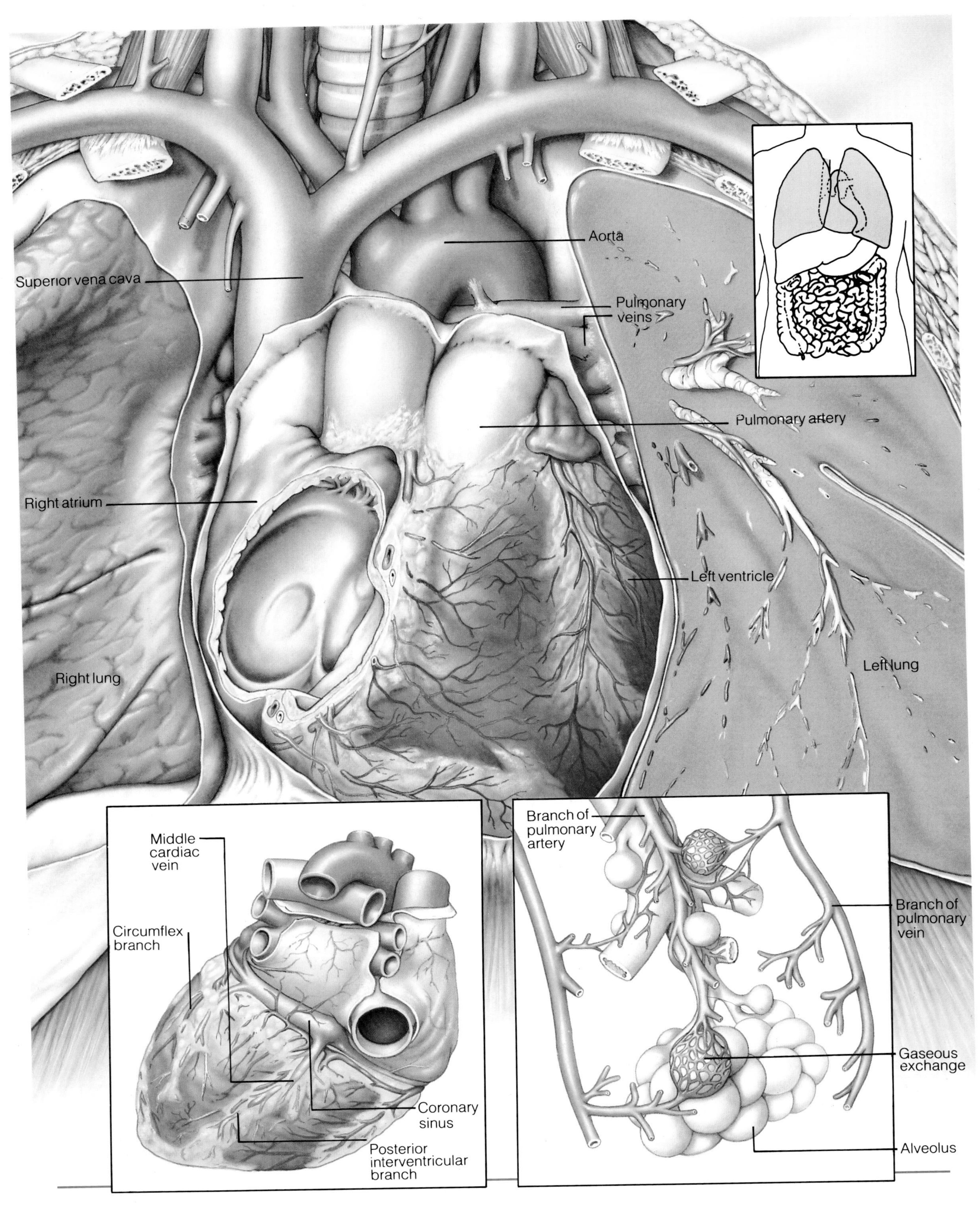
Aorta
Superior vena cava
Pulmonary veins
Pulmonary artery
Right atrium
Left ventricle
Right lung
Left lung
Middle cardiac vein
Circumflex branch
Coronary sinus
Posterior interventricular branch
Branch of pulmonary artery
Branch of pulmonary vein
Gaseous exchange
Alveolus

Two important secondary circulations concern the heart and lungs *(left)*, and the digestive and excretory organs *(right)*. The heart has its own blood supply, and in the lungs capillary veins and arteries meet at sites of gas exchange. The hepatic portal system (shown in purple) carries the products of digestion to the liver. Waste products in blood are carried along the renal artery to the kidney, which filters them out along with excess water to form urine.

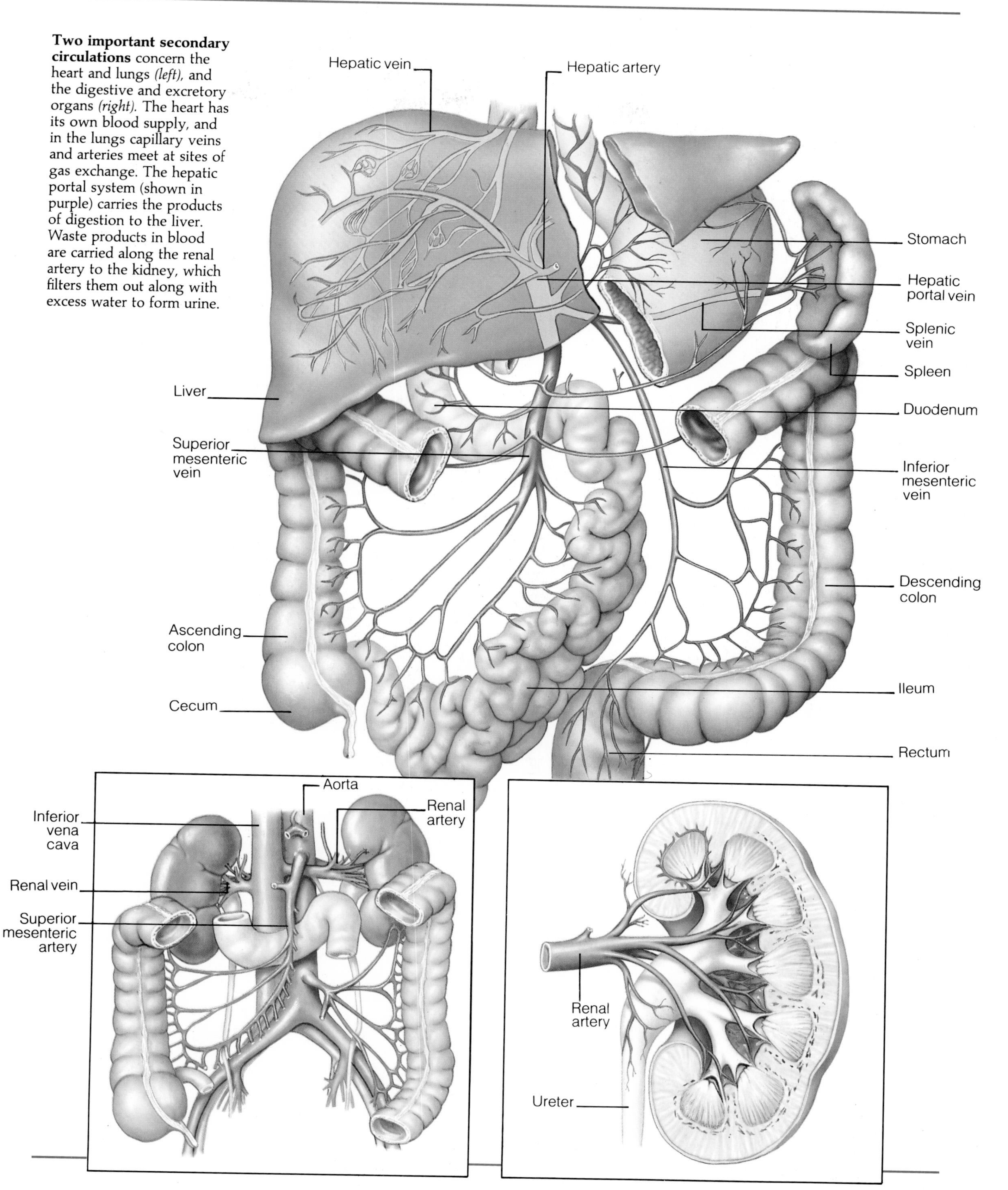

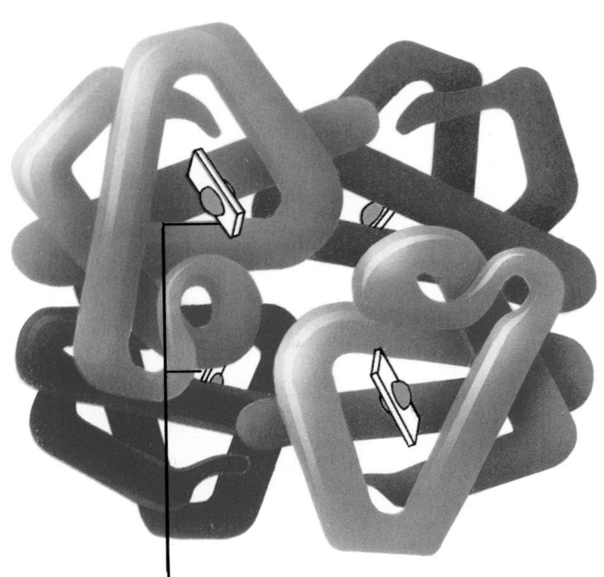

The giant hemoglobin molecule, shown modeled *(above)*, consists of four convoluted protein chains, each surrounding an atom of iron. Every red blood cell has 300 million hemoglobin molecules, each capable of holding four pairs of oxygen atoms. They pick up the oxygen in the lungs, and surrender it after traveling to the tissues in the blood. A lack of hemoglobin or red cells results in palor and other symptoms of anemia.

blood. It must be jettisoned before its levels creep up and become poisonous. So it is not a one-way street in the lungs, but two-way traffic. Carbon dioxide molecules on the way out trace a reverse path to oxygen molecules coming in. The carbon dioxide, now in gaseous form, is eventually breathed out of the lungs and into the atmosphere.

The numbers game

Oxygen does not simply float at random, as dissolved molecules washing about in the blood. At least, only around one percent of it does. The other 99 percent has a personal carrier: a large protein molecule known as hemoglobin. With a contorted molecular shape resembling a piece of modern, chunky jewelry, hemoglobin consists of some 10,000 atoms that make up four intertwined amino acid chains. Each chain cradles a ring of carbon, hydrogen and oxygen atoms known as a heme group. Nestling in each heme as a centerpiece is a single atom of iron.

Iron is an "oxygen magnet." It is strong enough to attract oxygen in the plentiful surroundings of the lungs, but not too powerful to keep hold of it in the oxygen-deprived environment of the tissues. On the return trip it returns the favor and transports some of the carbon dioxide. But most of the carbon dioxide molecules (about 70 percent) enter a chain of chemical reactions that help to maintain the balance of acidic and basic substances in the blood.

In this way each molecule of hemoglobin carries four molecules of oxygen. Again, hemoglobin does not float randomly in the bloodstream. Roughly 270 million molecules of it are jammed together and parceled up in a thin skin to make a scoop-centered, doughnut-shaped, pinched-disk of a cell. This is the red blood cell, or erythrocyte. It is one of the smallest and simplest cell types in the body, being not much more than a membrane-bounded thick soup of hemoglobin, without a nucleus (control center) and structures found in other cells.

Red cells are the most numerous cells in the body. In a tiny drop of blood, there are five million of them. They shuttle round and round, fetching and carrying. Like other cells, they are not immortal. After perhaps 75,000 trips between lungs and tissues, lasting some three months, the red cell dies. To meet the demands of this extraordinary turnover, about three million new red cells are "born," and the same number die, every second.

Meal on a tennis court

The other half of the "requirements equation," energy and raw materials, comes in via the small intestine. Blood on its round-body trip passes into another network of capillaries, this time in the microscopic "fingers," the villi, projecting from the lining of 23 feet of small intestine. Spread out flat, the villi easily outdo the alveoli – they could cover an area the size of a tennis court. Food, broken down by digestion into molecules, along with vital water, passes from the intestinal canal into the bloodstream at the villi, by the process of absorption.

We eat only occasionally, yet our cells need a constant supply of energy and materials. How is this achieved? Blood flows from the capillary network in the villi to a large vessel, the portal vein, that runs from the intestines to the liver. Inside the liver are up to 100,000 tiny hexagonal units, the lobules. They are literally drenched with blood, which is by now low in oxygen, but loaded with newly-absorbed nutrients.

Liver cells in the lobules work their metabolic magic as they convert, store, recycle and release: glucose, fats, proteins and other nutrients are dealt with, as and when needed. There is no shortage of supply. Some 30 percent of the blood pumped by the heart in one minute passes through this marvel of a chemical factory. Blood leaving the liver carries the processed nutrients, to be distributed to all tissues.

The liver, unique among organs, has a second blood supply. The hepatic artery brings bright red blood, fresh from the lungs, to deliver much-needed oxygen. This blood eventually mixes with the portal supply, drains into the main hepatic vein, and heads off back to the heart.

Living plumbing

Nestling just below the liver, in the upper abdomen, are the two kidneys. These intricate examples of living plumbing filter wastes from the blood, chiefly urea (which lends its name to the kidneys' product, urine), various salts, and water surplus to requirements. Every day, the kidneys filter all of the body's blood an equivalent of 60 times.

Each kidney contains more than a million delicate minifilters, the nephrons. By a series of twists, knots and loops, and some judicious exploitation of chemical concentration gradients, unwanted materials go one way into the bladder, while the cleaned and filtered blood returns to the circulatory system. More than any other organ, the kidneys determine the composition of blood and, in turn, the balance of the internal environment.

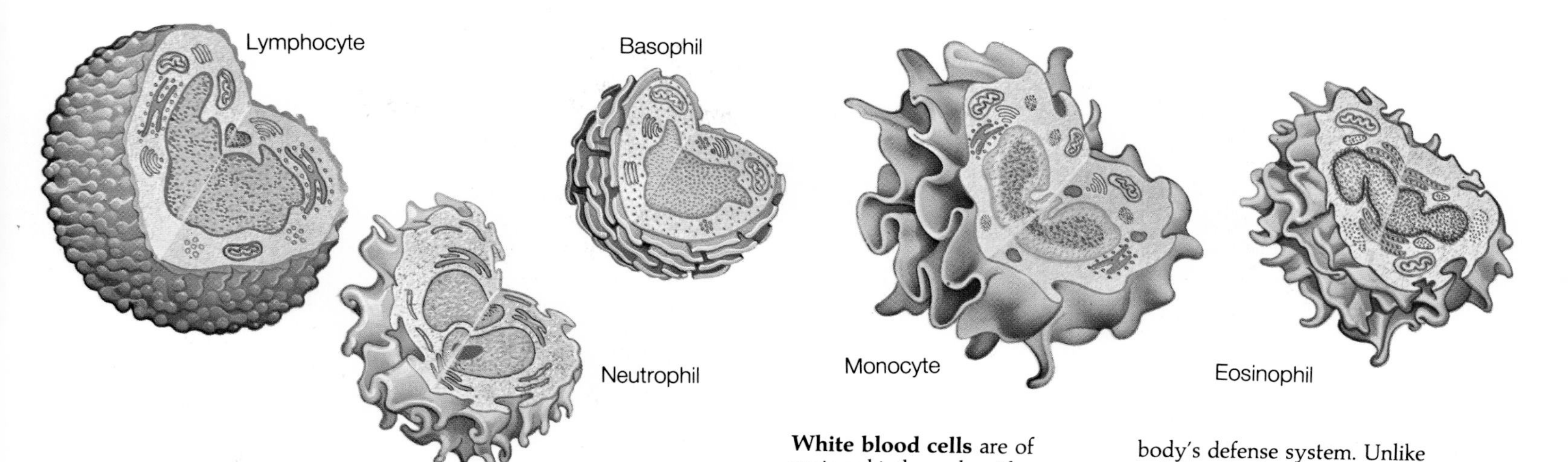

White blood cells are of various kinds, each with a particular job to do in the body's defense system. Unlike red blood cells, white cells have a central nucleus.

Blood under analysis
The chemical and particulate cocktail that is blood cannot be dissected by scalpel and forceps, like any ordinary tissue. But in the fourth century BC, the Greek physician Hippocrates observed that a clear flask of salt-treated blood, left to settle of its own accord, separated into three layers. The uppermost is clear and straw-colored; this is plasma The thin middle layer is made up of white cells. The heaviest layer, red in color and 45 percent by volume, consists of red cells.

Plasma is about 93 percent water – water being the body's solvent. Dissolved in it are any number of substances, from simple salts containing sodium and chloride, to small carbohydrates such as glucose, to lipids (fats), to complex proteins such as globulins (many of which are enzymes) and albumens.

At any one moment the vast, pervasive network of some 10 billion capillaries contains five percent of the body's blood. Here, in the capillaries, blood gives and takes. It is the great and silent exchange. Oxygen leaves the red blood cells, passes through the single-cell-thick capillary wall and diffuses through the watery interstitial fluid to reach the cells. It passes carbon dioxide diffusing in the opposite direction. Foodstuffs, too, pass from blood to tissues, while wastes and by-products such as urea do the reverse. Many other body chemicals, such as hormone messengers, join in the great exchange.

White knights of the battlefield
The thin white layer in the flask of settled blood contains an army ready to defend the body to the death – and die they do, in their billions, when an infection or other disease takes hold. The defenders in question are the white blood cells, or leukocytes, outnumbered 600 to 1 by red cells. But, unlike the reds, the whites are "complete" cells, with nuclei and other internal structures. Self-contained and self-sufficient, they are ready to go anywhere at a moment's notice.

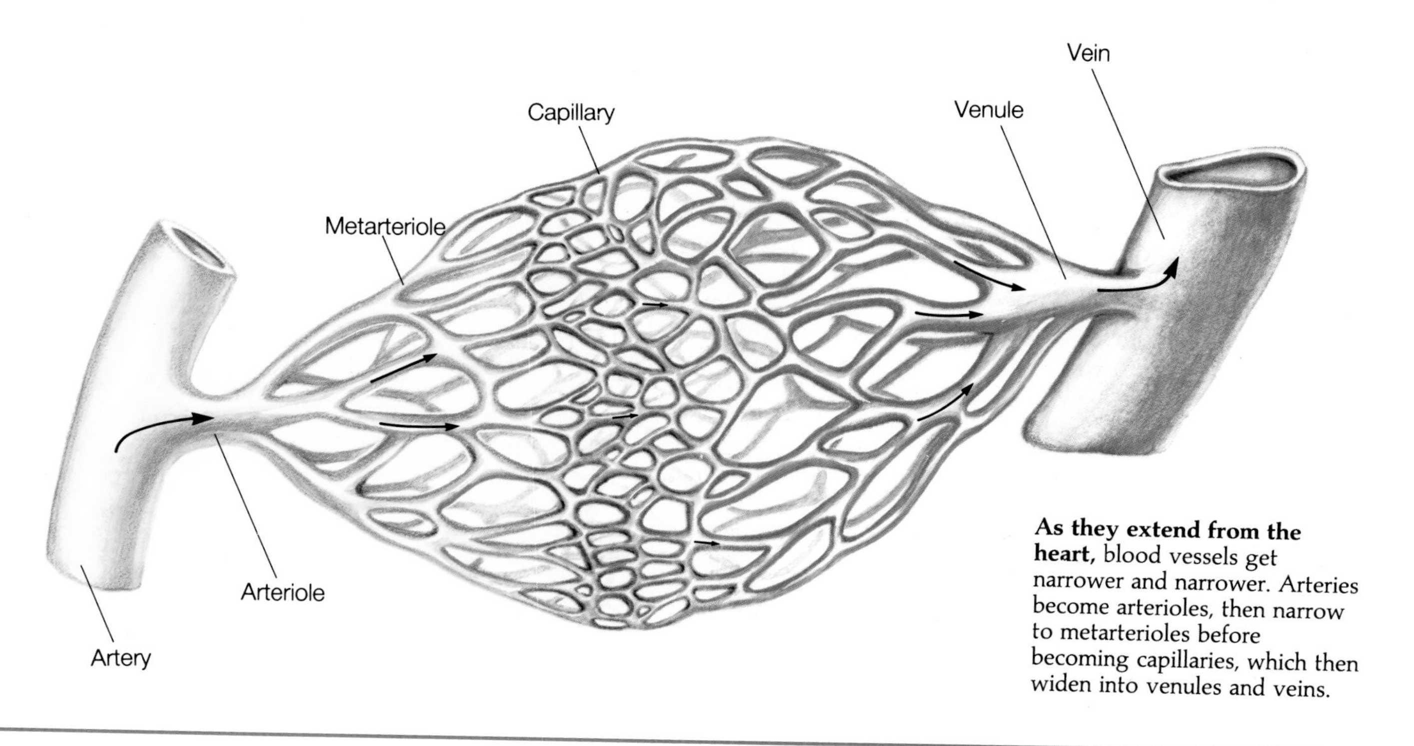

As they extend from the heart, blood vessels get narrower and narrower. Arteries become arterioles, then narrow to metarterioles before becoming capillaries, which then widen into venules and veins.

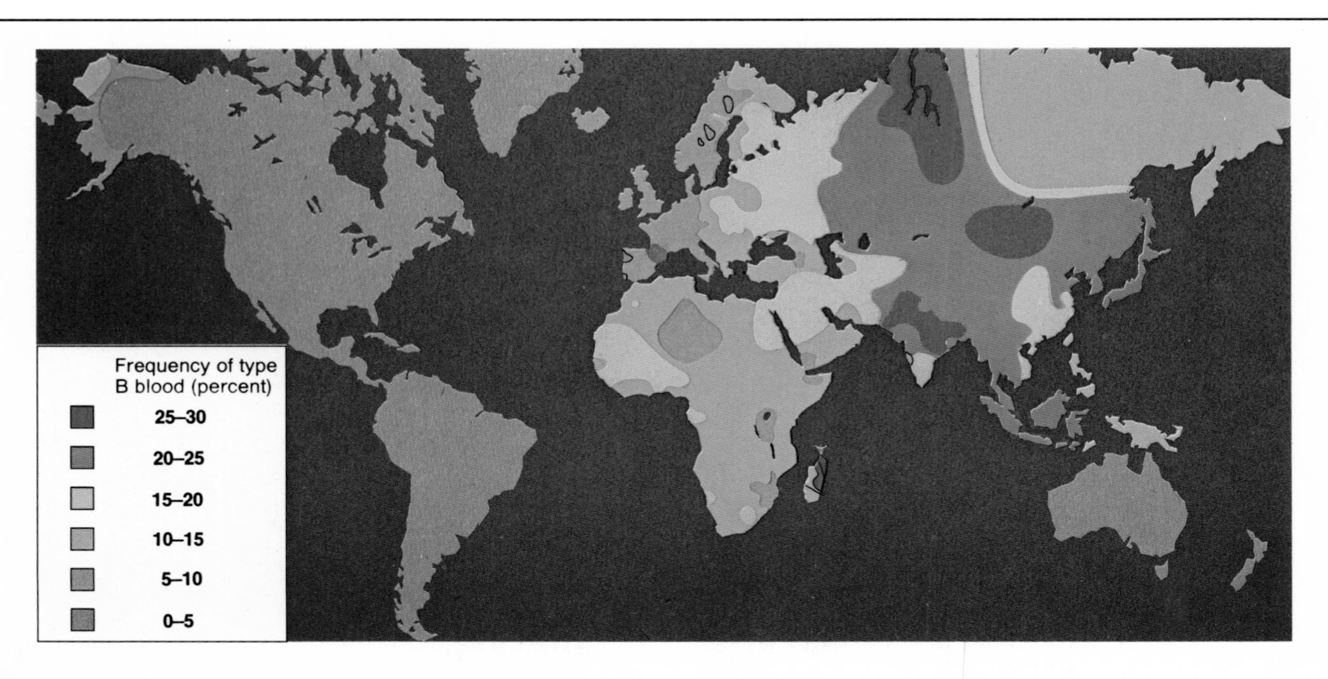

The world distribution of blood types provides anthropologists with clues about the migration of ancient mankind. The map shows the present-day distribution of type B, which is commonest in central China. Its incidence spreads westward through Europe, possibly reflecting the thirteenth-century Mongol invasions, and eastward across the Bering Strait into North America, indicating a common ancestry between American Indians and Mongolian peoples. The compatibility of blood types is of crucial importance in blood transfusions, and the types that match and can safely be mixed are shown in pink in the diagram *(right).* Donor types (D) are listed in the left-hand column, with recipient types (R) along the top of the chart. Any blood type can be mixed with the same type (A with A, B with B, and so on), either A or B can be mixed with type AB, and type O – often called the universal donor – can safely be mixed with any other type. But an attempted transfusion in which type B is donated to a recipient of blood type A, for example, would result in clumping and dangerous consequences for the recipient.

D↓ R→	A	B	AB	O
A				
B				
AB				
O				

There are five major types of leukocytes. When viewed under the microscope, three of them have a grainy, granular appearance. These are neutrophils (about 60 percent of the total), eosinophils (3 percent) and basophils (7 percent). Their names recall the way each type takes up a standard laboratory stain on a microscope slide. The other two smoother types of leukocyte are lymphocytes (25 percent) and monocytes (5 percent).

The leukocyte army is a team of specialists. Half the circulating army patrols in the blood, while the others are out and about in the tissues, checking that all is well. White cells are mobile. They can creep amebalike along capillaries, squeeze out of them through gaps between the cells of their walls, and pour themselves along spaces between the cells of the tissues.

Invaders are the enemy. They may be bacteria, viruses, fungi or parasites. Somehow they always seem to be getting in, through cuts in the skin or through the delicate linings of the respiratory and digestive tracts. When they appear, the army swings into action. Basophils and some lymphocytes act like mines, "blowing up" and releasing chemicals which trigger the disease-coping inflammatory processes.

Neutrophils, eosinophils or monocytes rush to the battlefield and literally gobble up the invaders, enveloping them in their cellular folds and absorbing them in a procedure termed phagocytosis (from Greek words meaning "cell eating"). The phagocytes crowd into the inflamed area, devouring as much as they can; they also set off through the body tissues to mop up invaders that have escaped the bloodstream.

Leukocytes are not the longest-lived of body cells. Most are in action for only a few hours. In healthy blood their numbers are low. Yet reserves are waiting, in the bone marrow, lymph nodes and spleen, where they live for a week. At the first sign of trouble, these reserves march to the battlefield. At the same time the white cell production line moves up a gear and hurriedly manufactures required new troops. It is this heated production that causes the fever of illness, while aching bones and a sore throat (from the swelling of the lymph nodes) shows that the body is busy fighting back.

The secret army

How do white cells distinguish friend and foe – "self" from "non-self"? Why do they not turn on each other, or attack red cells or some other part of the body? The answer lies in the almost mystical sub-world of immunology. We must descend below tissue and cell level, to the very molecules which make up cells – more specifically, to the molecules which link together to form the "skin", be it of a bacteria, virus, leukocyte, or any other speck of organic matter.

The white army recognizes its body's own cells. Their molecular coats are of known color and pattern, and signify allegiance. But the microbial invaders wear coats of different colors and patterns. The units of which these unfamiliar coats are constructed are known as antigenic determinants, and their wearers are called antigens. To the white army, antigens are the signal to attack.

The various types of phagocytes follow the signal and simply consume invaders, of any type, alive or dead. But the lymphocytes are altogether more cunning. They are the secret army within an army, the stealth force. There are two main divisions, T lymphocytes and B lymphocytes. Under a microscope they look alike, yet their roles are different, although complementary. They do not all enter the fray. These cells are individually preprogrammed to recognize a specific "fingerprint" of antigenic substances carried by one particular invader, from a common cold virus to the bacterium of tuberculosis.

T lymphocytes themselves come in three varieties. Some are killers; they attack the invaders directly with potent chemicals. Others are suppressors; they help to regulate the fightback, protecting the body from the excesses of its own defense. The third type are helpers; they prod the B lymphocytes into action. (The "T" stands for thymus, the gland behind the breastbone that processes these lymphocytes.)

The magic bullets

B lymphocytes make antibodies. These are not another cell type. Antibodies are the body's "magic bullets," each with the identity of a specific antigen engraved on it. They are protein molecules craftily shaped to stick onto, disrupt and disable intruders. They form the cornerstone of the immune response.

The scenario for war might run as follows. Suppose a virus dares to enter the body. The lymphocytes programmed to detect

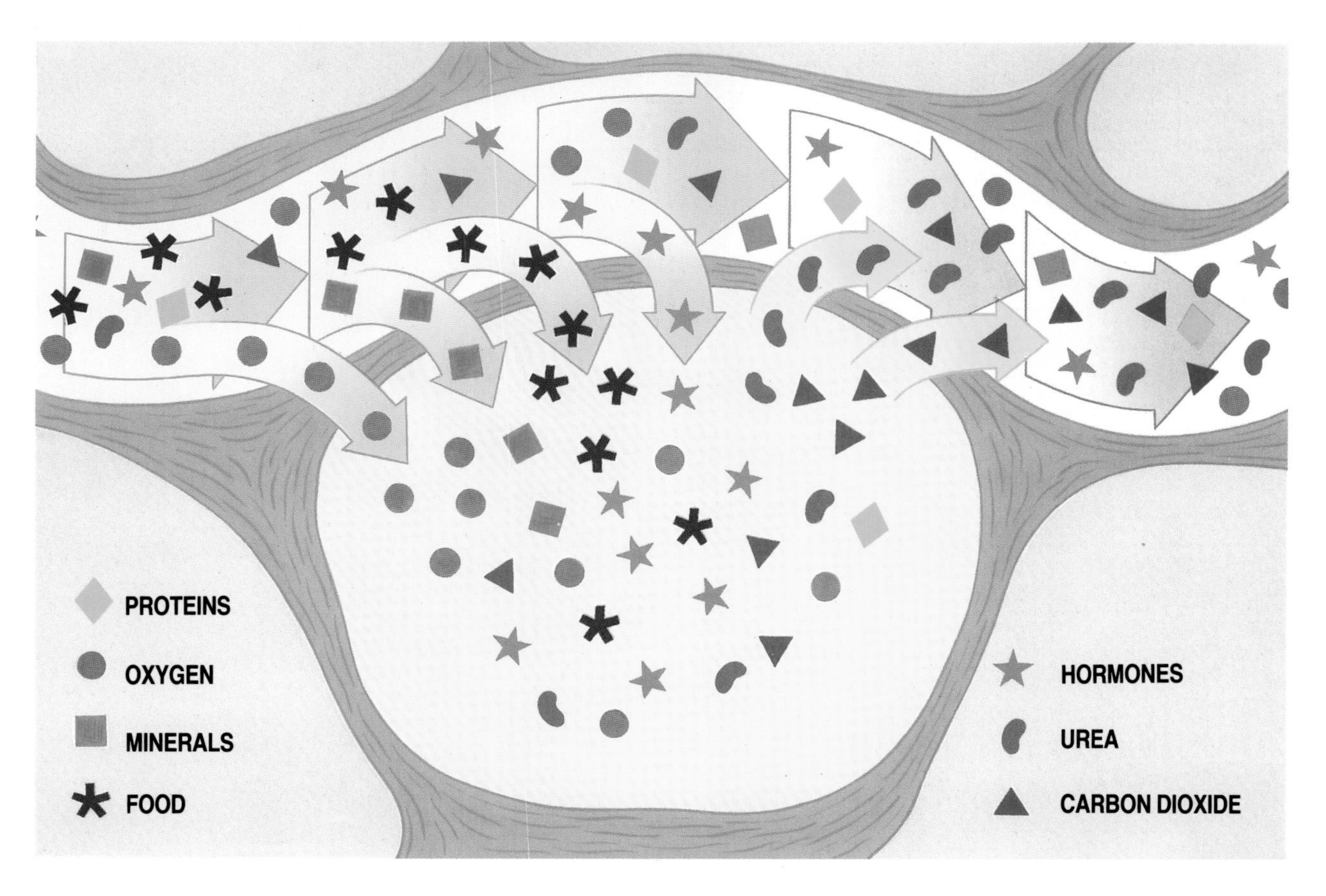

Blood is the body's principal transportation system, carrying vital materials (pink arrows) to the cells and transporting waste products (yellow arrows) away from them.

it are activated. They might number only a few dozen, in the millions of lymphocytes hanging around in a lymph node. When they meet their antigens, they stick or "bind" to them in mutual recognition. Immediately, the B cells are transformed. Many go on to clone from themselves the plasma cells, which are antibody factories. Within a few days, one lymphocyte has multiplied into hundreds of plasma cells, each of which mass-produces antibodies at the rate of around 2,000 every second.

The antibody is shaped like a Y, with the two arm tips as the business end. These lock onto antigenic determinants on the invaders – destroying them, and only them.

Agglutination helps. This is "clumping" of the antigens, facilitated by antibodies. Each arm of the Y antibody may attach to a different antigen, so linking the two together. Antibodies are generally sticky molecules, in any case, and sometimes glue themselves to other proteins or tissues. This process quickly snowballs. In no time, antigens too small to be noticed – some even small enough to "hide" by staying in solution – become knotted together and coated by antibodies. They are now bigger, easier targets for those hungry scavengers, the macrophages and neutrophils among the white cells.

A cellular memory

As the battle continues, other B lymphocytes, stimulated by their T colleagues, turn into "memory cells." They increase the size of the lymphocyte clone programmed to detect that particular invader. Then, next time the invader appears, the body is pre-prepared and can mount a more powerful defense more quickly. This is the basis of immunization. Once we have caught a certain infection, such as mumps, we will not suffer from it again. The corps of "mumps memory cells" sees to that. In contrast to other leukocytes, lymphocytes are long-lived cells, persisting for perhaps 20 years in some cases.

The body can be tricked – luckily. By giving it a weakened or disabled potion of microbial invaders, or only fragments thereof, the immune system reacts as though the fake invasion was real. We become immune, or resistant, without suffering from the infection. The "potion" is a vaccine, and the process by which we become resistant to that infection is termed immunization. It has saved millions of lives around the world.

Families of blood

Agglutination, and a sort of antigen, figure in another feature of blood, which has also saved countless lives. People have long had the notion of giving some of a healthy person's blood to one with greater need – a badly-bleeding accident victim, for example. Early on, such transfusions were hit-and-miss affairs. Sometimes they saved, but sometimes they worsened the situation, because the red cells inexplicably and unpredictably clumped together and clogged up the recipient's circulation.

At the beginning of this century a young Viennese researcher, Karl Landsteiner, set out to prove that there were individual differences in human blood. He succeeded, and in 1930 was awarded a Nobel Prize.

Landsteiner and his colleagues discovered blood groups, or blood types. They took samples of blood from different people (including themselves) and separated them into plasma and red cells. Then they mixed various red cells with different plasmas, and noted what happened: in some mixtures the red cells clumped together, in others they did not.

From this simple beginning, the intricacies of the ABO system

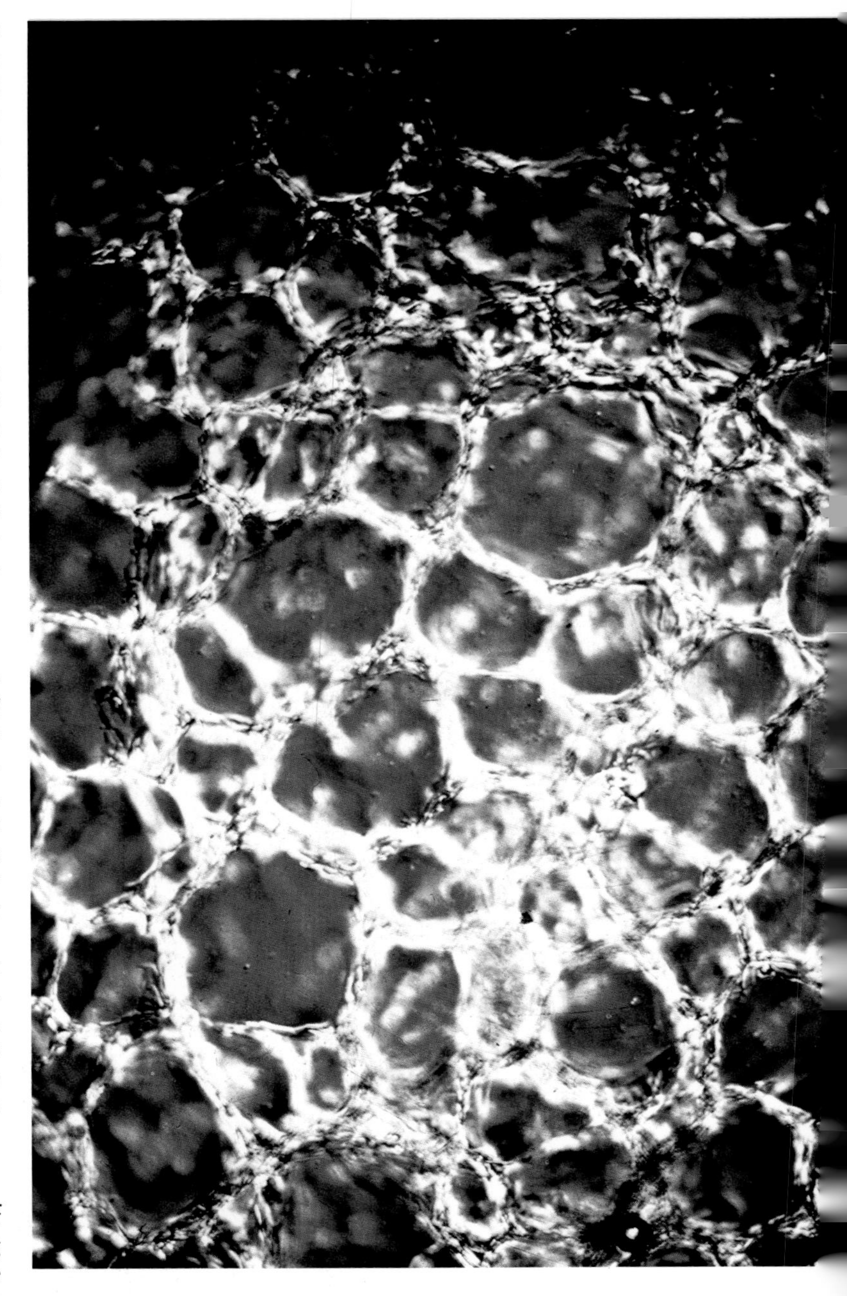

This delicate tracery is a close-up photograph of capillary blood vessels encircling an alveolus, the tiny grapelike structure in the lungs where gas exchange takes place. The total surface area of all the millions of alveoli is huge.

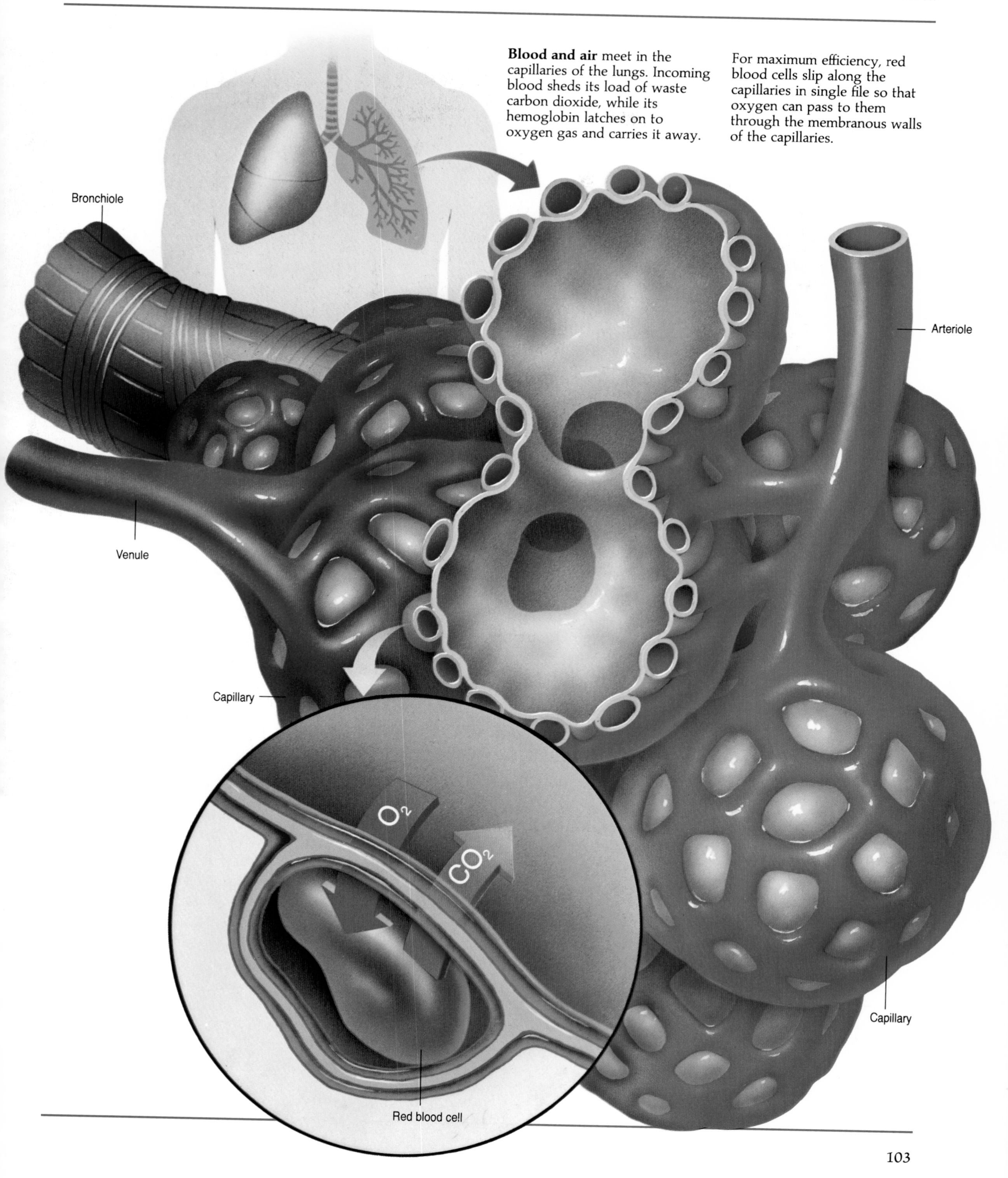

Blood and air meet in the capillaries of the lungs. Incoming blood sheds its load of waste carbon dioxide, while its hemoglobin latches on to oxygen gas and carries it away.

For maximum efficiency, red blood cells slip along the capillaries in single file so that oxygen can pass to them through the membranous walls of the capillaries.

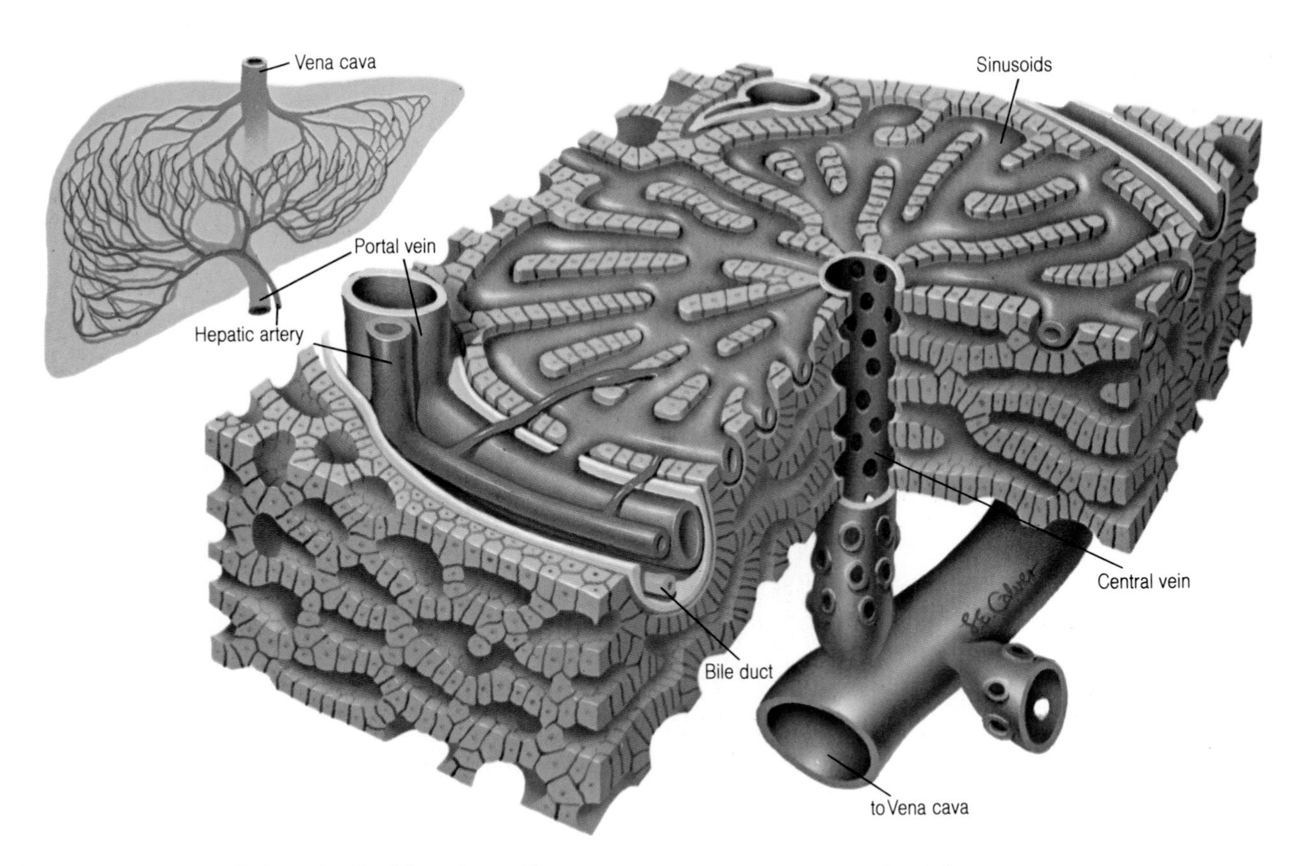

of blood types were unraveled. Red cells, like other cells, carry a specific molecular pattern on their surfaces. The pattern includes a type of molecule called an agglutinogen – an antigen that triggers agglutination, or clumping. There are two types of agglutinogen – A and B. Some people have only A; others have only B; some have both (AB); and some have none (O).

The plasma reciprocates. People with A agglutinogens on their red cells (that is, A type blood) have anti-B agglutinin, a type of protein, in their plasma. If red cells carrying B agglutinogens ever get into A type blood, the anti-B agglutinin in the A plasma reacts quickly with them and prompts them to clump, gluing them together like piles of sticky doughnuts.

People with B agglutinogens on their red cells have anti-A agglutinin in their plasma. The reverse applies to them.

Those with AB red cells have neither anti-A nor anti-B agglutinins in their plasma. Those with type O blood (neither A nor B agglutinogens) have both anti-A and anti-B agglutinins. O is the "universal donor" because, lacking agglutinogens, it cannot trigger clumping no matter what is in the plasma. Similarly, AB is the "universal recipient."

Landsteiner's original work has been much expanded. Today physicians recognize ABO, Rhesus and dozens of other systems of typing blood. It may be that a person's "bloodprint" is as unique as his or her fingerprint.

The liver receives blood from two different sources. Like other organs, it has its own supply of oxygenated blood, delivered along the hepatic artery. But it is also served by the portal vein, which carries the products of digestion from the intestines. Once the liver has converted these into nutrients that can be used by body cells, the blood drains into a central vein that joins with the vena cava, the vein leading back to the heart.

Solids out of the liquid

Blood functions in many ways once invaders are in the body. It is also in action as they try to gain access. Blood clots, or coagulates, to seal a leak, to save itself from losing itself, to retain vital body chemicals, and to help repel the ever-waiting invasion of body hijackers from without.

When a blood vessel is bruised, cut or otherwise injured, three

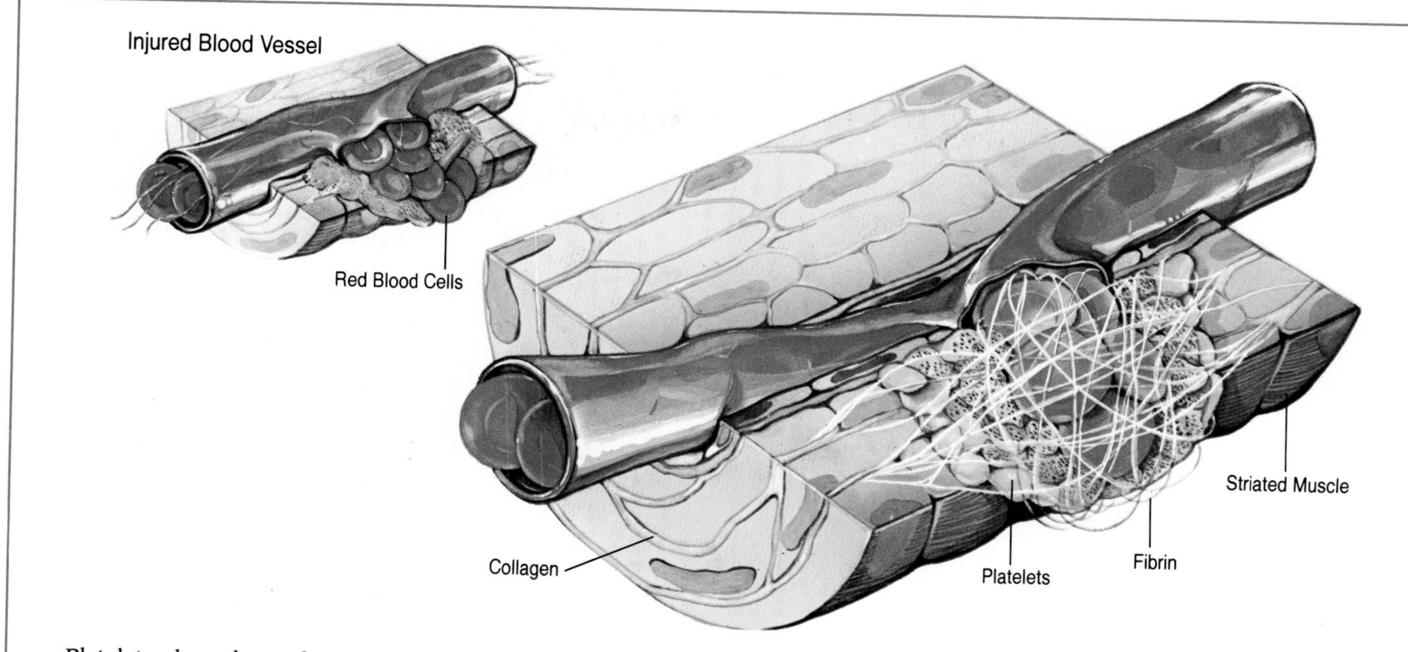

Platelets play a key role in stemming the flow of blood from an injured vessel. A soon as a blood vessel hemorrhages, platelets cling to collagen in the vessel wall and release adenosine diphosphate (ADP). This substance acts as a signal that calls other platelets to the wound. Another chemical from the damaged tissue, thromboplastin, converts the protein fibrinogen into strands of fibrin, which enmesh red cells and bind them into a plug.

mutually reinforcing processes move into action. The damaged vessel contracts, to restrict the flow of blood and therefore minimize any loss. Second is formation of a platelet plug. Platelets are "almost cells" – they resemble cell fragments, rather than whole cells. Only one-quarter the size of red cells, they float in the blood plasma, ever ready to seal a break. They live for about 10 days, and an average adult makes 200 billion new platelets every day.

As platelets gather at the scene they release serotonin, a substance that helps constrict the vessel and so slow bleeding. They also trigger the release of other chemicals that cause yet more platelets to arrive and clump. In less than a minute a loose plug of platelets is helping to block the wound.

For the dozens of small wounds that occur daily, in the normal course of wear and tear, platelet plugs are often sufficient to stem the blood flow. If not, a clot is what is needed. This is the third process. Clotting of blood is a complicated business, a cascade of chemical reactions involving almost all parts of blood. The early stages rely on circulating blood proteins called clotting factors. (Factor XIII, one of their number, is missing in people who have hemophilia.)

The clotting factors work to convert prothrombin, another circulating protein, into thrombin, an enzyme, which in turn transforms fibrinogen, yet another soluble plasma protein, into fibrin. The trick is that fibrin is not soluble. It rapidly "condenses" into a mesh of microscopic threads which trap platelets, red cells and anything else handy. Platelets play their part, releasing yet another substance, fibrin-stabilizing factor. Well named, this links the growing fibrin threads into a more durable meshwork. More cells are snared, and so the clot grows.

Although it looks firm, a newish clot is 99 percent water. But within minutes it starts to contract and squeezes out a pale fluid. This is serum – blood plasma without the fibrinogen, clotting factors and other clotting paraphernalia. Exposed to air, the clot hardens to a brittle lump which is the body's natural Band-Aid.

Lymph: blood's poor relation

Blood is not the only liquid tissue. Its "mirror-image", perhaps more a poor relation, is lymph. Lymph derives from blood, and returns to it. The fluid and white cells that leak from the capillaries must go somewhere. They drain from a nebulous network of ill-defined spaces into a more organized system of channels, which join together in a reflection of the capillary and venous sections of blood's circulatory system. The channels are lymphatic vessels, and the fluid they convey is lymph. It contains white cells galore, and also proteins, fats and other products of digestion – for the villi, lining the small intestine, are rich in lymph vessels called lacteals, and are especially important in the absorption of fats.

Lymph meanders along, vaguely driven, second-hand, by the far-off pressure of blood. The vessels converge and eventually empty back into the blood circulation via a large channel, the thoracic duct, into the thoracic vein near the heart.

Lymph has nutritive and waste-clearing functions, although they are less defined (perhaps less studied) than those of the blood. But lymph is most certainly both reservoir and major transport canal for white blood cells. Dotted around the system are lymph nodes, small nodules of lymphatic tissue. They are especially evident in the neck, armpits and groins. In routine matters, white cells congregate there and scavenge the blood, picking off the odd intruder or piece of debris. But there is frenzied activity during an infection, as the white cells multiply in the nodes and swarm into the bloodstream, ready to do battle. The nodes increase in size: we refer to them as "swollen glands."

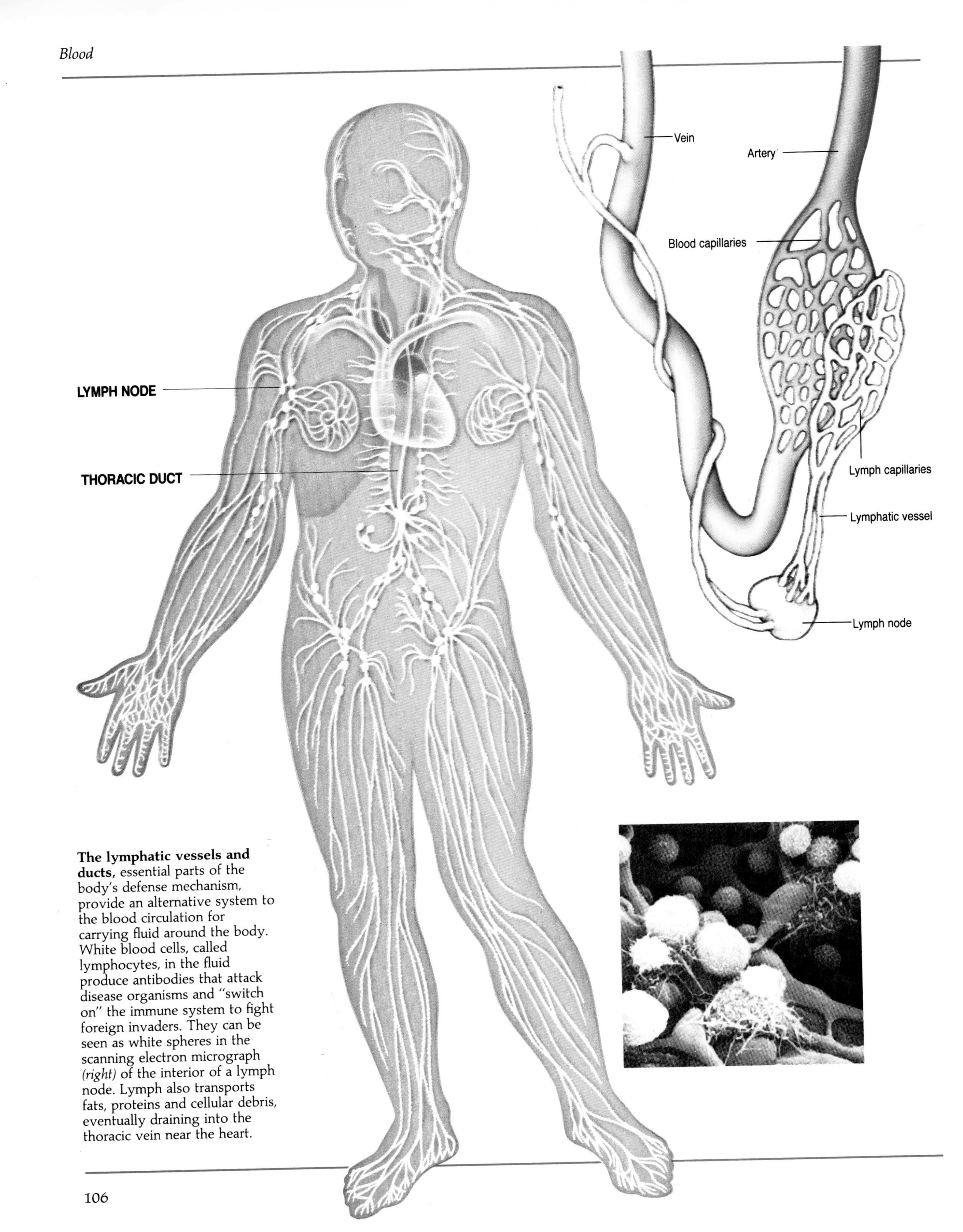

The lymphatic vessels and ducts, essential parts of the body's defense mechanism, provide an alternative system to the blood circulation for carrying fluid around the body. White blood cells, called lymphocytes, in the fluid produce antibodies that attack disease organisms and "switch on" the immune system to fight foreign invaders. They can be seen as white spheres in the scanning electron micrograph *(right)* of the interior of a lymph node. Lymph also transports fats, proteins and cellular debris, eventually draining into the thoracic vein near the heart.

Blood and bone

The dazzling array of blood cells turns over at an amazing rate. An average adult produces roughly 100 billion neutrophils each day, not to mention the hordes of basophils, eosinophils, other white cells, red cells and platelets. Where do they all come from? Paradoxically, the cells that live in blood, that most fluid of tissues, are born in the body's most solid substance: bone. Or rather, in the jellylike marrow contained in most bones in a child, and certain bones in an adult.

The red cells serve as an example. In the womb, during the first few weeks of life, they are made in the yolk sac of the embryo. Halfway through pregnancy, the liver, spleen and lymph nodes take over production of red cells. In the weeks before birth, manufacture switches to the bone marrow. In an adult, the marrow of the skull, ribs and spine makes most red cells, but any bone marrow in the body can be called upon to contribute in an emergency.

Red cells are descendants of "primitive cells," so named because they have yet to develop fully. Under hormonal influence the primitive cell becomes a rubriblast that splits into two, and again, and again, until there are 16 descendants. These offspring have already begun to fill themselves with hemoglobin. Less than a week later, the red cells – lacking nuclei they are sometimes called corpuscles – are ready to enter the circulation.

The platelets' ancestors, megakaryocytes, are also residents of the bone marrow. As the meaning of their name suggests, they are giant cells which literally fragment into thousands of much smaller platelets.

The various types of white cells originate from marrow, too. Their lines begin with "stem cells," which divide to multiply, producing immature leukocytes that gradually grow up. Mature lymphocytes, for instance, exit the marrow at the rate of some 200,000 each second.

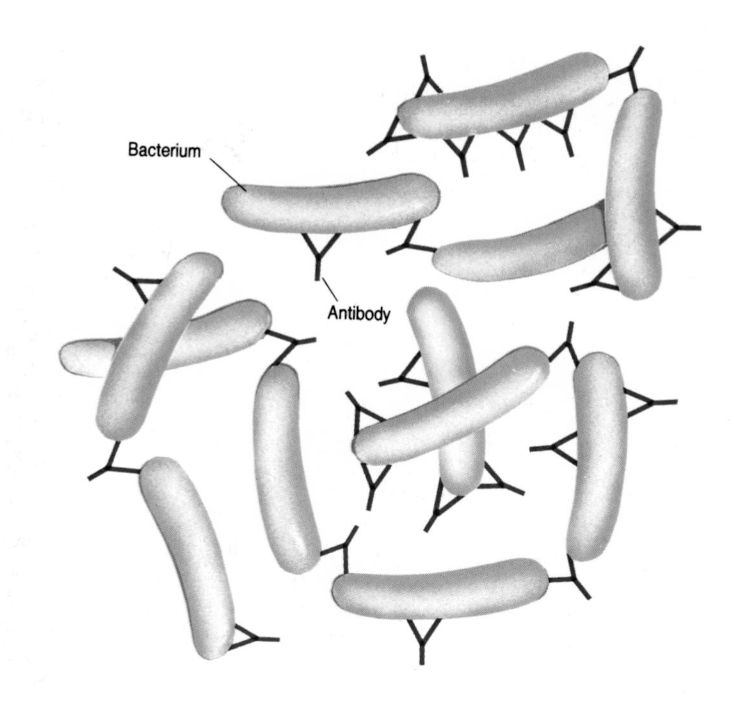

Antibodies are Y-shaped protein molecules that stick onto invading bacteria or form bridges between them so that they clump together. This clumping, or agglutination, makes the bacteria easy targets for phagocytes.

Other blood functions

In this account some of the main functions of blood have been summarized. A complete analysis of what blood does may never be possible. Every few years, scientists discover yet another hormone or enzyme that blood distributes, or another cell type to which it plays host, or another link in the vastly complex network of germ-fighting processes that sway to and fro on its "battlefield." We only have to think of the 1980s – the decade that brought us AIDS, a blood-borne disease – to realize that modern science has still to take a full inventory.

In some circumstances, blood itself becomes diseased. If there are too few red cells, or if the cells' hemoglobin content is reduced, anemia occurs. It can be caused by nutritional disorders, sudden or persistent bleeding, or a disease of the blood-producing bone marrow. Twenty percent of the people in the world suffer from iron-deficiency anemia (iron is a key component of hemoglobin). Pernicious anemia results from a deficiency of vitamin B_{12}, usually because the vitamin is not absorbed into the bloodstream from the intestines and not because there is insufficient in the diet.

One form of the disease, called sickle cell anemia, is a hereditary disorder that affects about one in 600 of the blacks in the United States. The red cells take on sicklelike shapes and the distorted cells clog veins and arteries, preventing the flow of healthy blood cells carrying oxygen and nutrients. One consolation is that people with sickle cell disease are immune from catching malaria.

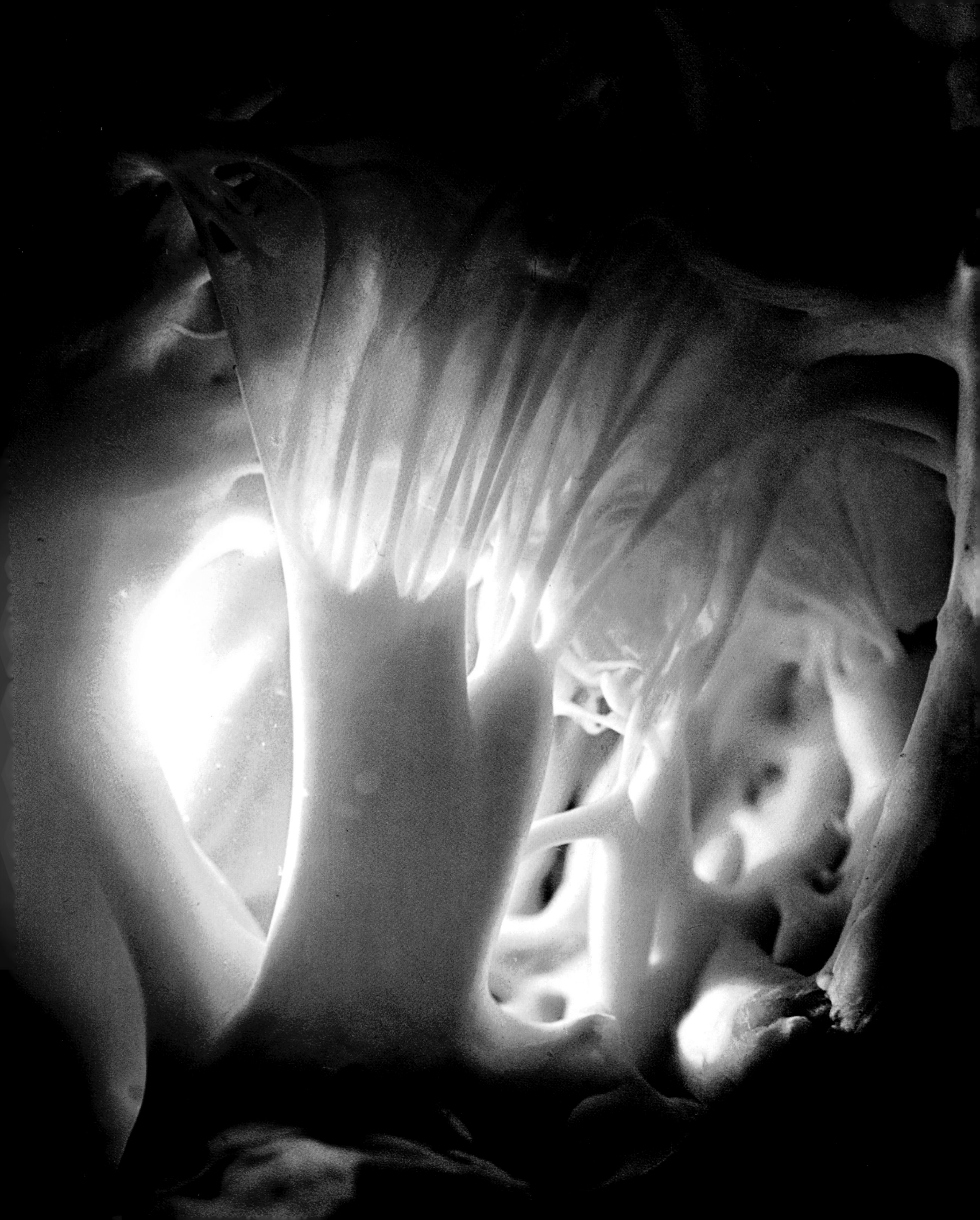

CHAPTER 8

The Heart

The fist-sized lump of muscle, valves and tubing that forces blood around your body is not the shape popularized on Valentine cards and representative of romantic interludes. Neither is it bright red, the color associated with such symbolism. And if it were to be run through by Cupid's arrow, the result would not be a star-struck lover – but almost certain death.

Recent times have seen the heart's place as a symbol of romance tempered by scientific reality. To the ancients, the heart contained the spiritual essence of the human being. The Egyptians weighed the hearts of the dead to measure truth. The Greeks saw the heart as a forge, burning impurities from the blood. Today, its disguise has been dispensed with, its cover has been blown. The heart is no more, no less, than a pump. But what a pump! Never resting, it thumps at least once each second (often more) from about the fourth week of conception until a couple of minutes before death. Each thump sends life-giving blood surging round the body. If each and every "lub-dub," the classic sound of a heartbeat, was written as a word in this book, you would have 10 million pages in your hands. Read each "lub-dub" in one second . . . and the book would last a lifetime.

Father of the heart

For 1,500 years, the heart remained shrouded in myth and mystery. Galen, the great physician of ancient Greece, assigned to it the role of a furnace, sucking in blood and combusting it to generate the heat that warms the body. Leonardo da Vinci correctly recognized the heart as made up mostly of muscle, but was also of the opinion that its job was to generate heat in some way or another.

Galen's teachings remained barely questioned until the early seventeenth century. An English physician, William Harvey, began to query the "Galenic religion" followed by doctors of the time. Through observation and experiment, on animals and human corpses, he followed a reasoned series of steps to argue that blood did not move as the ancients contended. The organ did not suck blood in, but squeezed it out. Blood did not flow back and forth in the vessels like a tide, but away from the heart through arteries and back to it via veins. Blood could not be combusted or "used" in some other way: while inside the heart the rate of flow in and out was far too great.

Harvey came to the now obvious conclusion that the circulatory system was by and large a closed circuit. The heart was the pump, and it drove the same blood round and round. In Frankfurt in 1628, after 12 years of experiments with more than 80 species of animals, as well as living studies and post-mortem examinations on human corpses, he published *Exercitatio Anatomica de Motu Cordis et Sanguinis in Animalibus* (Anatomical Treatise on the Movement of the Heart and Blood in Animals). It had an instant and profound effect.

The main gap in Harvey's proposals was that he could not explain how blood got from arteries to veins. This was the pre-

Like stalactites in a dimly lit cave, muscles and tendons span the cavern that forms the lower right chamber of the heart. Contraction of this part of the living pump forces blood through the pulmonary valve to the lungs, where it is recharged with oxygen. Continually about once every second, the never-tiring heart beats for a whole lifetime.

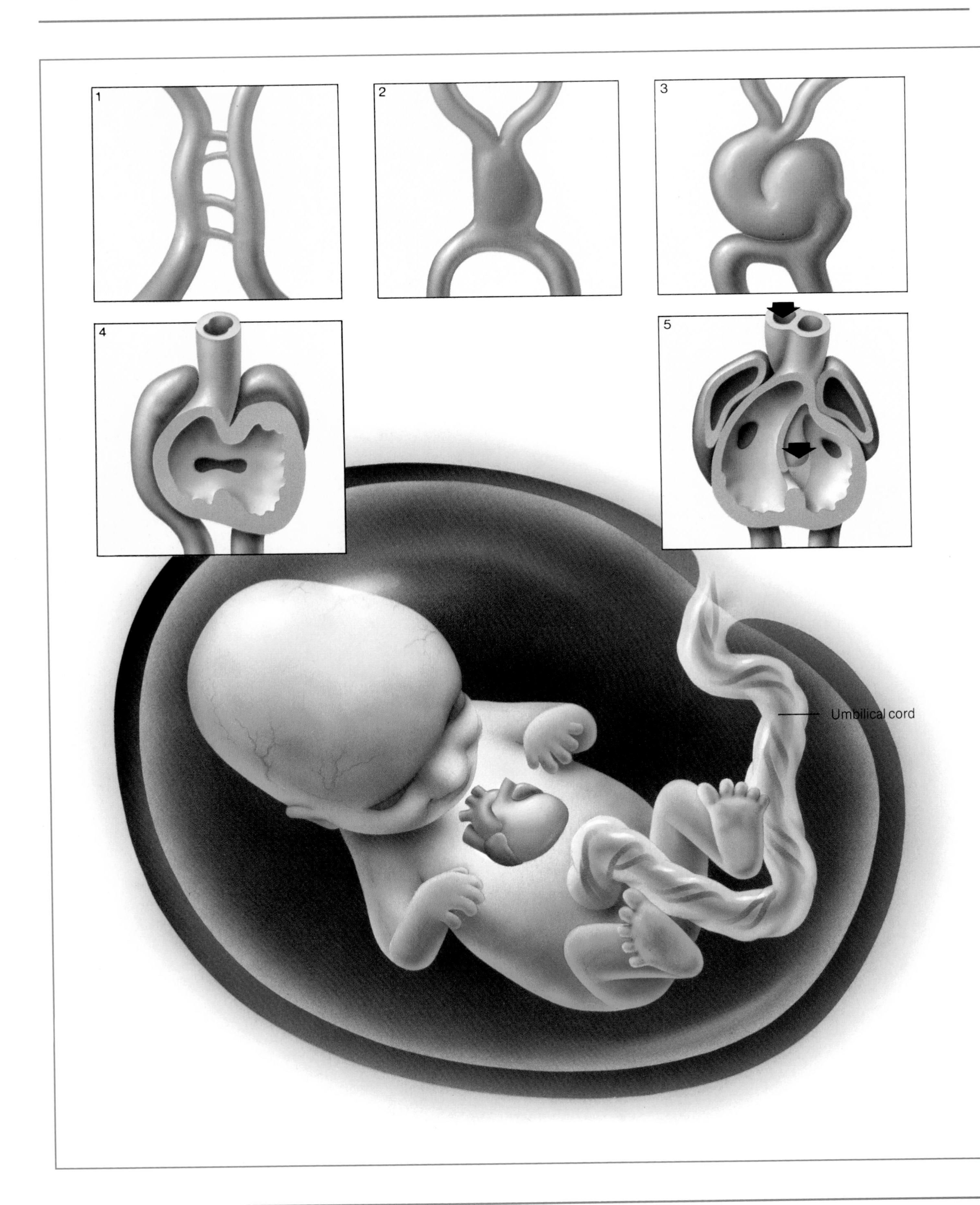
1
2
3
4
5
Umbilical cord

An unborn baby's heart develops in stages *(left)* by the fusion of the embryo's main vein and artery where they pass close to each other in the chest cavity. Three weeks after conception (1) the two tubes link together. They fuse to form a single chamber (2), before the upper atria and lower ventricles begin to take shape (3). Beginning in the fifth week (4), the atrioventricular canal is gradually split by growing ridges of tissue to provide two separate pathways for blood, one through each side of the heart. When the septum forms down the center (5), the heart has four separate chambers. By the time the fetus is eight weeks old, it possesses a tiny heart that is a miniature version of an adult's. It pumps blood to and from the placenta along the umbilical cord, collecting oxygen and getting rid of wastes.

Located in the middle of the chest and almost surrounded by the lungs *(below)*, an adult heart is a muscular organ a little larger than a man's fist and weighing nearly a pound. It rests on the domed diaphragm where it passes over the liver. Blood passes to the heart along two main veins: the inferior vena cava, carrying blood from the lower body, and the superior vena cava, which brings in blood from the arms and head. After being oxygenated in the lungs, blood leaves the heart along the aorta, the body's main artery, which branches to serve the head and arms before arching downward to carry blood to the rest of the body.

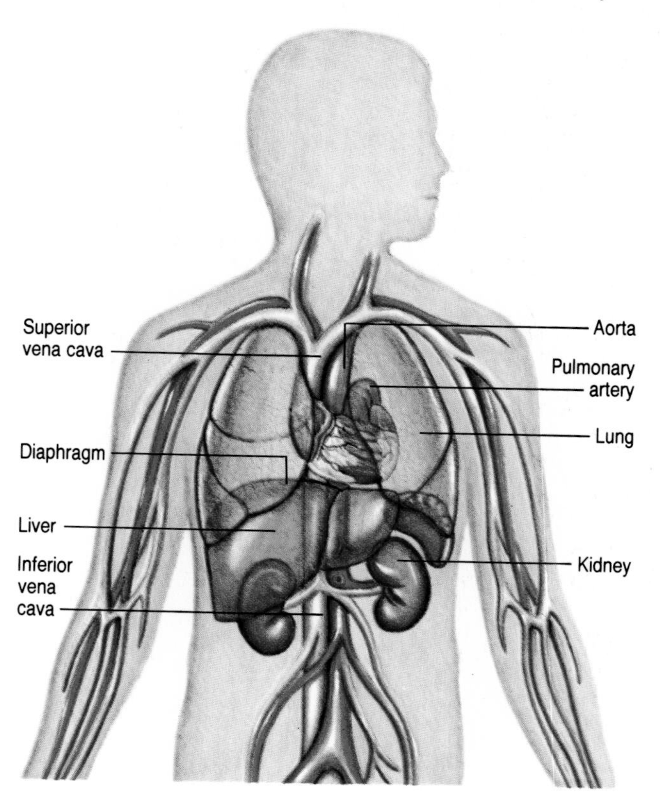

microscopic era, and without a microscope, the tiny capillaries that make this connection were still invisible.

The impact of William Harvey's work on medicine cannot be overemphasized. As with the Darwinian revolution in biology, it is difficult today for us to imagine how things were ever different. Within a generation, the notion of "Harvey's heart" had been absorbed into the mainstream of anatomy and physiology, and the scientific era of medicine had begun.

The heart in the womb

The importance of the heart to survival is indicated by the speed at which it develops after conception. Every cell in the body needs a continuing supply of oxygen, and of the nutrients that provide the energy and raw materials for cellular metabolism. This is as true of a tiny embryo, gradually evolving human form in the womb, as it is of a marathon runner. The oxygen and nutrients are brought by the blood, a physical, fluid conveyor belt driven around the body by heart power. In embryological development, therefore, the heart and its major blood vessels are formed and functioning long before any other organ.

The heart begins as two tiny tubes lying near each other in the primordial chest region, cross-connected by several even smaller vessels and surrounded by a sheath of muscle. Three weeks after conception, the tubes begin to fuse in the middle of the sheath. After a few days they have joined along their length to create a single, continuous chamber. By the fourth week of pregnancy this minute pocket of muscle, less than a twenty-fifth of an inch long, is folding and twisting – and has started to beat. Chambers form, marked on the heart's outer surface by grooves called sulci. The organ becomes S-shaped, and the single ventricle (lower chamber) splits into two. Until now the blood has (like the ancient teachings) ebbed and flowed, but rapidly it assumes a one-way thrust around the budding circulatory system.

By the fifth week, the squirming embryonic heart has almost taken on the basic U-shape of the adult heart. The upper chamber, the atrium, divides into two. Each atrium connects with the ventricle on the same side, to create not one pump, but two. Thus the double nature of the heart comes into being. During the sixth and seventh weeks the blood vessels coil and connect, and the valves that ensure a one-way blood flow are miraculously molded in living tissue.

It is the end of the second month of pregnancy. The tiny form is now nearly an inch long, and recognizably human. It is no longer termed an embryo: its correct name is a fetus. In the fetal chest, the heart pulses steadily. All four of its chambers, linked two-by-two to form adjacent left and right pumps, and separated by a tough dividing membrane known as the septum, are distinct. The valves are working. The beat is on.

Before birth, the blood circulation must be modified because the baby's respiratory and digestive systems do not work, while the placenta does. There is no breathing; the fluid-filled lungs cannot absorb oxygen from the air. The heart has a "hole" that allows communication between its left and right sides. Blood that would otherwise flow to the lungs is short-circuited back around the body, to the placenta, which is the source of oxygen (diffusing in from maternal blood) at this stage. By birth, the hole in the heart has closed. Stimulated by the newborn's first breaths, various other changes occur in the heart and circulation, the adult pattern is established and the infant is finally made physically independent of its mother by the cutting of the umbilical cord. This subject is dealt with in more detail in Chapter 6.

The heart muscle has to have its own blood supply in order to work, and this is provided by the coronary arteries, which branch off the aorta. Cardiac veins return the "used" blood to the vena cava.

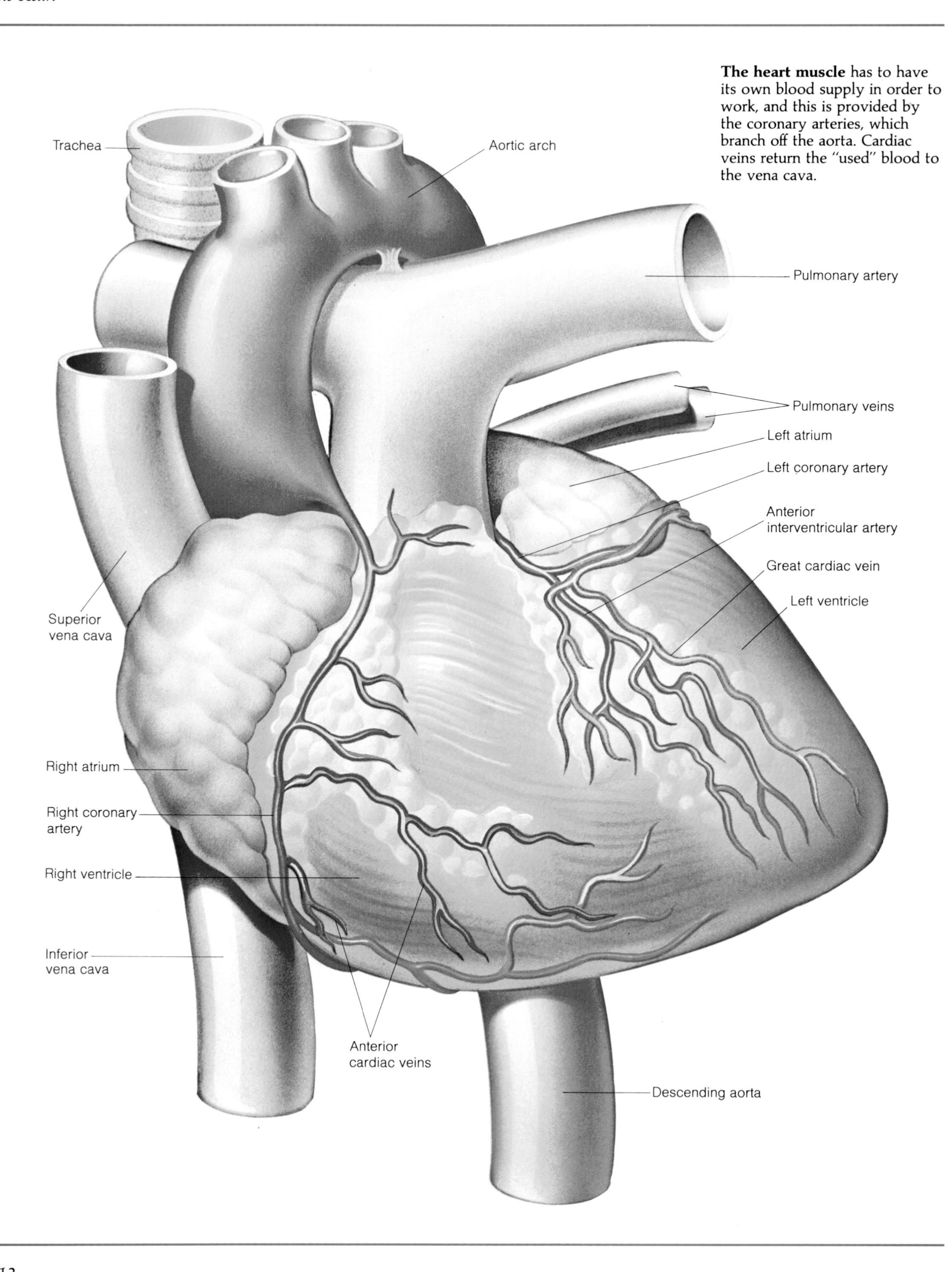

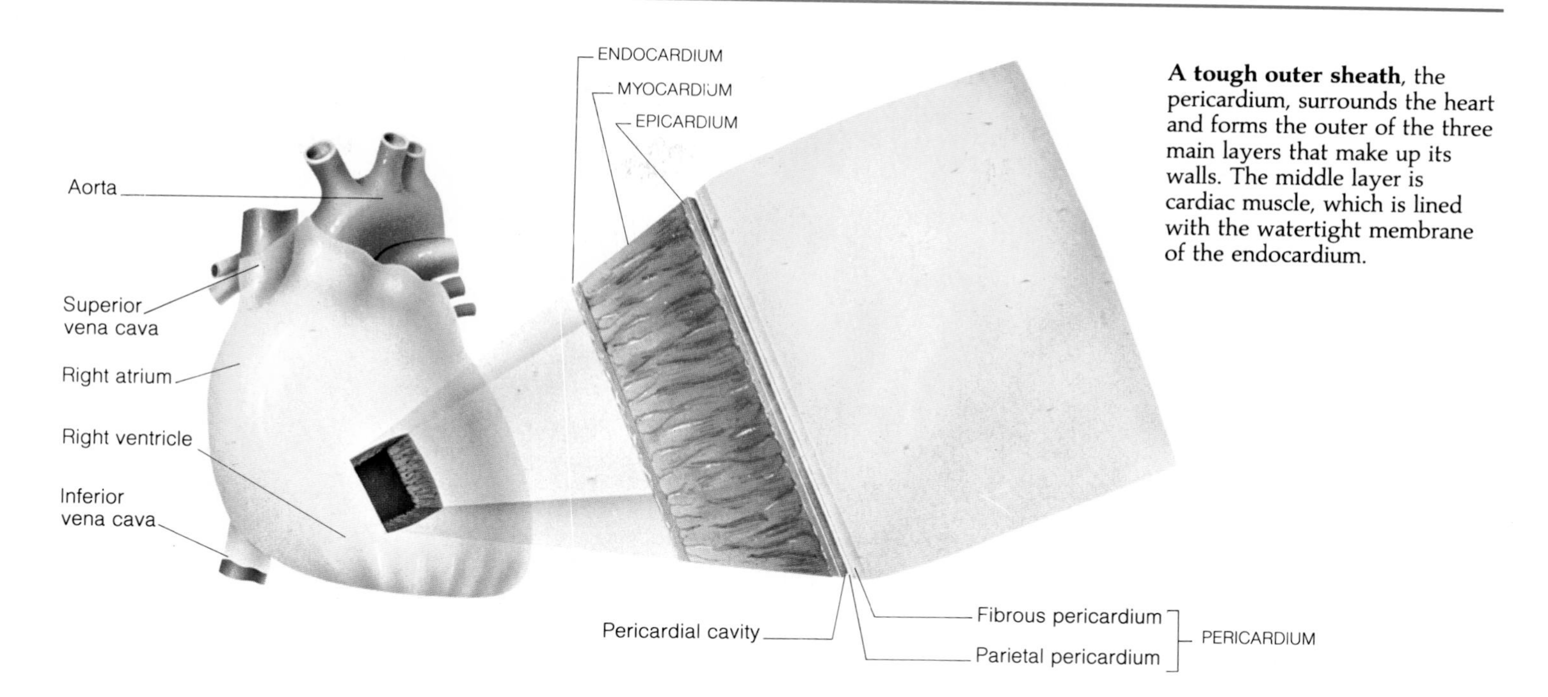

A tough outer sheath, the pericardium, surrounds the heart and forms the outer of the three main layers that make up its walls. The middle layer is cardiac muscle, which is lined with the watertight membrane of the endocardium.

Half a heart each

Each side of the heart is a self-contained pump. Its atrium (upper chamber) is thin-walled and distensible, and receives blood flowing in from the main veins. The blood passes from the atrium through a valve to the ventricle (lower chamber), which is the muscular, thick-walled, pumping part. This contracts to squeeze blood out through another valve into a main artery.

The right pump is like the right hand – on the left, as you look at the body from the front. It drives blood to the lungs for oxygenation. This "refreshed" blood returns to the left pump, which sends it around the rest of the body to deliver the oxygen. The blood returns to the right side, completing a figure-eight.

Tireless mover

Here are some of the heart's vital statistics. It beats on average 70 times per minute while at rest. This adds up to 100,000 heartbeats each day, and about 2.5 billion in a lifetime. But the true number is greater than this, because for a major part of each day the body is active, and the heart has to pump faster in order to increase the supply of blood and satisfy the demands of the muscles.

An average person's body contains eight to ten pints of blood. At rest, a typical heartbeat expels about two and a half fluid ounces. This represents a theoretical cardiac output of roughly eight to 16 pints of blood per minute, depending on the circumstances: up to 2,600 gallons each day – enough to fill a small road tanker. Again, the actual volume is greater since the body does not rest all day. After strenuous exercise, the cardiac output may be ten gallons per minute.

Cone in the chest

The average adult human heart is roughly cone- or pear-shaped. It is about the size of a man's fist, measuring nearly five inches long, about three inches from side to side, and two and a half inches front to back. It sits in the lower chest, slightly off-center, with two-thirds of its bulk to the left of the midline. Its pointed end, or apex, is directed to the left, slightly forward and downward, so that the heart is in fact almost sideways on, with its right half facing the front. The atria, often called "upper" chambers, are in reality around the back of the heart.

A shallow groove, the coronary sulcus, encircles the outer surface of the heart and marks the division between the atria and ventricles. On the front and back, two other channels, the interventricular sulci, designate the left and right ventricles. The sulci are padded by fat and guide blood vessels around the heart itself, so that overall the organ has smooth, rounded contours.

The heart is flanked by the lungs and major blood vessels, and its apex rests on the dome-shaped sheet of muscle, the diaphragm, which forms the chest floor below. It is well protected in a "cage" formed by the breastbone (sternum), ribs and spinal column, with their associated muscles and ligaments.

Coverings and linings

Wrapping the heart from base to tip is a covering known as the pericardium. Its tough outer sac, the fibrous pericardium, forms a protective sheath and is tethered by ligaments to the breastbone, spine and other parts of the chest cavity, firmly anchoring the heart in position.

Inside the fibrous pericardium is a thin but tough double-membrane. It covers the heart's surface as its outer layer, the epicardium, and folds back on and around itself to form the parietal pericardium, which lines the fibrous portion. There are a few drops of pericardial fluid in this "double-bag," in the narrow pericardial space between the two layers of membrane. Without this the heart's powerful pulsing movements would mean considerable friction and disruption, but the lubrication offered by the pericardial sac allows it to thump away in a kind of frictionless bath.

Inside the heart, lining its chambers and covering its valves, is the endocardium, another remarkable tissue. This glistening layer of cells must withstand the considerable internal hydraulic pressure of each heartbeat. It must also "blood-proof" the heart, so that it does not leak, by generating an ever-renewing barrier as blood rushes past within.

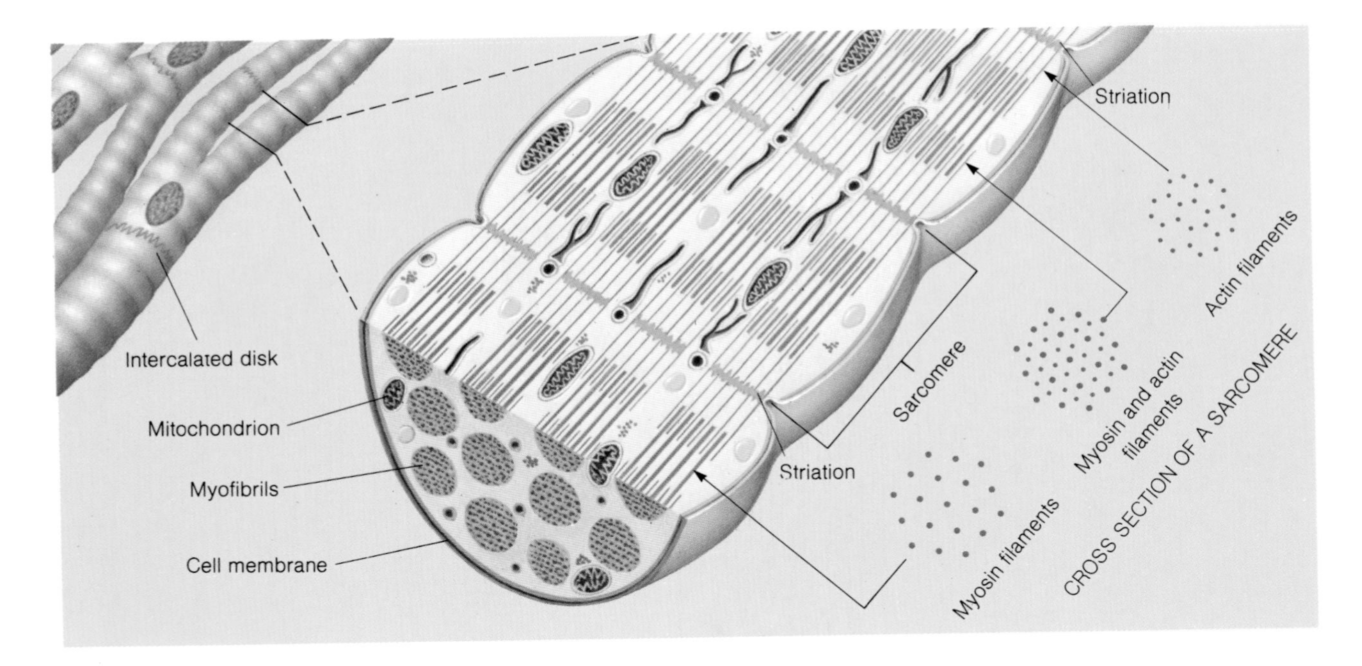

Dissected lengthwise, a cardiac muscle fiber reveals the origin of its banded structure. Parallel bundles of myofibrils are made up of sections called sarcomeres, which consist of overlapping filaments of the proteins actin and myosin. The filaments slide over each other to make the fiber contract, and the combined effect of thousands of such contractions maintains the pumping action of the heart.

The muscle that never fatigues

Between endocardium and epicardium is the heart's powerhouse: a layer of heart or cardiac muscle, technically termed myocardium. Spiralling bands of muscles wrap themselves around a framework of dense, fibrous tissue that forms the "skeleton" of the heart. (The fiber skeleton and the valves inside the heart make up about half of the organ's weight.) The muscle bands curve around each chamber to form its wall, and they also coalesce into loops which support the valves that separate the atria from the ventricles.

The atrial walls have much less cardiac muscle than the ventricles and so are quite thin, the left atrial wall being the more substantial. These chambers are not so much power pumps as expandable reception chambers for incoming blood (*atrium* is Latin for *entrance hall*). The left ventricular wall, which produces the pressure to drive blood all the way to the fingers and toes and back again, has the greatest muscular mass – it measures up to half an inch thick in places. The right ventricle's task of circulating blood through the lungs is less arduous; this chamber sits on the side of the left ventricle, and its walls are less than a quarter of an inch thick.

Cardiac muscle is a curious hybrid of the other two main muscle types in the body, skeletal and visceral muscle. Under the microscope, cardiac muscle is seen to have the stripes, or striations, of skeletal muscle, the type that moves the bones of the skeleton and is largely under voluntary or conscious control. Yet heart muscle is not under conscious control: it responds to the part of the nervous system called the autonomic or involuntary nervous system, as well as to its own internally generated electrical commands. In this respect cardiac muscle is more like the visceral, or smooth (stripe-less), muscle which lines the stomach and other internal organs.

The fibers of cardiac muscle are in effect enormous cells, tiny fractions of an inch long. Like skeletal and visceral muscle, they contain bundles of actin and myosin filaments, and in respect of the molecular basis of contraction, they resemble other muscle tissue in the body. They also have an abundance of mitochondria, the cellular power centers that convert food into energy. But they differ from other muscles in the way that electrical signals, constituting a nerve message, travel though the mass of fibers.

Where two cardiac fibers meet end to end, there are distinct dark bands known as intercalated disks. At certain points along each disk, the outer membranes of the two adjacent cells fuse, so that the two cardiac fibers share the same membrane at these so-called "tight junctions." Electrical signals flow almost unimpeded through the tight junctions. But at other points along the cell membrane the electrical resistance is hundreds of times greater. So the signals follow the line of least resistance and hop from one fiber to the next, leaving in their wake a chain of contractions. Structurally, cardiac muscle is a latticework of separate cells. Functionally, it behaves as a syncitium – a group of cells that have merged to act with common purpose, like a single giant cell.

Open and shut

Working within the heart's tireless muscles are four equally durable valves. They open and shut in a set order every time the heart contracts, ensuring blood flows in the correct direction around the system, rather than the same portion simply being sucked in and ejected with each beat. The four valves are in two

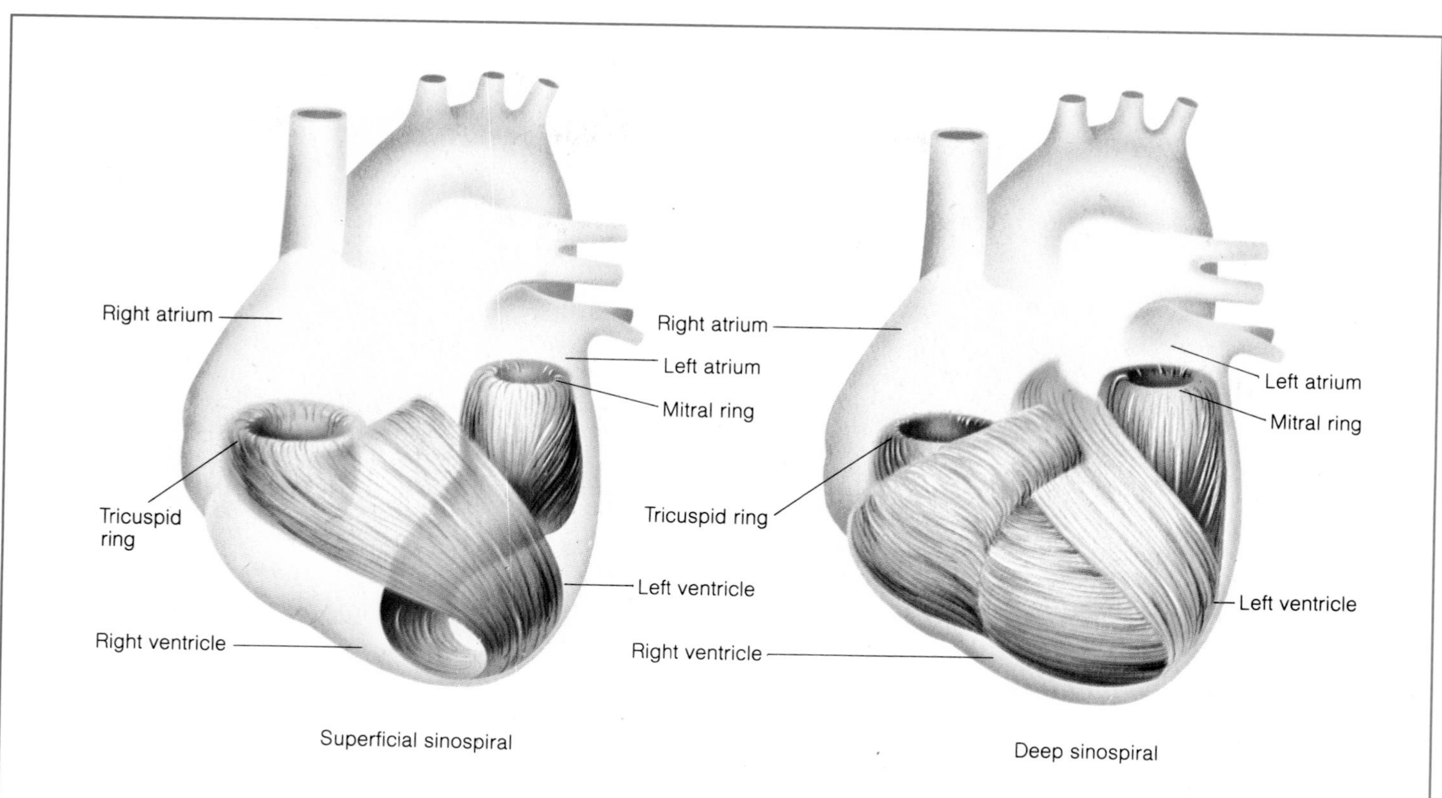

Superficial sinospiral

Deep sinospiral

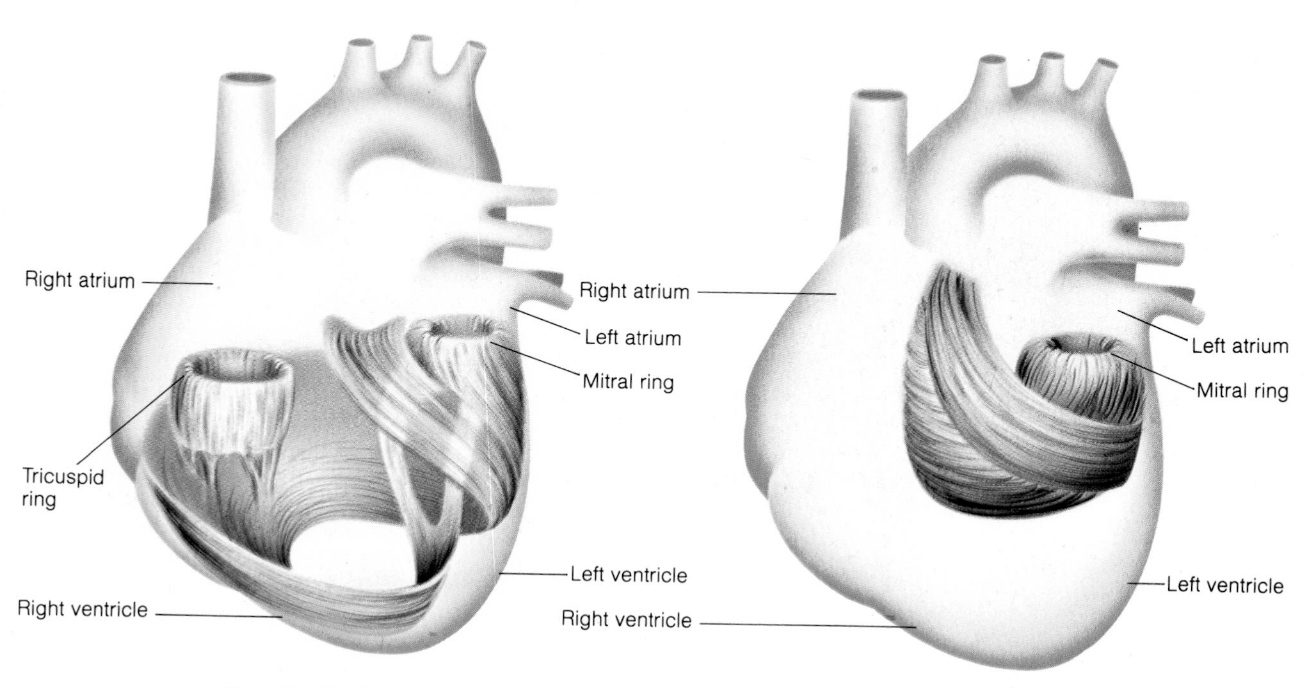

Superficial bulbo spiral

Deep bulbo spiral

The power stroke of the heart comes with the contraction of the ventricles, the thick-walled muscular lower chambers which pump blood either to the lungs or into the body's main circulation. The secret of their action is the way the muscle fibers spiral round the chambers, and include doughnut-shaped rings of muscle round the tricuspid and mitral inlet valves, to lock them tightly shut during the power stroke and prevent back flow.

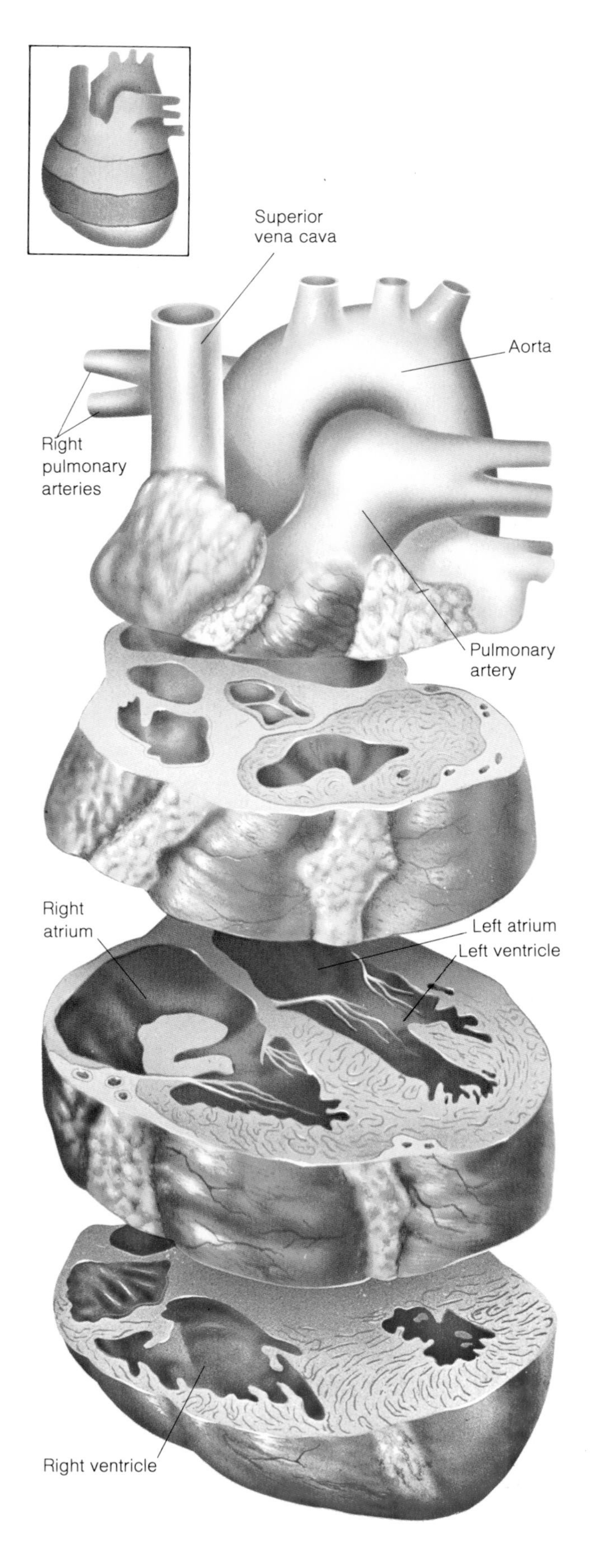

The heart's chambers vary in size and their walls vary in thickness, as shown by the cross sections *(left)*. Most noticeable is the thickness of the muscle that makes up the ventricle walls. The left ventricle (which contracts to pump arterial blood out of the heart) is more powerful than the right.

Tiny capillaries surrounding grapelike alveoli are sites of gas exchange in the lungs. Venous blood (blue), carrying carbon dioxide, is pumped to the lungs along the pulmonary arteries by the right side of the heart. After oxygenation, arterial blood (red) returns to the left side of the heart via the pulmonary veins.

pairs: atrioventricular valves between atria and ventricles, and semilunar valves between ventricles and main arteries. Each is sculpted from tough, rubbery flaps of fibrous tissue, the flap being sandwiched between two layers of endocardium and anchored to the tough rings of the muscle and fibrous skeleton.

The tricuspid valve guards the entrance to the right ventricle, ensuring that blood comes in from the right atrium but does not exit that way. As its name suggests, it has three tooth-shaped flaps, or cusps. The mitral valve does the same job for the left ventricle; it possesses two unequal cusps. It is named, rather less scientifically, from its resemblance to a bishop's miter.

Each of these atrioventricular valves works in a supremely simple way. As the ventricle contracts, the blood inside is pressurized and it pushes up on the undersides of the cusps, forcing their edges together and pinching them shut to create a seal. Long, thin tendons (chordae tendineae) lead from the edges of the cusps down through the ventricular chamber and are anchored to muscles in its sides; these prevent the cusps from being turned "inside out" and forced through into the atrium. As the ventricular muscle relaxes and the ventricle expands, the cusps swing back against its wall and so allow blood to surge through from the atrium.

While the atrioventricular valves are closing, the semilunars are opening. These valves were christened from their three crescent-shaped cusps, which are hollow and pouchlike. The pulmonary valve permits blood to flow from the right ventricle out into the main artery, the pulmonary artery, on its route to the lungs. The aortic valve fulfills the same function for the left ventricle, letting blood surge out into the body's main artery, which is called the aorta.

These two valves also operate in a simple way. Blood flowing the correct way pushes the cusps flat against the wall. Blood attempting to flow the wrong way fills the cusps and balloons them out so that their edges come together and make a seal.

As the valves slap shut to prevent blood's backflow, they make a noise. You have probably heard it: the "lub-dub" of a heartbeat, immortalized in literature, theater and romance. The "lub" represents the atrioventricular valves closing, while the "dub" is the sound of the semilunar valves shutting.

The crooked crown

All tissues of the body need a blood supply, especially active muscle. The heart is active muscle, and each of its fibers is paralleled by a capillary bringing bloodborne oxygen and nutrients. Paradoxically, the organ cannot use the blood coursing through its chambers. The rate of flow is too fast, and the internal pressure too great; they would rupture the delicate network of cardiac capillaries. In any case, blood in the right side of the heart is poverty-stricken as far as oxygen is concerned – it would be unable to supply this vital substance to the fibers there.

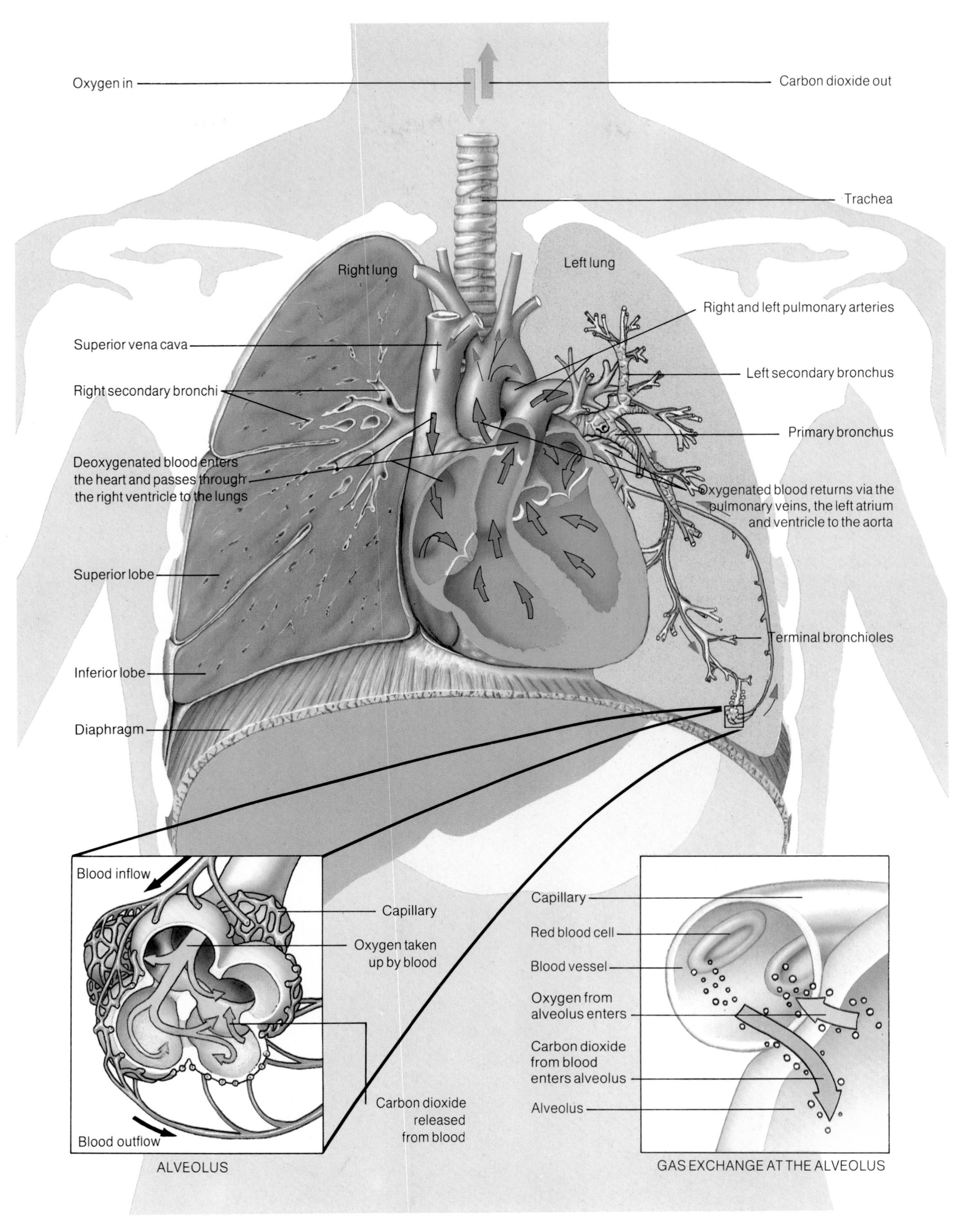
Oxygen in
Carbon dioxide out
Trachea
Right lung
Left lung
Right and left pulmonary arteries
Superior vena cava
Left secondary bronchus
Right secondary bronchi
Primary bronchus
Deoxygenated blood enters the heart and passes through the right ventricle to the lungs
Oxygenated blood returns via the pulmonary veins, the left atrium and ventricle to the aorta
Superior lobe
Terminal bronchioles
Inferior lobe
Diaphragm
Blood inflow
Capillary
Oxygen taken up by blood
Carbon dioxide released from blood
Blood outflow
ALVEOLUS
Capillary
Red blood cell
Blood vessel
Oxygen from alveolus enters
Carbon dioxide from blood enters alveolus
Alveolus
GAS EXCHANGE AT THE ALVEOLUS

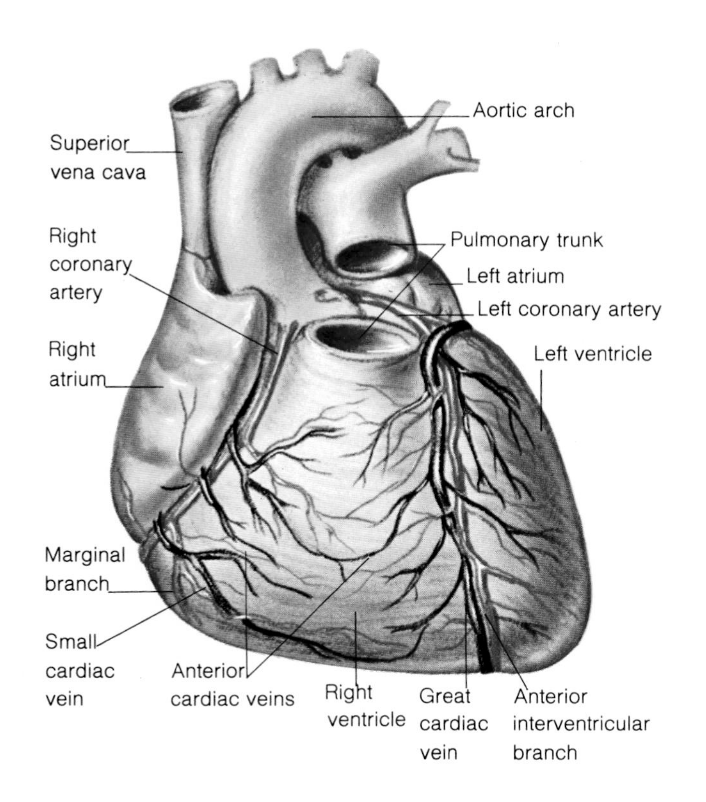

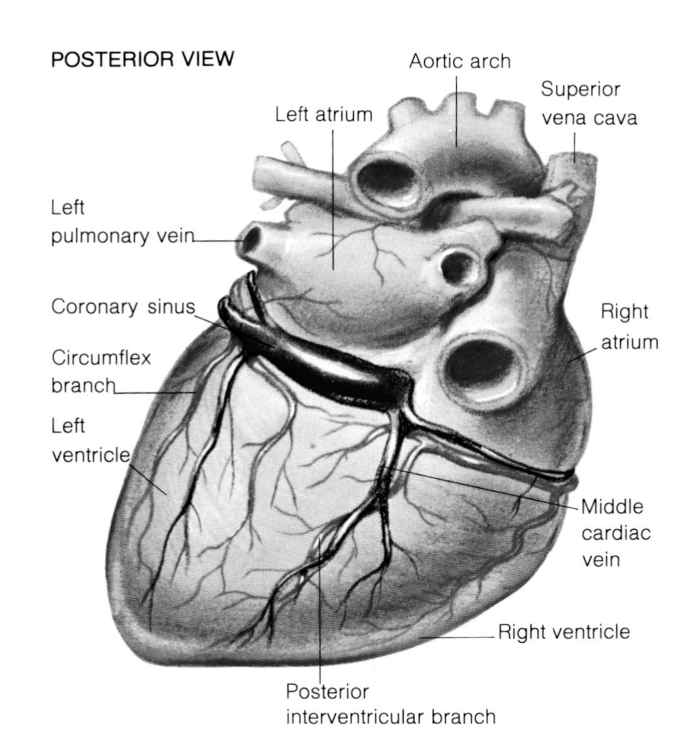

Encircling the heart like a crown, the coronary circulation *(above, left and right)* brings oxygen-carrying blood to the heart muscle itself. During times of great exertion, the heart beats faster. The heart muscle's oxygen demand rises and the coronary arteries have to deliver more blood to the very organ that pumps it there.

The "lub-dub" sound of a beating heart is caused by alternate diastolic and systolic phases *(below, left and right)*. The finely tuned rhythm of the cardiac cycle can be traced in this sequence of diagrams, in which nonoxygenated blood is shown in blue and oxygenated blood is shown in red. As the heart relaxes in diastole (1), both upper chambers (atria) fill with blood – nonoxygenated blood arrives in the right side from the body's main veins, and oxgenated blood returns to the left side after its trip to the lungs. The mitral and tricuspid valves open, and at systolic contraction (2) the atria force blood into the heart's lower chambers (ventricles). The ventricles then contract in their turn (3), pumping nonoxygenated blood through the pulmonary valve and on its way to the lungs, and forcing oxygenated blood through the aortic valve into the body's main circulation. The atria relax again as they reenter diastole (4), and fill with blood once again to restart the cycle.

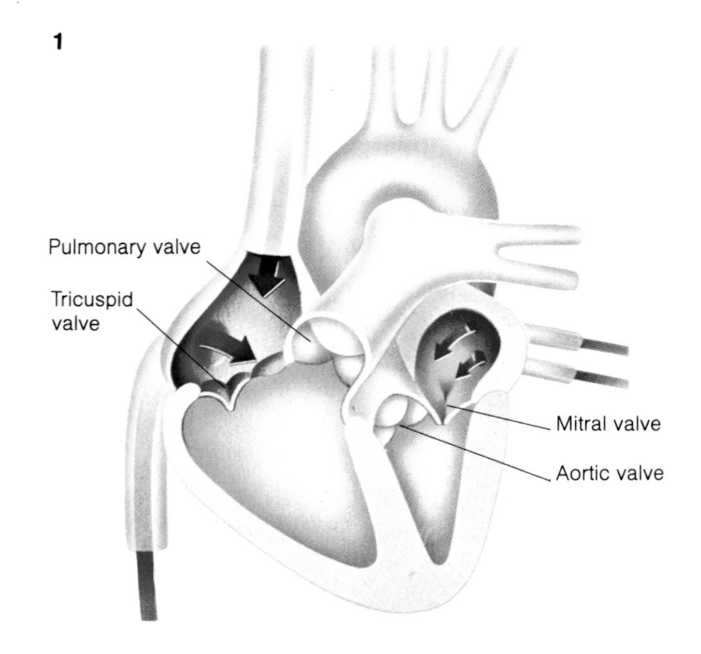

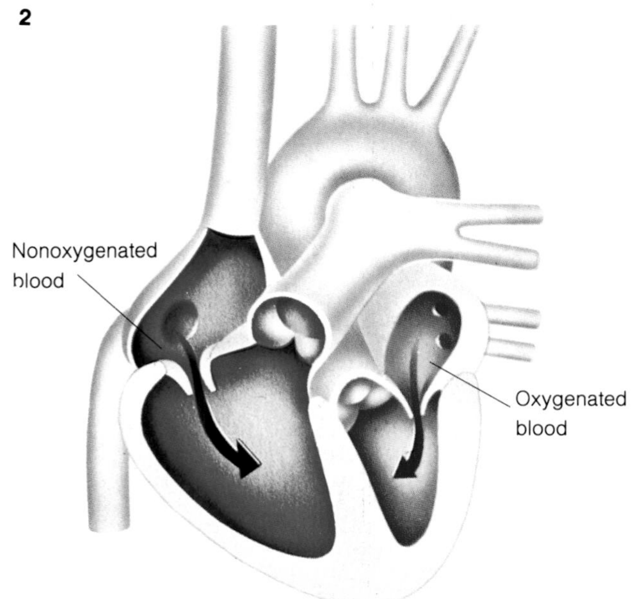

So the hollow muscle that is the heart keeps five percent of the blood it pumps (only the brain takes a greater supply). This organ, filled with blood, maintains its own supply from the outside. Two coronary arteries branch from the main aorta just above the aortic valve. No larger than drinking straws, they divide and encircle the heart to cover its surface with a lacy network that reminded physicians of a slightly crooked crown (*coronary* comes from the Latin *coronarius*, belonging to a crown or wreath). They carry about 130 gallons of blood through the heart muscle daily.

The left coronary artery divides into the anterior descending coronary artery, which carries blood down the front of the heart to both ventricles, and the circumflex artery, which winds around the back to nourish the left ventricle and atrium. The right coronary artery curves round and down to send one branch, the marginal artery, to the fronts of the right atrium and ventricle; a second branch, the posterior descending coronary artery, nourishes the backs of both ventricles.

The coronary arteries are mirrored by a system of cardiac veins which carry blood back from the muscles of the heart. These veins generally parallel the main arteries, to join at a small chamber, the coronary sinus, which empties into the right atrium.

Twice around the block

We can now trace blood's double circuit in more detail. Dark venous blood, low in oxygen and high in waste carbon dioxide, is squeezed out of the right ventricle, through the pulmonary valve and into the pulmonary artery. Shortly this branches right and left, one branch supplying each lung. In the lungs, the arteries continue to branch repeatedly until they become capillaries, microscopic blood vessels which encircle the millions of tiny air sacs (alveoli). Here the blood rids itself of carbon dioxide into the alveolar air, and recharges itself with oxygen from the same, turning bright reddish-scarlet in the process.

The oxygenated blood drains along the four pulmonary veins to the heart's left atrium, through the mitral valve. It then passes into the thickest-walled, most muscular chamber of the four, the left ventricle.

As this contracts, the oxygenated blood is forcefully expelled through the aortic valve into the aorta, the chief artery with branches to all parts of the body. In the vast capillary network that permeates all tissues, blood finally donates its oxygen and takes on board carbon dioxide, again turning dark reddish-blue. The capillaries unite, eventually to form the two main veins, the superior and inferior venae cavae, that guide the blood back to the right atrium, thence through the mitral valve to the right ventricle. The circuit is complete; it takes only half a minute.

The cardiac cycle

Heartbeat is a simple cadence of contraction and relaxation. A complete beat is termed the cardiac cycle. Its contractile phase, when blood squirts out of the heart into the arteries, is systole. Its relaxation phase, in which the heart momentarily rests and draws in another gulp of blood, is diastole.

The exact nature of the contraction – how it progresses through the various chambers – is elegantly ingenious. Seen from without, as the layers of cardiac muscle contract in systole, they seem to wring the heart of blood. The contractile wave begins near the top of the right atrium and crosses both upper chambers. Then another wave of contraction arises from the apex, at the bottom of the heart, and travels upward to envelop the ventricles.

A moment's consideration shows this must be so. The atria begin their contractions from the top because they funnel blood through to the ventricles; the latter squeeze from below, because blood must be ejected upward into the aorta and pulmonary artery. In systole the heart simultaneously shortens, flattens and twists about one-quarter of a turn, thrusting its left half toward the front of the chest.

During a typical heartbeat, with a duration of around four-fifths of a second, the atria contract for about 10 percent of the time and the ventricles for about 40 percent. Ventricular contraction begins with a 0.06-second phase, followed by the main phase of 0.11 seconds in which 60 percent of blood output is expelled. The third phase, lasting around the same time as the second, ejects the remainder of blood for that cycle.

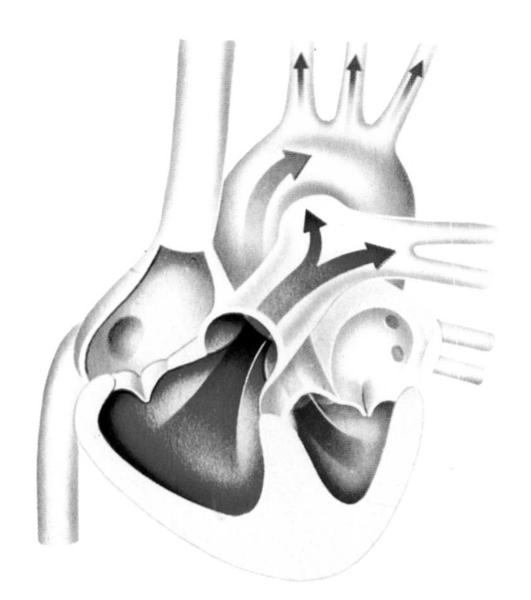

4

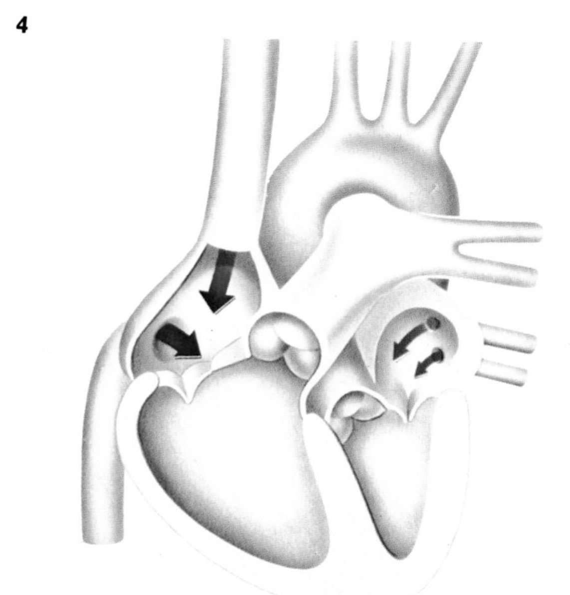

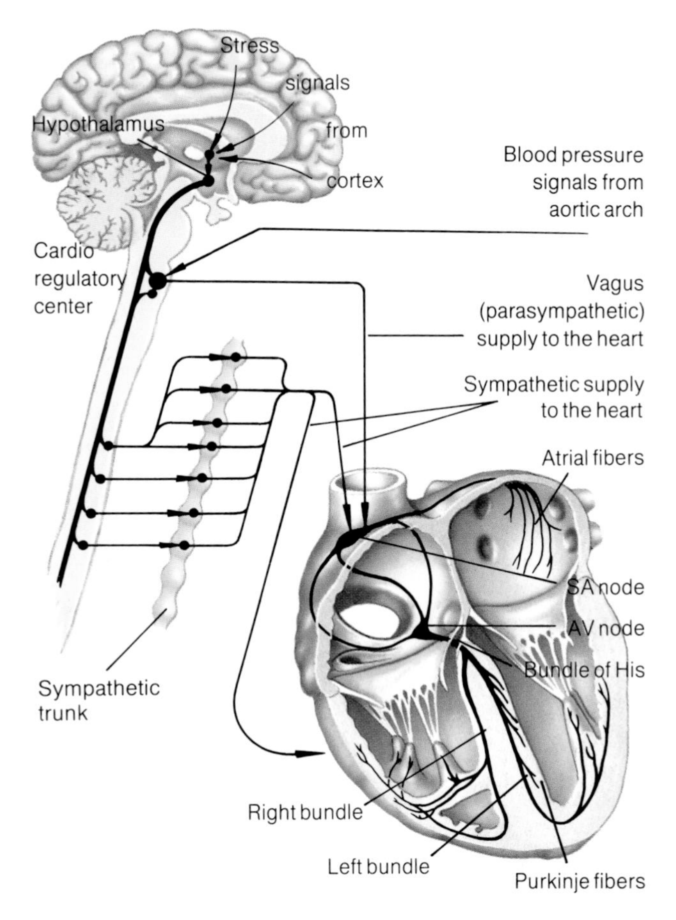

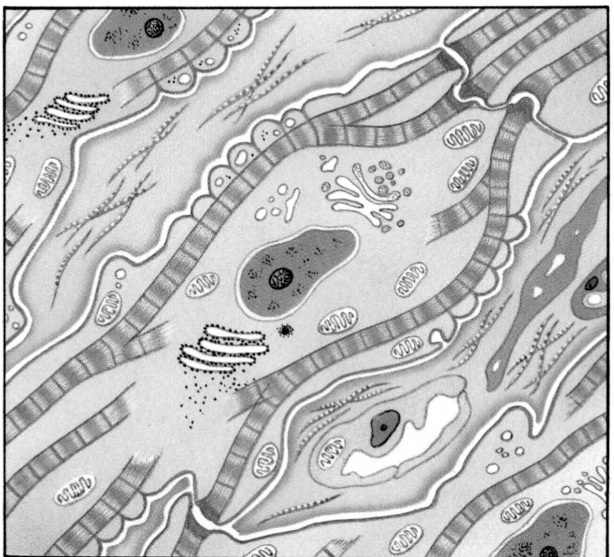

High power view of Purkinje fiber cell

Electrical signals along nerves spark the heart muscle into regular action. The sequence starts when the SA node, the heart's natural pacemaker, receives commands from the brain via the vagus and sympathetic nerves. It signals to the AV node and bundle of His, before impulses pass on to the Purkinje fibers that impregnate the walls of the ventricles. Thus stimulated, the ventricles contract. Purkinje fibers are shown in close-up *(bottom left)*, and can be located in the cutaway view of the heart *(right)*. Their nerve cells conduct impulses four times as fast as other nerves in the heart muscle.

The wiring within

The smoothly choreographed motion of a beating heart gives the impression of order within. Electricity is the key. Millions of tiny electrical nerve signals flash around the body every second, instructing and coordinating muscular activity. The heart too has its electrical signaling system. However, it is self-contained and based, not on nerves in the strict sense, but on muscle cells modified to transmit electrical signals. Clusters of these unique cells lie buried in the heart, and the results of their teamwork are manifested as the power to pump.

The heart's tempo is set by a bundle of specialized cells high in the wall of the right atrium. Equipped not only to transmit but also to generate, these cells make up the sinoatrial, or SA, node. This is the heart's own natural pacemaker. The cells create their own impulses; they "self-excite," and their excitement is infectious.

At the start of a beat, signals from the SA node flash around the atrial walls to stimulate cardiac muscle contraction. They also pass through to the septum, the muscular wall separating the two sides of the heart. Deep in the septum, close to where the heart's four chambers converge, lies a second group of cells, the atrioventricular, or AV, node.

The AV node is a relay station. It receives the signals from the SA node, which have been traveling along a highly conductive pathway in the atrial wall, at a speed of around 24 inches per second. In the node itself, the speed of conduction falls to only two inches per second. Then the AV node fires the signals onward along another electrical conduit, a tract of yet more specialized cardiac fibers named the bundle of His (for the German physiologist who discovered it). The signals speed up as the bundle of His forks into two branches, each spreading into a profusion of tendrils called Purkinje fibers – the "distribution nerves" of the heart. The fibers network their way through the ventricular walls, spreading the electrical messages at the rate of almost six feet per second, to the waiting cardiac muscle fibers.

A controlling influence

The SA node sets the pace of cardiac contraction, with its inherent firing rate of about 70 times per minute. Self-excitation seems to be a general built-in feature of cardiac muscle, however. The other nodes and fibers can also generate their own pulses of electricity, although in the intact heart their rates are subservient to the SA node. An isolated atrium, for example, will beat away by itself around 140 times each minute; a ventricle does so too, although at the slower rate of 30 to 40 beats per minute.

The SA node is not its own master. Nerves of the autonomic system run from the cardioregulatory center in the brain stem, bringing messages that modify the rate of cardiac contraction. The cardioregulatory center is sited in the medulla, where other "vital centers" control various automatic processes, such as

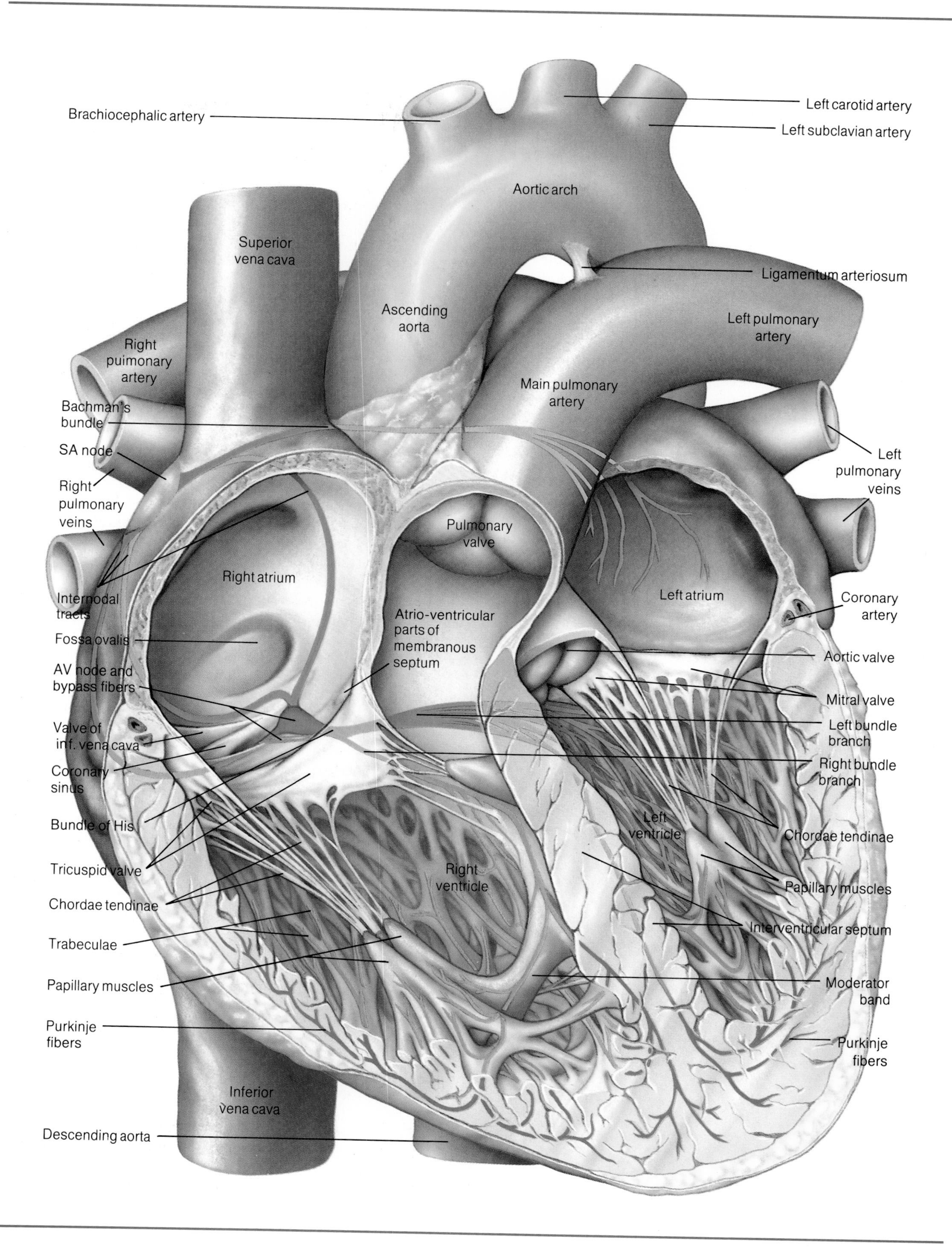
Brachiocephalic artery
Left carotid artery
Left subclavian artery
Aortic arch
Superior vena cava
Ligamentum arteriosum
Ascending aorta
Left pulmonary artery
Right puimonary artery
Main pulmonary artery
Bachman's bundle
SA node
Left pulmonary veins
Right pulmonary veins
Pulmonary valve
Right atrium
Left atrium
Internodal tracts
Coronary artery
Atrio-ventricular parts of membranous septum
Fossa ovalis
Aortic valve
AV node and bypass fibers
Mitral valve
Valve of inf. vena cava
Left bundle branch
Coronary sinus
Right bundle branch
Bundle of His
Left ventricle
Chordae tendinae
Tricuspid valve
Right ventricle
Papillary muscles
Chordae tendinae
Interventricular septum
Trabeculae
Papillary muscles
Moderator band
Purkinje fibers
Purkinje fibers
Inferior vena cava
Descending aorta

breathing rate and depth, and management of blood pressure, without the need for conscious intervention. Hormones and other chemicals also influence the heart rate, as explained later.

Pumping ion

In order to understand how the specialized cardiac tissue generates and transmits electrical impulses, it is necessary to descend to the level of atoms and their behavior at the membrane which surrounds each cell. An atom should be electrically neutral. Inherently it has an equal number of protons, each of which bears a positive charge, and electrons, which carry negative charges. Under certain conditions, and especially when in solution, some atoms tend to lose or gain a particle or two. The elements sodium and potassium, especially vital for nerve function, tend to lose an electron – and with it, one negative charge. This leaves a balance of one extra positive charge. An atom "charged" in this way is called an ion. Calcium ions also form, and due to the nature of their atoms, each bears two positive charges.

In the heart, the cardiac muscle-cell membrane acts as a pump. It drives sodium ions out of the cell, and does so faster than it pulls potassium ions in. An excess of positive charges soon accumulates on the outside of the membrane. Suddenly, when the threshold is reached, the flow reverses. Ionic "gates" open in the membrane and sodium gushes back into the cell. This rapid shift in positive charges is called depolarization, and it creates an electrical current. Quickly the ions are pumped back out again, during the stage termed repolarization.

Depolarization in the self-exciting SA node is the event that sparks off a heartbeat. Waves of depolarization, referred to previously as electrical signals, speed along the cell membranes of the conducting pathways, through the heart. When they reach the terminals of the Purkinje fibers they leap across to the cardiac muscle fibers and set in motion the molecular ratchets of actin and myosin, which slide past each other to shorten the fiber. All this happens in less than a second. What began as pumping ions ends up as pumping blood.

Electrical echoes

The activity of the heart sends tiny electrical "echoes" throughout the body. Waves of depolarization travel round and through the heart like a series of ripples from stones thrown quickly into a pond, one after another. The waves also "leak" into surrounding organs, because body tissues are good conductors of electricity. The strength, size and timing of the waves differ in different parts of the body. We are not aware of them, but they reach the skin and are measurable by apparatus on the outside.

In 1887 the British physiologist Augustus Waller made the first recording of the human heart's electrical signals without exposing the organ. He used a mercury column, which, being of high density, responded slowly and clumsily to the heart's nimble rhythm. Dutch physiologist William Einthoven refined the procedure by wiring the electrodes to a sensitive galvanometer he designed himself, and for which he received a Nobel Prize in 1924.

Einthoven's "string galvanometer" occupied two rooms, weighed nearly a third of a ton, and required five people to operate it. At the time it was a stunning advance. Today the same process can be performed by a device no larger than a briefcase. This is the electrocardiograph, an invaluable diagnostic tool for doctors. It records the sequence, length and strength of each beat as seen from the bioelectrician's viewpoint, giving a frame-by-frame account (either as a paper trace or on a TV monitor) that we have come to know as an EKG or ECG – an electrocardiogram.

The ups and downs of each heartbeat, as recorded on an EKG trace, have been formalized into three main groups, sequentially known as P, QRS and T. The P wave denotes the start of electrical contraction, the QRS complex signifies ventricular contraction, while the T wave represents repolarization – the "recovery" of the electrical conducting tissues in the heart, in readiness for the next beat. The shape and frequency of the waves tell physicians much about how well the heart is working.

Stroke volume and Starling's law

An EKG trace provides an electrical profile of the heart in action. The peaks and troughs of an "exercise EKG" are particularly revealing, showing how well the heart copes with the increasing demands of strenuous activity.

Busy muscles demand more oxygen and energy, so they "use" more blood. The heart must increase the circulation rate, sending more blood in a given time – both to the muscles, and to the lungs for extra supplies of oxygen. Cardiac output (the volume of blood pumped in one minute) is a product of the amount of blood pumped per beat and the number of beats in each minute.

The first of these variables is known as the stroke volume. Under restful conditions the left ventricle accepts from its atrium about one fluid ounce of blood during systole, to add to the reservoir of around the same amount it already contains. It outputs this input volume during diastole. But cardiac muscle possesses a strange property: the more it is stretched, the more it contracts. This feature was first noted by Ernest Henry Starling, of University College in London, in 1914, and it is now commonly known as Starling's law of the heart. It states that the heart's force of contraction is proportional to the rate at which the veins fill it and the resistance against which it pumps.

As the body moves into action, muscles process blood faster. The volume of blood in the veins rises, while muscles around the veins compress them; both factors combine to speed the return of blood to the heart. There, the surge in volume stretches the heart muscle during diastole, and (obeying Starling's law) the muscle responds with a stronger contraction at the next systole.

A healthy heart can raise stroke volume by about two-and-a-half times in this manner: the more blood it pumps with each stroke, the more room it makes for fresh blood on the next beat. The "post-systolic" reservoir (the amount of blood remaining in the heart after contraction) gradually shrinks as the stroke volume climbs. Indeed, high stroke volume is a sign of a well-muscled heart. But cardiac fibers have their limits, and the heart cannot build muscle indefinitely. The limit is set not so much by the cardiac muscles themselves, but more by their coronary blood supply. There inevitably comes a point when the bulk of myocardium outstrips the ability of the coronary circulation to supply it with oxygen and energy.

Rates and reactions

It is not necessary to monitor the heart itself in order to assess its rate of beating (or the heart rate, the second variable in the cardiac output equation). Simply feel the pulse in the wrist; the pulse rate is the heart rate, usually expressed in beats per minute. Each throb of the wrist pulse is a pressure wave that has passed along the artery in the arm, like the loop that travels along a rope

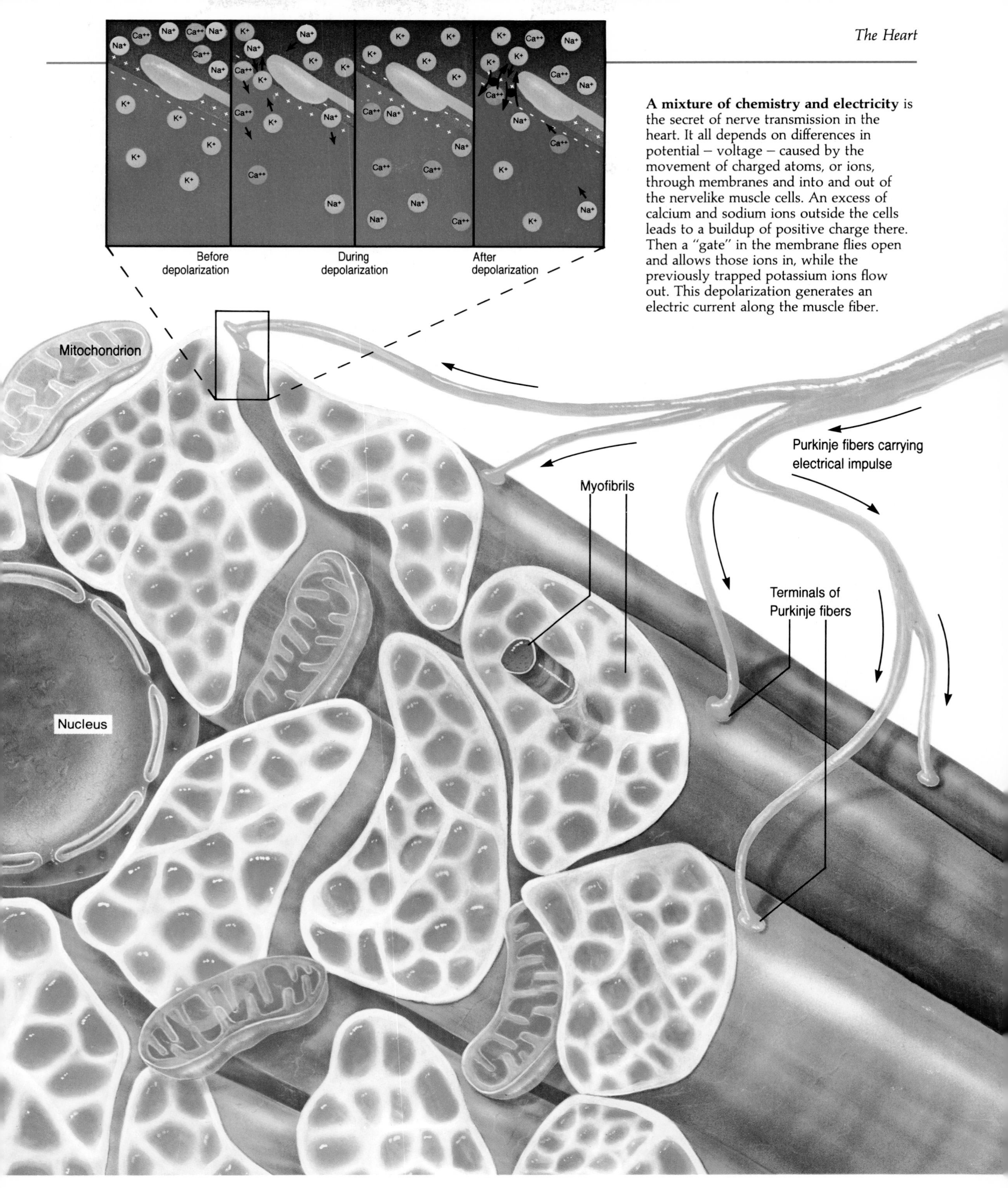

A mixture of chemistry and electricity is the secret of nerve transmission in the heart. It all depends on differences in potential – voltage – caused by the movement of charged atoms, or ions, through membranes and into and out of the nervelike muscle cells. An excess of calcium and sodium ions outside the cells leads to a buildup of positive charge there. Then a "gate" in the membrane flies open and allows those ions in, while the previously trapped potassium ions flow out. This depolarization generates an electric current along the muscle fiber.

MAPPING THE HEART'S SPARK

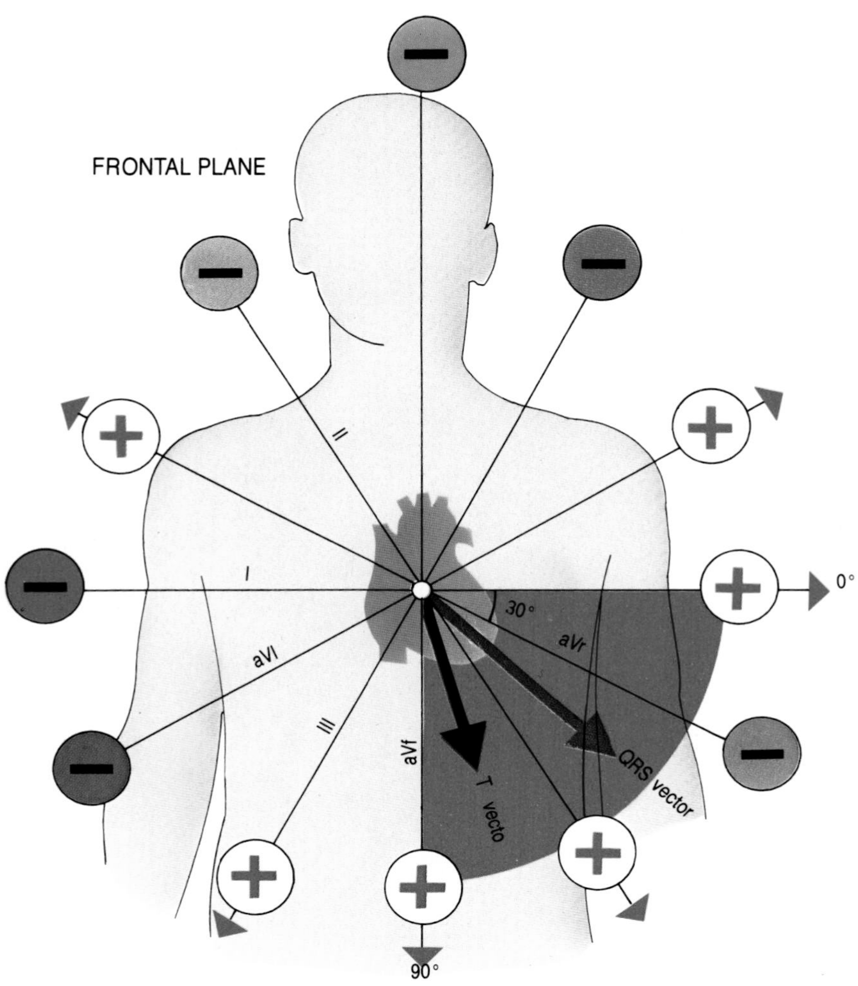

An electrocardiogram – an EKG or ECG trace – follows the electrical activity of nerves and muscles in the heart. There are various characteristic patterns or waves, shown in the band across the bottom of these pages. For ordinary diagnostic purposes, physicians use three waves, illustrated as segments of a typical trace *(opposite page, far center).*

The P wave charts current flow through the atria (shown in purple), from the SA node to the AV node. The QRS complex follows the spread of depolarization through the ventricles (shown in orange). Finally the T wave results from currents generated during ventricular repolarization.

Various waves can be picked up as electrical echoes from the heart by connecting the EKG machine's leads to the skin at different sites on the body. The blue arrows *(left and below)* represent the direction of depolarization as electrical impulses flash through the ventricles during the QRS wave. Normally, the course is somewhere between 0 and 90 degrees on the diagram *(left).* During the T wave, the ventricles repolarize and current flows roughly in the direction of the black arrows.

By fixing two or three electrodes to the wrists and ankles, physicians get a frontal view of the heart's activity by charting the direction of current flow in two dimensions. Together with horizontal-plane electrodes, explained on the opposite page, the EKG can give a three-dimensional picture of the heart's electrical activity, and hence its health.

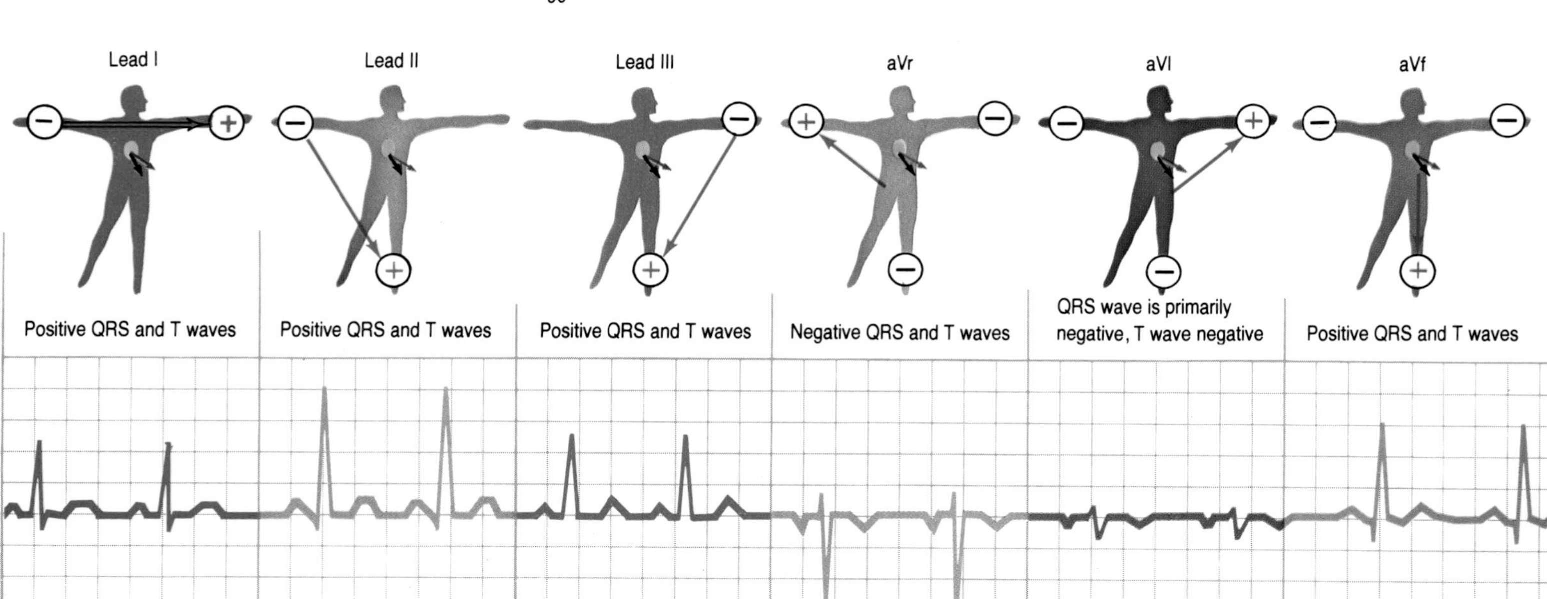

For a horizontal view of the heart's electrical activity, six EKG leads are strapped across the chest. Numbered V_1 through V_6, their responses to the QRS wave are diagrammized *below*. For instance the V_1 leads picks up negative spikes as the electric current races away from it, whereas the V_6 lead registers a strong positive pulse as the heart fires toward it.

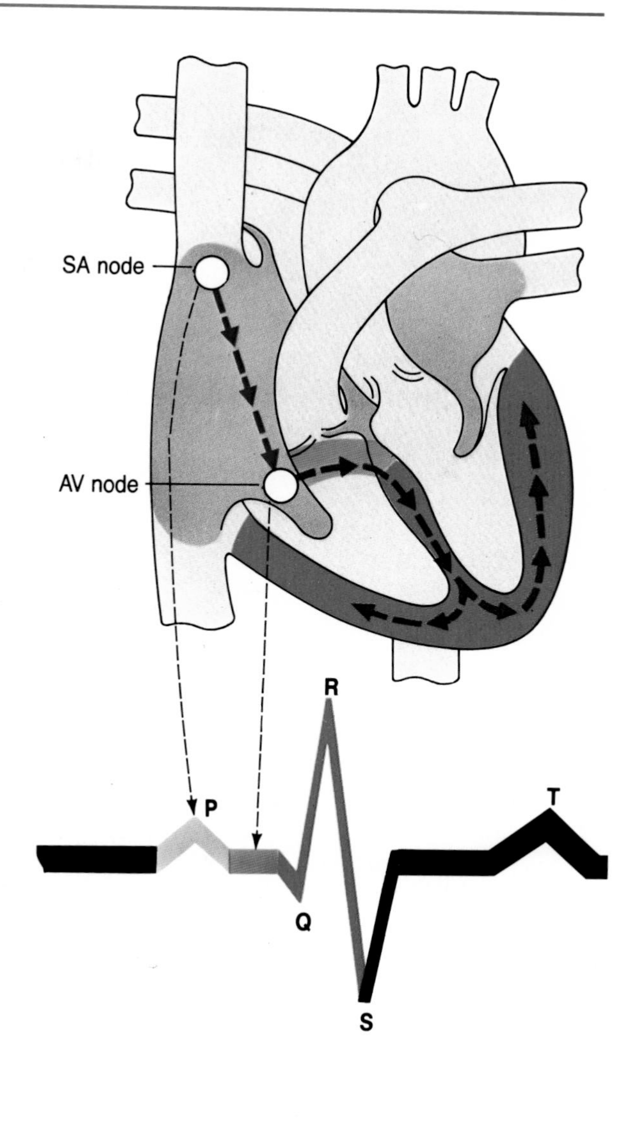

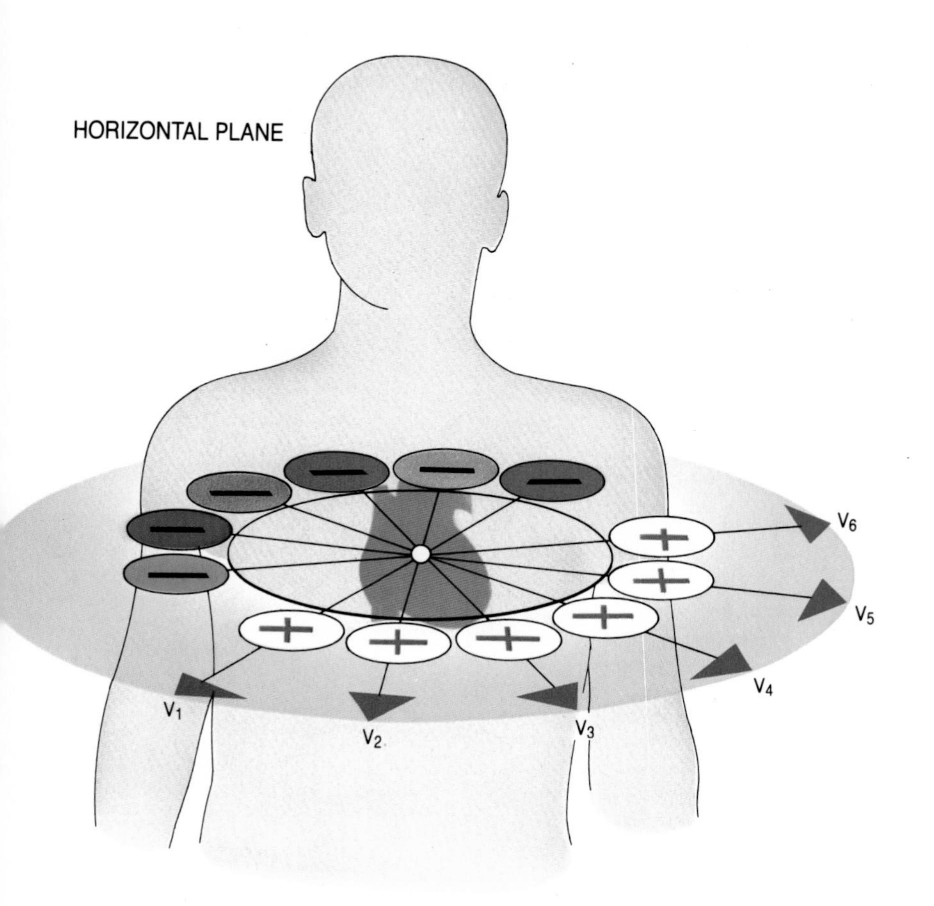

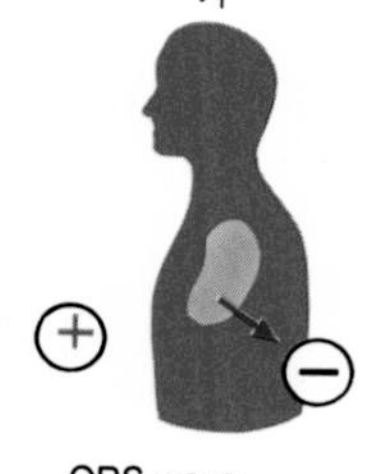

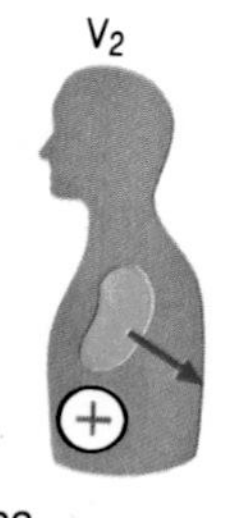

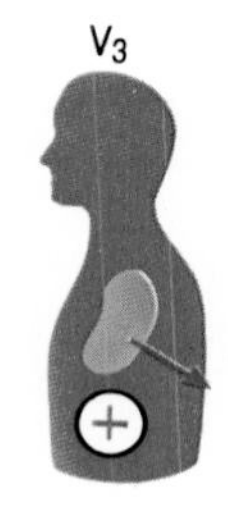

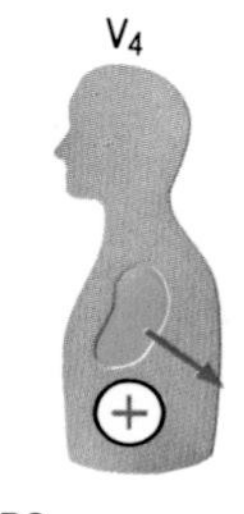

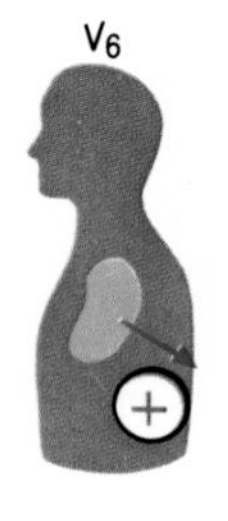

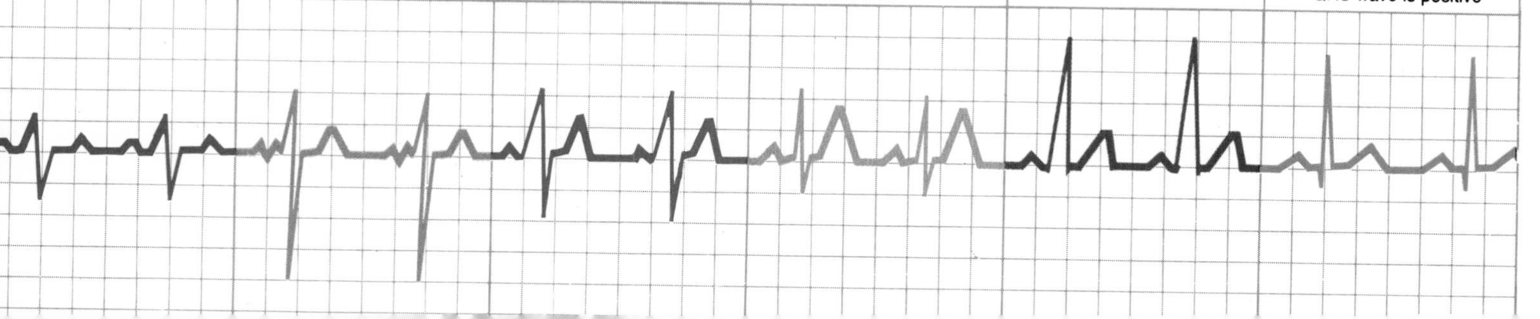

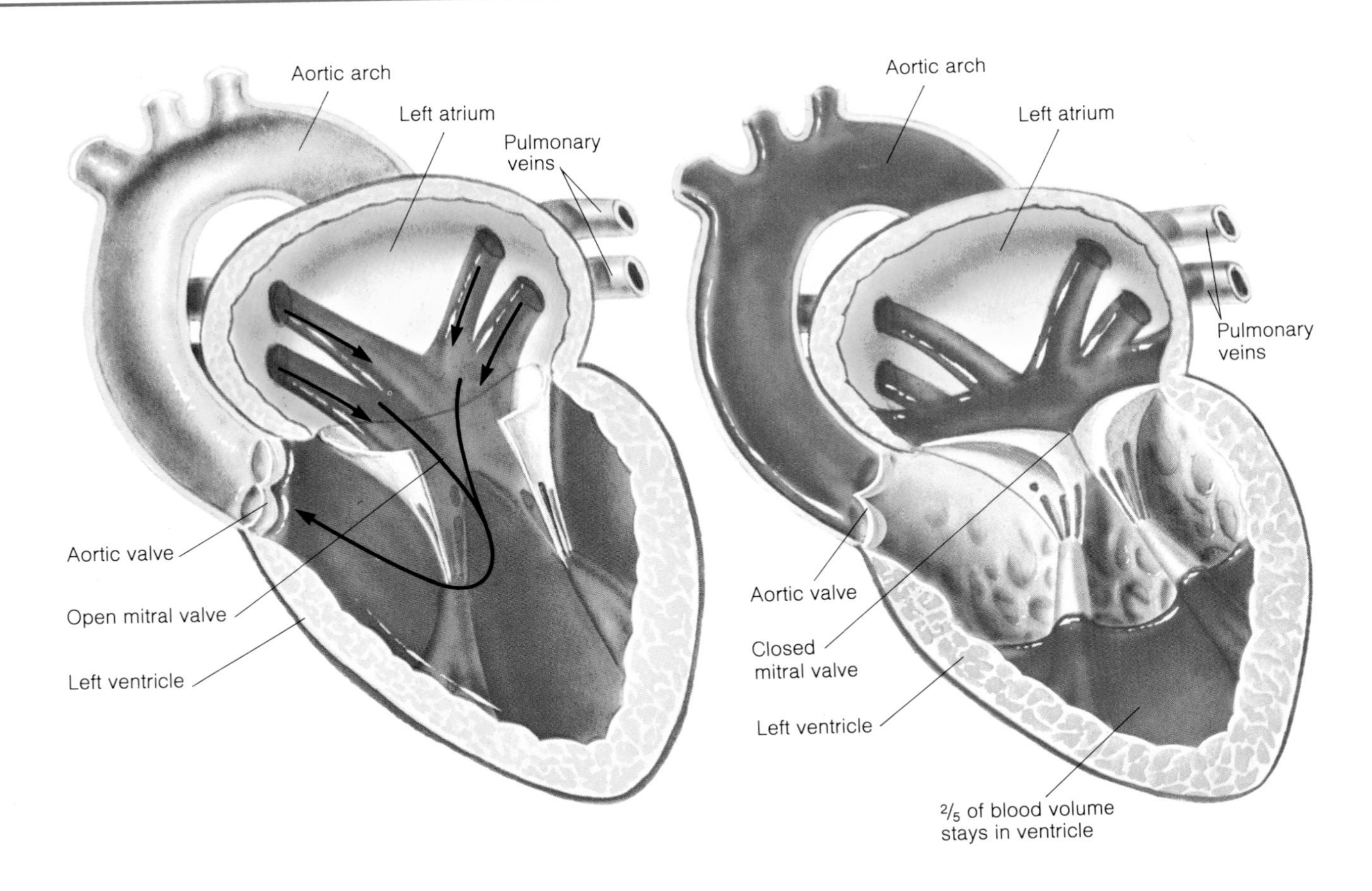

when you flick up one end. The pressure wave originates in the heart itself, at systole, and is transmitted as a "bulge" of blood that surges outward through the arterial tree.

A wide array of factors temper heart rate. They range from body temperature – each Fahrenheit degree of fever raises the heart rate by three to five beats per minute – to hormones and other chemicals. The adrenal glands, atop the kidneys, secrete the hormones epinephrine (adrenaline) and norepinephrine (noradrenaline). The latter strengthens the contraction of heart muscle and constricts the walls of blood vessels. Epinephrine relaxes the vessels that convey blood to skeletal muscles, thereby allowing them to receive a boosted supply – as well as stimulating the heart action, releasing energy-containing glucose into the bloodstream, and encouraging glucose uptake by cells. These hormones do not act instantly. Epinephrine takes about one minute to stoke up heart rate.

A quicker and more precise control involves the nervous system, as mentioned previously. Stimulation of the parasympathetic part of the autonomic system (along the vagus nerve) slows the heart rate, while signals from the sympathetic part of the system increase it. Heat, light, love, danger – such experiences feed from the "higher" conscious levels of brain activity, in the cerebral cortex, inward and downward to its more primitive vital centers, including the cardioregulatory center. Here they stimulate the sympathetic system to raise the heart rate – and a person feels the "bounding and pounding" of the organ in the chest as it picks up speed and power. Both systems reach right to the biological core. One counsels the heart to conserve energy; the other is sympathetic to the circumstances and encourages the heart to burn energy when the situation demands.

Like other muscles in the body, the heart thrives on exercise. Regular physical demands improve the strength of the heart muscle, increasing by more than twice the amount of blood pumped out with each beat – the so-called stroke volume – which can be an important asset when the body's muscles need an increased supply of oxygen.

The speed of the beating heart is controlled by two elements of the autonomic nervous system. Parasympathetic nerves, entering the heart with the vagus nerve, slow it down, whereas sympathetic nerves speed it up. Both sets of signals originate in the brain's medulla, which responds to the body's physical and emotional needs.

Feeding the hungry millions

The heart, marvel of stamina and power, is at the functional – and roughly anatomical – center of the circulatory, or cardiovascular, system. Pumping blood is all very well, but the blood needs to be delivered to where it is needed, as demands on various parts of the body change. The arteries and veins are the highways of the delivery network, while the capillaries are the byways and trading posts, feeding the millions of cells in the body's every nook and cranny.

Arteries divide repeatedly to form capillaries, which unite repeatedly to form veins. The entire length of the system has been estimated at a staggering 80,000 miles.

The functions of the circulation are many. Oxygen, of course, and nutrients are a regular delivery order for the tissues. Carbon dioxide and other molecules of metabolic refuse are just as regularly collected up for disposal. The circulatory system also distributes heat from "hot," metabolically active organs such as the liver to cooler, more quiescent areas like resting muscles.

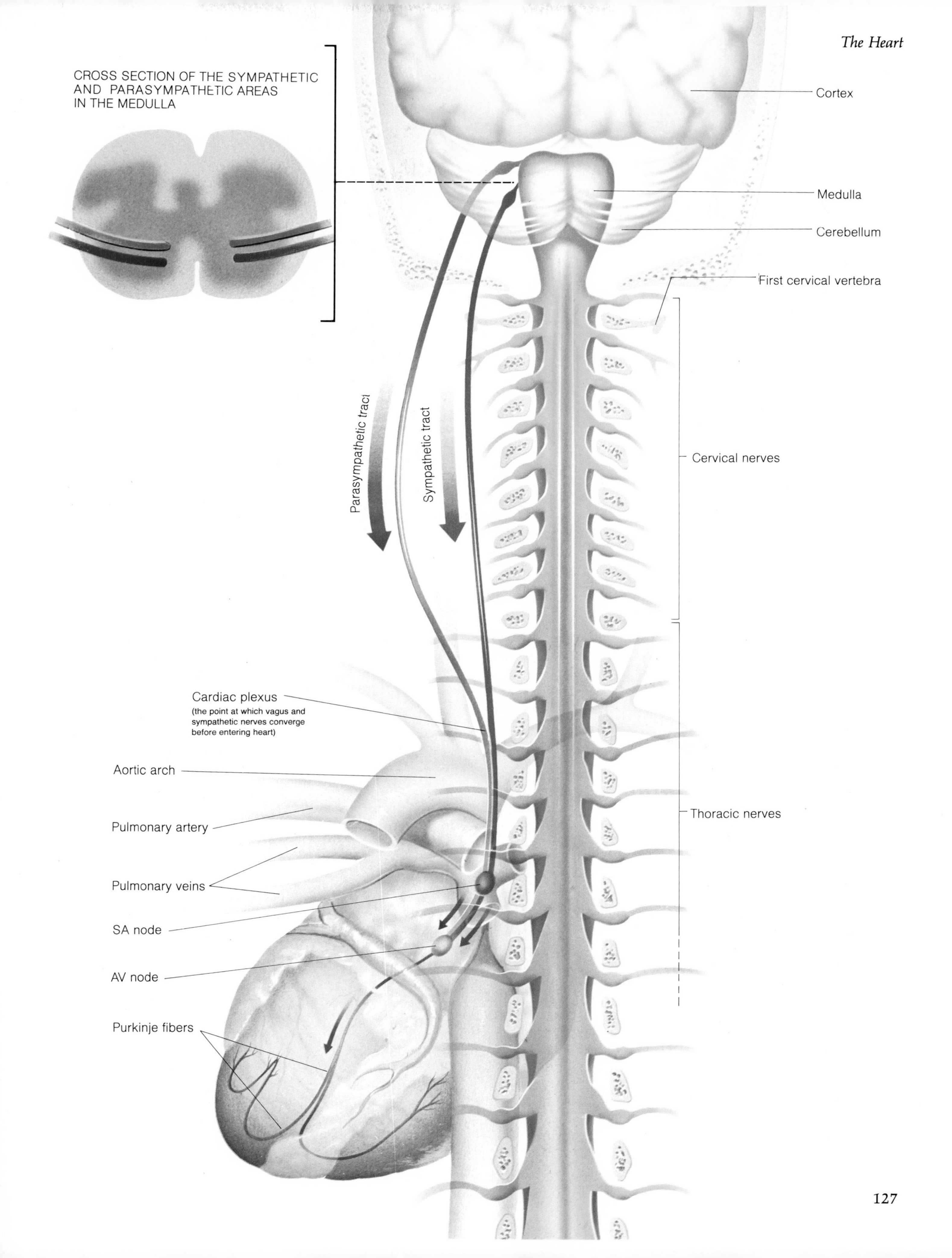
CROSS SECTION OF THE SYMPATHETIC AND PARASYMPATHETIC AREAS IN THE MEDULLA
Cortex
Medulla
Cerebellum
First cervical vertebra
Parasympathetic tract
Sympathetic tract
Cervical nerves
Cardiac plexus
(the point at which vagus and sympathetic nerves converge before entering heart)
Aortic arch
Pulmonary artery
Pulmonary veins
SA node
AV node
Purkinje fibers
Thoracic nerves

Integrated into this aspect is a thermostatic function: divert more blood to the skin and more heat is lost to the external environment, while reserving the bulk of blood flow for the internal organs instead helps to conserve heat.

The cardiovascular system also serves a longer-term regulatory function. A country's road transport network evolves with demand: areas of increased activity receive bigger and better roads, while lack of action allows routes to fall into disrepair and decay. So with the circulation. In response to lengthy periods of elevated oxygen demand from any organ, the blood vessels to and within that organ increase in size and number. Conversely, depressed oxygen demand shrinks the vessels in number and diameter.

From surge to ooze

Arteries are the vessels that guide blood away from the heart. You might think that all arteries contain bright red blood, under great pressure, which would spurt out if the artery were severed. In most cases, especially of arteries likely to be damaged in an accident, this is true. But the first assumption is incorrect. The pulmonary artery and its branches leading to the lungs carry dark reddish-blue blood, low in oxygen, and not the highly-oxygenated, bright red version. By the same token, the pulmonary veins transport oxygen-rich, bright scarlet blood and not the dark venous variety.

Arteries and veins have the same layers of tissues in their walls, but the proportions of these layers differ. Lining the bore of each is a thin endothelium, and covering each is a sheath of connective tissue. But an artery has thick intermediate layers of elastic and muscular fibers, whereas in a vein these are less developed. The thick, elastic arterial wall helps it withstand and absorb the pressure wave newly generated in the ventricles and hydraulically transmitted by the fluid medium of the blood. The wall expands with the swelling force of systole, then snaps back to urge the fluid on its way as the heart takes a rest during diastole; the semilunar valves prevent any backflow. In this way the pressure peaks are gradually flattened along the branches of the arterial tree.

As blood enters the capillary network, the pressure falls off. By the time it reaches the veins, the red fluid is not so much surging, as oozing. There is no need for strength or containment here, so the venous walls are thin to the point of floppiness. To compensate for this lack of dynamic force, many veins lie sheathed in skeletal muscle. The slightest shift of a limb squeezes the vein and drives the blood toward the heart. To ensure the right direction of flow, valves are again the order of the day. Veins, particularly those in the arms and legs, have many semilunar valves. These swing open for each faint pulse of blood, then flap shut to prevent flow reversal.

Pressure and tension

The cardiovascular system thrives in a state of high tension, a condition created by two opposing forces. One is pressure from within, the pressure exerted by the cardiac muscle as it compresses and squeezes out the blood on its multitude of routes around the body; this is transmitted through the fluid medium as force exerted on the vessel walls. The second is pressure from without, the constraining effects of the grasping muscles around arteries and veins, which constitutes the "peripheral resistance."

Blood pressure is measured in millimeters of mercury (mm Hg), the height to which the blood's internal force can raise a column of mercury. Barometric pressure (the lows and highs of weather) is often measured in the same way. A sphygmomanometer is an instrument that measures blood pressure, and like the information from an EKG device, the figure gives valuable information concerning the state of cardiovascular health. Yet in one sense, it matters little how well the heart is working; it is the good circulation of blood around the body that is of fundamental importance to health and fitness.

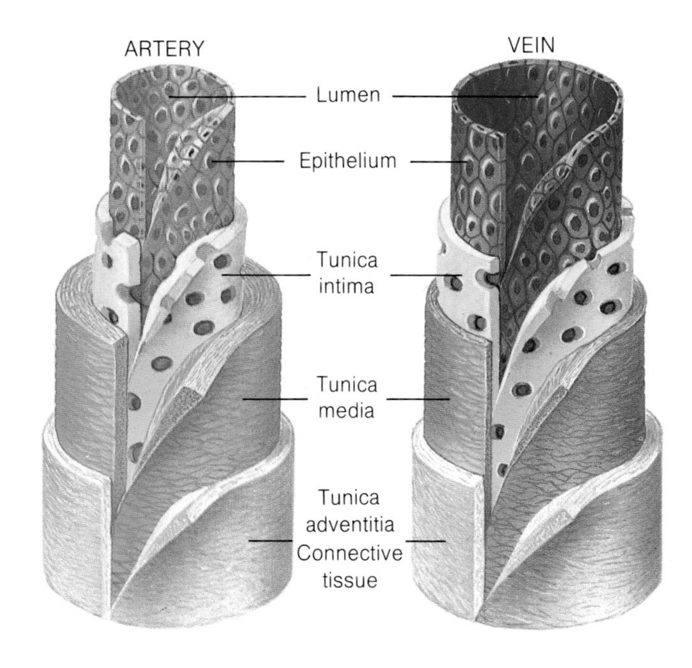

The heart pumps blood, and blood vessels carry it around the body. There are two main types of blood vessels: arteries and veins. Arteries carry oxygen-rich blood from the lungs and deliver it, along finer and finer vessels, to the tissues where it is needed. Veins pick up the deoxygenated blood, now bearing waste carbon dioxide, and carry it back to the heart and on to the lungs. Arteries work at high fluid pressure, and have muscular walls stiffened by elastic membrane. Low-pressure veins have a simpler structure, and are kneaded by muscles *(right)* to assist blood flow back to the heart. One-way valves in the veins of the legs prevent any backflow of blood.

English curate Stephen Hales made the first accurate measurements of blood pressure in animals, and he published his results in 1733. An early form of sphygmomanometer was developed by Italian physician Riva-Rocci in 1896. It used a band around the forearm to restrict blood flow through the arteries there, and a column of mercury to gauge the threshold pressure at which blood flow ceased when felt at the wrist's pulse.

Today in a typical blood pressure reading, a pneumatic cuff around the upper arm is inflated to squeeze the arm and halt temporarily the flow of blood. The operator listens with a stethoscope to an artery "downstream" at the elbow or forearm, and gradually deflates the cuff. Note is made of the maximum blood pressure, at which blood just starts to force its way past the cuff; this is the systolic pressure. The minimum, at which flow is full and unimpeded, is diastolic pressure. The two figures are written, for example, as 120/80, and spoken of as "one-twenty over eighty." The mercury column is still widely used as a pressure indicator, although electronic digital versions have

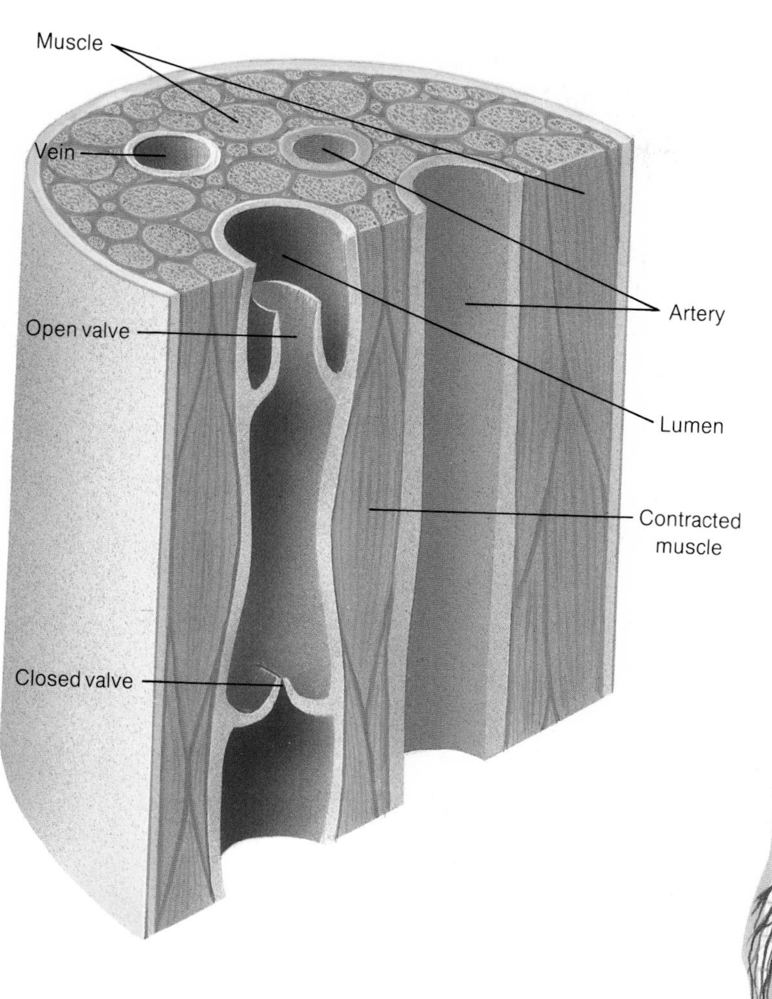

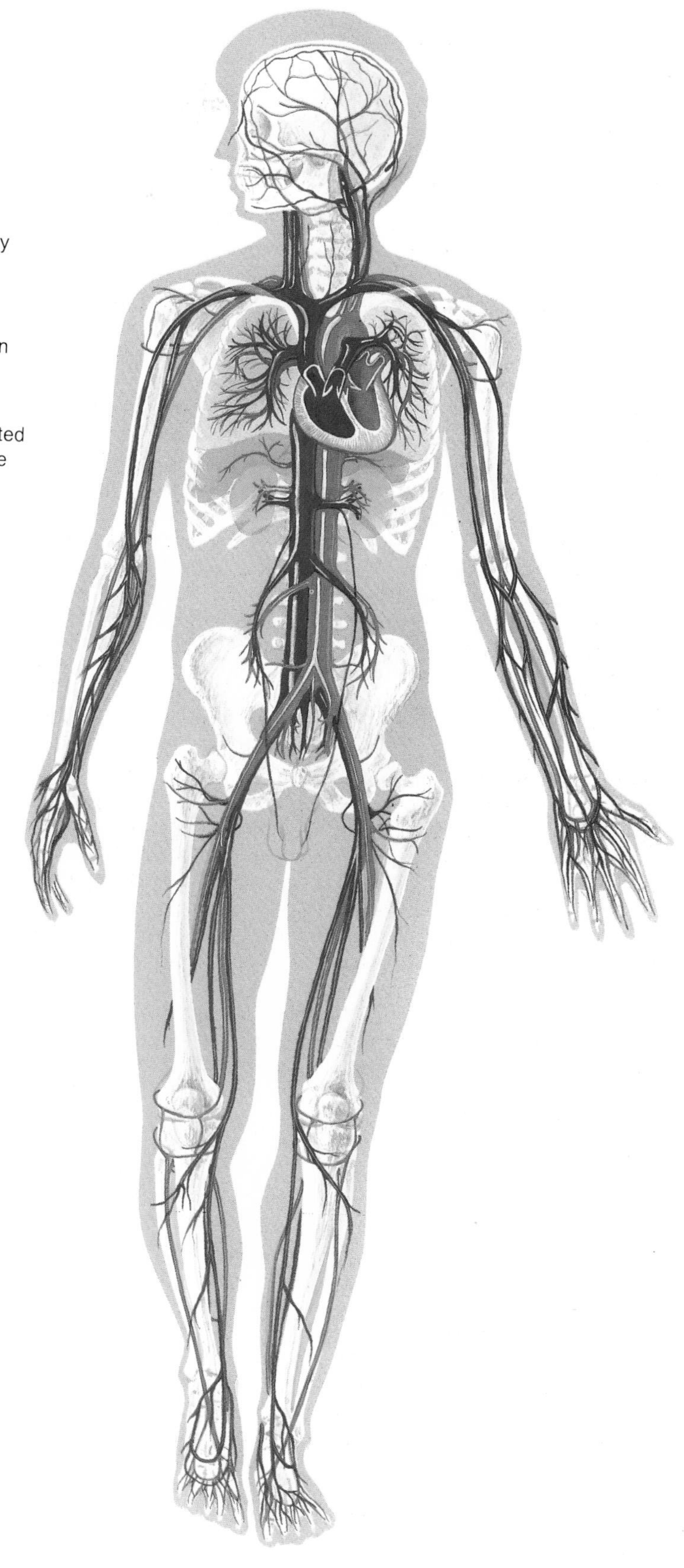

become more common in the past decade or two.

Blood flows down a "pressure gradient." Its pressure is highest in the left ventricle and aorta, where it averages around 100 millimeters of mercury, with systolic peaks higher. The level has dropped to around 35 millimeters by the time the smallest arteries, the arterioles, are beginning to branch into capillaries. It has fallen by another 20 millimeters when the capillaries are joining to form venules, the smallest veins. In the venae cavae, the main veins almost back at the heart, the blood pressure registers near zero.

Blood's flow rate follows a similar pattern. Urgently propelled by the contracting ventricle just around the corner, its speed in the aorta is around 12 inches per second. It courses down the arterial tree, losing velocity as it goes, so that by the time it reaches the small arterioles it is coasting at about an inch a second. In the capillaries it hardly moves at all. However, because the average capillary is only a twenty-fifth of an inch long, blood is inside it for only around one second – time enough to exchange oxygen for carbon dioxide, to trade fresh nutrients for stale wastes. Time is no problem on this microscopic scale. Macroscopically, neither is area: the combined surface of the capillaries totals 6,000 square yards, larger than a football field.

Piecing together the coronary puzzle

Four thousand times each day, the grim drama of a heart attack is replayed. It may strike with a sledgehammer blow, causing death in minutes; or it may pass almost silently, so that the victim is unaware of the damage to the pump within. Middle-aged men and the elderly are most at risk. The epidemic was noticed in the United States and other Western countries around the beginning of this century. It became associated with prosperity, modern lifestyle, affluence and the "good life."

Today, as a result of worldwide population studies, scientists have found and fitted many of the pieces in the heart attack jigsaw. It is a complicated one. Slowly, gradually, the epidemic is fading in some regions, as health education seeps through the populace. The peak incidence of death from heart attacks in the United States was during the mid-1960s; there has been an increasing decline in the death rate since then. Countries such as Britain lag behind, but the decreasing trend seems to be at last happening there, too.

Nearly all heart attacks result from an insidious disorder known variously as coronary heart disease, coronary artery disease, or simply coronary or heart disease. It involves the narrowing of the coronary arteries, which supply the precious, nourishing, oxygenated blood to the cardiac muscle.

Healthy arteries have smooth linings and a clear passageway inside. In coronary disease, the inner linings of the coronary arteries become lumpy and thickened. Deposits of cholesterol (a natural fatty substance), connective tissue and smooth muscle cells build up, beginning the deadly process known as atherosclerosis – furring-up of the arteries with these internal and infernal lumps, bumps, plates and veneers of deposits, generically termed atheroma. Hardening of the arteries (arteriosclerosis), where the atheroma in the arterial wall calcifies into a rigid chalky substance, is also involved.

Recent research indicates that the inner lining of the artery, the endothelium, is active in these processes and may initiate formation of the deposits after becoming diseased, damaged or "insulted" in some other way. The deposits, known as atherosclerotic plaques, can eventually clog the artery completely. Or they may narrow the arterial bore and disturb the smooth flow of blood, so that the blood responds by clotting; clots and plaque then conspire to block the artery.

Atherosclerosis can, and does, occur in almost any artery in the body. But in the heart its effects can be critical. The body depends on a strongly pumping heart to circulate life-giving blood – and this includes to the heart muscle itself. If the coronary arteries become blocked, the cardiac muscle begins to fail, and so the blood circulation decreases, which includes the circulation to the heart muscle itself.

A multitude of factors

Coronary disease is often said to be "multi-factorial" in origin. Thorough studies of family medical histories, diet and social habits have identified risk factors, which increase a person's chances of developing the condition. Some risk factors, such as sex, age, family history of heart conditions, or diabetes, cannot be altered, unless one can follow the doctors' pseudo-adage: "Choose your parents well." But other risk factors are within our power to modify. They include smoking, elevated levels of cholesterol and certain other fats in the blood, high blood pressure, obesity, lack of exercise, some types of stress, overconsumption of sugar or alcohol, and certain personality traits.

It is possible for anyone to put things right, or at least righter, and reduce his or her chances of developing coronary disease, thereby lessening the risk of a heart attack. The basic messages

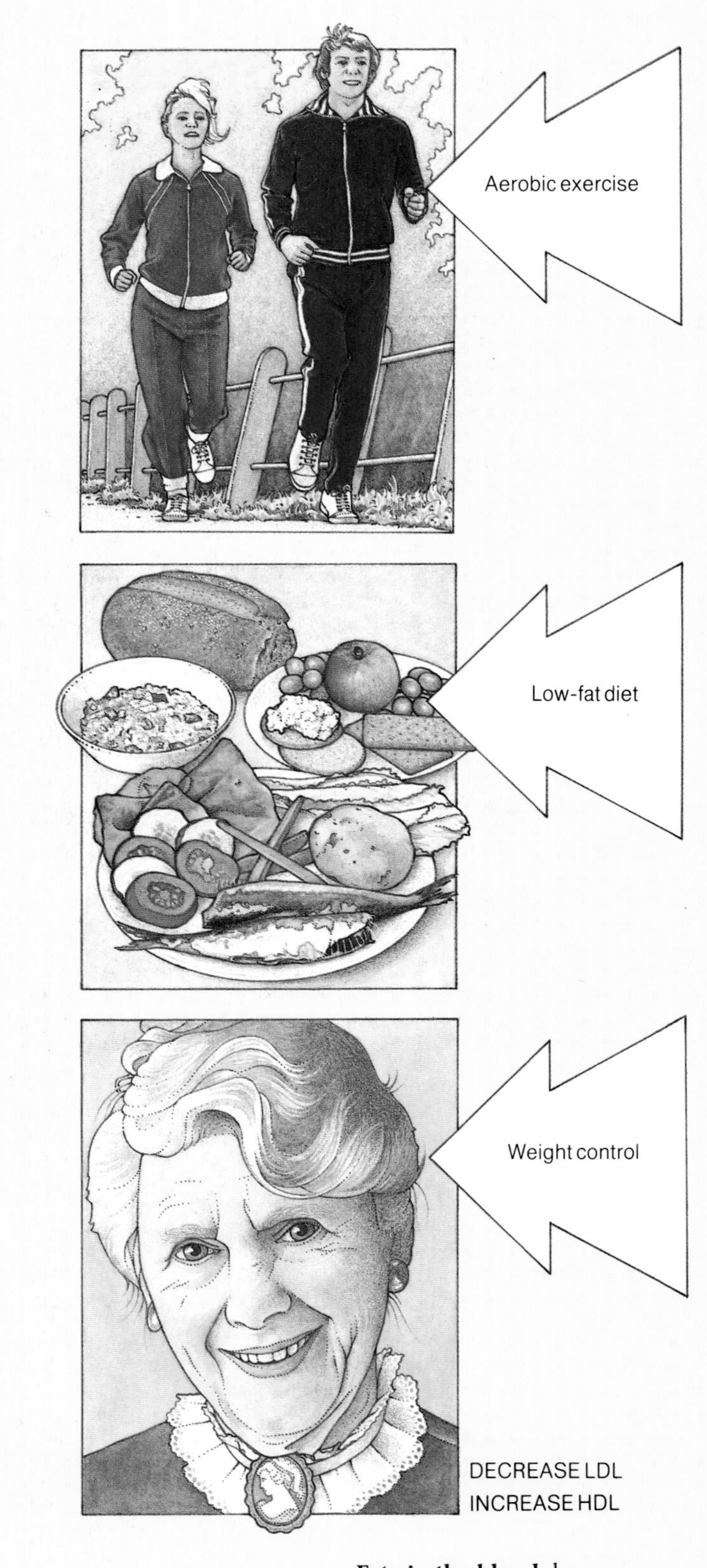

Fats in the blood, known medically as LDLs (low-density lipoproteins) and HDLs (high-density lipoproteins), give clues

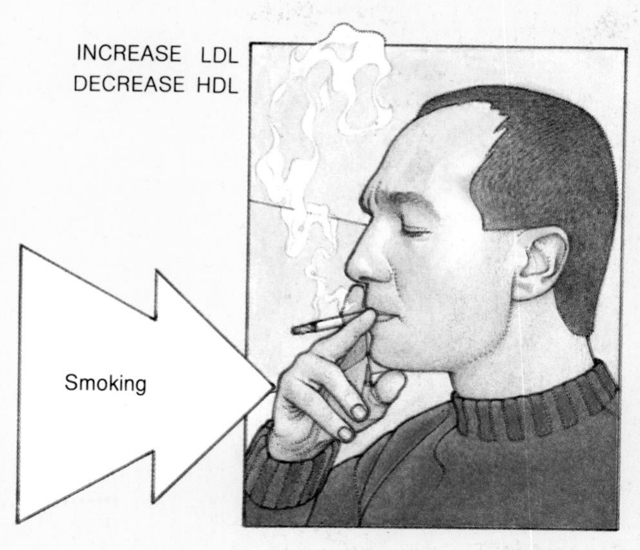

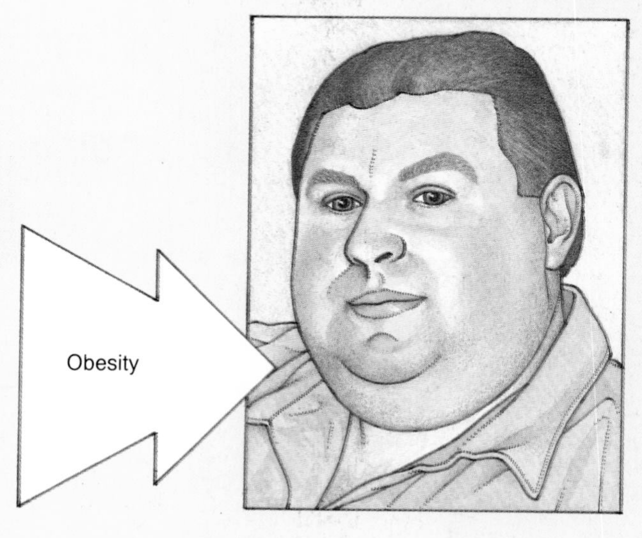

to health. Exercise and sensible eating keep good HDLs high, whereas an unhealthy lifestyle increases bad LDLs.

are: do not smoke, exercise regularly and keep fit, do not become overweight, eat less animal fats in food, and get treatment for high blood pressure without delay.

The cholesterol connection

Population studies have shown which factors, from genes to fatty foods, are associated with coronary disease and heart attacks. Work continues to unravel the biochemistry that underpins the statistics.

Cholesterol is a natural and vital body chemical. This soapy, waxy substance is essential for the healthy functioning of nerve tissue and to produce bile and several hormones. In addition to extracting this chemical from foods, chiefly animals fats and dairy produce, the body manufactures its own cholesterol in the liver. There is a feedback process to help regulate blood cholesterol levels; the more a person consumes, the less the body produces. But if the diet contains large amounts of cholesterol, or the body's metabolism of it is flawed in some way, the excess may contribute to the formation of plaques in the arteries.

Cholesterol is insoluble in the watery serum of blood, so it circulates chemically bound to lipoprotein molecules (those containing both lipids, or fats, and proteins). Research indicates there may be "good" and "bad" lipoproteins. The bad ones may be LDLs, or low-density lipoproteins, which are rich in cholesterol. High blood levels of LDLs have been closely linked with atherosclerosis. The good ones are the small, heavy HDLs, or high-density lipoproteins. Some researchers contend that HDLs clear fat from artery walls and return it to the liver for excretion.

Physicians consider that a simple blood test to measure the ratio of HDLs to LDLs can be an important pointer to the risk of developing heart disease. Medical studies have shown that the risk of heart disease goes down as blood HDL levels go up. This would seem to indicate that a person could lower the risk of heart disease by raising his or her HDL level. Would it work? The evidence leans towards a qualified "yes." Low HDLs have been found in people who smoke, and in those who are obese or sedentary. So changing such habits could boost HDL levels, and place the threat of coronary disease at a greater distance.

The sweet-heart

Many people who consume large quantities of sugar have high blood levels of triglycerides, another fat implicated in formation of plaques. However, as sometimes happens in such a complex field, the evidence from around the world is not consistent. Britons who consumed about four ounces or more of refined sugar each day suffered heart attacks at a rate five times greater than those eating only two ounces of sugar daily. Yet in countries such as Honduras and Costa Rica, where sugar consumption is very high, heart disease rates remain low.

In general, the coronary-conscious diet contains fewer fat meats, dairy foods and sweets, less salt, and more vegetables, fruits, whole grains, poultry and seafood. In a healthy diet, roughly 15 percent of calories should come from protein, 50 percent from carbohydrate and 35 percent from fats.

Cardiovascular conditioning

Hearts evolved for use, not abuse. The best conditioning for the cardiovascular system is aerobic, or dynamic, exercise such as sustained running (not short sprints), swimming and cycling. Sustained activity increases the muscles' needs for oxygen, which are met by an increase in heart rate and stroke volume, and

Regular exercise is good for you and good for your heart. Under instructions from the cardioregulatory center in the brain, the heart beats faster and more strongly during exercise, and breathing becomes deeper and more frequent. Blood flow is diverted from the liver, kidneys and intestines to the muscles, which have a greater need for the extra oxygen it carries. Regular training can also increase the number of blood capillaries in the muscles, and the volume of blood pumped out of the heart by each beat.

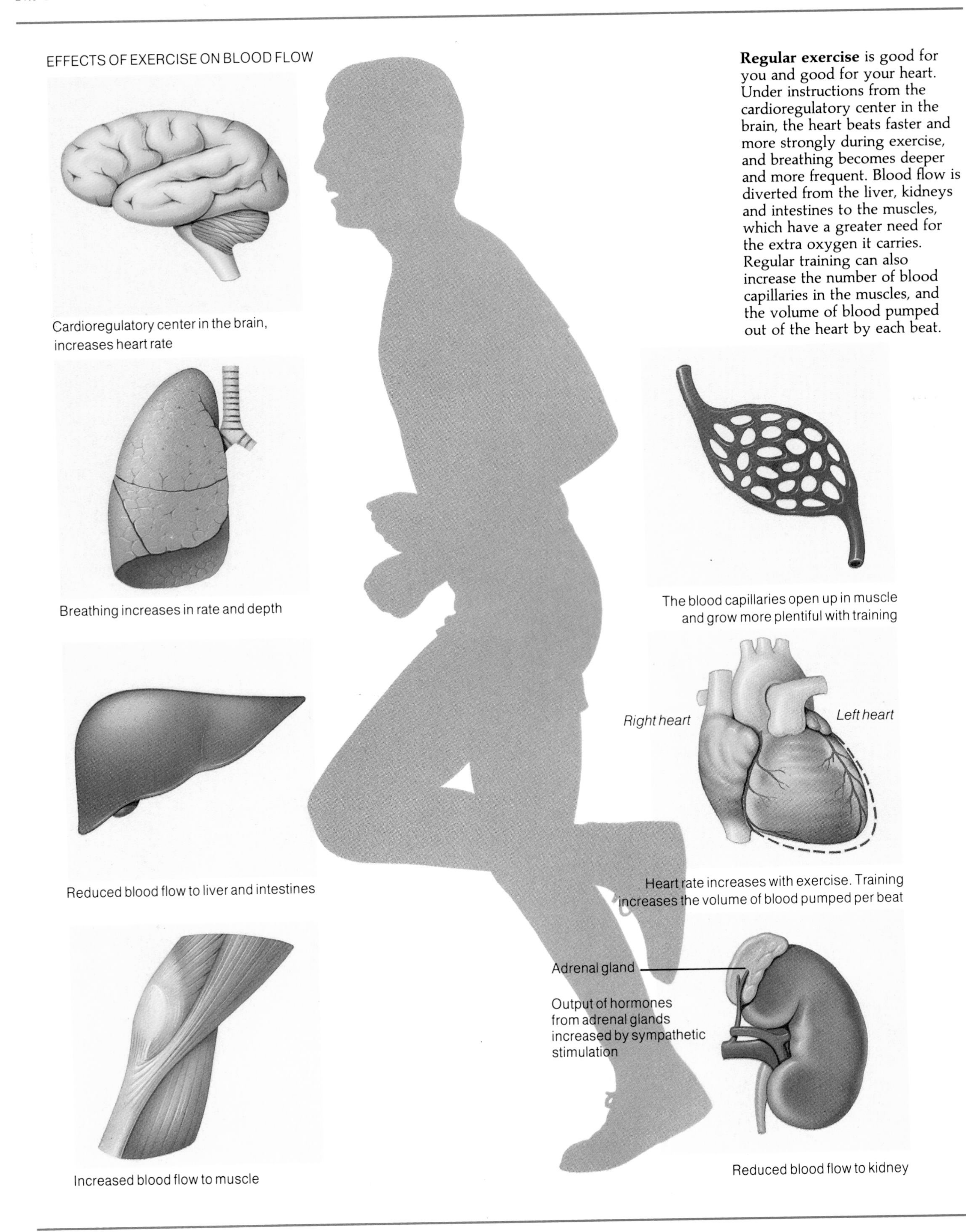

also deeper, faster breathing. Exercise improves the health of muscles, including cardiac muscle and those involved in respiration. With training, the heart and lungs grow stronger and more efficient, and the arteries become less rigid and more elastic. Possibly the coronary circulation itself is augmented by extra blood vessels supplying the cardiac fibers; these "collateral" vessels could provide another route for blood should one of the main coronary arteries become narrowed or blocked.

Exercise is not without risks. A person who has led a coronary-tempting lifestyle for many years and who then suddenly takes up strenuous exercise, such as a tough game of squash or vigorously working out with weights, is being less than sensible. The heart may protest or even fail at the strain, quite apart from the risk of sports injuries to the flabby body, such as torn muscles and sprained joints.

Most doctors recommend that new converts to exercise who could be at risk from heart disease should have a thorough check-up, including a fitness assessment, and perhaps an exercise EKG and respiratory function test. They should adopt a graded exercise program. It is important to begin by jogging, not sprinting, along the road to health.

Under too much pressure

Major medical studies of heart disease such as the Framingham project (begun in 1948, in a suburb of Boston) have implicated high blood pressure, or hypertension, as a risk factor. A blood pressure reading of around 120/80 is considered average for an adult. A person whose readings are consistently above 140/90 may be said to have high blood pressure, although other factors such as age will influence a physician's decision to impose such a label on a potential patient. In general, the diastolic (second) reading is considered more important. Framingham figures indicate that coronary disease is three to five times more common among people with hypertension than those with normal or near-normal blood pressure. Indeed, some of the credit for the declining mortality from heart disease in the United States has been given to the improved identification and treatment of people with raised blood pressure.

Hypertension is a classic "silent killer." Frequently there are no symptoms until the advanced stages; a person may feel healthy and vigorous for years, yet run an extremely high risk of stroke, heart disease and kidney conditions. High blood pressure is a result of the heart being forced to pump harder, usually in order to overcome an increased peripheral resistance set up by narrowed blood vessels. This overwork makes the vessels stiff and brittle, and the heart adds muscle fibers, which places additional burden on the circulation. Eventually, the overworked heart may fail. Before the advent of antihypertensive drugs, heart failure was the most frequent cause of death among people with high blood pressure.

So much for the effects; what of the causes? Here we are in less well charted territory. The great majority of cases are so-called "primary hypertension" – that is, the high pressure is not secondary, not a consequence of some other body problem. Often, there is no clear cause.

Natural regulation of blood pressure involves baroreceptors, specialized pressure-detecting clumps of nerve cells embedded in the walls of arteries such as the aorta and the carotid artery in the neck. When the blood pressure rises, the baroreceptors alert the brain. Impulses from the brain then pass through the autonomic nervous system to the arterioles, instructing the muscle in their

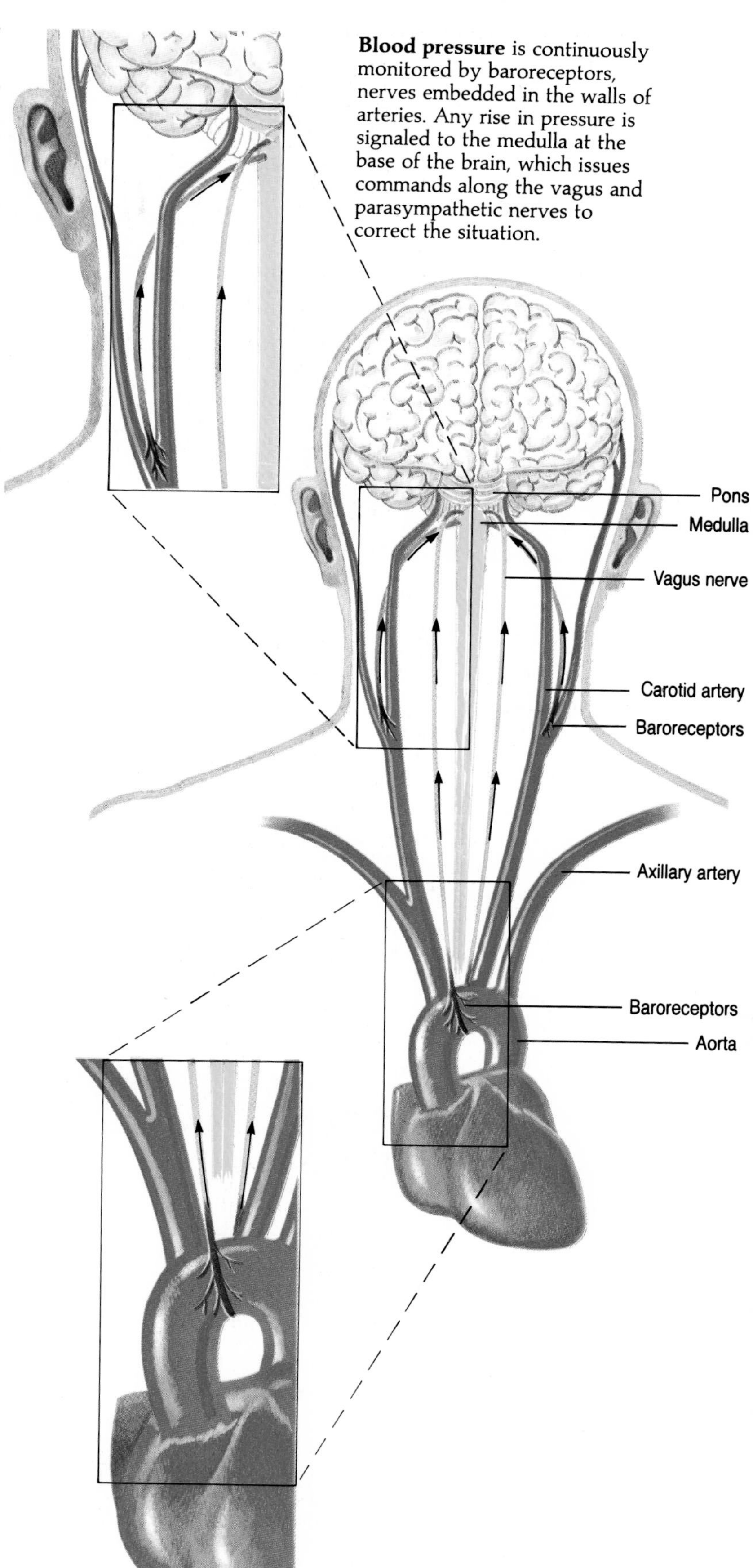

Blood pressure is continuously monitored by baroreceptors, nerves embedded in the walls of arteries. Any rise in pressure is signaled to the medulla at the base of the brain, which issues commands along the vagus and parasympathetic nerves to correct the situation.

walls to relax and widen the arterial bore. This permits more blood to flow and decreases the peripheral resistance. The result should be a drop in blood pressure.

Some experts theorize that the "barostat" mechanism may be flawed in hypertension. Turning up the thermostat in a home forces the heating system to work harder, driving up the average temperature. In high blood pressure, the barometric feedback system may become "reset" to a higher level somewhere along the line, resulting in persistently raised pressure.

In other cases the flaw might be chemical. Substances in the bloodstream may distort barostatic messages, causing the arterioles to respond improperly. A newer approach concentrates on the heart-brain connection. Animal studies have led to speculation that a rise in blood pressure dampens feelings of anxiety in stressful situations. When a person consistently denies angry, anxious feelings, barostatic messages could be eventually disregarded and so the blood pressure remains elevated.

The heart, smoking and drinking

In the public consciousness, cigarette smoking may be more associated with lung cancer than heart disease. However, smoking causes more overall illness via the heart than through its effects on the lungs and respiratory system, partly because heart disease is so widespread in the first instance.

Tobacco smoke affects the heart in several ways. The nicotine it contains causes the brain to instruct the adrenal glands to produce more epinephrine (adrenaline), the hormone that speeds heart rate. A faster heartbeat increases the heart's need for oxygen, and the pump must work harder to compensate. If the coronary arteries are already narrowed or blocked, the extra workload may be critical. Large amounts of epinephrine can also lead to irregularities in heartbeat. In addition, nicotine constricts blood vessels, which pushes up blood pressure.

Carbon monoxide, a second villain in tobacco smoke, competes with oxygen for a place in the bloodstream. It is a chemical "hijacker" because it combines preferentially with the molecules of hemoglobin, contained in red blood cells, which normally gather and ferry oxygen to the tissues. In this way carbon monoxide may reduce by one-fifth the amount of oxygen delivered to the body tissues.

A pregnant woman who smokes risks her baby's cardiovascular system as well as her own. The tobacco smoke contaminants have the same noxious effects on the baby's heart as on the mother's, increasing its heart rate and blood pressure while reducing its vital oxygen supply.

Alcohol's effects on the heart have been well characterized. While scientists have found no direct link between drinking and atherosclerosis, it is known that alcohol can bring on irregular heartbeats, raise blood pressure and elevate blood levels of certain fats. Alcoholics are at risk from an enlarged, failing heart.

Even so, the situation has become rather less clear-cut in recent years. In 1979 a study at Harvard Medical School identified a link between alcohol and heart disease, but surprisingly it was a positive association. It seemed that light drinkers (one or two average drinks per day) had a lower coronary risk that heavy drinkers – and than abstainers. But the statistics are complex, with many factors of diet and lifestyle intertwined. Since then, other surveys looking at possible connections between HDL levels and alcohol consumption have not confirmed the Harvard findings. Experts do not advocate drinking as a preventive measure against coronary disease and heart attacks.

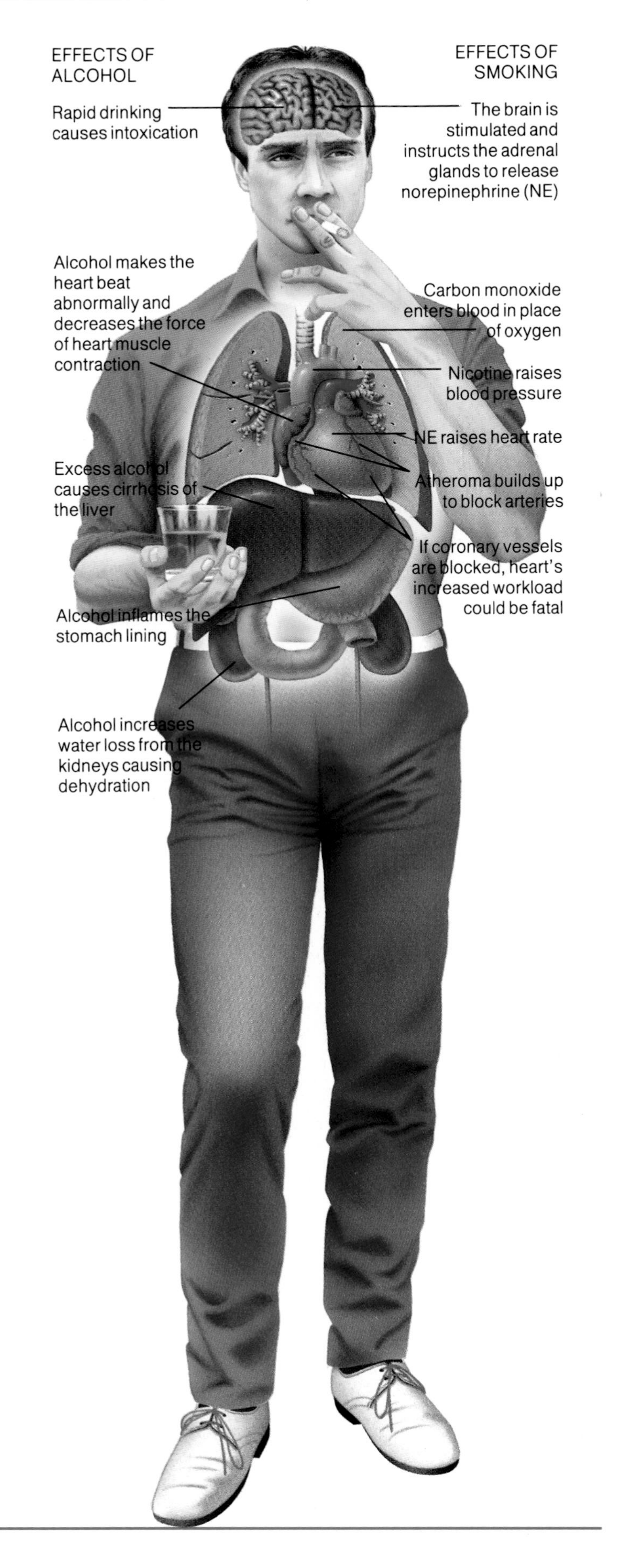

Abuse of the body with drugs invariably has a damaging effect on the heart. The effects of regular consumption of alcohol or tobacco can be wide ranging *(left)*. In addition to inflaming the stomach (gastritis) and scarring the liver (cirrhosis), alcohol affects the heart beat and eventually attacks the brain. Nicotine and carbon monoxide in tobacco smoke – both toxic substances – can have equally devastating effects on the heart and blood pressure. Deaths from heart disease in men *(below)* correlate with the number of cigarettes smoked per day.

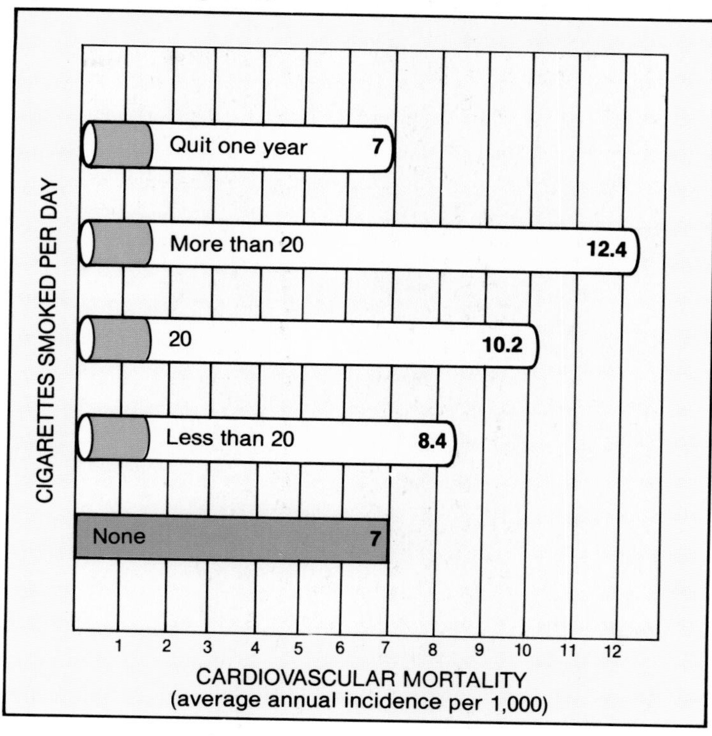

The A-to-B guide to stress

The relationship between stress and heart disease is more speculative than for other risk factors. This is partly a lexicographical problem: the word "stress" is ill-defined, overworked, tired and confused. It is difficult to identify, let alone quantify.

Physiological mechanisms linked with the responses to what most of us perceive as "stressful" events have been coined as the "fight-or-flight" reaction. The nervous and hormonal systems gear up the body to meet the threat by causing heart rate, blood pressure and respiration rate to soar, in preparation for physical activity. The large amounts of epinephrine and norepinephrine that pour into the bloodstream at this time encourage cardiac muscle to burn oxygen at a very high rate, quickly depleting the heart's oxygen supply. The resulting oxygen deficiency could lead to a heart attack in a person with pre-existing coronary disease. A Swedish study of 150 middle-aged men found that those who experienced a high degree of psychological stress were six times more likely to develop coronary disease within five years than those who reported relatively low stress levels.

In 1959, in an attempt to encapsulate the features of a high-pressure, stress-loaded life, cardiologists at Mount Zion Hospital, San Francisco, first described the "Type A" personality. The typical Type A person is competitive, achievement-orientated, impatient and quick to anger – everything the Type B individual is not. In a large survey involving 3,500 middle-aged men, long-term Type A behaviour was the best predictor of coronary disease. Such men were almost three times more likely to develop heart problems than their patient, less competitive Type B counterparts.

The Type A concept has taken its place among the many risk factors linked to heart disease, factors that themselves are interrelated, in ways as complicated as our modern lifestyle. They include smoking, diet, alcohol, exercise, blood pressure, weight control, personality, and the temptations for each of us to do what we enjoy. Trying to avoid coronary disease is for some like threading a way through a multidimensional maze, to its healthy heart. Yet increasingly the medical evidence shows that it can be done, and that each year more people are doing it.

Angina, thrombosis and infarction

Not all heart attacks strike without warning. A type of chest pain called angina pectoris ("angina" for short) often accompanies the build-up of fatty deposits in the coronary arteries. Angina may be vaguely discomforting, or a viselike pain crushing the chest and stabbing into the limbs. It usually lasts a few minutes and is brought on by exercise, emotional stress or heavy eating.

The pain is due to myocardial ischemia, or "cramp" in the heart muscle; the fibers temporarily run short of oxygen because the narrowed coronary arteries cannot match the demands of an increased blood supply. The typical heart attack is set in motion by coronary thrombosis.

The plaques in the arteries act as magnets for platelets, cellular fragments floating in blood, which are involved in the clotting process. The platelets stick to the plaques and may grow into full-scale clots, or thrombi. As the thrombus enlarges it reduces blood flow through the artery. In a coronary artery, this narrowing – and perhaps eventual total blockage – is known as coronary thrombosis, or a "coronary." Eventually, the cardiac muscle cannot go on. Its blood supply becomes very poor, or even ceases altogether, and the muscle tissue begins to "suffocate" through lack of oxygen. A heart attack is in progress.

A heart attack is a life-threatening emergency usually resulting from a myocardial infarction, which takes place when the blood supply to part of the heart muscle is cut off because a coronary artery is blocked. A healthy artery (1) has a smooth lining of epithelial cells. But in some people, perhaps because of a high level of cholesterol in their diet, fatty deposits called atheroma build up beneath the epithelium (2), narrowing the artery and reducing its flexibility. Eventually calcium salts become deposited in the artery walls (3). These build up until the vessel is almost completely blocked. Lack of blood supply to an area of muscle causes an infarct to occur and that area of the heart muscle dies and ceases to function.

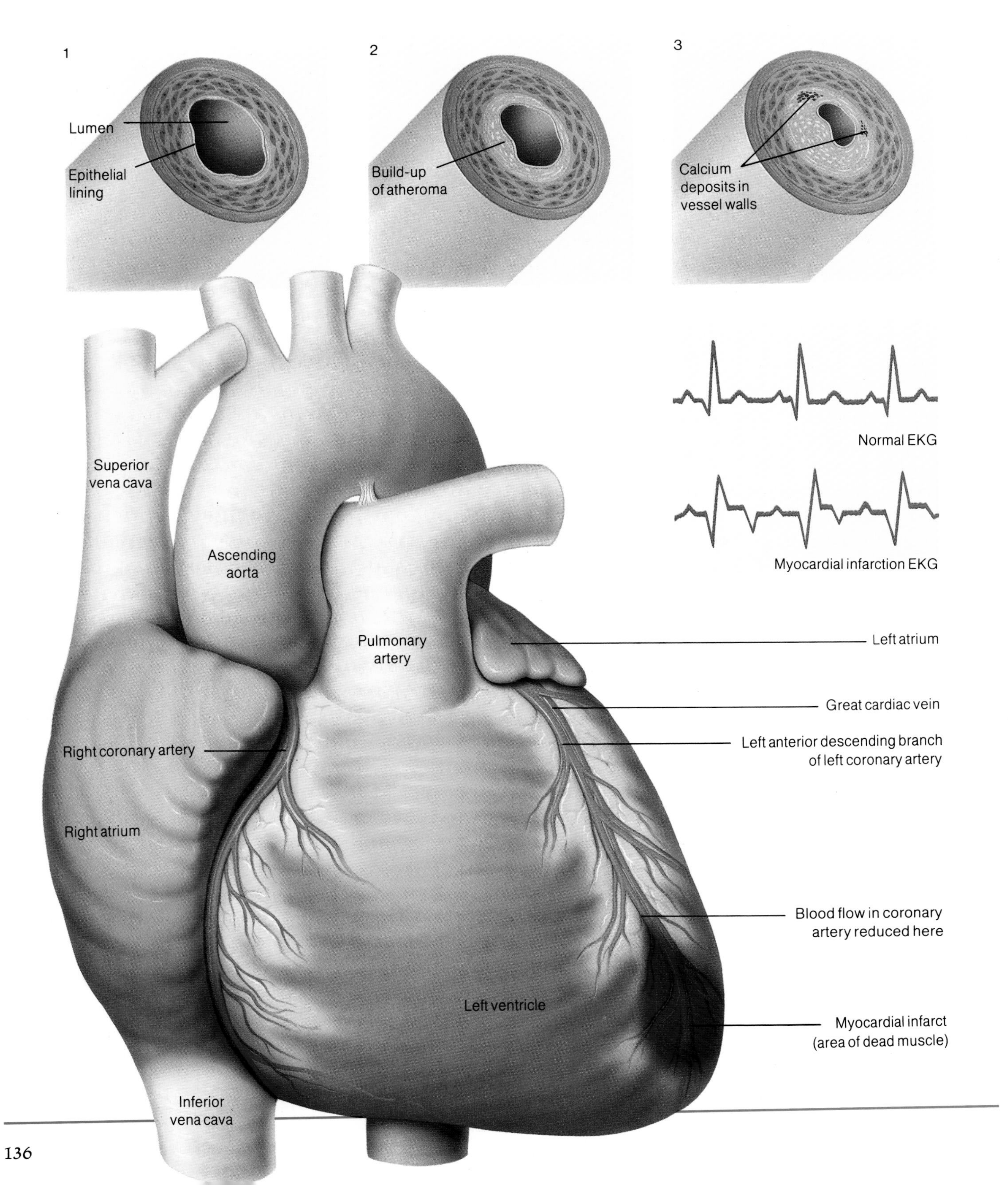

Unless action is taken, the cardiac fibers begin to die, an eventuality doctors refer to as infarction. Putting together the words for muscle, heart and tissue death creates "myocardial infarction," and this is used as one technical name for the common type of heart attack.

The heart may still be able to pump enough blood to keep the body alive. But there is chest pain from the afflicted heart tissue. The lungs fill with fluid; the kidneys cannot clear the bloodstream of waste; and the victim may be confused because the brain runs short of oxygen. These are symptoms of shock.

Advances in treatment are various and hopeful. Expert on-the-spot resuscitation and urgent administration of new drugs, such as the thrombolytics that dissolve away clots within minutes, and an old one aspirin are saving more and more lives. A long, thin tube (catheter) bearing a tiny blow-up sausage-shaped balloon can be passed into the coronary artery and judiciously inflated to massage away narrowings (coronary angioplasty). Coronary bypass surgery can graft in other body tubes (such as less essential veins from the legs) to carry blood around a coronary narrowing, as mentioned later. Nevertheless, the key to today's coronary epidemic sits in all our pockets; it unlocks the door by which we leave behind smoking, fatty foods, excess alcohol, obesity, inactivity and stress, to enter a new era of healthier diet, exercise and a sensible, relaxed approach to life.

The pump begins to fail

Coronary disease imposes great demands on the heart muscle. Cardiomyopathy is the generic term for disease of the cardiac muscle itself. It takes two principal forms: hypertrophic and congestive.

Hypertrophy, or "overgrowth," involves thickening of the cardiac fibers that make up the bulk of the heart wall. Sometimes the septum, the tough partition between the heart's left and right sides, swells and restricts blood flow out of the left ventricle to the body (the systemic circulation). The thickened wall may distort the cusps of the mitral valve so that it does not seal securely, allowing a leaky backflow of blood. Symptoms of this ailment include breathlessness, dizziness and irregular heartbeats noticed as palpitations.

In the more common congestive cardiomyopathy, the heart chambers enlarge while their muscular walls degenerate and fibrous, non-contractile tissue replaces the cardiac muscle fibers. As blood flow slackens through the enlarged heart, fluid builds up and pools in the lungs and other organs ("congestion") and the risk of blood clots increases.

As the heart's beat becomes weaker, congestive heart failure may develop. One or both ventricles are unable to stay the pace. In left ventricular failure, blood trying to enter the left pump from the lungs is forced to back up and accumulates in the lung tissue, "waterlogging" it and impairing respiration. As fluid accumulates around the system, one noticeable result is swelling of the ankles; the kidneys are also affected, reducing urinary flow. In right ventricular failure, blood that should be arriving on this side of the heart backs up in the main veins and raises the blood pressure in important internal organs such as the liver, spleen and kidneys. Various drugs can minimize these effects, although in severe cases another heart – a transplant – is the only answer.

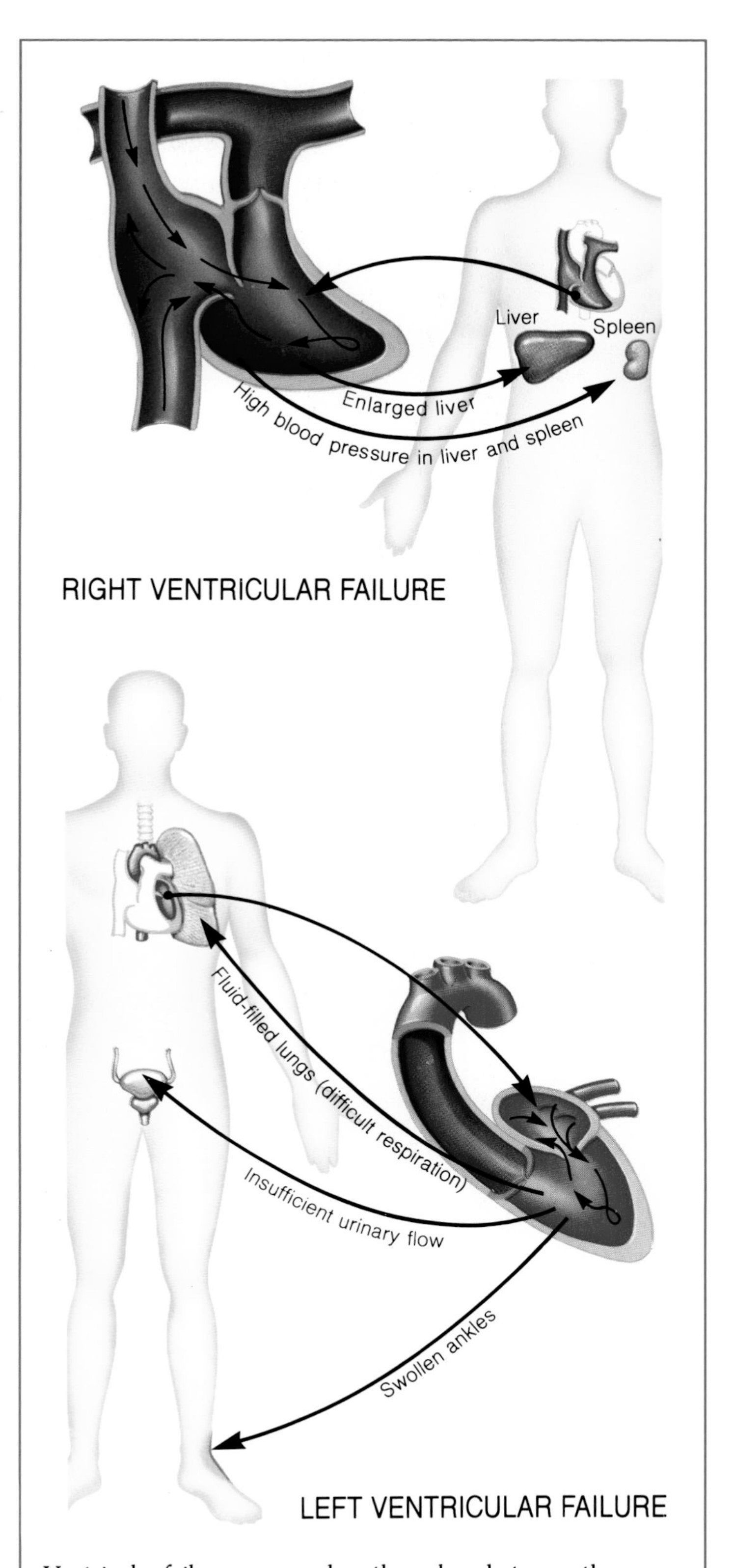

Ventricular failure occurs when the valves between the heart's upper chambers (the atria) do not close properly, so that as the lower chambers (the ventricles) contract, some blood is forced back through the atria toward the organs from which it came.

Infections in the heart

The heart can suffer ailments brought on by bacterial or viral infection. Pericarditis is inflammation of the pericardium, the baglike series of membranes around the heart. Infected by a virus, it becomes rough and thick, and excess fluid seeps into the pericardial space. The condition manifests itself as pain in the chest or left shoulder; most often occurring in young people, it is usually benign.

One of the most serious of past epidemics, now largely conquered by modern medicine, is rheumatic fever. Innocent enough to begin with, this bacterial infection (by a type of streptococcus) produces a sore throat or ear trouble. In some sufferers, however, the body fights the bacteria by manufacturing antibodies that, in a cruel twist of fate, may attack certain heart tissues.

Most vulnerable are the valves, particularly those that work hardest, the mitral and aortic valves on the heart's left side. Twinned disorders result. The valve cusps become inflamed and stiff, and their edges may partly fuse to narrow the opening and restrict blood flow in the correct direction; this is stenosis. The twin is regurgitation; on the backbeat, the scarred valve cannot seal fully, and so blood refluxes, or leaks back the wrong way.

Murmurs soft in the ear

The normal "lub-dub" sound that bounces up the stethoscope tube to the doctor's ear is accompanied by quiet swishes and swooshes of normally flowing blood. Valvular disease alters the usual pattern of sound, producing "murmurs." Some murmurs indicate a departure from the norm but do not necessarily signify a health problem; these are "innocent" murmurs. Others point to problems.

In aortic stenosis, for example, the left ventricle struggles to squirt blood through a narrowed, stiffened aortic valve. Internal ventricular pressure may rise to more than 300 millimeters of mercury (three times normal). As you might put your thumb over the end of a hosepipe to produce a narrow, powerful jet of water, so the stenosed aortic valve creates a "nozzle effect." Blood gushes at high pressure through the restricted opening into the aorta, setting up intense vibrations in its wall. The doctor, listening a couple of feet away, can hear the turbulent flow and vibrating vessel wall.

Murmurs can have four separate cadences, depending on the phase of the cardiac cycle and the valve or valves affected. Mitral regurgitation and aortic stenosis are audible during systole; mitral stenosis and aortic regurgitation are heard during diastole.

The severity of valve disease varies. Some people live for years with no pain, and perhaps only occasional discomfort. In most cases, however, the condition worsens with time. Fluid collects in the lungs (a condition known as pulmonary edema), leading to breathlessness, and the atrial chambers may stretch so much that the electrical signals sparking through their walls are disrupted, causing irregular heartbeats.

Flutter and fibrillation

If the wall of the left atrium is overstretched, the electrical impulses triggered by the SA node (the heart's own pacemaker) must travel a greater distance before they elicit contraction and fire onward to the AV node, bundle of His and Purkinje fibers. This opens up a new arena of heart problems: cardiac arrhythmias, the short-circuiting or interrruption of the electrical ripples that coordinate the heart's contraction.

In heart block, the ripples grind to a halt in the AV node, the relay station between atria and ventricles. Heart block is a question of degree. It can be intermittent, in which case every second or third wave of impulses passes through the AV node. Or it can be complete: when the ventricles tire of waiting for instructions from above, they take up their own, slower rhythm of contraction – termed ventricular escape.

An ectopic, or premature, heartbeat is often due to ischemia. Reduced blood flow to a patch of heart muscle gives it "cramp" and makes it impatient and irritable. It may depolarize on its own initiative, sending an extra and uncoordinated electrical ripple through the cardiac tissue, confusing its smooth rhythm of contraction. These impulses pose little danger provided they remain random and rare, perhaps in response to stimulant chemicals or excessive fatigue. The anxious heart simply skips the occasional beat. But should ectopic beats come in a lengthy sequence, interfering with the SA pacemaker and chronically confusing the pattern of conduction, there is cause for concern.

A flurry of such beats can sometimes build into arrhythmic frenzies called fibrillation or flutter. In atrial flutter, the upper chambers "tremble" at 250 to 350 times per minute, but maintain an orderly rhythm. In fibrillation, the regularity is lost and cells fire at random across the heart, producing chaotic shaking that may soon cause real trouble.

The malformed heart

In the womb, the intricate set of natural plumbing that is the heart emerges from twisting, turning, thickening and folding of vessels. In about one baby in 200, something goes awry in this process. Tubes link up incorrectly perhaps with serious consequences for blood flow. These problems are known as congenital heart defects.

The causes of congenital cardiac disorders remain uncertain. In a tiny fraction of cases, faulty genes inherited from ancestors give the wrong plumbing instructions. In other instances, there may be poor nutritional supply (such as shortage of an essential vitamin) during the critical two-month development period after conception. Infection by microbes such as rubella (German measles) virus or administration of teratogenic (embryo-harming) drugs may also have an impact.

The fetal blood circulation makes scant use of its pulmonary circuit, to the lungs. In the womb, the lungs are filled with amniotic fluid, not air, and so no oxygen enters the unborn baby by this route; the precious dissolved gas is obtained from the mother, through the placenta. Before birth, a short-circuiting and very short tube, the ductus arteriosus, channels most of the blood bound for the lungs into the aorta instead. At birth, this duct should close within a few days; if it persists as a throughway, termed a patent ductus arteriosus, the baby suffers respiratory difficulties.

The development of the septum, the tough wall separating right pump from left, may form in error. A passage between left and right atria (atrial septal defect, ASD) permits blood to drain from the left atrium (where the pressure is greater) to the right, leading to the latter's enlargement. Ventricular septal defect (VSD) mimics the effect of patent ductus arteriosus, allowing freshly oxygenated blood in the left chamber to mingle with deoxygenated blood in the right.

Some congenital heart defects are the cause of the so-called "blue baby" complexion, which results because wrongly circuited blood is not able to nourish all the body tissues with oxygen. Great strides have been made in the past quarter century in the field of surgically repairing such defects. Judicious combinations of drugs, blood exchanges (some of them intra-

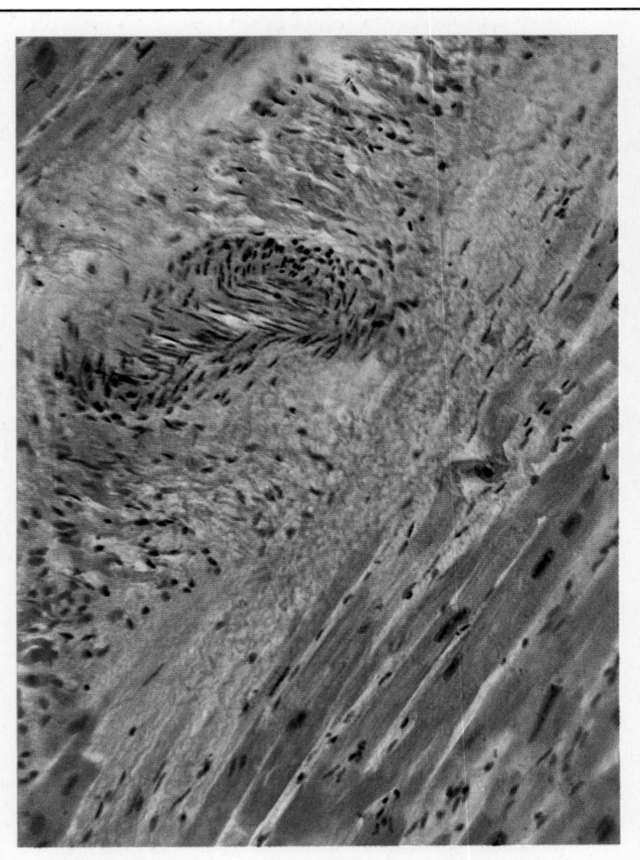

Rheumatic fever damages heart valves, which can become inflamed and scarred as a result. The microphotograph *(left)* reveals how inflammation in heart muscle disrupts its striated structure like a knot in a piece of straight-grained wood. Affected valves become hardened by scarring and may fuse together, so that they become less flexible and cannot close to form a watertight seal. This condition, known as stenosis, can affect the mitral or aortic valves *(below)* and is often first detected as a heart murmur, which is the sound of turbulent blood squirting through a narrow opening or sloshing back through a leaky valve.

A physician can hear a murmur using a stethoscope, or make a chart recording of it with a phonocardiograph. Not all murmurs are dangerous, but if valve disease is severe it may be treated by corrective surgery or by fitting an artificial replacement valve.

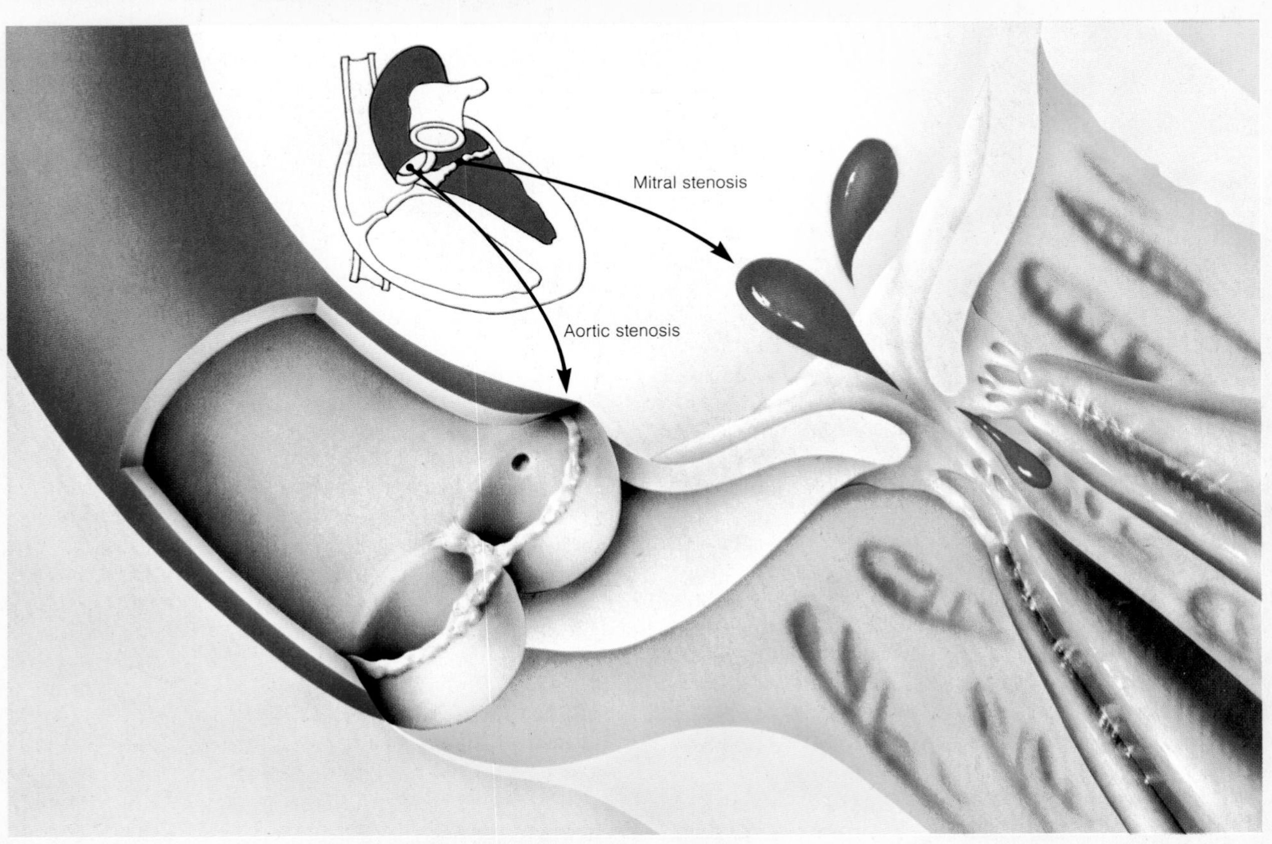

NORMAL CIRCULATION

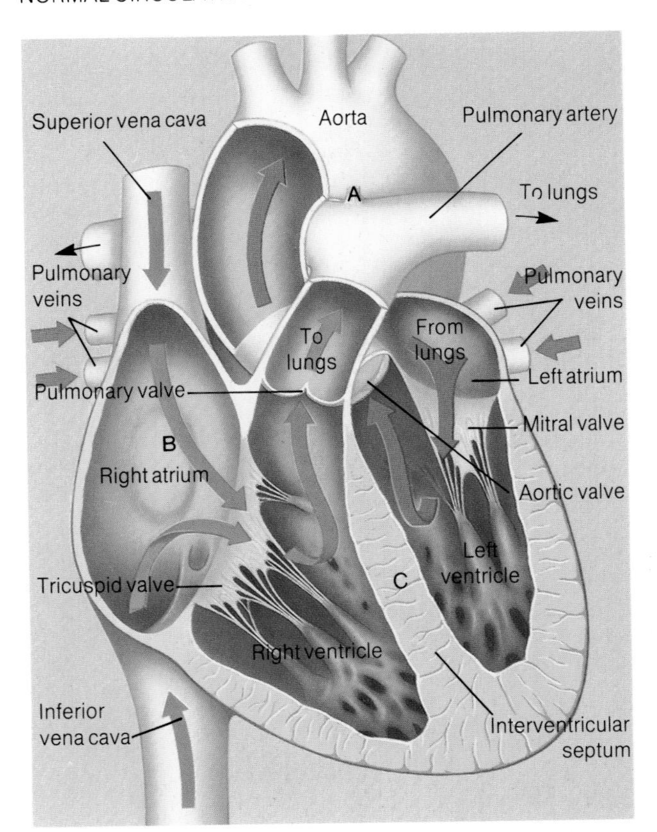

A PATENT DUCTUS ARTERIOSUS

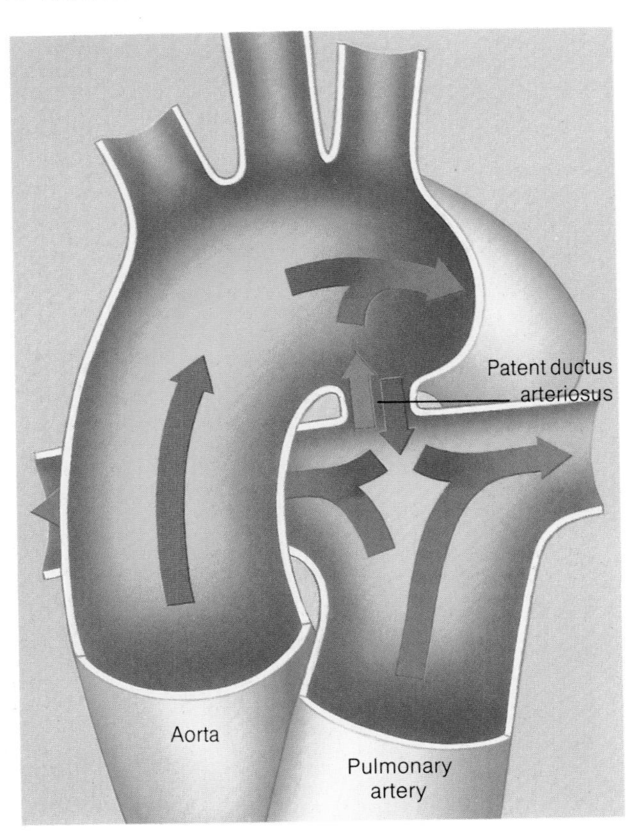

Effectively two pumps side by side, a normal heart *(above left)* ensures that venous and arterial circulations are kept entirely separate. Trouble occurs when abnormal connections allow the two types of blood to mix, so that the heart has to work harder to get enough oxygenated blood to the tissues. Many such heart defects are congenital, although they may not be obvious at birth.

Patent ductus arteriosus (A) occurs when the channel that allows blood to bypass the

uterine) and step-by-step operative repair of the faulty tubes, holes and connections have rendered congenital heart disorders a largely treatable complex of malformations.

Bypassing the heart

Aristotle once proclaimed: "The heart alone . . . cannot withstand injury." In the 1890s, British surgeon Stephen Paget declared: "Surgery of the heart has probably reached the limits set by Nature."

Today, surgeons treat the heart, if not with less respect, then perhaps with less reverence. In the 1950s cardiopulmonary bypass equipment (the "heart-lung" machine) freed the heart from its workload for an hour or two, so that it could be manipulated, incised and restructured with less difficulty. In 1952 the first artificial heart valve, a small ball inside a plastic tube, was successfully sewn in place. Improvements to valve design, including the silicone ball that bobs up and down inside a metal cage, came thick and fast. Today, an artifical heart valve is a routine piece of replacement equipment.

In the same year, the first artificial pacemaker sparked the heart into action every second or so. Intended as a temporary device, it was switched off after 52 hours, when the patient's heart began beating on its own once more. In 1960, the first fully implantable pacemaker was installed in a patient's chest wall. In 1967, the first coronary bypass operation utilized a length of saphenous vein from the patient's leg in order to replace a diseased section of coronary artery.

On 3 December 1967 the world caught its breath. South African heart surgeon Christian Barnard carried out the first heart transplant at Groote Schuur Hospital, Cape Town. The patient was Louis Washkansky, 54 years old and diabetic. His life was prolonged by 18 days. Today, heart transplants make only minor headlines, and patient survival is measured in years, not days. The spell of the heart, once worshipped and still romanticized, is fading at last.

Heartening news

Successful though heart transplants are, they still have to deal with the problem of rejection – the defensive action of the body's immune system which regards the transplanted heart as "foreign" tissue and attempts to destroy it. The problem is tackled by using a donor heart that is a good match to the tissue

B ATRIAL SEPTAL DEFECT

C VENTRICULAR SEPTAL DEFECT

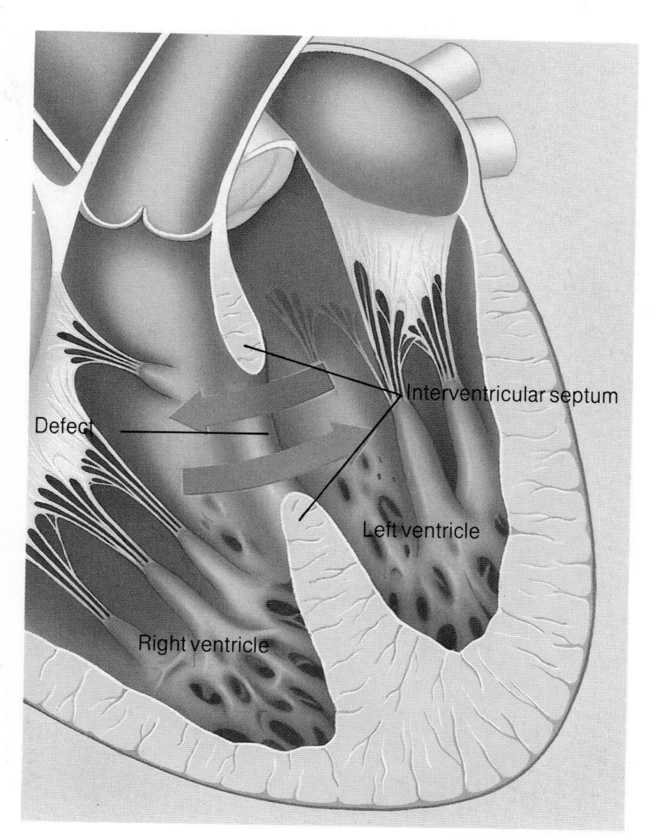

lungs in the fetus does not close when the baby is born. A septal defect is a hole in the partition, or septum, which separates the two sides of the heart. It occurs when the two ridges of tissue that grow together to form the septum do not meet properly in the developing fetal heart. In an atrial septal defect (B) the hole is between the upper chambers, and in a ventricular septal defect (C) it is between the lower chambers. Usually all of these conditions can be corrected by surgery.

type of the recipient, and by administering immunosuppressive drugs to the recipient.

An entirely different treatment for somebody with a failing heart is to provide an artificial replacement. Pioneered by American scientist Robert Jarvik, the first artificial heart was implanted by surgeon William de Vries into a 62-year-old dentist from Seattle, Barney Clark.

One problem with the Jarvik heart is that blood tends to stick to its inside surfaces, and then become detached and form small blood clots. Anticoagulant drugs minimize the effect, but even so there is always a risk that a clot will travel to and block an artery in the brain and cause a stroke. William Pierce overcame this problem with the Pennsylvania State Heart, which had seamless, smooth chambers for blood, and valves made entirely of plastic. It also incorporated an automatic blood-flow regulator, allowing the pumping rate and therefore the blood supply to vary to match the needs of the body.

An artificial heart, slightly larger than a natural one, consists of two separate, identical pumps, joined together by Velcro patches. It is implanted into the recipient's chest, and powered by compressed air, which is pulsed to move a membrane that forces blood through the device. One-way valves ensure that the blood flows only in the right direction. But there is one major drawback. The external drive system and timing device together weigh 375 pounds and are connected to the mechanical heart by two plastic tubes six feet long. Movement is severely limited because wherever the recipient goes, all this equipment must go too. Current research is being carried out into developing much smaller power units, which will be carried like a shoulder bag and give the recipient true mobility.

Dr Jarvik estimates that 25,000 people will be fitted with artificial hearts by the mid-1990s. At present, transplanted hearts have a better survival rate than mechanical ones. A way of making the best use of both methods is to use an artificial heart as a stopgap, to keep the patient alive while waiting for a well-matched natural donor heart for transplantation.

The first person to benefit from this approach was 25-year-old Michael Drummond, who in 1985 received a Jarvik heart – to replace his own which had been wrecked by a virus disease. He did not have long to wait for a good match to become available, and nine days later the artificial heart was replaced by transplanting a natural one from a donor.

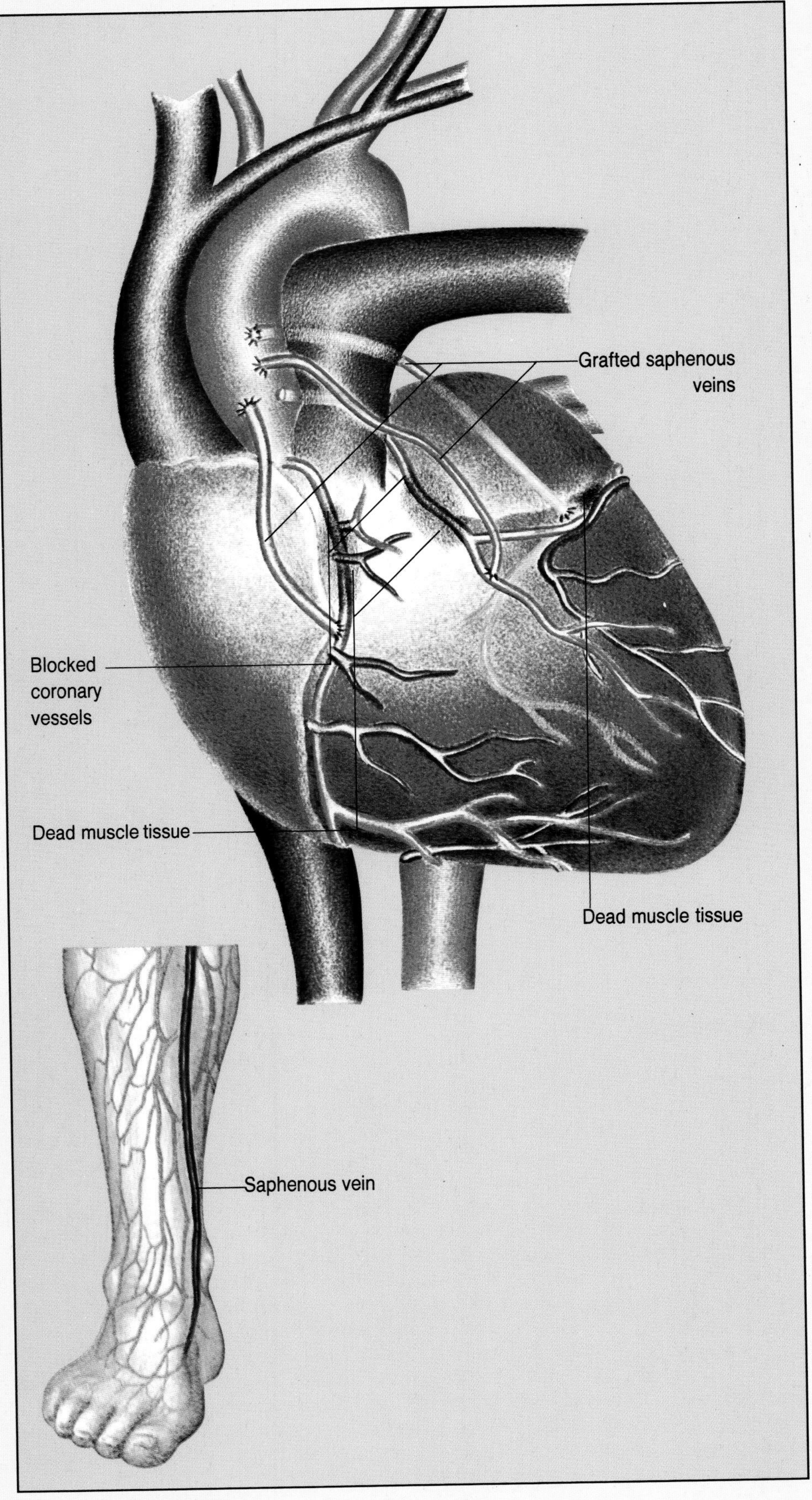

A coronary bypass operation
Millions of people suffer from coronary heart disease. A coronary artery blocked by a buildup of fatty deposits starves part of the heart muscle of blood and hence oxygen, causing the pain of angina or leading to coronary infarction and a heart attack. If drug treatment fails, the highly successful coronary bypass operation may be performed. Segments of the saphenous vein in the leg, where veins are plentiful, are removed and used to bypass blocked or damaged coronary arteries. One end of the graft is joined to the aorta (to pick up a supply of blood), and the other end is spliced into the coronary circulation beyond the point of obstruction. A cardiac surgeon performing the operation *(opposite page)* uses a low-power microscope, draped in sterile plastic, to give a magnified view of the coronary blood vessels. Since the operation was perfected in the late 1960s, it has been successfully performed more than half a million times in the United States alone.

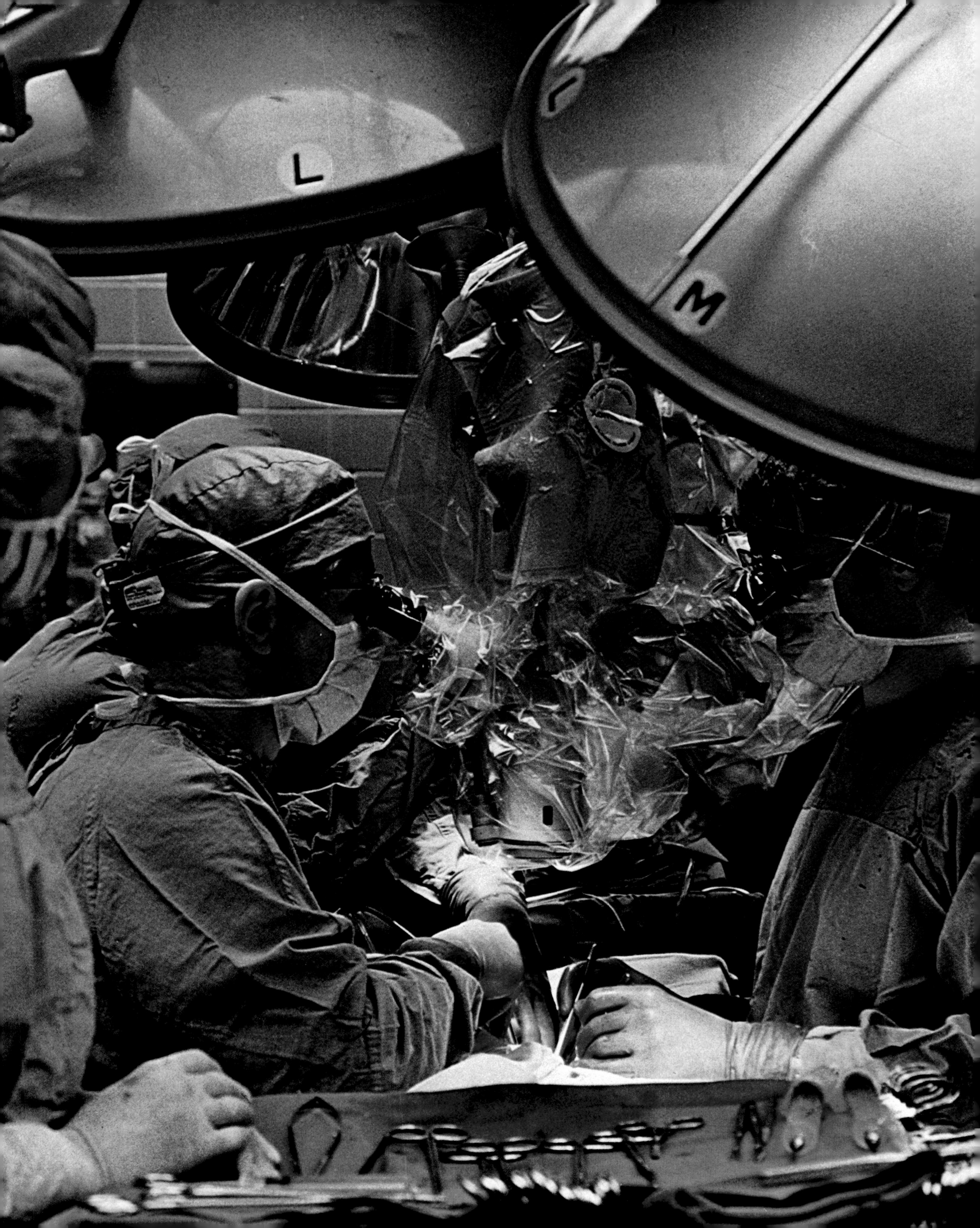
L
M

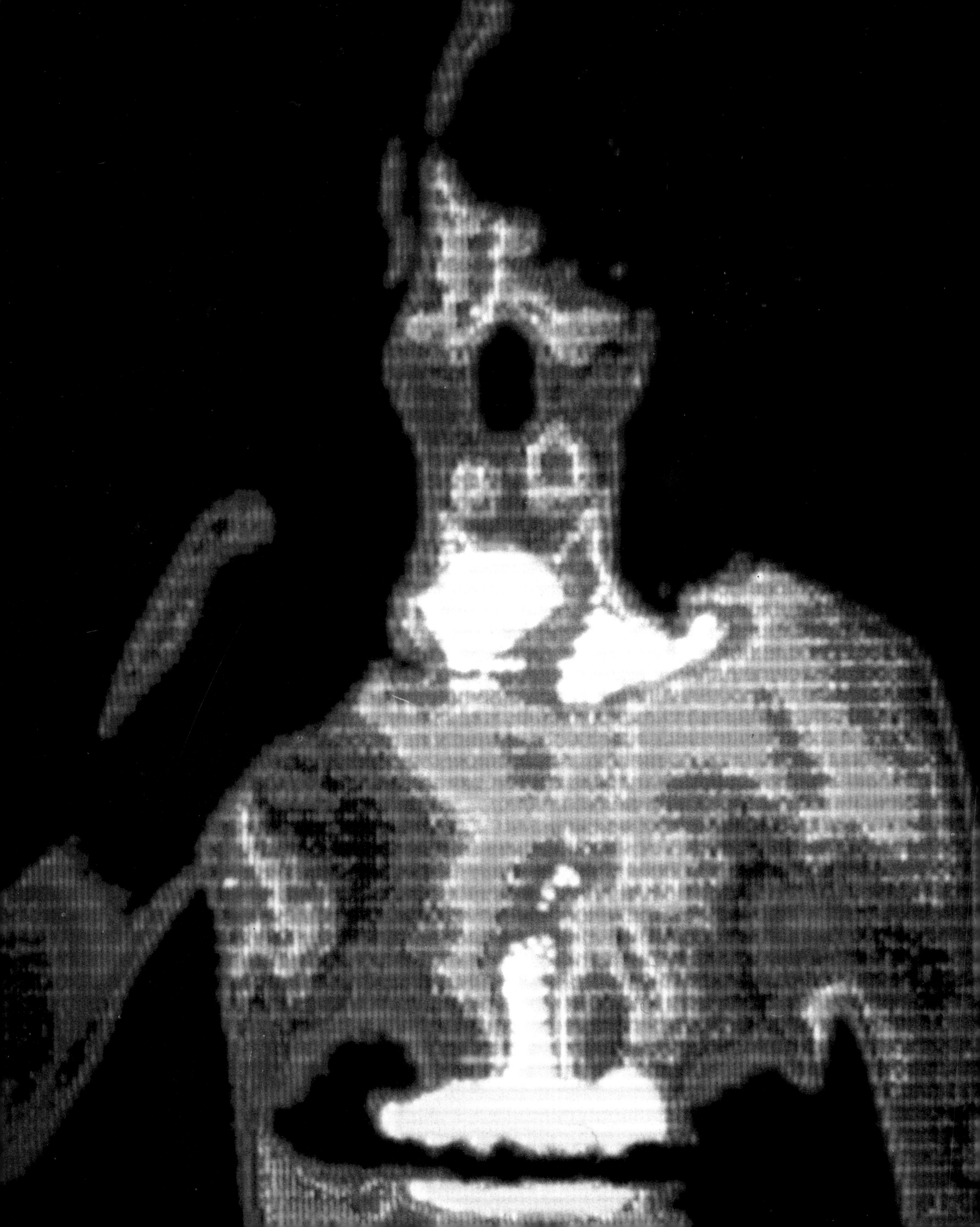

CHAPTER 9

Digestion

Most people enjoy eating, and food is a popular topic of conversation. It is also a popular topic with writers and philosophers. François Rabelais, 16th-century French cleric and renowned scholar, in his epic and satirical *Gargantua et Pantagruel*, penned in anticipation: "I drink for the thirst to come ... Appetite comes with eating ..." At the other end of the spectrum, it is difficult to argue with the deadpan realism of Ludwig Feuerbach, 19th-century German philosophical materialist: "A man is what he eats."

On the outside, we call it food. It may be a crust of bread or a lavish 15-course banquet, but the digestion shows no favors. A healthy human eats about half a ton of it each year. If this food is to be of value to the body, rather than merely pleasing to the taste buds, then it must be digested. The system that performs this complex cascade of chemistry is in essence a tube, more than 25 feet long.

The digestive tube is twisted, swollen and shaped along its length into several distinct regions: mouth, pharynx (throat), esophagus (gullet), stomach, ileum and colon (small and large intestines), rectum and anus. Food passing along the system is mixed with a staggering laboratory of chemicals, which break it down by means of a multitudinous series of interlocking reactions into small molecular units. These can be absorbed into the body, chiefly into the blood and lymph systems.

On the inside, the nutritive molecules succumb to a variety of possible fates, depending on the body's long-term needs and the necessities of the moment. Some molecules act as energy sources that power the biochemical gyrations of every cell; others become building-blocks, for construction of tissue during growth, or in reconstruction of repaired or worn-out tissues. In times of plenty, certain molecules may simply be set aside until times of shortage – one of the body's many forms of life insurance.

Not all we consume is digestible. Wastes remain, and the system must package and dispose of them in an efficient way. It may be the less glamorous end of the tale, but no less important for the system's functioning – and so just as vital for good health.

Drink a cup of hot coffee and feel the warmth. This computer-generated thermograph shows the warmth – white and red are hottest, blue and purple coolest. One mouthful of hot liquid has just arrived in the stomach, while the next is caught in the act of being swallowed. Thermography allows specialists to study the digestive system in action. Together with other techniques it provides valuable insight into the convoluted journey food makes as it is being transformed into the substances that fuel the body.

Regional specialization

The organization of the human digestive system – often more economically termed "the gut" – is perhaps best appreciated by following its formation in an embryo. In its earliest history, buddings from the primordial gut generate such diverse structures as the middle ear, tonsils, several glands – including the parathyroids, thymus and thyroid – as well as the lungs, liver and pancreas.

During the third week of embryonic life, sheeting, folding and curling of cell layers form the basis of the human gut. The embryo is hardly recognizable, but it does have a bulging head fold, a middle section and a tail fold. Inside the head is the foregut, a part of the yolk sac that will form the important

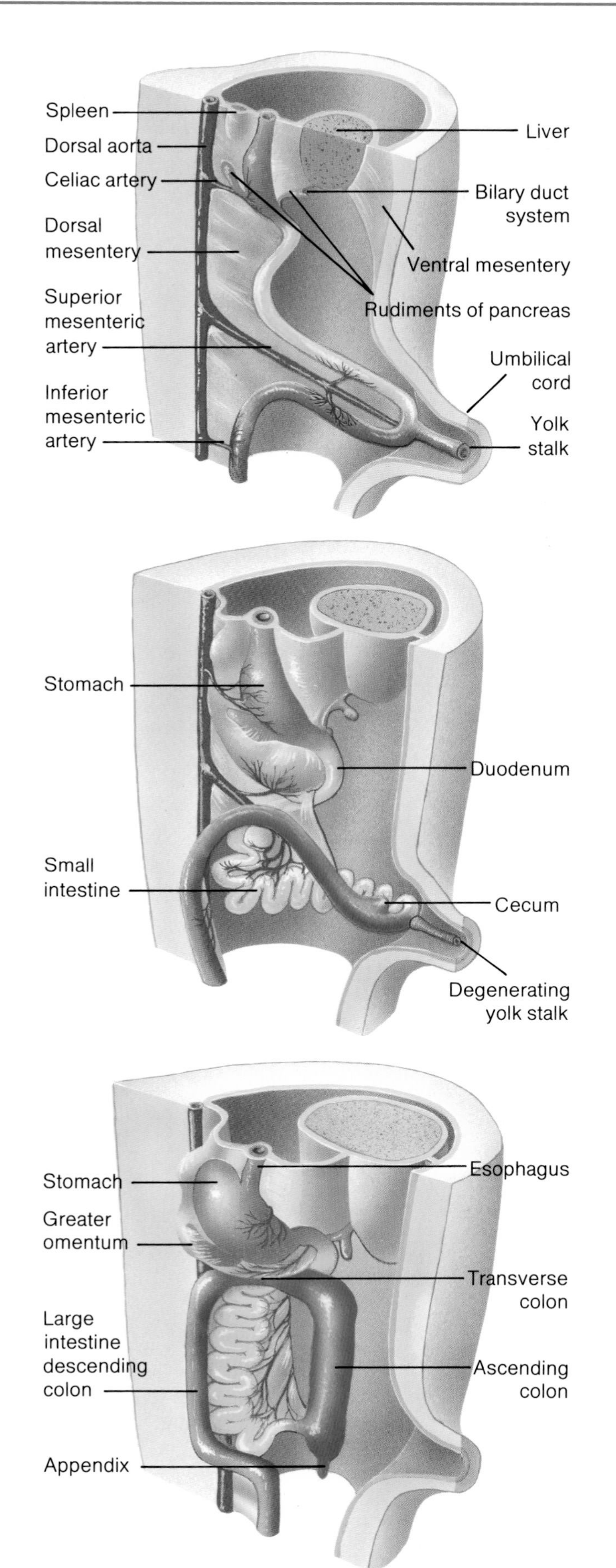

digestive organs from the mouth along to the liver. In the central section, the midgut takes over and as it develops it will shape the system from liver to colon. The tail fold encloses the hindgut, which in both senses brings up the rear.

Tunnels, pits and pouches

A couple of samples aptly demonstrate the minuscule contortions of the developing gut, which take place in an embryo that is only about the size of a grain of rice. There is no mouth – at first. Gradually, however, the anterior (front) cavity of the foregut, in the middle of the head, extends itself toward what will become the facial region. Simultaneously, a small pit forms in the central part of the head's surface. This dimple deepens and bores its way inward, to meet the foregut, which is excavating outward. Eventually the two join, and one of the gut's connections with the exterior is established.

During the fourth and fifth weeks of embryonic development, a set of four grooves, the pharyngeal pouches, appear along the side walls of the foregut, just behind the joint just described. At the same time, inpushings from the embryonic "neck," the pharyngeal clefts, deepen as though to meet the pouches. These strange structures are considered to be evolutionary echoes from our distant past, when our fishlike ancestors had gill slits which communicated from the interior of the throat to the outside, at the neck. Millions of years of gradual change have fashioned different fates for these pharyngeal structures.

The first pharyngeal pouch on each side is destined to form the middle ear. Adults retain a remembrance of this origin in the Eustachian tube, a tiny tunnel that connects the middle ear cavity to the back of the throat. Pouch number two makes the primordium of the tonsil. Pouch three lends itself to the development of the thymus and parathyroid glands. The front fork of subdivided pouch four links with pouch three to contribute to the parathyroids; its rear fork becomes integrated into the thyroid gland.

Meanwhile the midgut elongates dramatically, since its destiny is to form the long loops of the intestines. By ten weeks after conception a bulbous swelling near its front end foreshadows the stomach; the ileum has folded itself into a long series of curves, while the colon has increased its diameter and taken up an inverted U-shape, to frame the tiny abdomen.

The hindgut has the simplest series of changes during embryological growth, for it eventually forms the rear part of the colon, the rectum and the upper portion of the anal canal. In

During the second month of life in the womb, the tiny human embryo's digestive tract begins to take shape from a series of ingrowing folds and pouches *(left).* The stomach and small intestine are first to form, followed by the large intestine, which gradually regionalizes into its ascending, transverse and descending sections. The liver, pancreas and other accessory organs develop at the same time. The hindgut – the exit of the digestive tract – is last to be completed.

The structures of the mouth and organs of the throat and neck develop in an embryo from the clefts of the pharyngeal pouches *(right).* The very simple origins of the structures in a five-week embryo can be seen in the upper two diagrams, which show a vertical section and an oblique horizontal one (along the red dotted line). Over the next weeks, cell clusters migrate and grow to lay down the basis of the tonsils and the thyroid and thymus glands in the neck (lower two diagrams).

Five week embryo

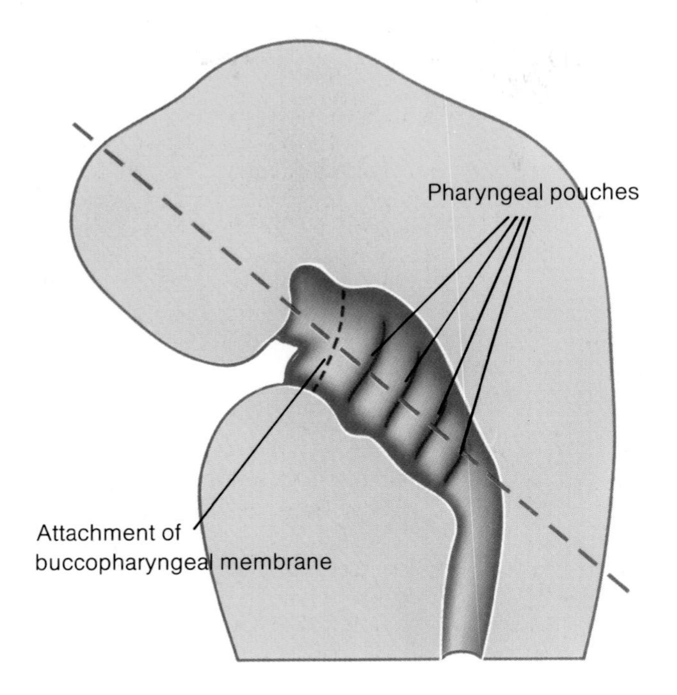

Schematic horizontal sections

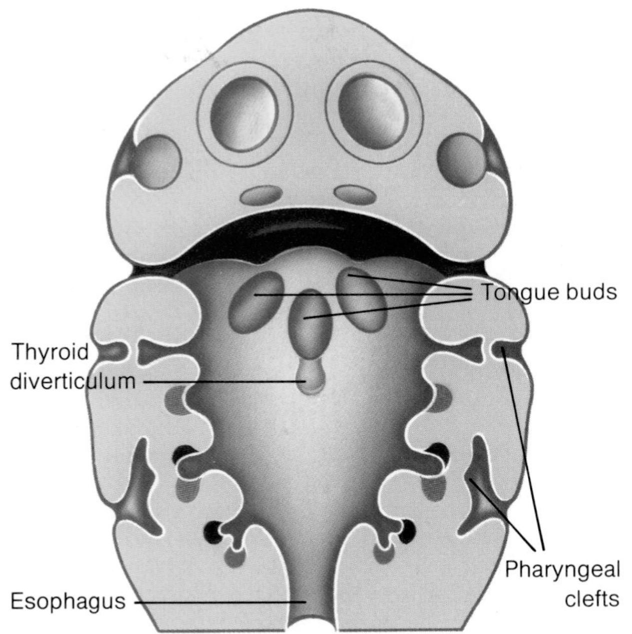

Six weeks

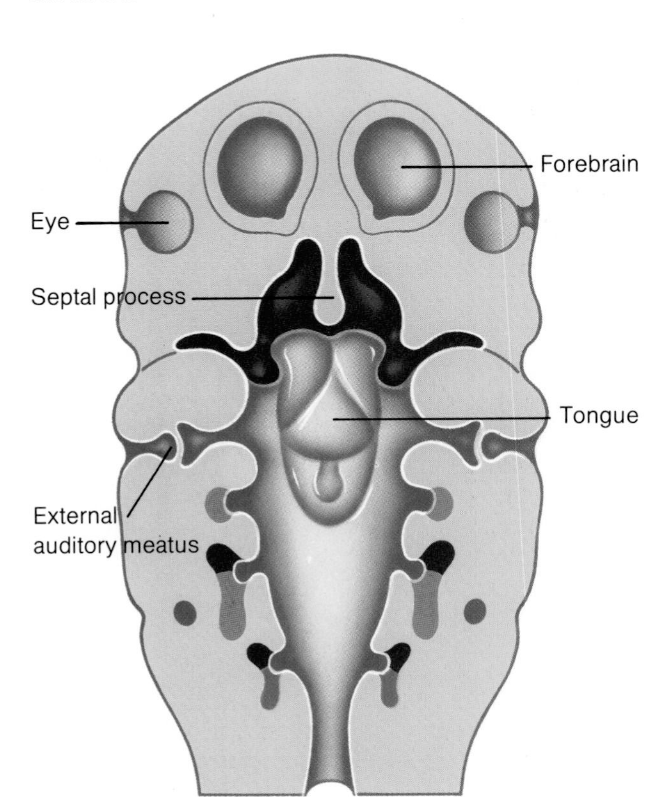

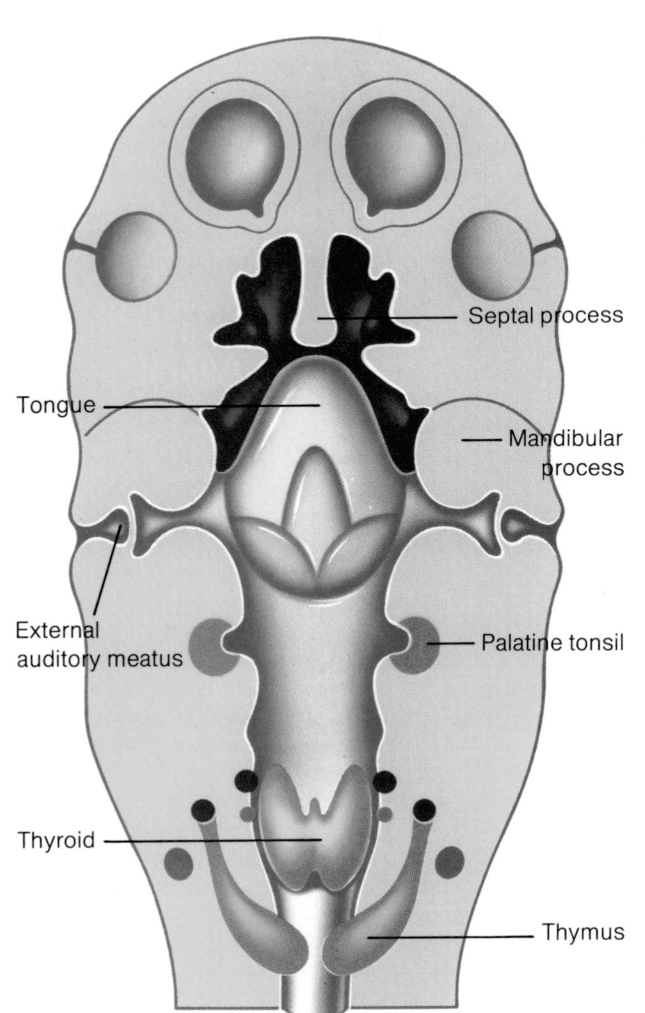

addition, it contributes to the formation of the urogenital organs.

Regional specialization is almost complete and the finishing touches are now in progress. The fetus is still less than three months old, and less than two inches long. Yet there is a wait of another five or six months before the system can be put to its first real test, when the newborn baby suckles at its mother's breast.

Essential fuels

A striking feature of human digestion is the wide range of foodstuffs on which it can survive and prosper. In zoological terms, humans are perhaps best described as omnivores with a definite leaning toward herbivory. Our adaptable guts, as well as our inventive brains, have allowed this species of large primate to spread across the globe and utilize for food whichever edible plants and animals were present in the region. Today, a significant proportion of the world's five billion human digestive systems exist on a relatively restricted, localized diet – most people eat what they grow. In the more affluent countries, people are able to sample foodstuffs imported from various places around the world.

The three principal constituents of food are proteins, carbohydrates and fats. The first group of substances is used essentially for construction and repair, while the second fuels the body's processes and functions; the third does both, building and fuelling. In addition the body must have vitamins, which are essential for normal growth and development, and minerals, which are substances required in tiny amounts to assist in many body processes such as nerve and muscle function.

Paradoxically, the diet as a whole should also contain material that is not actually digestible, such as the cellulose in plants. It may be called "fiber," "roughage," "bulk" and various other names. The usefulness of fiber in the diet is slowly becoming understood, in that it seems to contribute physically and chemically to a healthily efficient digestive system. Its absence is currently linked to a variety of intestinal ailments, from constipation to colonic cancer.

The final and central dietary requirement forms the medium in which all else happens. It contains no energy, no building blocks, no vitamins, no fiber. It is plain water.

Building with proteins

About one-sixth of your body weight consists of proteins, which make up the main part of tissues such as muscles, skin, hair and nails. All proteins contain the element nitrogen, as well as the carbon, hydrogen and oxygen found in many other organic molecules. They are big molecules: one protein molecule may contain thousands of atoms, and the larger ones can be seen individually using today's advanced electron microscopes. There are two main types of protein, structural and functional.

Structural proteins are the body's building blocks, scaffolding, girders, cladding plates and tiles. Functional proteins are the enzymes, which are the chemical handlers and regulators – the construction workers themselves. All the thousands of processes upon which human life depends are controlled by enzymes, and all enzymes are functional proteins. (Enzymes are especially prevalent in the chemistry of digestion, where they act as organic catalysts to break up large food molecules into ones small enough to be absorbed.)

Each protein eaten in food is made of a chain of subunits, called amino acids, strung like beads on a necklace. A small protein may have one or two dozen amino acids; a large one has many

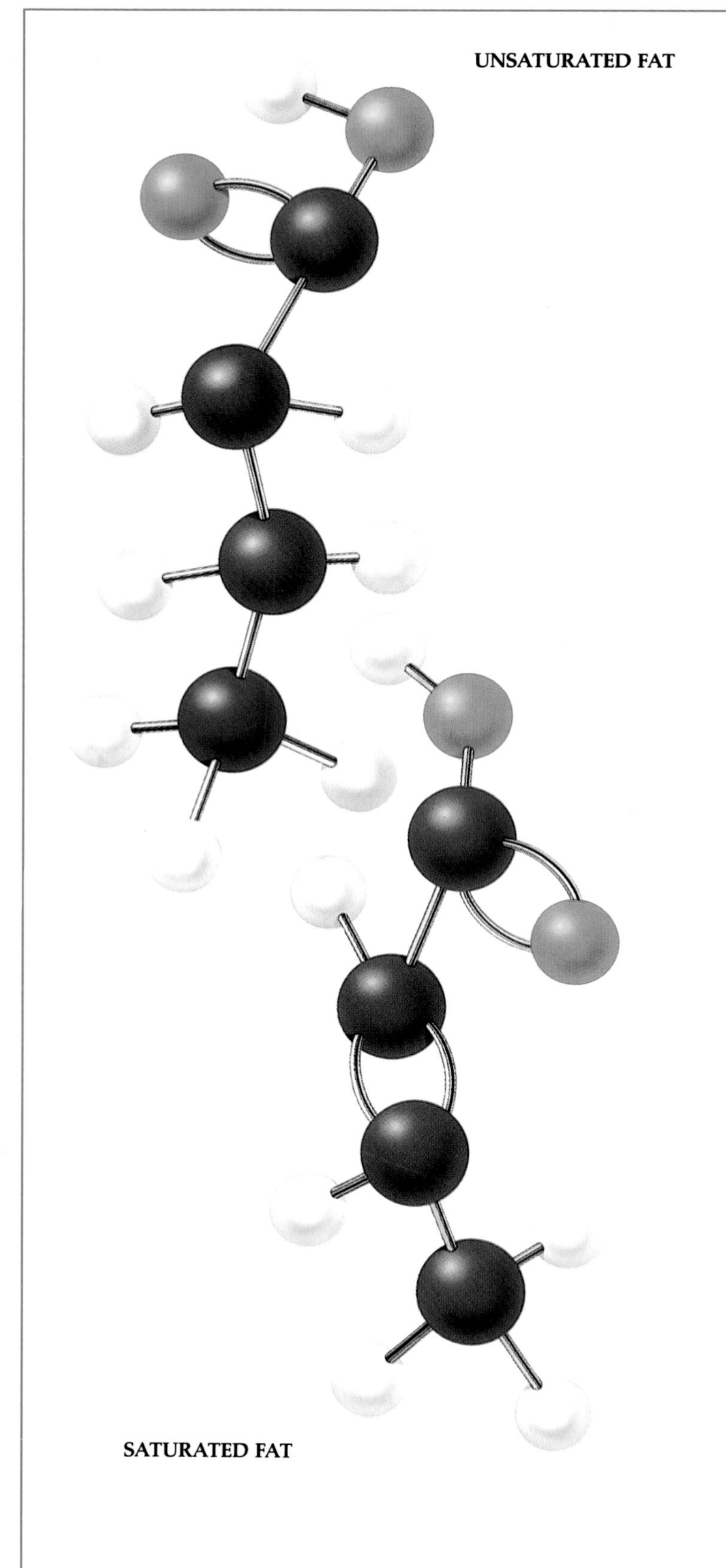

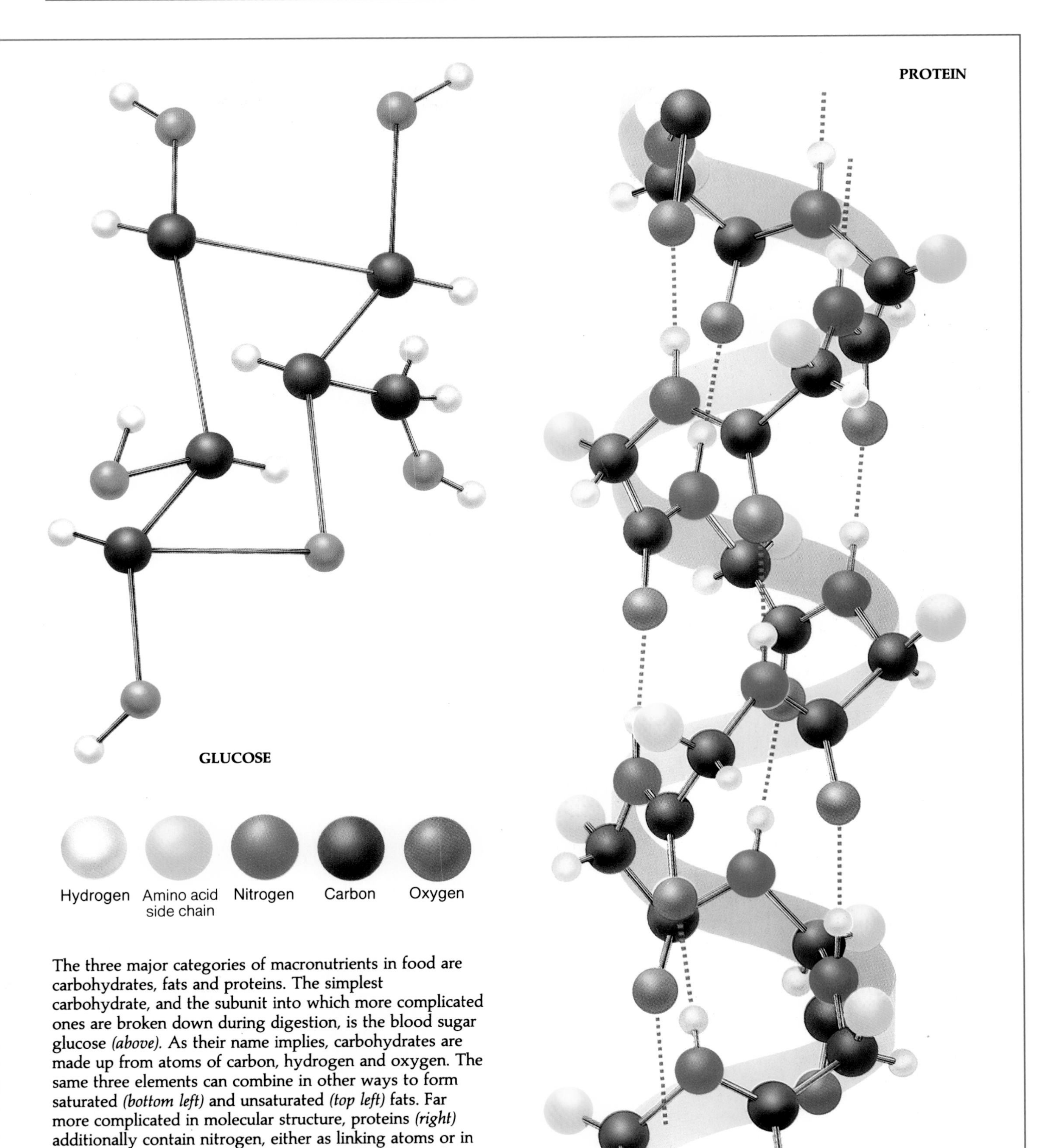

The three major categories of macronutrients in food are carbohydrates, fats and proteins. The simplest carbohydrate, and the subunit into which more complicated ones are broken down during digestion, is the blood sugar glucose *(above).* As their name implies, carbohydrates are made up from atoms of carbon, hydrogen and oxygen. The same three elements can combine in other ways to form saturated *(bottom left)* and unsaturated *(top left)* fats. Far more complicated in molecular structure, proteins *(right)* additionally contain nitrogen, either as linking atoms or in amino acid side chains. Typical protein molecules are long spiral chains, suited to the fibrous structures they form.

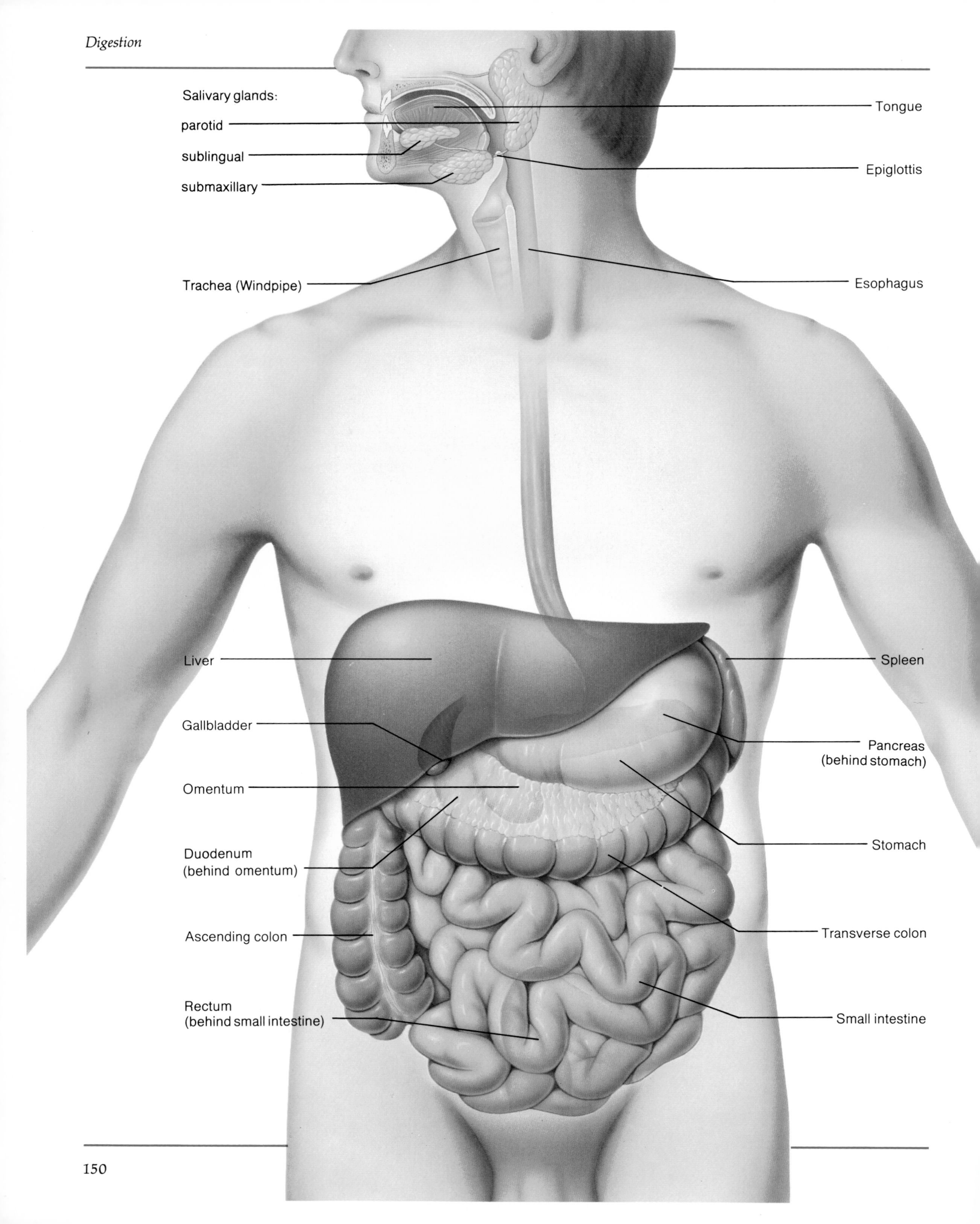
Salivary glands:
parotid
sublingual
submaxillary
Tongue
Epiglottis
Trachea (Windpipe)
Esophagus
Liver
Spleen
Gallbladder
Pancreas
(behind stomach)
Omentum
Stomach
Duodenum
(behind omentum)
Transverse colon
Ascending colon
Rectum
(behind small intestine)
Small intestine

The alimentary canal extends from the mouth to the anus *(left)*. The first part, as far as the stomach, is fairly straight; the rest has coils and loops.

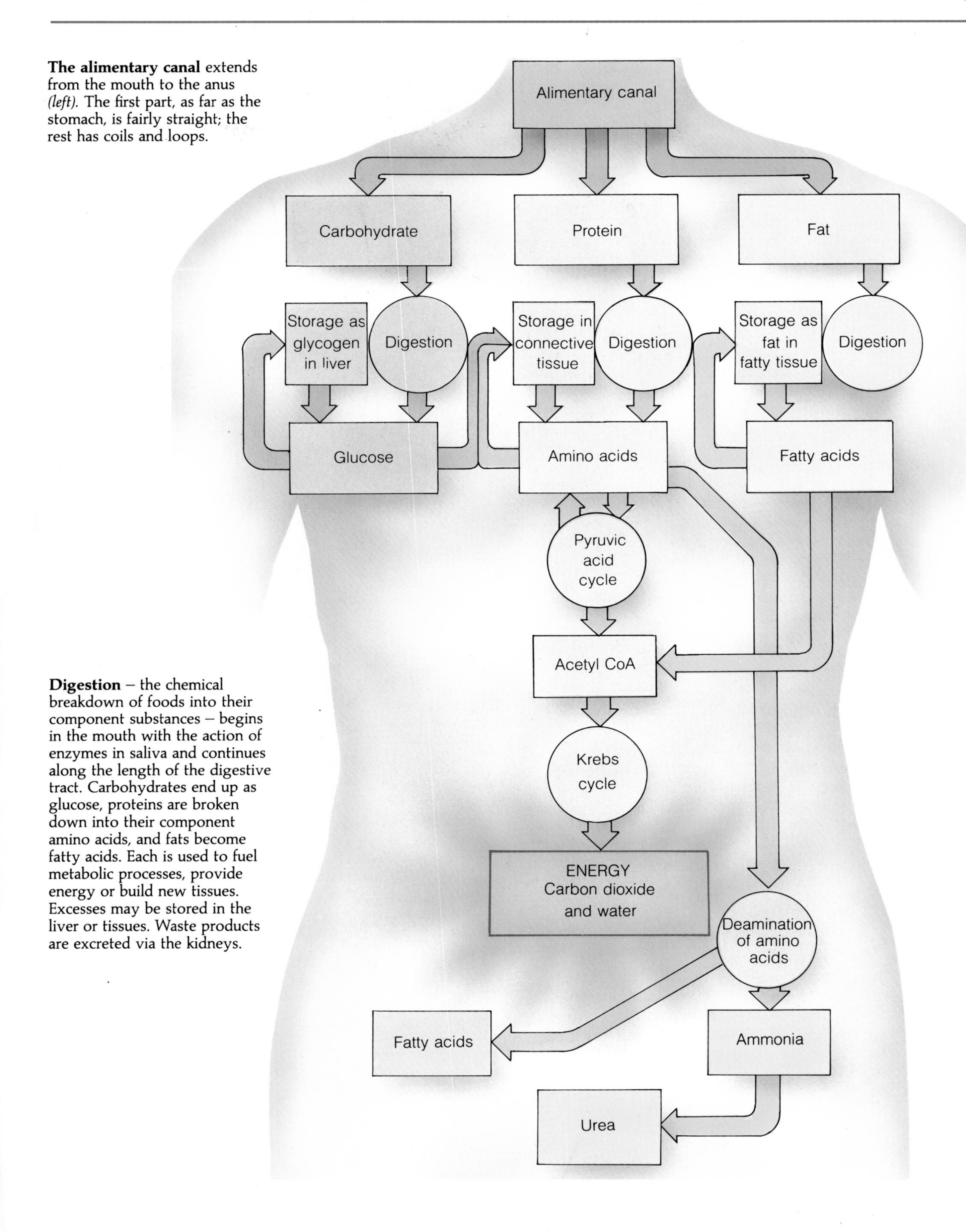

Digestion – the chemical breakdown of foods into their component substances – begins in the mouth with the action of enzymes in saliva and continues along the length of the digestive tract. Carbohydrates end up as glucose, proteins are broken down into their component amino acids, and fats become fatty acids. Each is used to fuel metabolic processes, provide energy or build new tissues. Excesses may be stored in the liver or tissues. Waste products are excreted via the kidneys.

hundreds. In the entire animal kingdom there are only about 20 different amino acids. Yet these can be linked and repeated in endless combinations, in the way that the 26 letters of the English alphabet may be shuffled to make the tens of thousands of words in common usage. Each protein, like each word, has individual "meaning" in its structure.

Fuel for the furnace

Sugar, and especially glucose sugar, is the body's preferred energy source. Chemically, carbohydrates are combinations of many individual sugar molecules, the smallest and simplest of which is glucose. The name "carbohydrate" derives from its constituents: carbon, hydrogen and oxygen.

Enormous chains of sugars congregate to make up those very large carbohydrates, the plant starches, which are eaten in the form of rice, potato, grains and many other plant parts. During digestion, these huge chains are snipped up into smaller ones that can be absorbed and either burned, to fuel cellular metabolism, or reassembled and stored in the energy closets in the form of glycogen "animal starch").

Fats, oils and lipids

Fats are the second form of fuel for the metabolic furnace. They are usually regarded (particularly in the kitchen) as semi-solid but easily meltable substances, such as butter and lard; oils are seen in a similar way, but they are more liquid – although still viscous – at room temperature. To resolve any confusion, the preferred scientific name for the group is "lipids."

Just as proteins are made up of amino acids, and carbohydrates consist of glucose subunits, so fats are combinations of subunits. Fatty acids are basically long strings of carbon atoms with hydrogen attached. There is usually a chemical "handle" at one end of a fatty acid, which is an organic acid (similar to that in an amino acid) hooked up to a small molecule called glycerol, or glycerine. A single glycerol can take in tow three fatty acids, and for this reason dieticians speak of the types of fats found in our food as being made up of triglyceride subunits.

Lipids are for building and burning. Every cell in the body depends on the lipid content of its cell membrane to preserve the structural and functional integrity of this cellular "skin." Second, fat provides the most concentrated possible energy source within the diet. Its nutritional calorie content (a measure of energy obtainable per unit weight) is twice that of carbohydrate or protein. But too much fat is too much to handle: excess of certain types, particularly animal-derived fats, is implicated in the furring up of arteries and the risk of heart attack.

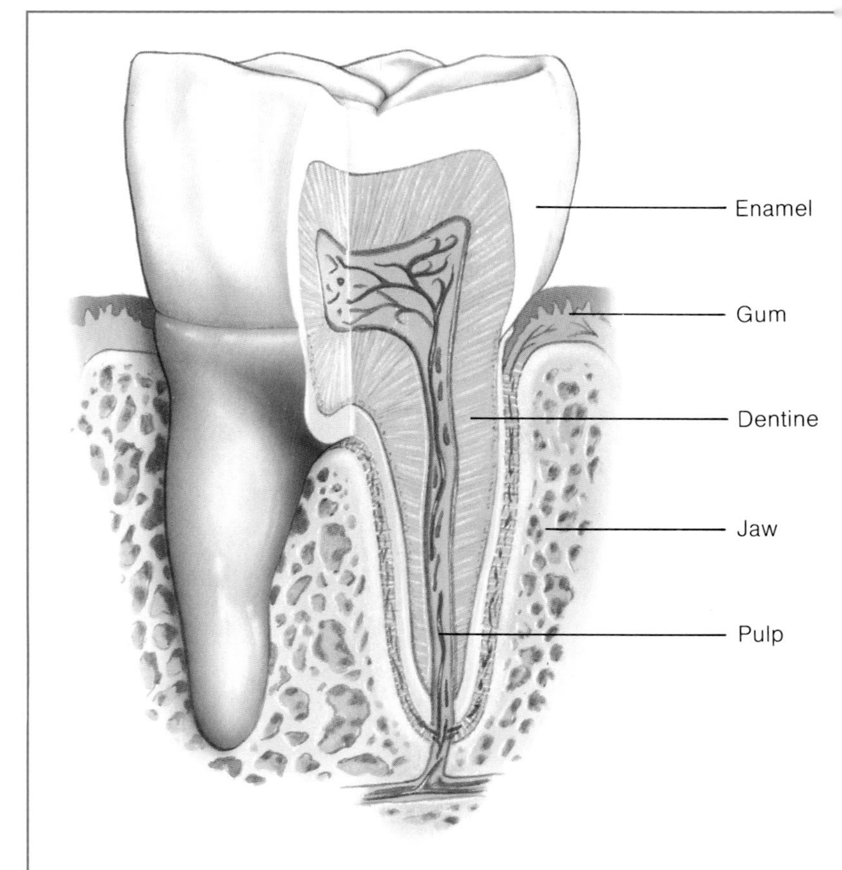

The enamel that forms the outer surface of a tooth is the hardest substance in the body. It overlays the dentine, which in turn surrounds the central pulp carrying nerves and blood vessels. A child's first, or milk, teeth (shown in white on the diagram) are easily displaced by the permanent teeth (shown in blue) as they push through the gums. The eight adult premolars – two in each side of each jaw – replace the child's molars, while the adult molars grow into previously unoccupied spaces at the back of the jaws. The dentition in the diagram includes eight adult molars. Four more, the so-called wisdom teeth, may erupt behind them. Incisors and canines, at the front of the jaw, are used for biting off and slicing food; the premolars and molars then crush it into pieces small enough to swallow.

The micronutrients

Vitamins and minerals are sometimes called micronutrients, because we have only to consume the equivalent of "pinches" of them in our food, rather than the great mouthfuls of the macronutrients which are proteins, carbohydrates and fats. Vitamins are a diverse group of chemicals, mostly smallish molecules, that serve as accessory ingredients in the many biochemical pathways of cellular metabolism. Their common feature is that, with a few exceptions, the body cannot manufacture them itself and so must take them ready-made from the diet.

So far, about 20 vitamins have been discovered. Their lack in the diet brings on the so-called deficiency diseases. For instance, the bleeding and ulceration of scurvy are due to lack of vitamin C (ascorbic acid), which is found in fresh fruits and vegetables, notably citrus fruits. The tiredness, weight loss and heart failure of beriberi are brought on by lack of vitamin B_1 (thiamine), contained in whole grains, nuts, beans and peas.

Minerals are elements such as calcium, sodium, iron, iodine and phosphorus. Some abound throughout the body, such as calcium, which is a major constituent of bones and teeth, and sodium, a vital functionary in the passage of electrical nerve impulses. Iron is essential for building the oxygen-carrying hemoglobin of red blood cells. Other minerals, the "trace elements," are needed in only small amounts. It is likely the body

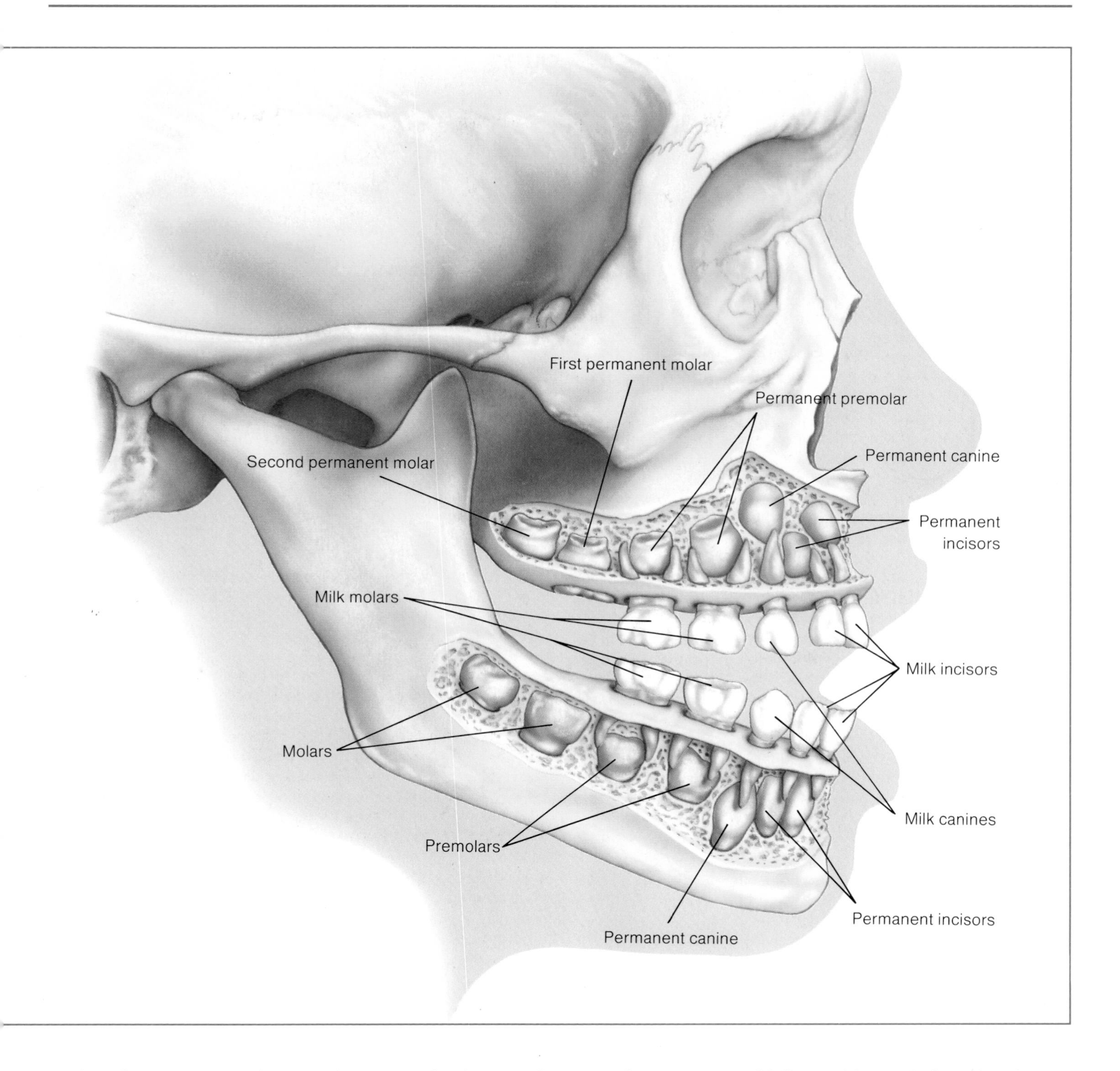

needs such tiny amounts, that some have yet to be discovered.

The list of dietary components, with their energy contents and chemical formulae, might make the task of achieving a balanced diet impossible. But the basic advice for a healthy diet is relatively simple. Be keen on variety, on fresh vegetables and fruits, wholegrains and some white meats. Avoid too much of anything, particularly meat (especially the rich, red types) and dairy produce. Imagine the kind of foods our forebears would gather and hunt for, on their prehistoric lifestyle. In the main, plucking fruit from a tree or digging up juicy roots was, for them, easier than trapping and killing wild animals for rich red meat. That is the food our bodies are still designed for.

Lip service

Part your lips and you open the entrance to the digestive system. Beneath the skin of the flexible and fleshy lips is a circular muscle, the orbicularis oris, which is tagged onto the complex network of muscles in the face, jaw and upper neck. Coordinated contractions of these muscles produce all lip and mouth movements, with their varied functions.

Some lip activities are communicative: both visual, from the merest hint of a smile to an enormous grin; and auditory, from the pursed intake of breath that signifies doubt, to the wide-open funneling of a scream of fright or rage. Some are respiratory: from the slight crack of quiet breathing, to open-mouthed gasping for air after a short sprint. Some are sexual (kissing is one example). And some are digestive, from the delicate sucking of liquid through a straw, to ensuring that what has gone into the mouth, to be chewed into a pulp, does not dribble out again.

A bite to eat

The digestive journey commences in the mouth. It is the start of a long campaign to break down food. The advance troops are teeth, tongue and saliva. Once in the stomach, the chemical infighting starts in earnest.

You may use knives, forks and spoons to cut and trim your food into dainty, mouth-sized chunks. But your teeth are just as capable as cutlery. Each tooth is firmly embedded in a special outgrowth of the jaw bone known as the alveolar process, and is cushioned by gum (periodontal) tissue and fibrous connective matter. The part of the tooth within the jaw is the root; the section exposed above the gum line is the crown.

At the tooth's core is its sensitive pulp, a tangled web of blood vessels that nourish the dental tissues, and nerve endings which warn of excessive pressure or pain. The pulp is shrouded in a layer of dentine, a fairly hard, bonelike material that makes up the bulk of the tooth's volume. Cloaking the dentine is enamel, the outermost whitish layer and the hardest substance made by the body. Odontoblast cells line the pulp cavity and manufacture the overlying dentine. Ameloblasts construct the enamel coating; they disappear once the enamel has been formed.

Like many of our mammalian cousins, we get through two sets of teeth in a lifetime (more if the dentist advises false ones). The dental primordia are laid down during fetal life, and so poor diet or disease in the mother can have serious and even permanent effects on the teeth of her children. If she takes the antibiotic tetracycline, for example, her child may have discolored teeth.

The first, deciduous or "milk" teeth, 20 in number, erupt from about the seventh month after birth. The age range is very wide, however; a few babies already have growing teeth at birth, while others still have none at their first birthday. The second or adult teeth begin to erupt from the age of about seven years.

Four by four

Human dentition is relatively unspecialized. True, we have the four main types of teeth seen among our mammalian relatives: incisors, canines, premolars and molars. But we lack impressive and distinguishing dental features, such as the huge pointed canines of a cat or dog, or the great crushing molars of a horse.

Starting at the front of the adult mouth, there are eight incisors, two on each side of the midline in both the upper and lower jaws. These chisel-shaped teeth are cutters and slicers. Bite into a crisp apple and see the groove pattern they leave. Behind the incisors the canines, one on each side of each jaw, making four in total. These are but a shadow of the great "eye teeth" found in the true carnivores, for which they seize and stab prey. Behind each canine are two premolars, the smaller of the crushing, grinding and chewing teeth. And behind these are three molars, the large, flat "cheek teeth" with the power to splinter bone and crack nuts.

These four types make up the 32 teeth of the full human dentition. In today's world, however, there are many departures. Premature decay, principally due to poor oral hygiene and an excessively sugary diet, can rot teeth and necessitate fillings or extractions. Sometimes a developmental abnormality means a tooth is missing, or two appear instead of one. In certain people the four rearmost molars, two upper and two lower in each jaw, never erupt. The appearance of these so-called wisdom teeth often occurs (when it does so at all) in the late teens or third decade of life, and was popularly linked with attaining the mature, experienced and balanced persona of adulthood.

A mobile muscle

Anchored to the floor of the mouth, and slung at the rear from muscles attached to a spiky outgrowth at the base of the skull, is the tongue – itself a powerful muscle. The tongue is covered by the lingual membrane, which is specialized in places to detect the flavor of food. Tiny onionlike sensory structures, the taste buds, enable you to savor the sensations of flavor, and warn you of food that is rotten and bad to eat. The muscle fibers of the tongue are richly supplied with nerves, enabling this organ to manipulate food in the mouth and place it between the teeth for chewing, without being bitten itself. The tongue also helps shape the sounds of speech, and its carefully coordinated movements are necessary in the initial stages of swallowing.

Down the slippery slope

It is difficult to place a solid item in the mouth and then swallow it at once. Think of taking a medicinal tablet: the natural inclination is to chew, unless the tablet is "disguised" in water. Food in the mouth is masticated with saliva, a watery fluid rich in the starch-breaking enzyme amylase, which lubricates the processes of chewing and swallowing. Up to three pints of saliva are secreted daily. Three glands on each side of the mouth release copious quantities when stimulated by the smell and taste of food (or even the thought of a meal) and by its physical presence in the mouth. These glands are the parotid (over the angle of the jaw, near the ear), the submandibular (tucked into the lower jaw) and the almond-shaped sublingual (under the tongue).

Gradually the food is chewed, pummeled and pulped, mixed with saliva into a moist mass called a bolus, and made ready for swallowing. At the start of a swallow, the tongue tip touches the roof of the mouth at the hard, front part of the palate – a shelflike plate dividing the nasal and oral cavities. Trapping the bolus behind it, the body of the tongue rises from the front to push the food back and down, into the pharynx (throat). The pharynx is an apparatus of muscles and specialized tissues used for accurate transport of the food to the esophagus (gullet) – most importantly past the entrance to the trachea (windpipe). A bolus entering the top of the trachea, rather than the esophagus, results in choking and possibly asphyxiation.

As the food reaches the pharynx it lifts the rear, soft part of the palate, which swings up like a one-way trapdoor to block its passage into the nasal cavity. Likewise, a flap on the lower front of the pharynx, the epiglottis, swings over and down to cover the entrance to the trachea; simultaneously the larynx raises itself to narrow its entrance and hide it beneath the epiglottis. As all this occurs, further muscular activity creates propulsive waves in the pharynx, massaging the bolus down into the esophagus.

The whole process of swallowing is controlled by a series of reflexes that ensure food and liquids pass smoothly into the next region of the digestive tract. We swallow hundreds of times each

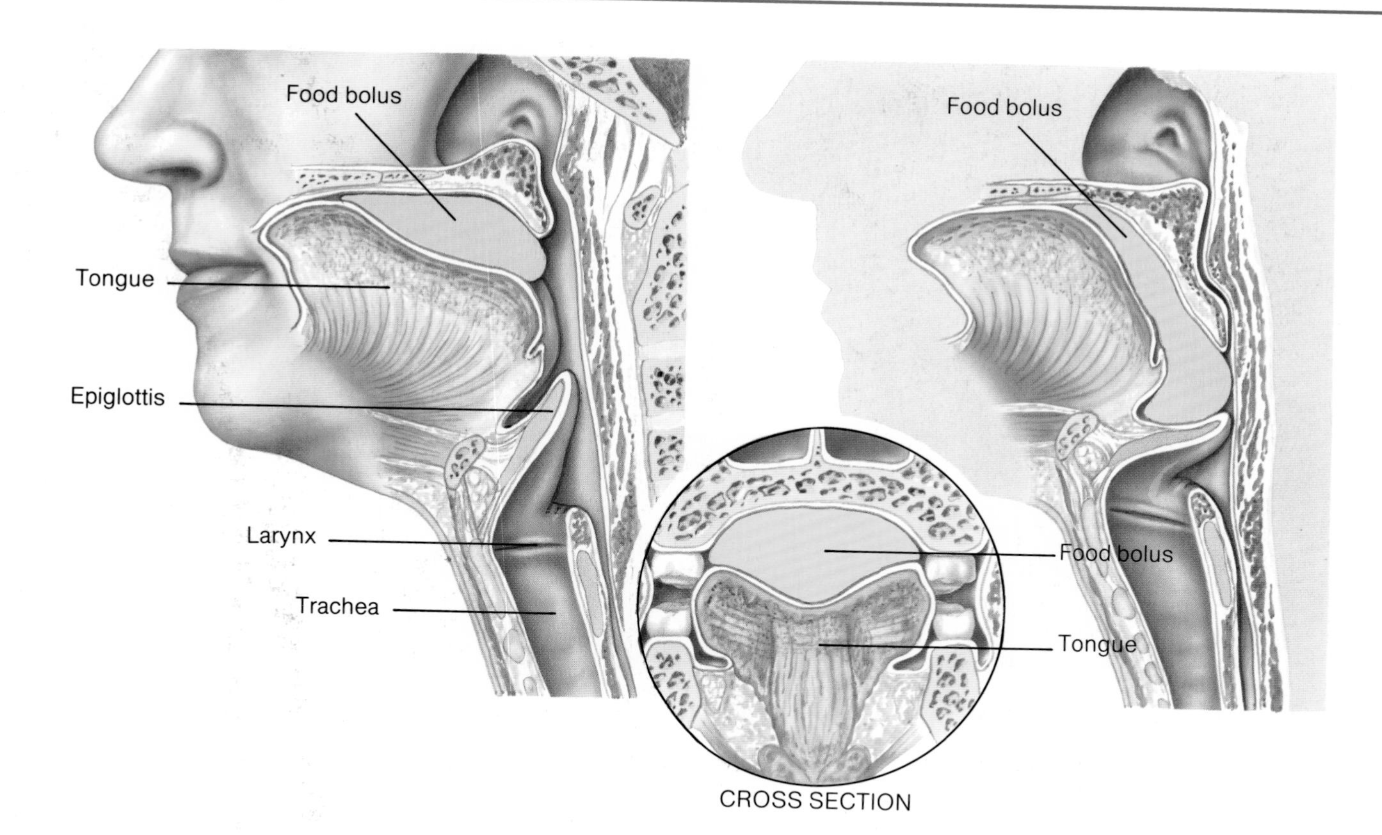

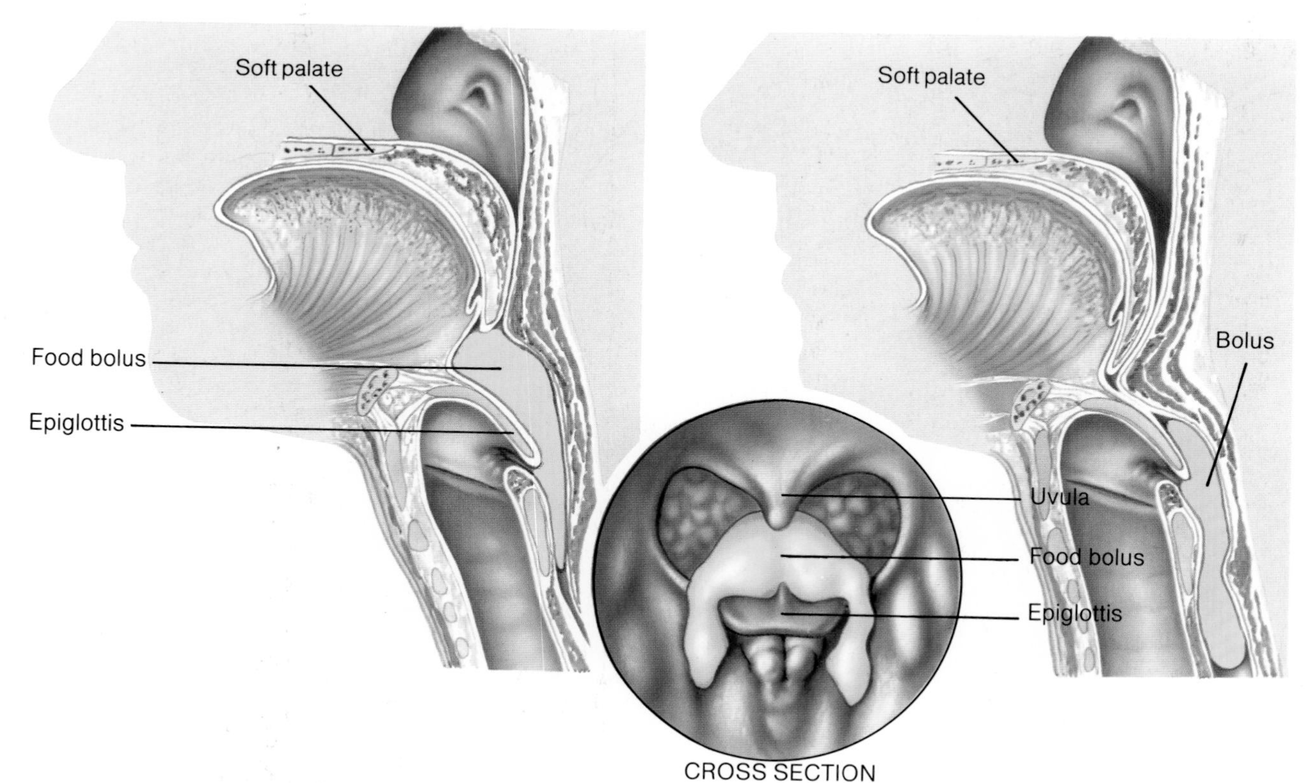

Chewing and mixing with saliva converts a mouthful of food into a soft ball called a bolus. Swallowing it involves a series of complex, but automatic, movements of the tongue and throat muscles. First the tongue rises at the front against the roof of the mouth to push the bolus into the pharynx. The soft palate swings upward to block off the exit from the nose, and the epiglottis closes the entrance to the trachea, or windpipe, to prevent food from "going down the wrong way." Once the bolus is in the esophagus, peristalsis takes over and squeezes it along the length of the esophagus to the entrance to the stomach.

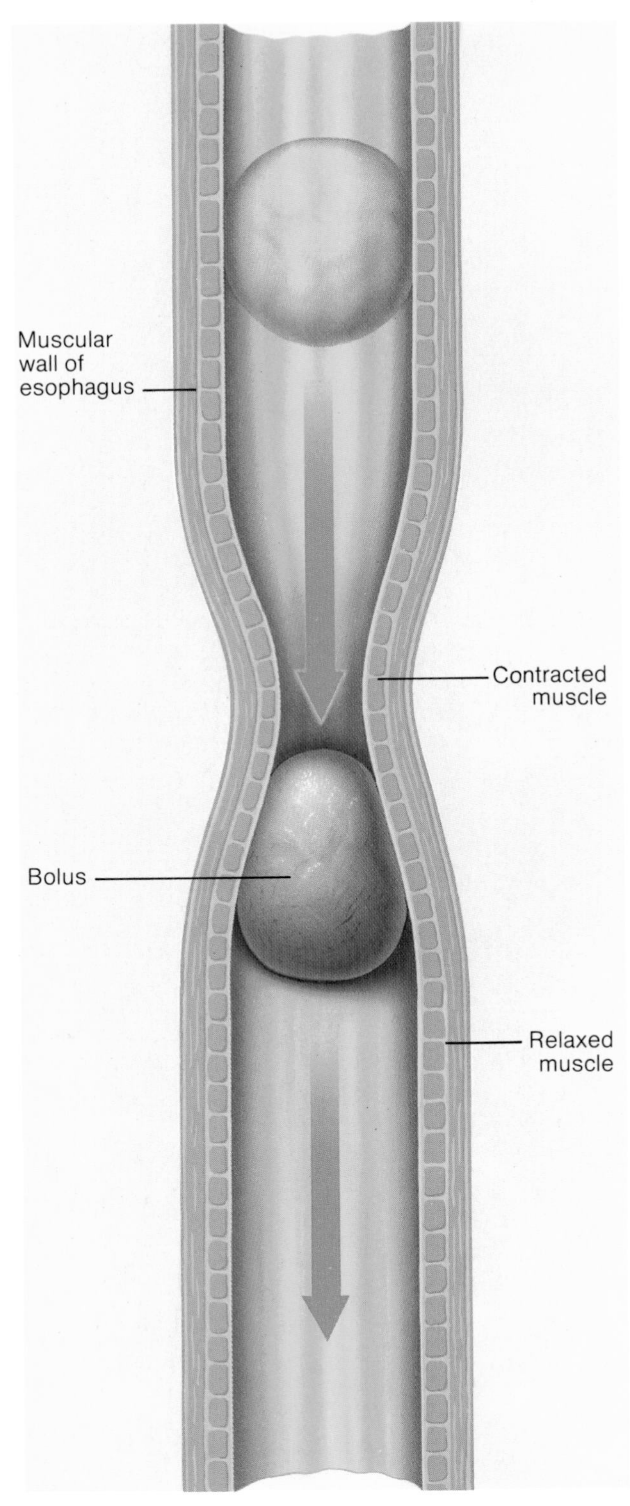

Peristalsis is a wave of involuntary muscular contractions that forces material along the digestive tract. Food is moved along the esophagus, and partly digested food and feces are squeezed along the intestines.

day, hardly ever bothering to think about it, yet each event is a potential disaster if food goes down "the wrong way."

The fate of the bolus

The whole of the gut, from the top of the esophagus to the rectum, can make snakelike writhing movements whereby the digestive contents are progressively pushed through the system. The principal components of this action, known as peristalsis, are the same throughout the gut, though their speed and force vary from one section to another.

Peristalsis is brought about by layers of muscles and nerves within the gut wall. In the outer layer, the muscle fibers are arranged lengthwise, pointing along the gut tube. The inner layer consists of fibers encircling the gut. Between the two layers is the myenteric plexus, a network of delicate nerve fibers. This is connected to another plexus, the submucus plexus, just next to the innermost mucous lining of the gut.

A lump of food or a bubble of gas in the central space – the lumen – of the gut stretches its wall. This initiates contraction of the muscle immediately behind the lump, which eases the lump forward into the next segment of the plexus, where the muscles are relaxed. Here the process repeats as this next segment of wall stretches. In this way waves of contraction flow along the gut, followed by waves of relaxation. Peristalsis in the pharynx and esophagus is so powerful that if necessary you can swallow food into your stomach even when upside-down. Reverse peristalsis can be initiated in times of trouble, causing the stomach contents to travel up the esophagus and be thrown out of the mouth with surprising force. We call it vomiting, and it is intended to rid the digestive system of bad foods, poisons or excessive contents.

Shutting off the system

Peristaltic waves travel down the ten inches of esophagus at about two or three inches per second. At the base of the esophagus is a ring of muscle, the esophageal sphincter. Like peristalsis, the actions of sphincters are useful features of the digestive tract. These rings of specialized muscle are capable of sustained contraction to close an orifice and seal off one section of gut from the next, while the contents are processed. Under automatic nervous control, the muscles relax occasionally and allow the contents to ooze through.

The esophageal sphincter is particularly important because the next region of the tract is the stomach – and this is filled with powerful, churning acid. If the sphincter weakens, or if there is high intra-abdominal pressure (as when bending forward or in obese people), stomach contents may well up into the lower part of the esophagus. The esophageal wall is not nearly so resistant as the stomach is to acidic contents, and the welling-up produces an unpleasant, bitter sensation deep in the throat. This is known as heartburn – although it has nothing to do with the heart.

The acid bath

The average adult stomach holds from two to three pints and manufactures the same volume of "gastric juices" every 24 hours. The stomach plays several roles in digestion.

It is a food hopper, or storage reservoir. The upper baglike portion, the fundus, holds a hurriedly-eaten meal and feeds it part at a time to the lower portion, the antrum.

It is a food-mixer. Strong muscles, principally in the antral wall, contract to squash and pulverize the contents into a sticky, slushy mass called chyme.

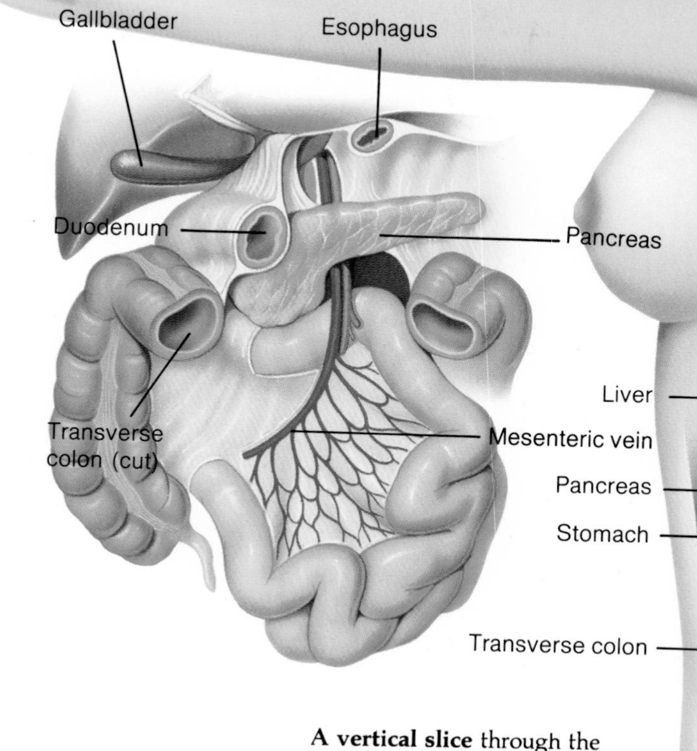

A vertical slice through the stomach and intestines shows how they are supported by the membranes of the mesentery and omentum. The smaller diagram *(above)* shows the arrangement of the organs.

It is a sterilizing unit. Parietal cells, in glands in the stomach lining, make powerful hydrochloric acid that kills many of the germs in unwisely consumed, contaminated food.

It is a digesting tub. The acid, along with the protein-splitting enzyme pepsin made by other cells in the stomach wall, sets to work to split and crack the chemicals in food.

And it is well protected from itself. A thick layer of mucus, secreted by yet other cells in its lining, coats the inside of the stomach and stops it digesting itself. If there is too much acid, or the mucous coat is deficient, the acidic contents erode raw spots in the stomach wall. These are gastric ulcers.

Getting the juices flowing

The control of acid and enzyme secretion in the stomach is crucial, because it must be coordinated with the appearance of the food on which these substances work. The organ is well supplied with nerves; most of which come from the autonomic nervous system – that part of the overall nervous system that controls automatic or involuntary movements, mostly involving the internal organs. The most important nerve is the vagus, which runs from the brain stem or medulla, at the base of the

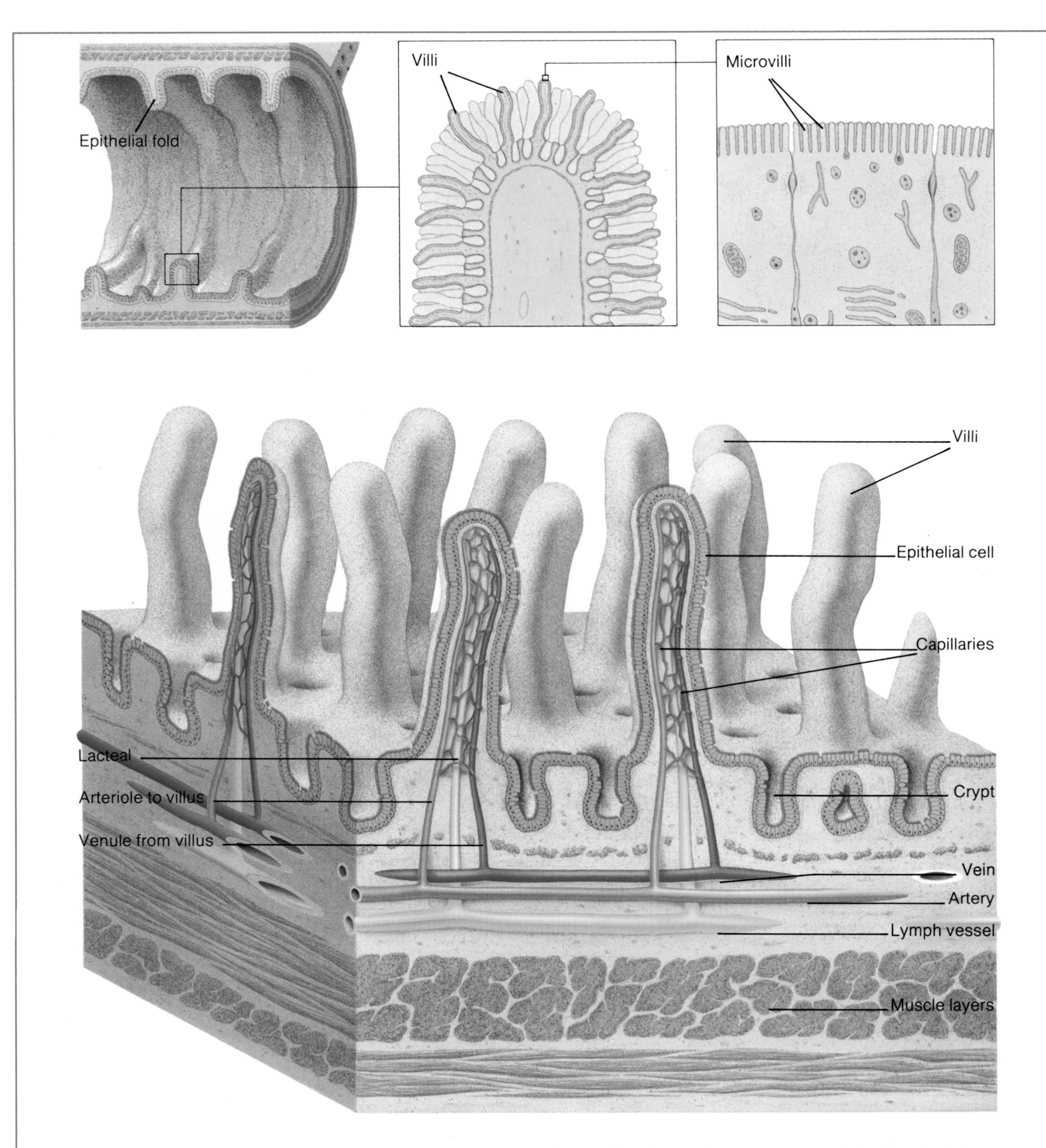

For efficient absorption of the products of digestion, the internal surface of the intestines is designed to have maximum area. There are three successive levels of maximization. First, the mucous membrane lining is gathered up into epithelial folds. Second, each fold bears thousands of fingerlike projections called villi. Third, the individual absorptive cells bristle with even smaller microvilli. Each villus is served by a tiny artery and a vein, whose capillary junctions enfold a lacteal tube which carries away the products of fat digestion.

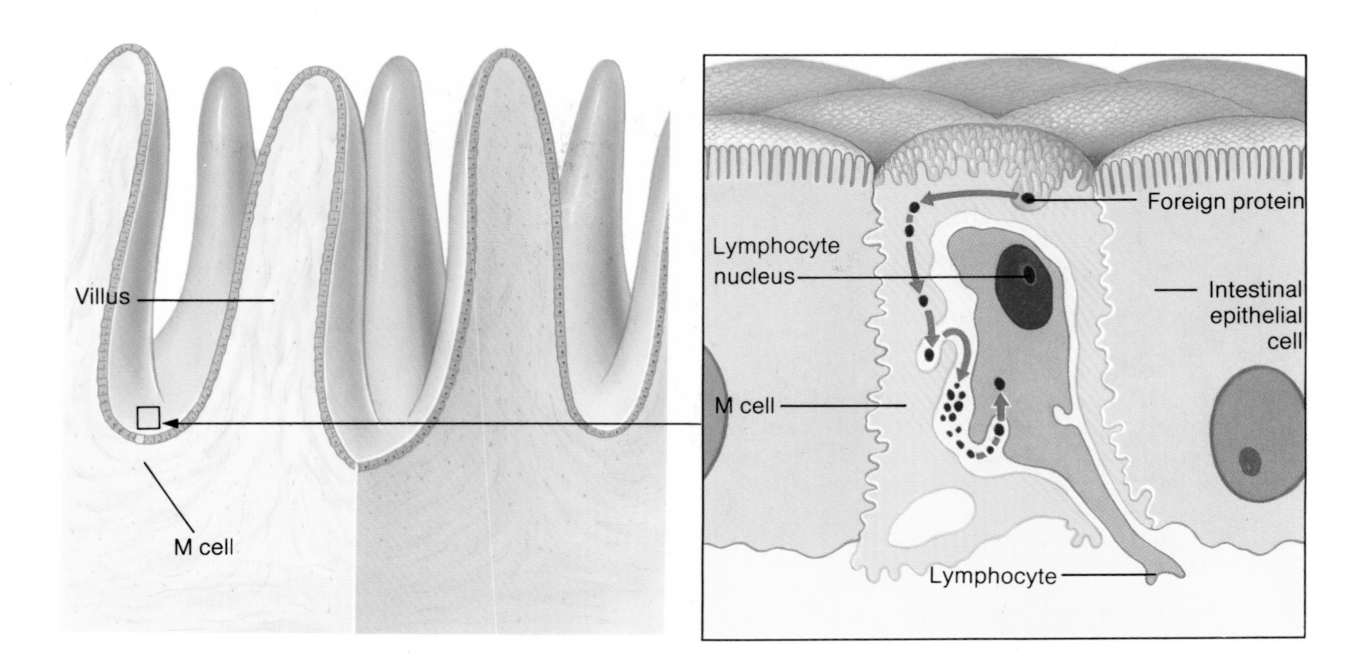

Bacteria and toxic substances in the intestines are dealt with by M, or microfold, cells in the surface epithelium of the villi. These defensive cells direct invaders – recognized as "foreign" proteins – toward a neighboring lymphocyte, which produces antibodies to neutralize the invaders.

brain, down through the chest and abdomen to control various automatic functions. The vagus carries nerve impulses directly from the brain to the groups of cells in the stomach lining which secrete gastric juices. The smell and taste of food, and its physical presence in the mouth, esophagus and stomach, all help to get the juices flowing.

Hormones play their part, too. Nervous stimulation of cells in the upper part of the stomach wall causes the cells to release the hormone gastrin, which spreads through the blood supply and encourages further release of juices into the stomach cavity. Therefore, soon after the chewed food enters the stomach, it is assailed by various chemicals designed to break it down.

So much for filling. What of emptying? The same dual nervous–hormonal control is in evidence. The pyloric sphincter at the stomach's exit relaxes occasionally to let squirts of chyme through into the duodenum, the first ten-inch section of the small intestine. In the enterogastric reflex, the squirts of chyme gradually fill the initial part of the duodenum and distend its wall. Sensors in the wall initiate the reflex, sending nerve signals along the vagus, and other small nerves of the celiac plexus network, back to the stomach. The signals dampen the stomach's movements so that the amount of chyme flowing from it decreases.

As chyme slops into the duodenum, it also triggers the release of the hormones secretin and cholecystokinin from the mucosal lining there. These hormones assist in limiting gastric motility and acid secretion, as well as stimulating organs such as the gallbladder and pancreas to prepare for the coming meal.

Chemical bombardment in the intestine

The small intestine has three main sections. First is the duodenum, already mentioned. Second is the jejunum, some eight feet long and coiled behind the area of the navel. Third is the ileum, three feet longer, which loops its way down to the lower right of the abdomen. Here it swells to form the cecum, the first part of the large intestine. Each section of small intestine is about an inch in internal diameter.

The initial attack by acid in the stomach does a great deal to break down food, but most "digestion" (in the chemical sense) and absorption of nutrient molecules takes place in the small intestine. The liquefied food that is chyme flows through the duodenum and there mixes with a battery of enzymes pouring in along a duct from the pancreas. Now the food is under heavy bombardment. Among the bigger guns are the pancreatic enzymes trypsin and chymotrypsin, which pound proteins into shorter and shorter segments until they are only a few amino acids long. Pancreatic amylase, another powerful enzyme, sets about the starches and other large carbohydrate molecules, and breaks them into smaller sugar molecules. The pancreatic juices also contain alkalis, which neutralize the corrosive acid of the stomach and allow these various enzymes to work at their peak efficiency in the now slightly alkaline environment.

Dietary fats present more of a problem. Another component of pancreatic juice, the lipid-splitting enzyme lipase, is ready to enter the fray. But its target is awkward. Because fats are not water-soluble, they tend to split up and regroup in the intestine as small globules, which lipase finds difficult to penetrate. The globules must be broken into much smaller droplets, a process chemists call emulsification. Enter bile, literally, along the bile duct leading from the gallbladder. Bile is an especially effective emulsifying agent and splinters fat droplets, so that the fragments can be more easily picked off by lipase.

The finger buffet

The small intestine faces a similar problem to the lungs. Each is in the absorption business – the lungs take in oxygen, while the intestines draw in nutrients. An advantage in this business is surface area: the wider you can spread yourself, the greater amounts you can absorb. Yet both systems must fit into the limits of the human torso. The lungs have spheres, in the shape of millions of tiny air-filled "bubbles" called alveoli, as a way of packing as much surface area into as small a space as possible. The design employed in the intestine is tiny "fingers."

The first factor which increases surface area for nutrient absorption is length: the small intestine is very long, coiled upon itself and packed neatly into the abdominal cavity. Examination of the intestinal lining reveals that it is far from smooth, which leads to factor two: its inner surface is folded and ridged, to increase further the surface area. Even closer study reveals factor three: the folds and ridges have a surface not unlike a tiny pile carpet, being thrown up into thousands of fingerlike projections called villi. Seen under a microscope, the surface membranes of the cells lining the villi display factor four: they are also thrown into even finer fingerlike projections, the microvilli. If the small intestine were a simple, featureless tube it would have an inside surface area of about four square yards. Factors two, three and four increase this to a staggering 250 square yards or more – greater in area than a tennis court.

The partly broken-down remnants of once-large food molecules drift along the microvilli. They are still under attack from various enzymes. For example, maltase, sucrase and lactase hack away at the carbohydrates, reducing them to the smallest and simplest sugar molecules. Dipeptidase and aminopeptidase ensure that once-huge proteins are rendered down to their individual amino acid subunits. Only at this size are the amino acids, sugars and most other products of digestion small enough to pass through the cells lining the villi, into the copious network of blood capillaries just behind them.

Triglycerides, the subunit products of digested fats, enter the intestinal cells rather like ghosts passing through a solid wall. They merge with the cell membrane itself, which has a lipid component, and are assimilated into the cell contents. Here they are parceled up in protein wrappers, which makes them soluble in water, to form tiny packets known as chylomicrons. They then tend to drift, not into the bloodstream, but into a microscopic channel called a lacteal that runs up the center of each villus and is filled with lymph fluid. Thus the lymphatic system (whose main role is defensive, fighting infection) becomes the main mode of entry for fats, although this system eventually links up with the blood circulation by draining into a main vein near the heart.

The intestine's defensive system

In 1677, Swiss doctor Johan Peyer described small patches of lymphatic tissue in the small intestine, which now bear his name. He theorized that they were involved in some sort of secretion. We now know that Peyer's patches are an important part of the way the small intestine protects itself from invasion by bacteria that have survived the rigors of stomach acid. The story is filling out, but there is still much research to be done into the complex way in which the intestine mounts a defense against infection.

Peyer's patches contain large numbers of lymphocytes, key cells in the immune system. They also contain isolated M or "microfold" cells, which seem to attract bacteria and other toxic substances to themselves and then funnel them inward to make contact with lymphocytes. This is a cunning system, because lymphocytes are responsible for defending the body against invading bacteria and other microorganisms. Large numbers of lymphocytes are manufactured within the Peyer's patches and in nearby lymph nodes, ready to repel microscopic invaders.

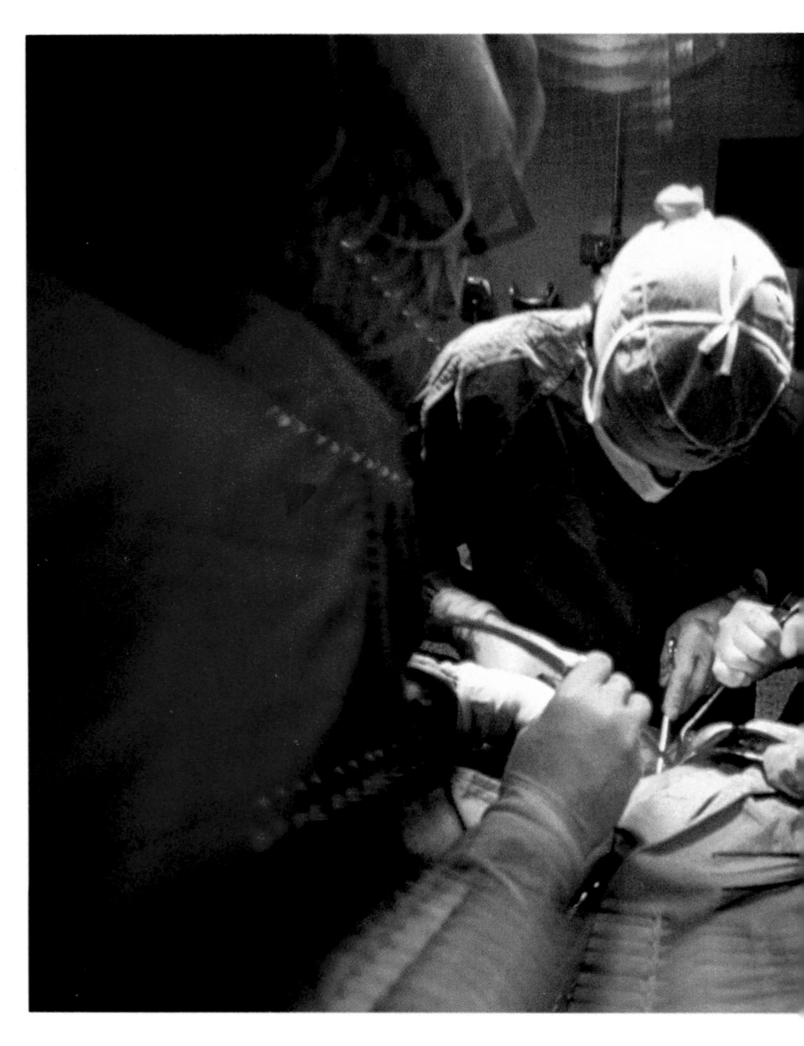

The mesenteric connection

The small intestine does not float freely inside the abdomen. If it did, there would be every chance that it could twist and knot, blocking the passage of food. Instead, its multitudinous loops are anchored to the rear wall of the abdomen by a tissue structure known as the mesentery. The appearance of this thin, membraneous sheet has been likened to an open fan, rooted by its "handle" to the back wall of the abdomen, and with its long, free edge supporting a great length of intestine. The mesentery not only holds the intestine in place, but also carries to it a blood supply (since, like any other organ, the intestine needs a supply of oxygen and suitable nutrients). In addition, the mesentery supports a vast network of blood and lymph vessels which transport the absorbed nutrients away from the intestine.

The abdominal cavity is lined by a thin membrane, the periosteum. This folds inside itself and extends to cover each digestive organ, in an almost impossibly complicated series of curves, folds and invaginations.

The battle is won, but the war is not over. Food popped into the mouth perhaps eight or ten hours previously is, at last, truly "inside" the body, broken into its smallest viable subunits and

Gallstones are a common disorder of middle-aged people, and if the gallbladder becomes inflamed it causes the pains of cholycystitis. Surgery to remove gallstones *(left)* is now routine, and if necessary the whole gallbladder can be removed.

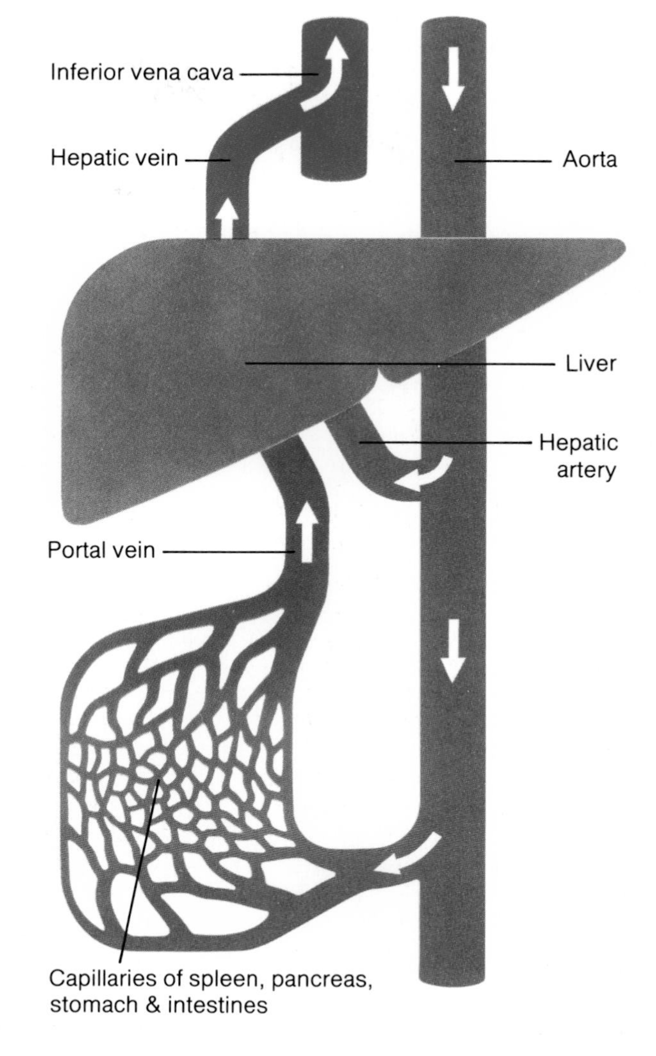

The liver has two blood supplies. Incoming blood from the hepatic artery, branching off the aorta, carries oxygen and nutrients for the tissues of the liver itself. Blood arriving along the portal vein carries the products of digestion from the stomach and intestines for processing by cells in the liver. "Used" blood, from both sources, leaves along the hepatic vein for return to the heart.

floating along in the bloodstream and lymphatic channels. Yet the story continues. The food molecules, now "prisoners of war," are shipped to the next port of call: the liver.

The chemical factory

A classical Chinese medical text, the Canon of Medicine, from about 2,300 years ago, labels the liver as the military leader of the body, the seat of anger and of the soul, and the source of tears and nasal secretions. Traditional Chinese medicine still places great emphasis on the liver and the way in which its congestion or blockage can lead to a wide variety of symptoms.

Some five centuries later the great Roman physician Galen came closer to appreciating the true functions of the liver. He believed that food substances absorbed from the intestine traveled, in the form of the substance chyle, along a special vessel directly to this deep-red, blood-rich organ. There they were converted into blood, with the addition of a natural spirit, pneuma, to help in growth and nutrition. The liver was thought constantly to generate blood, which was used up and disappeared in the tissues.

Galen was correct in one sense. The products of digestion do indeed travel along a special vessel directly to the liver. It is called the portal vein, and thus the liver is unique among body organs in being fed by two blood supplies – one along the portal vein, and the other along the usual incoming branch off the aorta, which in the liver's case is the hepatic artery.

The liver nestles in the top of the abdomen, just below the diaphragm, with its bulk on the right side. It is the body's largest gland and largest internal organ, over three pounds in weight, and at rest accepts a blood supply of two and a half pints each minute. Between meals, more than three-quarters of this supply comes from the intestines via the portal vein; the remainder arrives along the hepatic artery. When food is eaten, more blood is diverted to the intestines to deal with the tasks of digestion and absorption, and blood flow in the portal vein decreases.

The liver has two main lobes, the right lobe and a smaller left one which crosses the midline to lie above the stomach. These two lobes are further divided into lobules, at the center of which are small blood vessels, draining into the hepatic vein. Between the lobules (which are one twenty-fifth of an inch or so across) lie the portal canals, and they contain three structures. First, there are fine branches of the hepatic artery and portal vein; blood from these flows through the lobule to reach the second, central structure, the tributary of the hepatic vein. Finally there is a channel to collect bile, which flows out of the lobule in the opposite direction.

At the microanatomical level, the basic functioning units of the organ are its cells, the hepatocytes, which are arranged in a

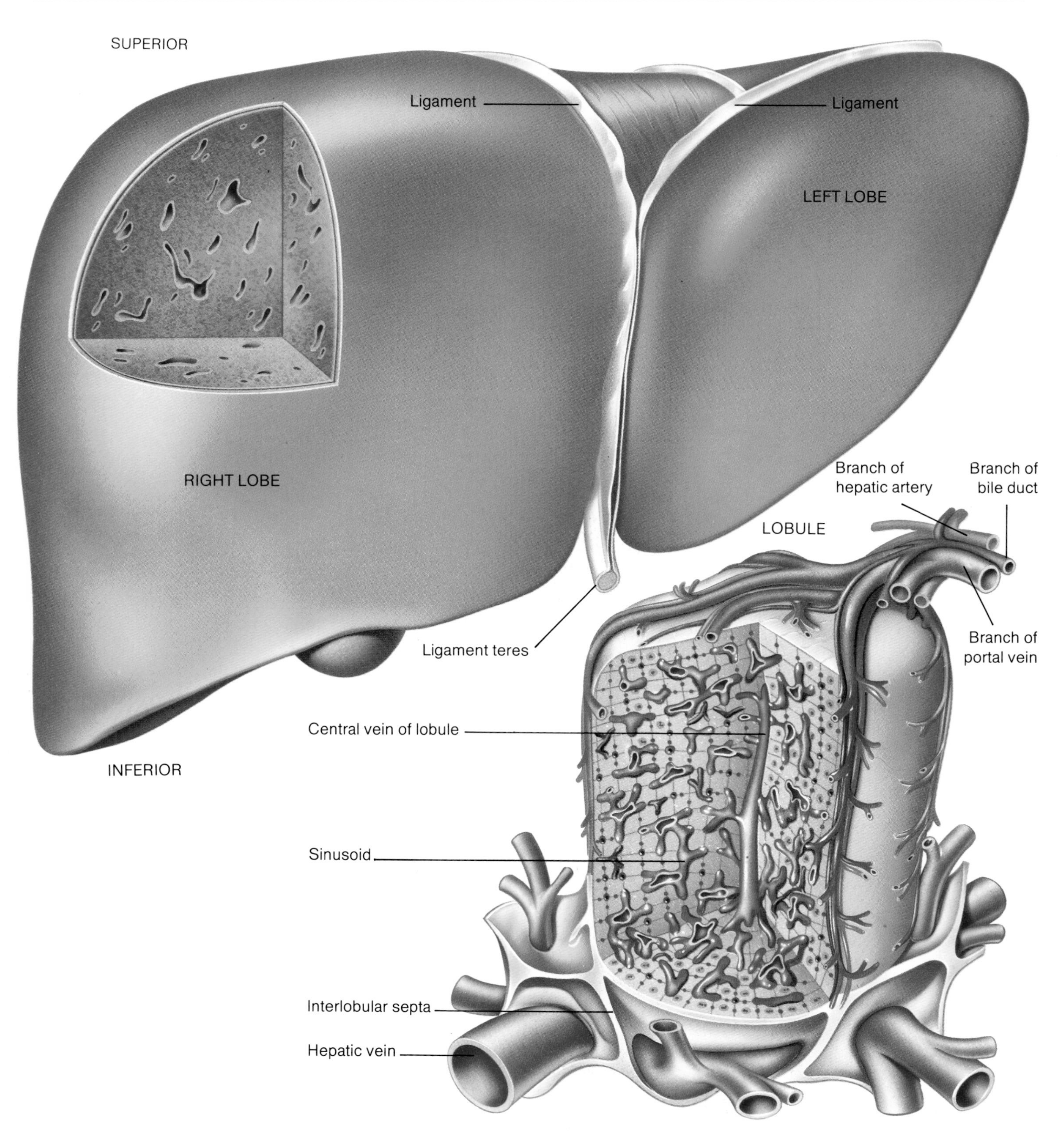

A complex biochemical factory, the liver is the body's largest gland. Viewed from the front (main image, *above*) or from below (main image, *opposite page*), it can be seen to consist of a large right lobe and a smaller left one. Each lobe is made up of thousands of lobules, penetrated by branches of the hepatic and portal veins. Together with tributaries of the hepatic artery and bile duct, these vessels split into finer and finer branches until they encircle individual liver cells. These cells, the

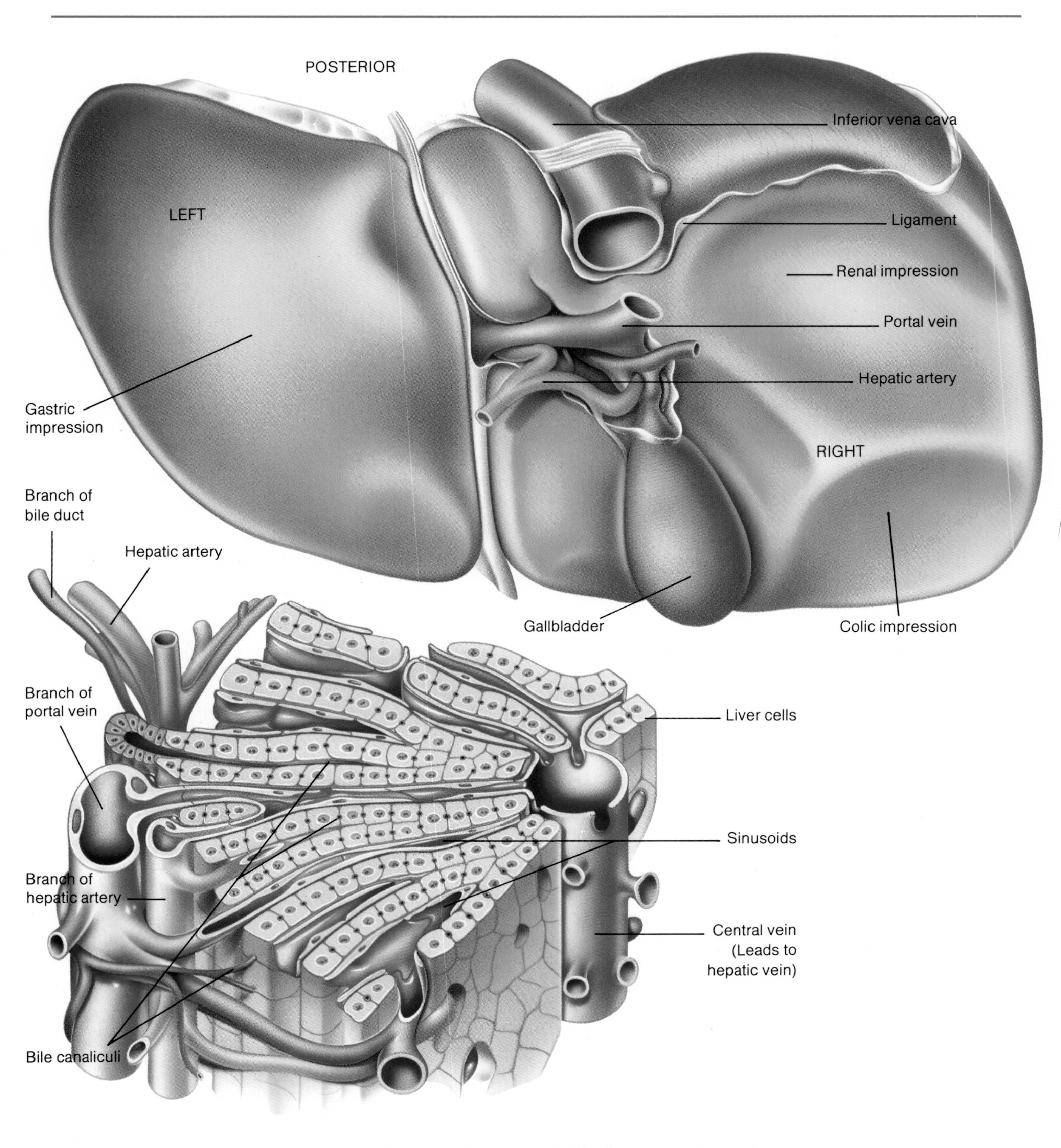

hepatocytes, have two main functions. On the one hand they deal with nutrients arriving from the intestines – for instance, by converting glucose to glycogen and storing it. On the other hand they receive toxic substances – such as alcohol in the bloodstream – and convert them into harmless waste products. Bile, produced in the liver, is stored in the gallbladder.

highly ordered way. The minuscule architecture of the liver is critical to its function, and any disorder of its structure can lead to disturbances in its working.

The ultimate food processor

The liver is the body's central "food processor." Chemically it is extremely active, playing host to more than 500 metabolic pathways – and doubtless there are still many more to be discovered. Some pathways involve the breakdown of complex chemicals (catabolism); others involve synthesis (anabolism), especially of protein molecules. The liver acts as a cleansing station for the blood, inactivating hormones and drugs. The Kuppfer cells that line the liver's blood vessels mop up unwanted elements and infectious organisms reaching it from the gut. In particular, the liver has three principal functions.

There is a key role in carbohydrate metabolism. Glucose from the intestines is chemically condensed to form glycogen ("animal starch"), which is laid down in the liver itself and in other organs. Excess glucose is converted and shunted off to adipose cells around the body, to be stored as fat.

There is a key role in fat metabolism. The lipid products of digestion arrive in the liver and certain of them are used to manufacture various vital fatty substances, notably cholesterol. This is an essential ingredient in the construction of some hormones and in nerve cell functioning, although it has become a villain in the story of atheroma and heart disease.

Finally there is a key role in protein metabolism. Unwanted proteins are disassembled into their amino acid subunits, which are broken down still further and rebuilt into other amino acids as the current requirements dictate. These amino acids are then reassembled in the correct sequence to make new, wanted proteins.

Proteins are distinguished from carbohydrates and fats by their nitrogen content (in addition to carbon, hydrogen and oxygen). While it is rearranging amino acids and proteins, the liver produces a certain amount of "free" nitrogen, surplus to requirements. This is rapidly converted into urea – the main waste product from the liver's chemical juggling of proteins – which then enters the bloodstream, to be filtered out by the kidneys and excreted in the urine.

Alcohol and the liver

If unfamiliar and noxious substances enter the body, accidentally or intentionally, the liver acts to "detoxify" them and render them harmless. Alcohol is the most infamous of these substances. Across the world, there is a fairly straightforward relationship between alcohol intake and the health of the liver. The more alcohol that is drunk, the more the liver suffers. The form in which this drug is taken does not seem to matter particularly: beers, spirits and wines are all much the same to the liver, which metabolizes 95 percent of the alcohol consumed.

The mildest form of liver damage produced by alcohol is an increased amount of fat in the hepatocytes. A fatty liver looks large and yellow. But this is not necessarily a permanent change. The liver, like only a few other human organs such as the skin, has remarkable powers of regeneration. If alcohol consumption ceases, it can usually return to good health at this stage.

The next stage of liver damage produced by alcohol is acute alcoholic hepatitis – inflammation of the liver cells. The liver becomes swollen and tender, and its owner feels ill and feverish and is often jaundiced.

The pancreas is a dual-purpose gland, tucked beneath the liver with its "head" in the curve of the duodenum. Cell clusters called acini produce alkaline digestive juices and enzymes, which flow along the pancreatic and common bile ducts to the duodenum. Other groups of cells, the islets of Langerhans, manufacture the hormone insulin, which enters the bloodstream and passes to all parts of the body to control the use of glucose by the cells.

This stage may progress to the next: cirrhosis. It may happen quietly, without any particular signs of ill health at first. Nevertheless a network of fine scar tissue (fibrosis) develops in the organ, which divides the cells into small islands, or nodules. Disruption of structure affects function, and the liver, having passed the threshold of its regenerative powers, begins to fail in some of its many metabolic tasks. It becomes small and shrunken, although the abdomen itself may become enlarged due to a build-up of fluid, a condition known as ascites.

There may be a whole range of other symptoms: the skin turns yellow in jaundice, the palms of the hands turn red, the nails become white and opaque, there are small, dilated blood vessels in the skin, and the sufferer becomes generally fatigued. About one in ten heavy drinkers – those who consume more than one-third of a bottle of spirits or its equivalent per day – are likely to scar their livers with alcoholic cirrhosis.

Bile and the gallbladder

The great Pythagoras of 6th-century BC Greece, who is perhaps better known for his mathematical prowess, developed a theory of life based on the four elements of earth, air, fire and water. These corresponded to the "four humors" of the body: blood (hot and moist), phlegm (cold and moist), and the two colors of bile, yellow (hot and dry) and black (cold and dry). The relative proportions of the humors were thought to determine health and intelligence. Our language today still echoes Pythagoras' theories, with terms such as melancholic ("like black bile") and phlegmatic.

Bile is a fluid formed in the bile canals, as mentioned previously in connection with the lobular construction of the liver. One of its goals is to emulsify fats in the intestine. The bile collects in branches of the hepatic duct, which lie in the portal canals between the liver lobules. Small hepatic ducts merge to form the main right and left hepatic ducts, which lead from the two lobes of the liver and join to make the common hepatic duct, through which all bile passes.

The common hepatic duct runs for nearly two inches before branching up the cystic duct to the gallbladder, a small pear-shaped pouch in a hollow below the liver's right lobe. There is great variation among individuals, but the average gallbladder, a temporary reservoir that concentrates the bile within, has a capacity of around one and a half fluid ounces.

The union of cystic and common hepatic ducts produces the common bile duct, which is nearly three inches long and one-fifth of an inch in diameter. It passes down behind a loop of

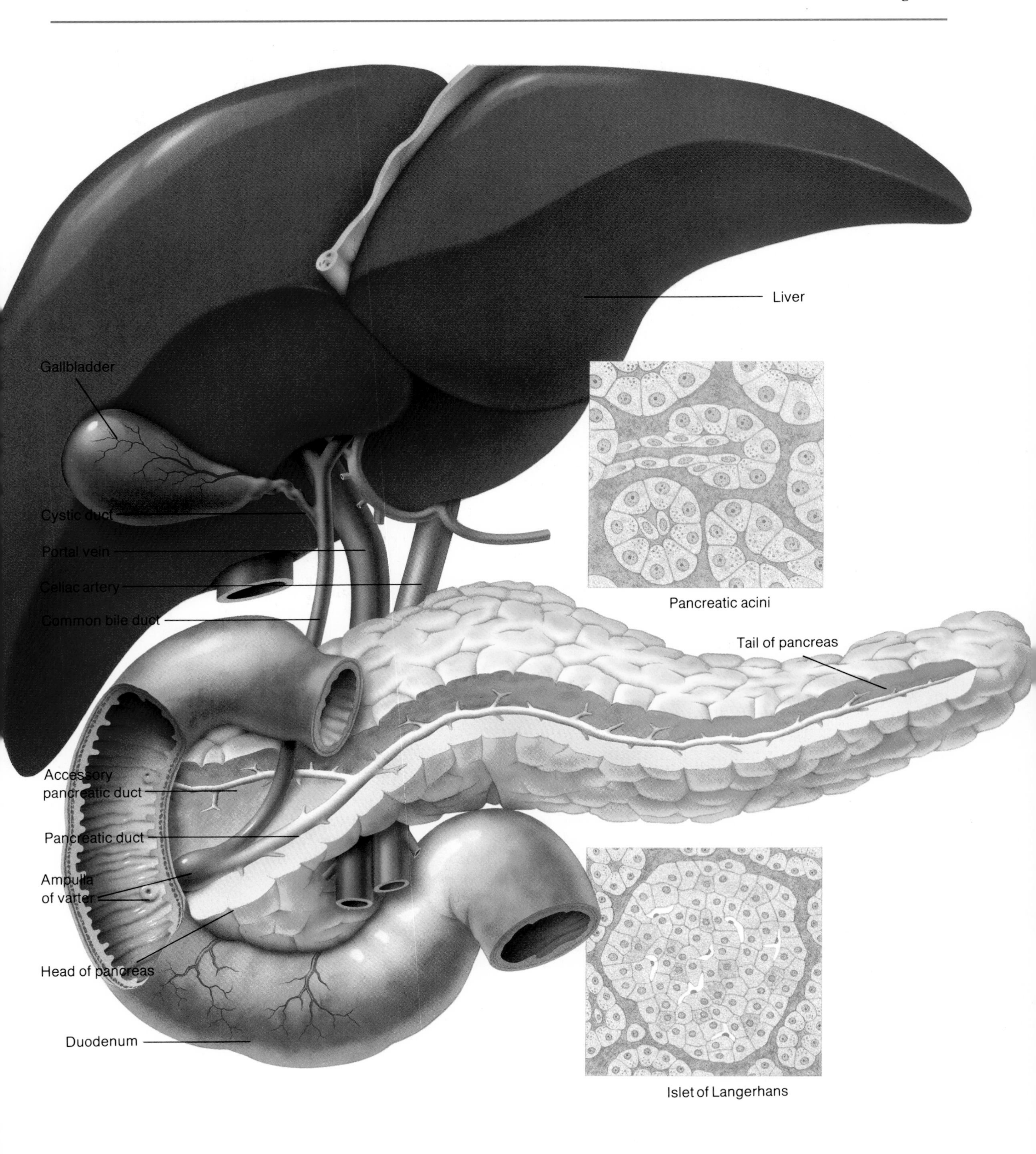
Liver
Gallbladder
Cystic duct
Portal vein
Celiac artery
Common bile duct
Pancreatic acini
Tail of pancreas
Accessory pancreatic duct
Pancreatic duct
Ampulla of varter
Head of pancreas
Duodenum
Islet of Langerhans

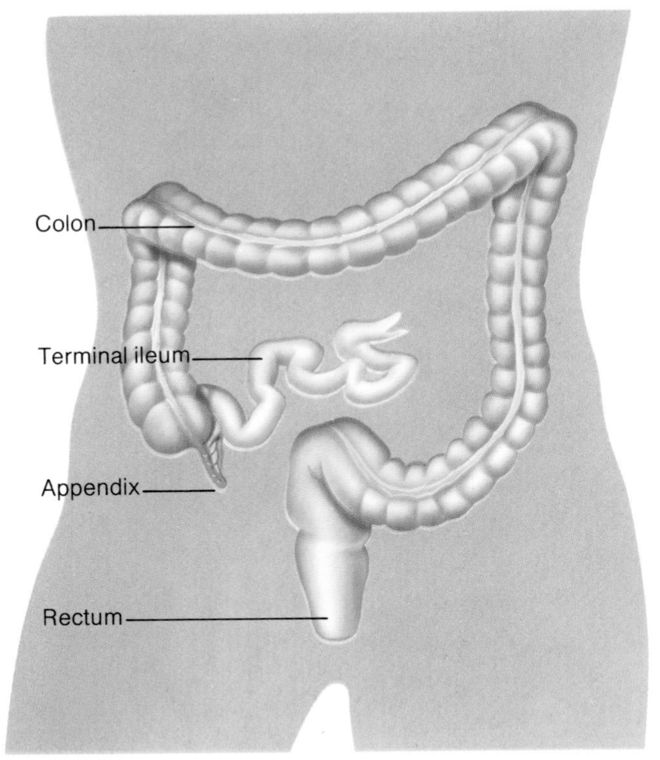

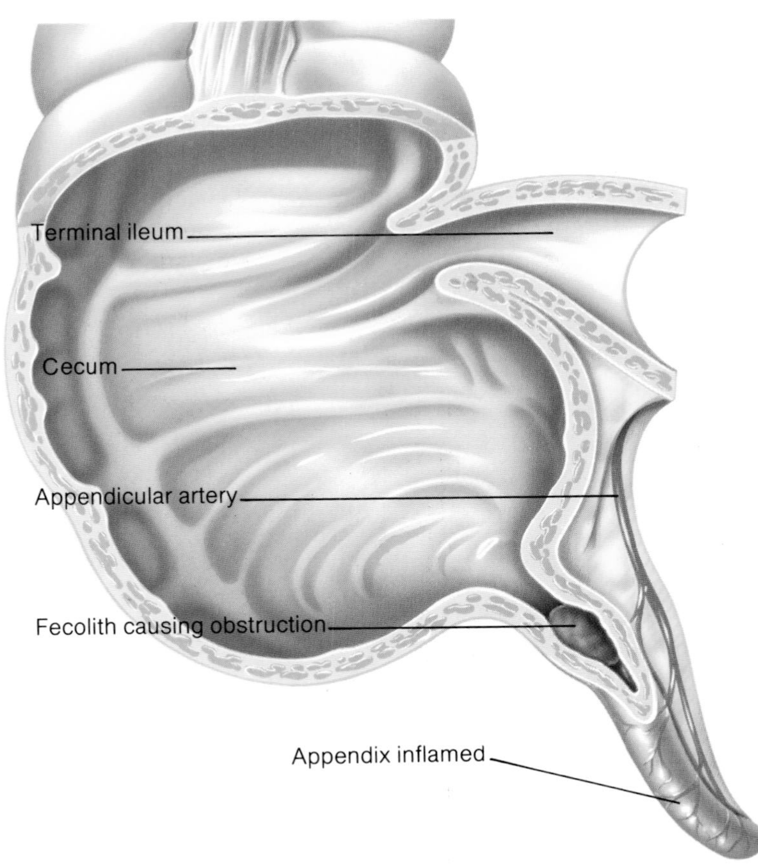

duodenum and through the larger, "head" end of the pancreas to discharge into the duodenum. Just before it enters the duodenum, the common bile duct is joined by the pancreatic duct carrying digestive juices from the pancreas. Smooth muscle in the gallbladder wall allows this sac to squeeze out bile when this is needed for digestion. And one of those useful digestive system devices, a sphincter (this one being the sphincter of Oddi), is sited at the junction with the duodenum, to control the flow of bile and pancreatic juices into the intestine.

Pigments and bile salts

Bile is a complex fluid. It contains a mixture of bile salts, lipids, cholesterol and assorted pigments, proteins and mineral salts such as sodium. Its yellow color is derived from bilirubin, mainly formed from the breakdown of red blood cells that have reached the end of their four-month lives. (A build-up of bilirubin in the blood, rather than its excretion in bile, is responsible for the yellow color typical of jaundice.)

The bile salts are essential for emulsification of fat globules in the upper small intestine, although they themselves are not absorbed through its lining into the body. However, they are not lost along with the undigested remains of food. The last part of the small intestine, the terminal ileum, resorbs most of the bile salts and cycles them back to the liver, to be incorporated again into bile. This system, the enterohepatic circulation, is so efficient that the body's entire pool of bile salts may cycle through it twice during the digestion of a single meal.

The liver manufactures around two and a half pints of bile daily. This is collected in the hepatic ducts and trickles along the common hepatic duct and then up the cystic duct to the gallbladder, to await a meal. But the gallbladder is no passive storage vessel. In it, there is such active resorption of mineral salts and water that the volume of bile decreases by nine-tenths to a thick, mucous consistency.

Release of bile is triggered by a hormone, cholecystokinin, which is in turn released by the duodenum when food is present in the stomach. The hormone reaches the gallbladder via the bloodstream and stimulates its contraction, causing the concentrated bile to be squeezed out along the bile duct.

Enzymes and insulin

The pancreas is a long, soft, glandular organ tucked behind the stomach, with its bulky head portion on the right fitting neatly into the C-shaped duodenum. It is composed of many tiny lobules that secrete its powerful digestive enzymes. Each lobule has a collecting duct that channels the juices toward the main pancreatic duct, and thence into the duodenum. There is a second, smaller, accessory pancreatic duct that enters the duodenum just before the main one. Cholecystokinin, the hormone that urges the gallbladder to expel bile, also initiates release of pancreatic juices.

Between the lobules are the islets of Langerhans, the source of the hormone insulin. This chemical messenger is released directly into the blood, not into the pancreatic ducts, and travels throughout the body to control the metabolism of glucose. Thus the pancreas is both an exocrine (ducted) and endocrine (ductless or hormone-secreting) gland. The islets make up less than one-hundredth of the pancreas by weight.

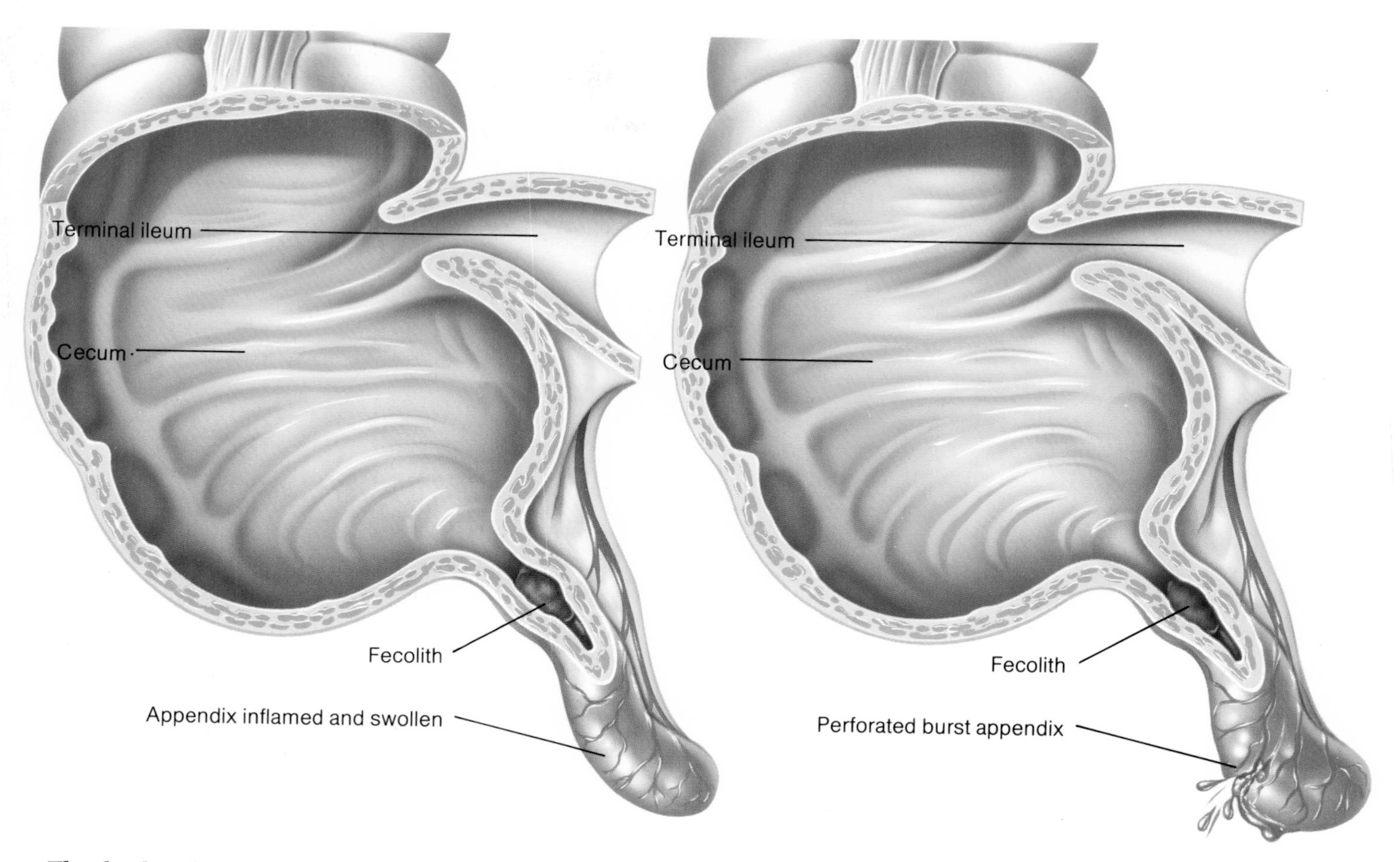

A useless nuisance, the appendix is a blind-ended tube off the lower part of the cecum. If it becomes blocked by a hardened piece of feces – a fecolith – the bacteria inside it multiply and cause inflammation. The inflamed appendix swells and fills with pus, and its owner suffers the abdominal pains of appendicitis. Untreated it may burst, causing life-threatening peritonitis. Antibiotic drugs can reduce the inflammation, but the only permanent cure is surgical removal of the appendix in an appendectomy operation.

The dead-end worm

Digestion in the small intestine takes place almost continuously, and in a slushy, watery environment. The colon (large intestine) resorbs water from the intestinal contents, making them firmer and more manageable. And it stores up the contents until it is convenient to void them from the body.

Although the colon performs these two useful functions, there is a particularly troublesome appendage attached to the first portion of it, which has no known function – the appendix. This dead-ended organ, resembling a fat, five-inch worm, is more correctly termed the vermiform ("worm-shaped") appendix. It branches from the cecum, the bulbous first portion of the colon which follows from the terminal ileum. It is made of muscle fibers sheathing a layer of lymphatic tissue, lined by epithelium resembling that lining the colon.

Many theories have been proposed to explain the function of the appendix. Some more recent versions revolve around its involvement in the immune system (because of the lymphatic tissue in the appendix). However, this organ has given no entirely satisfactory explanation of its existence, so it may be that it is indeed an evolutionary remnant of our prehistoric past. As though indignant at our ignorance, the appendix pays us back by occasionally becoming swollen and inflamed, in the acute condition known as appendicitis. The most frequent causes of appendicitis are blockage of the space within (its lumen) by a hard lump of some sort – a fruit stone, perhaps, or an unusually firm nodule of fecal matter (a fecolith) – or obstruction by the swelling of lymph tissue in its wall. The usual treatment is an appendectomy to remove the inflamed appendix.

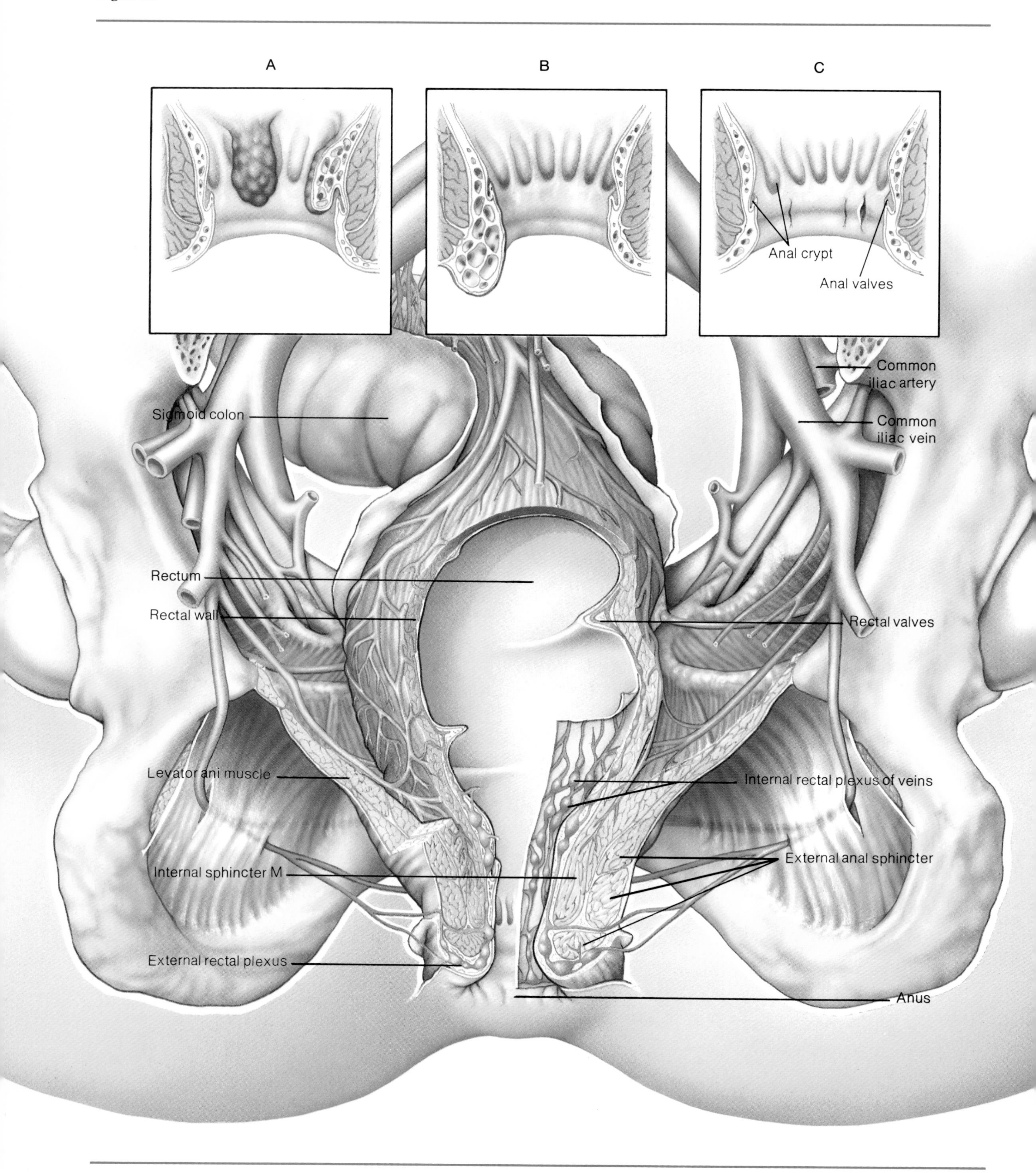
A
B
C
Anal crypt
Anal valves
Common iliac artery
Sigmoid colon
Common iliac vein
Rectum
Rectal wall
Rectal valves
Levator ani muscle
Internal rectal plexus of veins
External anal sphincter
Internal sphincter M
External rectal plexus
Anus

The contents of the rectum are confined by the internal and external sphincters of the anus. The area is well served by veins, and damage to them causes the bleeding of hemorrhoids. Internal hemorrhoids (A) occur near the entrance of the anal canal. External hemorrhoids (B) are accompanied by swelling and cause considerable pain during defecation. The consequent reluctance to defecate can, in turn, lead to constipation, which only makes the condition worse. An anal fissure (C) involves splitting of the skin that lines the anal canal, again resulting in bleeding and painful defecation. Severe hemorrhoids may need treating with surgery; anal fissures often heal spontaneously.

The penultimate stage

The colon, at just over two inches across, is considerably wider than the "small" intestine. It is also much shorter, at less than five feet long. Beginning with the ileocecal sphincter, a valve at the entrance to the cecum, it runs up the front right-hand side of the abdomen (ascending colon), across the top beneath the liver (transverse colon) and down the left side (descending colon), to turn back toward the midline in an S-shaped bend (sigmoid colon) before opening out into the next region, the rectum. It is anchored to the back wall of the abdomen by the mesentery.

We are what we eat – but not all that we eat. Some food remains undigested, notably the cellulose of dietary fiber. This, combined with leftover digestive juices and bile contents, sloughed-off cells from the gut lining, and dead and dying bacteria, makes up the brownish fecal matter. As most of the water is being absorbed, the now semisolid feces are pushed by peristalsis along the colon into its last portion and then to the rectum, a six-inch, distensible, baglike structure, where they remain until evacuated. Plenty of fiber and other roughage in the diet helps to give the feces a moist, bulky quality so that they pass swiftly and easily through the colon, as nature intended.

Bacteria make up almost one-third of the dry weight of feces. These microorganisms dwell in all regions of the gut, although the bacterial "profile" differs from one region to the next. Feces, held in the colon and rectum, are an especially favored breeding ground for bacteria.

A normal person plays host to millions of colonic bacteria. In their life processes, they help produce and facilitate the absorption of several important chemicals into the body, including vitamin K. Hence they are more than benign; some are positively helpful. However, one of their side-effects is to produce gas, mainly methane, as they break down the food residues to provide energy for themselves. The quantity and quality of this gas, or flatus ("wind"), depends partly on the food eaten. Beans have a justifiable reputation because they contain a number of trisaccharide sugars which cannot be broken down in the small intestine. When these sugars reach the colon, they encounter bacteria capable of metabolizing them – but with large amounts of gas produced as a result. Just as people vary greatly in the frequency of their bowel movements, so there is variation in the amount of flatus released each day.

Journey's end

The final region of the gut is the two-inch-long anal canal. Of all the sphincter mechanisms in the digestive tract, this is probably the best known. So are its possible disorders, such as hemorrhoids or "piles" (swollen veins), fissure, fistula and abscess.

In fact there are two anal sphincters, internal and external. The internal sphincter is normally in a contracted state to prevent any leakage of fecal material through the anus. As the rectum becomes stimulated by the accumulation of feces, nervous impulses pass from it to the anal area and trigger relaxation. Of course, in an adult this is rarely a simple reflex reaction. Rectal distension also produces a conscious desire to empty the bowels. The external anal sphincter has a large degree of voluntary nervous control, so that we can keep this contracted until we will it to relax, allowing defecation to take place.

There is no "normal frequency" of defecation, although each individual tends to have his or her own pattern, and a sustained departure from this pattern might provoke suspicion. Perhaps the old saying, "from once every three days to three times a day," represents something like the range across which a physician would not show concern. In some people, worry or anxiety can heighten the sensitivity of the system, necessitating frequent visits to the bathroom.

Food poisoning

Illness caused by eating bad food – that is, food contaminated with harmful bacteria and other pathogenic microorganisms – was once common before the days of canning, freezing and domestic refrigerators. Yet modern mass production of food, such as intensive farming of chickens for eggs, part-cooked TV dinners from supermarkets and other kinds of fast foods have meant that if a food bug does get a hold, it can infect thousands of people very rapidly.

Most notorious of food contaminants are microbes of the salmonella group, which cause typhoid fever as well as food poisoning. They are rapidly killed by heat, and so foods that are adequately cooked or pasteurized – for 20 minutes at a temperature of 140°F – are safe. The danger can still come from dirty cooking utensils or from warmed-over or insufficiently defrosted foods.

In North America in the early 1970s, an outbreak of salmonella poisoning was traced to a chocolate factory, which proves that even seemingly innocent foods can harbor the microbes (which originated in the raw cocoa beans). Other potential sources in food processing plants include the feces of rats (in which salmonella can survive for 150 days), cockroaches and flies.

Food poisoning may also be caused by the clostridium bacterium, whose spores can survive heating to the temperature of boiling water (212°F). If cooked meat containing the spores is allowed to cool slowly, they germinate and the bacteria multiply rapidly to produce a toxin which causes vomiting and diarrhea if the meat is eaten. For this reason, cooked meat that is not to be eaten immediately should be cooled rapidly after cooking and refrigerated. Strict hygiene and thorough cooking are the most effective measures against all forms of food poisoning, in food factories, commercial kitchens and in the home.

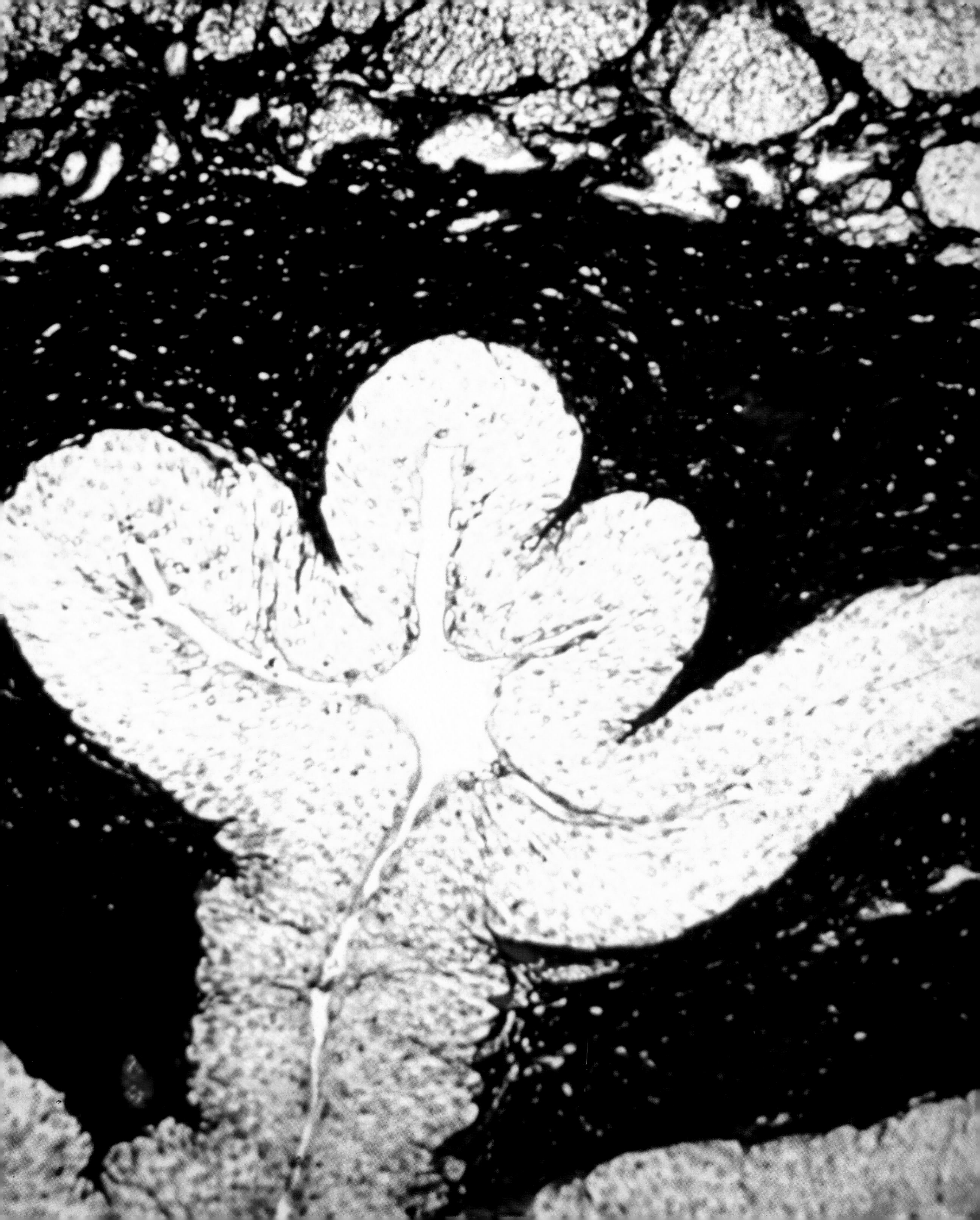

CHAPTER 10

The Kidneys

The contemplation of physical well-being tends to involve thinking mainly of inputs: getting enough oxygen, food and fluids. The less appealing matters of waste disposal do not receive such attention, but they are just as vital, for without them the body is liable to be poisoned by itself. The body has four routes for elimination of wastes. One is the skin, for removing certain salts and minerals. A second is the lungs, for getting rid of carbon dioxide and of water in vapor form. A third is the intestines, for removal of undigested leftovers from food – although in a sense, undigested remains have never been truly "inside" the body. The fourth involves the principal excretory organs, the kidneys.

The body's balancing act

The paired kidneys, in the upper part of the abdomen toward the back, perform a masterly balancing act. They balance the fluid levels in the body. They balance the body's acid/alkaline nature, and they balance concentrations of salts, minerals and other substances. In times of plenty, unwanted materials are removed; when times are hard, the kidneys conserve.

The kidneys work through the medium of the blood. Blood flows everywhere in the body, fetching and carrying, distributing and collecting. Part of its circulation is through the kidneys, and there it is balanced – filtered, purified, cleaned and adjusted. Like a commercial effluent treatment plant it is a continuous process, 24 hours each day.

The kidneys form part of a major body system, the excretory system. Its other structures include the two ureters, tubes which lead from the kidneys to the bladder, and the urethra, which leads from the bladder to the exterior of the body. Waste disposal by the system can be described relatively simply. The kidneys filter blood and produce a filtrate, urine, which contains unwanted substances. The urine is stored in the bladder until such time as it is convenient for it to be expelled during the act of urination (technically called micturition).

The simple description belies a marvel of metabolic juggling. Hundreds of pints of fluid and dozens of chemicals are involved. The renal arteries supply blood to the kidneys. The flow they carry is an astounding 750 to 1,000 pints daily, approaching one-quarter of the heart's output, and roughly equivalent to the body's entire blood volume circulating through the kidneys 20 times each hour. However, only one to two thousandths of the blood flow emerges at the other end, as urine. The average daily production of urine is just over two and a half pints, although it varies considerably according to how much we eat and drink, how active we are, the ambient temperature, and other factors.

Microscopic sieves

The Roman physician Galen, of the second century A.D., suggested that the interior of the kidney was a sieve which filtered out impurities into the urine. He imagined it as one large sieving surface with innumerable pores too small to see. In fact,

The star-shaped space at the center of this microphotograph is a section of a ureter, one of the two tubes that carry urine from the kidneys to the bladder. Less than one-eighth of an inch in diameter, the ureter has muscular walls that squeeze the urine along. Because of its narrowness, it is easily obstructed. A blockage – perhaps by a kidney stone – causes urine to back up to the kidney and gives rise to the excruciating pain of renal colic.

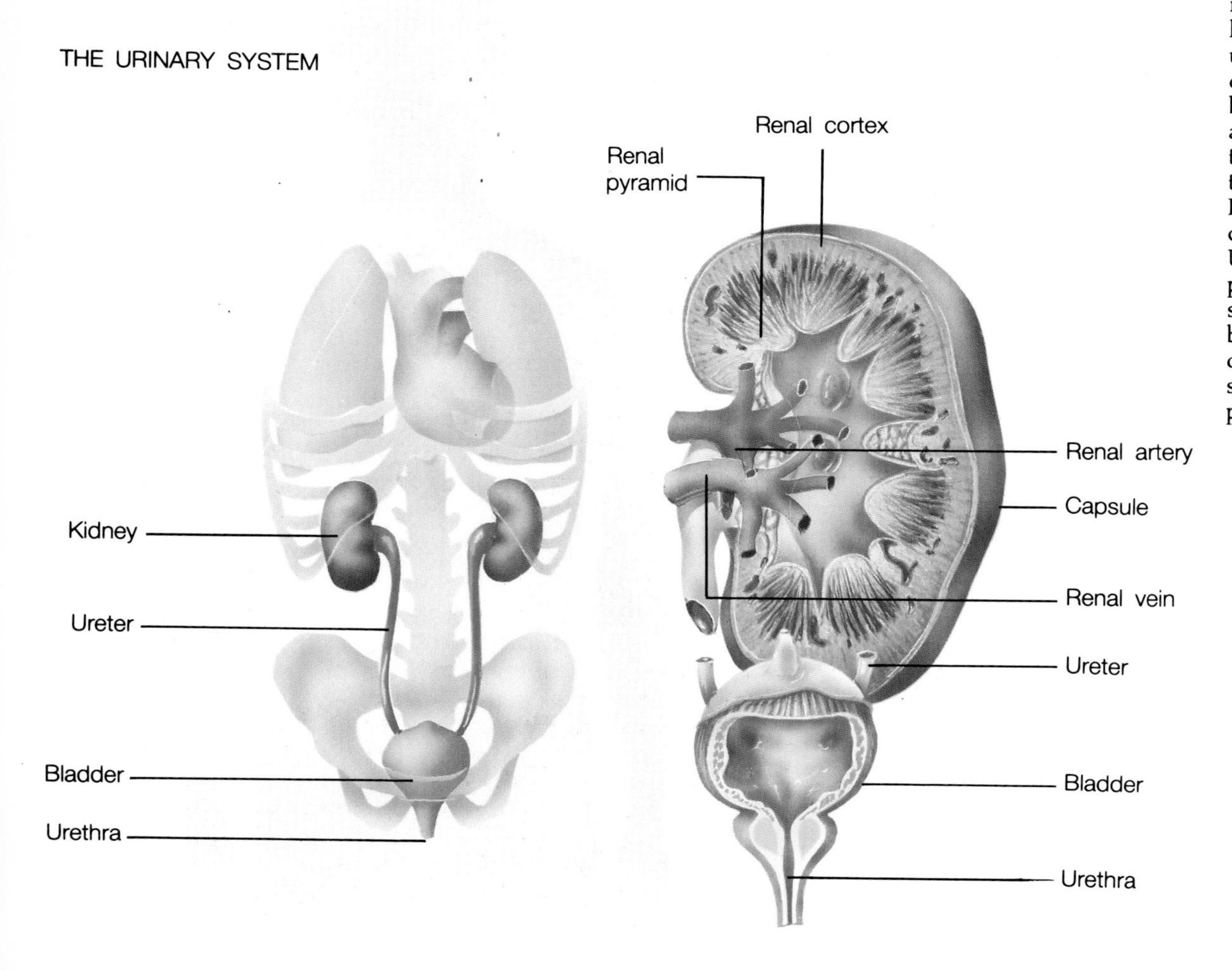

The urinary system filters the blood and stores waste products in urine. It consists of the kidneys, ureters, bladder and urethra. In addition to removing dissolved wastes from the blood, fed in along the renal arteries, the kidneys maintain the balance of salts, water and trace elements in the body. Purified blood is returned to the circulation along the renal veins. Urine drains from the renal pelvis into the ureter, to be stored in the bladder. In men the beginning of the urethra, the outlet from the bladder, is surrounded by the globular prostate gland.

his notion was not too wide of the mark, considering that the role of the heart and the circulation of the blood were not realized until 15 centuries later, by William Harvey.

However, the gross anatomy of the kidney reveals almost nothing of its finer workings. Slice one in half and you first encounter its tough external coat, the renal capsule. Inside the organ, the smooth outer layer is known as the cortex; the more stripy, textured inner layer is the medulla, and it is grouped into lobes termed pyramids. On its concave side there is a converging system of collecting spaces, the renal pelvis, which gradually narrows to form the thin tube of the ureter.

The hidden secrets of kidney function had to wait until the invention of the microscope. Celebrated Italian microscopist Marcello Malpighi studied its tissue and published his observations in 1659. But with no wider framework of knowledge, and with the mighty Galen's "sieve" still deeply entrenched in the medical beliefs of the time, Malpighi failed to make progress. He was denounced by his colleagues for imagining what he saw through the eyepiece.

In 1842, the English physician William Bowman wrote the treatise that proved a watershed in renal studies: "On the Structure and Use of the Malpighian Bodies of the Kidney With Observations on the Circulation through that Gland." Bowman had unraveled the secret of the first part of kidney function, describing how the maze of tiny tubes and blood vessels in the organ managed to filter wastes and water from the blood. Only two years later Carl Ludwig, in Vienna, showed that this initial filtrate was then concentrated in a secondary part of the tube network, as the body reabsorbed most of the water from it. We now know that the kidneys filter up to 350 pints of water and solutes from the blood each day, in the first part of the process; and that reabsorption of water from this filtrate brings down its volume to two and a half pints of actual urine per day.

Kidney beans

An average kidney is about four inches long, up to three inches wide, and nearly two inches deep. It weighs around five ounces and is deep red in color. Because it cannot informatively be described as kidney-shaped, it is often termed "bean-shaped."

Humans are symmetrical organisms. We have paired arms, paired lungs, and paired kidneys. The two kidneys are tucked in the upper abdomen, the left one being about half an inch higher than the right. Many people perceive their kidneys as being lower and more to the side than they actually are. A common misconception is that they are just under the skin of the "flanks," on the sides of the abdomen, broadly at waist level. In fact they are buried much more deeply, centrally and higher, being only two inches or so from the body's midline and directly behind the lower ribs. The top of each kidney is more rounded than the base, and tilts inward toward the midline. As you stand up from a lying position, your abdominal organs "sag" and the kidneys move down with them, about an inch lower than their recumbent pose. They also bob up and down during breathing.

Behind the kidneys is the musculature of the spine; the left kidney is flanked by the pancreas, jejunum (part of the small intestine) and the colon; the right kidney, displaced downward by the bulk of the liver, touches the duodenum and colon. This array of ribs, spine and other adjacent parts provide superior protection for such important organs. In addition, the kidneys are padded by a special type of surrounding fat which holds and cushions them firmly.

Atop each kidney sits an adrenal, or suprarenal, gland. These glands are not involved in excretion, being part of the body's endocrine or hormonal system.

This arrangement is the normal one. Some people have only one kidney, either through a developmental abnormality or following surgical removal of the second kidney because of disease. Or the two kidneys may develop fused together into a horseshoe shape. These departures should pose no problems, provided the existing kidney is not diseased in any way.

The problem of nitrogen

Why does the body need to excrete? Why is it not possible to take in exactly what we require, and no more? Because the human body is dynamic. There is turnover: cells die, and need replacing. The substances that make up these cells are built mainly from the elements carbon, hydrogen, oxygen and nitrogen. So are the by-products left in the body when proteins are broken down.

Carbon and oxygen can be lost as carbon dioxide (CO_2) in the breath during exhalation. Hydrogen and oxygen can be combined to make "metabolic water" (H_2O). But nitrogen poses a problem. Although in its elemental state it is an inert gas (the air we breathe is almost 80 percent nitrogen), the concentrated leftover chemicals produced by the body, known as nitrogenous waste, are toxic. The liver deals with most of the nitrogenous waste, converting it to a relatively simple chemical, urea. Urine is a solution of urea in water, along with various other substances, as the circumstances dictate.

Keeping the inside constant

The kidneys also cope with the variety of inputs and activities to which the body is subjected each day. You might overload your system with food one day but hardly eat the next. You might then have a very active day, moving energetically and sweating profusely without taking in fluid to compensate. The following day might be spent in restful repose, yet you might take in large amounts of fluid. Through the medium of the blood and other internal fluids, the kidneys iron out these fluctuations. They are the great levelers.

Body cells, left to their own devices, cannot tolerate much change. They are specialized and delicate devices, able to perform their singular and specific functions as part of the body's team, but unable to look after themselves in more basic ways – such as controlling their temperature and immediate chemical environment. They must be cared for and cossetted. The 19th-century French physiologist, Claude Bernard, first recognized the presence of the milieu interieur, or "internal environment."

The body is not a series of watertight compartments, each holding an organ. Cells contain their own, intracellular, fluid. They are bathed in extracellular fluid, which undergoes a slow interchange with the intracellular fluid and with both blood and the lymph fluid of the lymphatic system. Together these fluids must provide constancy of temperature, of sugar and salt concentration, and of acidity and alkalinity. Then the specialist workers, the cells, can get on with their jobs. The kidneys are the carers, keeping the inside constant.

A million microfilters

The kidney's sieves, or filters, can be seen only with the help of a high-power microscope. Their tiny tubes coil and intertwine, so that early researchers had much trouble trying to untangle the organ's microanatomy, which has been described as "raspberries and spaghetti." However, today we have a clear description of the nephron, the active site of filtration.

There are about one million nephrons in each human kidney. Each nephron consists of several parts. There is a tiny knot of capillaries, the glomerulus ("raspberry"). The glomerulus is cupped in the expanded end of a long tube. The expanded part is known as Bowman's capsule (in honor of William Bowman), and the long tube (the "spaghetti") is termed the renal tubule. It twists and turns in the vicinity of the capsule and then throws a long loop, the loop of Henle, from which it returns to the neighborhood of the capsule. Finally the tubule winds away from the glomerulus and capsule, and joins with other tubules to form a larger collecting duct.

The glomeruli, Bowman's capsules and their adjacent parts of the renal tubule are embedded in the kidney's outer cortex – a million sets of each. The long loops of Henle (many are nearly an inch long) and the collecting ducts are found in the medulla. The ducts come together and open into the renal pelvis, the kidney's central space. If the tubules of all nephrons were uncoiled and joined end to end, they would stretch for 50 miles.

A single nephron's structure is perhaps best detailed by following the fate of blood as it flows along the renal artery, heading for the kidney. In this way it is possible to trace, step by step, the conversion of certain elements of the blood into urine.

The vital blood supply

The renal artery leaves the aorta and divides into five branches, which fan out on reaching the renal pelvis in order to supply five distinct areas of the kidney. About one person in three has one or more of these branches coming directly off the aorta and entering the segments within the kidney. For them, a single renal artery as such does not exist.

The kidneys, like the heart and brain, receive a "privileged" blood supply in preference to other parts of the body. Even when total blood flow varies widely, because of dehydration or blood loss, the flow rate through the kidneys remains at roughly 120 pints per hour. This proves their importance to the body: their job is so vital that it must continue, even if this means putting other systems under stress.

Once inside the kidney, the arteries branch again and again as they travel up the pyramids of the central medulla, until they reach the edge of the cortex. There they fan out, and smaller branches lead off at right angles to feed every part of the cortex. These tiny interlobular arteries give rise to the even smaller arterioles, which go on to form the tangles of capillaries, the glomeruli. The initial filtration of the blood is about to begin.

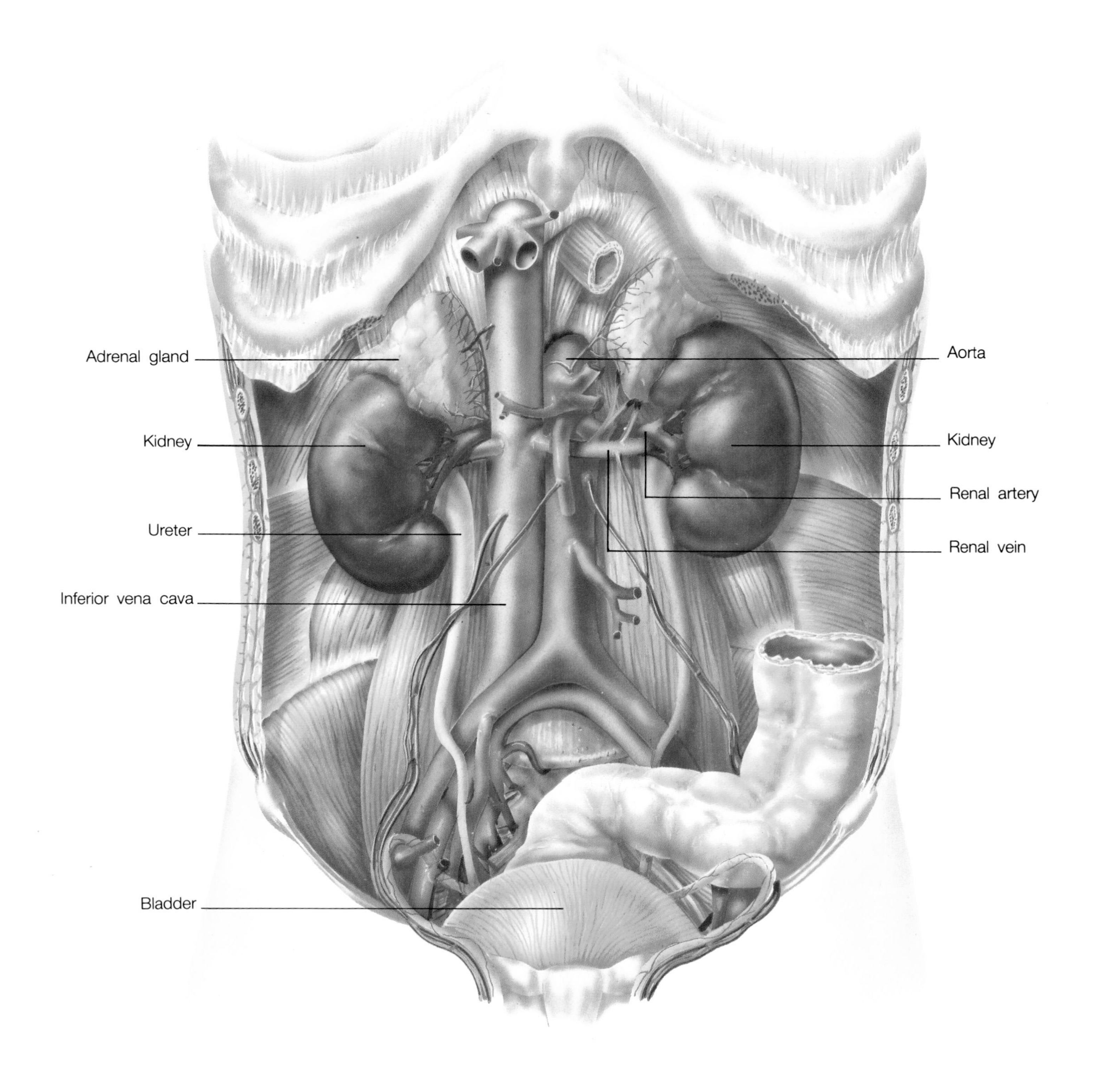

For maximum protection against injury, the kidneys are tucked up toward the rear of the abdomen, guarded by the lower ribs. Their never-failing blood supply flows along the renal arteries, which branch off the aorta, the body's main artery from the heart. Blood cleansed of wastes by the kidney's filters returns along the renal veins to the inferior vena cava, on the way to the heart to be pumped back into the circulation. The main structures within the kidney *(right)* are the outer cortex, and the inner medulla. Bunches of medulla form pyramids, from which urine drains into the pelvis of the kidney, then enters the ureter.

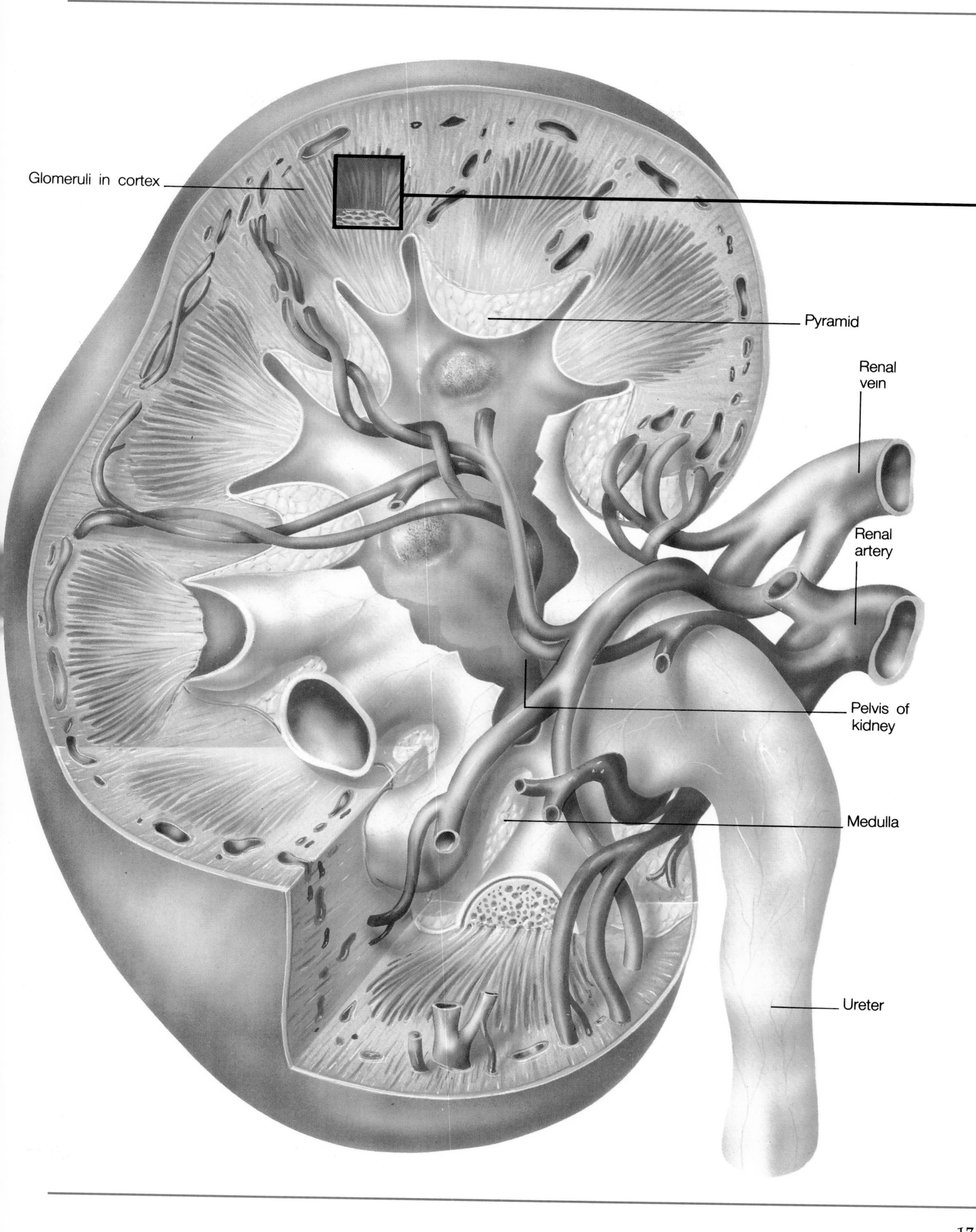
Glomeruli in cortex
Pyramid
Renal vein
Renal artery
Pelvis of kidney
Medulla
Ureter

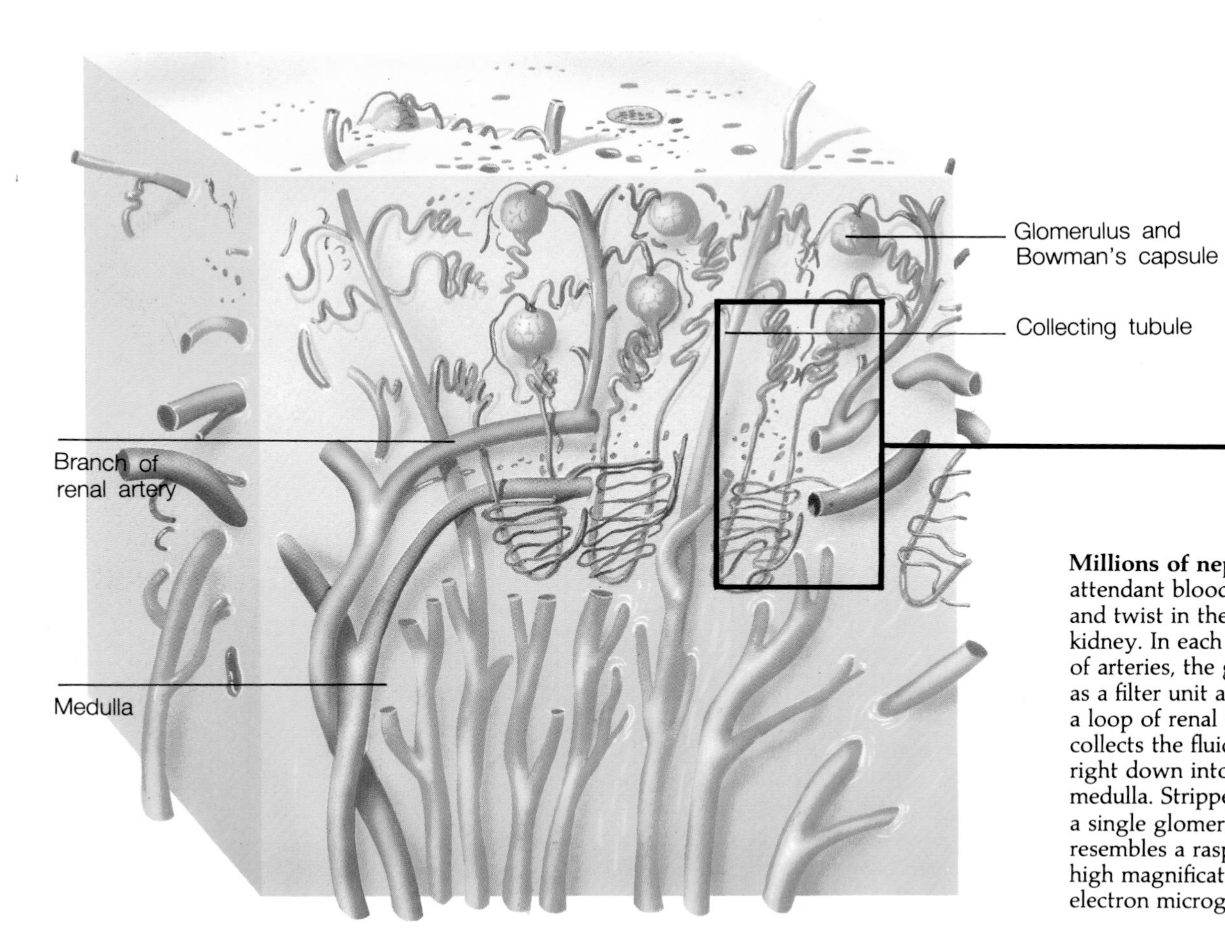

Millions of nephrons and their attendant blood vessels loop and twist in the cortex of the kidney. In each nephron a knot of arteries, the glomerulus, acts as a filter unit at the wide end of a loop of renal tubule that collects the fluid filtrate and dips right down into the kidney's medulla. Stripped of its capsule, a single glomerulus *(below)* resembles a raspberry in this high magnification scanning electron micrograph.

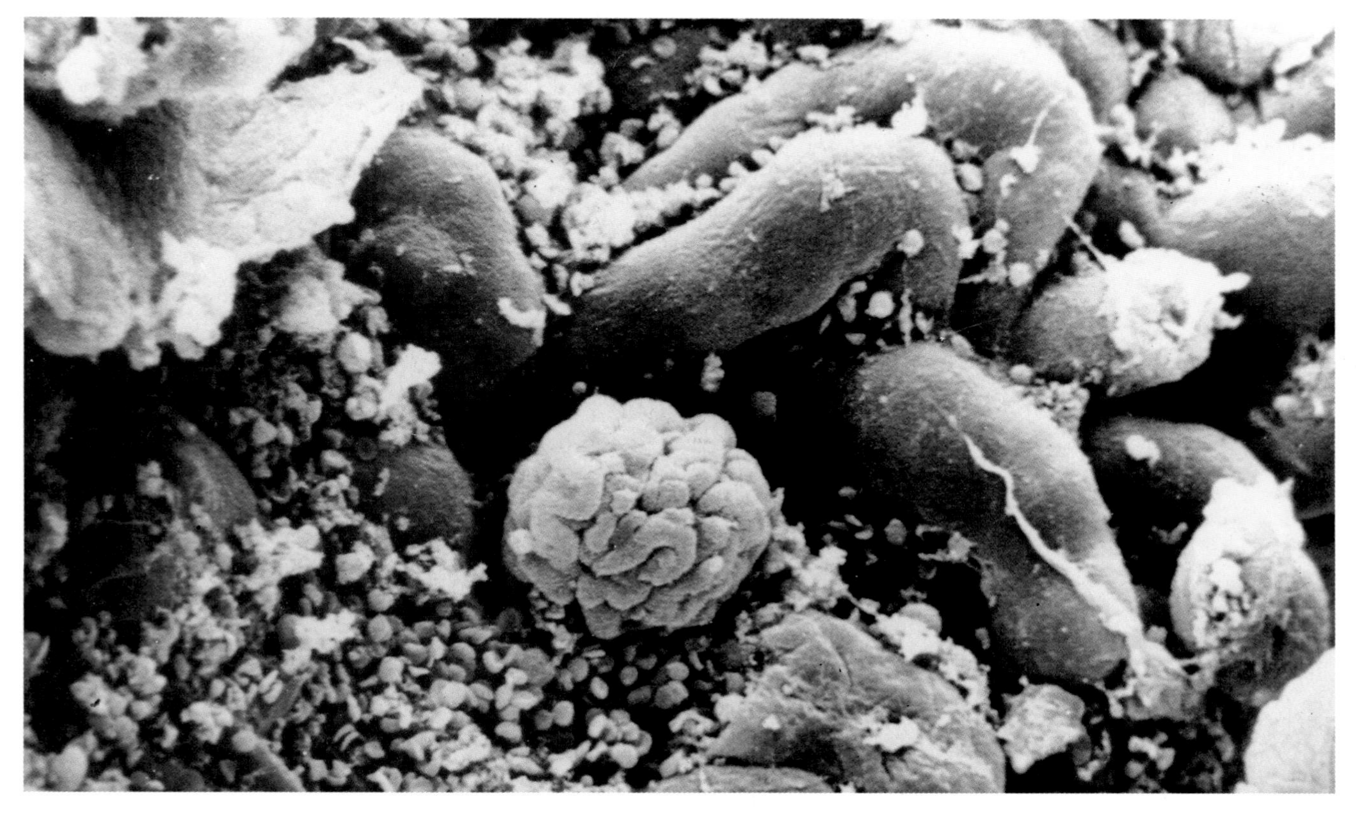

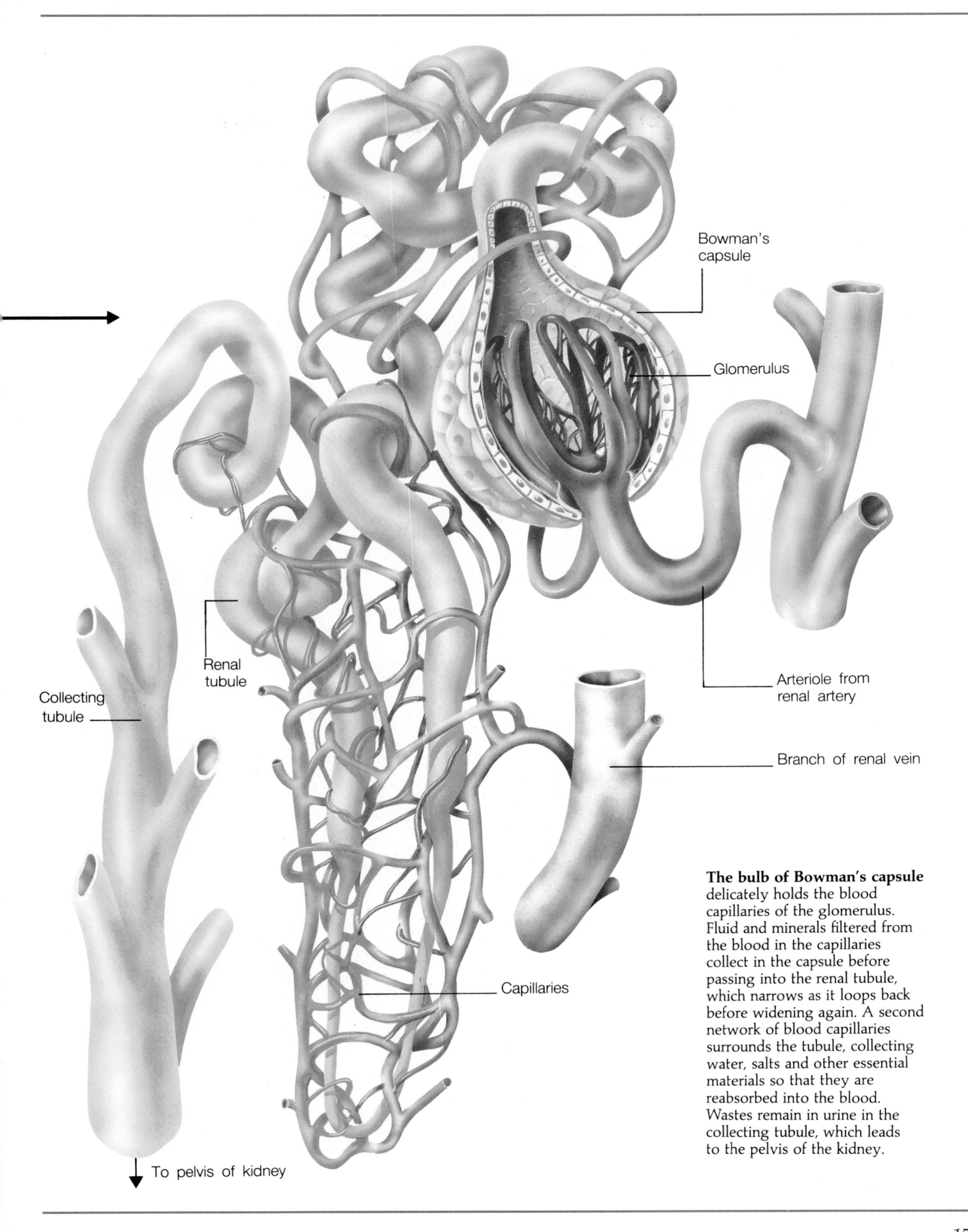

The bulb of Bowman's capsule delicately holds the blood capillaries of the glomerulus. Fluid and minerals filtered from the blood in the capillaries collect in the capsule before passing into the renal tubule, which narrows as it loops back before widening again. A second network of blood capillaries surrounds the tubule, collecting water, salts and other essential materials so that they are reabsorbed into the blood. Wastes remain in urine in the collecting tubule, which leads to the pelvis of the kidney.

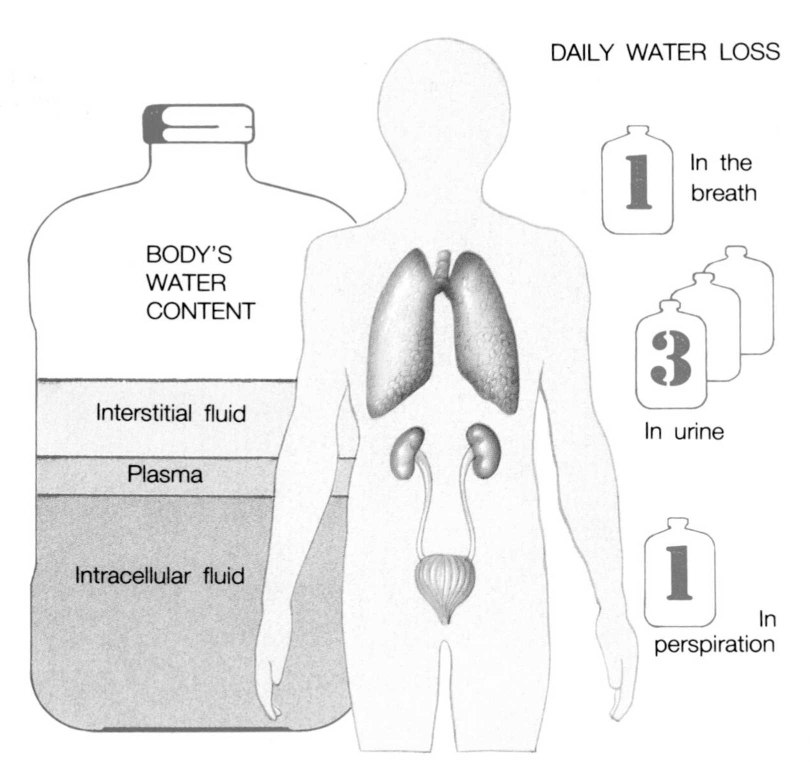

More than half the human bodyweight is water, contained in the blood, lymph, cells and intercellular spaces. Each day we drink an average of five pints of fluid, and to keep the body in balance we also lose five pints each day. One pint is lost as water vapor in the breath, one pint in perspiration, and three pints in urine as water filtered from the bloodstream by the kidneys.

The filtering surface

The actual filtering surface is the glomerular basement membrane, which acts as a selective barrier between the capillaries of the glomerulus itself and the beginning of the urinary space within the Bowman's capsule. The basement membrane is made of a mat of filaments which, on the capsule side, interlock with each other like the bristles of two hairbrushes pressed together. Gaps or "slit pores" between the bristles allow comparatively small molecules to pass through. Nothing as big as a red blood cell can normally traverse the membrane from blood to the urinary space.

The fluid pressure of blood in the capillaries is much higher than that of the fluid in the urinary space (the journey from the heart, where the pressure originates, to the kidney is very short and blood still retains much of its force). Plasma, the watery part of the blood which contains many dissolved substances, is pushed through the glomerular basement membrane by this irresistible pressure gradient. With it travel smaller molecules such as urea (the nitrogenous waste product), creatinine (a waste from muscle metabolism), glucose, mineral salts and some of the small blood proteins. It is, in effect, a pressure-filtration system. Bigger molecules and blood cells stay in the capillary.

The basement membranes offer a large surface area for filtration. In a normal day, for an average adult, about 350 pints of fluid from food and drink are filtered from the glomeruli of both kidneys into the tubules. This means that two and a half fluid ounces of liquid pass through the collective basement membranes of one kidney each minute.

Super selectivity

The liquid that comprises this initial filtrate contains many substances that the body cannot afford to lose, including glucose (its chief energy source), amino acids (the building-blocks of proteins), and mineral salts such as sodium, chlorides and phosphates. Water itself also passes through the filters in abundance, and the body cannot risk losing it at such a rate. So, after the initial "blunderbuss" of filtration, comes selective reabsorption. It is rather like trying to sort items in a large box. The renal method would be first to tip them all out onto the floor (initial filtrate production), then put back the ones to be kept (selective reabsorption), and throw away the unwanted ones (urine removal).

The Bowman's capsule gives way to a part of the winding renal tubule called the proximal tubule. The cells lining it are unlike those in any other area of the nephron. Instead of being flat and interlocking they have brush borders, with their outer membranes thrown up into thousands of fingerlike projections or microvilli. This greatly increases the surface area of this portion of tubule. (The small intestine also employs microvilli to increase its surface area, for absorption of nutrients.) And the presence in these cells of many mitochondria – the cellular "power plants" that supply energy – reveal that energy-consuming reactions are actively occurring here. In fact, 80 percent of the water filtered out of the blood is reabsorbed by this portion of the nephron, along with 65 to 70 percent of the sodium that got through the gomerular filter.

Strangely, urea is also reabsorbed – although it is one of the principal unwanted materials in urine. Its re-entry into the body is an unavoidable side-effect of renal chemistry. About half of the urea passing into the tubule at the capsule end is withdrawn back into the body. But successive cycles from blood to glomerular filtrate ensure an acceptable throughput of urea, with eventual expulsion.

Where do the reabsorbed substances go? Encircling the tubule are thousands of tiny blood vessels, ready to receive them via the wall of the tubule and their own walls, and transport them back through the venous system. The vessels are known as stellate veins because of their star shape. Blood from the glomerular capillaries flows into these tiny veins and becomes "rejuvenated" with the reclaimed water and salts. It then travels onward as the veins unite and grow, forming an enlarging network that heads toward the renal pelvis. At the pelvis, the veins come together to form the renal vein, which transports blood back to the heart.

Around the loop of Henle

Meanwhile, back at the nephron, the developing urine passes from the proximal tubule into the downward-sweeping loop of Henle. This delves into the kidney's medulla, becoming as it does so much thinner in diameter (down to one five-thousandth of an inch) and with thinner walls. Metabolically, this descending portion of the loop is passive. Water passes through its wall to the blood vessels beyond by simple osmosis – the natural passage of water through a membrane from a weak solution to a more concentrated one, until both solutions are the same strength. In this way, the urine becomes more concentrated.

On the upward part of the loop, the cells in its wall are thicker and fatter, with no brush border. These cells are thought actively to pump glucose, and sodium and other desirable mineral salts, from the urine and through themselves, out to the surrounding blood vessels. This part of the reclamation process, like that in the proximal tubule, requires energy because it has to run against the prevailing concentration gradient.

One staggering feature of the nephron is that, amid the tangles of capillaries, tubules and supporting cells within renal tissue, each nephron is virtually self-contained. The capillary of a glomerulus winds onward but sticks with its own tubule, encircling it intimately all the way around the loop of Henle and to its final portion, the distal tubule.

Regulating sodium

Back in the cortex, at the "far end" of the loop of Henle, is the portion of convoluted tubule known as the distal tubule. This is involved in the regulation of sodium, a valuable body commodity which is used to adjust the balance between extracellular and intracellular fluids.

In the urine, sodium is in the form of positively charged ions

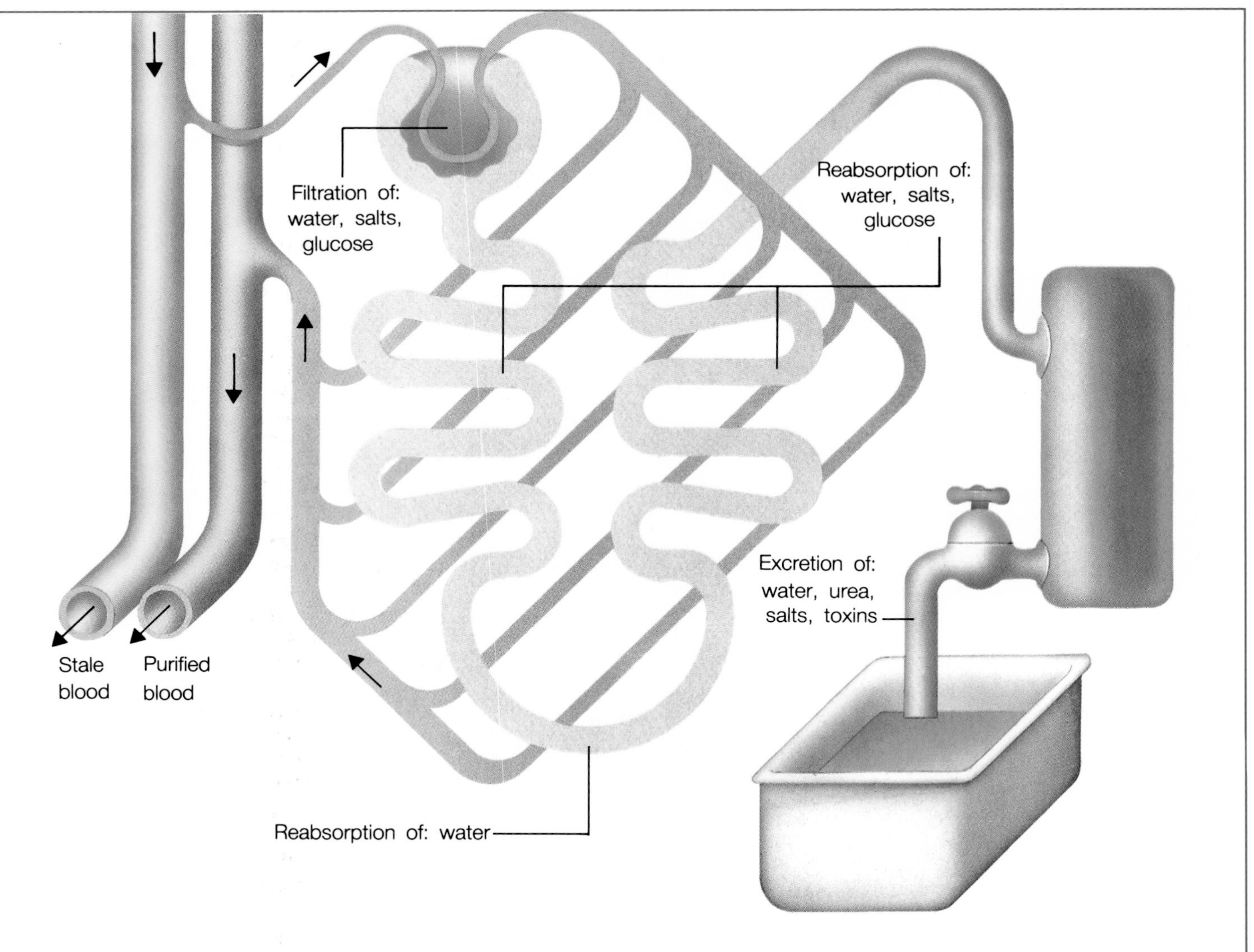

The whole of the urinary system is like an ingenious plumbing setup – with two separate circuits containing special pipes that selectively allow water and salts to leak through them. In engineering terms, the kidneys comprise a bypass filter, which every day siphons off 380 pints of fluid from the blood passing through them. Nearly 99 percent of this fluid is reabsorbed into the blood, leaving only two to four pints of waste-containing urine to be stored in the bladder and ultimately passed out of the body. But, overall, it is an extremely selective filter which allows the blood to retain glucose, salts and minerals but cleanses it of urea and poisonous substances.

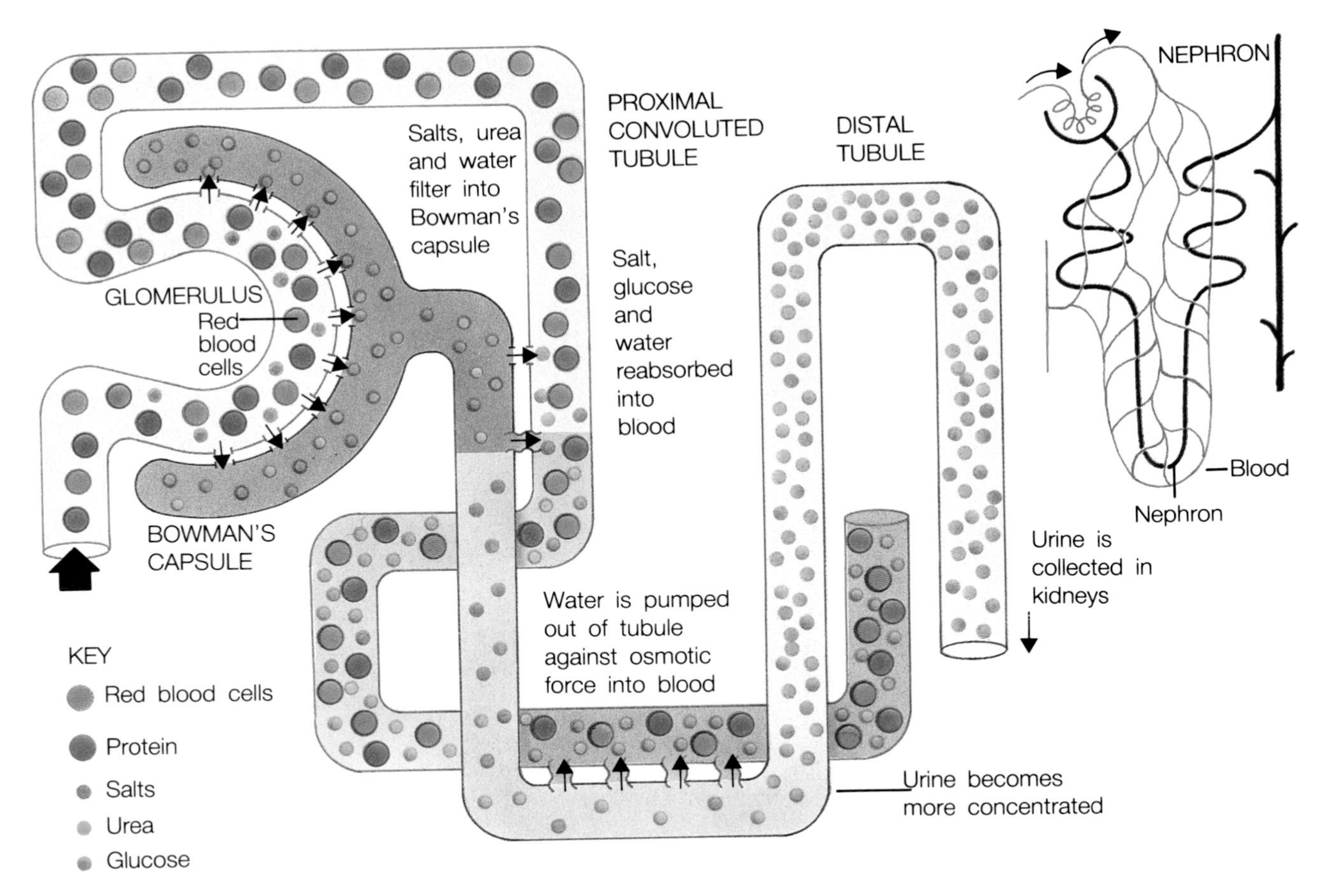

Different substances pass in and out of the blood at different stages of its journey around a kidney's nephron. Red blood cells and protein molecules are too big to pass through the filter of the glomerulus, and remain in the blood all the time. Salts, nitrogen-based urea and a great deal of water are filtered out into Bowman's capsule, and pass along the tubule. First salt, glucose and some water are reabsorbed by passing into the adjacent blood capillary. Then more water is pumped back into the blood, leaving mostly urea and some salts in the urine, which therefore becomes more concentrated along the loop.

(Na^+), along with accompanying negative ions of chloride (Cl^-) and bicarbonate (HCO_3^-). Most of the sodium (up to 70 percent) and other ions that pass through the glomerular pressure filter, into the Bowman's capsule, is reabsorbed in the proximal tubule. The distal tubule reclaims less, but its action can be varied by the hormone aldosterone, to "fine-tune" the body's sodium balance.

The first stage in hormonal regulation of sodium reabsorption begins in the kidney. When flow through the renal blood vessels increases, the vessel walls release the substance renin. Once in the bloodstream, renin stimulates a vital biochemical transformation – the conversion of a plasma "prehormone" substance into the near-hormone angiotensin I, which then undergoes a further chemical conversion in the blood to angiotensin II. The hormone angiotensin II acts on the two adrenal glands, on top of the kidneys, and causes them to release another hormone, aldosterone. (Angiotensin II also acts to raise blood pressure by increasing the resistance to blood flow.) The aldosterone is carried around in the blood and acts on the distal tubules of the kidneys to increase their rate of sodium reabsorption. So more aldosterone saves sodium, while less allows its escape.

Sodium and the other ions cannot be reabsorbed alone, however: they are in solution and because of this they "drag" water with them, from the tubules into the blood vessels.

Maintaining water balance

The distal tubules from different nephrons unite to form a combined or collecting tubule. Its wall gradually tapers, increasing in thickness as it unites with other collecting tubules and heads down to the renal pelvis.

Scientists believe that these collecting tubules show a unique response to antidiuretic hormone, ADH, one of the many blood-borne chemical messengers secreted by the pituitary gland at the base of the brain. When the concentration of solutes in the extracellular fluid rises, as happens when the body is dehydrated, special cells in the brain detect the increased concentration and switch on the release of this hormone. ADH travels in the blood to the kidney, "instructing" it to reabsorb more water.

Under the influence of the higher blood level of ADH, the cube-shaped epithelial cells that line the collecting tubules become permeable and allow water to enter, possibly because the hormone opens up little pores in their membranes. This allows water to pass from within the tubule to the surrounding tissue, so continuing the concentration process of urine. If the hormone level falls, the walls become "waterproof," and the urine in the tubules therefore remains dilute, so that little water is reabsorbed into the body. Depending on the prevailing internal conditions, urine can be four times as concentrated as the body fluids (it looks dark yellow when this happens), or if necessary up to three times as dilute.

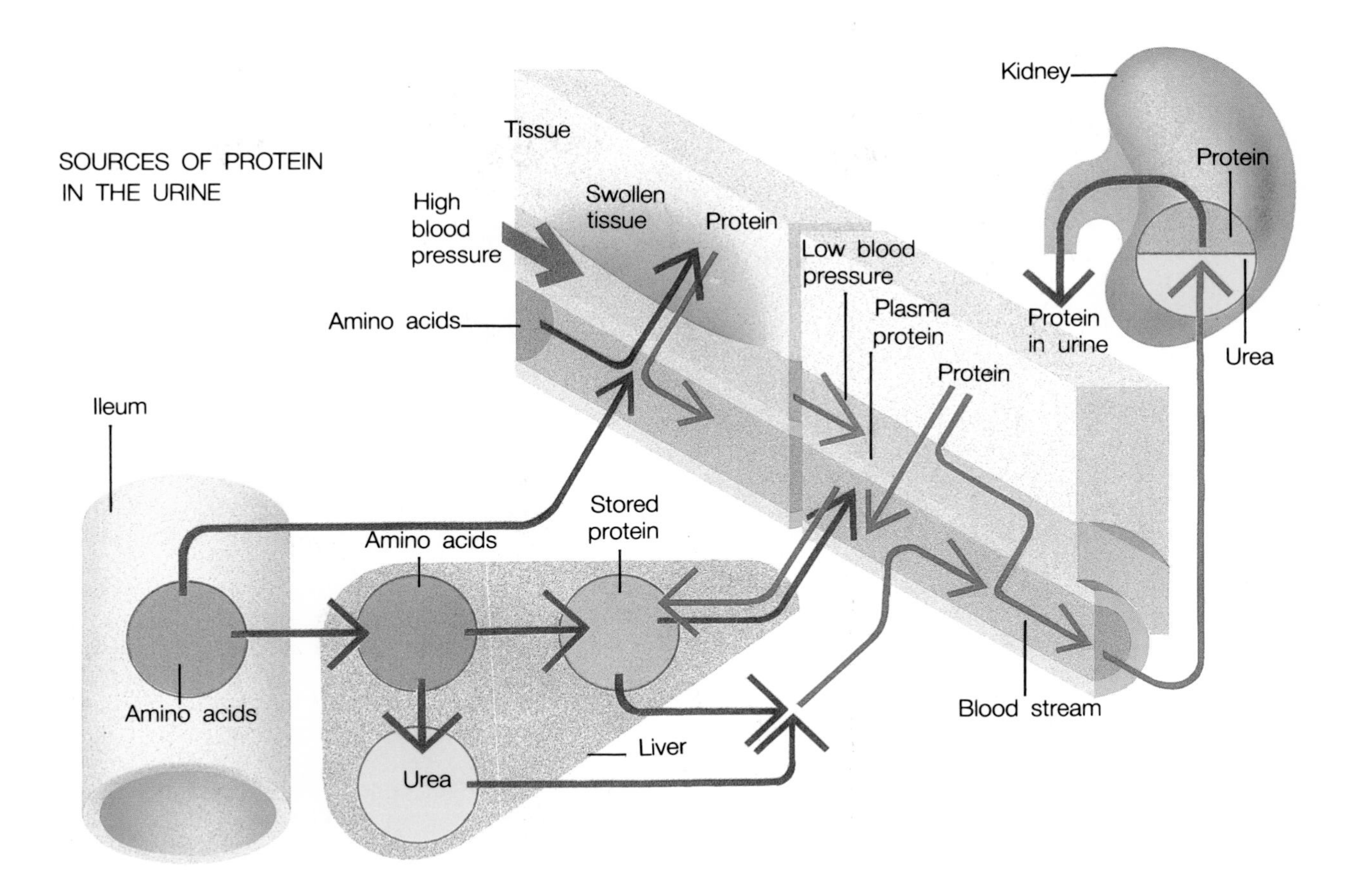

Most protein in the food we eat is broken down into amino acids during digestion, and stored or processed by the liver, which releases urea as a waste product. Some protein in the urine is normal, especially after exertion, but larger amounts are not. High blood pressure may cause swelling of tissue and release of protein, resulting in edema, and low blood pressure may cause an increase in plasma proteins. The excess may pass to the kidneys in the bloodstream and be filtered out, causing the presence of protein in the urine (proteinuria), which is usually symptomatic of metabolic disorder.

The kidneys and blood pressure

The hormones ADH and aldosterone both have another important job to do – both are involved in the long-term maintenance of blood pressure. The link between blood pressure and urine production is clearly shown by the following sequence of events.

There is an accident. The victim suffers severe wounds and begins to bleed profusely; this means a sudden fall in blood volume, and a precipitous drop in blood pressure. The body's immediate response is to shut down many of the blood vessels, except those supplying the most vital organs such as the heart and brain, as well as increasing the heartbeat rate.

In the longer term, as blood clotting seals the wounds, the body needs to restore its fluid loss. Enter the kidneys. The low blood volume stimulates release of ADH, which ensures the maximum amount of water is reabsorbed from the tubules back into the body's fluid systems. Meanwhile the low blood pressure initiates the release of renin, which, as described above, increases the angiotensin II level. This in turn increases the aldosterone level, which, via the distal tubules, pulls more sodium and other ions – as well as their accompanying solvent, water – from the urine into the body. This helps replace the fluid loss and restore the volume, and pressure, of the blood.

What happens when the situation is reversed and the plasma volume expands or blood pressure is raised? Renin release is inhibited, which means aldosterone secretion is reduced, and the release of ADH is minimal. The net outcome is that less sodium and water is reabsorbed, and the body rids itself of extra fluid. This helps reduce blood volume and lower blood pressure.

Hypertension and the kidneys

People with hypertension (persistently high blood pressure) may be put on a low-salt diet. The reasoning follows the link explained above between blood pressure and urine production. With low levels of salt in the body, very little renin and angiotensin II (which has blood pressure-raising effects) are free in the blood. Furthermore, if less salt needs to be reabsorbed by the kidneys, less water is reabsorbed too, which again helps to keep blood volume, and hence pressure, in check.

Many disorders of the kidneys, including either high or low blood pressure, can damage the nephrons and interfere with the kidney's filtering process. Normally nitrogen-containing wastes such as urea, from protein breakdown, are removed into the urine. An ill kidney cannot oblige, however. Protein-derived compounds such as urea build up in the body – and they are toxic. This is why people suffering from a kidney disorder may be put on a limited protein diet, to help keep down the level of urea in the blood. Proteins with a high "biological value," in eggs, milk and lean meat, help increase the utilization of urea, thereby lowering its level in the blood. A high calorie intake is also needed so that the body does not turn to proteins for energy, which would break down to more nitrogenous wastes.

Into the renal pelvis

It is time to return to the developing urine, flowing along the collecting tubules, eventually to leave the kidney and pass along the ureter to the bladder. The journey within the kidney is not quite over. Running up from the funnel-shaped end of the ureter and pelvis, to meet the medulla, are larger urine-collecting tubes or calyces, which branch and disperse into the body of the organ. The bulges of medulla between the calyces are known as papillae. One papilla collects urine from 70,000 nephrons.

Just before reaching the top of each papilla, the collecting tubules bearing urine from the nephrons join together to form the ducts of Bellini. The papillae also contain the lower portions of the hairpin-shaped loops of Henle, before the tubules sweep back up to the cortex again.

As a result, this area of the kidney has two separate structures available to reabsorb water and concentrate the urine. It is known as the counter-current mechanism. If the renal papillae are not functioning correctly, urine leaves the tubules overdiluted and full of sodium. This vulnerable area of the kidney bears the brunt of any infection or damage from chemicals or drugs.

The collecting tubules, therefore, join up to form the renal papilla, which sits in a calyx in the renal pelvis. Waves of muscular contraction (peristalsis) coax the urine from the calyces, through the renal pelvis and down the ureter. The muscles of the calyces are involuntary, and it is thought their contractions are triggered by distension in the tubules (a similar system occurs in the muscular wall of the heart). Once the peristaltic waves have been initiated, they travel ten inches along the full length of the ureter from the kidney to the bladder.

The storage organ

An empty bladder looks small and wrinkled, like a prune. This expandable bag is made from a basketwork of interlacing involuntary muscle fibers which run in whorls, sweeping up and down from the apex (top) to the neck (outlet) of the organ. Collectively, the fibers are known as the detrusor muscle. The detrusor's unique layout of fibers means that when the muscle contracts, the exit hole at the bladder's neck opens.

The two ureters track down the back of the abdomen and feed into the bladder via two small flaplike openings. Due to the angle at which the ureters enter the bladder, the openings – ureteric orifices – function as one-way valves. So, in health, when the bladder contracts to expel its contents, the urine cannot flow back up the ureters ("reflux") toward the kidneys. Developmental abnormalities or disease may result in faulty ureteric orifices into the bladder. They may need to be repaired by surgery, because the reflux of urine up to the kidneys can carry with it bacteria or viruses lurking in the bladder.

An average bladder can usually hold about a half pint of urine before it reminds its owner that it needs emptying. If ignored, it can continue to expand in capacity to nearly a pint, but by this time thoughts are of little else except finding a lavatory. Yet the bladder can continue to stretch if necessary, as happens in certain disorders involving retention of urine. In men, for example, the prostate gland is located at the exit from the bladder, all but encircling the urethra as it conveys urine from the body. Enlargement of the prostate, or disease may swell and stiffen the gland, so that it squeezes shut the urethra and blocks the exit of urine. In some cases the bladder swells to accommodate up to three and a half pints of urine.

The bladder's inner lining is formed from a folded, wrinkled membrane, known technically as the transitional cell epithelium. As the bladder expands, so the epithelium flattens out to maintain its integrity as a resistant, "urine-proof" barrier. However, a small inverted triangle of lining is unwrinkled. This is the trigone, which has at its two upper corners the ureteric orifices, and the urethral opening at its lower, central point.

The urge to "go"

The muscles in the bladder wall are not under voluntary control – but the muscles just below them, in the "pelvic floor" at the base of the abdomen, are. Thus it is possible to control bladder emptying. A band of muscle, the vesical sphincter, surrounds the neck of the bladder. A few seconds before urination the pelvic floor drops and the vesical sphincter relaxes, along with various other muscles in the base of the abdomen. Urine enters the urethra as the detrusor muscle in the bladder wall contracts. And away it goes, as out it flows.

Urine's final passageway is the urethra. In a woman it is about one and a half inches long and a quarter inch in diameter, and carries only urine. In men, on the other hand, the urethra is up to eight inches long and soon after leaving the bladder it becomes a dual-function tube, for the semen-carrying ejaculatory ducts join it inside the prostate gland. The male urethra also has a series of helical folds in its lining near its exit from the penis. These give the departing urine stream a narrow, corkscrewlike twist – otherwise it would emerge as an uncontrolled spray.

When the kidneys complain

Three main types of complaints affect the kidneys. They are stones, various degrees of inflammation (often connected to infection), and failing function.

Kidney disorders may be first noticed by some alteration in the nature of urine: in its smell, color, quantity or clarity. But renal complaints also have less specific, more generalized, effects on the body. Imbalances in body fluids, salts and blood composition have widespread effects. They range from vague feelings of fatigue and muscular cramps to blurred vision, nausea, fever and accumulations of fluid (edema) that produce puffiness around the ankles, eyes and other body parts.

The way such symptoms are combined gives physicians a clue to the underlying renal disease. Laboratory analysis of a urine sample also provides much valuable information. It is always important to report signs of ill health to a doctor promptly, but with kidney or urinary problems especially this advice is doubly sensible. Once a nephron is damaged, it cannot regenerate. A million nephrons per kidney may not be enough.

Rolling stones

Kidney stones, or renal calculi, are responsible for about one in every 1,000 hospitalizations in the United States. They are an old problem – stones have been found in Egyptian mummies. Some patients have a family history of stones, and sometimes their formation is due to an inherited metabolic defect.

More than 80 percent of kidney stones contain calcium, usually as calcium oxalate; other types contain phosphates or urates (as are found in people with gout). Stone formation is linked to the level of calcium in the blood. Overactivity of the parathyroid glands in the neck, erosion of bone tissue by cancer,

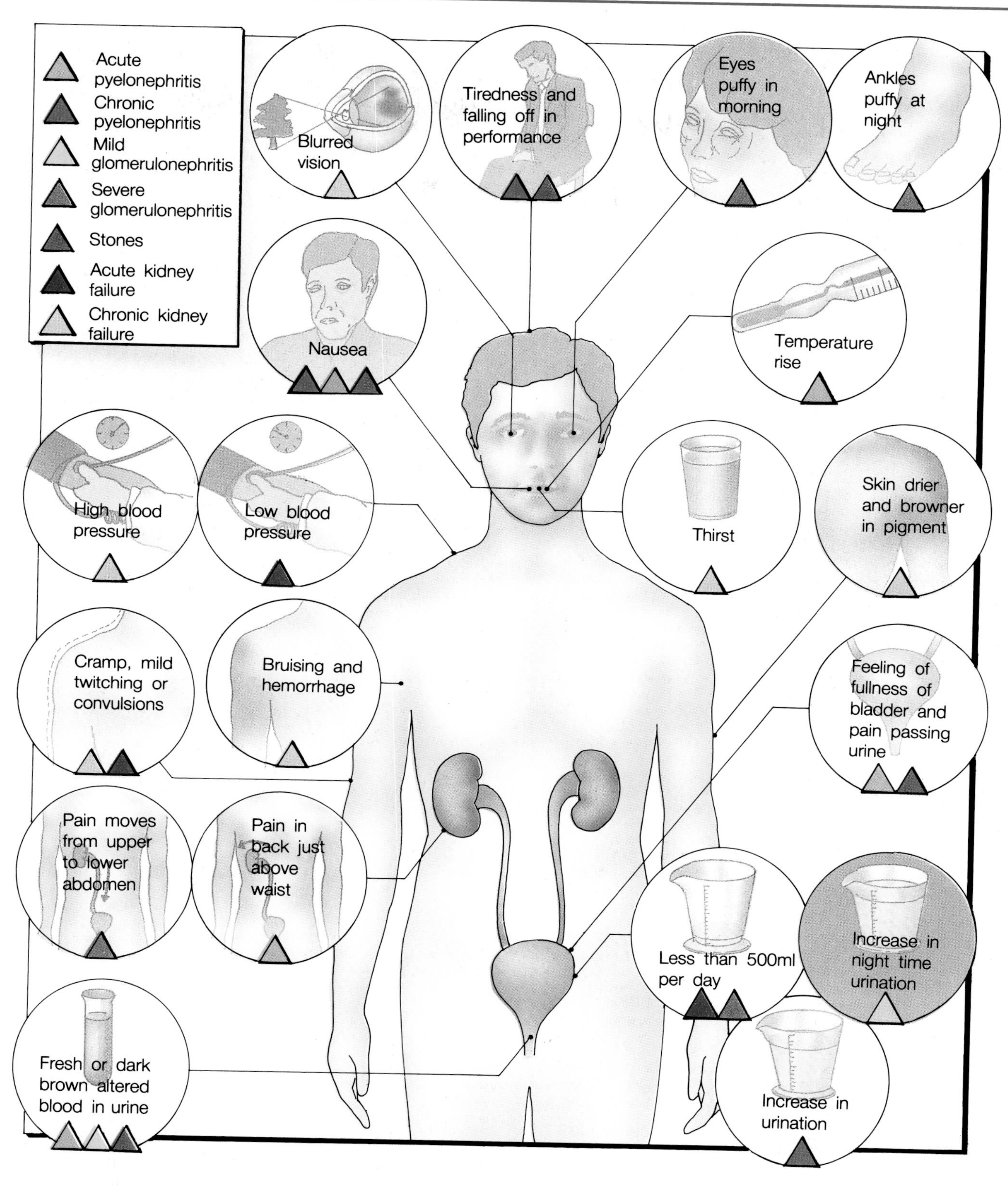

Poorly functioning kidneys can cause a host of varied symptoms affecting many different parts of the body. For instance, acute pyelonephritis – inflammation of the kidney substance and its pelvis – causes dark urine, back pain, nausea and fever, whereas in its severe form glomerulonephritis causes fluid retention, resulting in low blood pressure, a reduction in urination and ankles swollen by edema. Nausea and pain that moves from the upper to lower abdomen may be symptomatic of kidney stones. They may also cause dark or cloudy urine, or blood in the urine. Most serious of all is kidney failure, which in its chronic form causes a wide range of symptoms from cramp and high blood pressure to thirst and increased urination at night; vision may also be affected. None of these symptoms should be ignored, and all need medical attention and treatment.

Total kidney failure once led to death within a few days. Today patients with this condition – perhaps waiting for a kidney transplant – can be given a new lease of life with an artificial kidney machine. It uses the principle of selective filtering or dialysis to mimic the function of a real kidney and separate potentially toxic waste products from other components of the blood. The key part of the machine, the dialyzer, consists essentially of a container of pure water in which dialysate concentrate has been dissolved. Blood from an artery in the patient's arm is pumped into the dialyzer, where it passes through a network of tubes made from a semipermeable membrane. This membrane has the property of holding back blood cells, plasma proteins and other large molecules (see diagram, *right*), but it lets through urea and other waste products. The blood then passes through a trap to remove any bubbles that have formed in it, and continues on back into the patient's circulation. The dialysate is also circulated, and the part that runs to waste corresponds to the patient's urine.

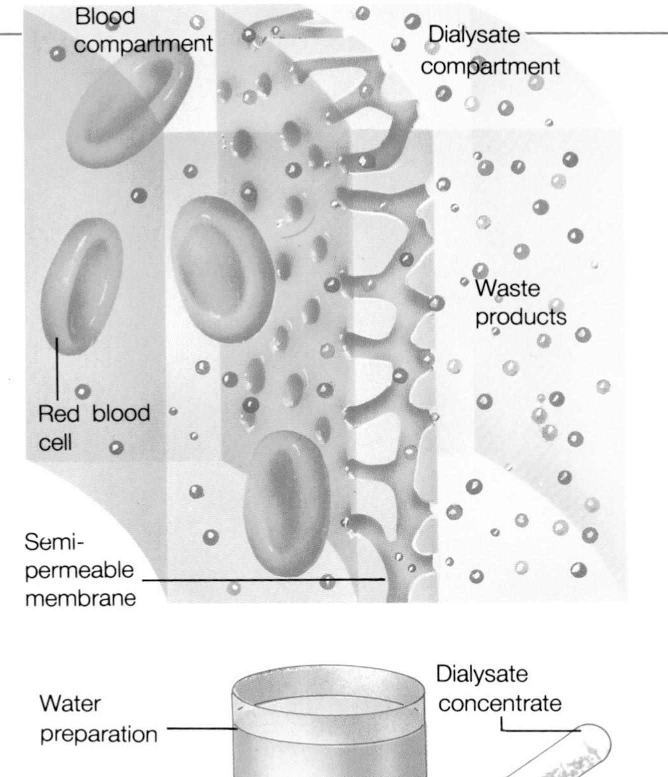

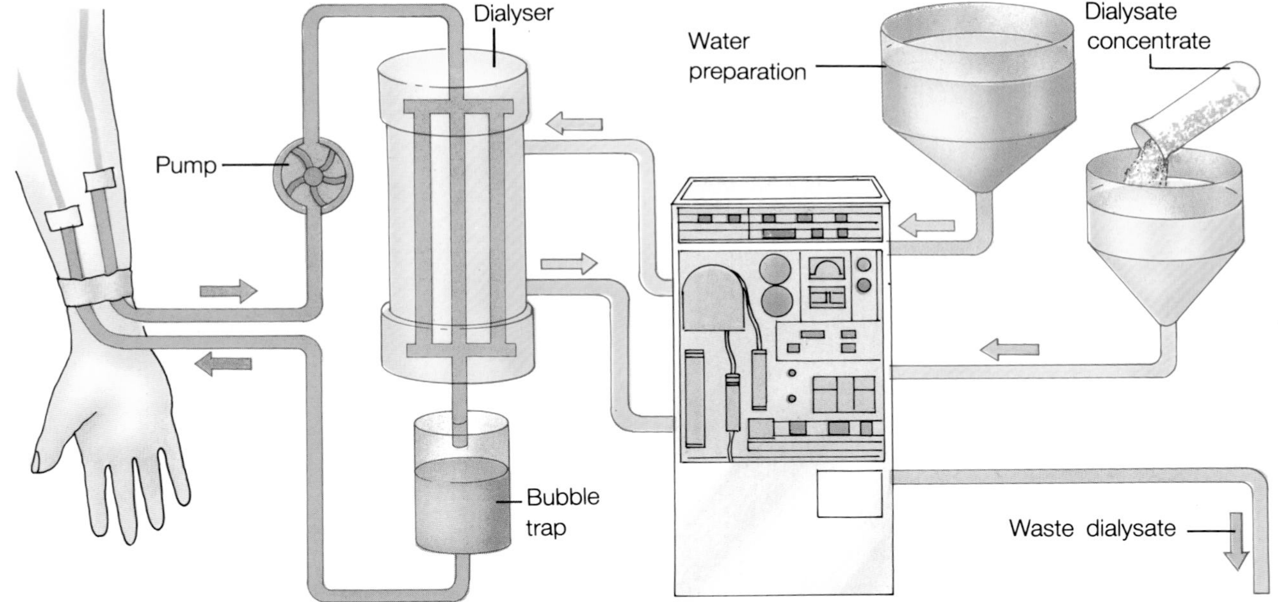

and dietary defects such as excessive milk intake or vitamin D overload are some of the culprits. Stones are also encouraged if the volume of urine is reduced, as in prolonged dehydration, or when its acidity alters, as when urine becomes infected. The most valuable, and simple, measure to prevent recurrent stone formation is to be sure of maintaining a high fluid intake.

The effects of a stone depend on its size. Very small ones often pass right through the urinary tract without producing symptoms. Larger than about a third of an inch, they are unable to enter the ureter and so remain in the renal pelvis – again, perhaps, without symptoms. Here they can continue to grow and yet remain largely anonymous, even when they almost fill the central urinary spaces and assume the branching shape of the pelvis and calyces, to become "staghorn calculi."

Stones between these two sizes tend to be most troublesome. They often enter the ureter but then become stuck. Urine backs up into the kidneys, while peristaltic movements of the ureteric wall try to massage the stones downward. The accompanying and excruciating pain, coming in waves that correspond with the peristaltic waves, is often called renal colic (more correctly, ureteric colic). Treatments range from drugs to surgical intervention. A thin, guidable probe known as a ureteroscope may be inserted up the urethra, through the bladder and on up the ureter, to grasp and withdraw the stone or crush it in place so that the urine stream washes away the fragments. Distintegration of kidney stones by bombardment with ultrasound waves has also proved successful.

Stones also grow in the bladder, where they are known as

vesical calculi. They can cause pain, urine retention and blood in the urine (hematuria). Benjamin Franklin, civil leader, scientist, writer and statesman of 18th-century America, had to travel to the Constitutional Convention in a sedan chair because of the pain from bladder stones. He resisted being put under the knife of the lithotomist ("stone-cutter") and spent his last two years in bed, reduced to a "skeleton covered by skin." Today, bladder stones can usually be blown apart by ultrasound or physically chopped up and removed by cystoscopy – probing the bladder with a telescopelike instrument inserted via the urethra.

Infection and inflammation

Women are at an anatomical disadvantage as regards the risk of germs invading the urinary tract. Because the female's urethra is much shorter than the male's, it provides an easy route for infection from the exterior. Cystitis, inflammation of the bladder, may be due to invasion by hostile microbes or to irritant substances in the urine.

Inflammation of the kidneys themselves goes by a number of names, depending on which parts of the organ succumb. Glomerulonephritis affects the filtering mechanisms, is associated with various underlying disorders, and accounts for up to 30 percent of cases of chronic kidney (renal) failure. Infections by some strains of *Streptococcus* bacteria, which cause sore throats and scarlet fever, sometimes lead to glomerulonephritis, which occurs a few weeks after the initial infection. If glomerular disease is severe enough to bring on sudden increase in blood pressure and hematuria, the condition is known as acute nephritis.

In acute pyelonephritis, the inflammation – often due to infection – is concentrated in the renal pelvis and tubules. Pregnancy, urinary obstruction, diabetes and certain drugs are predisposing factors, while the symptoms include fever, hematuria and shivering attacks (rigors). If the infection remains untreated, abscesses form in the kidney and permanently scar it.

Most patients with chronic pyelonephritis have obstruction and reflux, as well as possible infection. Unlike the acute infections, there may be no specific symptoms and the condition is revealed only during routine health screening, although some patients suffer from high blood pressure or general malaise. It accounts for 20 percent of cases of chronic renal failure.

Failure of the kidneys

When the kidneys can no longer cope, for any of more than a dozen reasons, the result is renal failure. It may be acute (sudden) or chronic (coming on slowly).

Acute renal failure develops over hours or days. Causes range from severe blood loss to severe glomerulonephritis, ingestion of a poison (such as antifreeze), to sudden and complete urinary obstruction. Provided medical help is obtained swiftly, possibly with intravenous feeding and short-term dialysis (with a "kidney machine") to tide the patient over the critical period, eventual recovery is almost certain. But it can be six weeks before the kidneys start producing urine again, and perhaps years before their function returns to normal – if it ever does.

Chronic renal failure is the major cause of death from kidney disease. Its symptoms are many and varied, often vague and minor at first. They result from a build-up of metabolic wastes in the bloodstream, combined with breakdown of the barriers that prevent leakage of substances from blood into urine, and a failure of other essential processes controlled by the kidneys. Analysis of urine samples tells the clinician how far the disorder has progressed, through the appearance in them of protein, red and white blood cells, and "casts" (protein-derived material dumped into the tubules and washed down into the bladder). When the kidneys fail there are two chief forms of treatment. One involves a machine, the other, someone else's kidney.

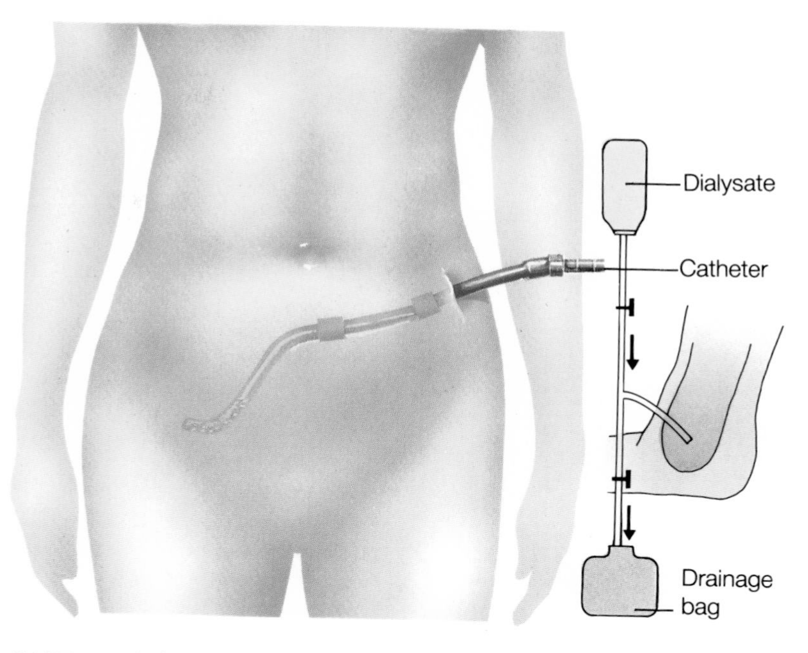

CAPD stands for continuous ambulatory peritoneal dialysis, an alternative to a kidney machine which gives the patient complete freedom of movement. It makes use of the peritoneum, the membrane that lines the abdominal cavity, as a natural semipermeable membrane to filter the blood. Dialysate fluid passes in through a catheter permanently inserted into the patient's abdomen, and once filtration has taken place the spent fluid runs out to be collected in a drainage bag.

The "kidney machine"

The kidney was the first major organ in the body whose chief function was duplicated by a machine, the renal dialyzer.

All artificial kidneys work on the scientific principle of dialysis – the passage of small molecules from one fluid to another through molecule-sized holes in a special semipermeable membrane. Large molecules such as proteins cannot pass through. In a dialysis (or hemodialysis) machine, blood – containing cells, plasma proteins, salts and wastes – is on one side of the semipermeable membrane. On the other side is a custom-designed salt solution which contains the right mixture and concentration of small molecules. Wastes and unwanted salts move from the blood through the membrane into the "salt bath," but blood cells and plasma proteins stay behind. The volume of fluid in the salt bath is large, so the wastes are rapidly diluted, and their levels in the blood become nearly normal.

The basic principle sounds simple, but the practicalities are not. Potential disasters include catastrophic bleeding upon a faulty link-up to the device, blood clotting in the machine, poisoning of the patient by the contents of the dialysis bath, and the introduction of infection. Dialysis may be needed for several hours each day, on several days each week.

Repeated "vascular access" – connecting the patient's blood circulation to the machine by inserting needles – soon damages the blood vessels. One solution is to make a relatively long-term artificial connection, termed a shunt, between an artery and a vein using a piece of silicone rubber tubing. Another is surgery to join an artery and vein to produce a fistula. Either structure can take a large number of hypodermic needle insertions. Some patients attend hospital, but the modern compact machines, prepackaged salt solutions and "foolproof" technical wizardry allow others to dialyze themselves at home.

A relatively recent development is peritoneal dialysis. This makes use of the body's own semipermeable membrane, the peritoneum, which lines the abdominal cavity and wraps itself over the organs within. Fluid is run into the abdomen through a rubber tube (catheter) and wastes are exchanged into the fluid through the peritoneum, just as in the dialyzer. The fluid is removed and fresh added, and so on, to "wash" the wastes from the body.

Two main types of peritoneal dialysis are available. In intermittent peritoneal dialysis (IPD), a machine pumps a large volume of fluid through the abdomen three or four times each week, over a period of 10 to 14 hours at a time. In the continuous type, fluid is run into the abdomen, left for several hours and removed, to be replaced by a new charge of fluid; three or four changes are made daily. Between fluid changes the patient is not connected to any machinery, even as dialysis is taking place, so this second method is sometimes referred to as continuous ambulatory peritoneal dialysis (CAPD).

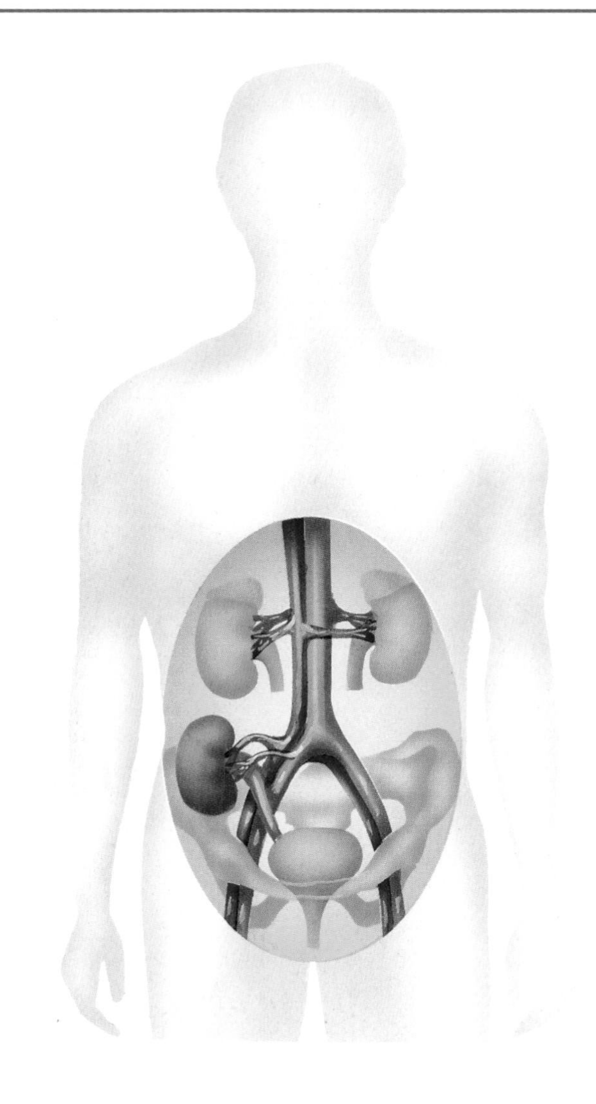

A permanent cure for nonfunctioning kidneys is provided by a kidney transplant *(above)*. Usually the patient's own kidneys are left in place, and the donor organ located lower down in one side of the pelvic cavity, where it is partly protected by the hip bone. To simplify making new connections to the femoral artery and vein, the donor kidney is positioned upside-down, and its ureter shortened to connect with the bladder.

Successful transplantation requires a good "match" between the tissue types of the donor kidney and the recipient. A poor match results in rejection, in which the recipient's immune system mobilizes to destroy what it regards as an "invasion" by foreign tissue. The problem can be overcome to some extent by the use of immunosuppressive drugs, many of which are steroids, whose crystal structure *(right)* resembles shimmering fronds of gold.

Kidney transplants

Today the kidney is still by far the most commonly transplanted organ. Since the 1940s, the methods developed for renal replacement therapy have helped advance transplant programs for many other organs, such as the heart and lungs.

Kidney transplantation does not hold the cure for every person suffering from renal failure. The great barrier is rejection of the "foreign" kidney tissue by the body's defensive immune system. However, the technique does have an advantage: a patient with failed kidneys can be kept alive by dialysis until a suitable, "closely matched" donor organ is available. Machinery cannot substitute for heart, lungs or liver, hence the relative lack of development in these transplant programs.

A kidney from a recently dead donor (cadaver kidney) may be used for the transplant, provided the appropriate permissions are obtained from relatives by the transplant staff. In other cases a living, healthy, close relative of the patient may decide to donate a kidney. He or she is selected to be a good tissue match. Removal of the kidney in a healthy person is a relatively simple and safe operation, and the remaining kidney has reserve enough to take the strain.

By the 1980s, the drug cyclosporin A had revolutionized the situation. It is an immunosuppressive: it damps down the recipient's immune system and so helps the transplanted kidney to "take." With careful tissue typing and matching, results are improving all the time.

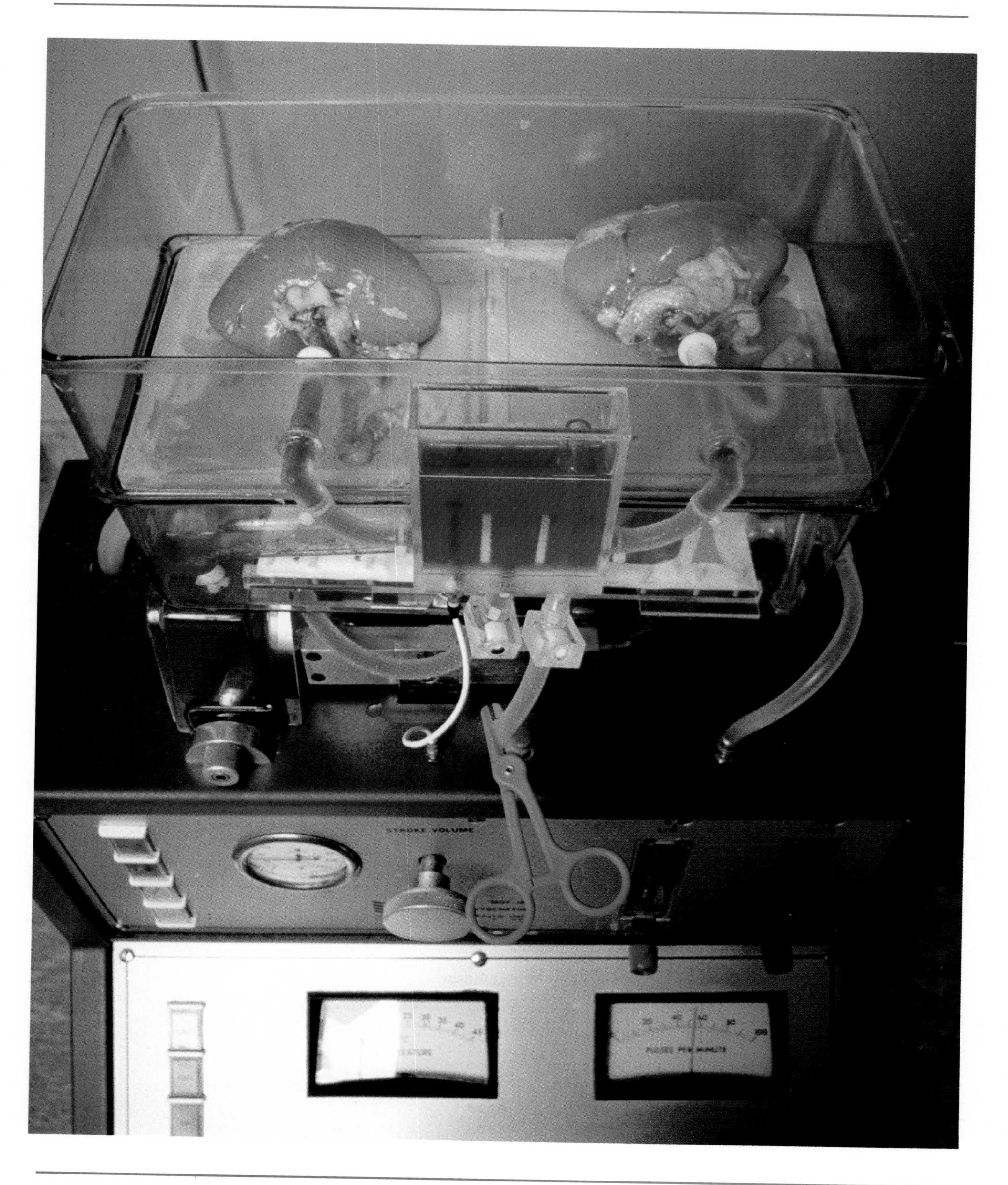
STROKE VOLUME
PULSES PER MINUTE

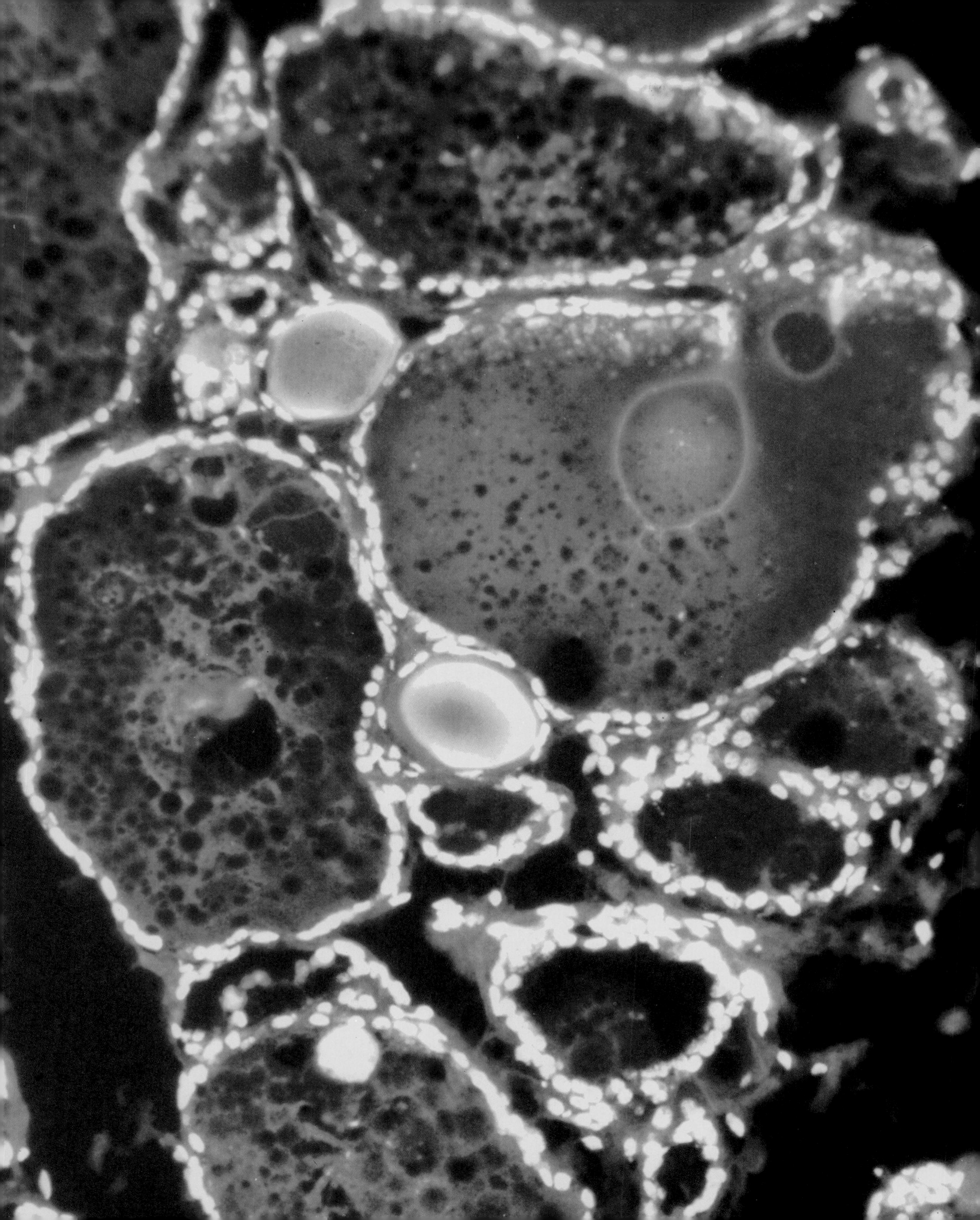

CHAPTER 11

The Endocrine System

The human body is a complex system of interrelated organs and tissues, all of which must work together if it is to function properly. A wide variety of physiological processes – ranging from growth and development to digestion and reproduction – have to be continuously monitored and controlled. Both monitoring and control have to take place without conscious intervention, and these are achieved by the endocrine system, which carries them out with the help of a highly ingenious array of chemical messengers called hormones.

Released into the bloodstream, the body's liquid communication system, hormones are carried to all parts of the body where they trigger the required action. But the effect of any particular hormone must be carefully controlled – if it goes on working too long or too vigorously it could disrupt the delicate balance of the body. The secret of this control mechanism is feedback, which is used in much the same way as it is in machines – the output of the process is sensed and fed back to modify the action of the initiating agent.

Many processes use negative feedback; the sensed output is fed back in such a way as to reduce or switch off the input, and as a result the system settles down to a stable condition. But if the output is disturbed by some external influence, this reacts on the input and the system responds so as to restore the stable state. A common example is a thermostat that controls a heating system by turning the heat on and off to maintain a particular temperature. Less common is positive feedback, in which the sensed output adds to the input, causing even more output that goes on growing until the limits of the system are reached.

Within the endocrine system negative feedback is normal. The organ responsible for outputting a hormone senses the level of hormone in the bloodstream, or the level of the substance being controlled by the hormone. When the level rises too high, the hormone output is reduced or switched off, or a counteracting hormone released to bring the level down. Conversely, if the level is too low the hormone output is stepped up or switched on. This delicate balancing trick ensures that the body always has the correct hormone levels it needs.

Hormone factories

Most hormones are proteins and they are manufactured in endocrine glands located in various parts of the body. The major ones include the pituitary (at the base of the brain), thyroid and parathyroids (in the neck), pancreas and adrenal glands (in the abdomen), and the gonads, or sex organs. The main feature of the glands is that the hormones they produce are secreted into the bloodstream (or sometimes other tissue fluids), in which they are carried through the body to act on various organs and tissues. Other glands in the body release their secretions into tubes or ducts which carry them to the site of action. These are the exocrine glands. Some glands – for example the pancreas – work in both ways and are known as mixed glands.

A hormone store in the thyroid gland is made to reveal its presence by the specialized technique of fluorescence microscopy. The scintillating membranes surround packets of hormones in an inactive form which can be quickly mobilized to meet the body's metabolic requirements.

The master gland

The most important endocrine gland is the pituitary. It is the master gland, and has a controlling effect on many of its colleague glands. About the size of a pea, it is suspended from the brain by a stalk so it sits just above the roof of the mouth, more or less in line with the bridge of the nose.

The thyroid produces hormones that control metabolism, the intricate biochemistry involved in the breakdown of food into useful substances and energy and the disposal of waste products of the body. It is a fleshy gland positioned in the neck just below the larynx. Four parathyroid glands are attached to the back of the thyroid. They control the body's levels of calcium and phosphorus, chemicals needed for healthy bones and teeth.

Located on the upper part of each kidney, the two adrenal glands are roughly triangular in shape. Their hormones control fluid and mineral balances in the body, the metabolism of glucose, and one of the body's most fundamental life-preserving functions – its response to stress and danger.

The pancreas is situated in an inaccessible position at the back of the abdomen, underneath the liver and in front of the vertebral column. It is mainly involved in the production of digestive enzymes, which pass along ducts to the duodenum at the exit of the stomach. But the pancreas also produces the important hormone insulin, which controls the levels of the energy-generating sugar glucose in the blood.

The gonads have different functions in men and women, though there are two in each sex. Men have external gonads, the testes, or testicles. Women have ovaries, almond-shaped organs about one and three-quarters of an inch long located within the pelvic girdle. Sex hormones – estrogens, progesterone, testosterone and androsterone – are produced by both males and females. But the end results of the actions of these hormones are very different in the two sexes because each produces significantly different proportions of each hormone. Predominance of testosterone and androsterone defines maleness, whereas estrogen and progesterone are the predominants in females.

Bodily health and balance depend vitally on the proper operation of the endocrine glands. Faulty behavior of just one of them can have profound effects on the whole body. In growing children malfunction of the pituitary, for example, may cause them to end up as dwarfs, or as giants. Similarly incorrect levels of the sex hormones result in inadequate or misdirected development. This may lead to infertility or the production of incorrect secondary sexual characteristics – such as the growth of facial hair in women or of breasts in young men.

Tumors can affect the endocrine glands, in the same way as they sometimes attack the rest of the body. But even non-cancerous tumors are liable to cause malfunction of the endocrine system and result in overproduction or underproduction of hormones. In such cases surgery or radiotherapy may be needed to deal with the tumor. Luckily it is also possible to treat many endocrine disorders by a simple process of hormone replacement therapy. The missing hormone is supplied in tablet or injectable form to reestablish the correct levels in the bloodstream. The best-known example of such therapy is the regular administration of the hormone insulin to people with diabetes mellitus, caused by lack of that hormone.

How it all works

It has taken much painstaking detective work from countless doctors and scientists to unravel the complex workings of the endocrine system. But even so they still have only a broad picture of many of its actions. Detailed understanding of how all the individual processes work is still incomplete and the subject of ongoing research. As knowledge grows about how hormones work, and about their effects, physicians are finding new ways to help maintain the body's balance and deal with the wide variety of endocrine disorders.

Chemical messengers

Using the bloodstream to carry messages between various organs is a comparatively slow but generally certain process, which can work locally or at a considerable distance from the place where the messages originated. But because all of the hormones circulating in the blood get carried to all the organs in the body, it is important that each of them has a very specific action. If the same hormone had two or more different functions, or the same function was performed by two different hormones, they would be bound to interfere with one another and it would be impossible for the vital regulatory feedback mechanisms to work.

To make sure that they do the correct jobs, the hormones act like keys. Each hormone is recognized by and fits, or unlocks, a specific target cell on the organ or tissue it is supposed to activate. The hormone can only affect the cells that recognize it, and the cells can only be affected by their specific hormones.

Chemical building blocks are molded together to make the individual hormones, which come in three main types. Most common are the protein and peptide hormones formed from amino acids. The substance cholesterol – abundant in animal fats and dairy products – is the basis of the steroid hormones. And there is a third group of miscellaneous hormones which do not really fit into either of the other two groups.

The actual production of hormones in the body is a simple manufacturing process. To begin with, blueprints for the protein and peptide hormones are produced in the nucleus of the relevant endocrine cell. From the "drawing office" the plans, in the form of ribonucleic acid (RNA), are passed out to an assembly plant – ribosomes in the body of the cell. In this assembly operation, known as transcription, the RNA controls the build-up of the protein and peptide molecules that make the hormones. Finally the hormone molecule is packaged into a pouchlike vesicle by the cell Golgi apparatus.

When needed, the molecules are expelled through the cell wall and effectively thrown into the bloodstream. Some hormones are made in a simpler way from "dummy" or precursor molecules. These molecules have no function of their own but can be stored in the cell and quickly transformed to give their corresponding hormone molecule.

Like burgers in a fast-food restaurant, steroid and miscellaneous hormones are made to order. When the body needs them, an enzyme in the cell is activated and sets off a series of reactions to build the hormone molecule from building blocks stored in the cell cytoplasm. Once synthesized, the steroid molecules pass easily through the cell membrane and straight into the bloodstream for distribution.

Not all of the hormones circulating in the blood are needed for immediate action. Some are banked to form a hormone savings account that can be drawn on as required. The banking works by binding the hormone to a protein so it is not directly active but can continue to circulate in the bloodstream. When the hormone is needed the binding is easily broken and the hormone released for use. Another of the system's devices is to make some of the

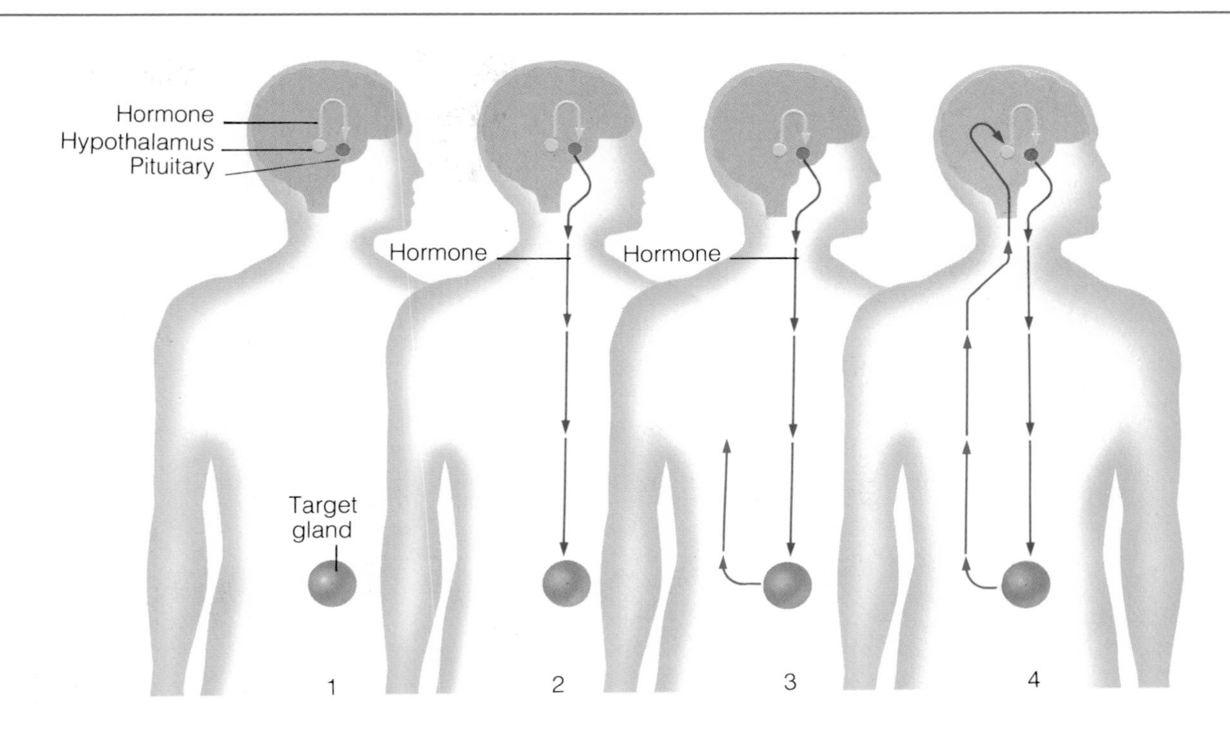

A chain of hormone reactions can be triggered by the hypothalamus in the brain, and controlled by a feedback mechanism. Initially (1) a hormone from the hypothalamus causes the pituitary to release one of its hormones (red arrows, 2). The pituitary hormone travels in the bloodstream to a target gland somewhere in the body, stimulating it to release a third hormone (blue arrows, 3), which is also carried in the blood to where it is needed. Some of it reaches the hypothalamus and "switches off" its pituitary-stimulating action (4).

hormones in one form and then process them by another body organ to make a different hormone with new functions.

Hormones in action

When a circulating hormone reaches its target cell it keys in, unlocks the system and then switches on the appropriate cell action. This switching process can take a number of different forms. In one it affects the membrane of the target cell so that it allows specific ions or molecules to pass through and start up the cell processes. In another the hormone binds onto receptors on the cell membrane and triggers the release of a secondary messenger that sets the cell to work. Steroid hormones can pass right through the cell wall and react directly with receptors in the cell cytoplasm.

After they have triggered the required cell action the hormones are of no further use and have to be cleared out of the system. Commonly they are transported to the liver to be broken down, or may be destroyed by the target tissues themselves. The water-soluble compounds that remain are efficiently disposed of by the body's excretory system.

Pituitary in charge

Overall control of the workings of the endocrine system is the responsibility of the pituitary gland. This conductor of the hormonal orchestra works in close cooperation with the hypothalamus, an organ which, like the orchestra leader, provides an all-important link between the brain, nervous system and endocrine system. The pituitary exercises its unique control by producing hormones that act on the other endocrine glands and stimulate them to release their own hormones.

Control of the individual glands involves both increasing and reducing their output, and many of these processes are regulated by a feedback loop involving the hypothalamus. Sensor cells in the brain, like the thermostat in a room heating system, monitor the levels of the various hormones in the blood and signal the results to the hypothalamus, which then modifies the activity of the pituitary. In turn this affects the action of the various glands, which reduce or increase their output so as to maintain the required balance.

The essential communication between hypothalamus and pituitary is also achieved by the use of chemical messengers, or hypothalamic hormones, produced by neurosecretory cells (nerve cells which make hormones) in the hypothalamus. They are released into the local bloodstream and pass to the pituitary, where they control its activity. Some of the hormones also pass into the general circulation, while similar products are manufactured in other areas of the brain and other organs of the body. The complex interactions of these hypothalamic hormones and the ways in which they help regulate the body's systems is even now not fully understood.

About the size of a pea, the pituitary is suspended from the brain by a slender stalk and sits in a bony pocket just above the back of the nose. Even though it is a small organ it is divided into two main parts, the anterior (front) and posterior (rear) lobes,

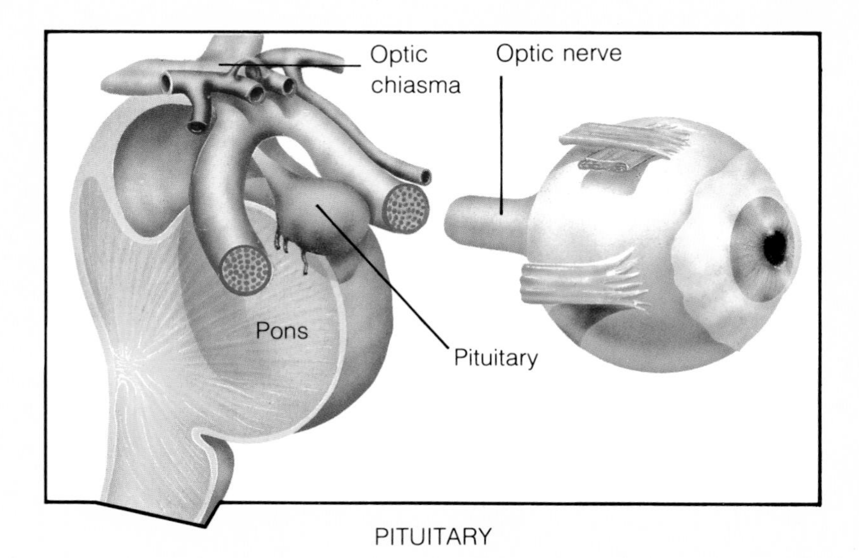

PITUITARY

The pituitary is the master gland whose hormones control most of the other endocrine glands in the body. This pea-sized structure hangs on a stalk from the base of the brain below the optic chiasma, the crossover point of the two optic nerves. The thyroid gland is located in the front of the neck, partly surrounding the larynx, with the four parathyroid glands embedded at its rear. Thyroid hormones control metabolism, and parathyroid hormone mediates calcium levels in the blood. Below the liver lies the pancreas. It produces insulin, which controls the way the body uses the blood sugar glucose.

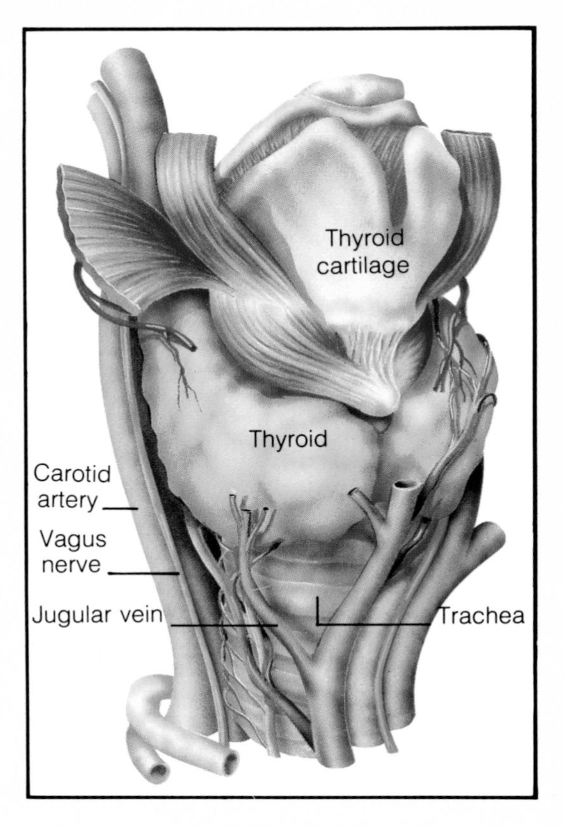

THYROID GLAND

which work in very different ways. The anterior lobe produces six different hormones. Two of these work directly on the body tissues, while the remaining four control the actions of other endocrine organs. In contrast the posterior lobe does not produce anything itself – it acts as a storage region for two hormones that are manufactured by the hypothalamus and carried to the lobe through a connecting duct.

Nearly all the body's tissues are affected by the action of somatotropin, also known as growth hormone, which is one of the direct-acting pituitary hormones. As the name suggests, it is one of the main hormones affecting growth, particularly of muscles, cartilage and bone. The correct level of somatotropin production is essential – both too little and too much have serious effects.

Children who have a shortage of growth hormone do not grow properly and suffer from dwarfism. Instead of developing normally they remain short and their faces stay childlike. All of the body is affected so everything is in proportion, it is just all small. Intelligence is not normally impaired. This type of dwarfism can be treated successfully if it is diagnosed early enough. With extra growth hormone supplied, the children develop normally and reach their full adult height.

On the other hand children whose bodies make too much growth hormone show the opposite effects and can suffer from giantism, growing rapidly, even to eight feet or more in height. In adults whose growth has already stopped, the sudden production of too much somatotropin has a different effect. It works on some of the other body tissues and causes the hands and feet to enlarge, an elongation of the face and the development of coarse features. This condition is known as acromegaly.

In a woman who has just given birth to a baby, breast-feeding makes the anterior part of the pituitary gland produce prolactin, the second pituitary hormone with a direct action. Prolactin stimulates the mother's breasts to produce milk, making sure that the action of the baby goes on ensuring that she has milk to provide for her infant. The prolactin has a restricting effect on the actions of the female sex hormones and so tends to reduce the woman's fertility – women who are breast-feeding rarely have menstrual periods. The reduced risk of pregnancy is nature's way of minimizing the demands made on the feeding mother.

Oxytocin from the posterior portion of the pituitary is also released by the stimulation of breast-feeding. This hormone makes the milk ducts in the breast contract so the milk literally squirts out of the nipple into the baby's mouth. A new mother trying to breast feed a baby may find it is very difficult to get her milk flowing. Worrying about her failure creates stress and, paradoxically, can inhibit the action of the oxytocin, so that milk does not flow as it should. Relaxation is the key – a calm approach lets the oxytocin get to work as it should and the baby gets its feed.

The sex glands are stimulated by the gonadotropins, two of the other hormones produced by the pituitary. These are luteinizing hormone (LH) and follicle-stimulating hormone (FSH), and their specific actions are dealt with in Chapter 15.

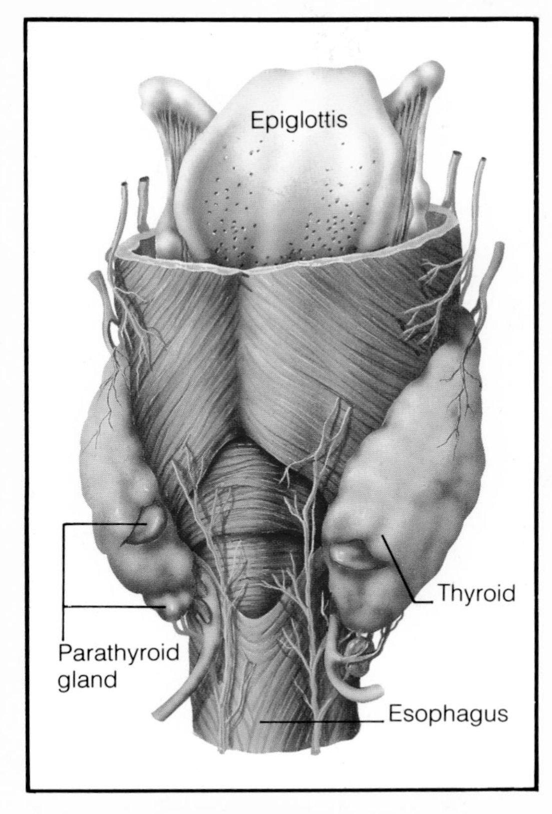

PARATHYROID GLAND

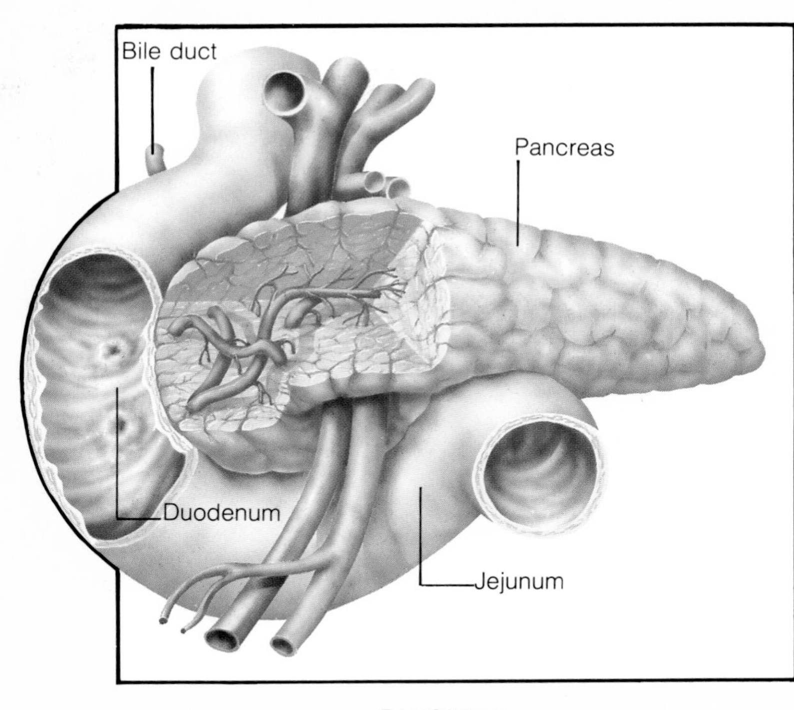

PANCREAS

The remaining two pituitary hormones are corticotropin (or adrenocorticotropin, ACTH) and thyrotropin (thyroid-stimulating hormone, TSH). As its name suggests, thyrotropin acts on the thyroid gland. Corticotropin influences the adrenal glands to make them increase the secretion of glucocorticoids. In turn, the glucocorticoids act on the pituitary to inhibit further production of corticotropin, giving a direct feedback loop that maintains the required glucocorticoid level.

Impotence and loss of sex drive in men is sometimes caused by the production of excess prolactin in the pituitary. The same fault causes infertility in women and may make their breasts start making milk even if there is no baby to be fed. These are typical examples of the complications that can arise when the operation of the pituitary is disturbed. Tumors are one of the more common causes of such disturbance, which often results in excess hormone production.

As well as acting directly, the excess may affect the operation of the other endocrine glands. An example is the excessive production of growth hormone, which causes giantism in children and acromegaly in adults. The usual treatment is to cut out the tumor by surgery or destroy it with radiotherapy. If surgery is undertaken the surgeon generally gets at the pituitary by working through the patient's nose. However, treatment of tumors may also damage the rest of the gland, and when this occurs, the patient may need continued hormone replacement therapy for life to compensate for the damage.

The posterior lobe of the pituitary is much smaller than the anterior lobe and its function is to store and release the hormones oxytocin and vasopressin, which are produced by the neighboring hypothalamus. As well as being vital to the release of milk from the breast of a feeding mother, oxytocin also stimulates the contraction of smooth muscle, most especially that of the uterus. Indeed it is one of the agents that causes contractions when a baby is being born and can in fact be used to induce labor at the end of a normal pregnancy. Oxytocin released when a newborn baby suckles on its mother's breast also causes contractions of the muscles of the womb and can help the ejection of the placenta after birth. Releases of more oxytocin with every breast-feed continues this muscular action and helps the womb return to its normal size.

Vasopressin is also known as antidiuretic hormone and is involved with the control of blood volume and salt concentration. If blood is lost, or there is an increase in the salt concentration, the condition is detected by receptors in the brain which trigger the release of vasopressin. The hormone then acts on the kidneys so that they absorb more water back into the blood and produce a more concentrated urine. This helps to dilute the blood salts and to increase the blood volume, so restoring the body to a healthy, well-balanced condition.

Thyroid gland

The rate at which the body's cells work – their metabolism – is controlled by hormones produced in the thyroid gland. The hormones concerned are thyroxine (T_4) and triiodothyronine

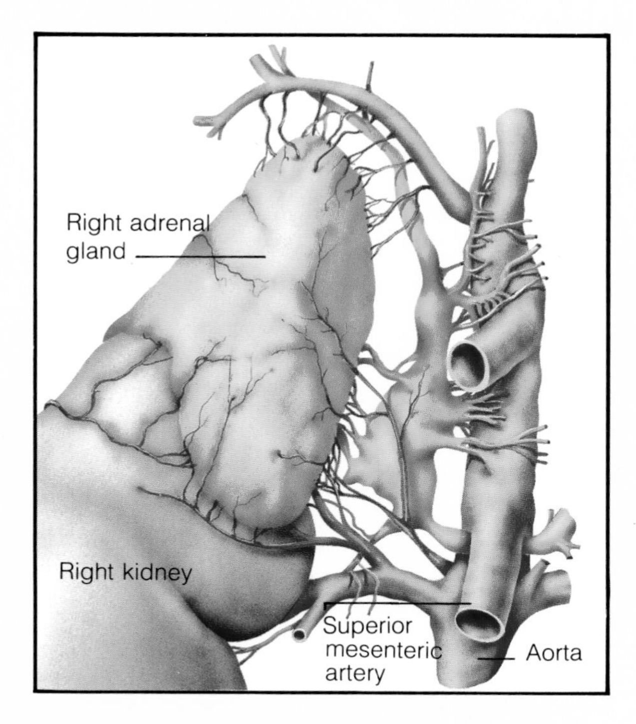

ADRENAL GLAND

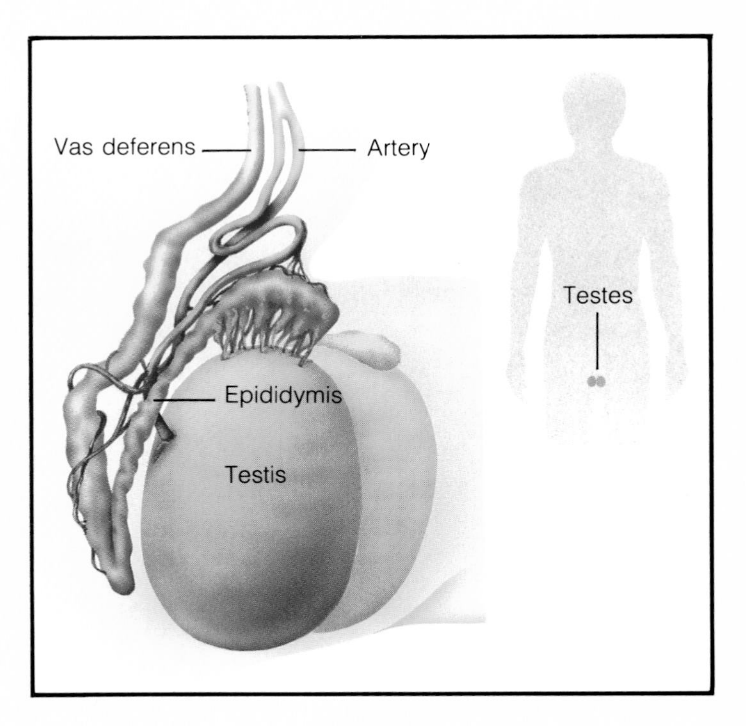

TESTES

The two triangular adrenal glands sit atop each kidney. Like all endrocrine glands, their hormones are carried around the body in the bloodstream and for this reason they have a plentiful supply of blood through direct connections with the main circulation. The principal adrenal hormone, epinephrine (adrenaline), prepares the body to meet stress or threat. Male sex hormones are produced mainly in the testes, a pair of egg-shaped glands located in the scrotum. The chief hormone, testosterone, brings about the development of secondary sexual characteristics in boys after puberty, and thereafter maintains a man's sex drive and maleness. The testes also produce and store sperm.

(T_3), and most of them are bound to proteins in the bloodstream. The small amounts of the hormones that remain free act to influence the rate of activity of the body cells, with the protein-bound hormones acting as a reserve supply which is released as the free hormone gets used up. Although the exact action is not certain it is believed that the cell action is actually triggered by T_3, and T_4 is converted to T_3 in other parts of the body to act as an additional reservoir.

Production of both these hormones in the thyroid is stimulated by the hormone thyrotropin (TSH) from the pituitary gland. Control of the process is another feedback operation carried out by the hypothalamus. In response to increasing thyroid hormone levels it acts on the anterior lobe of the pituitary, which in turn reduces the TSH level. This then cuts back the level of stimulation of the thyroid, and so diminishes the amounts of hormones produced. Calcitonin is a third hormone produced in the thyroid and it acts to help control the calcium levels in the body.

The key chemical element in the thyroid hormones is iodine. If there is a shortage of iodine in the diet, the thyroid gland becomes more and more enlarged in its attempts to gather sufficient iodine to maintain normal levels of the hormones. Such enlargement creates a characteristic swelling of the neck known as a goiter. To reduce the likelihood of this happening, iodine compounds are often added to table salt, to make sure there is a sufficient supply in the diet. Iodine deficiency is not, however, the only cause of goiters, which are also produced by other diseases of the thyroid.

Thyroid disease is the second most common endocrine disorder after diabetes. Excessive activity of the gland, with the production of too much of the hormones, results in hyperthyroidism or thyrotoxicosis. Graves' disease is a specific type of

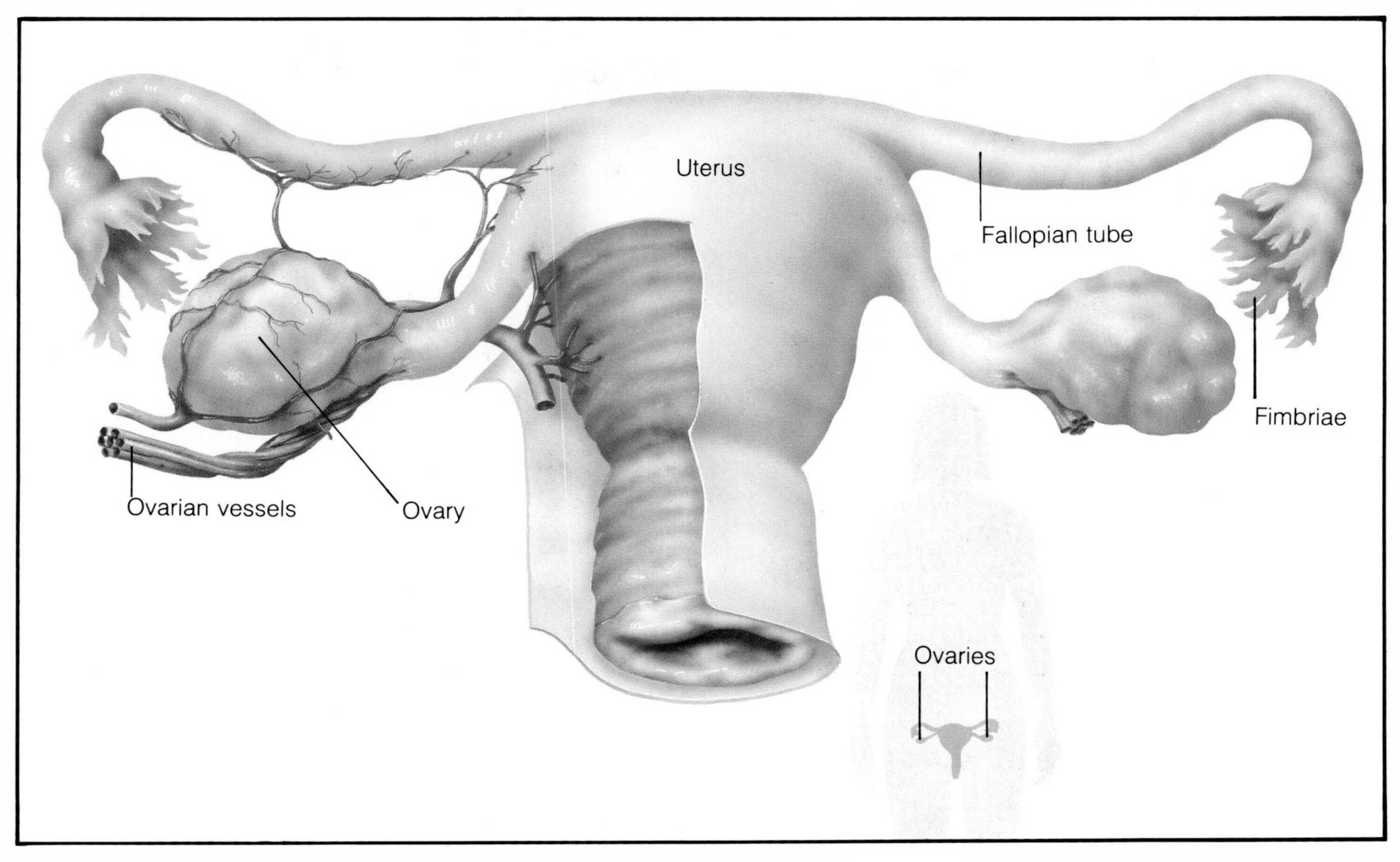

OVARIES

Female sex hormones are made in the ovaries, a pair of glands located in the abdomen on each side of the uterus. Initial production of these hormones, mainly estrogen and progesterone, brings about the development of secondary sexual characteristics in girls after puberty. During a woman's childbearing years they are also responsible for controlling the menstrual cycle, the approximately 28-day sequence of subtle bodily changes that includes, at about its midpoint, the release of an egg (ovum) from one of the ovaries. This event, called ovulation, can be prevented if a woman takes hormones in the form of the contraceptive pill; it ceases during pregnancy and at the menopause.

this condition which is characterized by swelling of the thyroid – leading to the development of a goiter in severe cases – and possibly bulging eyes that give the patient a typical staring expression. The high levels of thyroid hormones in the blood stimulate the body cells into increased action, resulting in a higher metabolic rate than normal. This causes excessive sweating, loss of weight and hunger. There may also be heart problems and emotional upsets.

Drugs that prevent the production of thyroid hormones are often used in the treatment of Graves' disease. Alternatively, surgery may be carried out to remove part of the gland. Another method consists of treating the patient with a radioactive isotope of iodine, which is taken up by the thyroid and acts to destroy part of it. Both of the treatments that destroy part of the gland tend to result in an underactive thyroid, which may then have to be treated by hormone replacement therapy.

Some people, typically middle-aged women, develop a goiter because of an autoimmune disease in which the body's defense mechanism is muddled. It treats its own tissue as "foreign" and produces antibodies to attack it. In Hashimoto's disease, the thyroid is infiltrated by large numbers of lymphocytes, or white blood cells produced by the body in its confusion.

As with the other endocrine glands, underactivity of the thyroid – or hypothyroidism – can have effects that are just as serious as those of overproduction. A baby suffering from a deficiency of thyroid hormones may stop growing and, if left untreated, the result can be the severe mental and physical retardation known as cretinism. To prevent this happening treatment has to start as early as possible – thus babies are tested for blood levels of the hormone soon after birth. The test involves measuring the level of thyroid stimulating hormone in a small sample of blood, with excessive levels showing that there

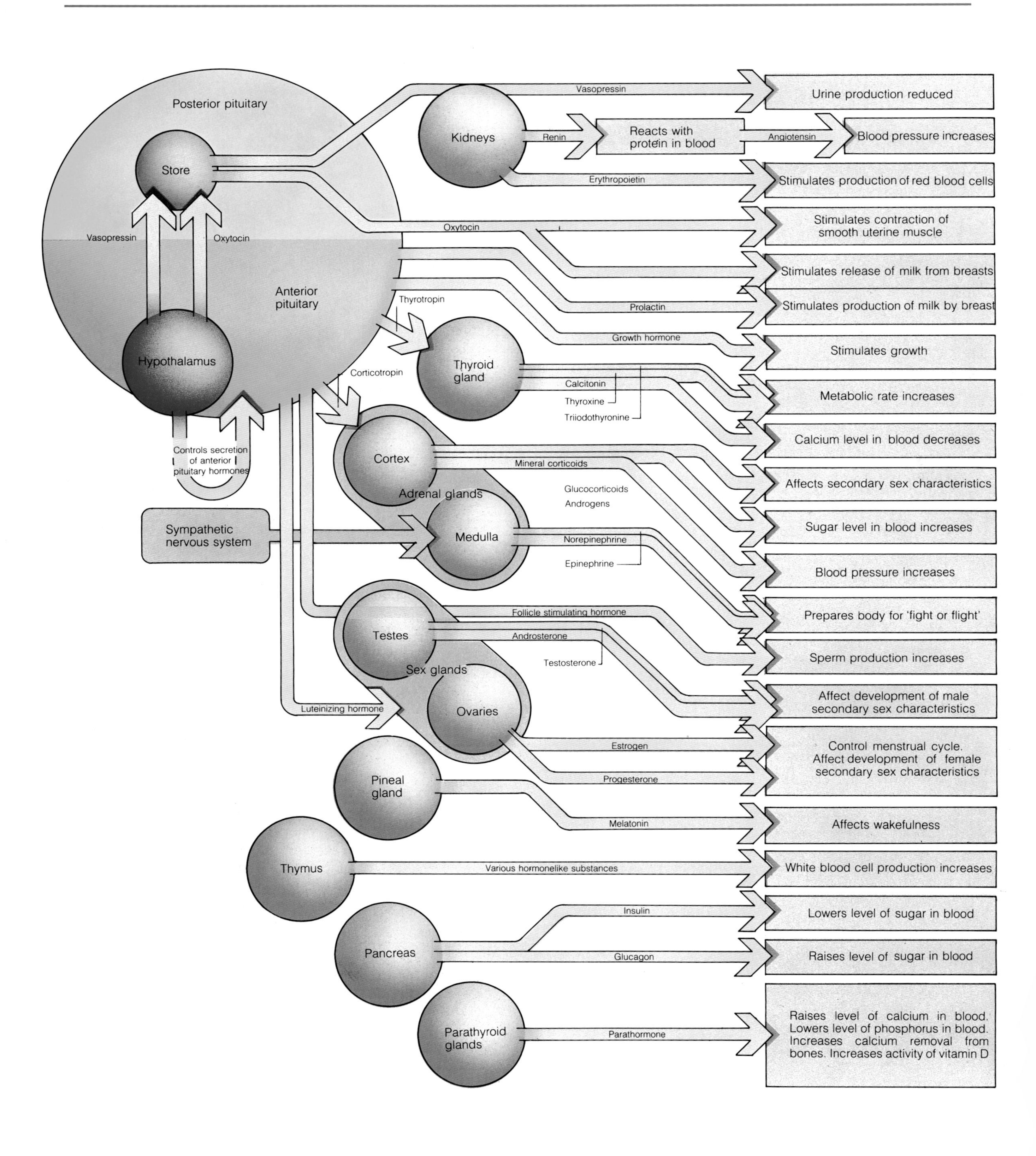

Posterior pituitary
Store
Vasopressin
Oxytocin
Anterior pituitary
Hypothalamus
Controls secretion of anterior pituitary hormones
Sympathetic nervous system
Kidneys
Vasopressin
Renin
Reacts with protein in blood
Angiotensin
Erythropoietin
Oxytocin
Prolactin
Thyrotropin
Growth hormone
Thyroid gland
Corticotropin
Calcitonin
Thyroxine
Triiodothyronine
Cortex
Adrenal glands
Medulla
Mineral corticoids
Glucocorticoids
Androgens
Norepinephrine
Epinephrine
Testes
Sex glands
Ovaries
Follicle stimulating hormone
Androsterone
Testosterone
Luteinizing hormone
Estrogen
Progesterone
Pineal gland
Melatonin
Thymus
Various hormonelike substances
Pancreas
Insulin
Glucagon
Parathyroid glands
Parathormone
Urine production reduced
Blood pressure increases
Stimulates production of red blood cells
Stimulates contraction of smooth uterine muscle
Stimulates release of milk from breasts
Stimulates production of milk by breast
Stimulates growth
Metabolic rate increases
Calcium level in blood decreases
Affects secondary sex characteristics
Sugar level in blood increases
Blood pressure increases
Prepares body for 'fight or flight'
Sperm production increases
Affect development of male secondary sex characteristics
Control menstrual cycle. Affect development of female secondary sex characteristics
Affects wakefulness
White blood cell production increases
Lowers level of sugar in blood
Raises level of sugar in blood
Raises level of calcium in blood. Lowers level of phosphorus in blood. Increases calcium removal from bones. Increases activity of vitamin D

Many bodily actions are brought about by a dozen endocrine glands and the main hormones they produce. The diagram *(left)* shows how the hypothalamus and pituitary (red and yellow) stimulate other glands (blue) to produce their hormones, which have the effects shown on the right.

may be a problem. Treatment with thyroxine gives a rapid improvement and means that the baby can develop normally.

In adults hypothyroidism causes myxedema, whose symptoms include tiredness, a hoarse voice, intolerance of cold, dry skin and loss of hair, depression, infertility and menstrual abnormalities in women. Diagnosis is carried out by analyzing the blood for levels of TSH (which will probably be high) and thyroxine and triiodothyronine (which will be low). The disease can be caused by various conditions, including iodine deficiency, damage to the thyroid, and malfunction of the pituitary. Treatment is by administering thyroxine orally.

Parathyroid glands

Calcium is one of the vital minerals. It is mainly used in the formation of bones and teeth, but there is also a small amount of free calcium in the body. It is needed for the brain, nerves and muscles to function properly and for blood clotting. It may also be involved in the everyday workings of cells. Vital control of the body calcium level is carried out by the hormone parathormone, produced by the parathyroid glands, pea-sized organs located on the back of the thyroid.

When the level of parathormone is increased, stored calcium is released from the bones. Because bones are mainly calcium phosphate, the release of calcium also causes a release of phosphate. To prevent excessive phosphate build-up, the parathormone also increases the rate at which the kidneys excrete phosphate from the body.

Overactivity of the parathyroids gives high levels of blood calcium while the consequential loss of calcium from the skeleton can soften the bones. Kidney stones can come about because of high blood calcium levels. Conversely, a lack of parathormone – possibly due to damage of the parathyroids – results in an abnormally low level of blood calcium. The main symptom is uncontrolled twitching and spasms of the body muscles, a condition known as tetany, with muscle cramps typifying less severe cases. Increasing calcium intake either in tablet form or by eating calcium-rich dairy products is the best treatment.

To get the calcium from food the body relies on vitamin D, which is processed in the liver and kidneys to make the hormone calcitriol. Output of calcitriol is stimulated by an increase in the level of parathormone and a decrease in the blood phosphate. Calcium levels are critical in growing children, and in women who are pregnant or breast-feeding their babies. To make sure that the calcium needs are met, the production of calcitriol is also stimulated by growth hormone, estrogen (produced during pregnancy) and prolactin (associated with breast feeding).

The pancreas – a dual-purpose gland

Active body cells need energy and they get it from carbohydrate food which is broken down into glucose and absorbed into the bloodstream. Glucose gets from the blood and into the cells with the help of the hormone insulin, which is produced in the pancreas. Most of the pancreas is engaged in producing enzymes needed by the digestive system, but dotted within the gland are thousands of small groups of cells. These are the islets of Langerhans, and it is here that the insulin is produced.

After you have eaten a meal, the blood glucose level starts to rise as your digestive system gets to work. Endocrine cells in the islets of Langerhans sense the rise and respond by increasing the rate at which they produce insulin. This lets body cells absorb the glucose, so reducing the amount in the blood back toward its normal level. Not all of the glucose taken up is needed at once, so the excess has to be preserved to meet future demands. To do this the cells convert the excess glucose into the carbohydrate glycogen, which is stored in the liver and muscles. When more energy is needed, the glycogen is released from store and converted back to glucose.

To tune the system to perfection the islets of Langerhans also produce glucagon, a second hormone with a complementary action to that of insulin. It comes into play when the glucose level in the blood starts to fall. To prevent the level falling too low, the glucagon stops the production of glycogen and starts to convert the glycogen in the liver back into glucose. Working in harmony, insulin and glucagon thus maintain the blood glucose at the level needed for the body to work normally.

But the system does not always operate with complete smoothness. The disease diabetes mellitus is the result of something going wrong with the making or use of insulin. Glucose builds up in the bloodstream but does not get into the cells. This causes tiredness and leads to weight loss because fat and muscle have to be burned up to provide the necessary energy. Blood glucose levels stay high and the glucose in the bloodstream spills over into the urine, making it characteristically sweet. The rate of urination goes up and is accompanied by extreme thirst.

Sufferers from diabetes may show a wide range of symptoms, and there are many possible complications, which may involve all parts of the body, including the eyes, kidneys and the blood vessels and heart. Most are commonest in diabetics who have had the disease for a long time.

For this reason diabetes is one of the most common causes of blindness, either by causing a cataract (in which the eye lens becomes opaque) or by causing actual damage to the retina. Similarly diabetes can make the arteries narrow and be a trigger of heart attacks. Sometimes the arteries supplying an extremity such as a hand or foot can close off so much that they become blocked. Because the limb gets no blood, gangrene may set in. At worst the affected limb has to be amputated.

In children diabetes is often caused when the pancreas fails to produce enough insulin. It can appear suddenly and cause severe disruption of the body's metabolism. In the past this juvenile diabetes was often fatal, but once insulin had been discovered and isolated, it became possible to treat juvenile diabetes by directly administering insulin to increase the hormone levels in the bloodstream. But insulin is a protein so you cannot take it as tablets – it would simply be digested and broken down along with other proteins in food. Instead it has to be injected directly into the body. This is a very effective treatment for diabetes, and the insulin used is obtained from the pancreas glands of cows or pigs, or from bacteria that have been genetically engineered so that they produce human insulin.

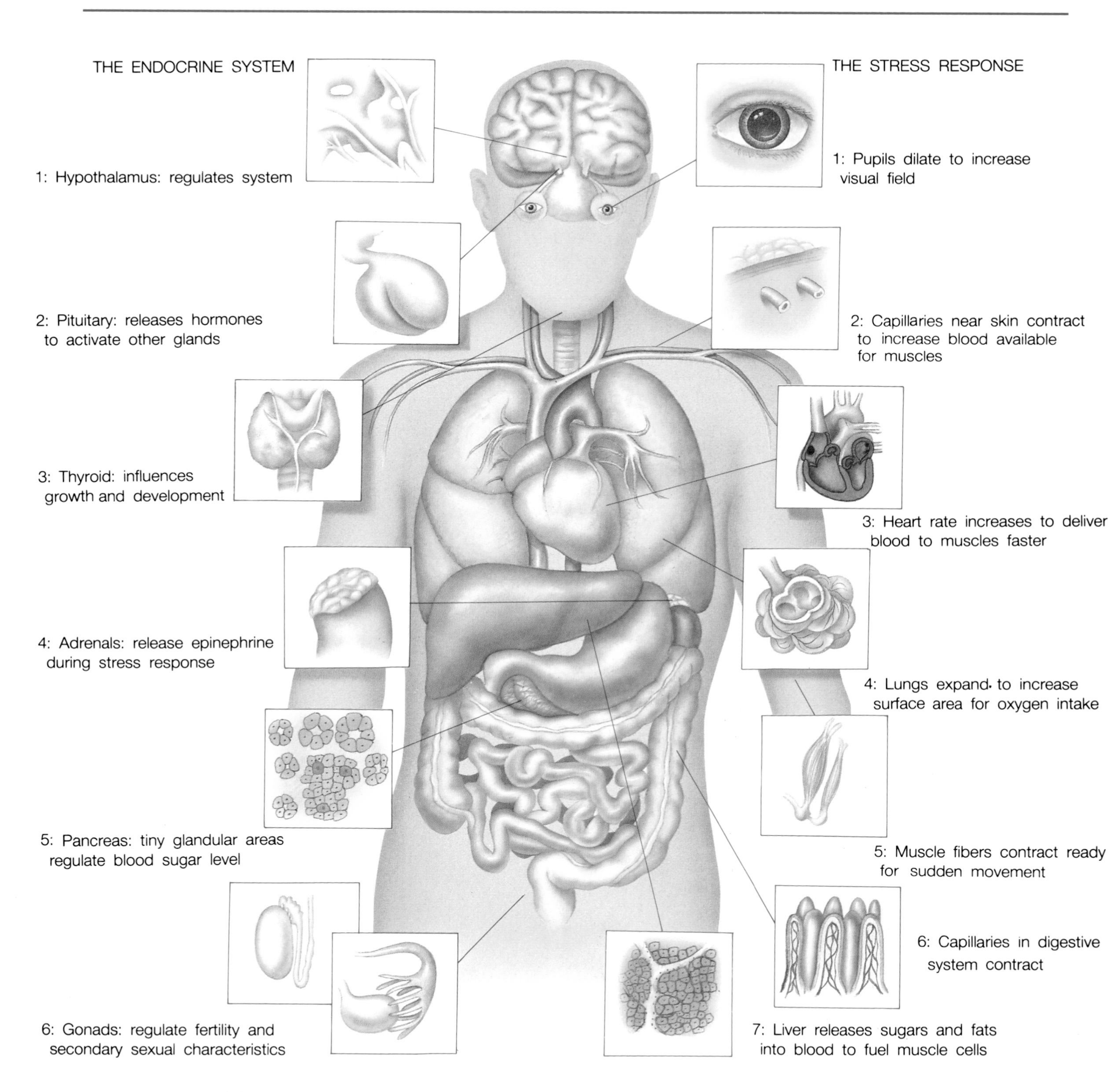

The chief glands of the endocrine system, and their actions, are shown on the left-hand side of the diagram. One of the fastest-acting reactions is caused by the release of epinephrine (adrenaline) from the adrenal glands in response to stress, reactions whose fundamental purpose is to prepare the body – particularly the muscles – for "fight or flight." Within seconds a host of changes take place, as shown on the right-hand side of the diagram: the pupils of the eyes widen to take in a better view of the surroundings; small blood vessels in the skin contract to divert blood away from the skin (making the person go pale) to the muscles, where it is needed; the heart speeds up to pump more blood to the muscles; the lungs expand to take in more oxygen (needed by the muscles); muscle fibers contract in readiness; small blood vessels in the intestines contract, again to divert blood to the muscles; blood sugar (glucose) and fats are released from the liver.

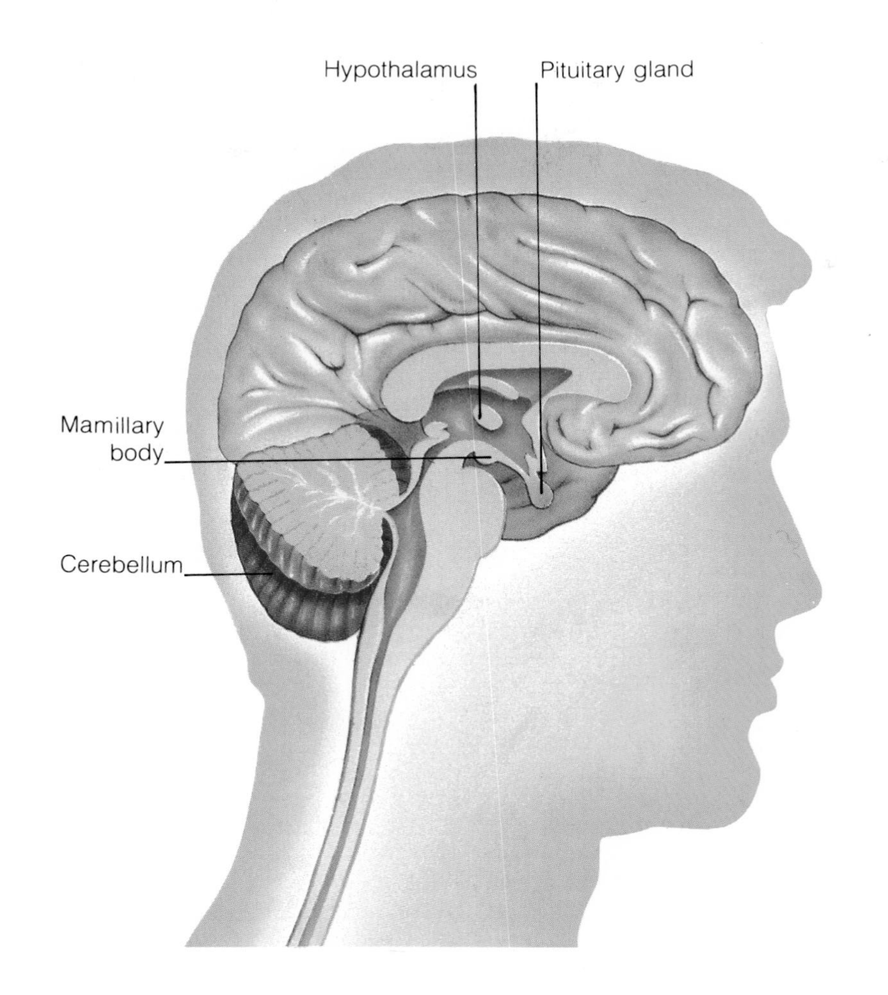

The hypothalamus and the pituitary lie close together and are linked to the brain by nerves and blood vessels. A message from the brain along nerves to the hypothalamus stimulates it to release hormones. Some, such as oxytocin and vasopressin, travel along neurosecretory tracts to be stored in the posterior lobe of the pituitary. They are released as required, oxytocin to stimulate contractions of the womb during labor and initiate the secretion of milk from the breasts, and vasopressin (also known as antidiuretic hormone) to prevent excessive water loss from the body. It does this by increasing the reabsorption of water from the blood by the kidneys.

Other hypothalamic hormones are carried along narrow capillary blood vessels into the anterior lobe of the pituitary, where they bring about immediate actions: prolactin, luteinizing hormone and follicle-stimulating hormone act on the breasts or ovaries, adrenocorticotropin (ACTH) stimulates the adrenal cortex to release its hormones, thyroid-stimulating hormone activates the thyroid gland, and growth hormone (somatotropin) stimulates growth.

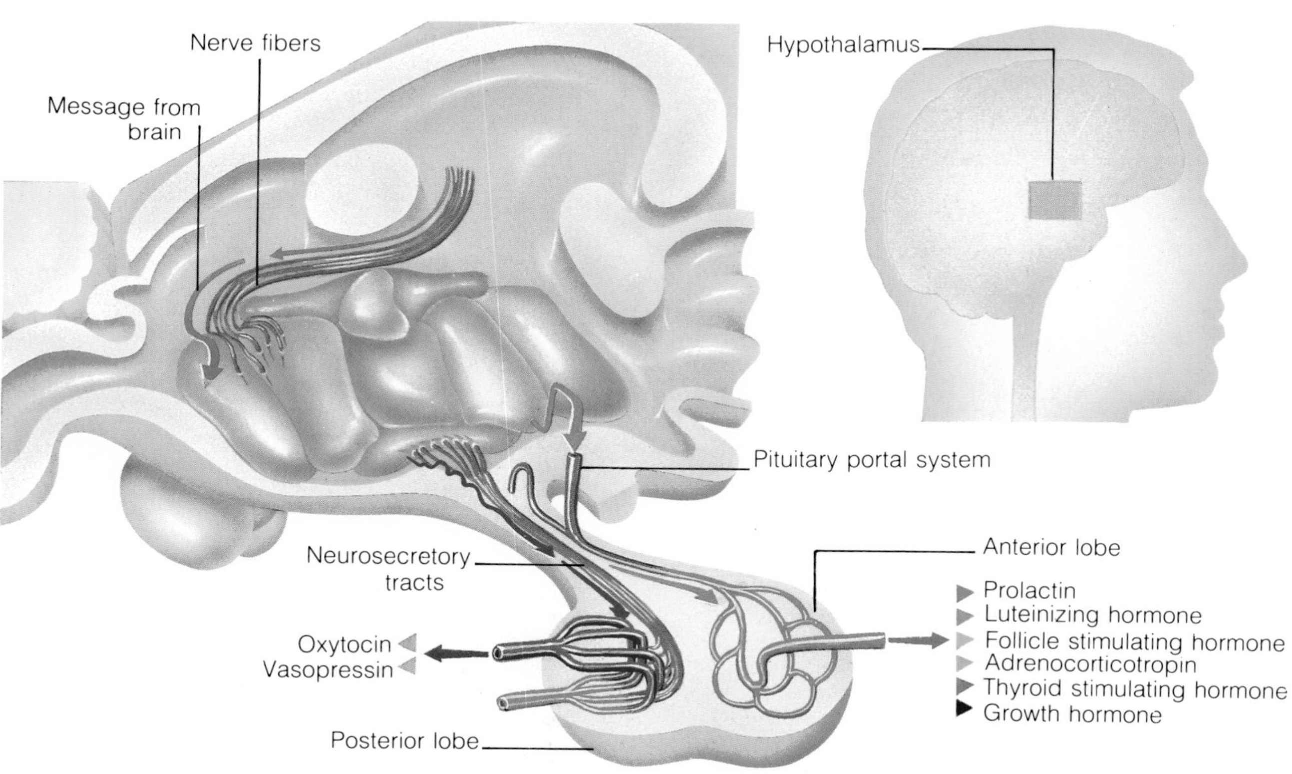

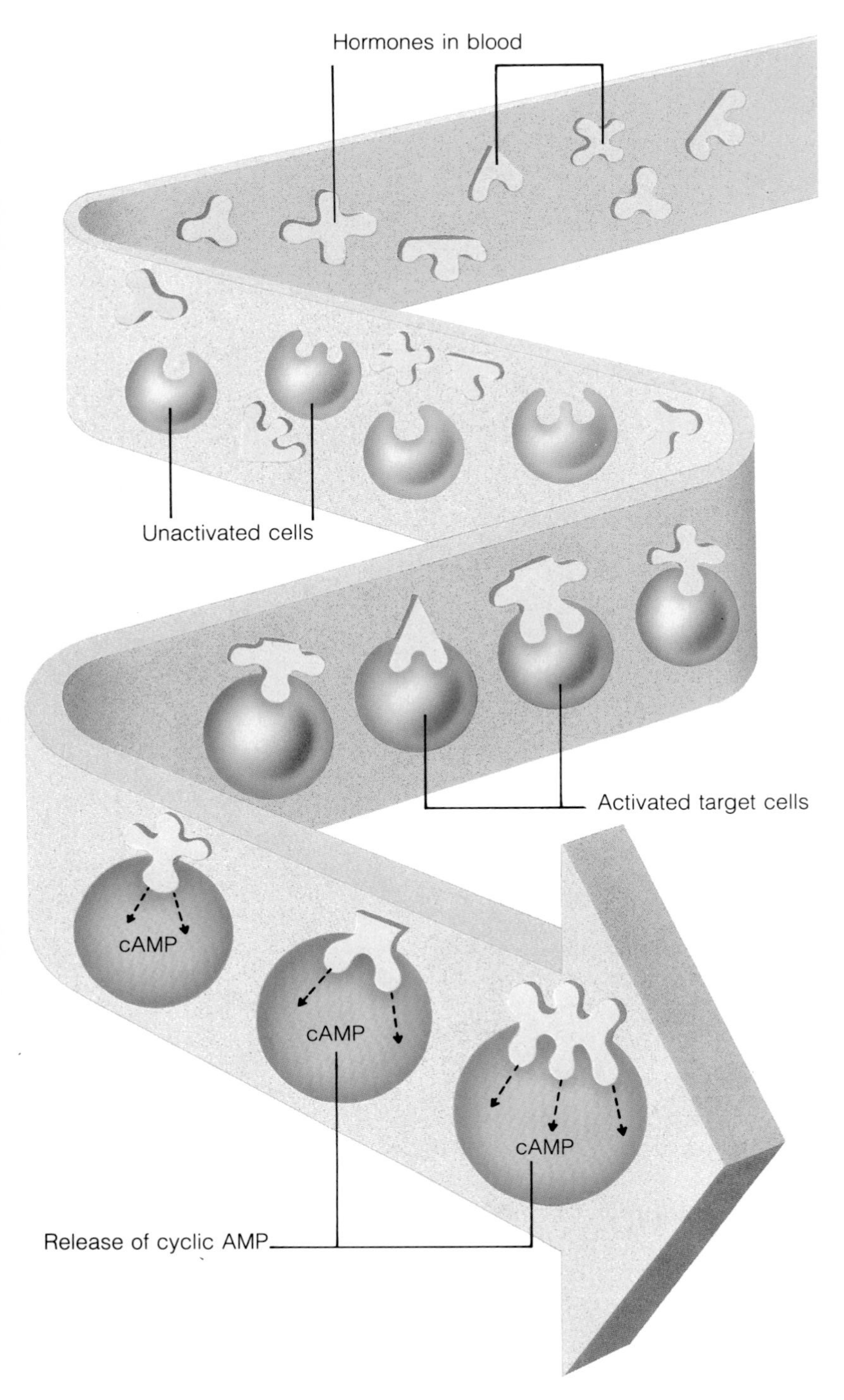

Every hormone is specific in its chemical activity and acts only on the cells in a particular target tissue. In order to achieve such fine discrimination, each hormone "fits" its target cell as a key fits a lock *(above).* Once locked on, it releases a chemical called cyclic AMP (cAMP). This in turn triggers a series of events that lead to the required action.

Blood volume is largely under hormonal control *(right).* When blood volume falls, the vagus nerve "tells" the pituitary, which releases two hormones: vasopressin and adrenocorticotropin (ACTH). Vasopressin slows down the release of fluid by the kidneys, and ACTH acts on the adrenals to produce aldosterone, making the kidneys retain salt.

A milder diabetes, more common in adults, results when the body fails to respond properly to insulin that is produced by the pancreas in the normal way. Sufferers are often overweight and the disease more commonly occurs after the age of 40. In a lot of cases all that is necessary to control the disorder is a change of diet to include less sugar.

The classic diabetes symptoms of extreme thirst and excessive urine production are also typical of another type of diabetes – diabetes insipidus – but this is not due to problems with the insulin level. It is a much rarer disease and caused by shortage of vasopressin, a hormone that is normally produced by the joint action of the hypothalamus and posterior pituitary. Lack of this hormone stops water being reabsorbed by the kidneys, so that too much fluid is lost from the body.

Adrenal glands

Each of the two adrenal glands, found on the upper part of the kidneys, has two main parts, an outer cortex and an inner medulla. These produce a number of different hormones that have a wide variety of functions. The outer part of the cortex produces the mineralocorticoids – steroid hormones that act to control the fluid and mineral balances of the body. Raised blood pressure can be caused by Conn's syndrome, in which overproduction of these hormones makes the body hold on to too much sodium and water.

Control of the way the body deals with carbohydrates, particularly glucose, is a vital function of glucocorticoids. They are produced by the two inner layers of the cortex. Male sex hormones – androgens – are also made by the adrenal cortex (even in women, although the main source in men is the testes). Genetic defects may make the adrenal glands produce an excess of androgens in early life, and when this happens to females it tends to generate masculine characteristics. These include enlargement of the clitoris, making it very difficult to establish a child's true sex. In extreme cases such girls can look like boys.

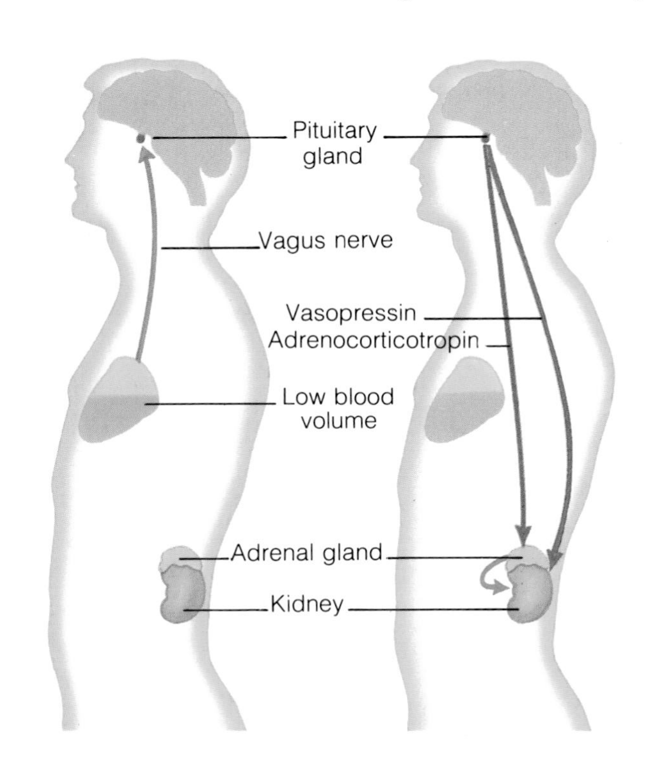

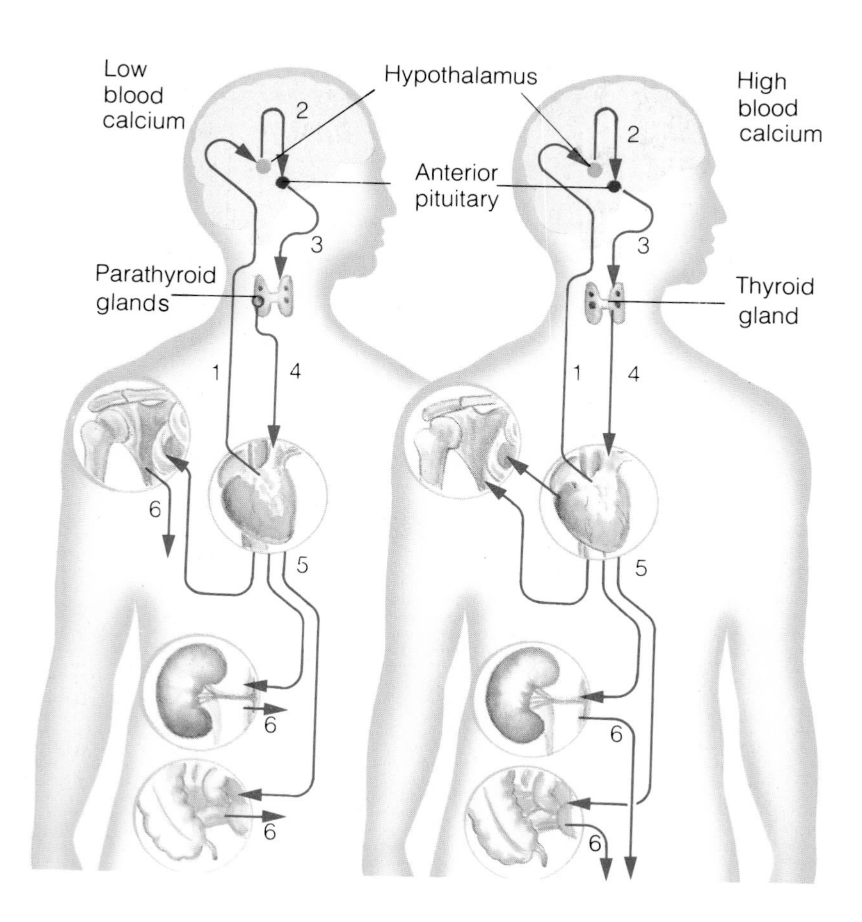

The bones are the chief store of calcium in the body but the maintenance of the correct calcium level in the bloodstream is controlled by the parathyroid glands. Low blood calcium is detected by the hypothalamus, which "informs" the pituitary. A hormone from the anterior lobe of the pituitary travels in the bloodstream to the parathyroids, which release parathormone. This has three effects: it speeds up the absorption of calcium from food by the intestines, reduces its absorption from the blood by the kidneys, and causes the release of calcium from the bones. If the level of calcium in the blood rises too high, the hypothalamus and pituitary cause the thyroid gland to produce the hormone calcitonin. This has the opposite effect to parathormone, reducing calcium absorption in the intestines, accelerating its retention by the kidneys and increasing the amount of calcium deposited in bones.

Appropriate treatment to block the production of androgen reduces the symptoms and allows the girls to become fertile.

The adrenal glands are also active at the moment of birth. Research has revealed that the birth process may be triggered by androgens made by the adrenal glands of the baby. In the last few weeks of pregnancy the level of output seems to increase and the androgens are converted into estrogens in the placenta. These estrogens upset the mother's hormone balance and lead to the onset of labor.

Addison's disease results from a shortage of adrenal hormones – its symptoms include dark pigmentation of the skin, weakness, weight loss and dizziness. It is treated by replacement hormone therapy using glucocorticoids. But this is complicated by the fact that the dosage needed varies considerably with the level of body activity. Overdosage of the hormone can cause the symptoms of Cushing's syndrome. With this condition the sufferer is usually obese and characteristically has a moon face, a hump at the neck, a protruding abdomen and relatively thin legs. There are many other possible symptoms, which may include a tendency to bleed, high blood pressure, weakness, stretch marks, diabetes mellitus, back pain, psychiatric disturbances and problems with the reproductive function.

In many cases Cushing's syndrome comes about because of problems with the pituitary gland. A tumor there may cause it to secrete excess ACTH, and this stimulates the adrenal cortex to manufacture excess corticosteroids. Other tumors, especially in the lungs, also produce ACTHlike substances which have the same effect. Another possible cause is a tumor of the adrenal cortex itself.

Inflammation occurs in a large number of diseases that afflict the body, including asthma and rheumatoid arthritis, and is often extremely painful. Treatment with corticosteroids is generally very effective in suppressing it, but considerable care with dosage is vital to avoid the complication of Cushing's syndrome. Athletes and bodybuilders sometimes use anabolic steroids to help build up their muscles, even though the drugs are officially banned. These hormones are closely related to the ones produced by the adrenal cortex, and excessive use of them can cause serious complications.

When we are frightened or subjected to sudden stress our bodies ready themselves for sudden action – either to fight or to flee. This is the job of the hormones associated with the adrenal medulla. Stress – either physical or emotional – triggers it to release epinephrine (adrenaline) and norepinephrine (noradrenaline), which muster the body's resources. The heart beats faster and more strongly, blood pressure rises and the blood flow is directed way from non-essential areas such as the skin to the muscles, so making them ready to spring into action.

Gonads – the sex glands

Sexual development and activity would be impossible without the action of the gonads – the testes in men and the ovaries in women. The different courses of development typical of a girl or a boy start in the womb when the embryo is about 40 days old. If the embryo is male the embryonic gonads start to develop as testes and begin producing the male sex hormone testosterone. This acts on the, as yet undeveloped, sex organs and they start to form a penis and scrotum. With a female embryo there is no male

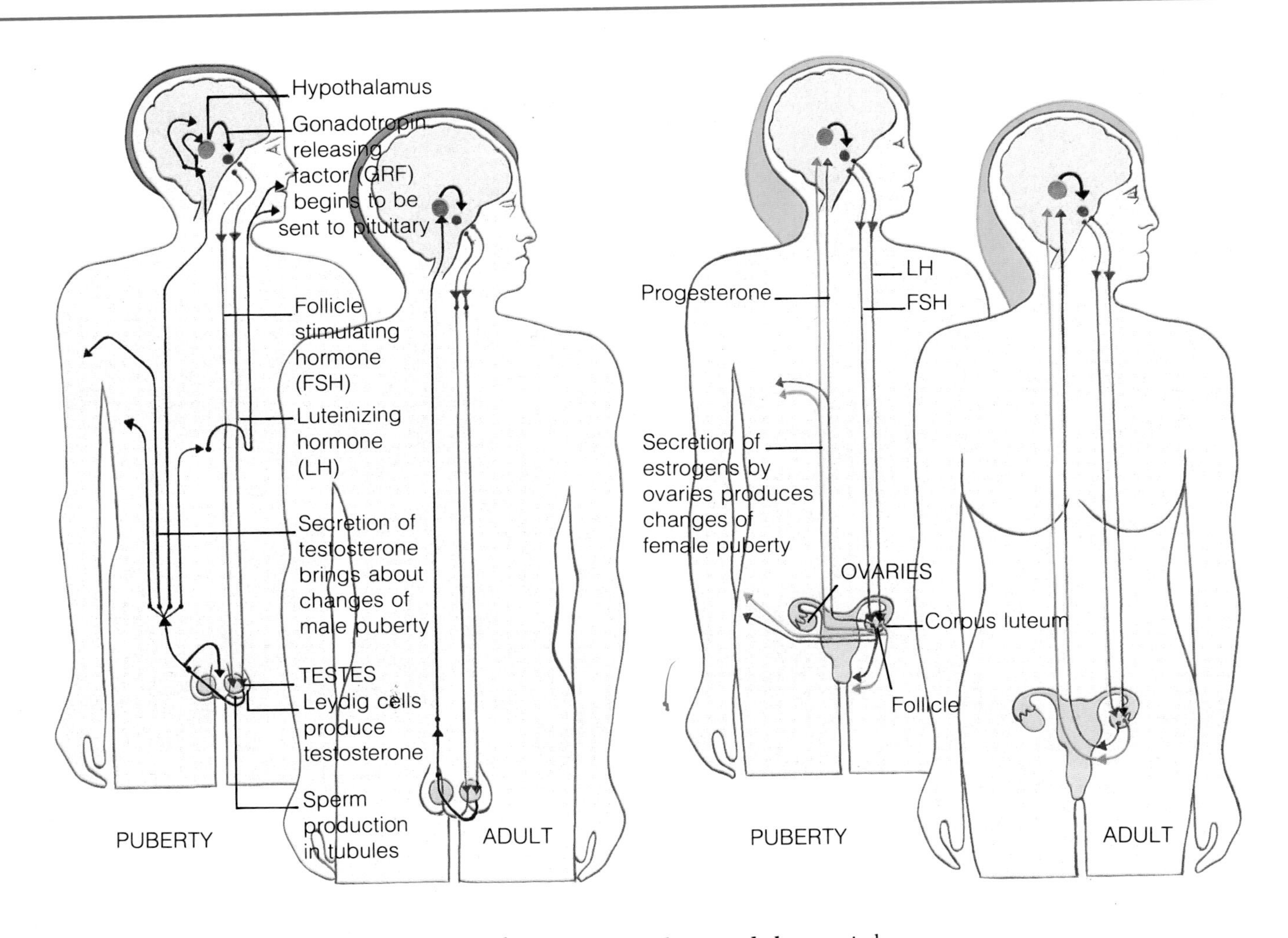

hormone and the sex organs form the Fallopian tubes, uterus, cervix and vagina, while the gonads become ovaries.

When a baby is born its ovaries or testes are not yet working at full capacity, but they do produce small amounts of hormones – mainly estrogens in girls and testosterone in boys. Similarly the pituitary gland produces minute amounts of its sex hormones but the system is really doing little more than ticking over. Then as puberty approaches the picture changes dramatically, with a sudden surge of sex hormones. Exactly what causes the changes is not certain but the results are obvious. Girls start to develop breasts and their hips widen as they turn into women; boys start to grow facial and body hair and their voices deepen as they become men; and in both sexes underarm and pubic hair starts to grow.

These secondary sexual characteristics start to show because of the action of the sex hormones – estrogens, progesterone, testosterone and androsterone. The first two are primarily associated with female sexual functions and the second two with male sexual functions, but all four are produced by both males and females. The different characteristics are the result of variations in the relative levels of the hormones.

Puberty normally occurs between the ages of 10 and 16 in girls, and 14 and 18 in boys, but some conditions can cause sexual development to occur before or after these ages. For example, if the hypothalamus or pituitary are damaged or affected by a tumor there may be a premature release of the gonadotropins. Abnormally early development of the gonads

The sexual changes in boys and girls at puberty are brought about by sex hormones from the gonads, which are in turn stimulated by hormones from the pituitary. As with most endocrine functions, there is a series of reactions beginning at the hypothalamus. In this case it produces gonadotropin-releasing factor (GRF), which acts on the pituitary to produce follicle stimulating-hormone (FSH) and luteinizing hormone (LH). These bring about the secretion of sex hormones by the gonads – the testes in boys *(above left)* and the ovaries in girls *(right)*.

takes place, with the children developing full sexual maturity and the appropriate secondary sexual characteristics. In contrast abnormalities in the testes or ovaries may restrict the production of the sex hormones, and delay the onset of puberty.

From their first menstrual period (menarche) onward, women have a regular, hormone-controlled, reproductive cycle that prepares their bodies for fertilization and childbearing. This menstrual cycle works through every 28 days or so, with ovulation – the release of an ovum – taking place about halfway through it. The womb lining is shed as the menstrual flow at the end of each cycle. All of the hormonal changes involved are controlled by the gonadotropins secreted by the pituitary.

During the first part of the menstrual cycle, follicle-stimulating hormone has a dominant effect. Its action causes one of the ovarian follicles to become dominant and fully develop into a Graafian follicle. Estrogen is released by the follicle and acts to prepare the lining of the womb (the endometrium) to accommodate a fertilized ovum. About 14 days into the cycle there is a surge of luteinizing hormone to provoke the release of the ovum by the ripe follicle, which becomes the corpus luteum. This organ then secretes progesterone, which acts with the estrogen to build up the lining of the womb.

Progress of the rest of the cycle now depends on what happens to the released ovum. If fertilization does not occur within 48 hours of release, the ovum dies. The corpus luteum degenerates, and secretion of progesterone decreases, from the seventh to the fourteenth day after ovulation. Reduction in the hormone levels makes the womb lining break down and it is shed by menstrual bleeding. The start of menstruation is taken as the beginning of a new cycle.

A different sequence of hormone production is set into operation if an ovum happens to be fertilized by a sperm from a man. About four days after fertilization the developing embryo burrows into and embeds itself into the lining of the womb and starts to form a placenta. The newly forming placenta releases a hormone that acts on the corpus luteum of the ovary so that it continues to produce progesterone and the womb lining is kept and maintained. When developed, the placenta takes over the functions of the corpus luteum to produce most of the steroid hormones needed, as well as being vital to the nourishment of the developing child.

Successful fertilization and implantation of the embryo involves a precise sequence of processes, all of which depend on the control exerted by hormones. As a result, it is comparatively easy – in theory at least – to interfere with it and prevent pregnancy occurring. This is the principle behind the contraceptive pill, which brought about a revolution in contraceptive practice.

Two main types of oral contraceptive have been developed, the combined pill containing both estrogen and progesterone,

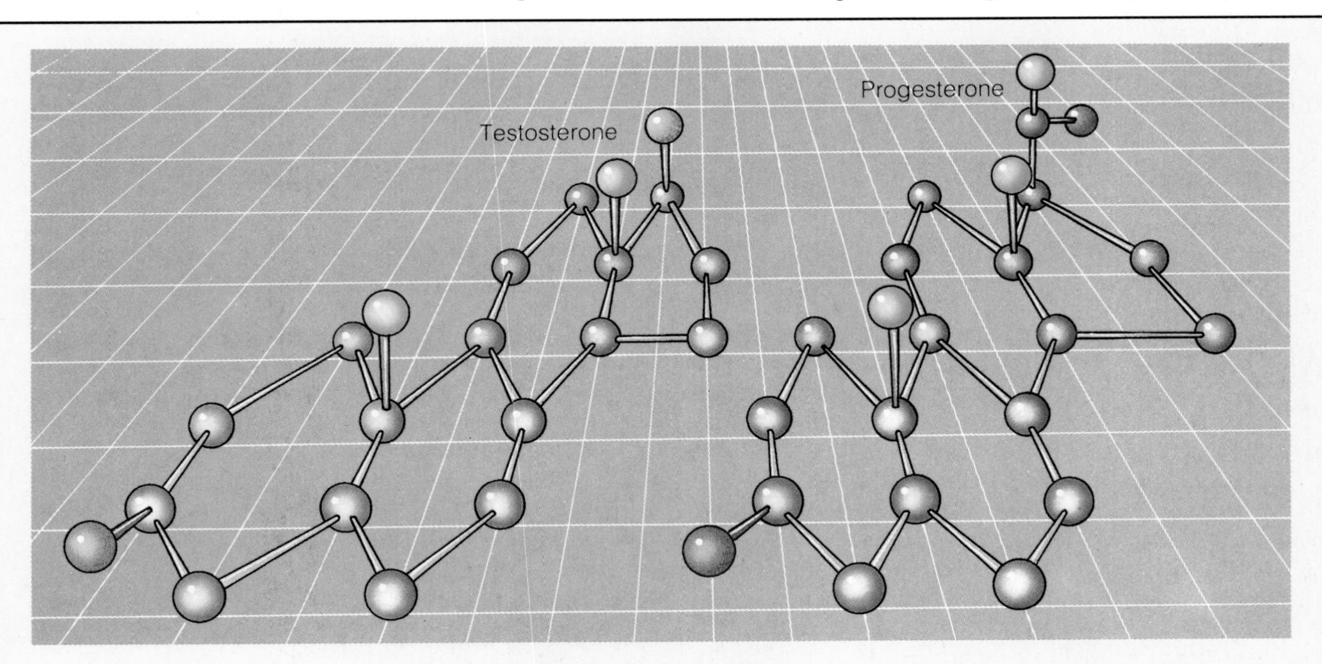

Most of the differences between males and females – established initially at the embryonic stage in the womb – result from the presence of sex hormones, chief of which are testosterone in males and estrogen and progesterone in females. But a comparison of the molecular structures of testosterone and progesterone *(above)* shows how little difference there is between the two. Both molecules consist of fused ring systems with atoms projecting from the rings, but at one corner progesterone has a three-atom grouping where testosterone has only one atom. It could therefore be argued that the chief difference between men and women is only two tiny atoms.

The regular daily rise and fall in the level of the adrenal hormone hydrocortisone (cortisol) in the bloodstream affects the body's metabolism, sleep and wakefulness and mental performance. Its secretion from the cortex of the adrenal glands is stimulated by the presence in the bloodstream of adrenocorticotropic hormone (ACTH), which is itself a product of the pituitary. Hydrocortisone is a multifunctional corticosteroid hormone important in the metabolism of energy-generating carbohydrates such as glucose and glycogen. Thyroxine from the thyroid gland influences the rate at

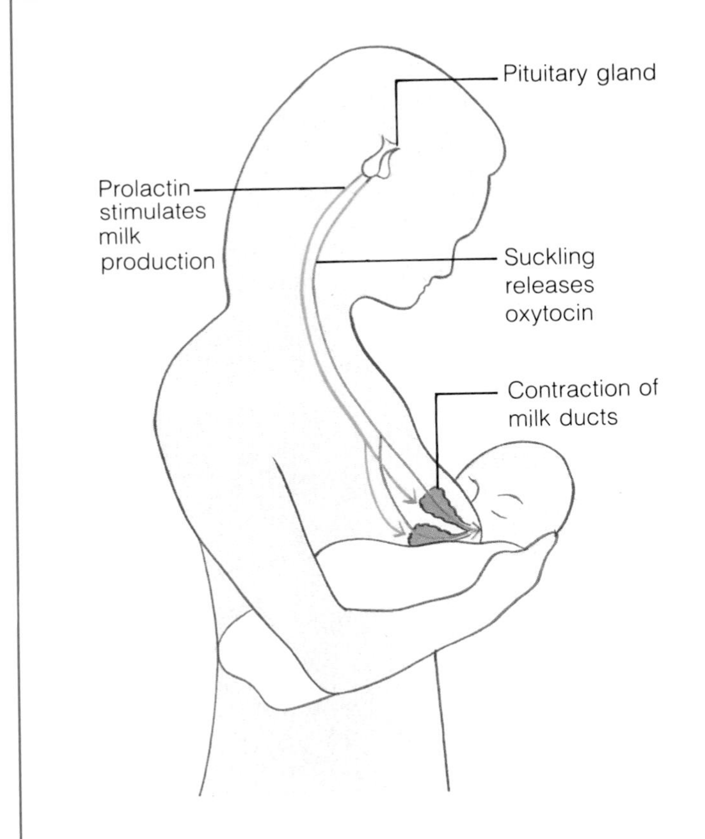

The production of milk after childbirth is triggered by the baby suckling at the mother's breast. This action makes the pituitary release the hormones oxytocin, which squeezes milk from the milk ducts, and prolactin, which maintains the production of milk.

and the progesterone-only type. When taken orally the hormones are absorbed by the woman's body so that the blood has relatively high levels of circulating hormones. These generate negative feedback on the hypothalamus and pituitary, which decrease their production of LH and FSH with the result that the processes of ovulation and implantation are both inhibited. Continued exposure of the body's organs to high hormone levels is believed to have adverse effects, however, such as increased risk of blood clots (thrombosis), and a lot of effort has been applied to developing pills with the lowest possible hormone concentrations.

Male sexual characteristics are similarly controlled by the action of the gonadotropins though there is, of course, no regular cyclic pattern as there is in women. Testosterone is the main male sex hormone and it is produced in the testes in response to luteinizing hormone (LH) from the pituitary gland. Aggression and sex drive are believed to be associated with this hormone though the exact influence is unclear. One possible pointer is the way castration dulls the libido, though it does not necessarily remove all sexual interest. Sperm production takes place in the testes under the joint influence of follicle-stimulating hormone (FSH) and testosterone. Feedback of the testosterone to the pituitary and hypothalamus helps to stabilize the action of male hormones in the testes.

Other endocrine tissues

The production and release of hormones is not the sole province of the main endocrine organs, and there are cells in various parts of the body that have an endocrine function. A noteworthy concentration occurs in the gastrointestinal tract, where they produce a number of hormones which have a local effect to stimulate the production of digestive juices. For example, the stomach secretes gastrin, which triggers the release of digestive acid from the gastric glands. Then as the part-digested food reaches the duodenum it causes the release of cholecystokinin, which makes the gall bladder contract to squeeze out its bile

which cells use up oxygen and produce heat, with the control of body temperature one of the major functions of the hypothalamus. Mental performance seems to be related to body temperature.

salts, and pancreozymin which brings about the secretion of enzymes from the pancreas.

Other organs that have an endocrine action added onto their main function include the kidneys and the thymus, together with the pineal gland. The main and crucial job of the kidneys is the excretion of excess fluids, minerals and the waste products of digestion and metabolism, but it also has several endocrine functions. One of the substances produced by the kidneys is renin, which helps control blood pressure by stimulating the production of angiotensin. This then causes a direct constriction of the blood vessels and stimulates the release of aldosterone (another hormone) from the adrenal cortex, with the end result being a rise in blood pressure. Another hormonal product of the kidneys is erythropoietin, which has a crucial role in the manufacture of red blood cells.

The thymus is central to the body's defense mechanisms, acting to help lymphocytes (white blood cells) mature. How this happens is unclear, but scientists believe it to involve the action of hormonelike substances such as thymusin. The exact function of the thymus is not known because the bodies of healthy adults appear to be able to function effectively without this gland, which shrivels into nonexistence in most adults. Yet in children the thymus is essential to the workings of the immune system (and patients with tumors of the thymus are liable to suffer from a range of infections.) Such tumors are also associated with various autoimmune diseases.

Yet another organ whose workings are not properly understood, but which seems to have at least some endocrine functions, is the pineal gland or "third" eye. Located at the front of the brain just behind the forehead, it is known to secrete a large number of active chemicals, one of the most important of which is melatonin. This substance is important in hibernating animals, and in humans its concentration in the blood increases during the night. Injections of melatonin make a person sleepy, so maybe it helps control the human sleep cycle, part of the natural daily rhythm of life.

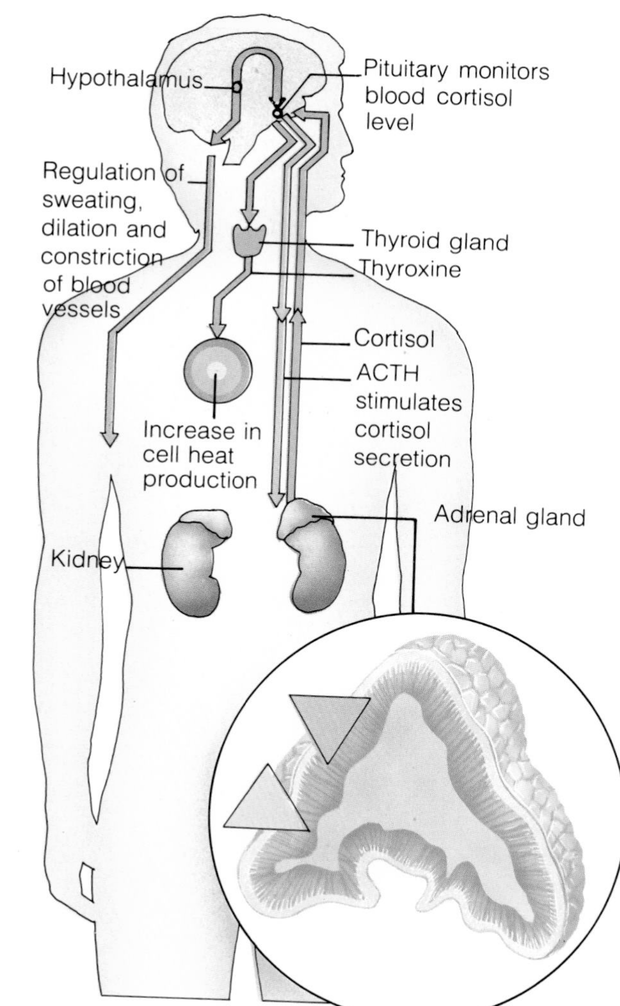

As well as being the pituitary's stimulator, the hypothalamus has a direct effect on body temperature by regulating sweating from the sweat glands and controlling the narrowing and widening of blood vessels in the skin. Through the pituitary, it influences the levels of thyroxine and hydrocortisone (cortisol), which control metabolic heat production.

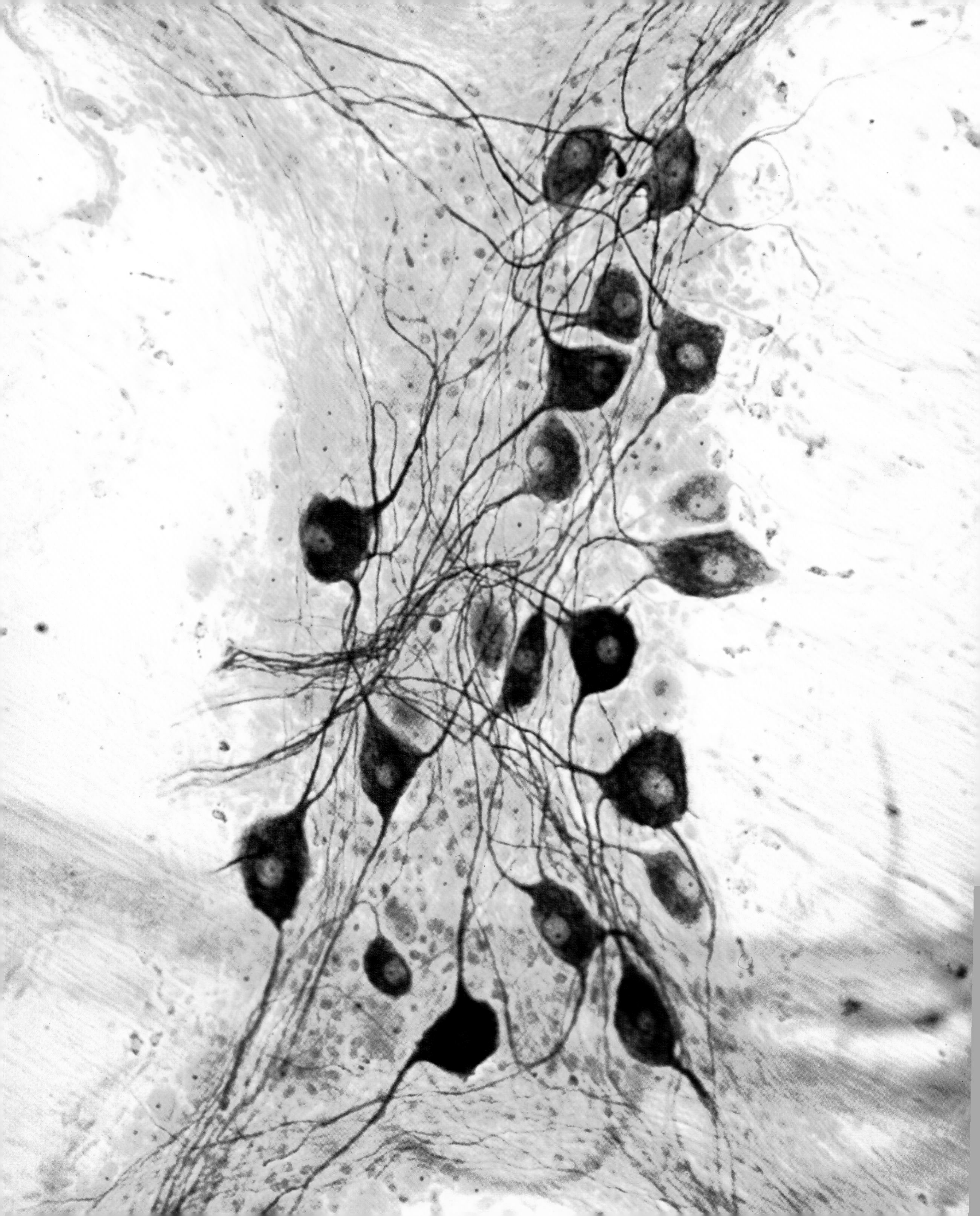

CHAPTER 12

The Nervous System

Silently and ceaselessly, at the height of human activity and during the depths of sleep, our nerves relentlessly transmit impulses to and from the brain. Every movement we make, from an unconscious twitch of an eyelid to the deliberate actions of driving a motor car, and every sensation we feel – even our dreams – depend on the workings of the nervous system. This complex and sophisticated network extends throughout the body. It is constantly gathering information of all kinds and transmitting commands to stimulate muscles and other organs into activity.

The main parts of this vital system are the brain, the site of sense and sensation; the spinal cord, the main trunk route of the communications network; and the nerves themselves, a branching network whose roots lead into and out of the cord. It is so intricate that the exact workings of each of these elements is still not fully known. Nevertheless it is possible to obtain a broad picture of how the complete system controls the body. Higher functions – those involving memory, comparison and decision making – take place in the brain, while the rest of the nervous system carries sensory information to the brain and commands from it to control the body's movements and reactions.

Systems within systems

Unlike the body's blood vessels or lymph canals, the nerves do not form a single system, rather there are several interrelated systems. Some of these are physically separate; others differ only in function. Together the brain and spinal cord make up the central nervous system. The rest of the network is the peripheral nervous system; physically part of it – but with its own specific functions – is the autonomic nervous system. This is responsible for those body functions not under conscious control – the unfailing beating of the heart and the automatic operation of the kidneys and the digestive system. The harmonious, smooth operation of these vital body functions is achieved by further dividing the autonomic nervous system into parasympathetic and sympathetic systems. These have broadly opposing actions and so provide checks on each other, working together to achieve a balance.

Like all other tissues in the body, from bone to blood, nerves are made up of cells. But nerve cells, termed neurons, are remarkable in many respects and include some of the largest – or at least longest – cells in the body, some being more than a foot long. The nervous system operates by means of minute electrical signals, or impulses, traveling along the length of the nerve cells. Acting at high speed it processes information from the sensory nerves, and initiates any required actions in a fraction of a second. Nerve impulses travel at up to 250 miles per hour. In contrast, the other main control mechanism in the body, the endocrine system, works by releasing hormone secretions into the bloodstream. This operates much more slowly, and can take many hours to achieve the desired effect.

Dark patches among a tangle of branching dendrites, in this microphotograph of part of the vagus nerve in the intestines, are the nerve cell nuclei. One of the vital nerves that arise directly from the brain, the vagus is among the longest nerves in the body. It carries messages from the brain to stimulate the secretion of digestive enzymes and activate the smooth muscles that propel partly digested food along the gut.

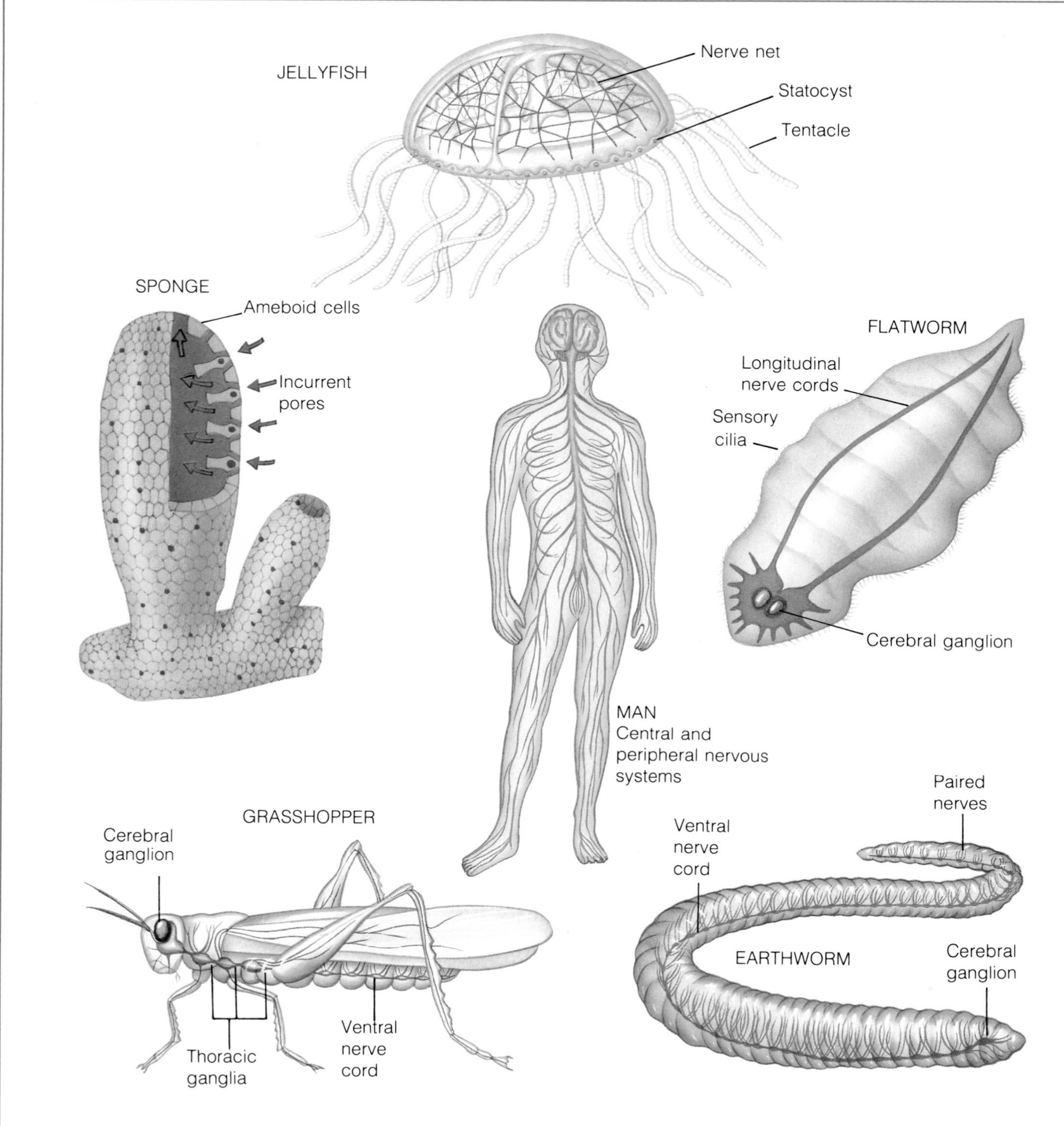

Human beings have the most highly evolved nervous system of any living creature, consisting of the central nervous system (brain and spinal cord) and peripheral nerves serving all parts of the body. On the other hand invertebrate animals have much simpler systems. A sponge, for example, has no nerves at all, and relies for information on "messenger" substances released by ameboid cells in its walls. Jellyfish have a diffuse network of nerve cells in contact with each other, whereas flatworms have a centralized system with cerebral ganglia and a pair of nerve cords. Earthworms have a similar arrangement, consisting of lateral nerves branching off in each segment of the body. In insects the cerebral ganglion forms a primitive brain that controls various activities.

A branching network

An anatomy teacher describing the nervous system has a similar problem to a wireman trying to explain how a telephone exchange works – it is extremely complex and has no real beginning or end. Within the body the nerves branch out like telephone lines from an exchange. They run to every extremity of the body, from the soles of the feet to the top of the scalp, and from the surface of the skin to the central organs of heart, liver and lungs. These nerves are actually single cells which have the highly specialized function of carrying information from one part of the body to another. Most are bunched together like the strands of a rope and form a cord. Individual nerves may be as small as one six-thousandth of an inch thick. They look creamy white, though this color fades as the nerve bundles split up.

Nerve cells have the same basic structure as all the other body cells, with a surrounding membrane containing the nucleus and cytoplasm. But they have a very special shape. Consider a typical motor nerve carrying instructions from the brain to a muscle. At one end there is a tuft of short rootlike projections or dendrites. At the other is a long thin projection, the axon, which may subdivide up to 150 times and be attached to 150 separate muscle fibers.

Although nerve cells may be thinner than a hair from your head, they may also be surprisingly long. For example, in a typical human the nerve running from the base of the spine to the tip of a toe is around three feet long. However, many other axons are only a fraction of an inch long. Most nerves act as links in a chain of nerve cells rather than connecting directly to a muscle. In such a chain each axon is in near contact with the dendrites of the next cell, but there is a tiny gap, known as a synapse, between them. Nerve impulses must somehow jump this gap and they do this with the help of chemical messengers, which are known as neurotransmitters.

A chain reaction

When an electrical nerve signal reaches the end of an axon, it triggers the production of a neurotransmitter chemical from tiny secretory cups, or vesicles. The neurotransmitter then quickly diffuses across the synapse and excites the dendrites of the next cell to induce a new electrical signal in their cell, and so on along the chain. This knock-on effect happens in a fraction of the time it takes to explain it. At each synapse there is some resistance to the transmission, so the nervous message gradually becomes weaker as it moves along the nerve chain.

Although the synaptic gaps appear to hold up the smooth flow of nerve signals, they have the vital function of allowing more than one nerve cell to influence the next. And cells can be subjected to many different influences, some of which may stimulate the cell while others "switch of" or inhibit it. The cell's response then depends on the balance between the various stimuli, so the multiple path action allows the system greater precision and control.

Nerve signals are one-way traffic: impulses can flow only from the axon of one cell to the dendrite of another, and reverse flows are impossible. So when you start walking, for example, the control impulses travel from your brain, down the spinal cord and along nerves to the leg muscles. These "command" impulses are transmitted along motor neurons, which are also known as efferent fibers because they carry messages away from the brain. But if you stub your toe, the sensations from your foot are carried back to the central nervous system by another set of nerve cells. These sensory neurons start from the sense organs and are described as afferent because their information travels toward the spine and brain. Both types of nerve may run in the same nerve bundle and they look much alike.

Automatic control

Many body actions are automatic – for example breathing, the beating of the heart, and the workings of the digestive system and kidneys. We do not have to keep reminding ourselves to breathe, or instruct our stomach to go into action after we have eaten a burger. These operations take place unknown to our conscious mind – but they are operating under the control of the nervous system.

Direction of these life-sustaining functions is the responsibility of the autonomic nervous system, but once again it is far from simple and more than one network of nerves is involved. The autonomic has two separate sections, called the parasympathetic and sympathetic systems, which differ anatomically in the routes they take in the body. The neurons for the parasympathetic system take a direct route. They start from the base of the brain and run directly to the target organ, where instructions are carried out.

Neurons for the sympathetic system follow a more tortuous path. Each nerve originates from the spinal column and runs to an intermediate synapse, where it meets a second nerve cell in a vertebral ganglion. The ganglion acts as a relay station, passing the impulse along the second nerve to the relevant organ. Pairs of ganglia are positioned under each of the 31 vertebrae from the base of the neck (the first thoracic vertebra) to the lower back (the fourth lumbar vertebra).

Most of the body's organs are connected to both sympathetic and parasympathetic nerves. However, the two types have opposing effects. Sympathetic nerves tend to "switch on" activity, possibly to deal with stress or an emergency. Parasympathetic nerves, on the other hand, usually "switch off" activity, and these nerves tend to restore the body to a state of calm. In this way the action of an organ can be carefully regulated in a balanced way.

Growing complexity

Because the human nervous system is so complex, it is difficult to analyze and understand. In contrast, many animals have much simpler nervous systems, and studying them can help give useful insights into how the human system works.

Very simple creatures such as amebas and sponges manage without any nervous system at all. The most primitive animals with nerves are sea anemones, jellyfish and corals. Their body cells are arranged in outer and inner layers, with a "nerve net" at the base of the outer layer. The net consists of interconnecting nerve cells, with two or three axons from each cell linking to the axons of other cells. As with vertebrate animals the axons do not fuse with each other – a tiny synapse is left that the nerve impulse has to jump across.

But with these simple animals all the axons are similar and there are no branching dendrites. Impulses can travel in either direction along the nerve cell, so any stimulation radiates across the entire body through the network. There are faint traces of nerve specialization in some species of sea anemone; they have one or two clearly marked conduction paths that let the impulses travel more quickly.

In a sea anemone the nerve cells are concentrated around the

The intricate network of the nervous system extends from the brain and spinal cord to reach every part of the body. The nerves of the peripheral system are involved in two-way traffic. Stimulation of sensory organs and receptors – which provide us with our senses of sight, hearing, taste, smell and touch – triggers them to send messages along nerves to the brain, where they are interpreted. The brain sends back instructions to the organs and tissues along the cranial nerves or via the spinal cord, enabling us to respond almost simultaneously.

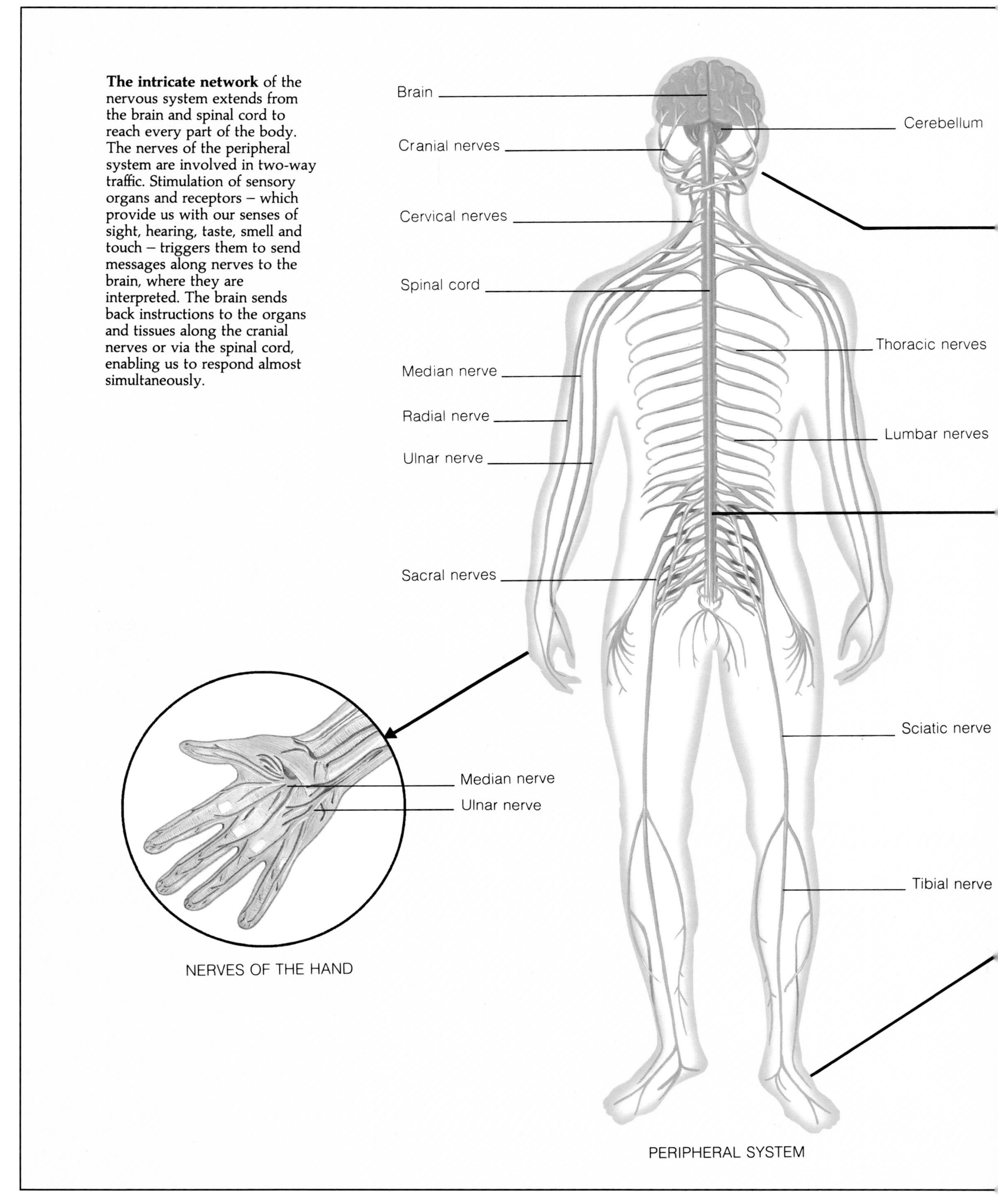

NERVES OF THE HAND

PERIPHERAL SYSTEM

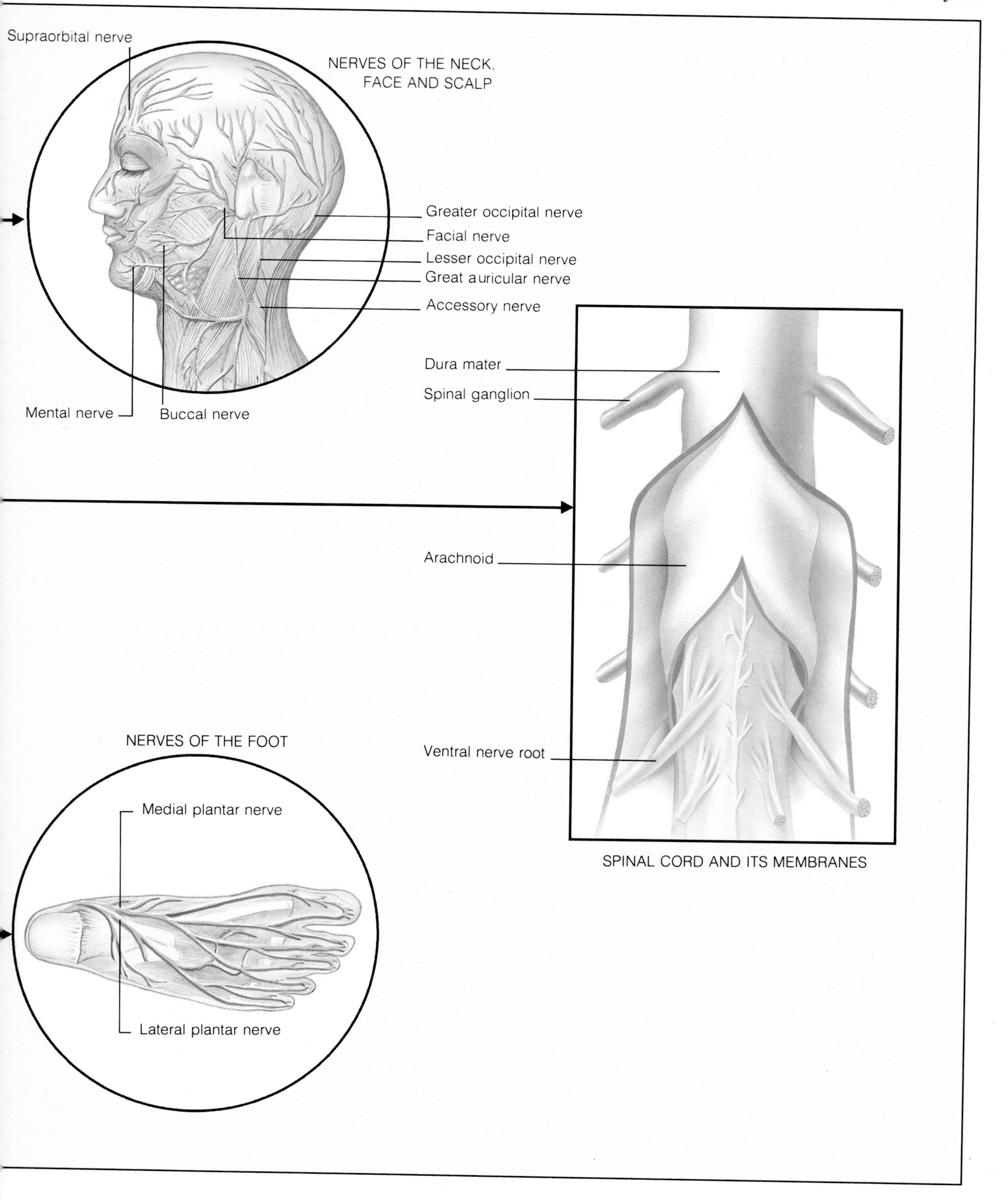
Supraorbital nerve
NERVES OF THE NECK, FACE AND SCALP
Greater occipital nerve
Facial nerve
Lesser occipital nerve
Great auricular nerve
Accessory nerve
Mental nerve
Buccal nerve
Dura mater
Spinal ganglion
Arachnoid
Ventral nerve root
SPINAL CORD AND ITS MEMBRANES
NERVES OF THE FOOT
Medial plantar nerve
Lateral plantar nerve

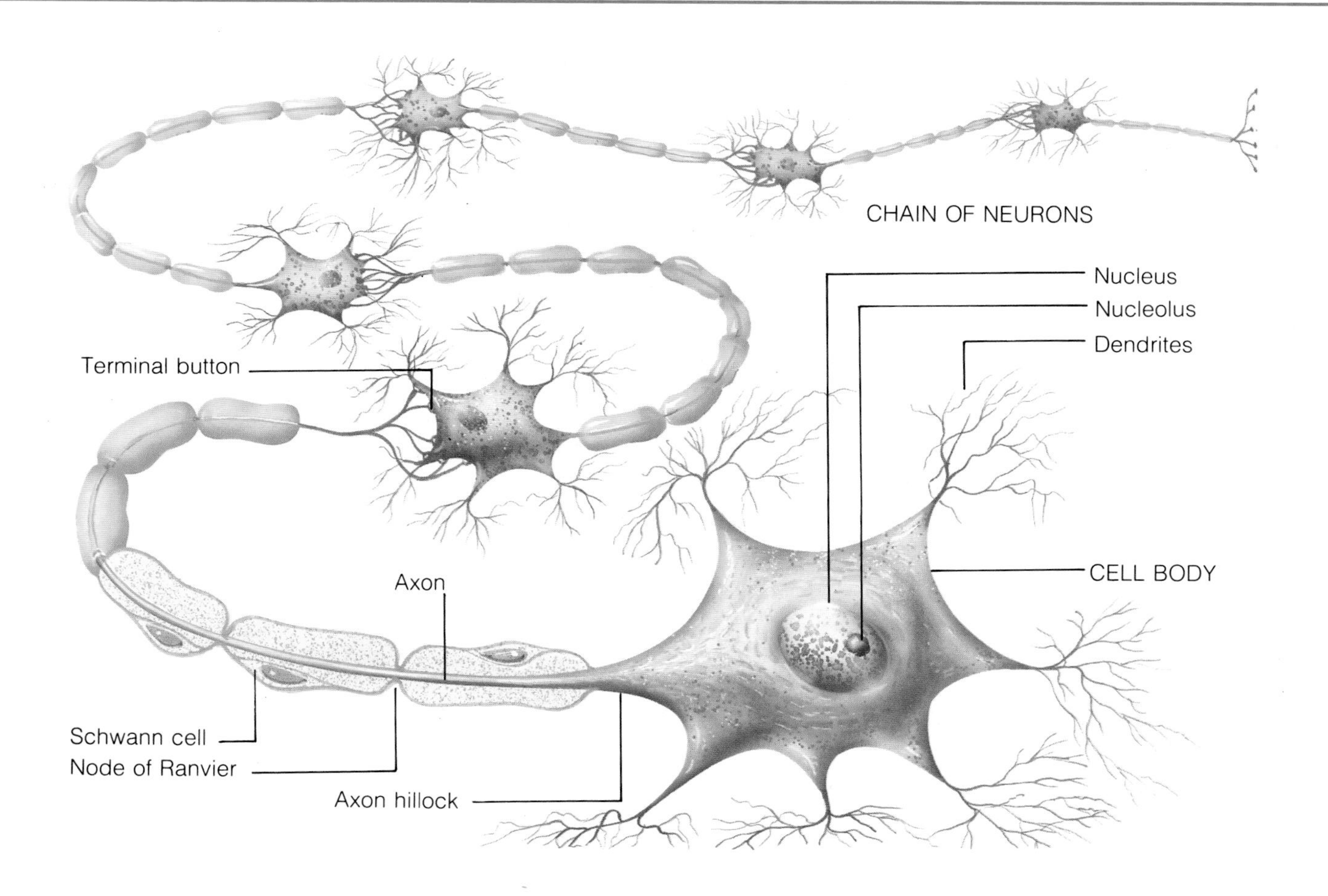

A single nerve fiber consists of a chain of neurons, or nerve cells. Each has a cell body containing a nucleus and bearing rootlike dendrites. An axon extends from a "hillock" on the cell body, and is wrapped around with Schwann cells separated by gaps at the nodes of Ranvier. Terminal buttons at the end of the axon link it to other nerve cells.

mouth region, but there is no brain. All higher animals have a distinct head region and most of them can move around. Movement makes it easier for an animal to exploit its environment but raises a different set of problems. Having a head – which leads the way during movement – allows the development of specialized sense organs to gather the information that helps solve these problems.

Flatworms show early stages in the development of such a system with rudimentary eyes near one end of their bodies. Although the eyes do little more than respond to light (which the flatworms shun), there is a definite concentration of nerve cells around them. These cells have a simple brainlike function that is recognized by calling them the cerebral ganglia. Two thick bands of nerve cells run back from the cerebral ganglia to form distinct nerve cords. This is an arrangement that is used by almost all of the more advanced animal groups.

The bonus of a brain

Further evolution resulted in "modular" animals made up of a number of repeating sections or segments, like beads on a string. Earthworms and leeches show this clearly – the segments can be seen on the outside of the body. Each segment contains a basic set of vital organs, and each contains a nerve cell ganglion and paired lateral nerves, with the ganglia of adjacent segments connected to each other. In addition there are three giant fibers that run the whole length of the worm. In the event of a violent stimulus these fibers – which bypass the ganglia – give rapid transmission of nerve impulses, and enable a rapid reaction.

In the head of an earthworm there is a fairly well-formed cerebral ganglion, and together with another ganglion under the gullet this forms a workable brain. But the brain is not in total command of the body; its main function is to integrate the activities of the separate ganglia and of the body as a whole. If an earthworm is cut into two pieces the rear piece – no longer connected to the brain – remains as active as the front portion.

Insects, crustaceans and spiders look more complex than segmented worms, but their nervous systems are little more developed. Typically they have a cerebral ganglion lying in the head between the eyes, and another ganglion on the side below their gut. From here a pair of nerves runs to the rear of the animal, with short nerves supplying each segment of the body emerging from swellings in the nerves. The brain acts to integrate the actions of the various segmental units, which are otherwise uncoordinated.

Mollusks – snails, squids, cuttlefish and octopuses – have the most highly developed nervous systems of all invertebrate creatures. Squids have a genuine brain protected by a cranium of gristle or cartilage. They also have a series of giant nerve cells that runs down the body. Indeed many recent laboratory studies of nerve action have been carried out using such cells from the giant squid. Sea squirts, animals that are intermediate between the invertebrates and vertebrates, show further development and have a hollow nerve cord running along the back of the animal – this is in fact a rudimentary spinal cord.

Backbone and spinal cord

With backboned animals – the true vertebrates – which all have distinct heads, the only remaining sign of segmentation is the row of vertebrae that make up the spine. Nerve cells from the primitive segments visible in the embryo form the brain and cranial nerves, some of them becoming the parasympathetic nervous system.

In all vertebrates – fishes, amphibians, reptiles, birds and mammals – the nervous system contains the same elements of brain, dorsal nerve cord and segmental ganglia. In vertebrates, therefore, the brain becomes vital to the body's functions, because the other nerve networks cannot support life on their own. So when the brain dies the whole animal dies with it.

The extensive nerve network

The lungs, liver, digestive organs and most of the other vital systems are concentrated in a particular part of the body. But the peripheral nervous system is widespread and diffuse. Nerves travel out from the spinal cord, branching repeatedly to form a network that extends throughout the body. Ultrasensitive regions such as the fingertips and lips have a high concentration of nerve endings, whereas less sensitive regions such as the back have fewer nerves. Depending on their location, the sensory nerves can end in receptors that respond to touch, temperature or certain chemicals. Some particularly sensitive regions, such as the tongue, have large numbers of all three types of receptors.

Most of the nerves of the peripheral system branch out

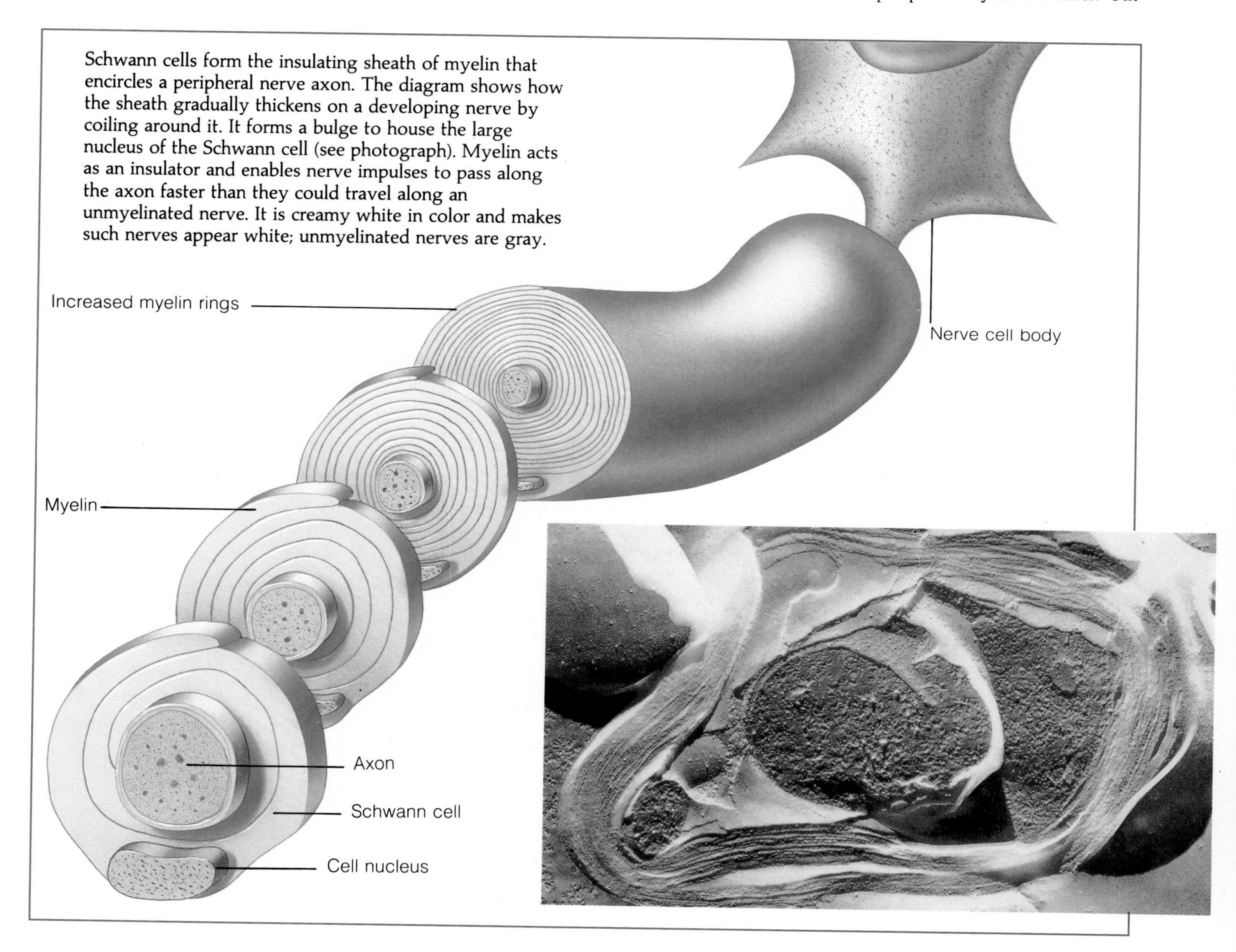

Schwann cells form the insulating sheath of myelin that encircles a peripheral nerve axon. The diagram shows how the sheath gradually thickens on a developing nerve by coiling around it. It forms a bulge to house the large nucleus of the Schwann cell (see photograph). Myelin acts as an insulator and enables nerve impulses to pass along the axon faster than they could travel along an unmyelinated nerve. It is creamy white in color and makes such nerves appear white; unmyelinated nerves are gray.

from the spinal cord like branches from the trunk of a tree. Collections of nerve cells (the ganglia) just outside the spinal cord act as sorting centers and help channel the information flow and any responses. Spinal nerves control all bodily movements, and this is why serious back injury that damages the spinal cord can result in paralysis. A few nerves, the cranial nerves, run directly from the brain to the face, teeth and mouth. Below the neck the only major cranial nerve is the vagus – which belongs to the parasympathetic part of the autonomic nervous system – extending all the way to the intestines.

Most nerve cells have a single axon and a large number of dendrites. These are around one twentieth of an inch long and branch repeatedly to give the neuron a structure resembling a bush. The tips of the dendrites terminate in thousands of smaller projections known as dendritic spines. Impulses arrive from other nerve cells through the spines, so that each neuron can receive messages from a number of others. All the messages are integrated in the cell body; if they result in the creation of an additional impulse, this is transmitted along the axon.

An axon generally emerges from the cell body at a raised region called the axon hillock. In humans the axon varies from about one ten-thousandth to one thousandth of an inch in diameter, and may be as short as one hundredth of an inch or as long as three feet.

Usually the axon does not branch along its length, but where it does the collaterals (branches) normally run off at right angles. Branching does occur at the ends, but not to the same extent as with dendrites, and each branch normally ends in a knoblike swelling – the terminal bouton or button. The swellings make contact with the dendrites, cell body or axon of other cells, or alternatively they end at muscle fibers.

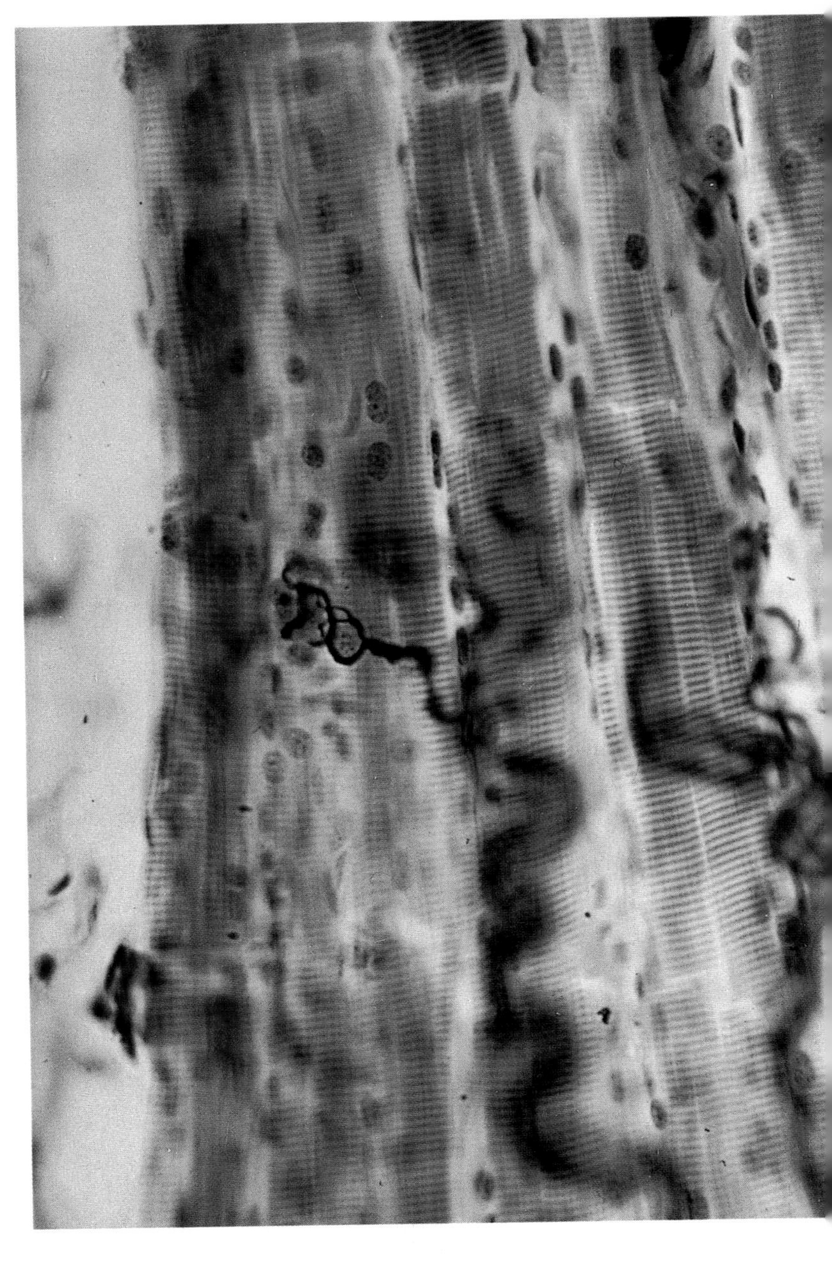

A nerve impulse arrives at a muscle through a motor end plate, the branched structure near the center of the photograph. The end plate then releases the neurotransmitter acetylcholine, which passes to the muscle and causes its filaments to slide over each other so that the muscle contracts. Meanwhile an enzyme, acetylcholinesterase, breaks down the acetylcholine and makes it inactive. This resets the mechanism in time for the arrival of the next nerve impulse, needed to keep the muscle contracting.

Supporting glue

Individual neurons are supported and held together by a mass of neuroglia (literally "nerve glue"), which is made up of large numbers of separate cells. The glial cells also form a barrier between the bloodstream and the nervous tissue, only letting through essential substances such as oxygen, water and some sugars. There is only one sort of glial cell – a Schwann cell – in the peripheral nervous system.

A single small axon, or a small group of axons, may be embedded in a single Schwann cell. But with larger axons the structure is more complicated. During axon development the Schwann cell wraps itself around the axon to form a double spiral with up to 80 turns. The Schwann cell membrane has a fatty layer and binds together like sticky tape to produce a thick sheath which insulates the axon from its neighbours. This sheath is myelin, and myelinated axons (found only in vertebrate animals) transmit nerve signals more quickly than ones without such protection.

But the first part of an axon from the axon hillock – the initial segment – remains uncovered with myelin, and even the subsequent sheath is not continuous. The Schwann cells are between one seventy-fifth and three fiftieths of an inch long and are separated from each other by a small gap – they look like a long link of sausages. As a result, minute sections of the axon are left uninsulated, and these are known as the nodes of Ranvier. The axon is exposed to the surrounding tissue at a node, and this is the only place in an axon where branching can arise.

A single nerve fiber consists simply of an axon with its myelin sheath, but individual nerves are made up of thousands of such

A motor end plate terminates in a series of bulges called synaptic knobs, which are separated from the muscle tissue by a small synaptic space. Within each knob are "packets," or vesicles, containing the chemical neutrotransmitter acetylcholine.

The arrival of a nerve impulse makes the vesicles release acetylcholine into the synaptic space. This, in turn, brings about the release of sodium ions, which act on the muscle filaments and make them slide over each other in muscular contraction.

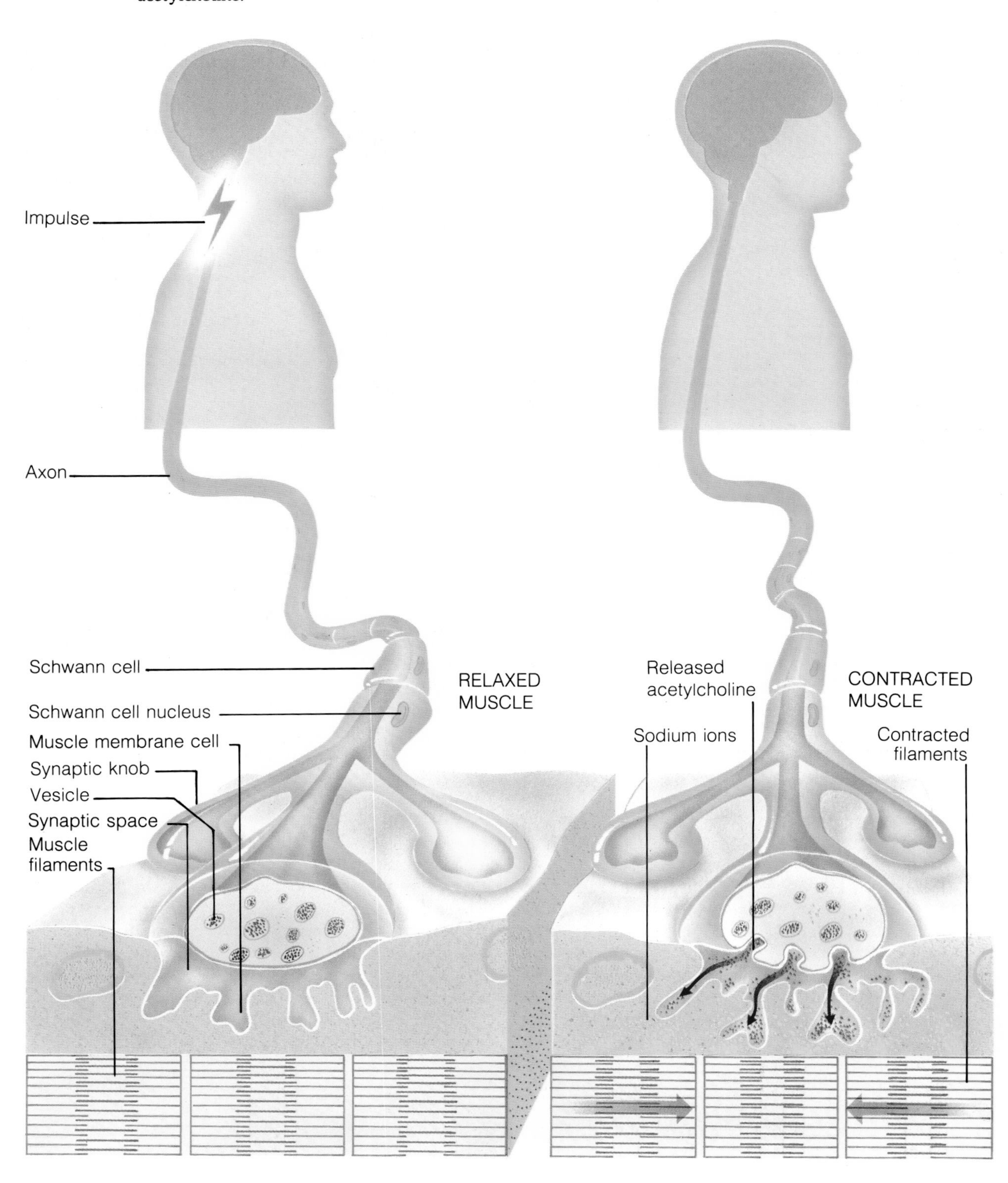

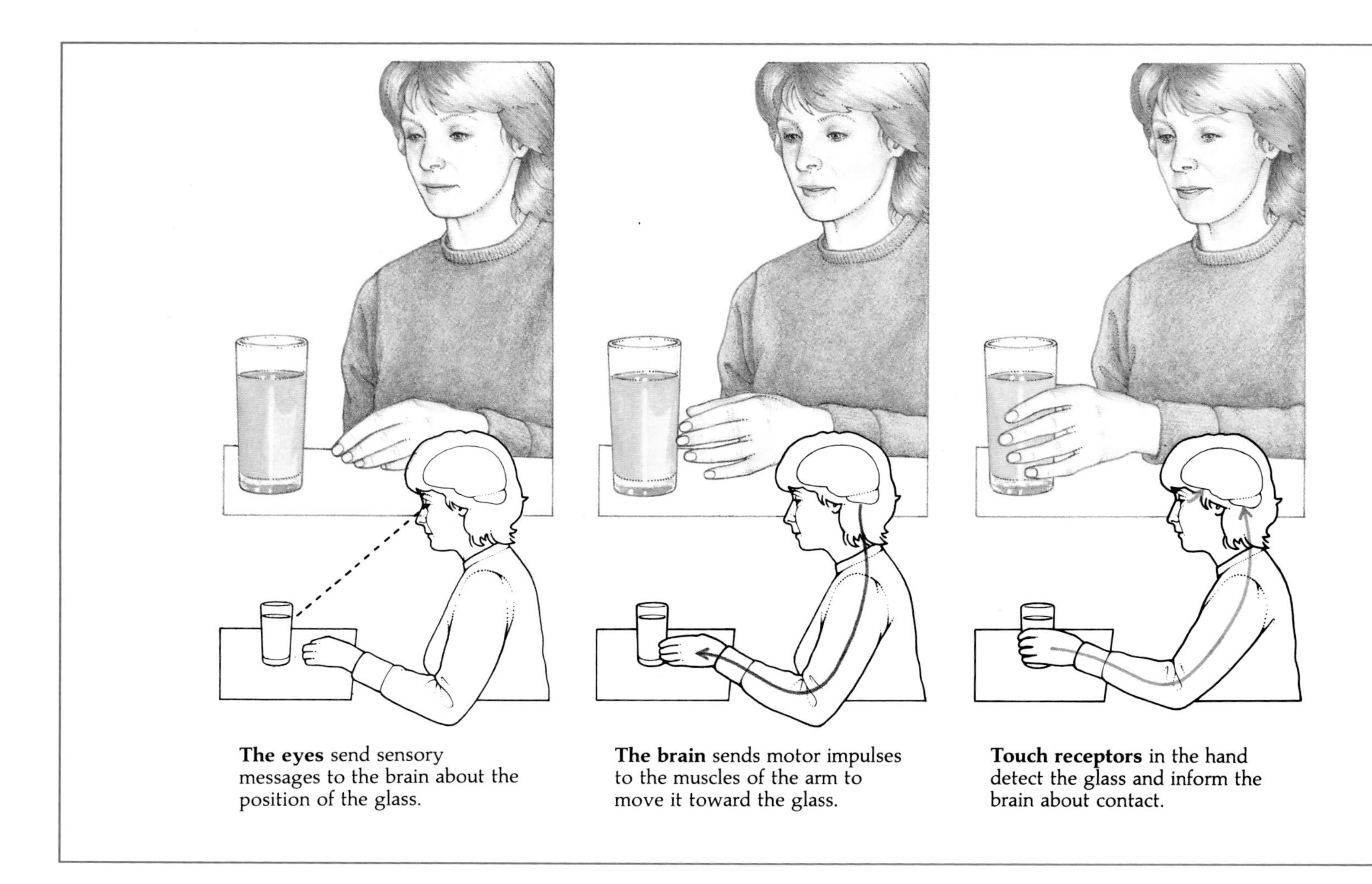

The eyes send sensory messages to the brain about the position of the glass.

The brain sends motor impulses to the muscles of the arm to move it toward the glass.

Touch receptors in the hand detect the glass and inform the brain about contact.

fibers, just as a telephone cable is made up of hundreds of individual insulated wires. Nerves contain a mixture of axons – of large and small diameters, myelinated and unmyelinated – some of them sensory axons carrying information to the central nervous system and other, motor, axons carrying commands back to the muscles and glands. Myelin sheaths around axons make the nerves look white, whereas unmyelinated nerves appear gray. For this reason, areas that consist mainly of unmyelinated cells, such as the central areas of the cerebral cortex in the brain and the spinal cord, are known as gray matter.

The potential for change

Comparing the nervous system with a telephone network is particularly apt because both involve electricity. Nerve impulses are very small electric currents. They are carried along an axon by charged particles called ions, in a complex process that combines electricity and chemistry. Passage of a nerve impulse is accompanied by an electrical voltage peak called an action potential. Its passage depends on the potential difference (voltage) that exists between the inside and outside of a nerve cell, across the cell membrane. When the nerve is at rest – when no impulse is traveling along it – the inside is electrically negative with respect to the outside.

The ions are not stationary, they can move, and the nerve potential is produced by this very movement of ions into and out of the cell. Inside the axon the concentration of potassium ions is 20 to 100 times that of the fluid outside the cell. Conversely the concentration of sodium ions is 5 to 15 times greater outside than inside the axon.

Natural processes try to even up these differences, and potassium ions diffuse out through the cell membrane, building up a potential across it until equilibrium is established.

Passing on the message

Information carried along one nerve is transferred to the next neuron, or to a muscle, through the synapses. The nerve impulse jumps the gap with the help of a "ferryboat" in the form of a chemical messenger substance. A typical synapse is made up of several discrete parts: it has a "before" presynaptic region, such as the terminal bouton of an axon of one neuron; the synaptic cleft or gap itself; and an "after" or postsynaptic region such as the dendritic spines of another neuron. Motor nerves end on muscle fibers in elongated structures called motor end plates. These plates are longer than the presynaptic boutons, but the general structure of the synapse is much the same.

The presynaptic region may be swollen to form a presynaptic knob containing many "packages" of one or more different neurotransmitter chemicals. The synaptic cleft itself is very small and is filled with a sticky material which glues the presynaptic and postsynaptic regions together.

Motor impulses from the brain make muscles grip the glass and raise the arm.

Sensory impulses from the eyes and arm inform the brain about the arm's position.

The lips send sensory impulses to inform the brain that the glass has reached the mouth.

When the impulse in an axon reaches the presynaptic terminal, it triggers the release of the neurotransmitter acetylcholine. This chemical diffuses across the synaptic gap and joins with receptor molecules on the postsynaptic membrane. There it sets off electrochemical reaction (depolarization) which excites the next nerve cell or the muscle. The process is not quite instantaneous because it takes time for the neurotransmitter to diffuse across the synapse. There is a delay of around half a thousandth of a second between the impulse reaching the presynaptic terminal of one axon and the corresponding action potential starting in the next nerve cell or muscle.

After release, the acetylcholine remains active for only one or two thousandths of a second before it is broken down by enzyme action, so ensuring that its effect is short lived. But the inactivated chemical is not wasted; it is taken up by the presynaptic membrane, where it is converted back into reusable acetylcholine in the presynaptic knobs.

Signals work together

When they trigger the next nerve cell, which they do by depolarizing the postsynaptic membrane, neurotransmitters have a positive effect. But synapses can also have a negative, inhibiting effect when the arrival of the neurotransmitter causes an increase in the synaptic membrane polarization. This makes subsequent depolarization more difficult, and it may slow down the impulse rate if the nerve cell is already active. As a result, a synapse may be either excitory or inhibitory, depending on where it is and on the neurotransmitter substance. For instance, acetylcholine is excitory at nerve-muscle junctions but inhibitory in some parts of the autonomic nervous system.

A single nerve cell, such as a motor cell in the ventral horn of the spinal cord, may have many hundreds of thousands of synaptic contacts – all of which may have excitory or inhibitory effects. Even when the arrival of a single impulse at an excitory synapse is not powerful enough to depolarize the cell enough to generate an impulse a following impulse may have an add-on effect and do so; this cumulative effect is called temporal summation. The phenomenon known as spatial summation is similar, but by contrast to temporal summation, it occurs when two synapses that are separated in space (on the same nerve cell) are excited at the same time.

Impulses at synapses with opposite effects can, however, cancel out each other. Or a strong impulse may override a weaker one of an opposite type. The overall response of a nerve cell to the many signals it receives therefore depends on the integrated effect of all its active synapses. This means that the response of a single nerve cell may depend on the input from hundreds of thousands of other neurons. The end result of all this activity may be, for example, a tiny, precisely controlled body movement.

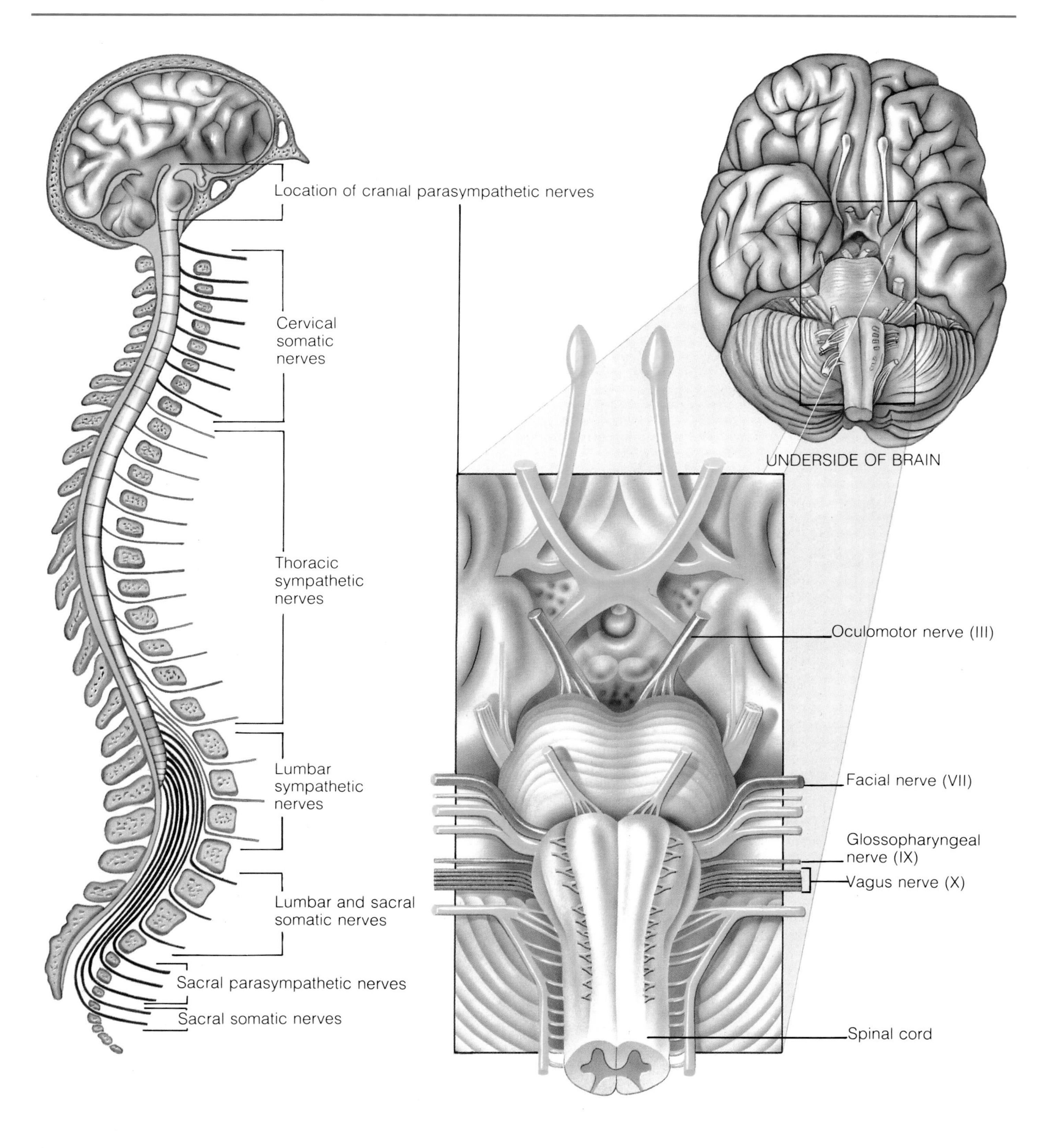

The autonomic nervous system consists of twin parts, known as the sympathetic system (red) and the parasympathetic system (blue). All the sympathetic nerves arise from the spinal cord, from 15 segments that make up the thoracic and upper lumbar regions. The parasympathetic nerves include four cranial nerves, directly from the brain, and eight spinal segments from the lower lumbar and sacral regions. The other eight cranial nerves are shown in yellow on the diagram, and the somatic nerves, from the cervical segments of the spine, are shown in black.

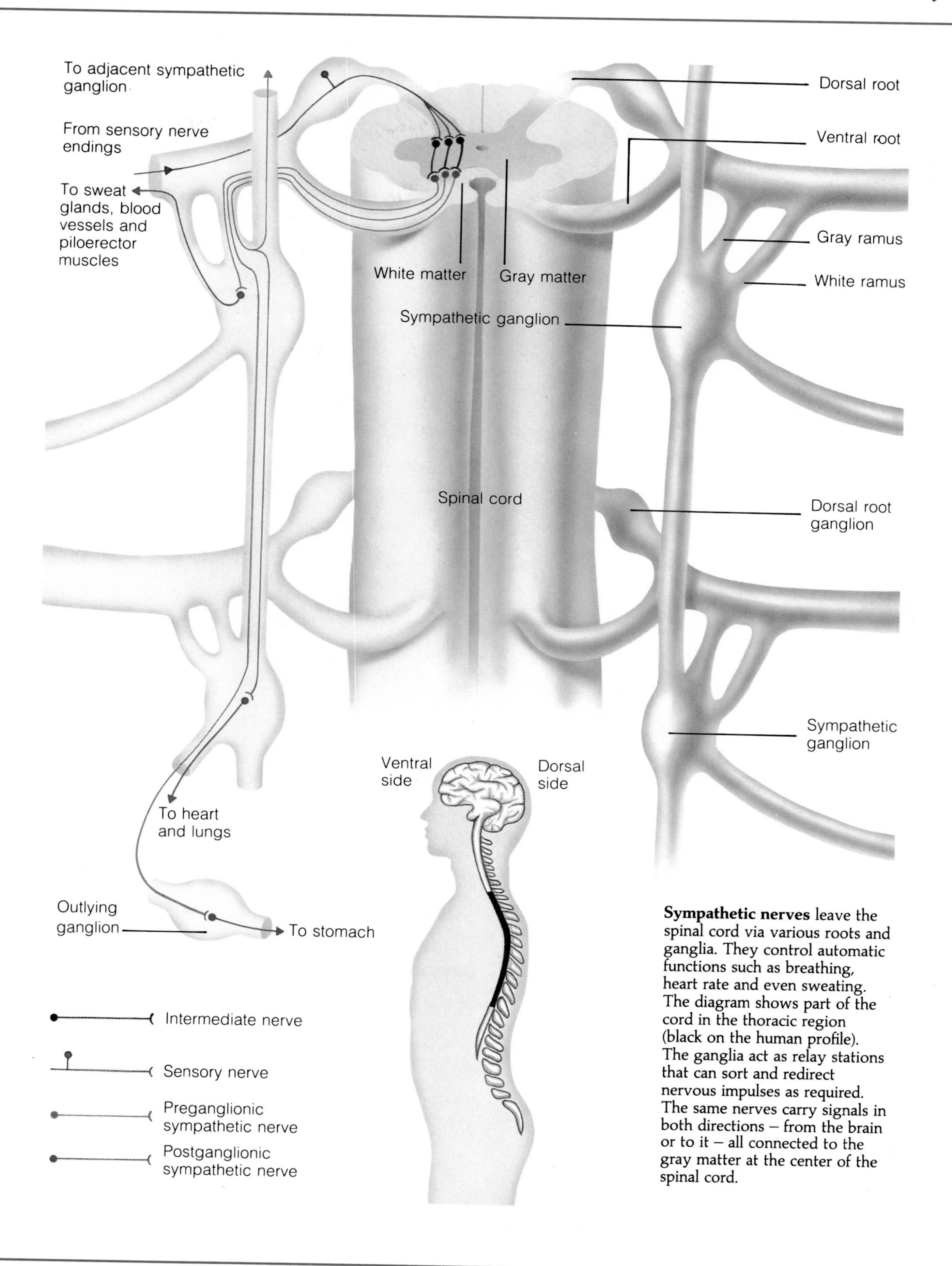

Sympathetic nerves leave the spinal cord via various roots and ganglia. They control automatic functions such as breathing, heart rate and even sweating. The diagram shows part of the cord in the thoracic region (black on the human profile). The ganglia act as relay stations that can sort and redirect nervous impulses as required. The same nerves carry signals in both directions – from the brain or to it – all connected to the gray matter at the center of the spinal cord.

Sensory messages and motor impulses

The brain is the main control system of the body and together with the spinal cord forms the all-important central nervous system. To be in control, the brain has to know what is happening in the rest of the body. It also needs a way of sending operating commands back to the body, and these two functions are carried out by the peripheral nervous system.

When you look at an object, say a glass of chilled milk, a sensory message – also known as an afferent impulse – travels from the eye to the brain and you see the glass. If you then decide to reach for the glass, motor – or afferent – impulses are transmitted from your brain to the muscles of your arm for it to make the necessary voluntary movement. Then as you grasp the glass, sensory messages travel back to the brain so that you know the glass feels cold. Other sensory messages are sent from special receptors, the proprioreceptors, in the arm joints. These messages tell your brain the exact position of your arm and let you lift the glass to your lips without having to look or feel where it is. And when you drink the milk, the autonomic nervous system takes over to trigger a series of automatic responses including salivation, contractions of the esophagus to move the milk down to your stomach, and the secretion of digestive juices.

In practice many of the neural messages concerned with voluntary actions actually take place without conscious thought – and the nervous system keeps working when you are asleep. For example, you do not ordinarily have to think about taking each breath in turn, but you can easily make yourself hold your breath, or breathe very deeply rapidly if you so desire. Similarly during sleep the skin receptors trigger muscle movements to turn your body in order to relieve the pressure on the areas of skin you have been lying on.

Rapid response system

The brain does not control all of the actions of the peripheral nervous system. When a rapid reaction is necessary, say in an emergency situation, the spinal cord can trigger the appropriate response. A typical example is the way you jerk your hand back after touching a hot surface. Sensory messages from the fingers travel up the afferent fibers to let the brain known that something is causing pain. The brain responds at once by sending messages to move the fingers. But at the same time the messages act directly through a reflex arc in the spinal cord to instantly trigger motor impulses that jerk the hand away. This reflex action moves the hand a few thousandths of a second before brain withdrawal impulses arrive, and so helps minimize any damage.

A physician can make a simple check on how some of the spinal reflexes are working by using the simple stretch reflex. This safeguards muscles from rupture by making them suddenly contract when they are stretched. Impulses from the muscle travel up the sensory fibers to the spinal cord and pass across it to produce a motor impulse that contracts the muscle. In a typical test the physician taps the tendon below the knee to stretch the muscle, and the leg gives a sharp jerk as the muscle contracts.

The stretch reflex is the simplest of the spinal reflexes because the impulse passes directly from the sensory to the motor nerve. All the other spinal reflexes use additional nerve cells – known as interneurons – to transfer the signal. And in some reflexes the interneurons can also come into play to balance other kinds of muscle contractions.

The brain in charge

The spinal reflexes are also influenced by the brain, which calms and modifies them. Without the brain's intervention, the reflex becomes exaggerated. This is what happens in spastic palsy or paralysis, in which damage to the spinal cord stops brain messages reaching spinal nerves in the area below the injury. Reflex actions are faster and limbs bend at the joints because the flexing muscles are stronger than the extensors. But if the spinal nerve is injured, there is a loss of sensation. Paralysis still affects the muscles but they also lose their tone and become flaccid.

Nerve fibers running from the peripheral nervous system join up to form larger nerves which connect with the spinal cord and the brain. There are 12 pairs of cranial nerves, which connect directly with the brain, and the spinal cord issues 31 pairs of spinal nerves. The cranial nerves carry either motor or sensory messages and supply the muscle functions and sensations of the head and neck. One of their most important tasks is the transmission of messages to and from the organs that provide the vital senses of sight, hearing, balance, taste and smell.

As their name suggests, the optic nerves carry messages from the eyes to the brain, while the oculomotor, trochlear and abducens nerves supply the muscles that help to move the eyeball. The trigeminal nerves transmit the face's sensations, and smell is dealt with by the olfactory nerves. The facial nerves connect with some of the tongue's taste glands as well as the muscles on the face. Other taste sensations from the tongue are transmitted to the brain by the glossopharyngeal nerves. These same nerves also help to supply the salivary glands through their autonomic fibers.

Tongue muscles are controlled through the hypoglossal nerves, while the two large muscles in front of and behind the neck which support the head are the responsibility of the accessory nerves. Sound impulses from the inner ear are carried to the brain by the cochlear or auditory nerves, and the vestibular portions of the auditory nerves convey the sense of balance from the semicircular canals in the inner ear. The vagus nerve is the tenth cranial nerve, and the longest one, connecting to all the major organs in the chest and abdomen.

Nerve damage

All of the cranial nerves pass into or out of the skull through small holes in the bone, and they are particularly vulnerable to injury at this point. An injury to the head can have unexpected effects if it damages any of these nerves. For example, damage to the olfactory nerve could result in a loss of the sense of smell, even though the nose itself is totally unaffected. Other confusing results arise from damage to the optic nerve – a tumor pressing on the optic nerve could, depending on its position, blind one eye or cause a loss of sight in both eyes.

Multiple sclerosis, as yet an incurable disease, destroys the myelin sheaths of nerve fibers in the brain. This disrupts the transmission of nerve impulses, and commonly occurs in waves; first one part of the nervous system and then another is affected and then recovers. Attacks may clear up completely or there may be repeated attacks and gradual deterioration.

Communication cord

The spinal cord is the major trunk route of the nervous system. It lies within – and is protected by – the vertebral column, which consists of vertebrae stacked atop each other and separated by fibrous pads (intervertebral disks) which have a cushioning effect.

The vagus nerve, otherwise known as the tenth cranial nerve, arises directly from the brain, but unlike the other cranial nerves extends well beyond the head. At its farthest extension it reaches the lower parts of the intestines. Disorders involving the vagus nerve – sometimes of psychosomatic origin – include problems with digestion caused by sluggish peristalsis, which slows the movement of partly digested food through the gut, and lowered secretion of digestive enzymes, which can give rise to gastric ulcers.

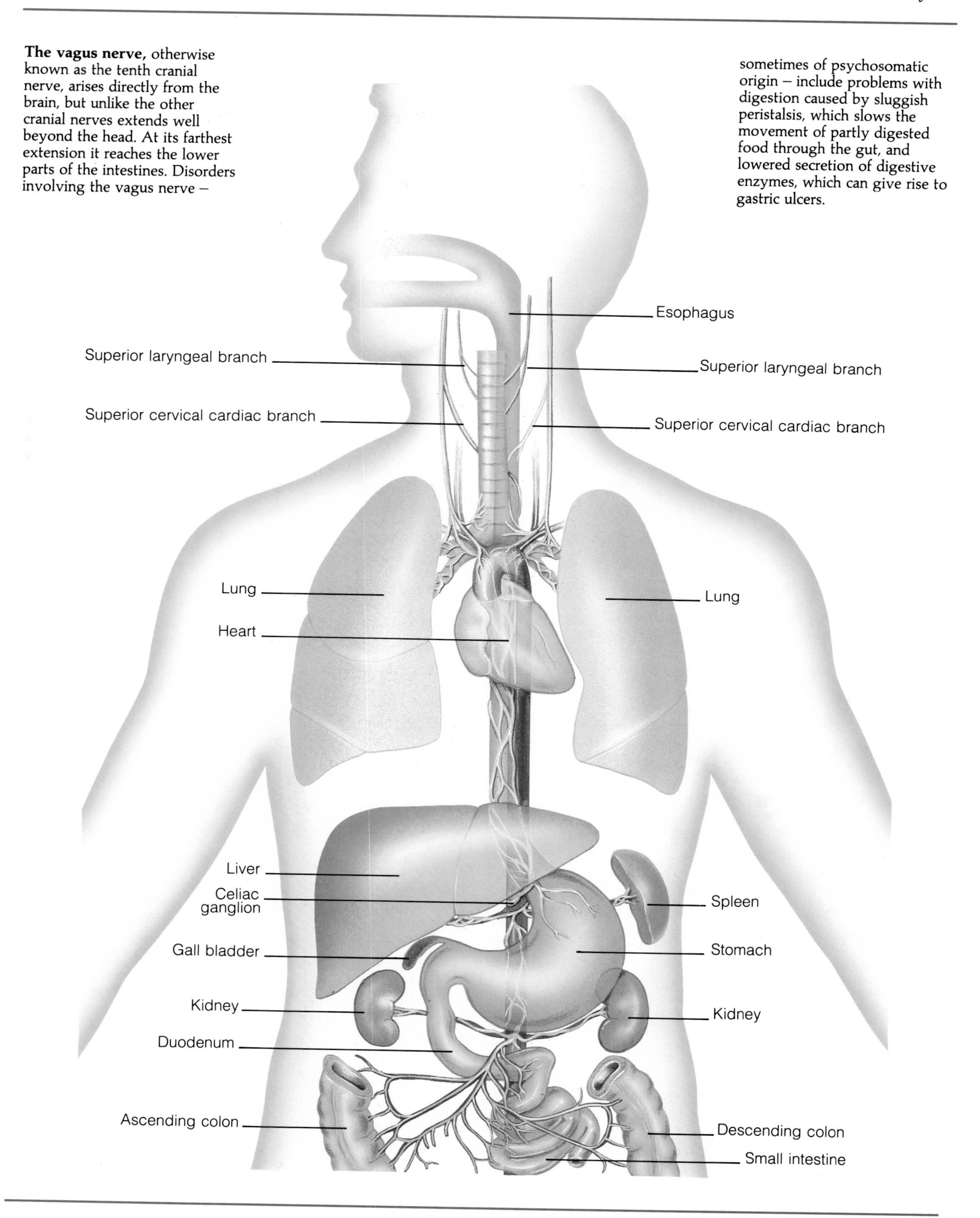

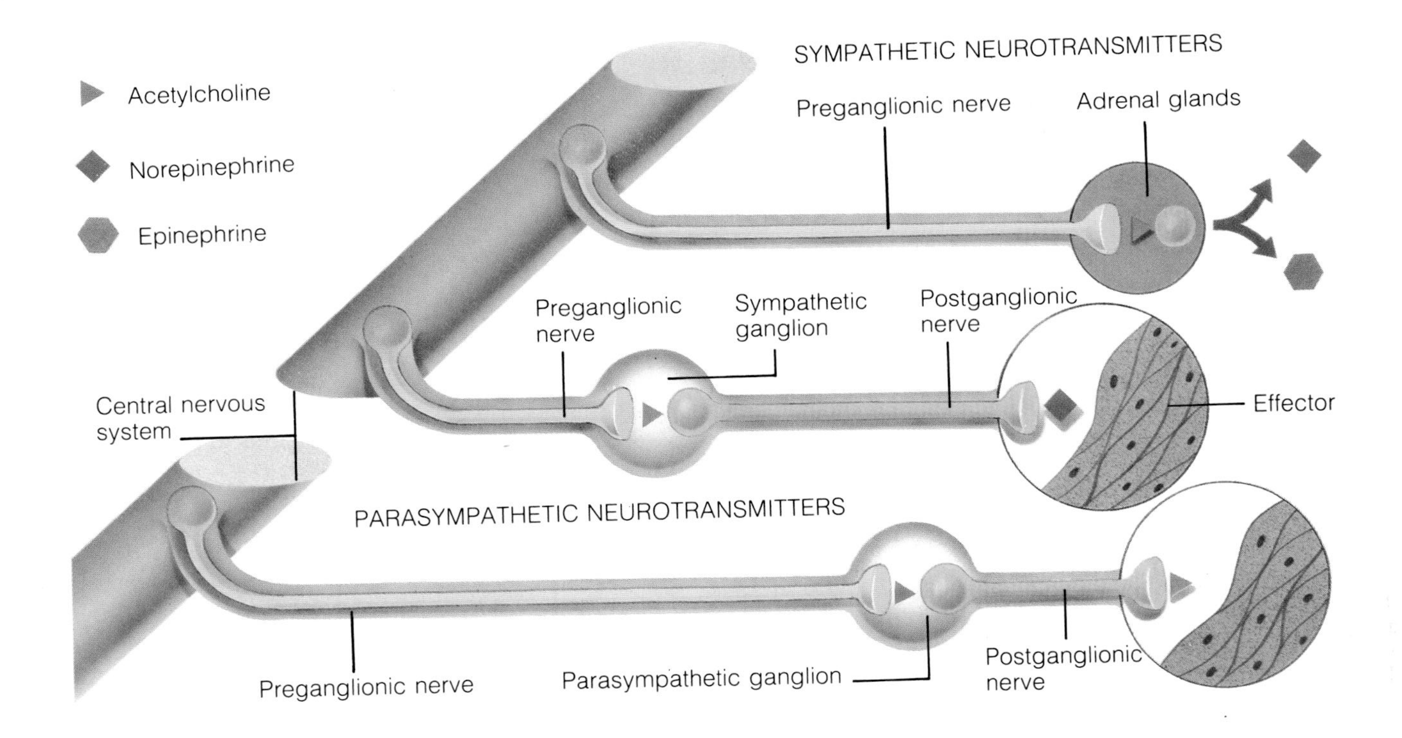

Neurotransmitters are the keys that unlock autonomic nerve action. These chemical substances are released at the end of a nerve to stimulate a particular organ or tissue into bringing about a particular effect. The choice of neurotransmitter depends on the function of the nerve – for example, all preganglionic nerves make use of acetylcholine. Those from postganglionic nerves differ between sympathetic and parasympathetic systems: the choice is from acetylcholine, epinephrine (adrenaline) and norepinephrine (noradrenaline).

Inside the column the spinal nerves branch off from the spinal cord in pairs. Each of the nerves has two roots, one in front of the other: the front one is the ventral root and carries only motor impulses, while the rear one is the dorsal root and carries sensation impulses.

Because they are made up of bunches of nerve cell axons, spinal nerves have a characteristic stranded appearance. In the ventral root the nuclei of the nerve cells lie on one of the bulges, or horns, of the spinal cord and are called anterior horn cells. But in the dorsal root the cell nuclei gather together in a clump – the dorsal root ganglion – which sticks up above the dorsal root.

The ventral and anterior roots of spinal nerves fuse together and wind down within the spinal column, alongside the cord, until the coalesced nerve reaches its exit hole, or intervertebral foramen. The spinal cord is shorter than the spinal column. Nerves from the top of the cord come out from the column almost at a right angle, whereas nerves that leave the cord farther down have to continue parallel to the cord until they reach their exit foramen. The nerves that emerge at the lumbar, sacral and coccygeal levels of the column travel down the spinal canal as a brushlike mass, the cauda equina, whose name literally means "the tail of a horse."

Once it has left the spinal column, the nerve again splits into two but this time each strand, or ramus, contains both sensory and motor fibers. The smaller strand, at the back, is the posterior primary ramus and carries impulses to and from the skin and muscles of the back. The larger front strand is the anterior primary ramus and supplies most of the nerve plexuses (networks of nerves) as well as the skin and muscles of the front of the body.

Many of the anterior primary rami run around the front of the body parallel to the ribs. The ramus at each side supplies the muscle and also a single strip of skin (a dermatome) which dips down lower at the front than at the back. These dermatomes overlap, so that the loss of one spinal nerve leaves an area of skin that is less sensitive but not completely numb. At the limbs the pattern changes from horizontal hoops to vertical stripes that run along the limbs. The legs, with their large muscles, have a generous supply of motor fibers, while the arms – and particularly the hands – also need a lot of nerve connections to accommodate the huge amount of sensory information they receive and to control delicate movements.

Secondary sorting stations

To meet the special demands of the limbs, the spinal nerves join up into complex networks of nerve fibers, or plexuses, before branching off along the limbs. They act as sorting stations and there are three major plexuses – cervical, brachial and lumbosacral – made up from the front spinal nerves. The cervical plexus in the neck collects and integrates fibers from two spinal nerves, sending the sorted fibers to the skin of the head and neck. The

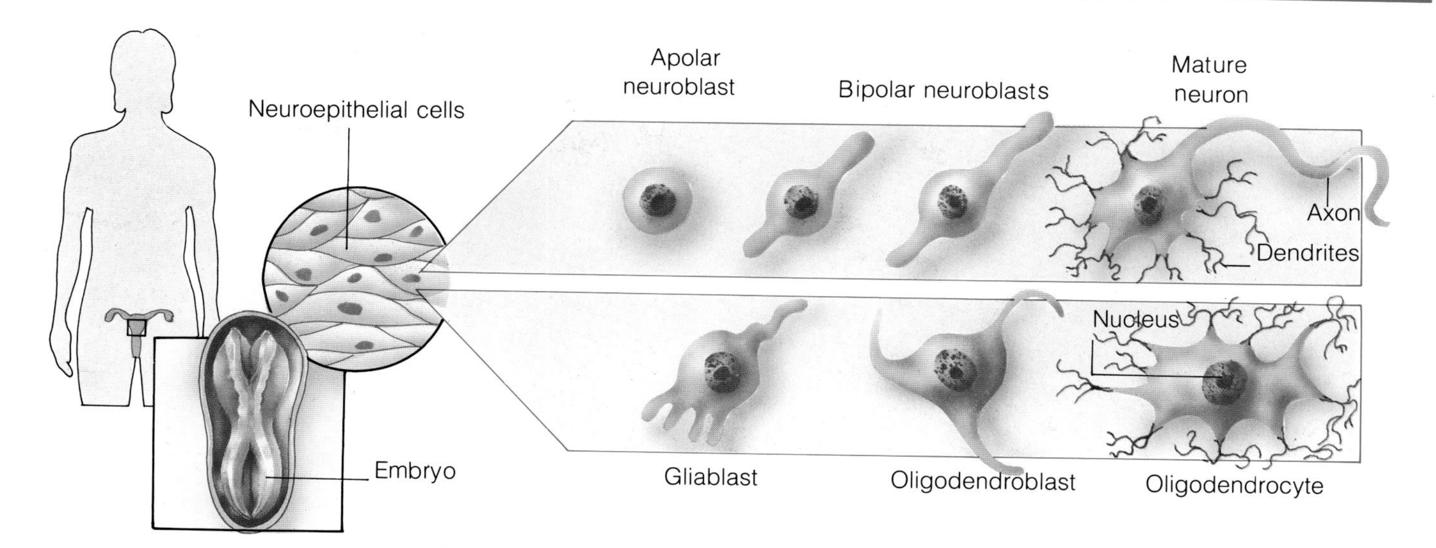

Formation of nerves and the nervous system begins very early in an embryo's development. Neuroepithelial cells around the closing cleft of the neural tube generate neuroblasts and gliablasts, the progenitors of the two types of nerve cells: neurons and oligodendrocytes.

phrenic nerve, which travels down to the diaphragm and is the main nerve used in breathing, also comes from this plexus.

Lying at the root of the neck, behind the collar bone (clavicle), the brachial plexus is made up of three or four spinal nerves. It crosses and intermingles the nerves and sends them down the arm, wrapped around the main blood vessels. Arm nerves are in virtually identical positions on everybody, so injury to a specific nerve always produces the same signs. For example, dislocation of the shoulder may damage the axillary nerve and create a numb area over the shoulder, while a fracture of the humerus in the upper arm can damage the radial nerve (which is wrapped around it) and cause a weakness of the wrist extension. Movement and sensation in the fingers and hands depend on the large median nerve, which has to pass through a tight band of ligament called the carpal tunnel in the wrist. If the nerve gets trapped there, the fingers tingle and the thumb weakens, a condition known as carpal tunnel syndrome.

Nerves for the legs come from the lumbosacral plexus, which gathers together eight spinal nerves. Many of the nerve fibers join up to form the sciatic nerve, which is about as thick as an adult's little finger and is the largest in the body. This huge nerve passes down through the buttock, making some sensory and motor connections to the hamstring at the back of the leg, then divides and thins out to supply the calf and foot. Pain caused by injury to this nerve is called sciatica, and often results from undue pressure being exerted on one of the nerves as it leaves the spinal column – though the pain is generally felt down the affected leg. One reason for the pressure is a "slipped disk" (prolapsed invertebral disk) – the abnormal protrusion of one of the centilaginous plates between the vertebrae in the spine.

Nerve tracts

Within a spinal nerve there is two-way traffic, since it carries both sensory and motor impulses for the part of the body it serves. At any given time there can be a mixture of different sensory messages passing along it. But once the nerve feeds the signals into the spinal cord, they are sorted and separated into groups of nerve fibers or tracts. One tract may carry sensations of pain and temperature, another those of touch or the messages that establish the body's posture. Yet another tract handles the instructional efferent motor impulses coming from the brain. This is the fundamental difference between the peripheral and central nervous systems – all the messages in the central system are sorted and streamed whereas those in the peripheral system are all mixed up.

Like the brain, the spinal cord consists of white matter – myelin-covered nerve fibers in their tracts – and gray matter – the bodies of the nerve cells. Toward the brain end of the cord there is a high concentration of white matter, while at the other end there is a large area of gray matter with just a thin coat of white matter. Some banding also occurs in the tracts, with fibers for the cervical spinal nerves lying next to those for the thoracic spinal nerves and peeling off as they leave the cord.

Destruction by disease

Physical damage or infection can have disastrous effects on nerves and on the parts of the body they serve, both in the short term and in the long term. Neurologists and other researchers have striven over the years to treat or prevent nervous disorders. One of the most notable successes was the development of a vaccine to counteract poliomyelitis. This once dread disease is caused by a group of viruses which attack motor neurons in the anterior lobes of the spinal cord and brainstem. Infected cells are destroyed, and in severe cases the body can become paralyzed; if the muscles that control breathing become affected, then a life may be at risk.

Rabies, another viral disease and potential killer, also affects the nervous system; it attacks the nerve cells of the spinal cord and the brain. Hydrophobia – a violent reaction on drinking water – is a common symptom of the infection in dogs, but in humans the incubation period is between 20 and 60 days and vaccination during this period, following a bite from an infected animal, will prevent the disease developing. Less life-threatening is the shingles, an infection resulting from the same virus that causes chickenpox in children but which may remain latent in the nervous system for many years until reactivated.

Tetanus, or lockjaw, is caused by a bacterium which multiplies in deep wounds. There it releases a toxin which attacks motor nerves and anterior horn cells, resulting in muscle rigidity and spasm. Syphilis is also caused by a bacterial infection and can attack the nervous system in the fully developed tertiary stage – the normal treatment is with the antibiotic penicillin in the early stages of the disease.

Leprosy is a disease that affects large numbers of people in tropical and subtropical countries. The bacterium affects individual peripheral nerves and makes them thicken. As a result, sensation is lost in the parts of the body supplied by these nerves, and the skin and bone often becomes damaged. Contrary to popular belief, leprosy is often curable.

Sequences of sense signals

Sensory receptors respond to a wide variety of stimuli, including sound, light, heat and pressure. But they all transmit their information to the brain as electrical impulses carried in the peripheral and central nervous systems. Sometimes the receptor stops producing impulses if the stimulation is constant – for example, the itch of a stiff shirt. It starts off by being very uncomfortable, but the sensation fades after half an hour or so.

When one of the receptors in the skin is stimulated – say when a finger touches something – it sets off an impulse which travels up the axon of the nerve cell. This cell is a first order neuron and the impulse travels along the axon from its source, past the nerve cell body in the dorsal root ganglion, and on to the spinal cord. There is transmits the nerve impulse to a second-order neuron, which lies in the substantia gelatinosa at the tip of the butterfly-shaped spinal cord.

The axon of this second nerve cell passes across the spinal cord and joins the lateral spinothalmic tract running up to the brain in the white matter of the cord. The transmitted nerve impulse zooms up the tract until it reaches the thalamus. This is the brain's main sensory relay station, and it passes the impulse over for the last part of its journey to the appropriate part of the cerebral cortex.

Having done its work in processing the information, the cortex may respond by producing a motor impulse – perhaps to move the finger away from the surface it has just touched. The impulse travels from the motor cortex to the motor nerves of the spinal cord, across to the other side of the cord and down the corticospinal tract, finally jumping a synapse to a spinal nerve, where it completes the journey.

Cross linked signals

At some stage in their travels all the impulses from one side of the body cross over to the other side. This explains why the right side of the brain is responsible for the sensory and motor functions of the left side of the body, and the left side of the brain for similar functions of the body's right-hand side. It also makes sense of the fact that a stroke in the right-hand side of the brain causes paralysis to the left side of the body.

The speed at which impulses move along the nerves depends on the type and size of nerve fiber in which they travel. The fastest fibers are alpha types, which carry information about the body's position and posture from the proprioreceptors of the joints. Touch, pressure, temperature and pain impulses are transmitted more slowly. Indeed some painful sensations travel in very fine nerve fibers so that the impulse moves very slowly.

Motor impulses coming from the brain are carried by the spinal motor neurons to end up at muscles. Each of the motor nerves ends in a group of muscle fibers together known as a motor unit. The number of muscle fibers in a motor unit depends on the part of the body it moves; eyeball muscles have five or six muscle fibers supplied by a nerve cell, whereas the large power muscles in the thigh may have as many as 150 muscle fibers or more connected to a single nerve cell.

Automatic actions

Twin control systems regulate and control all of the body's internal functions. But each works on a different principle. While the endocrine system uses chemical messengers, the autonomic nervous system employs electric nerve signals. The autonomic system has two parts, the sympathetic and parasympathetic systems. These originate in different parts of the central nervous system. Their actions tend to oppose each other and each uses a different neurotransmitter to transmit its signals to the various target glands and organs.

A pair of fibers makes up each of the autonomic pathways – a preganglionic neuron and a postganglionic neuron, and these fibers generally connect to the smooth (involuntary) muscle of internal organs and tissues. In the sympathetic part of the autonomic nervous system all of the preganglionic neurons are located in the spinal cord, originating from the 12 thoracic and first three lumbar segments of the cord. In contrast the parasympathetic fibers leave the central nervous system along with several of the cranial nerves and from the second, third and fourth sacral segments at the base of the spinal column.

Chain ganglia

Small nodules of nervous tissue, the sympathetic ganglia, are strung together along each side of the spinal column from the neck to the hip. Connections between neighboring nodules form the sympathetic chains. Before they enter the ganglia, the sympathetic nerves emerge from the ventral roots of the spinal cord, separate from the spinal nerves. They can then take one of three routes: they can connect with the postganglionic neuron in that ganglia; they can travel up or down the sympathetic chain to connect with the neuron in another ganglion; or they can pass on through to connect with outlying ganglia.

All of the small unmyelinated fibers that control the size of blood vessels, sweat glands and the tiny muscles that make the hair stand on end go back from the ganglia into the spinal nerves. Some postganglionic fibers, such as those that go to the heart, run directly to their targets without rejoining the spinal nerves. The postganglionic fibers which regulate the stomach, intestines, liver, kidneys, bladder and sex organs come from the outlying celiac and mesenteric ganglia.

One sympathetic pathway that does not conform to this two-neuron arrangement is the one that activates the all-important adrenal glands. Here preganglionic fibers run directly from the spinal cord, through the sympathetic chain, and on to the adrenal medulla. When they reach their destination they trigger the release of the sympathetic horomones epinephrine (adrenaline) and norepinephrine (noradrenaline).

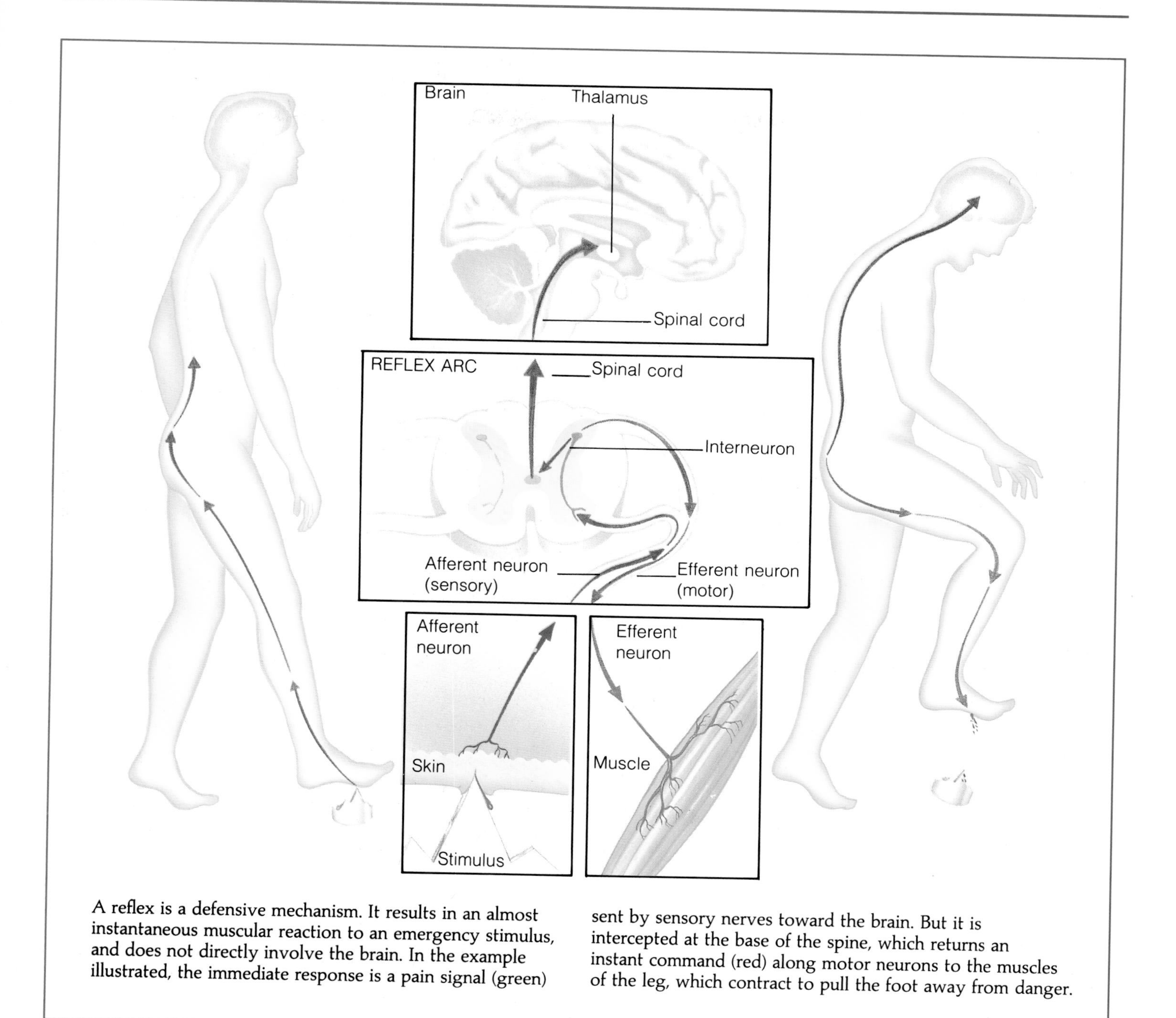

A reflex is a defensive mechanism. It results in an almost instantaneous muscular reaction to an emergency stimulus, and does not directly involve the brain. In the example illustrated, the immediate response is a pain signal (green) sent by sensory nerves toward the brain. But it is intercepted at the base of the spine, which returns an instant command (red) along motor neurons to the muscles of the leg, which contract to pull the foot away from danger.

Ever-branching nerves

The parasympathetic nervous system uses a different arrangement, with the ganglia situated very close to, or even within, the target organs. As a result, the preganglionic fibers are relatively long, while the postganglionic fibers may have a length of no more than a fraction of an inch. Approximately three quarters of the parasympathetic fibers are in the two vagus nerves, which branch repeatedly to supply nerves to all the major organs in the chest and abdomen.

Other parasympathetic cranial nerves go to the muscles of the eyes and to the nasal, tear and salivary glands. And at the lower end of the spinal cord the remaining parasympathetic fibers emerge from three sacral segments and congregate in the pelvic nerves. The branches of these nerves are distributed to the intestines, rectum, bladder, the kidney ureters, and the external genitals, where they are involved in various sexual responses – such as the male erection.

Many of the body's organs receive nerve impulses through both sympathetic and parasympathetic fibers. These two sets of signals often have a balancing effect, but there is no easy way to tell if a sympathetic or parasympathetic stimulation will "turn on" or "turn off" an organ. But as a general rule, sympathetic stimulation prepares the body for action – for instance by increasing the heart rate and blood pressure and diverting blood

Mild stimulation of the skin, such as that caused by a minor pinprick, activates large diameter nerve fibers. When their impulses arrive at the spinal cord, SG cells partly close the pain "gate" and limit the amount of stimulus passed on to the T cells and hence to the brain. As a result, little pain is felt.

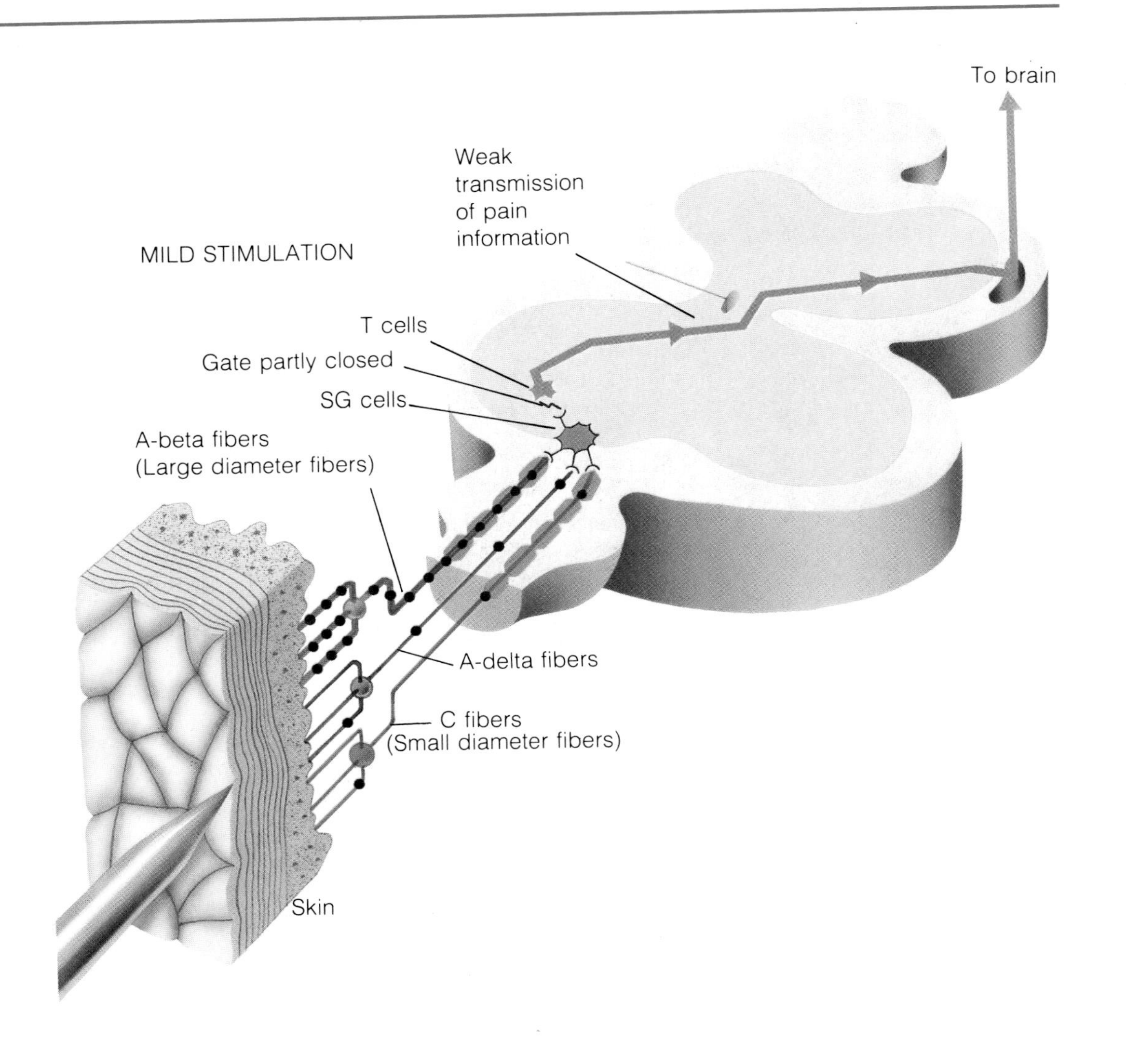

to the muscles – while parasympathetic stimulation has the opposite inhibiting effect, calming the body.

Nervous system first

In an embryo the nervous system is the first major structure to become visible during growth, and starts to develop at the beginning of the third week after conception. When this happens, a groove takes shape in the embryonic disk, and gradually gets deeper while the two long edges fold together, touch, and then close to form an open-ended cylinder. This cylinder is the neural tube, and ridges called neural crests form on each side of it. At the same time regular segments, called somites, are patterned along the embryo before gradually developing into other organs. One end enlarges into three swellings – the beginning of the brain – and a cavity within the swellings spreads down into the developing spinal cord.

This period is a critical one; if the tube does not close properly the result is the congenital malformation spina bifida. In its mildest form, spina bifida occulta, the bones of the lower back fail to make a complete circle around the spinal cord. The break may be no more than a hairline crack, and may never be detected. But if the gap is a wide one, nerve tissue pushes up under the skin and there is a serious risk of damage to the nerves or spinal cord. Many such cases can be treated surgically after birth, though this may not produce a complete cure.

Only four and a half weeks after conception the spinal cord consists of three layers arranged around a central tube. Within the tube primitive nerve cells, or neuroblasts, start to grow, produced by neuroepithelial cells surrounding the tube. As the number of neuroblasts increases, the cord gradually takes on its characteristic butterfly-wing shape.

Once its basic plan has been laid down, nervous system development continues at a rapid pace. Other neuroblasts generate the neurons for the central nervous system, and once sufficient neuroblasts have been produced, the neuroepithelial cells start to make glial cells for the central nervous system. The neural crests produce Schwann cells to sheathe the nerve fibers, meninges to form the membranous covering for the brain and spinal cord, and cells for teeth, skin colour and parts of the adrenal glands.

Around this time the embryo also starts to develop arms and legs, and as the limbs grow they take their segmental nerves with them. By the third month in the womb the fetus looks like a tiny baby and the neural tube has been replaced by a backbone with a spinal cord and nerves. Neuroblasts made by the neural crests form the sympathetic nervous system, while those for the

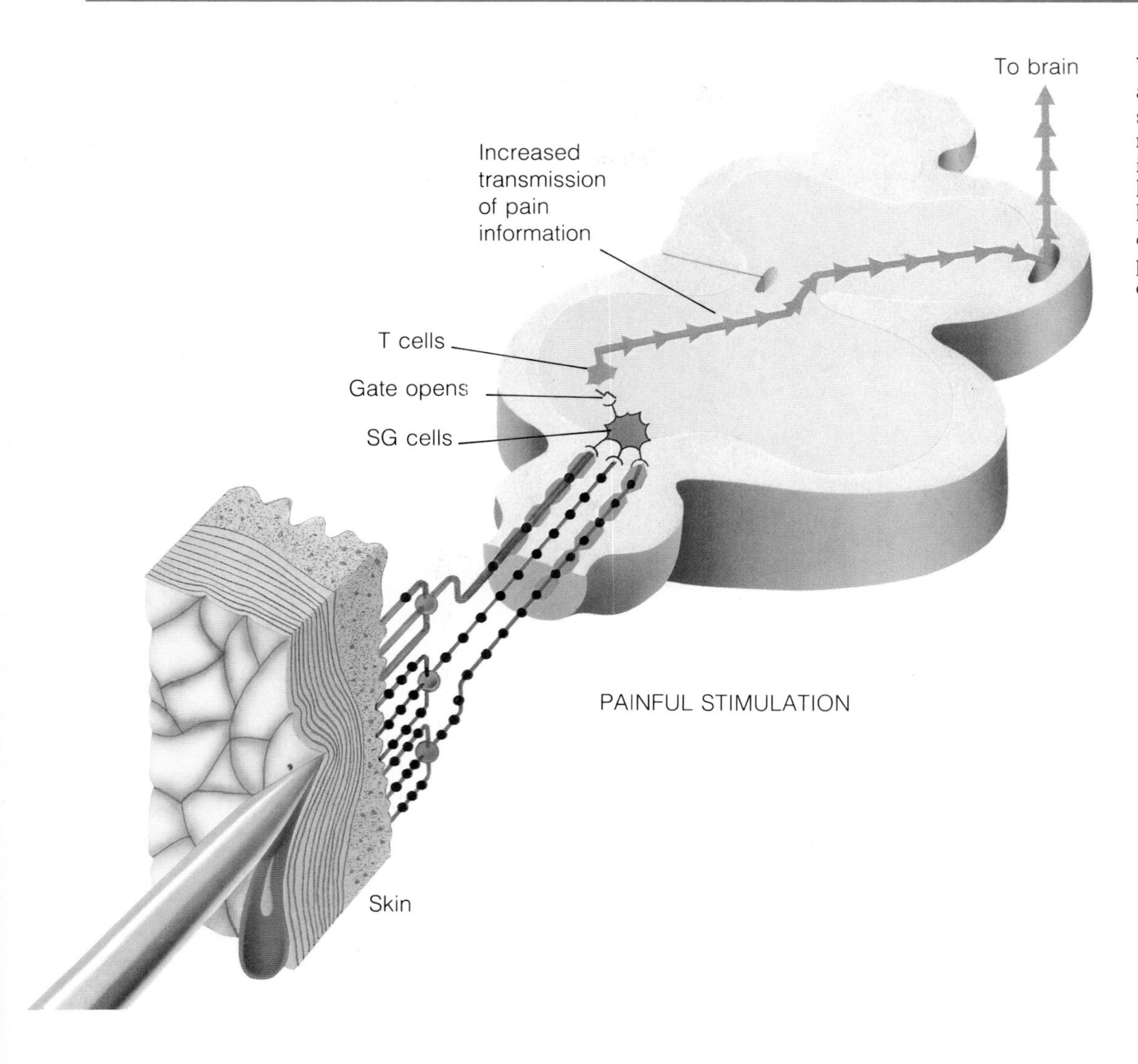

Vigorous stimulation, as when a sharp point penetrates the skin, activates the small diameter nerve fibers. When their impulses overcome those of the large fibers, SG cells can no longer prevent the "gate" from opening and allowing a strong pain message to reach the T cells and pass on to the brain.

parasympathetic system are produced in the neuroepithelium.

Meanwhile the brain takes shape from the three initial swellings, which grow to form the hindbrain (nearest the spinal cord), the midbrain and the forebrain at the top. The forebrain rapidly grows larger than the other areas and the thalamus and hypothalamus form within it. By 13 weeks the forebrain has grown into two large cerebral hemispheres with smooth surfaces, but as they continue to develop within the confined space of the skull the surfaces become crumpled into hills and valleys, giving a vastly increased surface area.

Development after birth

A baby's nervous system carries on developing even after it has been born; the brain goes on enlarging and many of the nerve fibers get their myelin sheaths. Growth of the myelin coverings seems to coincide with increases in the baby's motor and sensory abilities – possibly as a result of the stimulation of the nerves as they respond increasingly to the baby's world.

In a very young baby the reflexes predominate; the whole body is used to express fear, hunger and often basic emotions. But as the brain develops so it starts to override the reflexes. By around three months babies have their bodies under more control and are starting to make voluntary responses. At six months there is good control of the upper body and limbs, with control of the lower limbs following a few months later as the baby crawls, then stands and walks.

By the time a child is two years old, all the structures of the spinal cord, brainstem and cerebellum are myelinated. The child can walk and run, and carry out simple manipulations of objects. Mental processes are equally well advanced. The two-year-old can speak in simple sentences and solve problems based on previous experience. This is the end of babyhood and, although the nervous system continues to grow, it adapts to suit new needs rather than developing fresh abilities – a process of development that can continue throughout life.

Stimulation and pain

Pain varies from the nagging, dull ache of a grumbling tooth to the sharp stab of a finger punctured by a rose thorn. It is a vital warning that protects the body from injury. But pain becomes unbearable when it is intense and prolonged, which is when the painkillers come to the rescue.

When you hurt yourself the sharp, or acute, pain sensations are transmitted to the brain through peripheral nerves and the spinal cord. Initially the first-order peripheral neurons at the point of injury transmit a signal along A-type nerve fibers to the

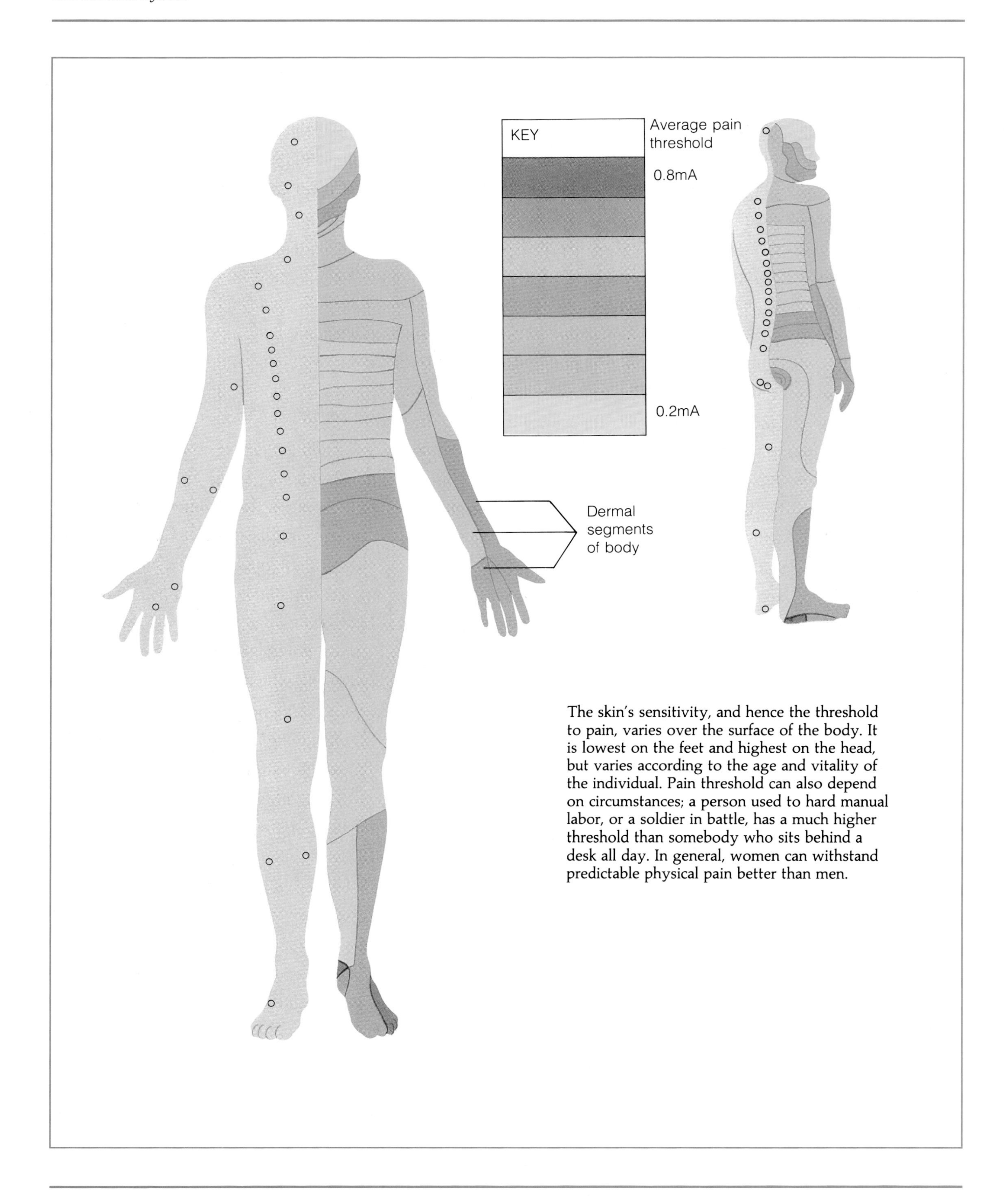

The skin's sensitivity, and hence the threshold to pain, varies over the surface of the body. It is lowest on the feet and highest on the head, but varies according to the age and vitality of the individual. Pain threshold can also depend on circumstances; a person used to hard manual labor, or a soldier in battle, has a much higher threshold than somebody who sits behind a desk all day. In general, women can withstand predictable physical pain better than men.

dorsal horns of the spinal cord. Here the second-order neurons take over, transfer the signal to the other side of the spinal cord, and pass it through the spinothalmic tracts to the thalamus of the brain. Duller and more persistent pain travels by another – slower – route made up from a chain of interconnected neurons, which run up the spinal cord to connect with the brainstem, the thalamus and finally the cerebral cortex. This system uses neurons with unmyelinated C fibers.

The autonomic nervous system also senses pain and transmits signals to the brain using a similar route to that for dull pain. Because the internal organs handled by the autonomic nervous system do not have very many sensory nerves, the pain sensation tends to be poorly localized, dull and persistent. Pains from the crucially important heart and kidneys are exceptions, and can be clearly identified by the sufferer.

Levels of discrimination

Nerve sensations register as pain only when the stimulation rises above a specific level. This "threshold" effect is due to the action of an ingenious gate mechanism, which modulates the signals travelling up the spinal cord. Pain information is transmitted by spinal cord transmission (or T) cells, activated by large or small nerve fibers running from the source of the stimulation. Within the substantia gelatinosa (SG) of the dorsal spinal cord there are other cells that act to inhibit the T cell transmission. These cells are activated by large-fiber input to the spinal cord, but they are also inhibited by the C fibers – the small pain fibers.

When no pain is present, the fibers concerned with touch and position sensations are activated together with the SG cell. This inhibits the T cell, closes the gate, and there is no pain perception. As soon as a painful stimulus occurs – say you step on a sharp rock – the activity of the small C fibers increases to a level where it overcomes the inhibition of the large-fiber activity on the SG cell. Once this happens the gate is open and a pain stimulus travels up the spinal cord to tell your brain that your foot is injured. The immediate reaction is to rub the painful area. The rubbing action increases the level of touch sensation, which augments large-fiber activity and so, in turn, activates the SG cells to close the gate. So rubbing a bruise does not make it better, but it does make it hurt less.

The conquest of pain

The sensation of pain is a natural warning that tells you something is wrong with the body: perhaps your hand is too close to a fire or the food you have just put into your mouth is bad. But once the warning has been heeded and action taken – you move your arm away from danger or spit out the offending mouthful – the persistence of pain is an ongoing problem, particularly if, like the sufferer from a headache or rheumatism, there is little or nothing you can do about it. The conquest of pain therefore ranks among the greatest achievements of modern medicine.

There are two main types of drugs that eliminate pain: anelgesics, which "switch off" the feeling of pain by the brain, and anesthetics, which temporarily deaden all sensation from part or all of the body. Analgesics range from mild painkillers such as aspirin and acetaminophen (paracetamol) to powerful sedatives – also called narcotic analgesics – such as barbiturates and morphine. Drugs in the first category are also referred to medically as NSAIDs (non-steroidal anti-inflammatory drugs) because, in addition to their analgesic effects, many of them have a valuable anti-inflammatory action in the treatment of conditions such as rheumatoid arthritis (a property shared with corticosteroid drugs). Aspirinlike drugs are also prescribed to lower the temperature of a patient with a fever, and are described as antipyretic.

Aspirin and its derivatives have an undesirable side-effect of irritating the stomach lining, and may cause gastrointestinal upsets such as dyspepsia and, in severe cases, bleeding (hemorrhage) from the stomach wall. Also aspirin is no longer considered safe for children under 12 years old since it has been associated with a disorder called Reye's syndrome. Mild analgesics are often used in combination with each other or with other drugs such as caffeine. Many are available in soluble form, which is less irritating to the stomach.

Narcotic analgesics act on the central nervous system to alter the perception of pain. The principal disadvantage of this class of analgesics is the risk of drug dependence, or addiction, and for this reason they are usually given only under strict medical supervision. One possibility for the future is that pharmacologists may be able to use the techniques of molecular modeling to produce drugs that mimic substances called endorphins and enkephalins, which are the natural painkilling chemicals formed in the brain.

Anesthetics are drugs with a much longer history than analgesics. Opium (the source of heroin) and Indian hemp (cannabis) have been known since ancient times. Tincture of opium – laudanum – was used to relieve pain in the middle ages, but by the 18th century there were no other drugs to conquer pain for a patient undergoing surgery, except perhaps strong alcoholic drinks.

The breakthrough came in 1800 when the British chemist Humphrey Davy described the painkilling effects of the gas nitrous oxide. It also had an exhilerating action, hence its popular name of laughing gas. But the gas was not employed as an anesthetic until the 1840s, when American dentist Horace Wells from Hartford, Connecticut, began using it when extracting patients' teeth.

At about the same time another American, Crawford Long, introduced the use of ether as an anesthetic in surgery. In 1846 a former assistant of Wells, Dr William Morton, used ether for dentistry and surgery. A year later Scottish obstetrician John Simpson employed chloroform to relieve the pain of childbirth. By the early years of the 20th century, barbiturates had been added to the number of general anesthetics in medical use. Nitrous oxide is still used, but other early anesthetic drugs have been replaced by safer gases such as halothane.

General anesthetics cause loss of sensation over the whole body as the patient loses consciousness. Local anesthetics affect only a specific area, and are usually injected or absorbed at the site where they are intended to act. One of the earliest, cocaine, was used as long ago as 1884, followed by cyclopropane, novocaine and lignocaine. When filling or extracting teeth, dentists still use lignocaine, which, along with bupivacaine, is also a valuable epidural anesthetic.

Epidural anesthesia is produced by injecting an anesthetic into the epidural space between the membranes that surround the spinal cord. It has the effect of deadening all sensation in parts of the body from the site of the injection downward – including the lower abdomen and legs. "Epidurals," as they are called, are therefore used mainly for abdominal surgery and during childbirth, including Caesarian sections.

CHAPTER 13

The Brain

The human brain is what sets us apart from other animals. It gives us the ability to reason, to communicate with others, to learn and to remember. These skills have made human beings the dominant species on planet Earth, able to tame and manipulate the environment, build ships and cities, travel to the moon and eventually reach other planets and perhaps, one day, the stars.

The brain is also the seat of the "human" qualities of love, compassion, mercy and forgiveness. Painting, poetry, music and drama – all of humankind's artistic achievements emanate from an irregular blob of jelly that makes up about two percent of the average person's body weight.

A source of mystery during the entire span of human history, the brain has recently yielded up many of its secrets – but not all. For many years to come, the brain will continue to puzzle scientist and philosopher alike.

Management from the top

Ballooning from the top of the spinal column like some weird, science fiction flower, the brain is the body's central computer which controls every thought and most movement. Information from all parts of the body is carried by sensory nerves to the brain, where it is integrated with direct input from the external senses. After the correct decision has been made, instructions for action are sent down motor nerves to the muscles. Not all of the action is conscious and voluntary; the brain also controls the "automatic" processes of the body, such as breathing, heart rate and digestion.

The adult brain weighs about three pounds, has a highly wrinkled appearance, and contains some ten billion nerve cells. By far the largest part is the ovoid mass of the cerebrum, where thinking takes place. Hanging beneath the back of the cerebrum is the cerebellum (the "little brain"), and jutting out from the middle is the brainstem, which merges into the spinal cord.

The brain and spinal cord consist largely of gray and white matter. Gray matter is made up of nerve cell bodies, or neurons; white matter is nerve fibers, or axons, in their insulating myelin sheaths. In the spinal cord the gray matter is contained within the white matter, in the brain the gray matter is on the outside.

Modern scanning techniques are gradually making the brain reveal the secrets of its structure – all the major component parts have now been located and named. Neurologists also now know a great deal about what the brain does. But the same cannot be said about how the brain works. A bewilderingly complex network of billions of nerve connections control all conscious and unconscious thoughts and actions. Exactly how they do so remains the subject of intensive study.

Care and protection

The tissues of the brain are protected from external knocks and damage by the strong, bony dome of the skull. Inside the cranial cavity, further protection is provided by a cushioning layer of soothing cerebrospinal fluid, a liquid secreted in the ventricles, or cavities, deep within the brain.

The brain is at once delicate and resilient. Large parts of it can be removed with no apparent effect, yet slight damage to other parts can result in total disruption of body functions and death. Much of our basic knowledge of the brain's operation comes from observing the effects of accidental damage or tumors. Electrical stimulation during surgery has also provided many

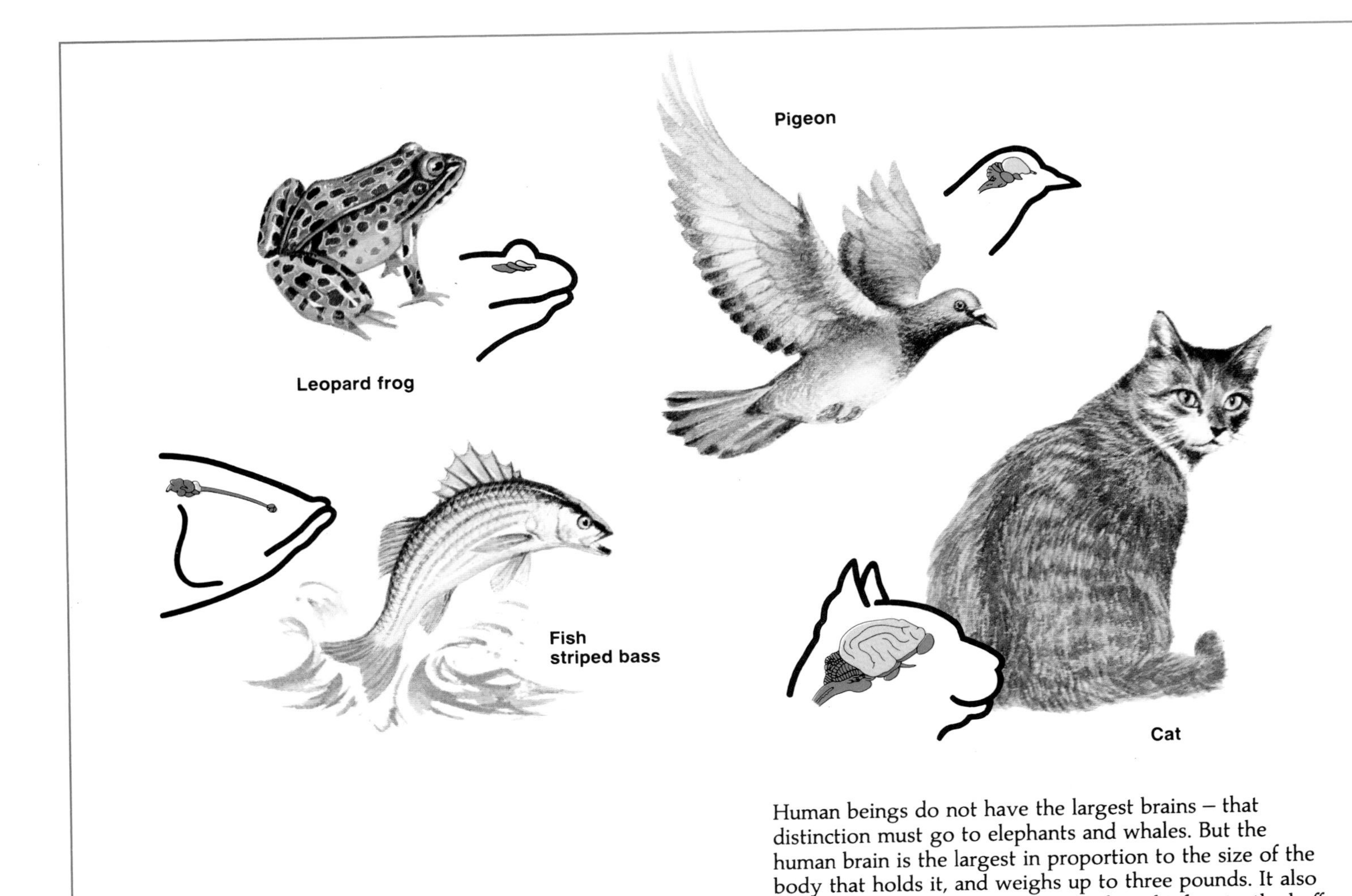

Human beings do not have the largest brains – that distinction must go to elephants and whales. But the human brain is the largest in proportion to the size of the body that holds it, and weighs up to three pounds. It also differs in the relative proportions of its chief parts, the buff-colored cerebrum and the evolutionarily older hindbrain.

clues. However, even the finest microscopic probes seem clumsy when applied to tissue in which a piece the size of a pinhead may contain up to five million cells.

The greatest source of information about the human brain is through experiments on animals. Because all mammals share the same basic brain structure, results of experiments on cats or monkeys can be applied to humans. However, scientists must be aware that it is the greater development of the brain, especially the forebrain, that makes us human, and not an animal.

Up the chain of command

The division of the brain into three basic parts – hindbrain, midbrain and forebrain – is based on their relative positions ascending from the top of the spinal column. An alternative concept is based upon three stages of the evolution of life on planet Earth. The first is reptilian, followed by paleomammalian (the very first mammals), then neomammalian (the later mammals, including humans). Although there is considerable overlap between the two concepts, the one based on location provides a more accessible description of the structures that make up the brain.

At the top of the spinal cord, the hindbrain was the first part of the brain to evolve. Comprising the cerebellum and most of the brainstem, it is sometimes referred to as the primitive brain.

In the region where the spinal cord gradually merges into the brainstem lies the medulla oblongata. This fibrous first inch of brainstem is the body's reflex center, the evolutionary core and primitive site of survival in human or dinosaur. Here are the small clusters of gray matter that control respiration, heartbeat, blood pressure and gastric movements. From the medulla also flash the involuntary commands to swallow, sneeze or cough.

Nervous disposition

As well as being the body's "automatic pilot," the medulla is also a major traffic intersection. Two thick bundles of nerve fibers, known as the pyramids, run down the medulla linking the brain to the muscles of the limbs. At a point known as the decusation (meaning "X shape") of the pyramids, most of the fibers cross over so that the left side of the brain controls the right side of the body, and vice versa.

Branching out from the medulla and adjacent areas above the decusation are the 12 pairs of cranial nerves, which serve the

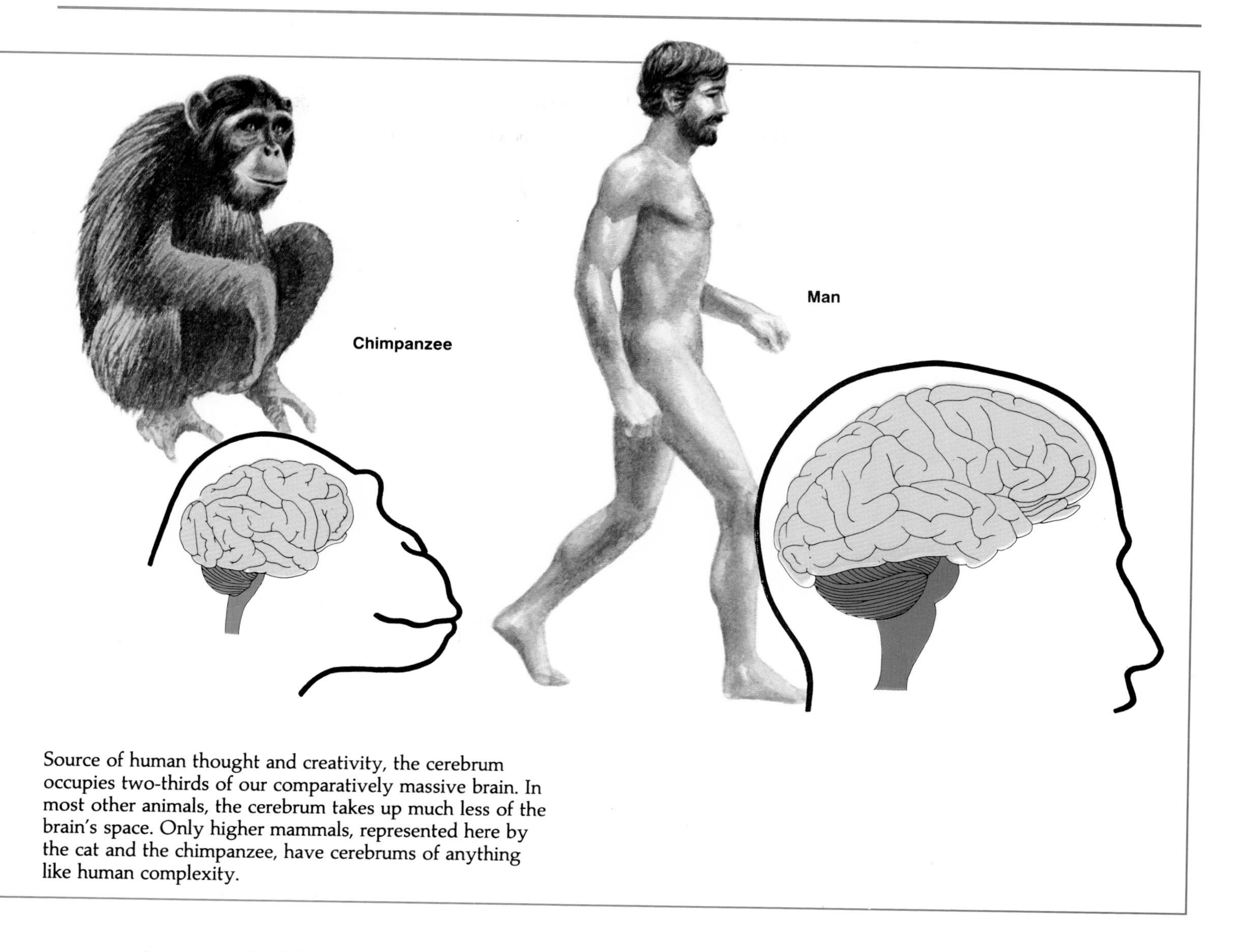

Source of human thought and creativity, the cerebrum occupies two-thirds of our comparatively massive brain. In most other animals, the cerebrum takes up much less of the brain's space. Only higher mammals, represented here by the cat and the chimpanzee, have cerebrums of anything like human complexity.

sensory and motor needs of the head, neck, chest and abdomen. Counting from front to back, and top to bottom, they are known by both name and Roman numeral: olfactory (I), optic (II), oculomotor (III), trochlear (IV), trigeminal (V), abducens (VI), facial (VII), acoustic (VIII), glossopharyngeal (IX), vagus (X), spinal accessory (XI) and hypoglossal (XII). To remember their initials in correct order, anatomy students recite, "On old Olympus's tiny tops, a Fin and German viewed some hops."

The view from the bridge

Arching over the medulla is the pons (named from the Latin for "bridge"), an inch-wide span of white matter which, in both function and appearance, links the cerebrum of the forebrain with the cerebellum of the hindbrain. From the pons itself extend a third of the cranial nerves, including the largest, the trigeminal. As its name suggests, this divides into three, with fibers fanning out to the jaws, face and scalp. Also located in the pons is a small piece of gray matter that controls the glands that produce saliva and tears.

Scattered throughout the pons and medulla, and extending up into the midbrain, are areas of closely interwoven gray and white matter called the reticular formations. These are the brain's chief watchdogs. Operating continuously during periods of wakefulness, the reticular formations are in essence the physical basis of consciousness, sending out the impulses that keep you awake and alert. When these impulses slacken, you fall asleep; if the formations are damaged, then prolonged unconsciousness, or coma, may result. And if they do not repair, coma persists.

Alarming development

Serving as the gatekeepers of consciousness, the recticular formations sort out the hundred million nerve impulses that assault the brain every second, separating the important from the inconsequential. The brain does not need a constant report of exactly how your shoes feel upon your feet; therefore, only about a hundred impulses per second are permitted through to the regions above the brainstem. Of these, the conscious mind heeds but a few. While you may be vaguely aware of nearby sights and sounds, concentration is limited to one at a time.

Together with their associated nerve pathways, characterized by short fibers enabling impulses to travel at high speed, the reticular formations also provide the body's alarm system: the reticular activating system (RAS). A nerve impulse requiring urgent attention, such as the smell of smoke, is detected by the RAS and sent straight to the cerebrum for action, temporarily overriding all other messages.

This mechanism also works in reverse. The conscious mind can instruct the RAS to block out everything else, and focus attention on one particular action that requires the highest skill and concentration. Without the RAS, a tennis champion could not ignore the murmurs from the crowd while serving for the match at Wimbledon.

The RAS also regulates a process known as habituation. An unusual sensation, such as the sound of a jack hammer in the street outside, immediately sets off as RAS alert. As soon as the sound has been recognized and the novelty has worn off, the brain instructs the RAS to halt its alerting action.

Smooth performer

Attached to the pons, and bulging out behind it, the cerebellum makes up about one-eighth of the brain's total mass. Its twin lobes are extremely wrinkled, and 85 percent of the surface area is hidden in the numerous deep folds. Working closely with the organs of balance in the inner ear, your cerebellum controls posture and governs your every movement. Apparently initiating nothing itself, it monitors impulses from motor centers in the brain and from nerve endings in the muscles.

Modifying and coordinating commands to swing and sway, the cerebellum grooves a golfer's tee shot, smoothes a dancer's footwork, and lets you lift a glass of water to your lips without spilling a drop. Learning a sequence of actions, like riding a bicycle or touch typing, takes both time and trouble, and many parts of the brain are involved. Once the sequence has been learned, however, the cerebellum is able to take over and ensure that it is repeated without conscious effort. There is evidence that the cerebellum plays a part in a person's emotional development, modulating sensations of pleasure and anger.

Well connected

A section through the two lobes of the cerebellum reveals matching leaflike patterns caused by the branching network of white nerve fibers. This prompted early anatomists to name it "arbor vitae," or tree of life. The exterior covering, or cortex, of gray matter is made up of three distinct layers. The middle layer consists of Purkinje cells, which are among the largest and most complex nerve cells in the human body. Each Purkinje cell can interconnect to up to a hundred thousand others, and no other cells in the brain have so many potential connections; such is the organization required for precise muscle control.

Between the cerebellar lobes lies a small structure known, for its wormlike shape, as the vermis. From it protrude three pairs of nerve fiber bundles known as peduncles. The lower two pairs, the inferior and middle peduncles, connect to the medulla oblongata and pons, respectively. These carry incoming, or afferent, impulses to the cerebellum. The superior peduncles above carry outgoing, or efferent, impulses to the red nucleus of the midbrain and thence via the thalamus to the cerebrum. Despite its crucial role in regulating our movements, the cerebellum does not exercise any direct control.

Relic of the past

Above the bridge of the pons lies the midbrain, the last inch of

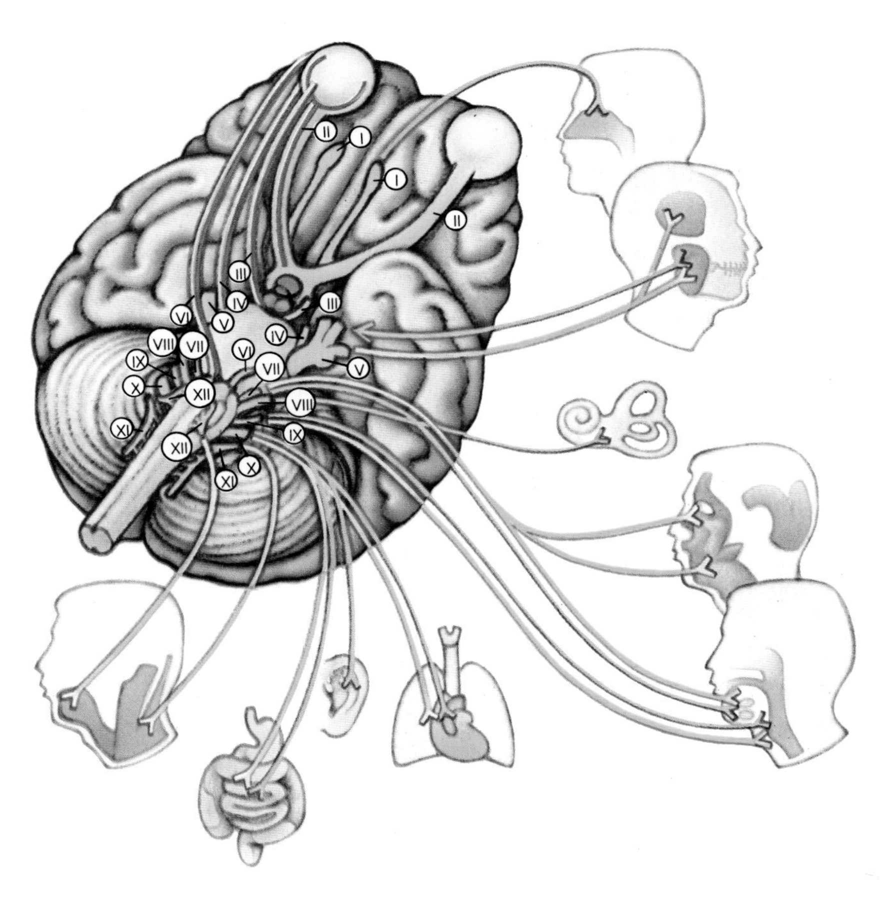

Conventionally identified with Roman numerals, the 12 pairs of cranial nerves branch out like power cables from the top of the brainstem. On the diagram, purple represents sensory nerves, which carry messages to the brain from sense organs and other receptors around the body. Orange is the color given to motor nerves, which transmit instructions from the brain to muscles in the face. An exception is the cranial nerve X, the vagus, whose motor fibers connect with the heart and digestive system.

the brainstem. Before the evolution of the cerebrum, the midbrain was responsible for the higher functions of the brain, such as processing the senses of sight and sound. These have now migrated to the forebrain, and in consequence the structure of the midbrain is much simpler than either the forebrain or hindbrain. Some evidence of its former eminence is still visible in the four bumps of the colliculi, which serve as relay stations for nerve impulses from the eyes and ears.

Two further peduncles rise to connect the brainstem to the hemispheres of the cerebrum. Each contains a dark core of cellular matter stained with the pigment melanin. This is the substantia nigra, motherlode of the biochemical agent dopamine, which guards against muscle rigidity and tremor. Down the center of the midbrain runs the cerebral canal, which carries cerebrospinal fluid, by now drained of nutrients and loaded with metabolic waste, down into the space surrounding the spinal cord where it is absorbed into the blood.

Primitive passion

The largest and most recently evolved part of the forebrain is the cerebrum, which is divided into the two cerebral hemispheres. In fact these are quarter spheres – the whole cerebrum makes a hemisphere – but the name has stuck. Within each cerebral hemisphere is a wishbone-shaped cavity where cerebrospinal fluid is produced, replenishing the supply three times per day. Known as the lateral ventricles, these drain into the centrally located third ventricle, down through the small fourth ventricle, and into the cerebral canal.

A complex of small structures, notably the thalamus and hypothalamus, cluster around the third ventricle and loop over the top of the brainstem; they make up the remainder of the forebrain. This is sometimes known as the "interbrain." Together with parts of the cerebrum and the pathways associated with a human's almost redundant sense of smell, this "interbrain" forms the limbic system – the most primitive part of the forebrain and the seat of human passions and basic drives.

Balancing act

Here at the very core of the brain is one of its most subtle and little-understood regions. Hidden beneath the enveloping folds of the cerebral hemispheres, the limbic system has been termed the "emotional brain," striving to maintain an equilibrium between fear and desire. It also attends to everyday wants and needs. The nerve cell masses in the system are interwoven with myriad pathways that convey not only the extremes of terror and ecstasy, but also more mundane electrochemical messages such as when to sweat and when to shiver, what to remember and what to forget. The limbic system is the body's balancing act on the tightrope of survival.

While the full workings of the system remain obscure, the basic role of the major components is understood. The thalami, twin eggs of gray matter perched above the top of the brainstem, are relay stations connected to a very large number of information channels that bring input from the main sensory systems, the cerebral hemispheres, the cerebellum and the reticular formations in the brainstem. Within the thalamus this mixed input is integrated and information patterns of greater complexity are sent back. The thalamus appears to be concerned with emotional shading, subjective feeling states, and awareness of identity and self.

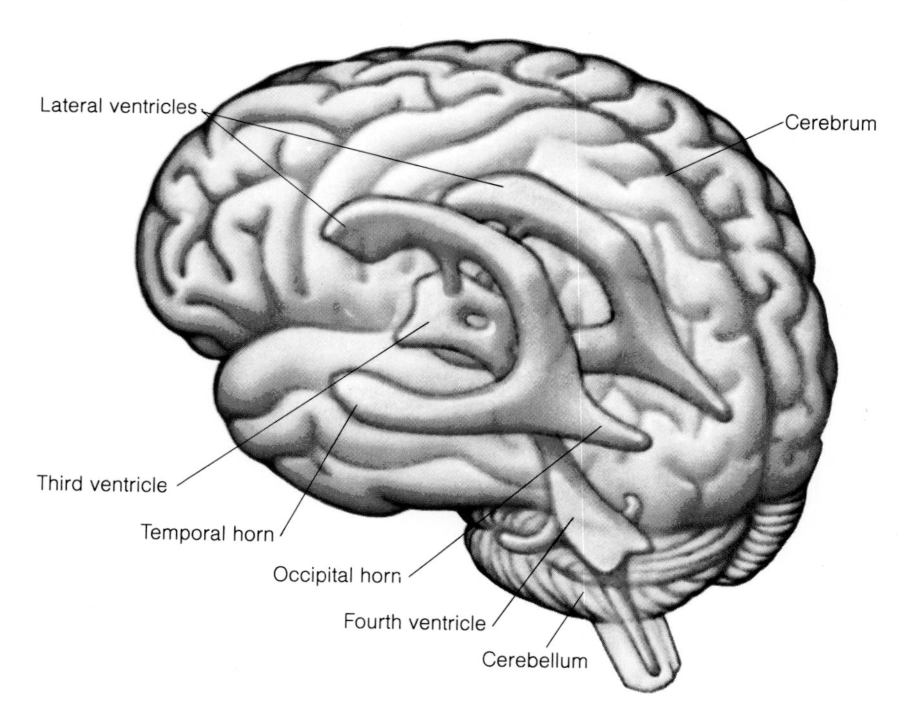

Internal shock absorbtion is provided by the pools of cerebrospinal fluid which bathe the brain and cushion its membranes. The plasmalike liquid is produced in choroid tissue and fills the ventricles, four cavities within the cranium (shown in blue). From the lowest fourth ventricle the fluid circulates down the spinal cord.

The little dictator

The size of a thumb nail and with one of the richest blood supplies in the body, the hypothalamus is the body's great regulator, maintaining internal temperature and fluid balance. Tiny receptors measure the amounts of glucose and salt in the blood, generating feelings of hunger or thirst when levels become low. Through its neighbor, the pea-sized pituitary gland, the hypothalamus organizes the release of hormones which affect growth and sexual behavior. It also initiates the "fight or flight" response to stress, sending chemical messages via the pituitary to the adrenal glands above the kidneys. Chemical control alone is not enough for the brain's little dictator, and cell clusters specializing in involuntary muscle control crackle with electrical energy as they fire off signals.

Although signs of crude rage and feelings of raw pleasure can be obtained by experimentally stimulating parts of the hypothalamus, more complex emotions, and "instinctive" behavior patterns such as finding a mate or rearing young, reside in the interaction of the limbic system as a whole. Above the hypothalamus, the amygdala is thought to modify rage and aggression to suit rapidly changing circumstances. To the rear, projecting from the roof of the third ventricle, is the pea-sized pineal gland. Once believed to be a vestigial third eye, it is probably a light-sensitive biological clock, which regulates sex-gland activity. Above the thalami a fibrous web, the fornix, terminates at the bulbous hippocampal formations. The hippocampus converts information from short-term memory to long-term memory, and constantly compares signals from the senses with stored experience.

Convoluted thinking

The cerebrum, arching beneath the curvature of the skull, is the seat of reason, imagination and creativity. Its surface is molded into a series of convolutions called gyri, separated by fissures, or sulci. The most pronounced fissure, the longitudinal cerebral fissure, separates the two cerebral hemispheres.

The convolutions of the cerebrum are not as dense and intricate as those of the cerebellum, and their arrangement is fairly constant in all brains. For ease of description, certain deeper fissures are used to divide each hemisphere into four lobes – frontal, temporal, parietal and occipital – although this is purely convenience and these areas of the brain do not coincide exactly with the cranial bones for which they are named.

The gray matter of the cerebrum, "the little gray cells" of a famous fictional detective, is mostly contained within a thin layer covering the gyri and lining the sulci. This surface layer, the cerebral cortex, is on average about one-quarter of an inch thick. Only about 30 percent of its area is externally visible; the rest is hidden within the fissures. Underneath the cortex is a mass of white matter, the connecting fibers.

Tying it all together

Hundreds of millions of microscopic threads form an array of connections between the cortex's nerve centers and distant parts of the brain. Although only a tiny fraction of the connections have been traced, their general organization is understood and they are divided into three different types.

The projection fibers squeeze into a compact band, the internal capsule, near the top of the brainstem. Both incoming

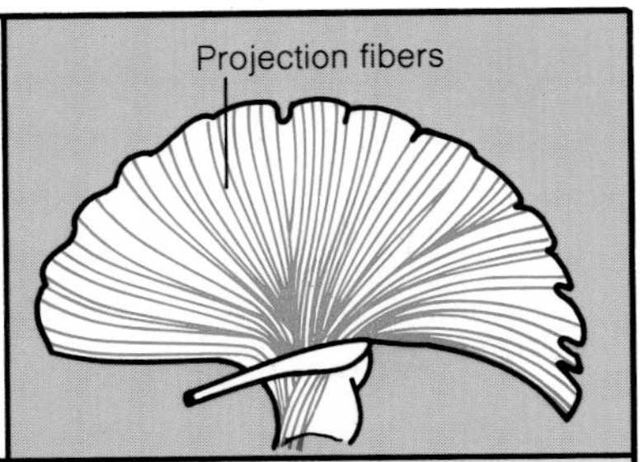

Projection fibers Forming the *corona radiata,* these fibers fan out from the brainstem. They relay impulses to and from the cortex.

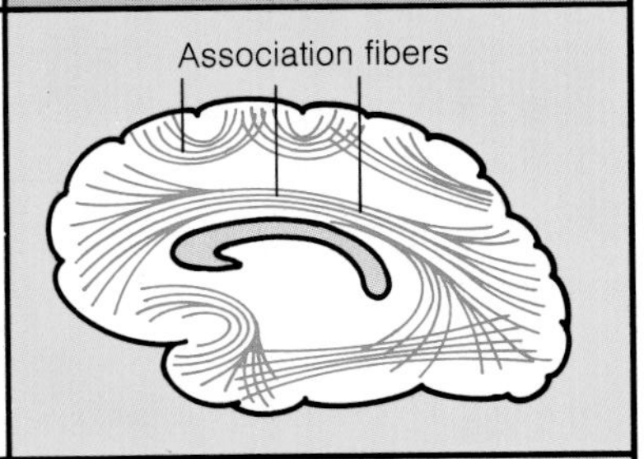

Association fibers Looping strands link different sections of the same hemisphere. This web subtly modulates the cerebral cortex.

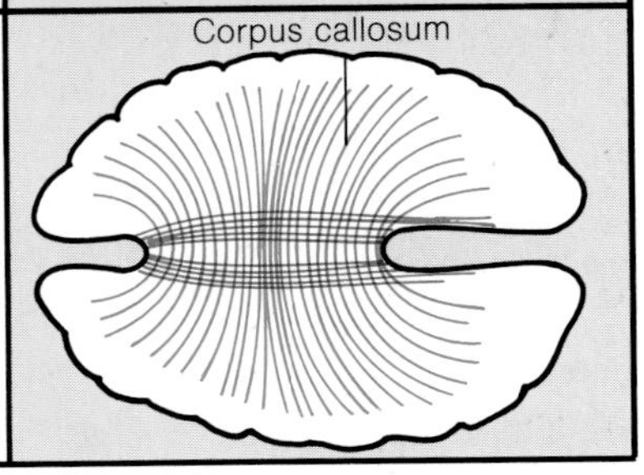

Corpus callosum This thicket of fibers joins the hemispheres, permitting the two sides of the brain to communicate with each other.

Cross connections within the brain are provided by various bundles of nerve fibers. A fan of projection fibers connects the brainstem with the cortex *(top)*. Within each hemisphere association fibers, shown in the sideways section *(middle)*, interconnect various parts of the cortex. To prevent the two hemispheres working in isolation, they are joined by the fibers of the corpus collosum, depicted in the horizontal section *(bottom)*.

Logical or creative? These two aspects of human intellectual activity are controlled by different sides of the brain. In most people, logical and scientific aspects are the responsibility of the left hemisphere. Imaginative, spatial thinking – and appreciation of art and music – is the province of the right hemisphere. Dominance of one side or the other may predetermine a person's aptitudes and preferences.

The powers of the human brain are made possible by the billions of interconnections between the nerve cells of which the brain is composed. This photomicrograph, in which the cells have been stained to reveal the nerve cells, show just how complex is the vast network for information transfer in the brain.

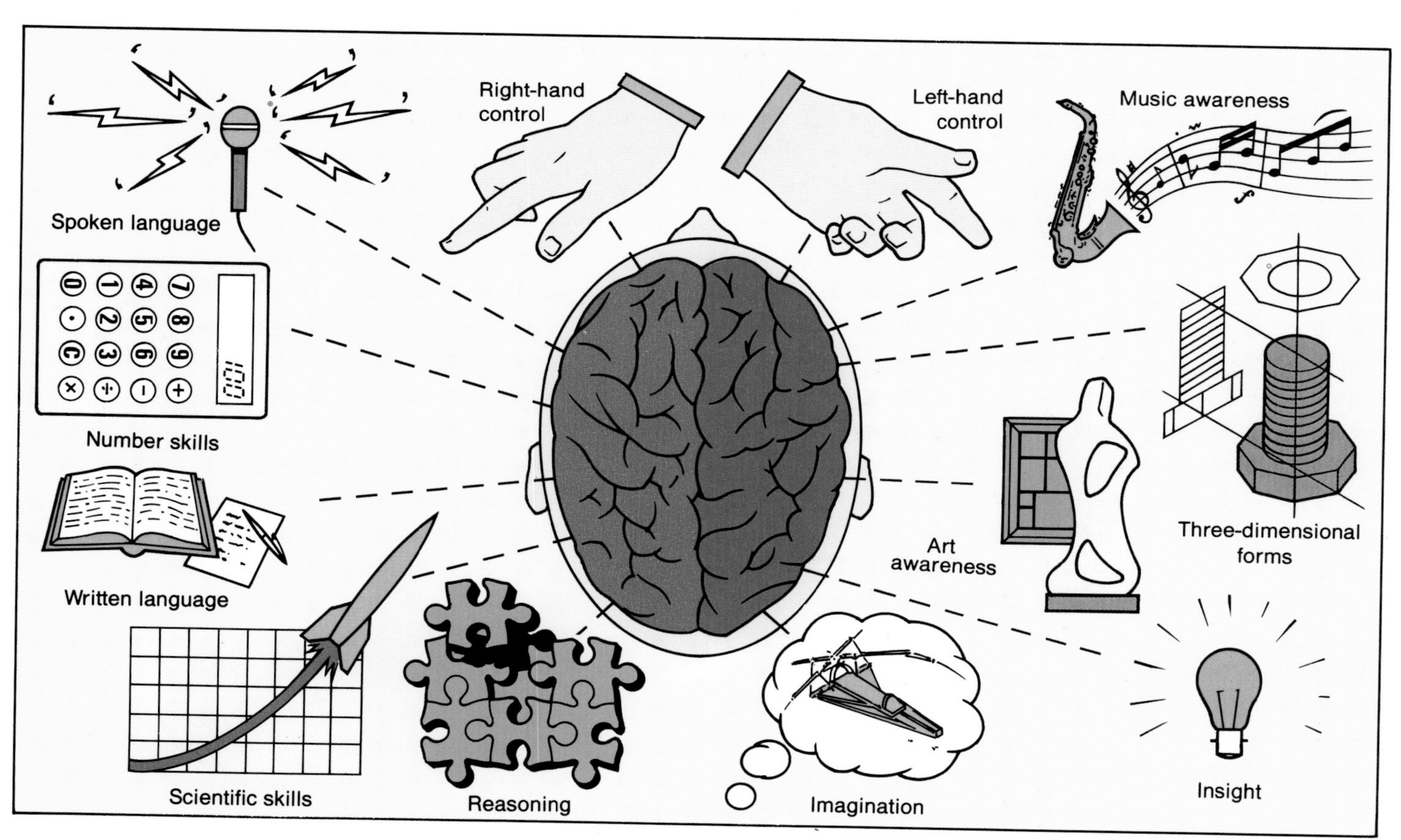

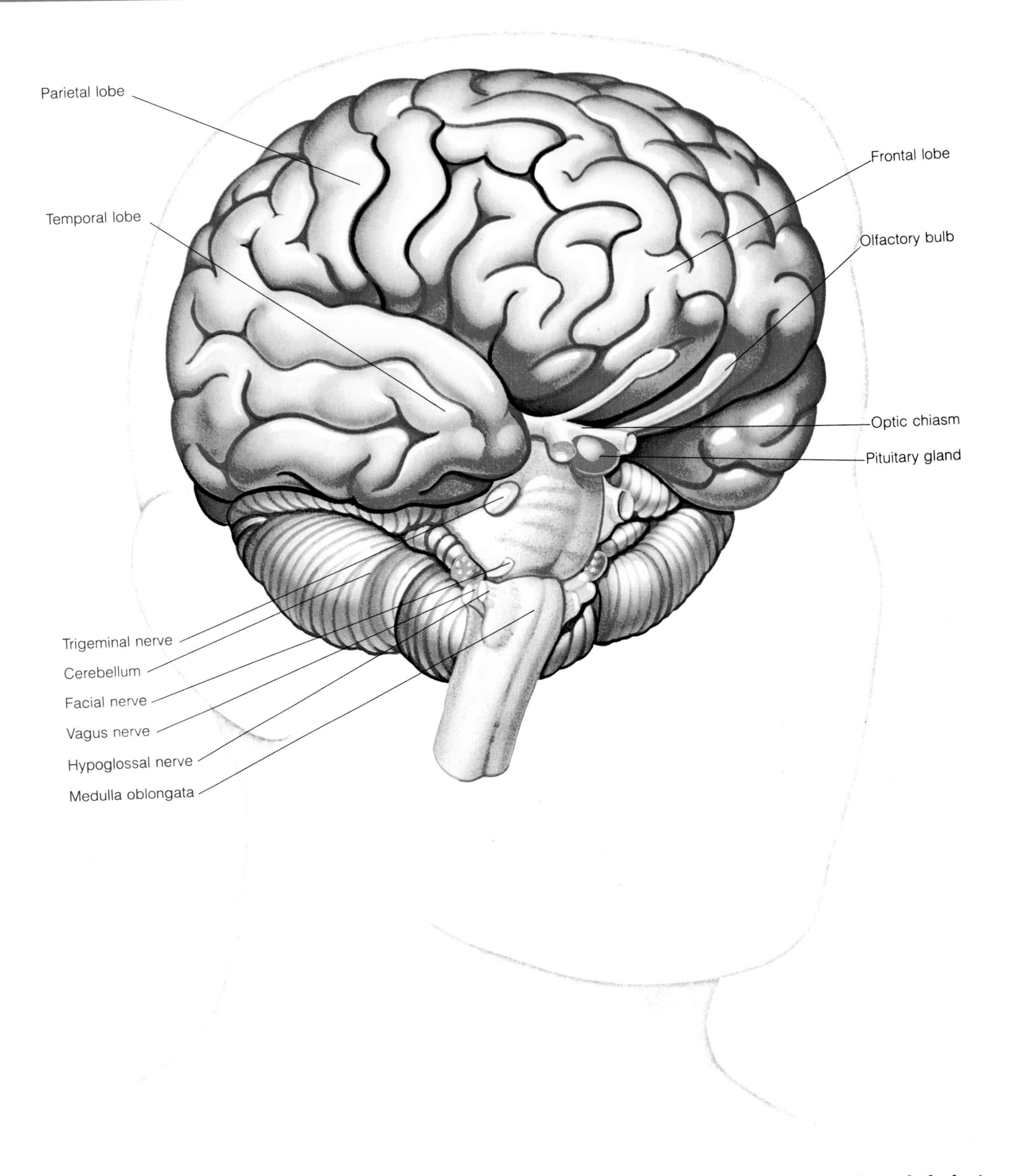

A section through the brain reveals its convoluted anatomy. This view shows the lobes of the cerebral cortex and vital nerve connections.

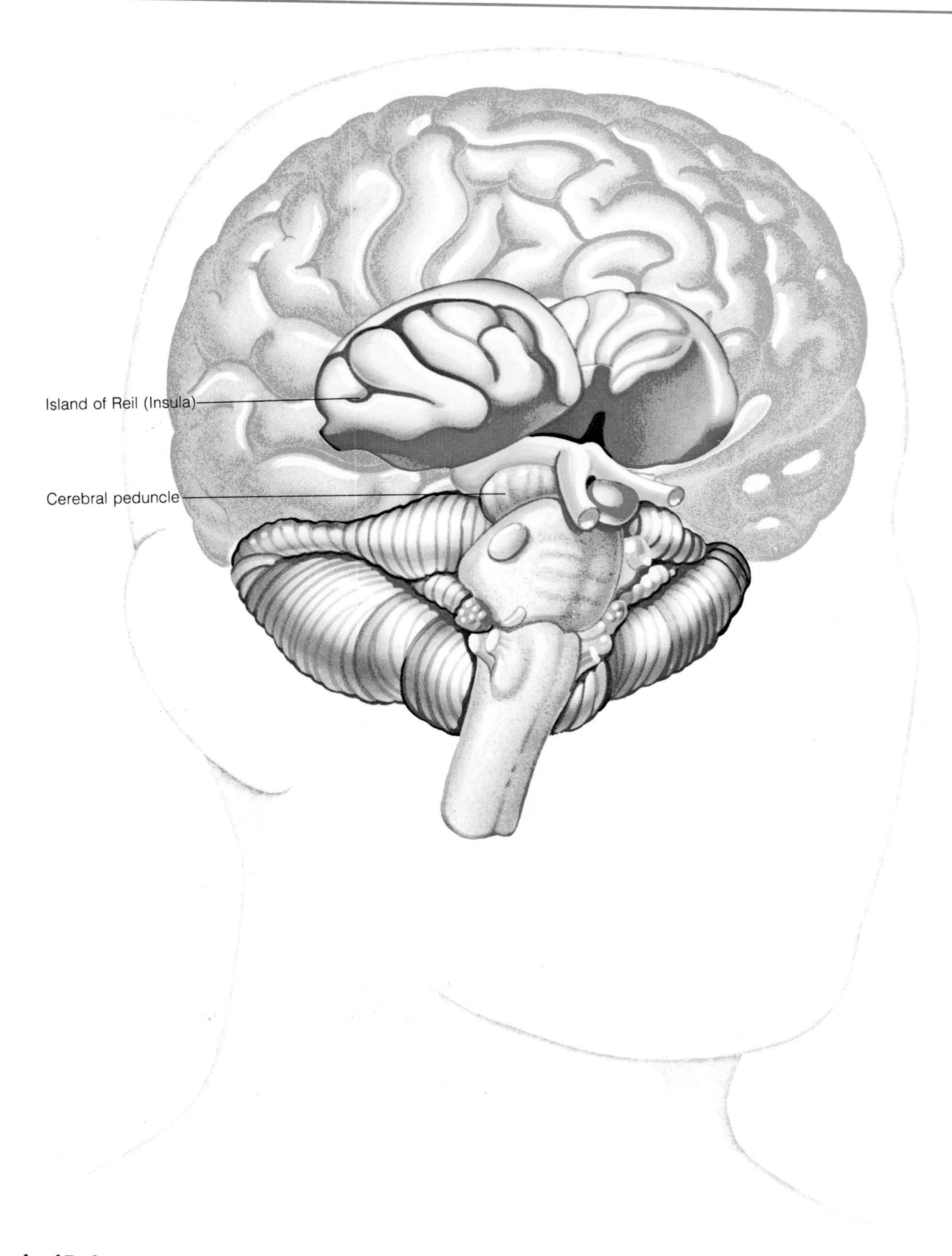

The islands of Reil, situated deep in the brain, perceive sensations from abdominal organs. The cerebral peduncle is part of the midbrain.

and outgoing fibers then flare upward and outward to all parts of the cortex, creating a pattern that anatomists have dubbed the corona radiata. The most numerous fibers link different parts of the same hemisphere, modulating the activities of the cortex and providing built-in self-sufficiency. These are aptly named association fibers.

One of the skeins of association fibers, the cingulum, is covered with a layer of cortex to form the cingulate gyrus, a major part of the "limbic lobe" of the cerebral hemisphere. This fifth lobe is a relative newcomer in terms of the brain's known geography, identified only quite recently, and its full extent has still to be agreed. Other components are the smaller dentate gyrus and the hippocampal formations, so named because in section they resemble the shape of a sea-horse.

The "limbic lobe" is the cerebral part of the limbic system, and it is believed that here the raw energies of aggression and sex drive produced by the hypothalamus are integrated into the rest of the brain, producing courtship and other socially acceptable behavior and behavior necessary to the continuance of human communities.

The third category of cerebral fibers are known as commissural fibers. These are gathered into a densely packed body, the corpus callosum, located at the bottom of the longitudinal cerebral fissure. Extending beyond the brain's midline and mingling with association and projection fibers, the corpus callosum links the cerebral hemispheres and enables the two sides of the brain to intercommunicate.

Crossing the great divide

Because of its size (it is more than four inches long) and central position, early scientists believed that the corpus callosum was essential to the brain's proper functioning. They saw it as unifying the two cerebral hemispheres into a single brain. However, in the middle of the last century it was noticed during autopsies that some people had lived apparently normal lives with half a brain, having only one cerebral hemisphere.

The mystery deepened in the 1930s when doctors discovered that epileptic seizures could be significantly moderated by surgically severing the patient's corpus callosum, and without any noticeable impairment of normal behavior. Further research, especially by a team at the California Institute of Technology led by Roger Sperry, revealed that although the two cerebral hemispheres are virtual mirror-images of each other, their mental functions are in fact very different.

The most basic distinction between the two is that the right hemisphere controls the left side of the body, and vice versa. This at least is symmetrical, and the crossover of nerve pathways inside the medulla oblongata is readily observable. Less apparent is the division of mental abilities – the geography of the brain merges into the geography of the mind.

Two brains in one

The CalTech experimental studies of split-brain people (with the corpus callosum severed) showed that the two sides of the brain process information differently. One side specializes in symbols and logic, while the other is adept at pattern and space perception.

Left hemisphere thinking appears to be analytical (taking ideas apart), linear (one step after another), and verbal (both written and spoken). It builds sentences and solves equations. Right hemisphere thinking is synthetic (putting ideas together), holistic (grasping relationships in a single step), and imagistic (visual thinking with the "mind's eye"). It listens to music and appreciates three-dimensional objects. The left side of the brain has given man science and technology; the right side is responsible for art and imagination.

Thus the proper function of the corpus callosum is now understood, it exists largely to unify awareness and attention and allow the two hemispheres to share learning and memory. For his work, Sperry was awarded a Nobel prize in 1981.

Panning for gold

Although it is possible to chart many specific brain functions, which side of the brain contains intelligence, and where is the elusive spark of creativity to be found? The answers are not easy to come by, not least because these highest human qualities are difficult to measure and almost impossible to define.

Traditional intelligence tests, with their sequences of numbers, missing words and logical puzzles, are heavily geared toward left hemisphere thinking. As such they have proved statistically reliable indicators of scholastic achievement. Outside of school, they are much less useful in predicting success in later life. How can qualities such as dealing with people, or susceptibility to gambling be measured; and to what extent are these and other important factors a function of intelligence?

In a search for a better alternative some experts have urged measuring reaction times, on the basis that the quicker you think, the cleverer you are. Others have suggested using "real-life" problems such as selecting the most cost-effective route from an airline schedule. None of the tests yet devised is particularly good at measuring divergent thinking, the ability to discover new answers, which psychologists believe to be crucial to creativity.

Superficial logic suggests that creativity, the opposite of reason, ought to be lodged in the right hemisphere of the brain alongside music and art appreciation. Certainly the ability to think in images or sounds is vital to the creative process. Mozart wrote entire concertos in his head – writing them down afterward was something of an anticlimax. However, artists need more than just inspiration and creativity to survive; they must be rational and self-critical masters of their discipline. It is the interplay between the two sides of the brain that is essential for all types of productive thought. Albert Einstein observed that his gift for fantasy meant more than acquiring knowledge.

Areas of responsibility

Established only by experimental study, none of the right–left separation of the brain's mental activity is visible to the anatomist's eye. Both cerebral hemispheres look the same, with irregular folds of cortex covering the whole of the surface. But the division of the hemispheres into four lobes, while partly for convenience, is not completely arbitrary because each lobe has a particular area of responsibility in the brain's organization.

The frontal lobe extends from behind the forehead up to the central sulcus that bisects the hemispheres at right angles to the main longitudinal fissure. The area directly behind the brow is confusingly known as the prefrontal lobe, and some researchers believe that powers of planning and choice are located here. Adjacent to the central sulcus is an area of cortex that controls every voluntary movement of the body – the motor cortex.

On the other side of the central sulcus, within the parietal lobe which caps the top of the cerebral cortex, is an area loaded with

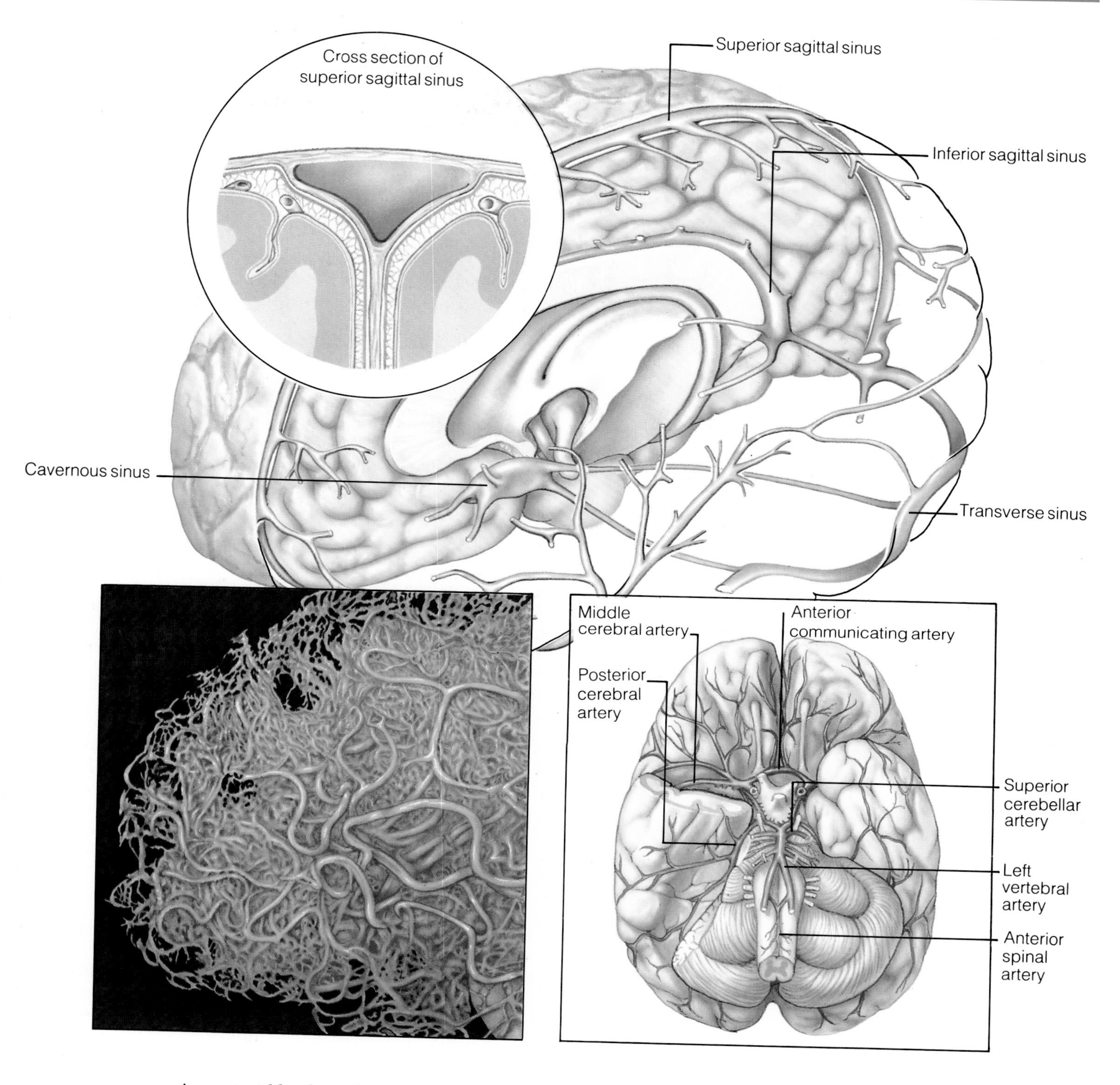

A constant blood supply is essential for proper functioning of the brain, and to carry this vital fluid there are hundreds of branching blood vessels *(above left)*. A built-in fail-safe feature gives the brain the best chance of receiving an uninterrupted supply of oxygenated blood, for without oxygen brain cells die within a few minutes, and can never be replaced. The arteries that lead to the cortex *(top and above, right)* branch off a closed loop. As a result, if an obstruction should occur in one place on the circuit the blood can flow around the other way to the arterial exit and on to the cortex. To remove carbon dioxide waste, produced by internal respiration of the brain's many cells, there is a network of veins on the surface of the brain that carry away the deoxygenated blood. They include internal drainage reservoirs in the form of enlarged cavities known as sinuses.

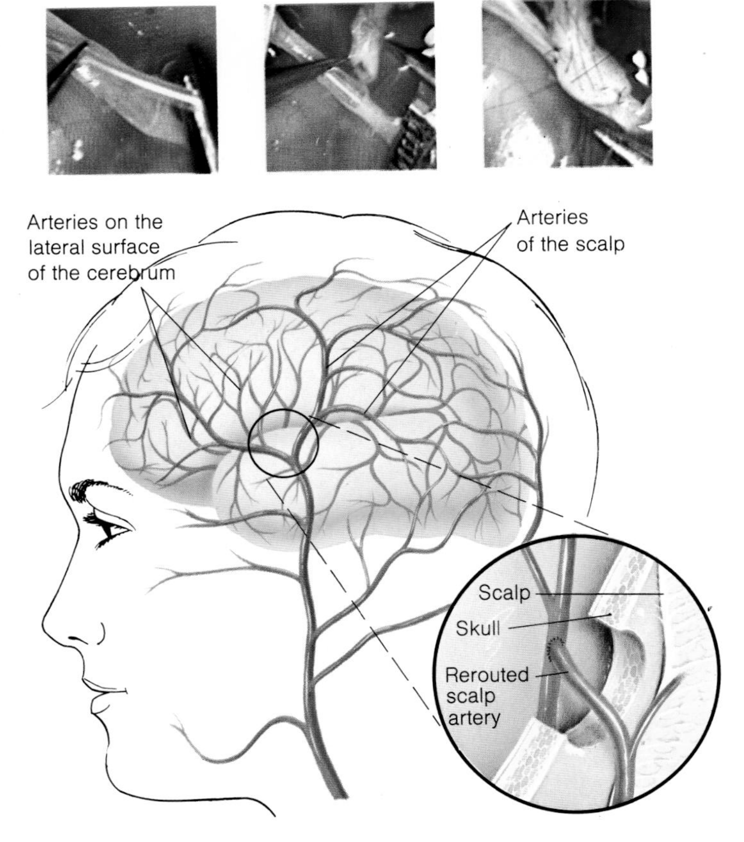

Bypassing a blockage restores the vital supply of blood and oxygen to a starving brain. With a piece of plastic tubing to keep it in shape, a brain artery *(top left)* receives a donor artery from the scalp *(center)* and the two are sewn together with microstitches *(right)*. The diagram pinpoints the location of the bypass, and shows how the artery from the scalp is routed through a hole bored through the skull at precisely the correct place.

cells responsive to touch, heat, pain, and body position. This area is known as the sensory cortex.

At the back of the skull, as far from the eyes as is anatomically possible, the visual cortex occupies an area of the occipital lobe. Impulses from the retinas race down the optic nerve, half of them crossing at the junction of the optic chiasma in front of the brainstem. They then fan out through clusters of gray matter called lateral geniculate bodies before hitting the visual cortex at speeds up to 400 feet per second. Not quite the speed of light, but fast enough.

In contrast, hearing is processed in the area of the brain closest to the ear. This is the temporal lobe, which is separated from the frontal lobe by the lateral sulcus at sideburn level.

These named areas of cortex have been identified because they are easily stimulated and the results can be observed. The rest of the cerebral cortex appears to lack direct connections with the central nervous system and has been dubbed the association cortex. These "silent" areas of the cortex are believed to be concerned with further elaboration of sensory information. They may also constitute the brain's memory banks.

Plumbing the depths

Deep in the floor of each lateral sulcus, completely overgrown by the rest of the brain, is the insula – a strange, vaguely pyramid-shaped structure. Also known as the islands of Reil, this has the same convoluted appearance as the cerebral hemispheres and a covering of cortex like the rest of the cerebrum. What the islands do – even what other parts of the brain they are connected to – is still very obscure. Almost the sum of our knowledge is that stimulating the insula sometimes produces increased salivation. Consequently, some scientists have decided that the insula is the area that processes our sense of taste.

Underlying the insula within each cerebral hemisphere is a series of masses of gray matter collectively referred to as the basal nuclei. These are located within the curve formed by the horns of the lateral ventricle. The two main structures, the lentiform nucleus and the caudate nucleus, are collectively known as the corpus striatum.

The two nuclei are separated from the insula by a sheet of gray matter covered in a layer of fiber called the claustrum. Alongside it is the biconvex shape of the lentiform nucleus (so called because it looks like a lens), which is divided into two parts. The darker and larger portion is the putamen; the smaller, lighter area is the globus pallidus (often shortened to the pallidum). The innermost basal structure is the caudate nucleus, which forms part of the floor of the lateral ventricle. The amygdala is also sometimes included, but is more properly seen as part of the limbic system.

The basal nuclei, and their associated fiber systems, form an extremely complex set of interconnections – they are another of the brain's relay centers. Information converges on the corpus striatum from all parts of the cerebral cortex, and from the thalamus and some areas of the midbrain. The resultant output, which is almost entirely routed through the pallidum, appears to exert a strong influence on both muscle tone and motor control. This is borne out by detectable concentrations of the neurotransmitter dopamine within the basal nuclei, possibly originating in the substantia nigra of the midbrain.

The outer limits

The entire brain is sheathed by a series of three membranes known as the meninges, which as well as serving to protect the delicate tissues of the brain, also provide a secure environment for the essential network of blood vessels.

The outer layer, the cerebral dura mater, is a thick, inelastic membrane which adheres to the inside of the skull, thus securing the brain. Within the dura mater run the venous sinuses, which drain blood from the brain.

The middle layer, the arachnoid, is thin and transparent, and separates the dura mater from the subarachnoid space, which contains cerebrospinal fluid and the larger blood vessels of the brain. The fluid is tapped off from the main supply at the fourth ventricle and maintains a uniform pressure on the brain, supporting it and distributing the weight evenly. The brain literally floats; a brain that weighs three pounds in air, weighs

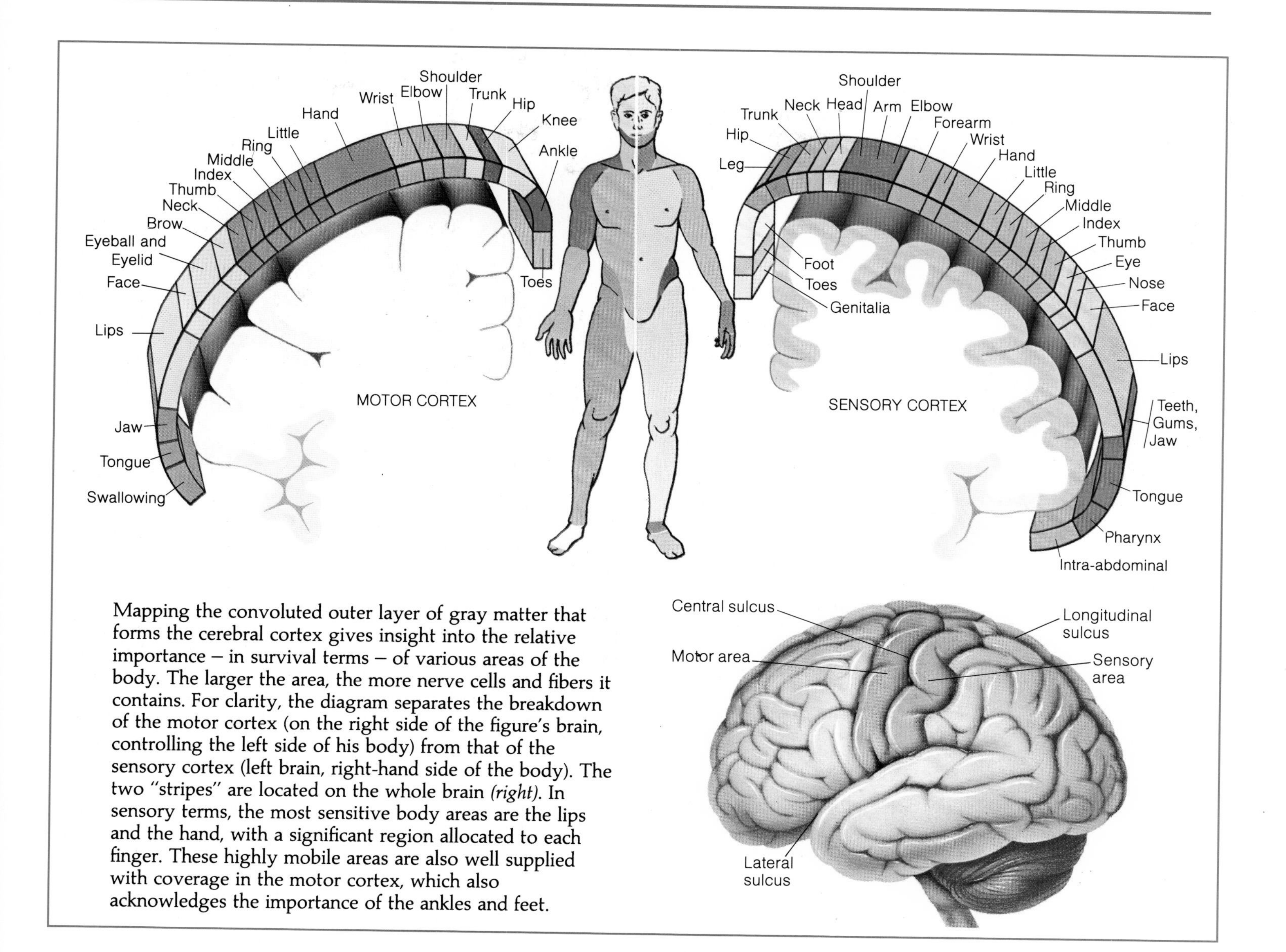

Mapping the convoluted outer layer of gray matter that forms the cerebral cortex gives insight into the relative importance – in survival terms – of various areas of the body. The larger the area, the more nerve cells and fibers it contains. For clarity, the diagram separates the breakdown of the motor cortex (on the right side of the figure's brain, controlling the left side of his body) from that of the sensory cortex (left brain, right-hand side of the body). The two "stripes" are located on the whole brain *(right).* In sensory terms, the most sensitive body areas are the lips and the hand, with a significant region allocated to each finger. These highly mobile areas are also well supplied with coverage in the motor cortex, which also acknowledges the importance of the ankles and feet.

about two ounces when suspended in cerebrospinal fluid.

The pia mater, the innermost layer, fits very closely around the brain following every contour and dipping deeply into the sulci. It consists of a dense plexus of minute blood vessels held together by extremely fine tissue.

Food for thought

The billions of cells in the brain, from the twinkling galaxies of the cortex to the deep recesses of the basal nuclei, whether they are complex and highly specialized or just links in the chain of command, all have one thing in common. They need blood to function, and without it they die never to be replaced. Blood brings the essential oxygen and carries away the unwanted waste product, carbon dioxide. Without a constant and uninterrupted supply of blood to the brain, a human being stops moving, stops thinking, and very soon stops living.

Blood is brought to the brain by the internal carotid and vertebral arteries. The vertebral arteries join at the base of the pons, forming the basilar artery, then diverge again, forming the two posterior cerebral arteries. Each carotid artery terminates by dividing into the anterior and middle cerebral arteries. These six cerebral arteries, which supply the forebrain and midbrain, are connected by communicating arteries that form an arterial loop – the circle of Willis – at the base of the brain.

The cortex is supplied by the cortical branches of these arteries, which divide within the pia mater into smaller vessels that drop vertically into the gray matter. The underlying white matter is served by the medullary arteries; these pass through the cortex and penetrate up to two inches into the brain. The cerebellum, pons and medulla oblongata are supplied with blood mainly from the cerebellar arteries.

Bringing it all back home

Within the brain, blood is delivered to the cells by a fine mesh of capillaries, then begins its journey back to the heart. The veins of the brain, many of which run alongside the arterial vessels, are thin walled and contain no valves. They drain into a network of venous sinuses contained within the thickness of the dura mater.

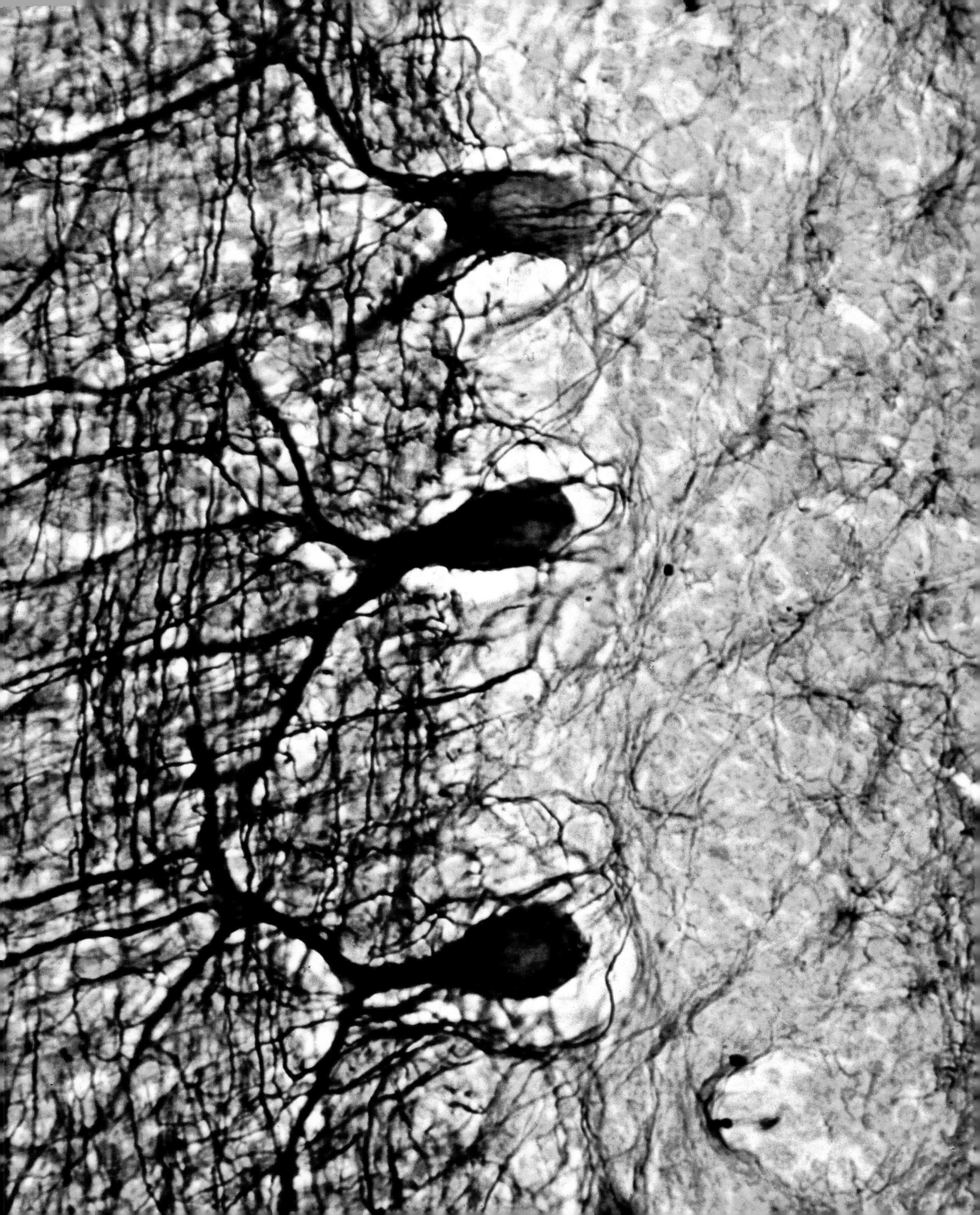

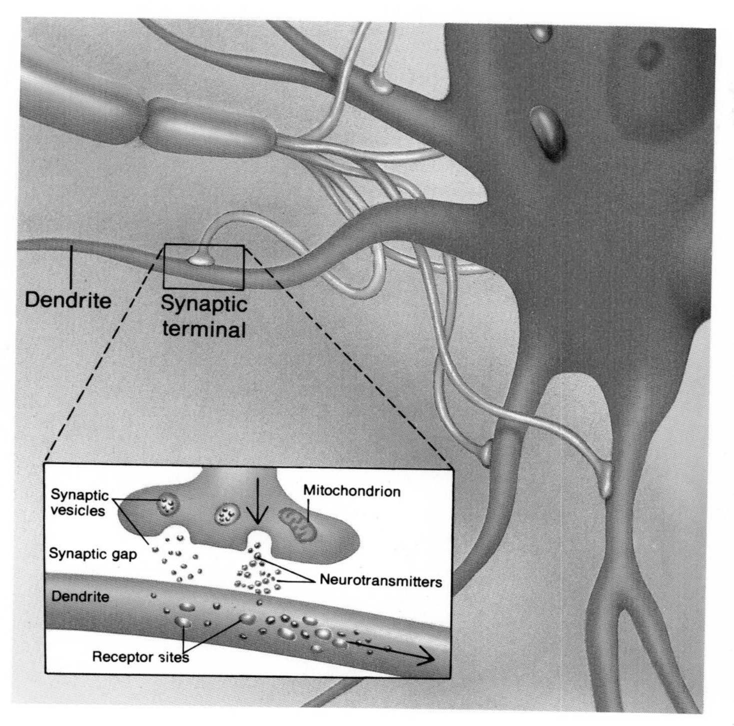

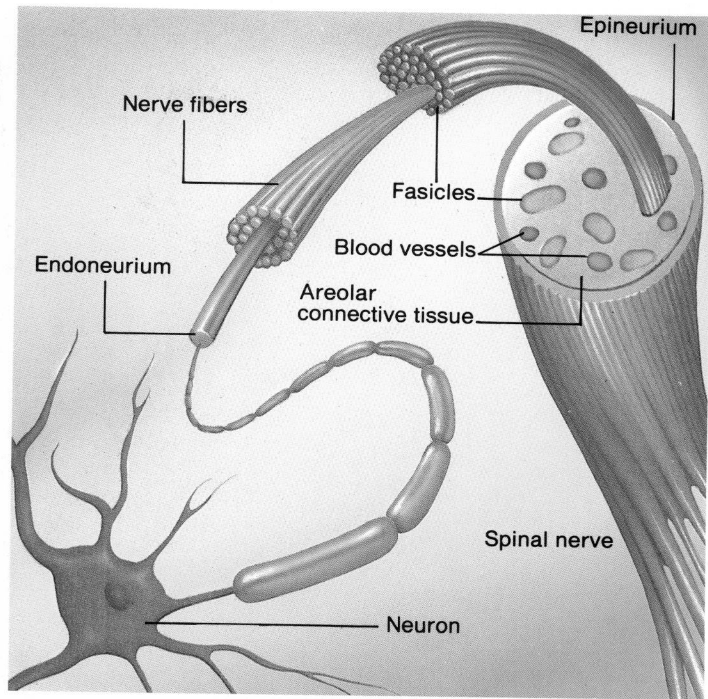

Like interlocking roots of ancient forest trees, the network of neurons in the brain *(left)* form billions of alternative conductive pathways for nerve impulses. They gather and interpret incoming information from sensory receptors, and send out commands that the muscles have to obey. Less well understood is how some connections, once established, provide short-term and long-term memory.

Fine tendrils from the end of a nerve axon *(above left)* latch on to the dendrites of neighboring nerve cells with their terminal buttons. In the synaptic cleft, at the point of contact, vesicles release neurotransmitters, which stimulate the receiving cell and pass on the nerve impulse. Bundles of axons *(above right)* form nerve fibers, which may be sheathed in a layer of myelin to form the major nerves linking body and brain.

The superior sagittal sinus, which grooves the bones of the skull along its centerline, also receives blood from the nose and scalp. The inferior sagittal sinus runs along the fissure dividing the two cerebral hemispheres. Both drain down into the transverse sinuses, which continue downward to become the jugular veins. The cavernous sinuses, above the nose on each side of the head, are criss-crossed by tiny filaments and have a spongy structure. There is no lymphatic system in the brain because the tissues do not require this form of drainage.

Small wonder

Interruption of the blood supply to any part of the brain can cause serious damage. A blood clot may block an artery, or a weak-walled section may balloon out into an aneurysm, which may then burst and cause a stroke. If not immediately fatal, a stroke can cause lasting impairment and the victim may be left without speech and minus motor functions on the side of the body opposite to the side of the brain that is affected.

Thanks to advances in microsurgery, doctors can now restore lost blood supply through bypass techniques which involve rerouting another artery. The exact location of the blockage can be determined by injecting the blood with special dyes that show up on X rays. The neurosurgeon then selects a suitable artery in the outer scalp, close to the blockage, and which has the same diameter as the blocked vessel. Cutting through the skull with a miniature cylindrical saw, the brain artery can be exposed by cutting and peeling back the dura mater. The bloodflow in both arteries can then be stopped by applying non-crushing microclips.

Peering through a 20-power microscope, the surgeon snips through the scalp artery, and makes an oval hole in the brain artery. The supply end of the scalp artery is fed through the hole in the skull. Using a needle as fine as a baby's eyelash, it is then sewn to the brain artery with about 20 minute stitches. When the clips are removed the supply of blood is restored.

Similar techniques can be used on aneurysms, enabling the swollen arterial wall to be collapsed gently before it bursts.

Quarter-inch miracle

The threat of blood starvation thus averted, the brain can continue working its normal everyday miracles. Thousands of tiny messages flashing across the cerebral cortex enable you to pick up this book, turn the pages and read.

If the quarter-inch thick cortex were lifted off the cerebral hemispheres, unfolded and spread out, it would have an area of about two and a half square feet and weigh 20 ounces or so. Like other gray matter, the cortex does not consist solely of neurons. It is in fact an intricate blending of nerve cells and fibers, neuroglia (non-excitable nerve cells that form a protective coating around nerve fibers and perform other specialist functions) and blood vessels.

Nor it is homogeneous; even without a magnifying glass the eye can discern several horizontal layers (or lamina) in a freshly

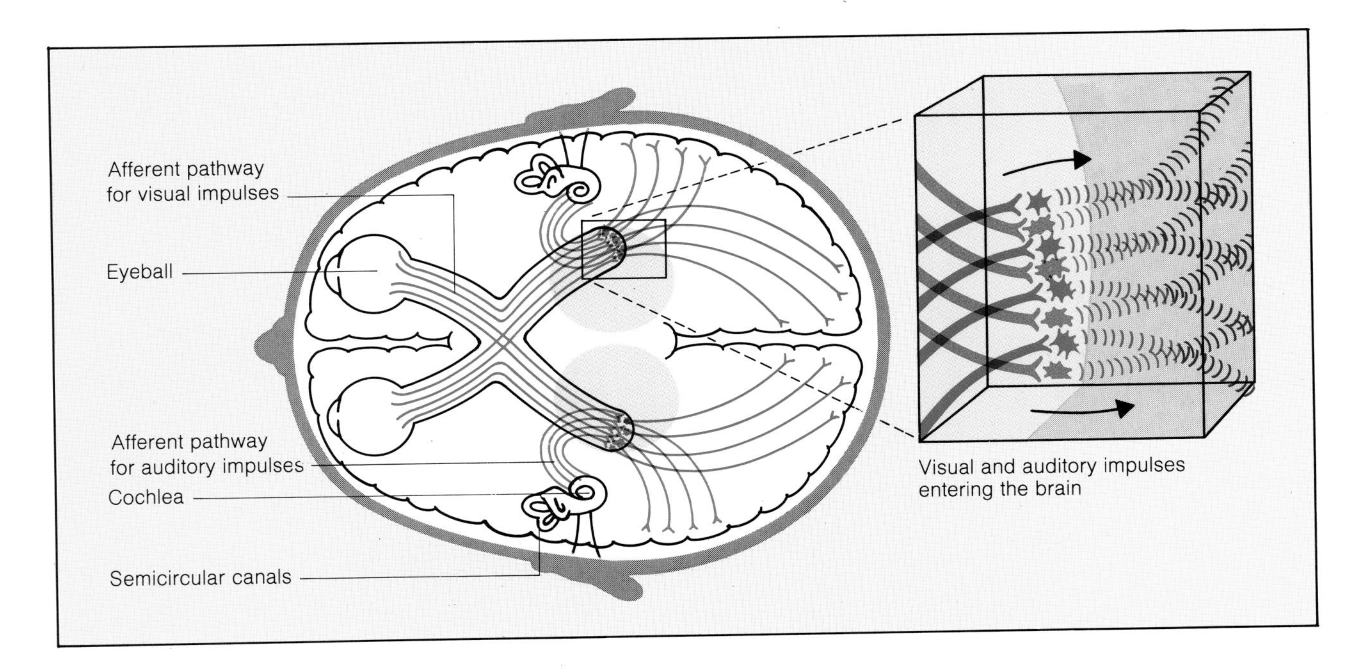

cut sample. Under the microscope there are six layers, and their relative thicknesses vary from one part of the cortex to another. In total, no less than 52 different cortical areas have been distinguished on the basis of laminar pattern.

The various types of neuron in the cortex – the most common are pyramidal, stellate and spindle cells – are generally found within their own characteristic layer, although this is by no means a fixed rule. Some idea of the microscopic density of the cortex can be gained from the fact that a sample just one twenty-fifth of an inch square, and a tenth of an inch thick, contains about 60,000 neurons. The accumulated length of the fibers totals an astonishing 10,000 miles per cubic inch.

Establishment of memories has been likened to a hologram's ability to three-dimensionally store millions of items of information. For instance, sights and sounds are received and interpreted in different parts of the brain. But when an event is recalled, the images and smells are remembered simultaneously, suggesting that their memories are somehow stored together.

Mapping human feeling

Although scientists can count the numbers, and describe the types, of neurons in the cortex, they cannot yet be linked in a coordinated scheme of interaction. Away from the microscope, understanding of the various surface regions of the cortex is much more precise.

The sensory cortex, spanning the front of the parietal lobe, receives information about temperature, pressure and pain from all parts of the body. Because this information is confined to the uniform sense of touch (even from inside the body), and excludes input from specialized sense organs, it is also referred to as the "somasensory" region (from the Greek word "soma" meaning "body"). Painstaking experimentation has shown that information from particular parts of the body is received at very localized areas of the sensory cortex. Mapping these areas has also shown that the most sensitive parts of the body (containing the most nerve endings), such as the lips and fingers, are allocated a proportionally greater area of cortex than, say, the arms or legs.

Running across the frontal lobe, parallel to the sensory cortex, is the motor cortex. Massed within its folds are batteries of motor neurons ready to fire, initiating every movement from a handshake to a shrug. The motor cortex also exhibits the same high degree of localization and, not surprisingly, the distribution of body areas closely follows that of the sensory cortex. The area devoted to the hands and fingers reflects man's high degree of manual dexterity, and the proportion controlling the mouth and lips indicates the importance of the spoken word. Because the motor cortex must also take account of information from other parts of the brain, especially from the visual cortex at the back of the occipital lobe, it is served by a correspondingly high number of fiber tracts.

There are approximately ten billion neurons in the brain and at least a million miles of fibers. A single incoming projection fiber may serve an area of cortex containing up to 5,000 neurons, each of which can potentially interconnect with up to 4,000 others. The number of possible interconnections thus runs into trillions.

The simple task of threading a needle, which involves precise motor coordination between two hands and constant monitoring by the eyes, probably involves most parts of the brain. A detailed description of exactly what happens lies many years in the future, if ever. At present knowledge is limited to an understanding of how neurons communicate, and how a single pathway may be selected from the potential billions.

Within the tangled web

The many types of brain cell, from the relatively massive pyramidal cells to the tiny stellate cells, generally share the same features as other nerve cells: a cell body, dendrites and an axon. The dendrites branch out from the cell body and form the cell's

"What is your name?" is a common enough question, and although you do not have to think about it, providing the answer involves a sequence of almost instantaneous interconnected events in the brain. First, the question is registered in the primary auditory area and structured in the neighboring Wernicke's area. The signals then pass forward, via the arcuate fasciculus, to Broca's area, which finally passes them to the motor area, which issues instructions to the muscles of the mouth and throat to speak the answer.

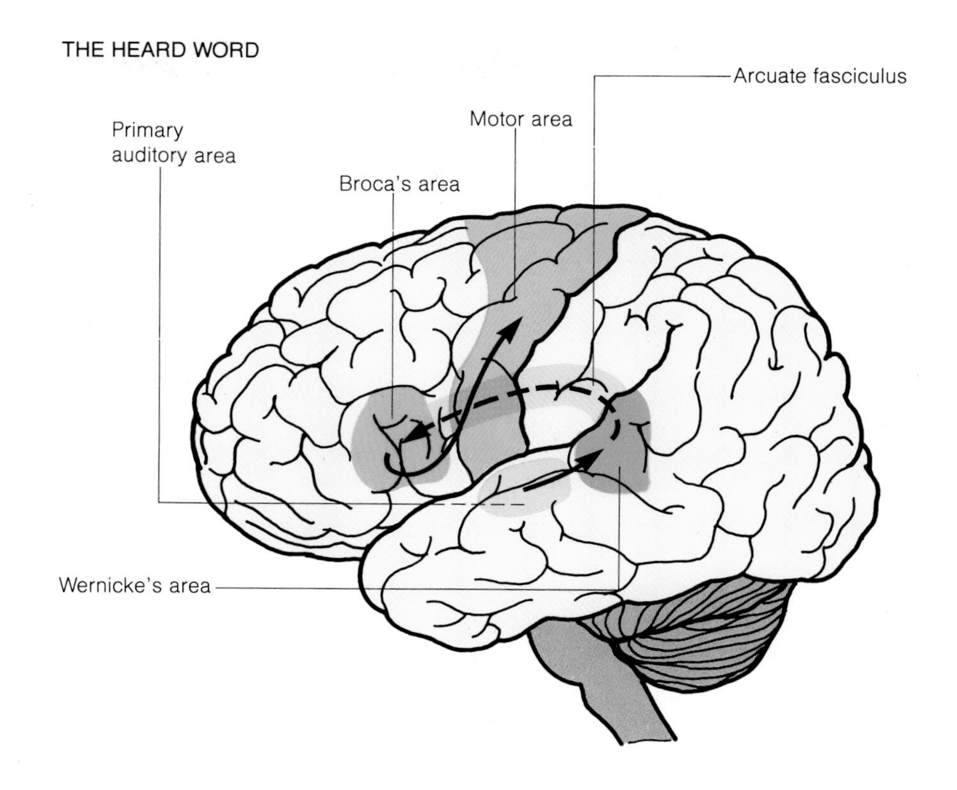

"Read this caption out loud" is a request that initiates a different sequence of events. As you scan the words, the nerve impulses triggered off the retinas in your eyes pass along the optic nerves to the primary visual area at the very back of the brain. Signals then pass for organization to the angular gyrus, before following a similar route as the auditory pathway through the arcuate fasciculus and Broca's area to the motor cortex. Once again the speech muscles are activated as you begin to speak the words.

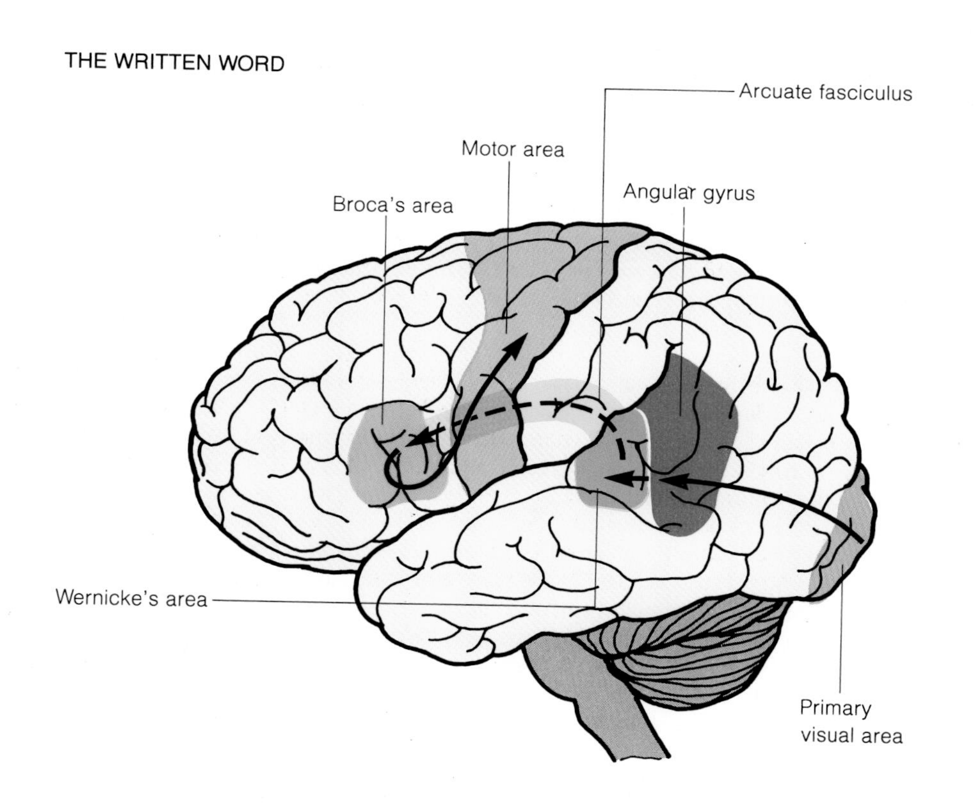

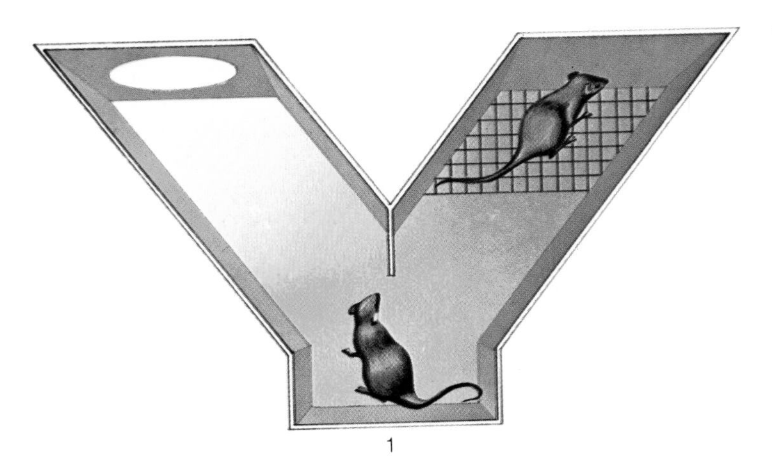

1

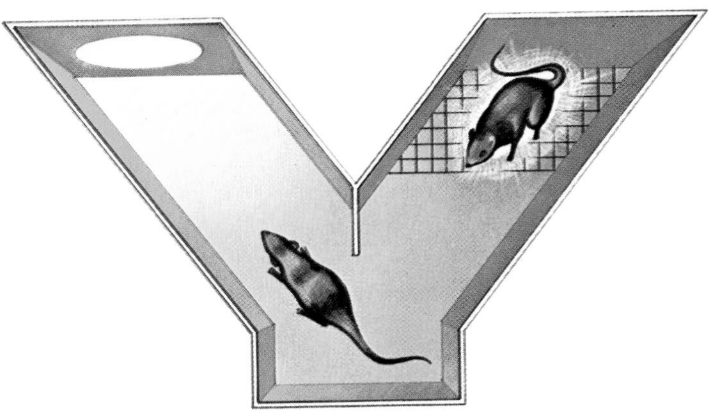

2

Memory chemicals have been proposed to account for remarkable experiments with rats. A rat is released into a Y-shaped box with light and dark corners (1), and the nocturnal rat naturally chooses the dark one. But given a discouraging electric shock (2), it soon learns to prefer the light corner (3). The learning experience is linked to the presence of a particular chemical in the brain. If an untrained rat is injected with a synthetic version of the same chemical (4), it shows an immediate preference for the light corner. This suggests that memories can be stored as chemicals.

receiving apparatus; the axon, a single long fiber, serves as a transmitter. The junction between the axon of one cell and a dendrite of another, across which nerve impulses travel, is known as a synapse. A typical brain cell has between one thousand and ten thousand synapses, tiny crossroads for the electrochemical traffic.

When one of a cell's dendrites receives an impulse, it causes a tiny wave of electricity to wash through the cell, causing it to "fire." The electrical charge flows down the axon, which terminates in a number of filaments, each tipped with a rounded terminal button. These buttons contain sacs called synaptic vesicles. Triggered by the electrical charge, they burst open and release chemical messengers, or neurotransmitters, which flow across the synapse and are picked up by receptors on a dendrite. This sparks electrical activity in the receiving cell, and so on, until the nerve impulse reaches its destination. After passing on the message, the neurotransmitter is either destroyed by enzymes or recaptured by the synaptic vesicles and stored for further use.

Not every cell contacted by neutrotransmitters fires in this way. To avoid a tangled web of confusion in the brain, many synapses are inhibitory – they prevent the receiving cell from firing. Synapses that cause the cell to fire are termed excitatory, and a constant interplay between excitatory and inhibitory synapses determines whether a particular neuron will fire.

Brain without pain

In recent years, researchers have discovered nearly 30 different chemicals that act as neurotransmitters, each designed to convey a different type of information. Perhaps the most exciting discovery may someday solve the problem of pain.

The juice of the opium poppy, especially when processed into morphine and heroin, is a powerful painkiller known from ancient times. In 1973, scientists at Johns Hopkins University discovered that opium binds to specific sites, known as opiate receptors, on nerves in the brain and spinal column. Once locked in, the drug slows down the rate at which the cell fires. As fewer signals pass along the nerves, the brain feels less pain.

But why does the central nervous system possess specific receptors for the juice of a particular flower? The answer came two years later with the isolation of a brain chemical called enkephalin, which is concentrated in parts of the brain's limbic system. When a pain impulse enters the spinal cord, special neurons release the enkephalin, which attaches to the opiate receptors and inhibits the release of the neurotransmitters that would pass the pain signal along. Opium's painkilling action is just coincidence, and the study of enkephalin and related brain chemicals may produce powerful non-addictive painkillers.

Circuits of the mind

While the brain's chemistry is gradually yielding up its secrets, the wiring of the brain's circuits remains a matter of conjecture and theory. Advances in other branches of science and technology are constantly providing new models for explaining the movement and processing of nerve impulses within the brain.

One of the most intriguing suggests that the brain may use the principles of holography. A hologram is a three-dimensional image produced by splitting the beam of a laser with a mirror. Half the laser light is aimed directly at a photographic film; the other half bounces off the object to be recorded before hitting the same film. The film does not record the image itself, but stores the interference patterns produced by the overlap of the two sets of light waves.

To reconstruct the image, a laser is directed at the film at the same angle as the original laser. By rotating the film, so that the laser strikes at different angles, many separate images can be recorded on the same plate. Even more remarkable is the fact that a corner broken off the plate can be used to reconstruct all the complete images stored on the plate.

Neurophysiologist Karl Pribram of Stanford University realized that the hologram's ability to reconstruct the whole image from a small part could explain some puzzling experimental results left over from the 1920s. It had been demonstrated that rats trained to run mazes and seek rewards could continue to perform these complex tasks even after major nerve pathways in their brains had been cut and large portions of cortex removed. This suggested that individual memories were not confined to specific points in the brain, but were dispersed throughout the

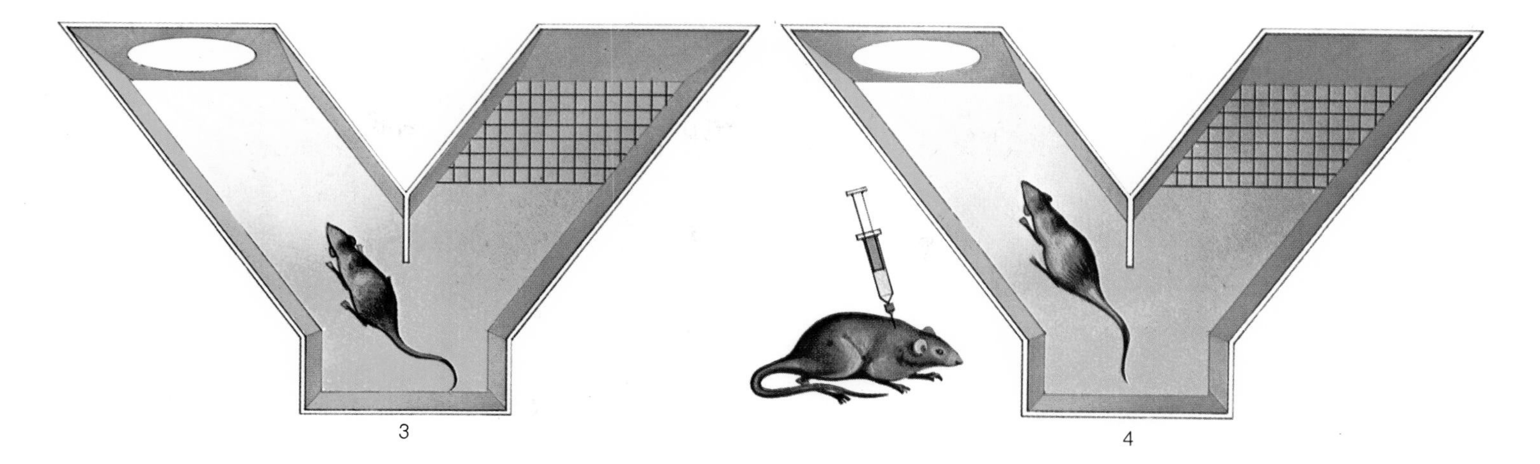

whole. Applying the new science of holography, Pribram theorized that individual memories are not encoded in patterns of neurons, but in chemical patterns that cause changes in billions of synapses. Each synapse may, therefore, retain millions of memories.

Holographic organization of the brain could explain how both the sights and sounds of an event, recorded by separate sense organs, can be recalled simultaneously. Similarly, two separate events can be recalled while all the intervening memories are ignored. However, it is just one theory among several.

Lost for words

Some functions of the brain are definitely located in specific regions. The left hemisphere skills of language and speech are controlled by three distinct centers located in the cortex of that hemisphere. Two of these centers are named after the 19th-century scientists who discovered them.

In 1861, Parisian surgeon Paul Broca performed an autopsy on a patient whose single utterance for 21 years had been the meaningless sound "tan." He could understand what was said to him, but could reply only with gestures and "tan, tan." The autopsy revealed that Tan (as the man was inevitably named) was missing a piece of tissue about the size of a hen's egg from the lower left region of the frontal lobe. This is now known as Broca's area. Situated next to the part of the motor cortex that controls the lips, tongue and vocal cords, Broca's area controls the flow of words from the brain to the mouth. Every minute up to 300 perfectly synchronized words can be spoken.

When Broca's area is damaged, typically by a stroke, a person is literally at a loss for words, though generally not as badly as poor Tan. He or she speaks in a characteristic telegraphic manner, as though some words remain trapped in the brain while others manage to squeeze out: "some . . . trapped . . . while . . . squeeze."

Listening in

Located at the top of the temporal lobe, Wernicke's area enables the human brain to comprehend speech. When words are heard, the sounds pass to the auditory area of the cortex. They then flash in neurological code to the adjacent Wernicke's area, where they are unscrambled into understandable patterns of words. If the words are to be repeated out loud, the patterns are transmitted to Broca's area along a bundle of nerve fibers called the arcuate fasciculus. Damage to Wernicke's area results in language that has the normal rhythms of speech but which contains meaningless syllables: "I'm a demaploze, I know my tugaloys."

The third language center in the left hemisphere is the angular gyrus, located between Wernicke's area and the visual cortex at the back of the occipital lobe. The angular gyrus bridges the gap between the speech you hear and the language you read and write. It transforms the visual image of the words that are read into sounds which the brain interprets, and vice versa when you write down a piece of dictated speech.

The language centers of the brain coordinate the skills of speaking, reading and writing; the words you use, their meanings and the rules of grammar that govern their use, are stored inside your memory. Of all the brain's marvels it is memory which enables you to learn, and which is responsible for human intelligence and awareness. The ability to compare past experience with the present, to retain information from one day to the next, gives humans their sense of time and provides each individual with a unique sense of identity based on his or her own particular memories.

Human memory remains an enduring mystery. Each successive technological advance in extrasomatic (outside the body) information storage has been used as an explanatory metaphor, but none of these analogies has been entirely satisfactory. Memory has been compared to a library, to a computer's magnetic disk store, and to the holographic storage systems of the future. One cubic centimeter of a holographic plate (an area about a quarter inch square) can hold ten billion bits of information. The only known system for storing information that is more sophisticated than a hologram is the human brain.

Sifting and sorting

Before an event can be stored in memory, it must be experienced by the brain. All the information flowing in from the senses is retained very briefly by a mechanism known as the sensory register. The image of a painting briefly haunts the retina after the eye is closed; the sound of a musical note echoes in the mind whan the musician pauses for breath.

These impressions are held in the register for about two seconds while the brain decides whether it wants to keep them or not. Most are allowed to drift away like smoke on the breeze to make room for the continuing flood of new information; those that the brain chooses to retain are diverted into the short-term memory. This is the brain's notepad where information is jotted down while the brain decides what to do with it. Short-term

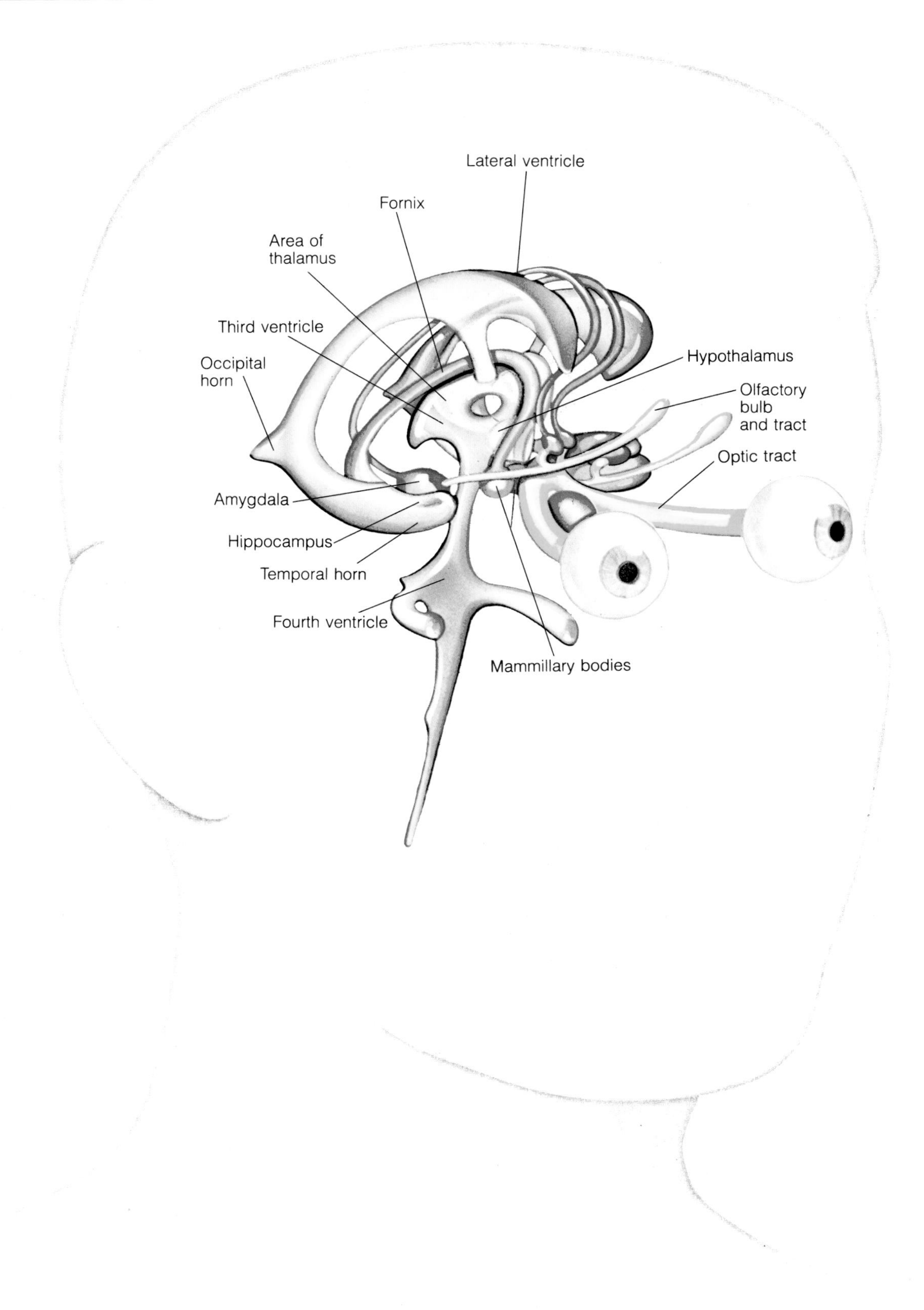

The inner structures of the brain curve around the central third ventricle, and are reception centers for the key senses of sight and smell.

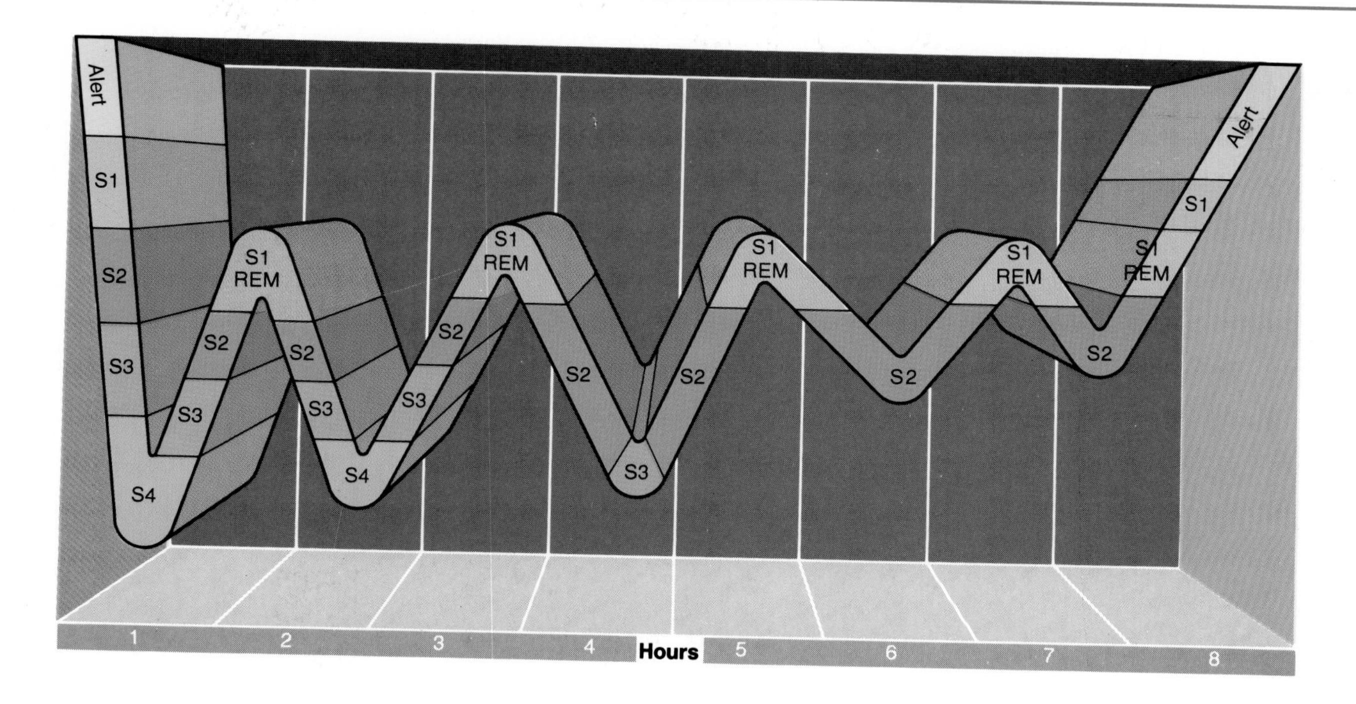

The ups and downs of sleep vary on a 90-minute cycle during the night. Of the four levels shown, S4 is the deepest sleep. S1 corresponds to shallow sleep, during which rapid eye movements (REM) accompany periods of dreaming.

memory is quite strictly limited and, unless a conscious effort is made to remember, information is held for only about a minute.

On average, just seven separate items can be retained at any given time. If a new piece of information comes in, the oldest entry is erased to make room. After looking up a telephone number, we say it to ourselves before dialing. This rehearsal concentrates our awareness and stops other information from entering the short-term memory and erasing the number before we can make the call. Short-term memory appears to be predominantly verbal, it is the sound of the numbers that we remember, not the visual symbols.

Picking up clues

Long-term memory, which stores everything we know about ourselves and our world, seems to employ a more sophisticated system of storage based on an elaborate network of cross references, enabling the reconstruction of a memory from a single clue. Thus the scent of the perfume our grandmother used to wear can trigger a cascading kaleidoscope of sights and sounds from childhood.

The question of whether short-term and long-term memory are interconnected, or form two separate systems, has still to be resolved. Many long-term memories, where we can recall an event in its entirety – what we were wearing, what we ate, what the scenery looked like – seem to be far too elaborate and visual to have passed through the limited abilities of the verbally orientated short-term memory. One theory that links the two suggests that all memory is based on patterns of neurons.

Every sensory impulse received by the brain causes a closed loop, or circuit, of neurons to fire. For an instant of time this circuit is frozen in the "on" position and the brain can perceive it. Before it fades, making way for successive loops of illumination, the circuit may create a structural trace or engram which can form the basis of long-term memory.

Searching for the invisible

Numerous experiments have shown that these engrams of memory do not reside in any particular structure or part of the brain; memory is everywhere. The association cortex, the silent majority that has no discernible function, has been put forward as a convenient location, but one that merely conceals our current ignorance.

Nor do we yet understand the nature of the engrams, although progress is being made. In the 1970s, Georges Ungar of the Baylor College of Medicine in Houston discovered that certain types of memory may be stored as chemicals inside the brain. Rats, which are normally nocturnal, were trained to avoid the dark corners of a specially designed box. Ungar then analyzed extracts from their brains and found a new chemical which he named scotophobin, meaning "fear of darkness." Untrained rats injected with a synthetic form of scotophobin showed the same aversion to the dark corners. These findings suggested that a specific chemical might be linked to every learned skill.

However, chemistry does not explain how we remember what color blouse Aunt Aggie wore last birthday. A more influential theory proposes that memories are stored in the nerve pathways themselves. The electrical activity caused by a loop of neurons firing off in response to a sensory impulse may alter the synaptic connections of those neurons. When the short-lived electrical activity dies down, the new connections remain, creating a nerve network that stores a specific memory. Activating one or two neurons in the chain triggers the loop back into illumination, thereby bringing the memory back to mind.

CHAPTER 14

The Senses

You can only know of the existence of an object – we can only *prove* it is there – because we can see it, hear it, taste it, smell it or feel it. You would be totally unaware of the presence of other people – or indeed of anything else in the world around you – if it were not for your senses.

Ceaselessly gathering information; your eyes, ears, tongue, nose and sense of touch provide a rich flow of signals to the brain, which interprets them and gives reality to the objects that stimulate them. And what powerful and emotive sensations they can be! Aided by memory, the sound of a songbird and the smell of a rose can conjure up a pleasant image of a country garden. On the other hand, even just the sight of blood might make you faint or vomit.

Sight and hearing are generally regarded as the most vital of the senses – the loss of either is considered a real calamity, placing significant restrictions on any person's experiences and way of life. The senses of taste and smell are usually allocated an inferior role, though this perception grossly understates their true importance.

Sensors for the basic sensations of touch and temperature are distributed all over the body, so that you can feel a fur rug with your toes or test the temperature of the baby's bath water with your elbow. The more specialized sense organs for sight, hearing, taste and smell are concentrated in the head. This comparatively exposed position is not without its risks, so most of the actual sensor cells are well protected by the bony structure of the skull – which also protects the all-important brain. Also, having the eyes, ears, nose and mouth on the head brings with it the advantages of mobility, making it easy to direct sensory attention to anything of special interest – or toward any potential danger. It also allows short, fast, nervous connections between the sensors themselves and the controlling and analyzing brain.

In fact, the sense organs are, in reality, elaborate and highly sensitive extensions of the central nervous system. All of the sensations detected by them trigger minute electrical impulses which travel along direct nervous pathways to the brain. Once there they are coordinated and processed to give an ever changing mental image of the world around us, stimulating a wide variety of conscious and unconscious responses.

A window on the world, the human eye collects the images that keep you in touch with reality. This view from inside the eye shows the lens and the adjustable aperture of the pupil. Light is focused by the lens onto the retina, where light-sensitive receptors trigger electrical nerve impulses to the brain. There the impulses are sorted and perceived as images. For correct functioning, therefore, the sense of vision needs the eyes and the brain.

The richness of vision

Sight is by far the richest of our senses, and it accounts for around three-quarters of our perceptions. With our eyes open, sensations come flooding in and are carried directly to the brain along the optic nerve. In principle the eye is like a camera: there is a lens system at the front to collect and focus light rays; the iris acts like a camera's aperture control; and the retina corresponds to the film which captures the images. There are even lens caps – the eyelids! But there is one big difference – unlike photographic film the retina can be used time and time again, continually capturing

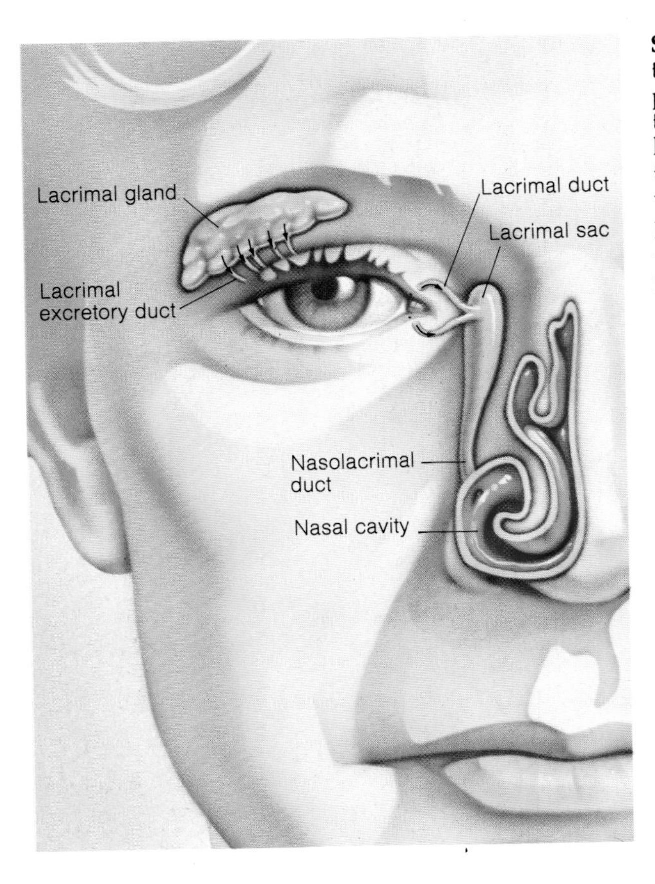

Stimulated by sorrow, joy or the emotionless accident of a piece of grit under the eyelid, tears are produced by the lacrimal gland. They flow over the surface of the eyeball, washing away germs and particles of dirt. Tears then pass along the lacrimal ducts and drain into the lacrimal sac, before draining into the nasal cavity by the nasolacrimal duct. When the eyes are watering copiously, the salty tears may be tasted in the throat.

Like well-disciplined troops on parade, the rods and cones line up in serried ranks along the inner layer of the retina. These are the receptor cells that trigger nerve impulses to the brain. But before light can reach them, it has to pass through layers of ganglion and bipolar cells. Each ganglion cell links with one or two bipolars which connect, in turn, to the photoreceptors. The inset shows how a lack of vitamin A in the diet causes receptors to wither and die.

images at a rate of ten per second throughout its working life.

For safety the eyes are cradled deep inside bony bowls in the skull. Lining the eye sockets there is a fatty layer that cushions shocks and gives a highly lubricated surface for the continual swiveling of the globular eye. Six muscles direct this movement and anchor the eyeball securely in place.

Automatic responses provide further protection for the eye and the vulnerable associated nervous system. Sudden flashes of bright light, loud noises, rapid movements near the eye, and even a particle of grit trapped by the eyelashes trigger an immediate response as the eyelids slam shut. When closed the tough fibrous plates of the eyelids form a waterproof and airtight shield over each eye.

Sealing the eye opening from lid to lid, and lining the eyelids themselves, is the conjunctiva, a transparent membrane that catches anything that gets past the first line of defenses. It is kept moist by a thin layer of slightly oily tears produced by the lachrymal glands above the eye and spread by the blinking action of the eyelids. Irritation, or the presence of a foreign body in the eye, makes tears flow – a reaction that can also be stimulated by mirth or misery. Tears contain a mild antibacterial agent, lysozyme, which provides the eyes with additional protection against harmful bacteria in the air.

Focusing the image

The white outer layer of the eyeball, the sclera, has a transparent circular segment that bulges out at the front to let light in. This is the cornea, which bends the incoming light and directs it toward the center of the eye. The bulging shape of the cornea is a precise curve which flattens out at the edges to minimize aberrations that would otherwise distort the image.

From the outside the sclera appears an opaque white, but its inner surface has a large number of blood vessels; this choroid layer is the main way in which essential nutrients are supplied to the eye. The choroid also contains a layer of the dark pigment melanin, which absorbs excess light. It also ensures that the interior of the eye remains dark, enabling the incoming light rays to give a good image on the retina – exactly as in a camera.

The cornea is also living tissue and to survive and work properly has to have a continuous supply of oxygen and nutrients. Other body tissues obtain these vital requirements through the blood. But in the cornea, blood vessels would interfere with the passage of light. So instead it gets most of its oxygen directly from the air, absorbing it through the tears. Nourishment comes via the aqueous humor, a watery transparent fluid that fills the space behind the cornea.

Controlling light intensity

Immediately behind the cornea, the choroid layer takes on an additional function; it forms the iris, a colored muscular diaphragm that surrounds the black central opening of the pupil. This precise ring of muscle constantly contracts and expands, narrowing and enlarging the size of the pupil to control the amount of light entering the eye. In bright conditions it can narrow to as small as six hundredths of an inch in diameter, while in the dark it may open up to as much as a third of an inch, gathering in whatever light is available. The action is fully automatic and triggered by the intensity of the light reaching the retina, but powerful emotions such as anger and fear, and certain drugs, can also affect the degree of iris opening.

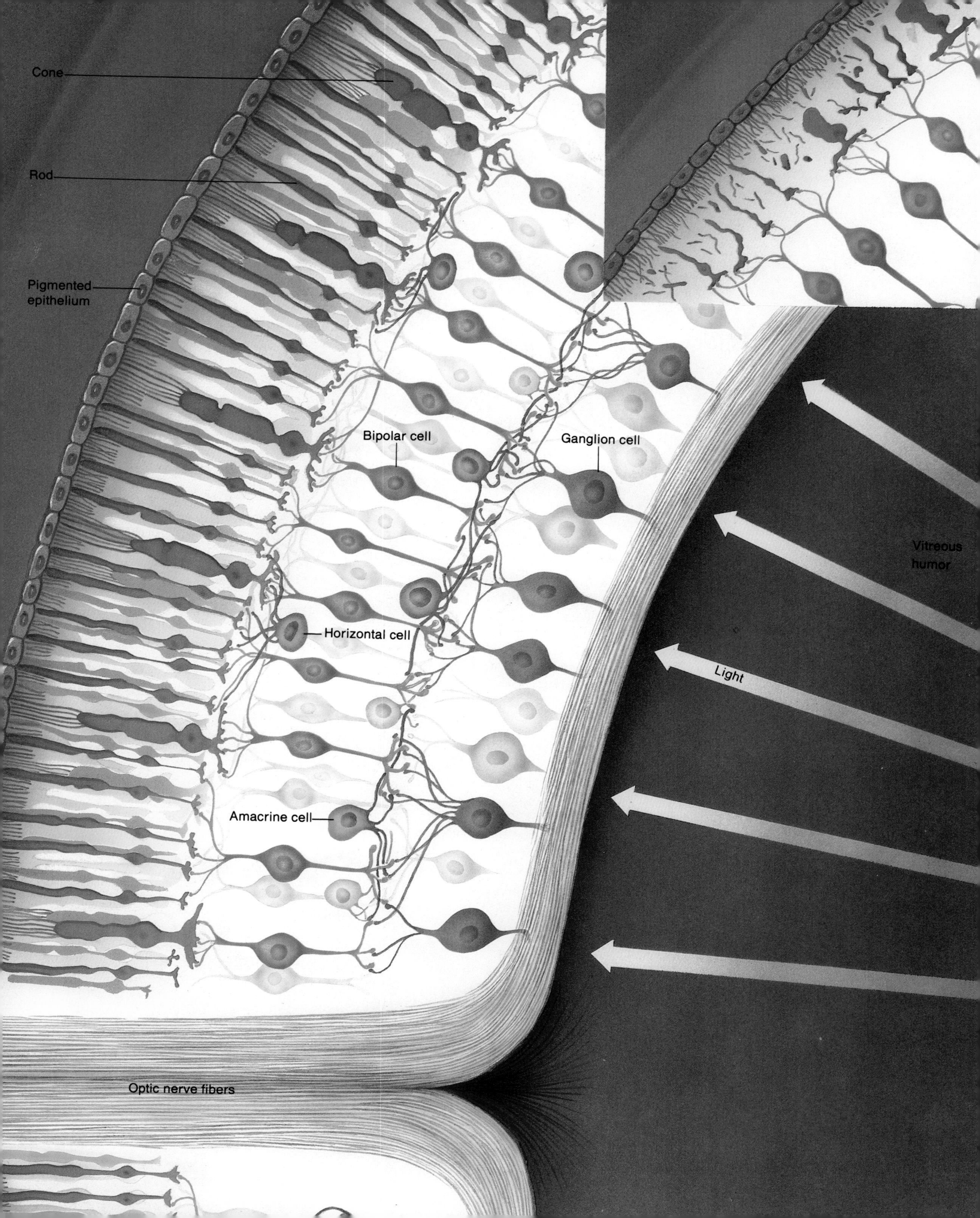
Cone
Rod
Pigmented epithelium
Bipolar cell
Ganglion cell
Vitreous humor
Light
Horizontal cell
Amacrine cell
Optic nerve fibers

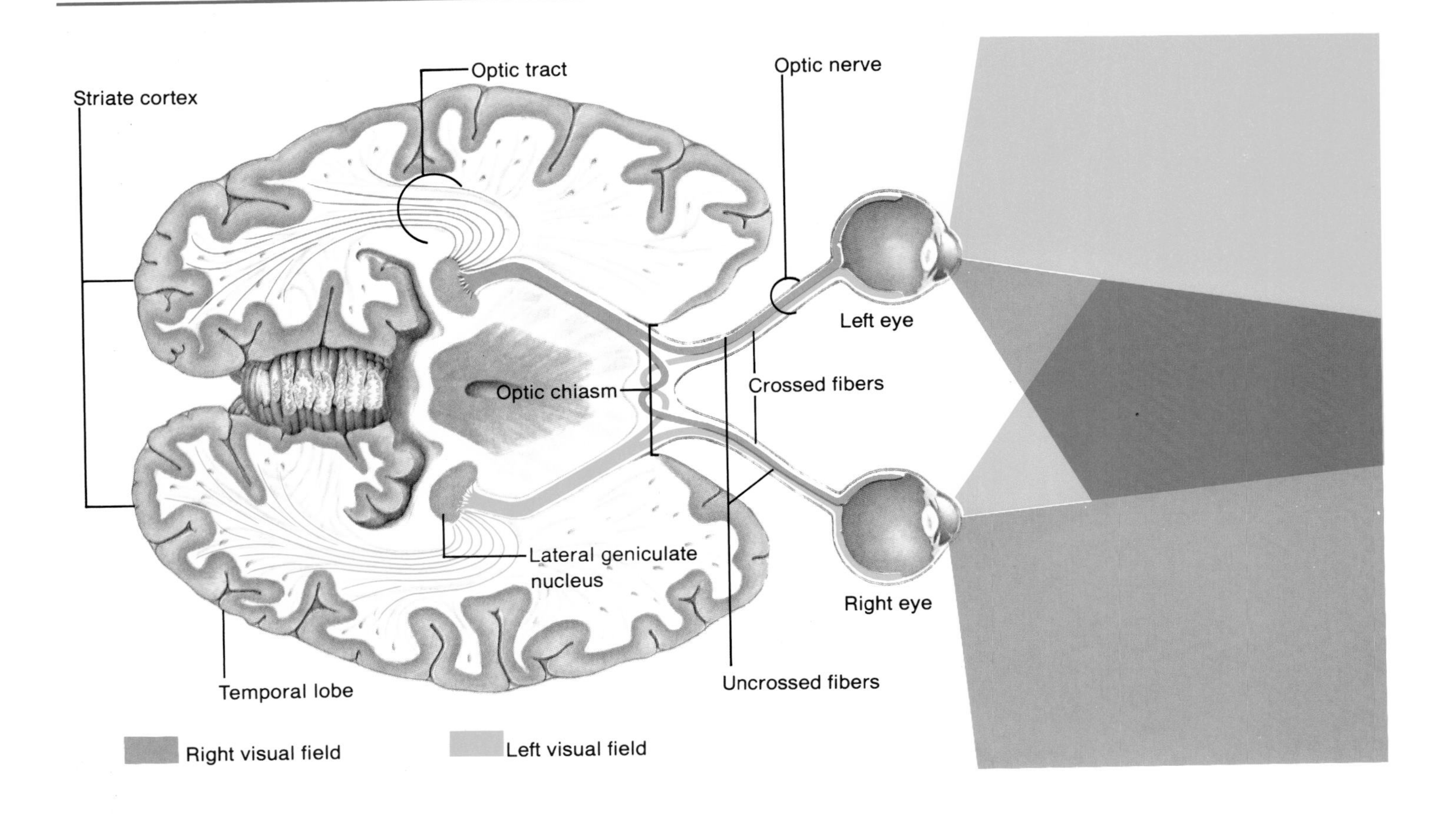

The iris is the colored part of the eye, and the complete range of colors found in people's eyes is produced by different amounts of the same pigment – the melanin found in the rest of the choroid layer. Blue and green eyes have a minimal amount of melanin, with the amount increasing to give the other colors. Most newborn babies have blue eyes because at first the melanin is concealed in the iris; after a few months it moves to the surface and the eyes take on their final color.

Fine image adjustment

The pupil is the front surface of the eye's lens; it always looks black because it is an opening that leads into the dark center of the eye. Incoming light is bent toward the center by the cornea, and then finely focused by the lens to give a sharp image on the light-sensitive surface of the retina. With a camera, focusing is achieved by moving the lens elements backward or forward, but the eye achieves the same result by the much more elegant process of changing the shape of the lens. This is the process of accommodation.

The lens is held in place by 70 fine ligaments – the ciliary zonule – which radiate from its edges like the gossamer strands that support a spider's web. The ligaments are, in their turn, attached to the choroid by the circular ciliary muscle. When the eye is at rest, the ciliary zonule pulls on the lens and flattens it. In this state the lens brings into sharp focus all objects that are more than about 20 feet away.

For nearer objects the lens has to take up a more rounded shape, and to achieve this the ciliary muscle contracts inward, relieving the tension on the ligaments so that the lens bulges into a more spherical shape. The amount of accommodation is

The sensory organs for vision – the eyes – are at the front of the head *(above)* but the actual visual sense is provided by areas of the brain at the back and sides *(below)*. Nerve impulses generated by rods and cones in the retinas of the eyes travel along the optic nerves to the optic chiasma, where they partially cross over. "Mixed" impulses from both eyes then pass via the optic tracts to the striate cortex at the back of the brain, before terminating in the temporal lobe vision area. In this way, right and left halves of the visual field merge. A blow to the back of the head jolts the striate cortex, sending false signals to the temporal lobes and making us "see stars."

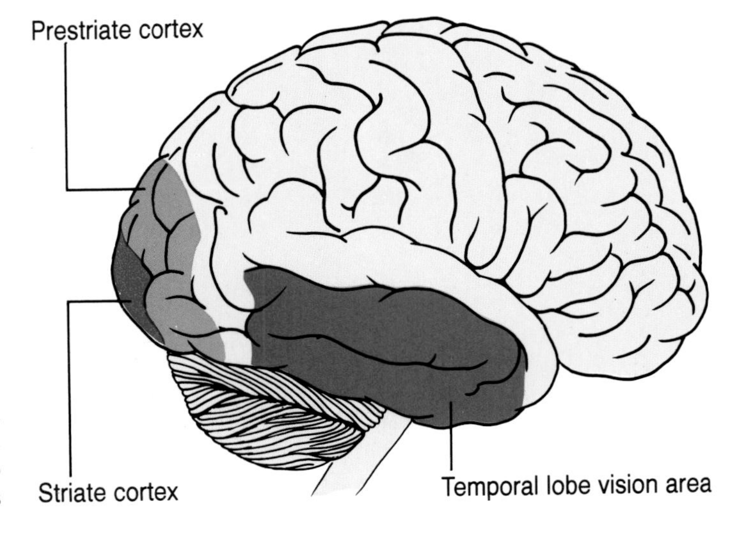

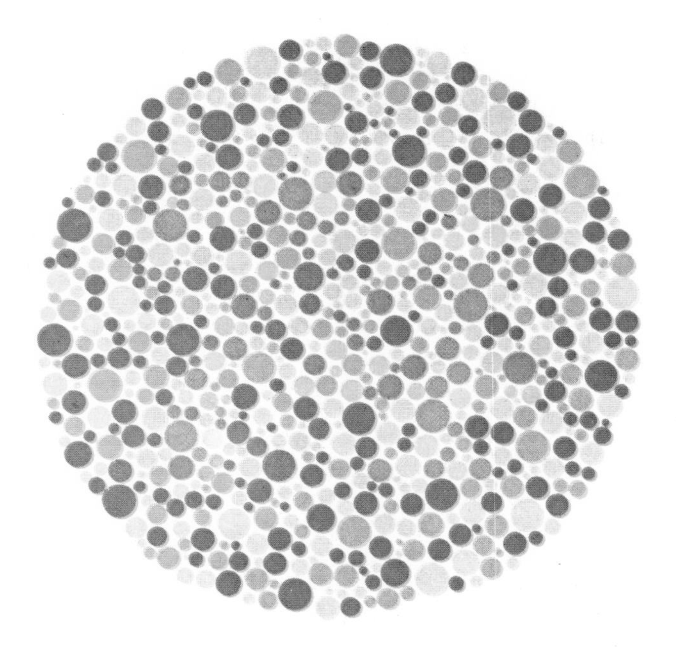

One person in 20 is color blind. In the most common form, red–green color blindness, the retina lacks cones that detect red or green. Somebody without green cones has difficulty in seeing the number 74 among the dots in the above chart.

limited, and a normal eye is unable to focus much closer than six or seven inches in front of the lens, a distance known as the eye's near point.

The lens is made up of a series of layers, like those in an onion. Each layer refracts the light slightly to give a smooth and gradual overall effect. More layers are added throughout life, and as they accumulate the older ones pack together to form a hard core. This rigid core reduces the flexibility of the lens, so that the amount of accommodation gradually reduces and the near point moves farther from the eye as we grow older.

The back of the lens is around one third of the way to the retina; the light rays make the rest of their journey through the jellylike vitreous humor filling the inside of the eyeball. Like the air in a football, the vitreous humor keeps the eyeball under pressure and so maintains its spherical shape.

Cones and rods for energy conversion

Finally the light rays reach the retina – the film of the eye's camera – and it is here that light energy is converted into electrical nerve signals. Crisscrossed with blood vessels, the retina has three distinct layers of nerve cells, all of them microscopically thin. Nearest to the lens is a layer of ganglion cells, then a layer of bipolar cells, finally backed up by the photoreceptors.

It is the photoreceptors that actually process the packets of light energy or photons that impact on the retina. As a result, the light has to pass through the ganglions and bipolar cells to get to the photoreceptors. The output connections to the fibers of the optic nerve are made by the innermost layer of ganglion cells. Each ganglion connects to one or more bipolar cells of the middle layer, which transfer information from the outer photoreceptors.

There are just two types of photoreceptor cells which, because of their distinctive shapes, are known as rods and cones. Rods are sensitive enough to respond to a single photon, the basic unit of light, but collectively create only a coarse gray image, which is nevertheless adequate for seeing in poor light. Fine detail and color vision come from the cones, but they need a lot more light and work best in bright daylight. Within the human eye there are about 18 times more rods than cones, specially arranged to give the best possible combination of night and day vision.

Zone of maximum sensitivity

Right in the middle of the retina – in the direct line from the pupil – there is a yellow spot called the macula lutea. Forming a shallow hollow, it is tightly packed with small cones and right at its center is a pit – the fovea – taking up about a quarter of the size. Over this small area the ganglion and bipolar cells have been moved aside, allowing light unrestricted passage to the receptors. This makes it the area of sharpest vision – to inspect something closely we focus its image directly onto the fovea.

Away from the fovea the relative proportion of cones declines, until at the edges of the retina most of the receptors are rods. This gives good peripheral vision of faint objects but no real detail – you can see something moving in the edges of your field of vision but you must look directly at the object to see what it is. The combination of rods and cones gives the eye a tremendous range of sensitivity, allowing it to adapt to widely varying light levels. The retina needs 40 minutes or so to adjust to poor lighting, but a fully dark-adapted eye requires 30,000 times less light to activate the retina compared to its normal daylight sensitivity. The best that film manufacturers can achieve is a series of photographic emulsions that together span a sensitivity range of about 20 times.

Adjoining the macula is the area where the nerve fibers fuse together to form the optic nerve and pass out of the eyeball. Where the nerve passes through the retina there are no sensors. The result is a blind spot at which no image can be formed. Normally, however, the brain simply fills in the image for the missing area so there is no "hole" in the visual field.

The chemistry of vision

Conversion of light photons into electrical signals in the nerve cells is a chemical process, in which the crucial substance is vitamin A. Molecules of this vitamin (obtained in food from green vegetables, dairy produce and liver) are carried to the choroid of the eye and absorbed by the photoreceptors. Inside a rod, a modified form of the vitamin combines with the colorless protein molecule opsin to form rhodopsin, which spreads through the tip of the receptor cell. Rhodopsin is light sensitive, like the silver salts in photographic film. When struck by a single light photon, a molecule of rhodopsin suddenly breaks down, generating an electrical impulse – a nerve signal – in the cell.

The sensitivity of the eyes seems to depend on the amount of rhodopsin present – maximum sensitivity is achieved with high rhodopsin concentrations. This chemical technique for detecting light is almost perfectly suited to its application. Indeed the same basic process is used by all animals with image-forming eyes, even though their organs evolved in different ways.

Cone receptor cells use a similar mechanism, but in their case the vitamin A is combined with three different opsins to give three light-sensitive pigments. Each pigment is sensitive to one of the three primary colors of light – red, green and blue. Depending on the concentration of pigment, each cone is therefore particularly sensitive to light of one of the three primary colors, and any color of the rainbow can be produced by various combinations of the primary ones.

Scanning system

Each eye is held in place by three pairs of taut, elastic muscles constantly balancing each other's pull. The superior rectus acts to roll the eyeball back and up, but is opposed by the inferior rectus. Similarly, the lateral rectus pulls to the side while the medial rectus pulls toward the nose, and the two oblique muscles act to roll the eye clockwise or counterclockwise. The muscles of each eye work together to move the eyes in unison. Because of the constant tension in the muscles, they can move the eye very quickly – much faster than any other body movement.

Working together, the eye muscles can carry out no less than seven coordinated movements, which between them allow the eye to track many different kinds of moving object. The first three movements – tremor, drift and flick – are the result of the constant, opposing, muscle tension. Tremor causes an almost imperceptible trembling of a point image (like a spot of light in a dark room), while drift makes the image move slowly off-center. But before the movement becomes really noticeable there is a quick flick to bring the image central again. Though these involuntary movements might seem distracting, they make sure that the image constantly moves over fresh parts of the retina. As a result, the receptors at any spot do not get overloaded with input and effective vision is maintained.

Smooth pursuit movements are similar to drift but on a larger scale; they are used to follow moving objects. High-speed saccadic movements move the eye in sharp steps – for example from word to word when reading and then back to the beginning of the next line. Each step appears to be pre-planned by the brain, possibly so that it can blank out the blurred image produced during the rapid movement from point to point. Although most saccadic movements are voluntary, they are also triggered automatically by sudden movements close to the eye, concentrating vision on the potential threat. The rapid eye movement associated with dreaming, during so-called REM sleep, is also a saccadic action.

Not only can the eyes see in color and scan moving objects, they also see in three dimensions. Binocular vision is created by the separation of the eyes, and each eye has a slightly different view of the same object. To prevent this resulting in double vision, particularly with near objects, the sixth eye movement, vergence, comes into play. The eyes turn inward to direct the images directly onto their respective fovea. During this movement the brain assesses the amount of muscle tension and uses it to estimate the distance of the object.

Most complex of all eye movements is the vestibulo-ocular system; it works to keep the image of an object on the high-definition fovea while the head and body are moving. This action is assisted by the vestibular apparatus in the inner ear, which provides the brain with a continuous flow of information about the way in which the head is moving.

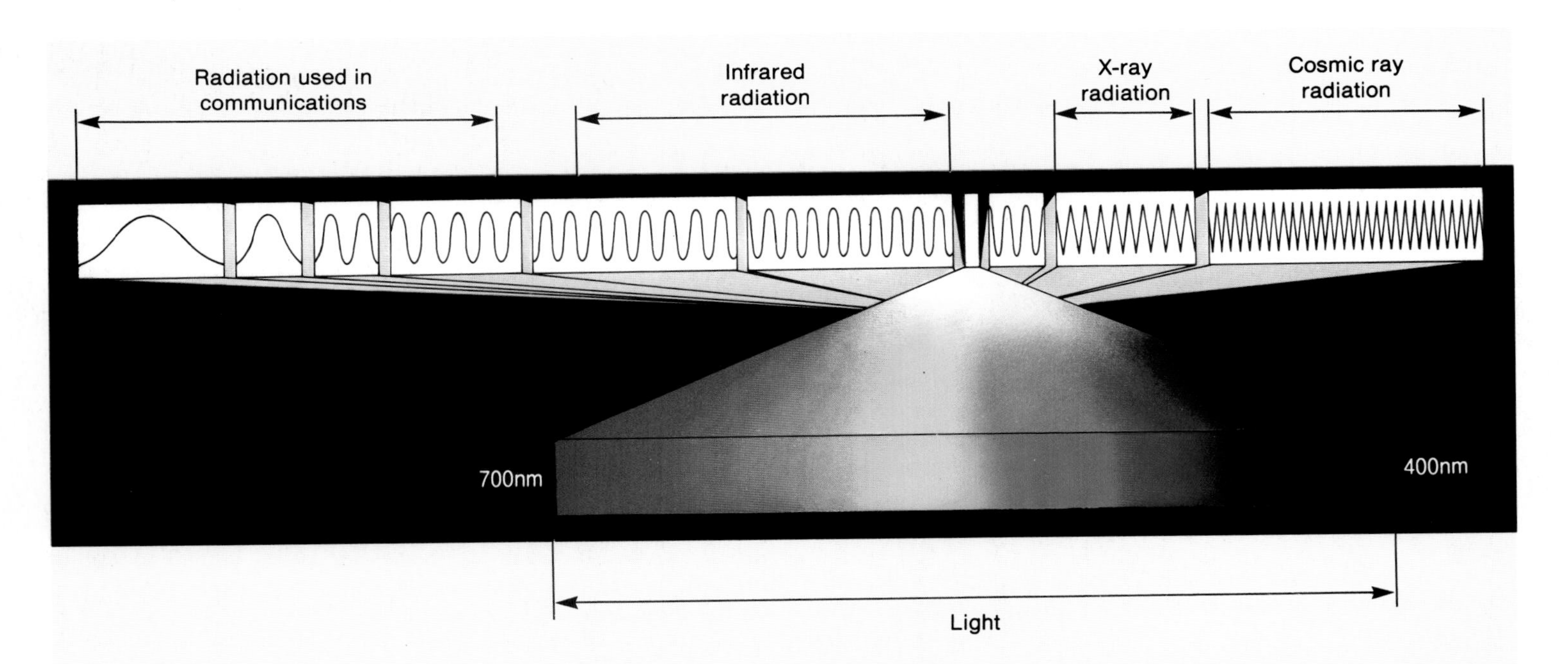

Light – essential for human vision – is just one of the various kinds of radiation that go to make up the electromagnetic spectrum. At the long wavelength end of the spectrum are radio waves and microwaves, used for radio and television broadcasting and in radar. Next comes infra-red, the invisible radiation given off by hot objects. The visible spectrum consists of the colors of the rainbow from red (with a wavelength of about 700 nanometers) to violet (400 nanometers). Beyond violet comes ultraviolet radiation, also invisible to the human eye, followed by X rays, gamma rays and finally very short wavelength cosmic rays. Some animals can "see" in infra-red or ultraviolet light.

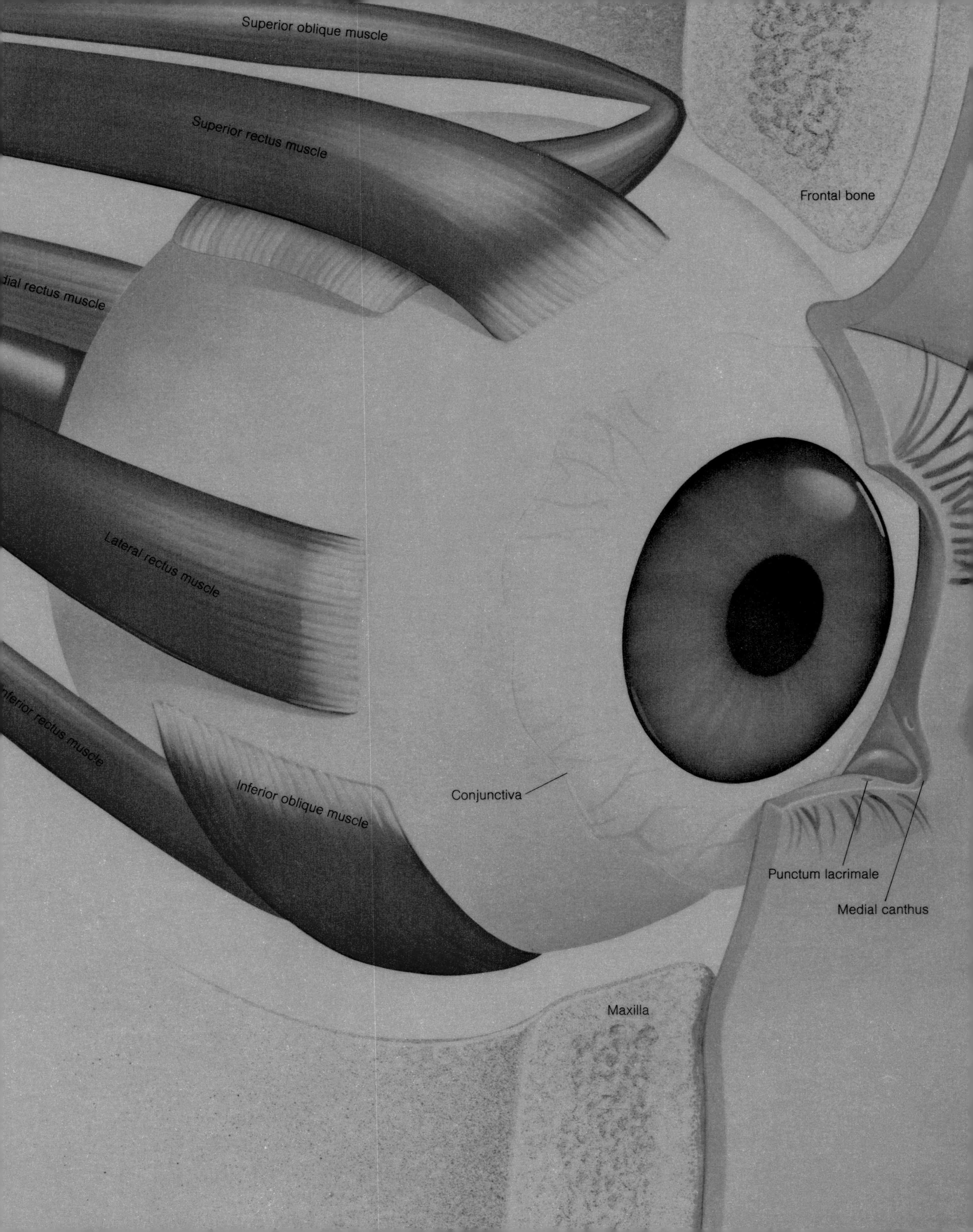

Superior oblique muscle

Superior rectus muscle

Frontal bone

dial rectus muscle

Lateral rectus muscle

nferior rectus muscle

Inferior oblique muscle

Conjunctiva

Punctum lacrimale

Medial canthus

Maxilla

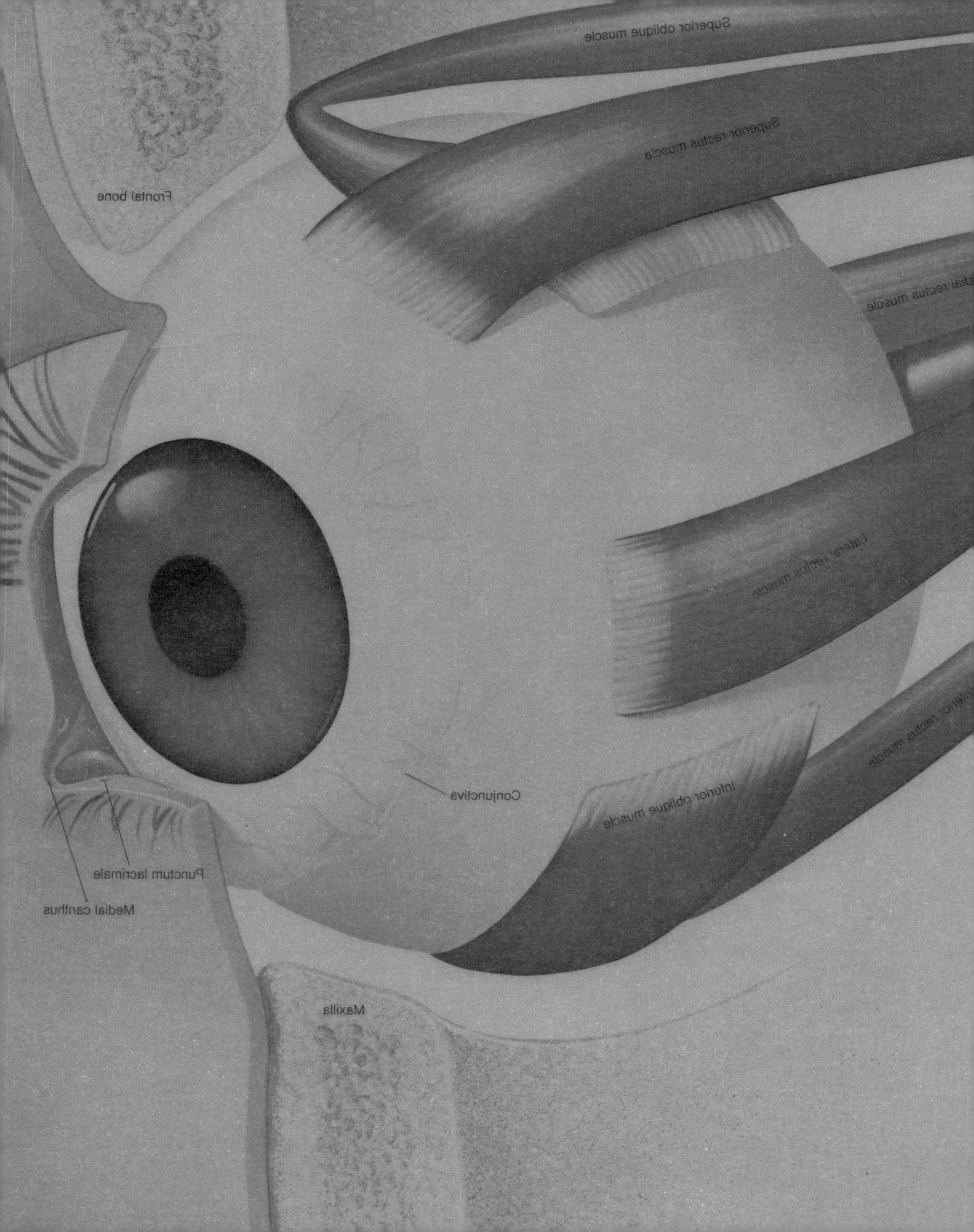
Superior oblique muscle
Superior rectus muscle
Frontal bone
dial rectus muscle
Lateral rectus muscle
nferior rectus muscle
Conjunctiva
Inferior oblique muscle
Punctum lacrimale
Medial canthus
Maxilla

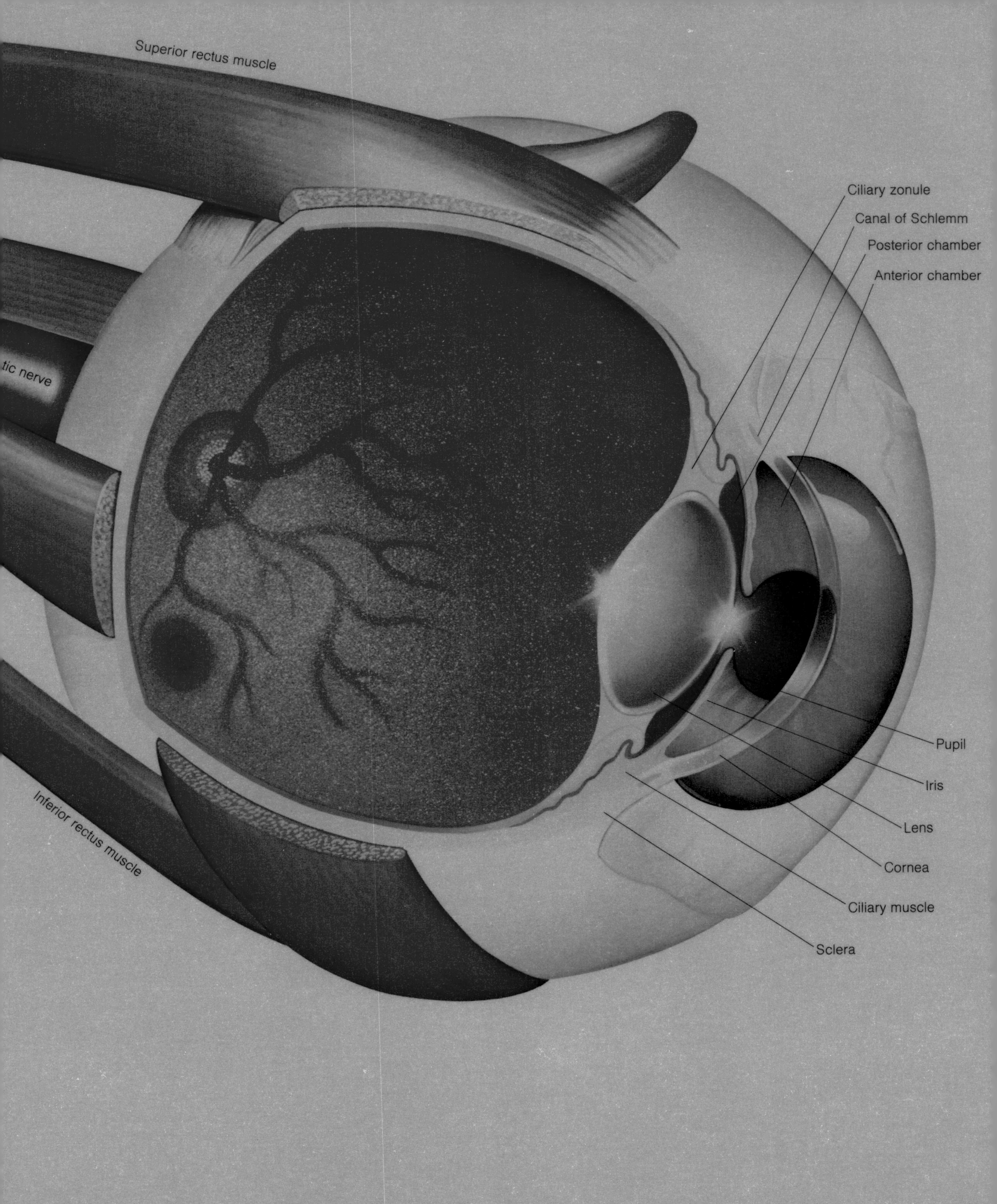
Superior rectus muscle
Ciliary zonule
Canal of Schlemm
Posterior chamber
Anterior chamber
tic nerve
Pupil
Iris
Lens
Cornea
Ciliary muscle
Sclera
Inferior rectus muscle

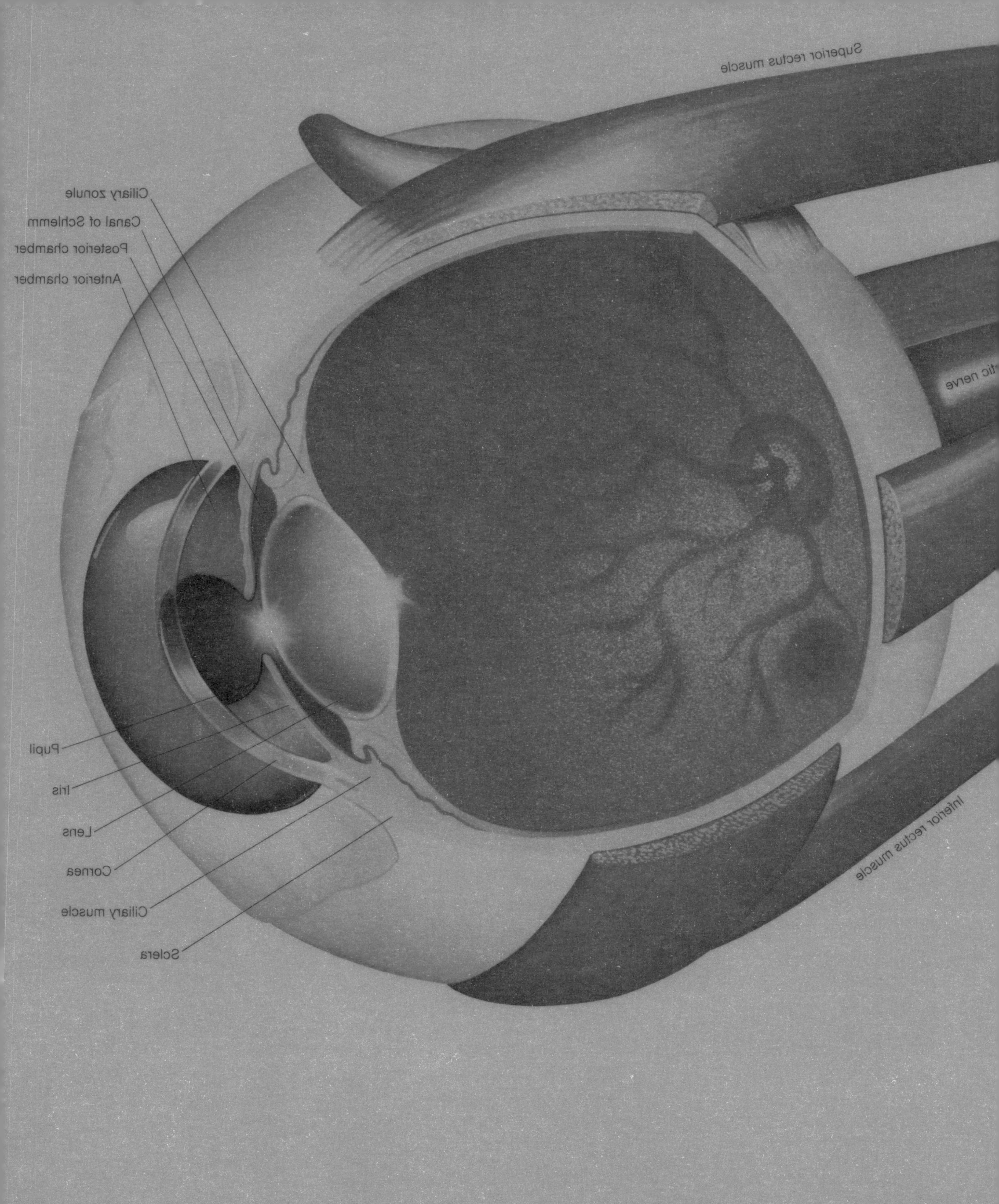
Superior rectus muscle
Ciliary zonule
Canal of Schlemm
Posterior chamber
Anterior chamber
tic nerve
Pupil
Iris
Lens
Cornea
Ciliary muscle
Sclera
Inferior rectus muscle

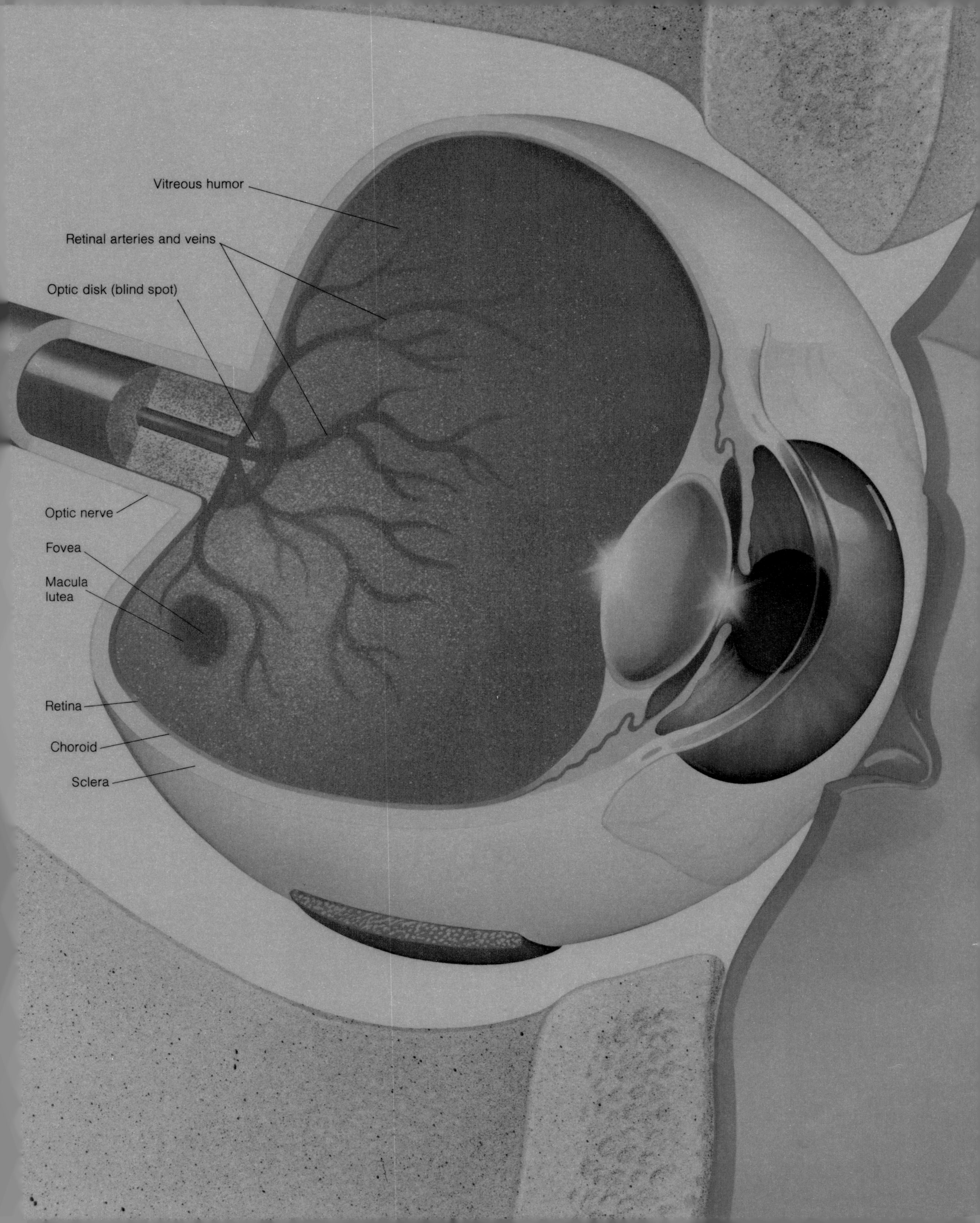

Vitreous humor
Retinal arteries and veins
Optic disk (blind spot)
Optic nerve
Fovea
Macula
lutea
Retina
Choroid
Sclera

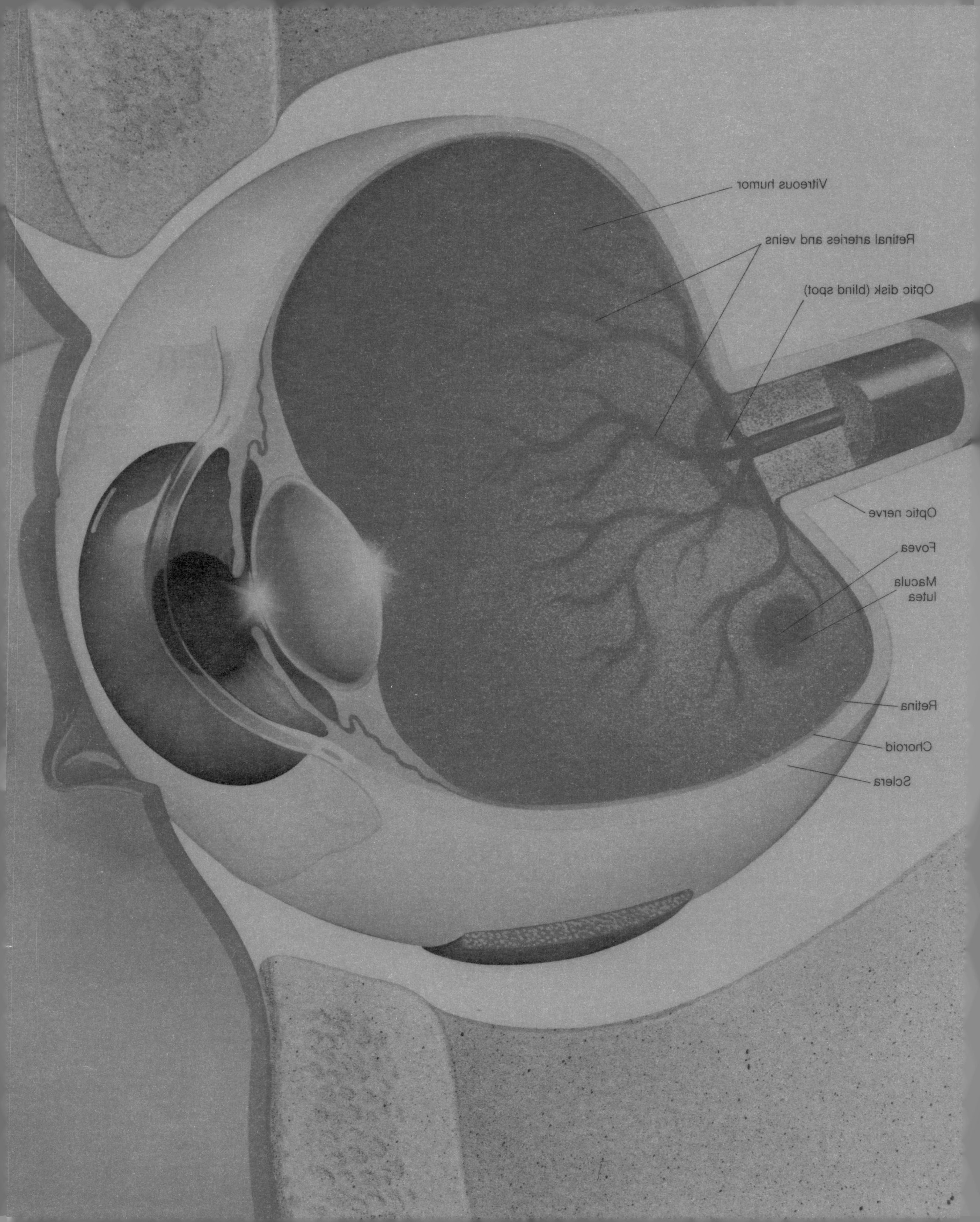

Vitreous humor
Retinal arteries and veins
Optic disk (blind spot)
Optic nerve
Fovea
Macula lutea
Retina
Choroid
Sclera

Processing the image

Capturing an image of an object – be it static or moving – and converting it to nerve signals is only part of the story. In order for us to see, the brain must analyze the signals and extract meaningful information from them. When they are stimulated, receptor cells in the retina release a chemical neurotransmitter that activates the intermediate bipolar cells. Groups of rods are connected to individual bipolar cells and it requires impulses from at least five rods to make a cell react and pass on a message to the associated ganglion. In contrast, each cone has its own bipolar and ganglion cell. For this reason, the eye takes slightly longer to respond to dim images as the system waits for sufficient impulses to arrive from the rods.

Within the paper-thin layers of the retina there are multiple nerve connections that rival the complexity of a major telephone exchange. Each ganglion is connected to one or more bipolar cells, while horizontal and amacrine cells cross link impulses to help integrate the visual messages. Nerve fibers from the ganglia collect together to form the trunk route of the optic nerve, which passes out of the eyeball and connects directly with the brain.

The two optic nerves – one from each eye – meet in the brain at a crossroads called the optic chiasma, where each nerve splits in half to form two tracts. Fibers from the inner part of each retina cross over and travel to the opposite hemisphere of the brain. As a result, signals from the right visual fields of both eyes go to the left hemisphere, and signals from the left visual fields combine and go to the right hemisphere. Although each part of the brain gets images from just half of the visual world, the brain merges them into an integrated whole. The partial crossing of the nerves at the optic chiasma also helps three-dimensional vision.

Into the visual cortex

Within the brain the impulses pass through a relay station, the lateral geniculate nucleus, then travel on to the visual cortex. This is where the signals are converted into mental images.

First stop is the primary visual cortex, or striate cortex, which recognizes the arrangement of objects in space, together with their shape and brightness. From there nerve signals project to the prestriate, or secondary visual, cortex. This region probably handles higher functions such as pattern recognition – certainly any damage to this part of the brain can make it impossible to perceive patterns. Individual segments in the cortex correspond directly to specific parts of the retina, to give point-for-point mapping of an image. The all-important images formed by the fovea get a more detailed treatment to increase visual acuity.

Feature recognition

"Seeing" an object, and recognizing it for what it is, involves image processing by cells in the retina and brain. Retinal ganglions respond best to specific contrasting features – such as a dark disk on a light background or a light disk on a dark background – and provide a first level of discrimination. Cells in the cortex also respond to light features, but in a more complex way. Three main types of cell are involved; simple, complex and hypercomplex.

Simple cells respond best to a clearly defined slit of light (or dark) or to edges between dark and light areas. However, the images have to be precisely located in the receptive field of the cell and in the proper orientation. Complex cells also respond to slits and edges, but generally react whatever the position or orientation of the image. Finally the hypercomplex cells respond best to even more specific features such as corners, angles and lines of a specific length, orientation and location. Together these cells form a hierarchy, building up from simple to more complex levels of image processing.

Some signals from the prestriate cortex pass on to the temporal lobes, near the sides of the head, which contribute to the processes of visual recognition and memory. Another section of the brain, the superior colliculus, acts to control visual attention. Objects on the periphery of vision are detected and the colliculus instructs the head and eyes to turn and so bring the object into focus inside the visual field. Another set of nerves carries signals to the cerebellum, the part of the brain responsible for muscular coordination. In this way, input from the eyes is used to regulate activities requiring keen muscular coordination, such as driving a car or catching a ball.

All possible hues can be charted on a three-dimensional color "tree." There are nine basic colors, called hues, on each level, increasing in saturation toward the edges. Brightness varies vertically, being greatest toward the top and least at the bottom. The central "trunk" of the tree represents gray, the color that results when hue and saturation are at a minimum, with light gray at the top and dark gray at the bottom.

Three-dimensional vision

We live in a solid, three-dimensional world, and our visual processes faithfully acknowledge this. But the images on the retinas are two-dimensional and virtually flat, so the brain has to translate these into a three-dimensional version. It uses a combination of different processes to achieve this effect and to work out spatial locations.

Some depth information can be obtained from the view as seen by a single eye, just as it can from an ordinary photograph. For instance, if one object appears to overlap another and obscures part of it, the brain simply assumes that the blocking object is nearer. This is the principle of superimposition. The mechanics of the eye provide another source of information. When you look at near objects, the lens of your eye has to be more curved and the amount of curvature is sensed and interpreted to give an estimate of distance. It works only for distances up to 20 feet or so, because the lens needs little or no adjustment for objects that are farther away.

For full depth perception the stereoscopic vision provided by two eyes is needed. Each eye views a scene from a slightly different position, and fusion of the two images gives the three-dimensional effect. With objects less than 200 feet away, the mechanism of vergence comes into play. Each eye is angled in slightly to focus directly on the object – and the bigger the angle the nearer the object. With moving objects another cue, motion parallax, can be used. Close objects pass more rapidly than distant ones, and the brain can compare apparent speeds.

The spectrum

To a scientist, light is a form of electromagnetic radiation. It

Saturation
Hue

occupies only a small part of the wide electromagnetic spectrum, which includes also many other forms of radiation from radio waves to gamma rays. Radio wavelengths can be as long as several miles, whereas gamma rays have a wavelength of no more than a six-quadrillionth of an inch (0.000000000000006 inch). Measured in nanometers (one nanometer is one billionth of a meter), visible light occupies a band stretching from 400 to 700 nanometers.

Within this visible spectrum the longest wavelengths (550 to 700 nanometers) correspond to yellow, orange and red colors; the mid region corresponds to green; and the shorter wavelengths (down to 400 nanometers) to blue and violet. Usually all of the colors are intimately mixed to give white light, although they can be separated, as they are in a rainbow or when white light passes through a glass prism.

Surprisingly the retina is able to respond to colors that fall outside the normal visible spectrum in the ultraviolet region. Normally this is not apparent because the lens of the eye filters out ultraviolet light before it can reach the retina. But patients who have had an eye lens surgically removed and replaced with a glass or plastic one (possibly as a treatment for cataracts) are able to see objects illuminated with ultraviolet light, which is invisible to eyes that retain their natural lenses.

Seeing the spectrum

Colored objects appear colored because of the presence of pigments. These selectively absorb some wavelengths of light and reflect or transmit others, giving an unbalanced mixture of wavelengths which the eye discriminates as a particular color. The effect is echoed in the color-sensing cones of the eye's retina as the three different types of cones generate nerve impulses corresponding to each of the primary colors. The same impulses also inhibit sensations for the opposing colors. Thus red light stimulates the red-sensitive cones to produce signals that register in the brain as red, but also tend to reduce sensitivity to green. Similarly blue signals inhibit both the red and green signals (which together make yellow, blue's opponent color).

Anything that goes wrong with the distribution or operation of the cones results in color blindness, a hereditary disorder that affects males to a much greater extent than females. True color blindness occurs when only the rods in the retina are active so that the sufferer sees everything in shades of black and white, but is a very rare condition. Much more common is dichromacy, in which one type of cone is missing from the retina. In most cases it is the red or green cones that are lacking to give red–green color blindness.

If the red cones are missing, red light reaching the retina cannot generate red color sensations nor have its normal inhibitory effect on green sensitivity. The end result is an abnormally strong green sensation. Conversely if the green-sensitive cones are missing, the result is a predominantly red or blue color sensation. Another cause of dichromacy is fusing of the red and green nerve channels, due to faulty neurological connections or mixing of the pigments in the cones.

Color blindness can also occur when the retina contains all three types of cone, but they are not in the correct proportions. This condition is called anomalous trichromacy and causes the unbalanced eye to see colors incorrectly, with diminished sensitivity in some parts of the spectrum. For example a bright red color might appear to be orange, or blue may look violet, depending on the precise defect.

Correcting lens faults

Color blindness is an inherent fault in the retina and cannot be cured. There are also a number of structural anomalies that can result in eye problems. Some of these are inherent, but others are caused by damage or simple old age. A complex battery of ophthalmic tests has been developed to assess how well the eyes are working, and to identify any problems so that they can be treated. By far the most common causes of poor sight are minor flaws in the construction of the eye which prevent it bringing an image to a focus on the retina. The usual errors are nearsightedness, farsightedness, astigmatism and presbyopia.

Nearsightedness, or myopia, is a condition in which a person sees near objects more clearly than distant ones. The near vision is not necessarily good, it is just better than the far vision. The defect is caused by a cornea that is too curved, an elongated eyeball, or a combination of both. They cause parallel light rays – rays from distant objects – to be focused some distance in front of the retina. And by the time the light has finished its journey it has started to spread out, giving a blurred spot of light instead of a sharp image.

As the name suggests, farsightedness, or hyperopia, is the opposite of myopia, with distant vision being sharper than close vision. In this case the cornea and lens do not bend the light rays enough, so that they reach the retina before they come into sharp focus, again giving a blurred image. Often the sufferer can bring distant objects into focus by using the accommodation of the lens system, but this does not work well enough to bring near objects into focus as well. Presbyopia has similar effects to farsightedness, making near objects look blurred, but it is a separate condition usually caused by a gradual age-related failing in the eye's powers of accommodation.

In a normal eye the cornea has an even, spherical surface, which uniformly bends all the light rays entering it to produce a properly focused image across the whole retina. But in astigmatism the cornea has an irregular shape, which makes the image distorted and out of focus.

Lenses for bending light

If the eye's optical system is not working properly it can be helped by using supplementary external lenses, either in the form of eyeglasses or contact lenses. For instance a convex lens makes parallel light rays converge, and can be used to correct farsightedness. Conversely concave lenses make rays diverge and are used to correct a myopic eye.

A more radical method of correcting vision involves surgery on the eye itself. Parts of the cornea can be cut free, reshaped to improve the optics, and stitched back into position. Alternatively sections of donor cornea are shaped to suit and slid into position under the original cornea. Another technique involves making a series of radial cuts in the cornea so that it takes up a better shape after healing.

In most Western countries, about half the population wears eyeglasses or contact lenses to correct common eye defects such as nearsightedness, farsightedness, astigmatism and presbyopia. All result from flaws that affect the way light is diffracted in the eye.

Nearsightedness

The nearsighted eye focuses rays of light before they reach the retina; from the focal point, the rays begin to diverge, resulting in a blurred image.

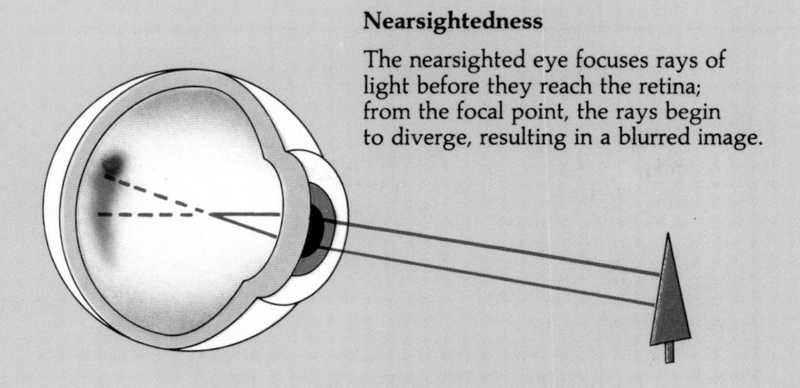

Farsightedness

The farsighted eye does not refract rays of light sharply enough to bring them into focus on the retina.

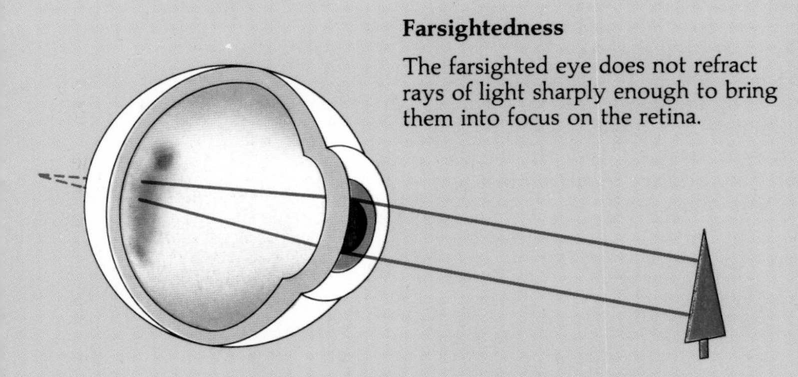

Astigmatism

The irregular shape of the astigmatic cornea bends rays of light unevenly, leaving a blurred image on the retina.

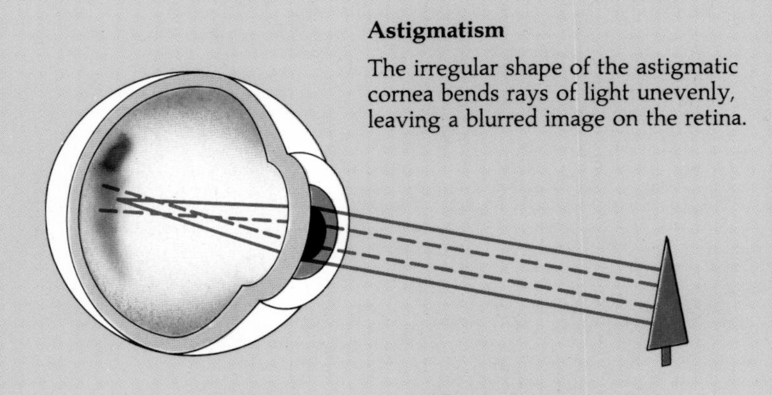

Presbyopia

As the eye ages, it loses much of its power to bring near objects into focus by changing the shape of its lens, a process called accommodation.

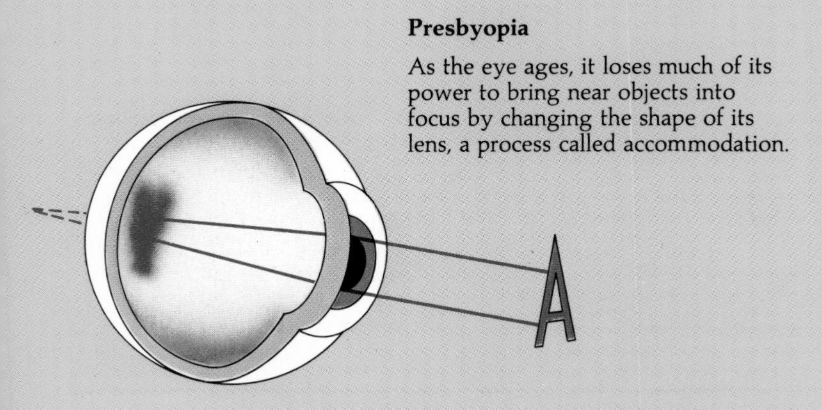

Concave lens

A concave lens spreads rays of light slightly before they reach the cornea, correcting nearsightedness.

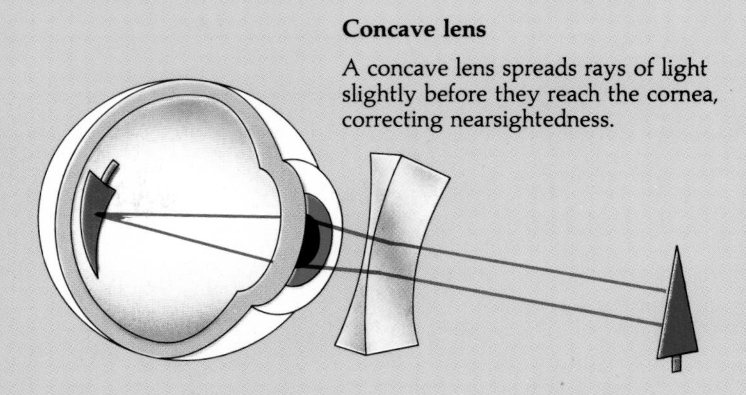

Convex lens

A convex lens bends rays of light inward before they reach the eye, correcting farsightedness.

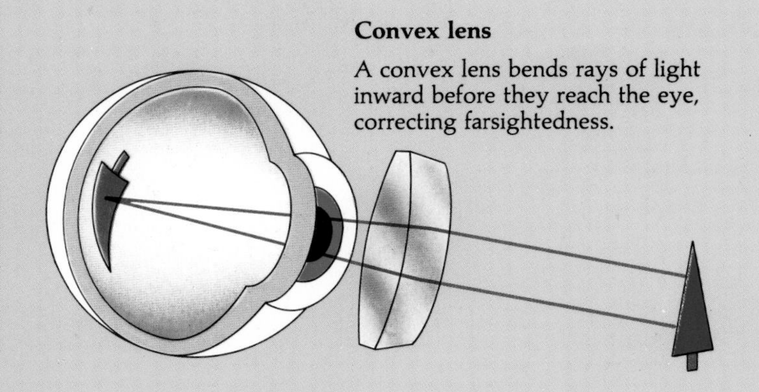

Cylindrical lens

A cylindrical lens bends only certain light rays, compensating for the oval shape of the astigmatic cornea.

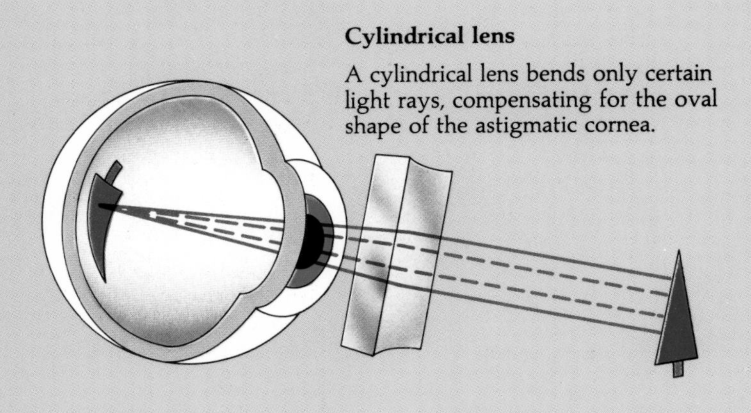

Biofocals

A convex lens, often in the bottom of bifocals, bends light inward and helps the presbyopic eye focus on near objects, like the print on a page.

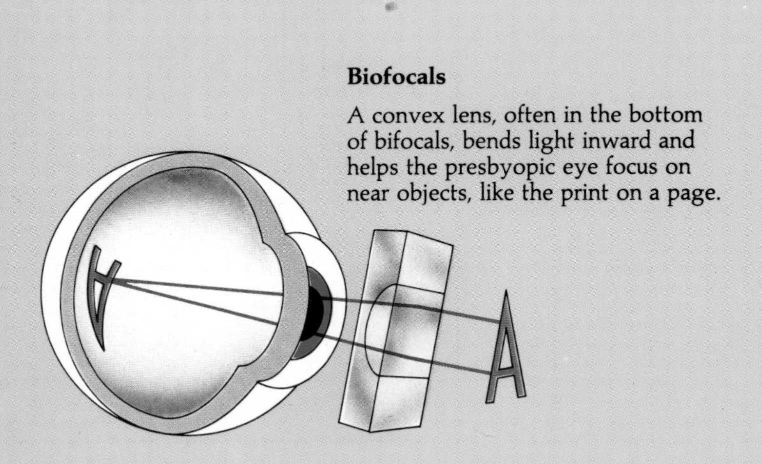

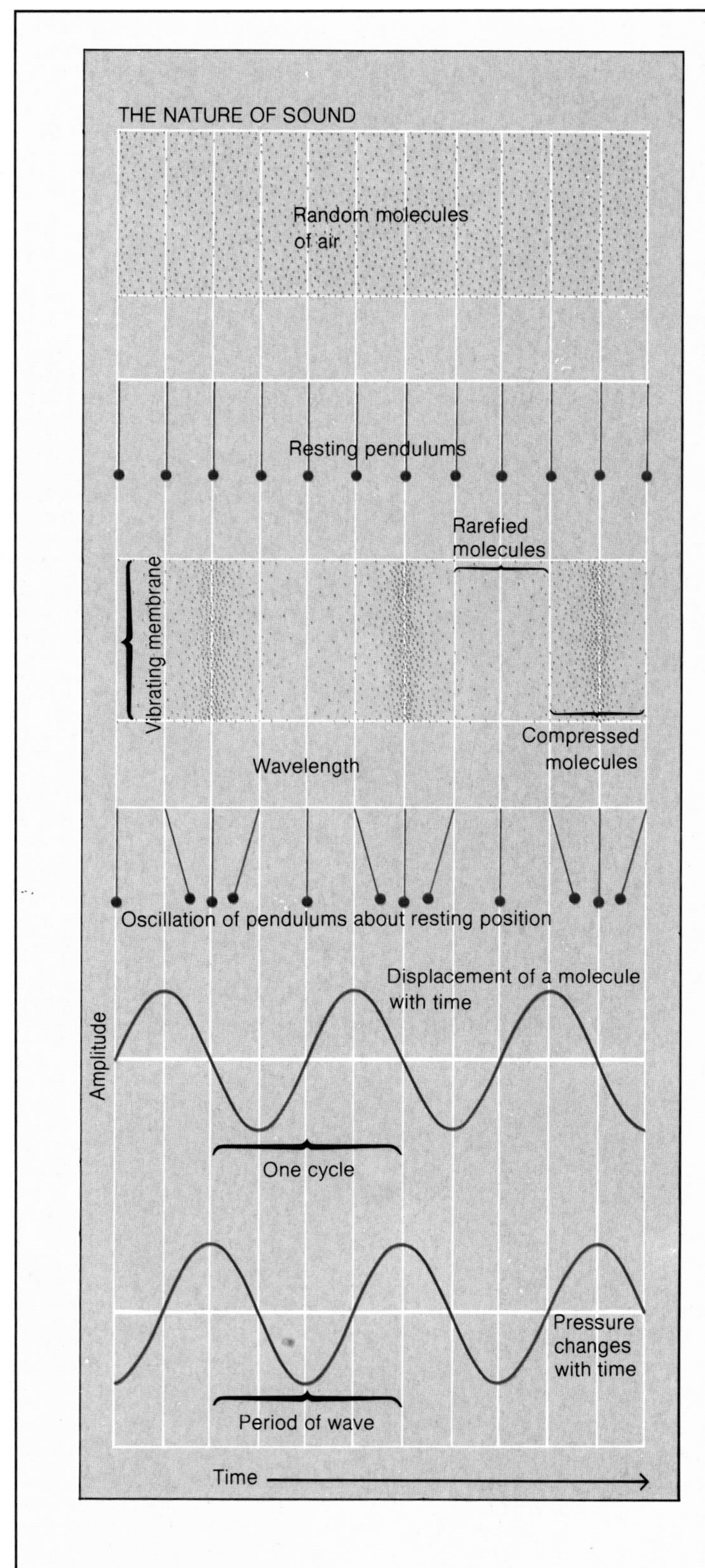

Sound is a form of energy that needs a medium in which to travel. When it travels in air it takes the form of regularly varying changes in pressure which behave as a wave motion, with a wavelength, amplitude (wave height) and period or frequency (number of waves per second). In the diagrams *(left)* air molecules are compared with a line of pendulums. Stationary air molecules can be set oscillating by sound waves vibrating a diaphragm. Like water molecules in a sea wave, the air molecules oscillate about their average position and are not carried along by the wave. The amplitude of a sound wave corresponds to its loudness or intensity, and the frequency is a measure of its pitch.

Rows of sound-sensitive hair cells form a colonnade along the organ of Corti, as revealed in this scanning electromicrograph. Each cell ends in a brushlike tuft of finer hairs, which detects the subtle vibrations of the underlying basilar membrane.

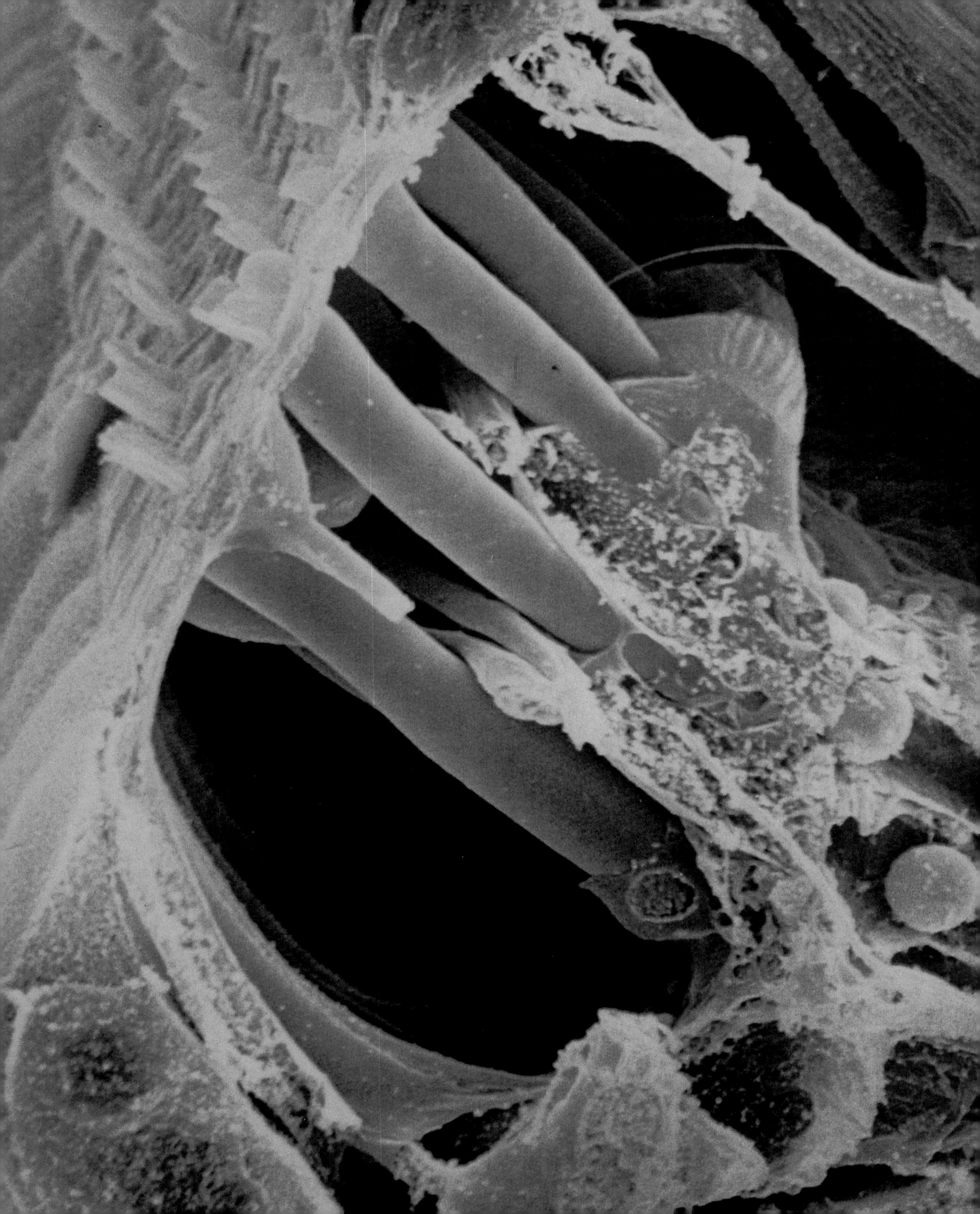

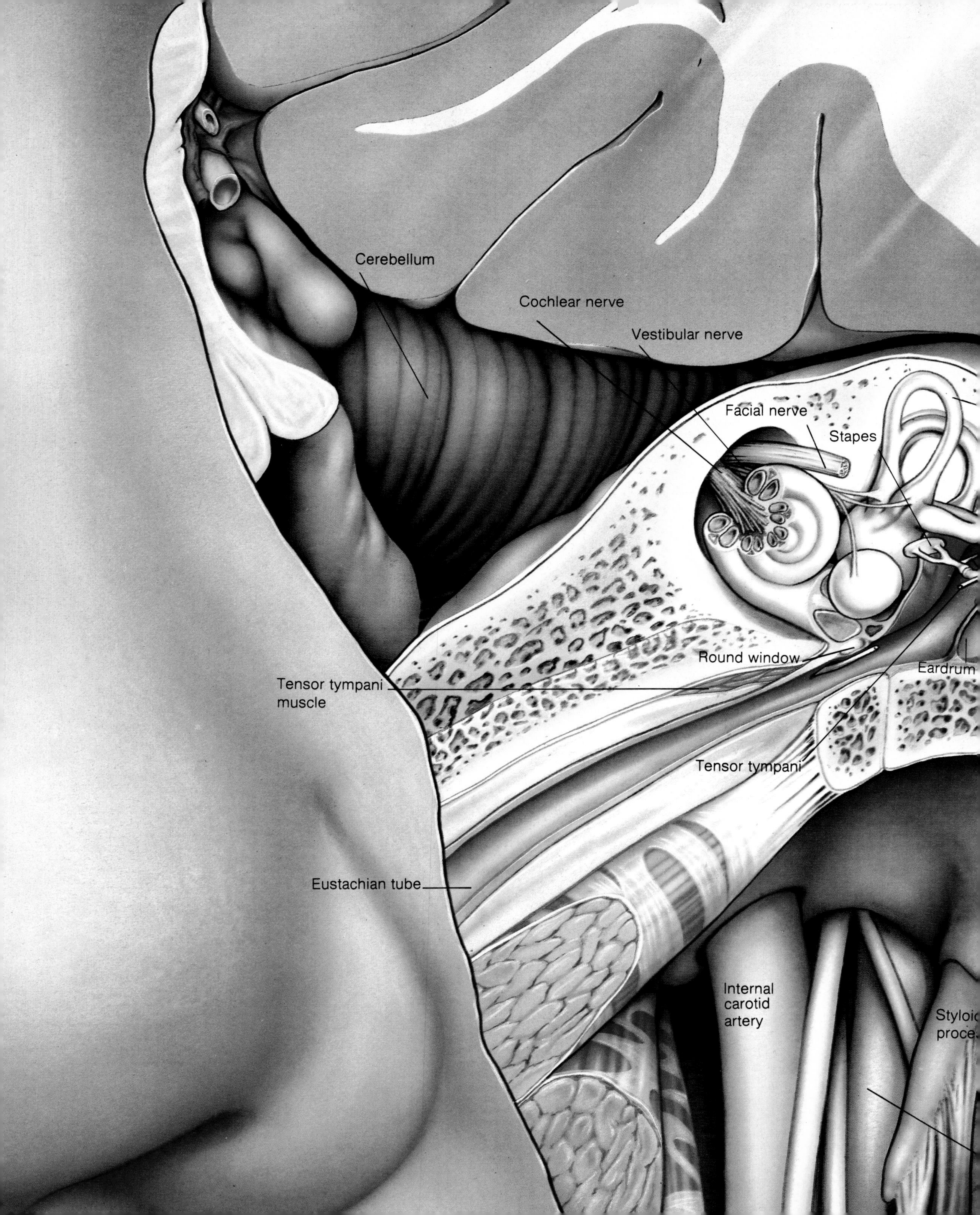

Cerebellum
Cochlear nerve
Vestibular nerve
Facial nerve
Stapes
Round window
Eardrum
Tensor tympani muscle
Tensor tympani
Eustachian tube
Internal carotid artery

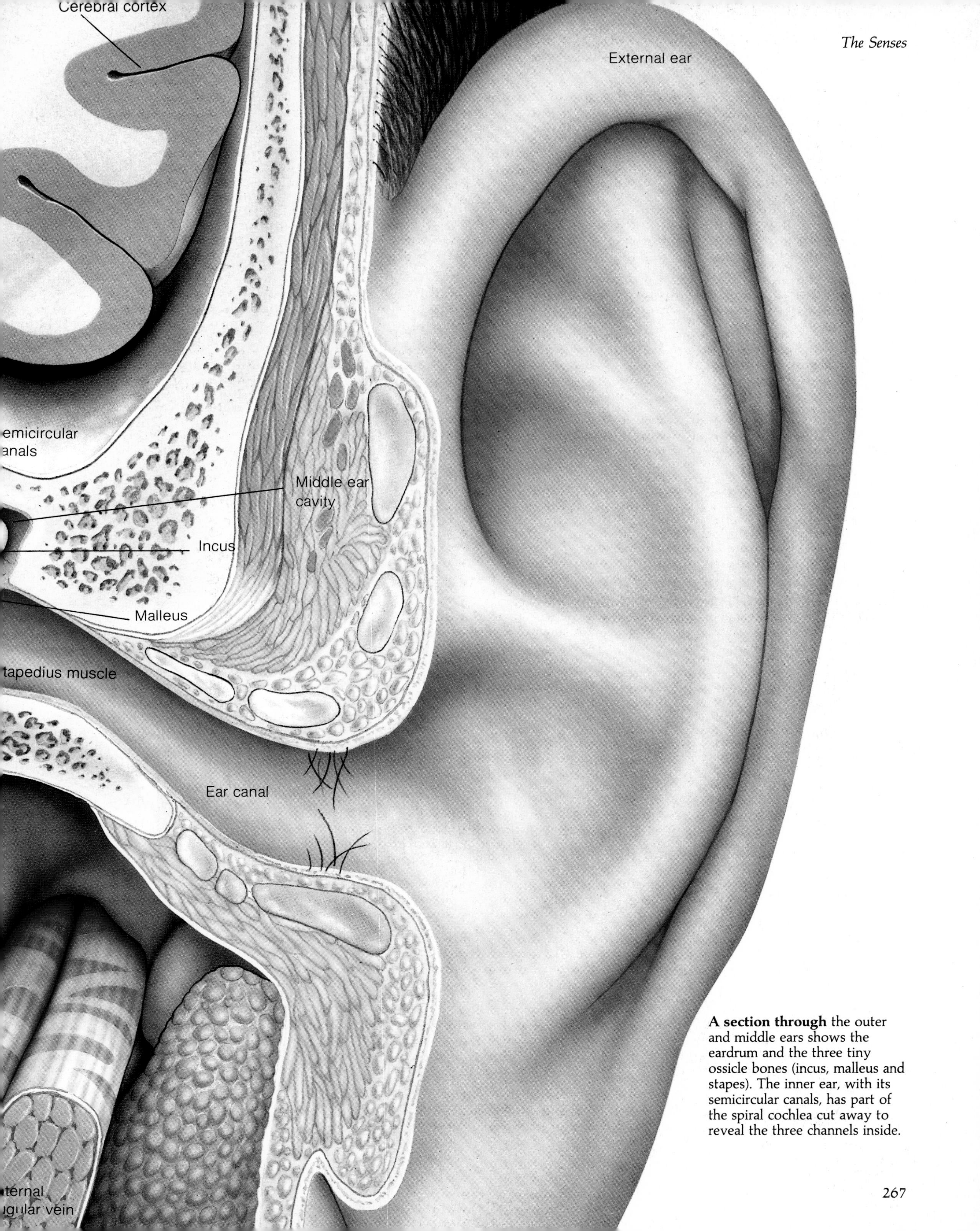

A section through the outer and middle ears shows the eardrum and the three tiny ossicle bones (incus, malleus and stapes). The inner ear, with its semicircular canals, has part of the spiral cochlea cut away to reveal the three channels inside.

Double images

Stereoscopic vision relies on the two eyes working together, and this ability is developed during the first year of a baby's life. But sometimes control is imperfect, or lost, and one eye starts to stray while the other maintains focused vision. This is squinting, or strabismus. Surgery can sometimes correct the faulty eye movements, but does not necessarily restore stereoscopic vision.

Another problem affecting children's sight is amblyopia, sometimes known as lazy eye, in which one of the eyes does not develop properly. The brain then rejects the image from the faulty eye. Provided it is diagnosed early enough the condition can often be cured by prescribing eyeglasses and periodically putting a patch over the good eye. This simple measure forces the child to exercise the amblyopic eye.

Damage, disease and defect

The worst disorder of the visual sense is blindness, which can be congenital or result from injury or disease. A common cause is damage to the cornea (by scarring or attack by a virus or bacterium), but fortunately it is often possible to restore normal sight by replacing the cornea. The damaged region is cut out and a replacement cornea – taken from a recently dead donor – stitched in place. As long as the donor cornea is in good condition it blends in with the remaining part of the original cornea, restoring the eye to good working condition.

Injuries and disease are not the main causes of loss of sight. Far more significant, in terms of numbers of cases, are the main enemies of the eyes: cataracts, glaucoma and retinal defects.

Starting as minute cloudy spots in the lens, cataracts spread gradually and insidiously to turn the whole lens a milky yellow white that scatters the incoming light instead of allowing it through to the retina. Vision gradually becomes less and less distinct – the effect is similar to looking through a waterfall, which gives the disorder its name – and blindness eventually results. Most common are senile cataracts in which the progressive clouding accompanies advancing age.

Surgery is the only solution to a cataract. The opaque lens is removed, to re-open a passage for light rays to reach the retina. Dramatic as it appears this is generally a highly successful treatment, and clear vision is achieved with the aid of powerful eyeglasses or contact lenses to substitute for the missing eye lens. Alternatively it is possible to implant a replacement plastic intraocular lens into the eye during the cataract operation.

Glaucoma is the name given to a group of diseases that causes a gradual buildup of pressure within the eyeball. The increasing pressure affects the eye lens and the nerve fibers of the retina, and if untreated the result is blindness. The gradual nature of glaucoma means that the symptoms are often not noticed until there has been a serious loss of vision, by which time treatment can do little more than save what is left. If the condition is detected early enough, however, drugs can be prescribed that will control it for life.

Repairing the retina

Some eye injuries and diseases cause the retina to become detached from the underlying layers, resulting in partial loss of vision. Prompt treatment often makes it possible to repair the damage by welding the retina back into place, typically by the use of laser beams fired in through the pupil.

Many diseases threaten the retina – indeed studying the eye can give valuable clues to problems in the rest of the body. However, there are two afflictions that stand out as major causes of blindness: diabetic retinopathy and senile macular deterioration. Diabetes attacks the vital capillary network supplying blood to the retina. The capillaries are destroyed, fluids leak out of the damaged blood vessels, and nerve tissue is affected. Provided the disease is restricted to its nonproliferative form it may stop naturally at this stage, with adequate vision being retained. But in the proliferative form of diabetic retinopathy, progressive deterioration continues, often made worse by an extensive growth of new blood vessels in an attempt to keep the retina supplied.

The hardest worked part of the retina is the macula – the sensitive central region of acute vision – and in many people it gradually deteriorates with age. For a variety of causes the supply of nutrients and oxygen gradually diminishes and the macula tissue starts to degenerate. Eventually the condition causes a loss in the powers of central vision.

The second sense

If vision is the primary and most important human sense, second place must go to hearing. The sounds you hear, and make, play a vital role in communication through the reception of speech, or the pleasures of music, as well as warning you of impending danger. Virtually all sounds are made up of vibrations in the air, and the ears have the task of detecting them and transforming them into electrical nerve impulses ready for the brain to analyze.

Sound vibrations correspond to pressure waves in the air, which spread out from their source like ripples on a pond. With simple sounds the vibrations have a regular waveform, rising smoothly to a positive maximum, falling away to a negative minimum of the same magnitude, and rising again to the maximum in regularly repeated cycles. Complex sounds are merely built up from a series of superimposed simple waveforms.

The frequency of a sound is the number of cycles it completes in a second and is measured in hertz (Hz), where 1Hz is one cycle per second. Frequency is recognized as the pitch of a sound; high-pitched notes have high frequencies and low tones have low frequencies. For humans the hearing range runs from about 20Hz to 20,000Hz, though the range varies a lot with age and among different individuals.

Loudness is just as important as pitch, and the ear is able to respond to an incredible loudness range, from the faintest whisper to the roar of a jet engine. Indeed the range is so large that scientists have to use a logarithmic scale to measure it, comparing any given sound to an agreed reference level. For practical purposes the unit used to measure sound levels is the decibel. Because the scale is logarithmic, a doubling in sound intensity is equal to an increase of 3 decibels. Thus a sound of 93 decibels is twice as loud as one of 90 decibels.

Inside the ear

Working out how sound waves are converted to electric nerve signals is simplified by the fact that the ear mechanism has three clearly identified and separate parts. These interconnected sections are the outer, middle and inner ear, which each have their own specific functions, processing the sound in sequence.

The outer ear collects the sounds, which are transferred through the middle ear to the inner ear, where they are converted into nervous signals. The only noticeable component of this complex system is the fleshy external ear, known as the pinna, which leads into the external ear canal. Together they

One of the main functions of the ear is to change sound waves into mechanical vibrations that can stimulate nerve cells. The key to the system are the three small ossicle bones, aptly named the malleus (hammer), incus (anvil) and stapes (stirrup). These bones are hinged together and fit inside the middle ear cavity (A).

For ease of explanation, the cochlea can be pictured as an uncoiled, tapering tube (B), with the basilar membrane and its rows of hair-cell receptors running down the center. Incoming sound waves are funneled along the ear canal (C), where they strike the eardrum and make it vibrate. The vibrations are picked up and amplified by the ossicles and passed on to the oval window in the cochlea. Vibrations of this window set up a traveling wave in the cochlear fluid, distorting the basilar membrane and triggering the hair cells to send nerve impulses to the brain.

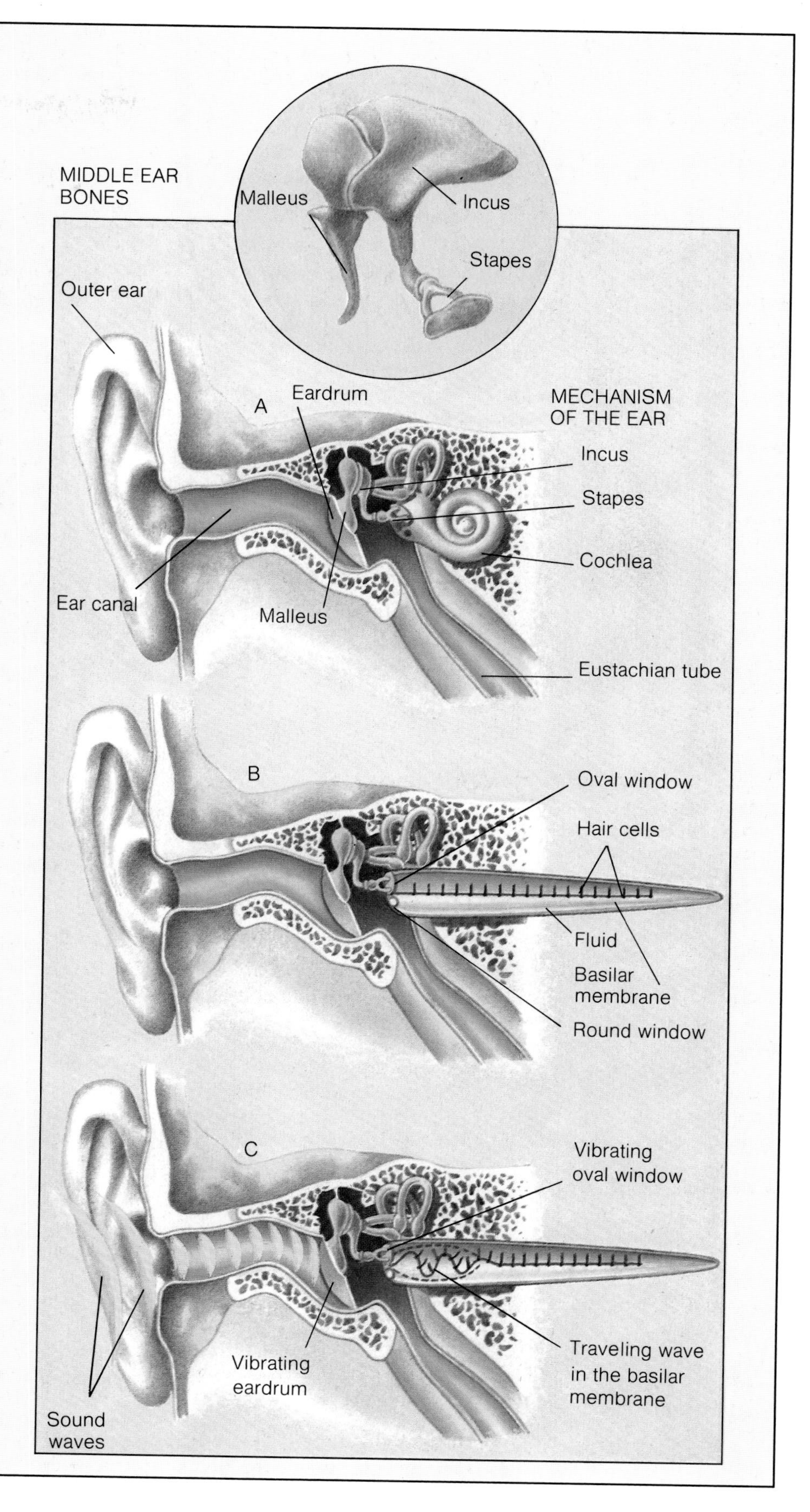

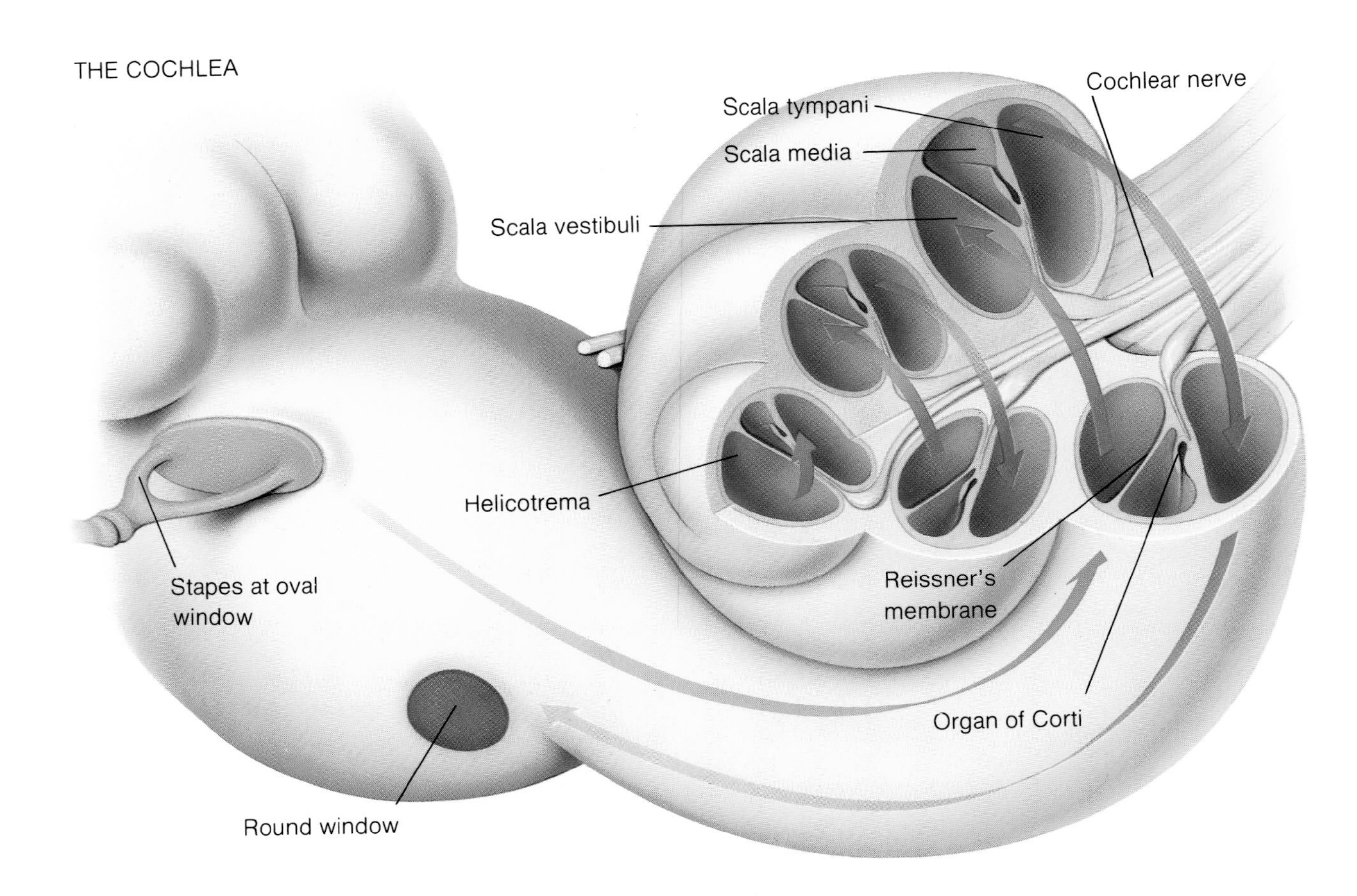

In ever decreasing circles, vibrations in the cochlear fluid (red arrows) spiral around channels to stimulate the organ of Corti to send nerve impulses to the brain. Initially the fluid vibrations are created as the oval window is vibrated by rapid movements of the baseplate of the stapes. On the outward trip, the vibrations travel along the scala vestibuli. Then, at the helicotrema (the apex of the cochlea), they change direction and spiral back (blue arrows) along the scala tympani. They finally arrive at the base of the cochlea, where they are dissipated by vibrations of the round window.

form the outer ear, which has the important functions of tuning in sounds and locating them. The outer ear increases the sensitivity of the system to the particular sound frequencies used in speech, and orienting the trumpet-shaped ear canal helps in locating the direction a sound is coming from.

Many of the sounds of speech lie in a comparatively narrow frequency range of 2,000 to 5,000Hz, and the pinna and ear canal form a tuned system that is particularly responsive to these frequencies. The conical opening linking the pinna to the ear canal resonates at 5,000Hz, so that quiet sounds at around this frequency build up into larger vibrations. Similarly the ear canal itself resonates at around 2,500Hz, and working together these effects amplify sounds that lie in the critical speech band.

Sound location

Identifying where a sound is coming from is rather more difficult. It depends largely on sensing differences in the intensity and timing with which sounds from a single source reach the two ears. For example, a sound coming from the left-hand side reaches the left-hand ear before it gets to the right-hand one, and it is also more intense at the left ear.

This discrimination does not work for sounds coming from directly in front or behind, and locating them relies on the shape of the pinna. Some of the sound waves from a source behind the head are scattered by the edge of the pinna and interfere with unscattered waves. The result is a reduction in the intensities of frequencies in the range 3,000 to 6,000Hz before they reach the eardrum. These interference effects change as the sound source moves around the head, and the complex variations in frequency and intensity can be analyzed by the brain to give the required position information. Similar effects are probably used to give height information about sounds.

The smallest bones in the body

Sounds captured by the outer ear end up at the eardrum, which lies at the end of the ear canal and forms an airtight seal between the outer and middle ears. The middle ear is, however, still an air-filled cavity; it is connected to the outside by the Eustachian tube. This is a narrow channel about one and a half inches long which runs from an opening at the back of the nasal cavity, to emerge through the floor of the middle ear. This connection ensures that the air pressure is the same on both sides of the eardrum – otherwise it would bulge under the pressure difference and so not be able to vibrate properly, or even burst. If the Eustachian tube becomes blocked – say by a head cold – the pressures become unequal and there is a temporary mild hearing loss.

Running across the cavity of the middle ear is a linked chain of the three smallest bones in the body, the auditory ossicles. Called the malleus, incus and stapes, these tiny bones provide the connection between the eardrum and another membrane, the oval window, which forms the boundary with the inner ear. The snail-like coils of the inner ear are filled with fluid, and the task of the middle ear is to transform the air pressure changes caused by sounds in the outer ear into fluid pressure changes within the inner ear.

The middle ear mechanism is a miracle of living engineering. It is needed because the direct transfer of vibrations between air and fluid is very inefficient; in such a set-up, most of the vibrational energy would be reflected away from the fluid rather than absorbed by it. Using a mechanical linkage overcomes this problem and allows efficient energy transfer to take place.

Set at an angle across the ear canal, the eardrum is an oval membrane with a stiff rim and a concave outer surface divided into regions that are relatively taut and flexible. Sounds reaching the eardrum make it vibrate in a complex pattern that ensures very efficient transfer of acoustic energy to the middle ear for frequencies below 10,000kHz, though the transfer becomes increasingly less efficient at higher frequencies.

A vibrating bone chain

The first of the ear bones, the malleus or hammer, is directly attached to the eardrum, and vibrations of the eardrum make the malleus rock. This movement is transferred through the incus (anvil) to the stapes or stirrup, and the geometry of the ossicle chain means that a force applied at the malleus is increased by about a third by the time it reaches the stapes end.

THE ORGAN OF CORTI

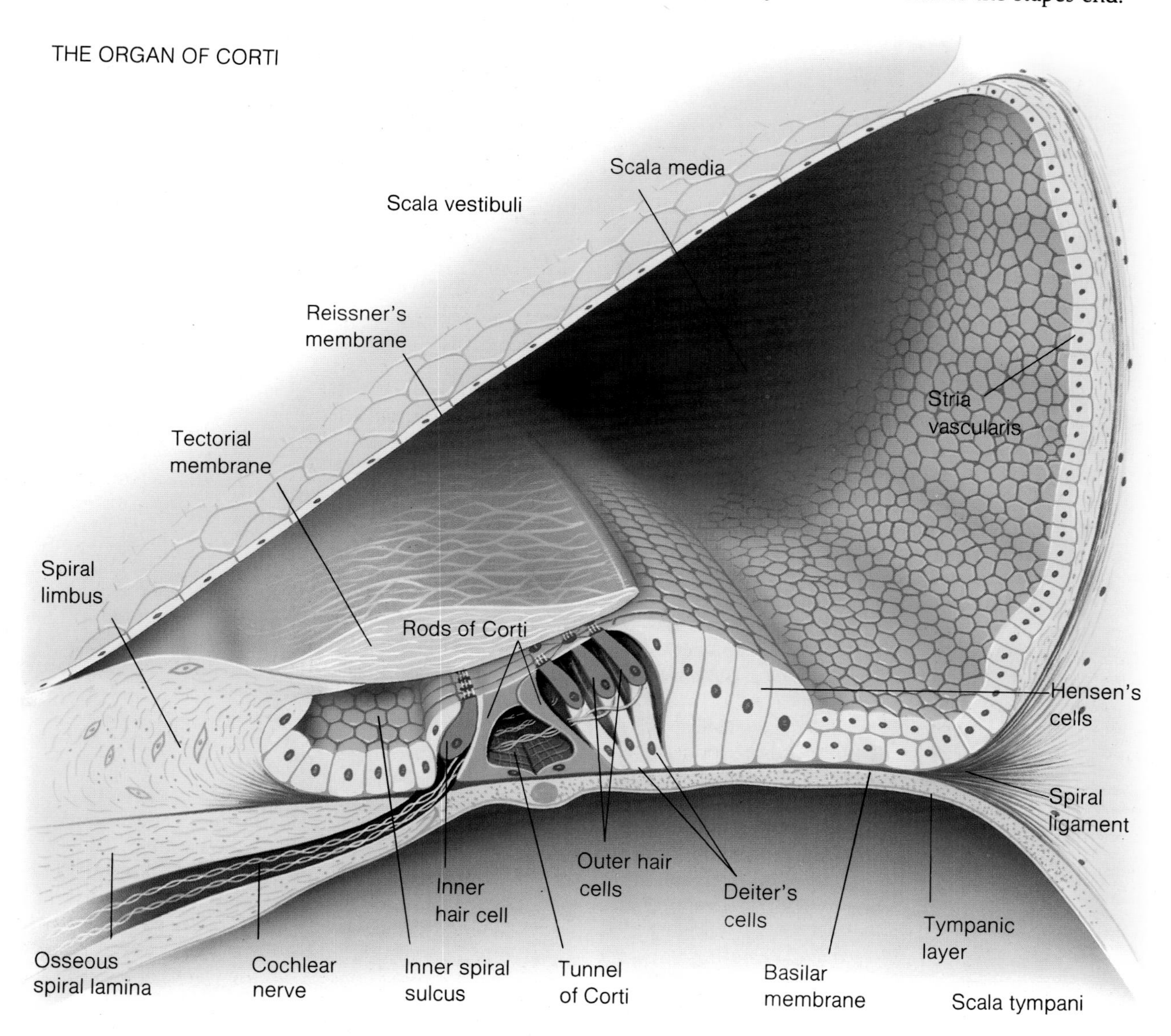

The wedge-shaped scala media, the narrowest and central channel, which spirals round the cochlea, has the basilar membrane along one edge. The membrane supports the organ of Corti, with its rows of hairlike nerve cells, the cells that are stimulated by pressure waves pulsing along the channel. As the hair cells vibrate in tune with the passing wave, they fire off nerve impulses which pass along fibers to the cochlear nerve and then on to the brain. Different parts of the basilar membrane are tuned to different frequencies, and so can "identify" and respond to tones of varying pitch.

Amplification of the signal pressure is also produced by differences between the sizes of the eardrum at the input end and the oval window at the output end. The eardrum is elliptical, and has an effective vibrating area of about one tenth of a square inch. At the other end of the ossicle chain, the footplate of the stapes, which bears on the oval window, has an area of about one two-hundredth of a square inch. Since the force from the eardrum is concentrated in a much smaller area at the end of the stapes, there is significant amplification of about 20 times.

Overload protection

Like a fuse in an electric circuit, the ear also has a device to protect it from sudden loud noises that could overload the system and permanently damage the ear. Transmission of energy through the middle ear can be modified by some small but quick-acting ear muscles. The tensor tympani muscle pulls on the malleus and increases the tension in the eardrum, and the stapedius muscle pulls on the neck of the stapes, acting to restrict the movement of the stapes footplate. Both of these actions increase the stiffness of the ossicle chain and reduce the efficiency of energy transfer through the middle ear.

With an ear that is working properly the ear muscles contract automatically when the sound exceeds 80 decibels. The response is very rapid; it starts within 15 to 150 thousandths of a second of the sound starting, and maximum protection is produced after 100 to 500 thousandths of a second. This is fast enough to protect against loud sounds at frequencies under 2,000Hz, but not fast enough to protect against the gross overload produced by an explosion with its rapid and intense sound peaks.

Inside the inner ear

Together the outer and middle ears convert pressure waves into a mechanical vibration. Final conversion of the sound into electric nerve signals takes place in the highly intricate inner ear. It consists of a maze of bony canals and membrane tubes filled with fluid, which contain myriad tiny hairs that sense the sound vibrations. But hearing is not the only function of the inner ear; it also contains a series of sensors that deal with head movement and body balance. These sensors are housed in the vestibular portion of the inner ear; the sound sensors are contained in the cochlea.

Shaped rather like a snail shell, the cochlea is formed from three ducts running in parallel: the scala media, which contains sound-sensing hairs; the scala vestibuli running from the oval window; and the scala tympani. Coiled up the ducts make two and a half turns. The scala vestibuli and scala tympani meet at the apex of the cochlea (the helicotrema) and are filled with a fluid called perilymph. The scala media is sandwiched between the other two ducts and contains a different fluid – endolymph.

Vibrations of the stapes at the oval window cause pressure waves in the perilymph of the scala vestibuli. The pressure changes pass along the duct to the helicotrema, and then spiral back along the scala tympani. This duct ends at the round window, a membrane that faces into the middle ear cavity.

Sound-sensing hair cells

A thin membrane – Reissner's membrane – is all that separates the scala media, or cochlear duct, from the scala vestibuli. As a result, any pressure changes in the perilymph are immediately transmitted through the membrane to generate corresponding pressure fluctuations in the endolymph. And it is here that the final conversion from vibration to electric nerve signal takes place. At the bottom of the cochlear duct is the basilar membrane, with its organ of Corti and the sound-sensitive hair cells. There are 12,000 outer hair cells and 3,500 inner ones. The tips of the outer cells are embedded in a flap – the tectorial membrane – sticking out into the duct. Pressure changes in the cochlear duct make the basilar membrane vibrate, transmitting bending and shearing movements to the hair cells. This stimulates them to produce a nerve signal, which is carried to the brain by the cochlear nerve.

But the cochlea does not simply respond to sounds. Its operation is much more sophisticated, acting to both identify pitch and to make an assessment of loudness. How this happens depends on the way the basilar membrane vibrates. As the cochlear duct runs up from the base to the apex of the cochlea, the basilar membrane gradually becomes more flexible and wider. The membrane therefore vibrates in different ways along its length: high-frequency vibrations tend to have the greatest effect at the base, whereas low-frequency vibrations cause more vibration near the apex. As a result, sound at any particular frequency makes some parts of the membrane vibrate more than others, stimulating a specific group of hair cells so that the note can be recognized. At the same time, the extent of the basilar movement depends on the loudness of the sound.

Keeping your balance

The second, larger, part of the inner ear is the vestibular apparatus, and has nothing to do with hearing. A series of fluid-filled chambers and ducts, it contains more clusters of hair cells. When they are activated, signals from these sensors pass to the brain along the vestibular branch of the eighth cranial nerve. The cells in the semicircular canals are the main sensors for detecting and responding to turning movements. When you turn your head, the signals from these cells are processed in the brain, which then activates the eye and body muscles so that they are ready to compensate for the tuning action.

Other receptor cells, the otolith cells, within the vestibular labyrinth sense the direction of gravity and straight line movements. Their main purpose is the control of balance – because humans have only two legs our balance is inherently unstable, and continual corrections have to be applied to our stance when standing so that we do not lose our balance and fall over. In addition the otolith signals have an essential role in coordinating eye and head movements.

Loss of hearing

Our ears are amazing instruments, able to identify and analyze sound with extreme precision, but their very complexity means that they are prone to failures minor and major. Impaired hearing is one of the most common features of old age, but problems can also occur in the young. One simple cause of hearing loss is blockage of the ear canal by the wax released from the ceruminous glands within the canal. Treatment is equally simple; the ears are syringed with warm water or treated with wax-softening compounds to dislodge it.

Sudden, intense noises or physical shocks may tear the membrane and perforate the eardrum. Small holes usually repair themselves without further treatment. More serious is the scarring that can be caused by middle ear disease. The scars reduce the sensitivity of the membrane, and the only effective treatment is plastic surgery to form a new membrane.

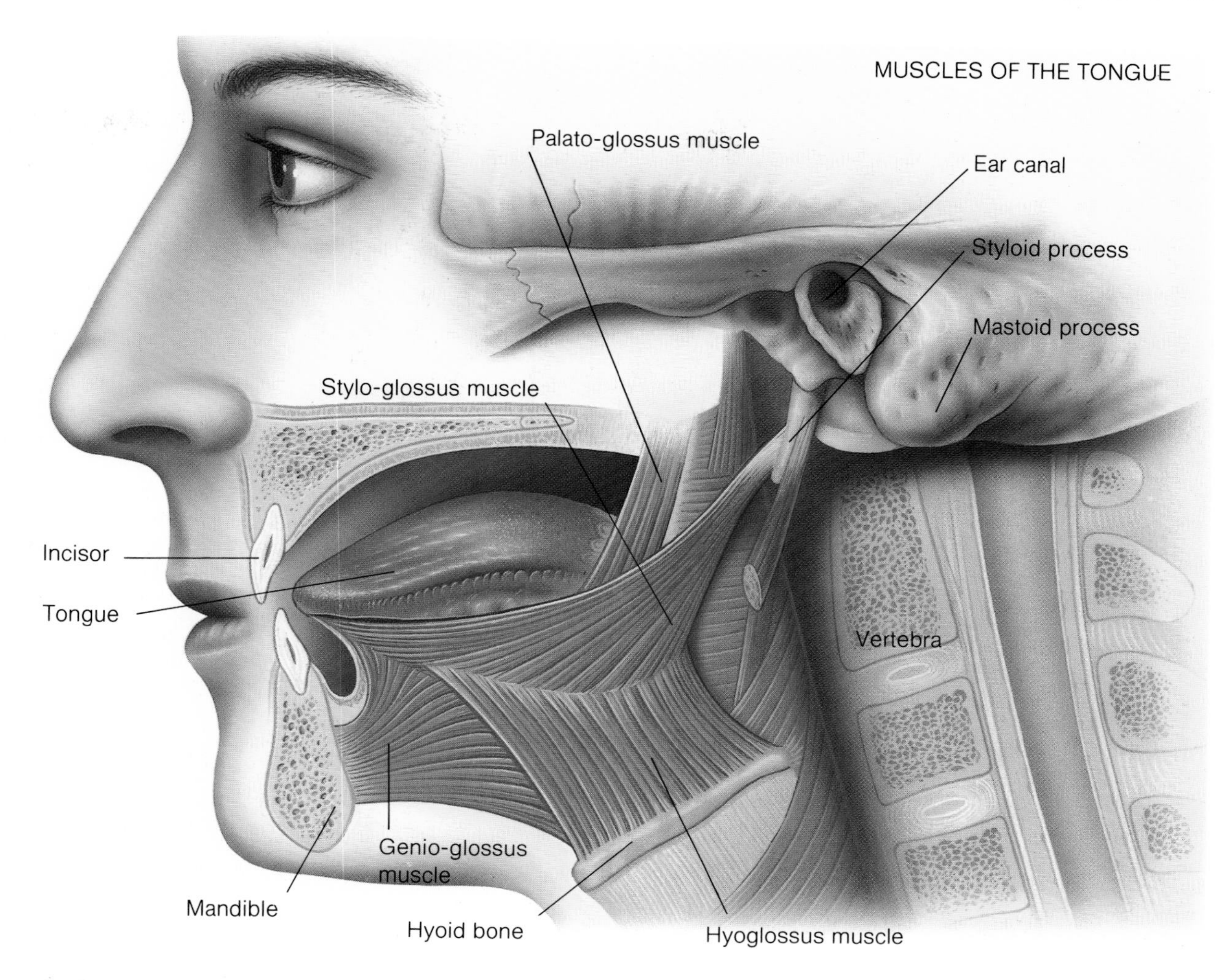

The tongue is a flexible organ whose change of shape is brought about by the intrinsic muscles inside it. Gross movements of the whole tongue are controlled by the extrinsic muscles, which take up a surprisingly large amount of space within the lower jaw (mandible). There are sets of four muscles on each side of the head, which between them can pull the tongue in almost any direction.

The Eustachian tube has an important function in balancing the pressures on each side of the eardrum, but this open channel also provides a ready route for bacteria and viruses to reach the middle ear. Various infections, known collectively as otitis media, may be caused by these microorganisms. The swelling that results can impede the action of the ear ossicles and distort the transmission of sound. Severe infections may even damage the eardrum and can spread to the mastoid bone, causing mastoiditis. Treatment with antibiotics is generally successful in controlling such infections.

Middle ear infections can result in permanent damage to the middle ear bone chain (ossicles), while the amount of movement these bones can make often diminishes with age. Conductive hearing loss ensues, and the only remedy is to make the sounds louder. Old-fashioned ear trumpets achieved this by simply trapping more sound energy and directing it into the ear canal. Modern hearing aids are much less obtrusive and amplify the sound electronically. Surgical techniques can also offer a cure by freeing or bypassing the faulty ossicles.

Prolonged exposure to high sound levels – either at work, in discos or through the headphones of a personal stereo – causes progressive deterioration in hearing because of degeneration of the hair cells in the cochlea. This same effect also occurs with age, and as a side-effect of some antibiotic drug treatments.

Tinnitus, or ringing in the ears when there is no external sound, is a curious phenomenon that may be caused by disease or

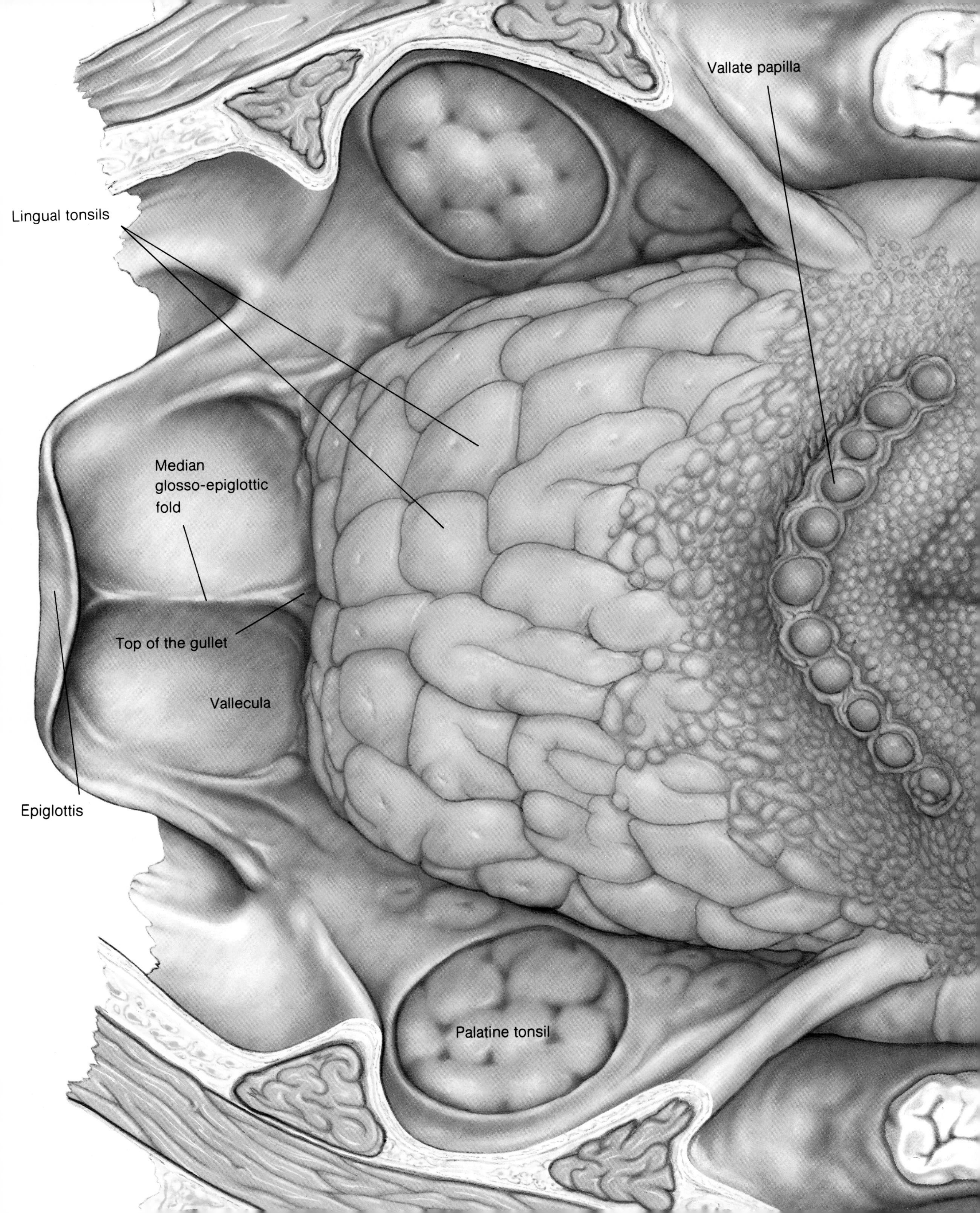

Vallate papilla
Lingual tonsils
Median glosso-epiglottic fold
Top of the gullet
Vallecula
Epiglottis
Palatine tonsil

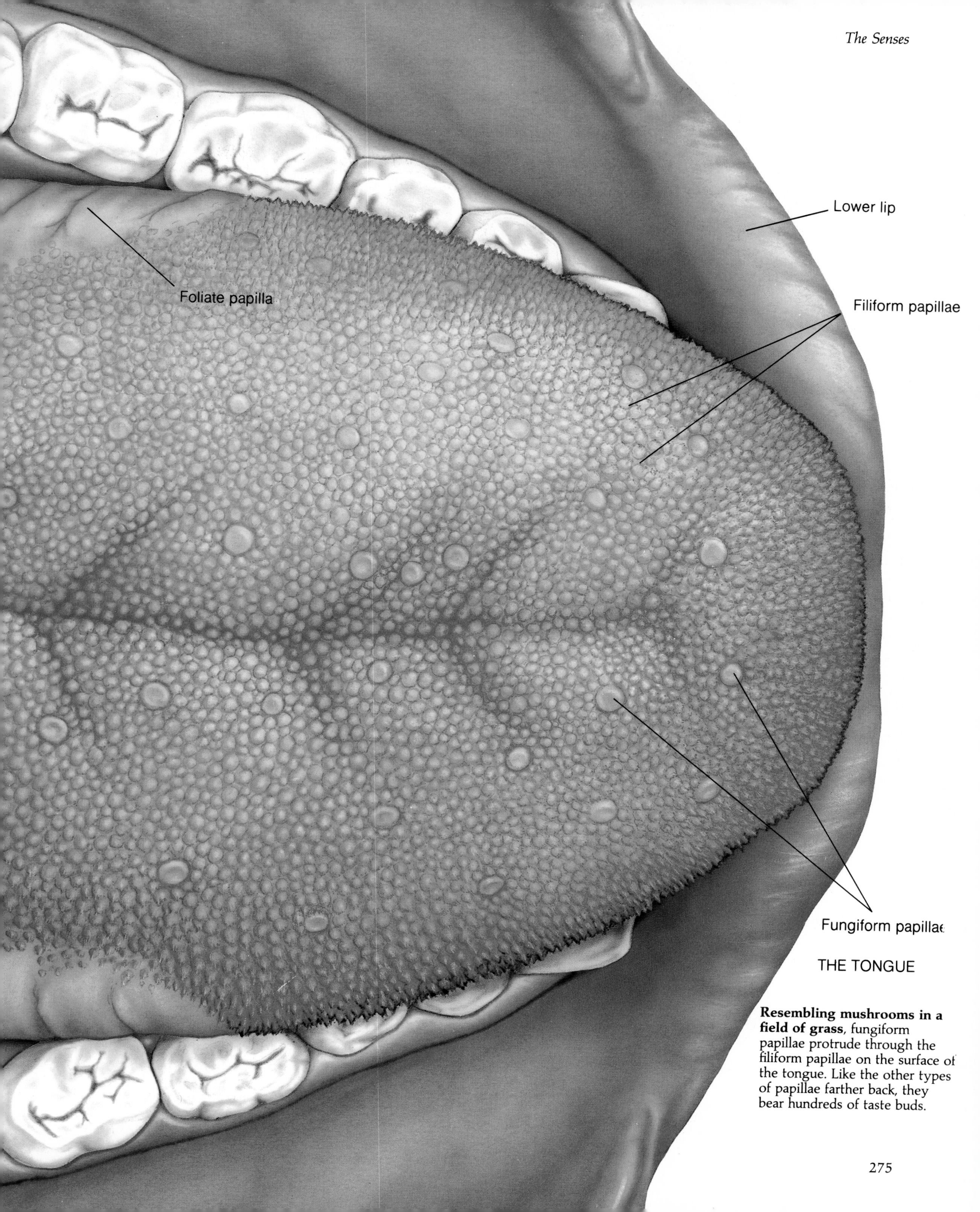

THE TONGUE

Resembling mushrooms in a field of grass, fungiform papillae protrude through the filiform papillae on the surface of the tongue. Like the other types of papillae farther back, they bear hundreds of taste buds.

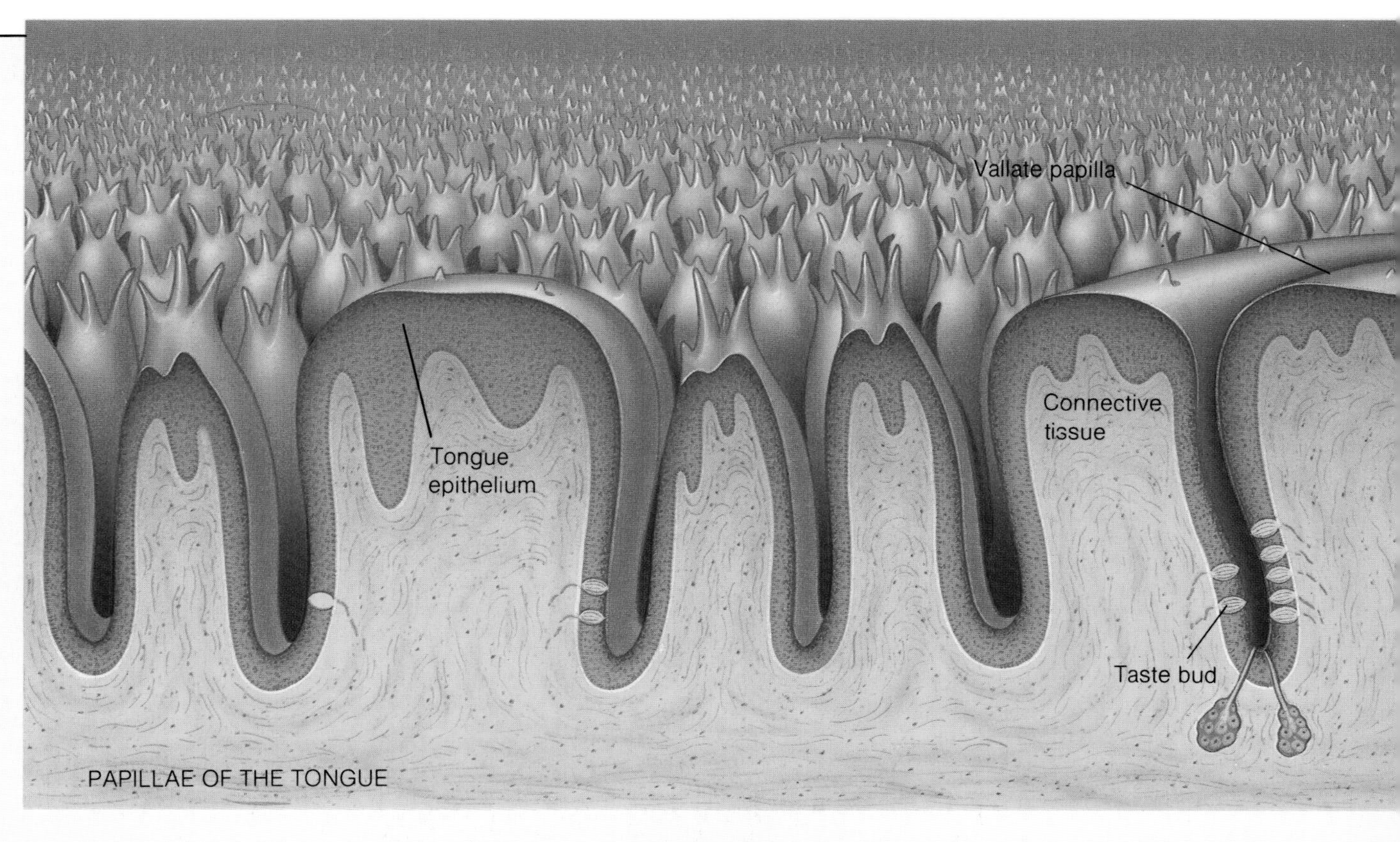

In a cross section of the central part of the tongue's surface *(above)*, three kinds of papillae reveal their shapes. Most numerous are the conical filiform papillae, whose taste buds are sensitive to sour flavors. The buds of the flat-topped fungiform papillae respond to both sweet and sour tastes,

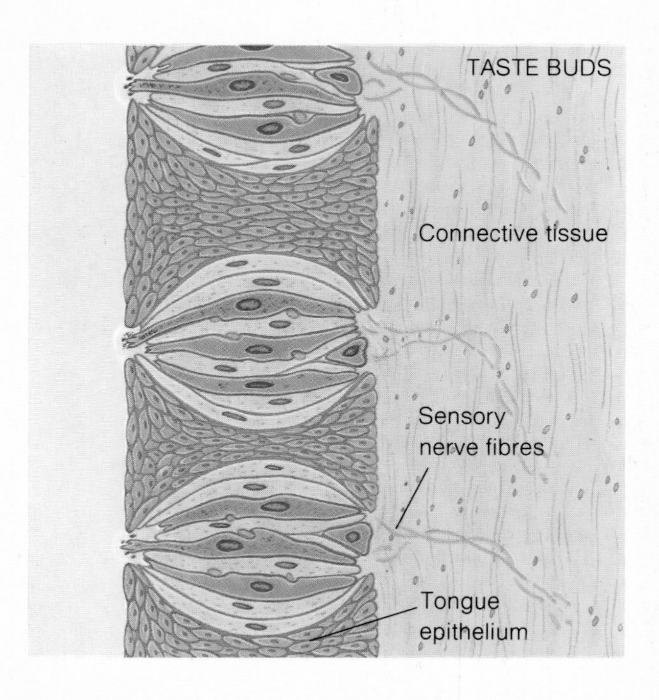

Individual taste buds can be seen *(left)* embedded in a papilla wall, which makes up the outermost layer or epithelium of the tongue. They respond to the presence of dissolved foods by stimulating nearby sensory nerve fibers. The impulses generated in these fibers travel along two of the cranial nerves directly to the brain.

A single taste bud in section *(right)* is revealed as consisting of different kinds of elongated receptor cells. A gap in the surrounding tissue, an apical pore, allows food molecules in solution to make contact with fingerlike microvilli projecting toward the pore. The receptor cell responds by releasing a neurotransmitter chemical from a synapse that surrounds a nerve fiber. Thus stimulated, the fiber depolarizes and fires off a nerve impulse to the brain, which interprets it as a taste sensation.

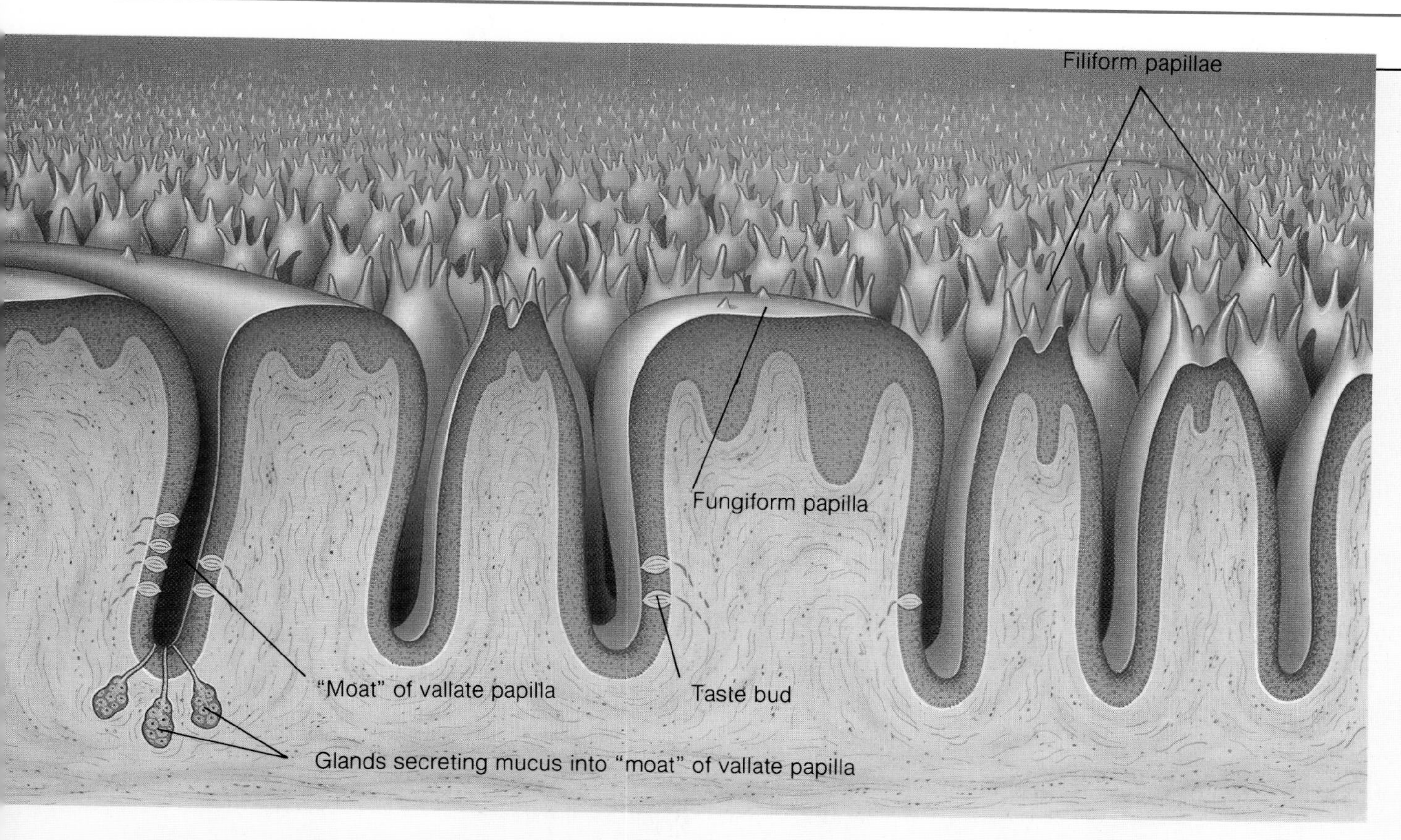

whereas those of the vallate papillae – concentrated in the "moat" surrounding its central mound – react to bitter flavors. The gaps between papillae are kept moist with mucus secreted from glands at the base of the gaps. Flavor molecules must dissolve for taste buds to detect them.

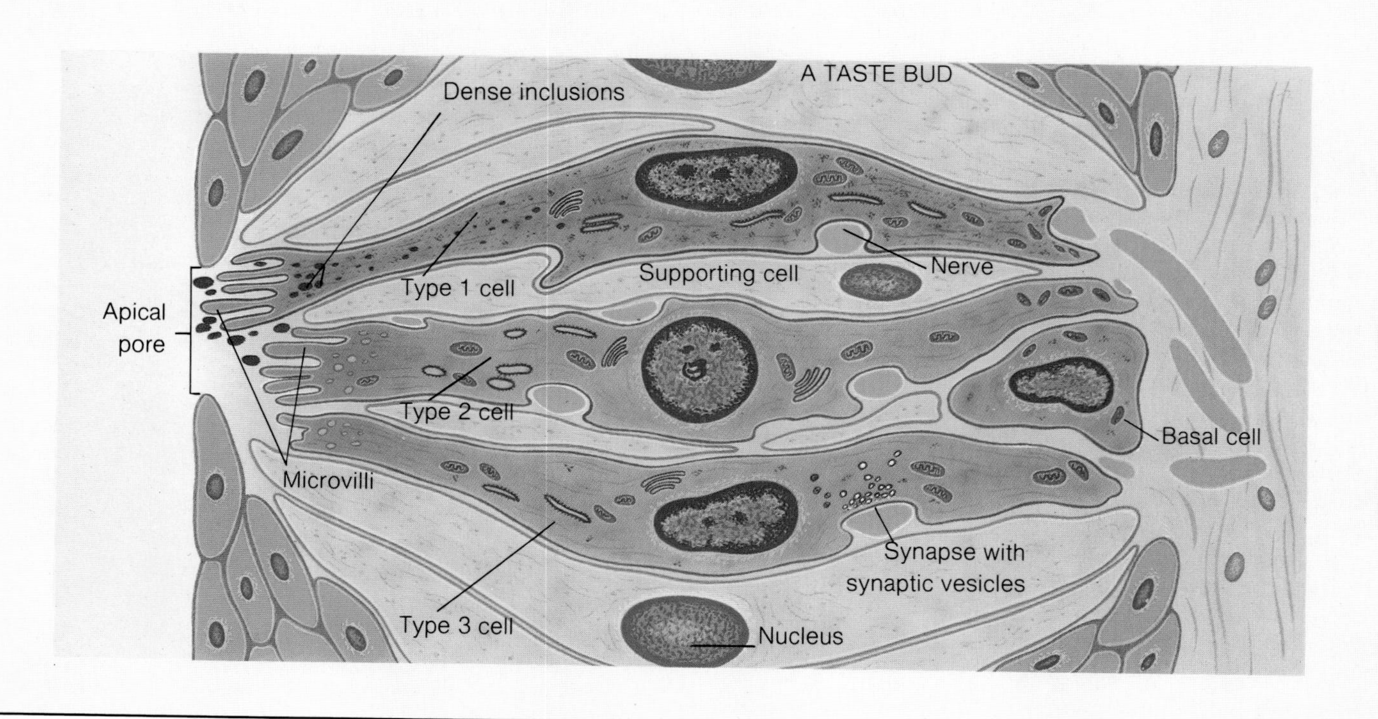

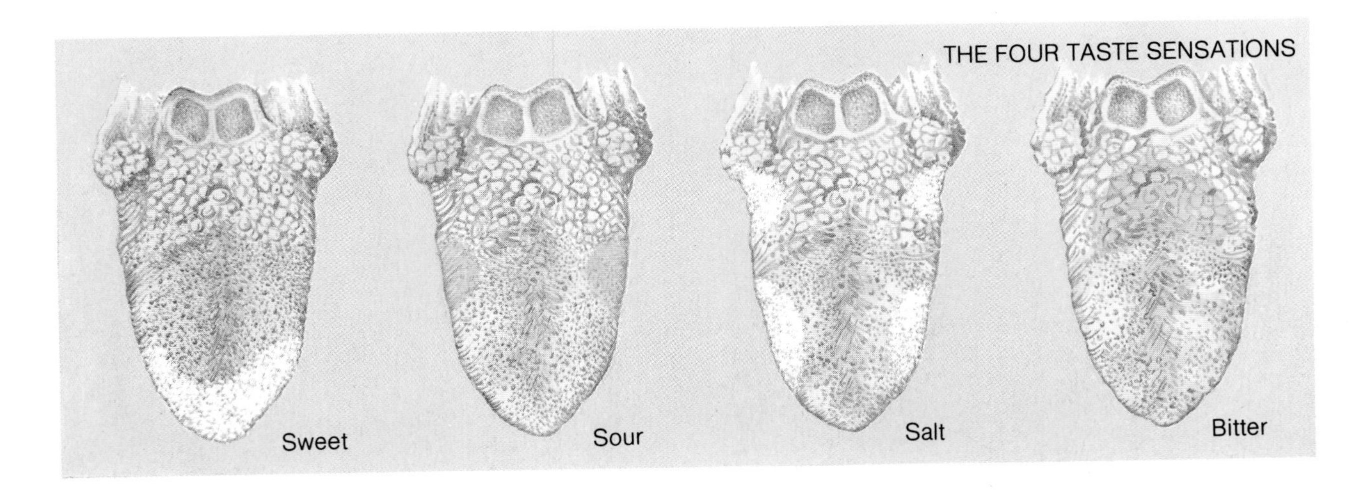

damage to the middle ear or cochlea. In itself it does not cause any problems, unless it interferes with normal hearing. Tinnitus is a symptom of Ménières disease, in which an increase in pressure of the cochlear fluid causes vertigo and hearing loss.

A versatile organ

The tongue is one of the most versatile organs in the human body. Its extensive complement of muscles allows a range of complex movements for chewing, sucking and swallowing, together with the vital function of modulating sounds to produce speech. But over and above these muscular functions it has a major role as the preeminent taste organ, and is also a highly sensitive touch sensor.

Most of the tongue consists of intimately interlaced muscles, arranged as paired blocks so that the left and right sides have independent sets of muscles. The two halves are divided by a fibrous septum running down the middle – externally this division is indicated by a central groove, or median furrow, running lengthwise along the tongue. Each block of muscles is formed from two arrays: a set of intrinsic muscles within the tongue for fine movements, and a set of extrinsic muscles connecting the tongue to the surrounding bones and used for large-scale movements.

Although it may seem to be floating freely in the floor of the mouth, your tongue is actually anchored in all directions by the four extrinsic muscle sets – the genio-glossus, hyo-glossus, stylo-glossus and palato-glossus. The genio-glossus runs from the front of the lower jaw into the tongue from tip to base. Contraction of these muscles (on either side) makes the tongue stick out as its whole foundation is pulled forward. A flat straplike muscle, the hyo-glossus passes from the side of the tongue down to one arm of the wishbone-shaped hyoid bone in the throat. Movement of these muscles pulls the sides of the tongue downward.

Linking the sides of the tongue to the base of the skull through the bony styloid process are the stylo-glossals. They act to pull the tongue backward and upward. Connected to the sides and rear of the tongue the palato-glossals run to the rear of the palate and lift the sides of the tongue when they are contracted.

Working together these extrinsic muscles have the flexibility to move the tongue in virtually any direction. But the movements they produce are relatively coarse, and fine shape changes are the province of the intrinsic tongue muscles. Again these are arranged in four groups: two run from the front to the back of the tongue; one runs transversely; and the fourth runs vertically.

Only four basic tastes can be detected by the tongue, though combinations of these give rise to a whole range of subtle flavors. Each type of taste is picked up by a different area on the tongue's surface – for instance, sweet flavors are detected by the tip of the tongue, whereas sour ones stimulate taste buds at the sides. The vallate papillae, which occupy a "V" shaped area toward the back of the tongue, detect bitter flavors.

The fine discrimination of the sense of taste involves complex nervous connections to various areas of the brain. This simplified diagram shows three pairs of cranial nerves carrying impulses from three parts of the tongue initially to the brain stem. From there links are made to subcortical regions such as the hypothalamus (shown only on the right) and the thalamus, as well as to higher centers in the sensory areas of the cortex (shown on the left).

Shaping sounds

In former times, a cruel punishment to deprive a transgressor of the power of speech was to cut out the tongue. In other words, even if the vocal cords are intact, speech is impossible without the fine movements of the tongue, which modify the basic vocal sounds. Acting in cooperation with the lips, the tongue forms a series of resonating passages within the cavity of the mouth. Sounds generated by the vocal cords travel upward and interact with the passages, so that they are modulated to produce the specific speech sounds, or phonemes. The precise movements and shape changes differ for every individual sound.

Typical movements include an up and down motion within the mouth cavity, making major changes to the characteristics of the air passage. At the same time the tip of the tongue can be rapidly applied to any region of the palate, inner surfaces of the teeth or the lips. The rear part of the tongue may be extended to the walls of the mouth cavity and to the pharynx. In combination these varied actions can momentarily check the outward-moving sound vibrations. They are basic to the production of the sounds for several consonants. For example, touching the tip of the tongue to the tips of the upper incisor teeth produces a "th" sound, whereas the "k" consonant sound is generated by momentarily blocking expiration by the back of the tongue.

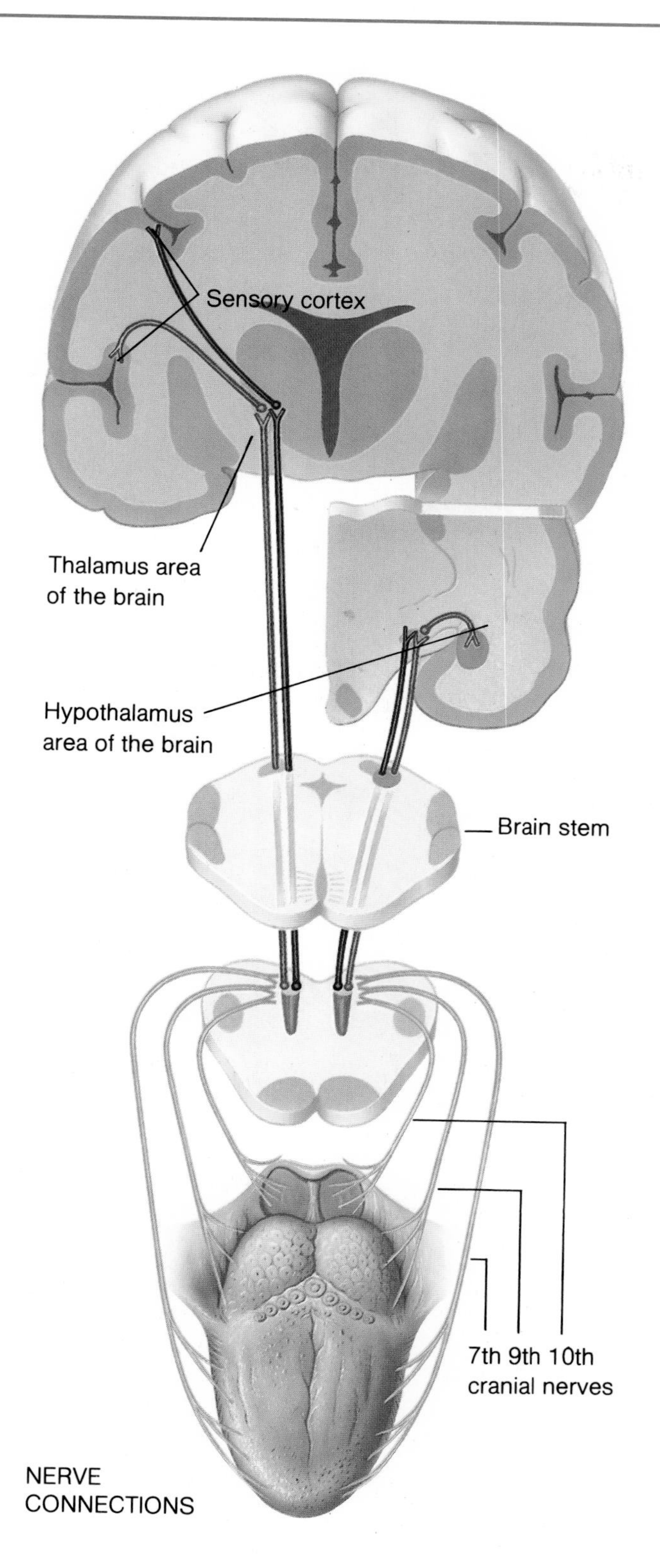

A rough surface
If you stick your tongue out and look at it in a mirror you will see only the front two-thirds of it. This part is known as the palatine section. The remaining third, at the back, is the pharyngeal section. The part you see has a rough velvety surface; and it is here that the taste receptor cells are massed. In contrast the underside of the tongue is covered with a thin, smooth mucous membrane loosely attached to the underlying muscles. At the midline there is a fold of tissue, the frenulum linguae. (Sometimes this fold is extensively attached to the floor of the mouth – in babies or young children – and they are literally tongue-tied, but the condition is easily treated by minor surgery.)

Close packed, short projections – or papillae – cover the upper surface of the palatine section of the tongue, and are what makes the tongue's surface rough. A sideways groove at the back of this section separates it from the pharyngeal section, which has a totally different surface configuration. Here the surface takes the form of a series of low, rounded hillocks caused by tiny tonsil-like lymphoid nodules and clusters of mucus-producing glands. Other similar glands are found right at the tip of the tongue, mixed up with the muscle strands. There is also a shallow central depression in the rear section of the tongue which marks the spot where the thyroid gland starts its development early in the life of the embryo.

Taste bud distribution
The papillae on the upper surface of the tongue carry a high concentration of taste buds, but this is not the only part of the mouth where taste buds are found. There is a thin scattering in many other parts of the mouth's mucous membrane, including the epiglottis (a small cartilage flap that helps seal the windpipe, or trachea), the larynx (voice box), the soft palate and the uvula (at the back of the mouth). There are even taste buds on the mucous membrane lining the upper third of the esophagus, which means that you carry on tasting food as you swallow it.

Types of tongue papillae
The furry surface of the tongue houses four main types of papillae: filiform, foliate, fungiform and vallate. Most numerous are the filiform, or conical, papillae, which are arranged in fairly regular rows running parallel to the central groove of the tongue. Some of these papillae have a simple conical form, while others have frilled tips, with each branch of the peak roughly conical. A surface coat of papillae follows all the irregularities of the tongue surface, and the outer cell layers are continuously and progressively converted into dead, hard scales. White discoloration or furring of the tongue – sometimes a symptom of disease elsewhere in the body – is caused by a buildup of filiform scales together with white blood cells.

The main job of the filiform papillae is to act as an abrasive coating, which helps give the tongue a cleaning and rasping action; this accounts for the details of their construction. When a sore patch or minor skin wound is licked, the filiform papillae have a cleaning effect on the surface. This action is then complemented by the antibacterial action of some of the components of saliva, such as the enzyme lysozyme.

Sensations of taste
All of the other three types of papillae are involved in the sensations of taste, and all have taste buds embedded in their surfaces. Lying near the back of the palatine section of the

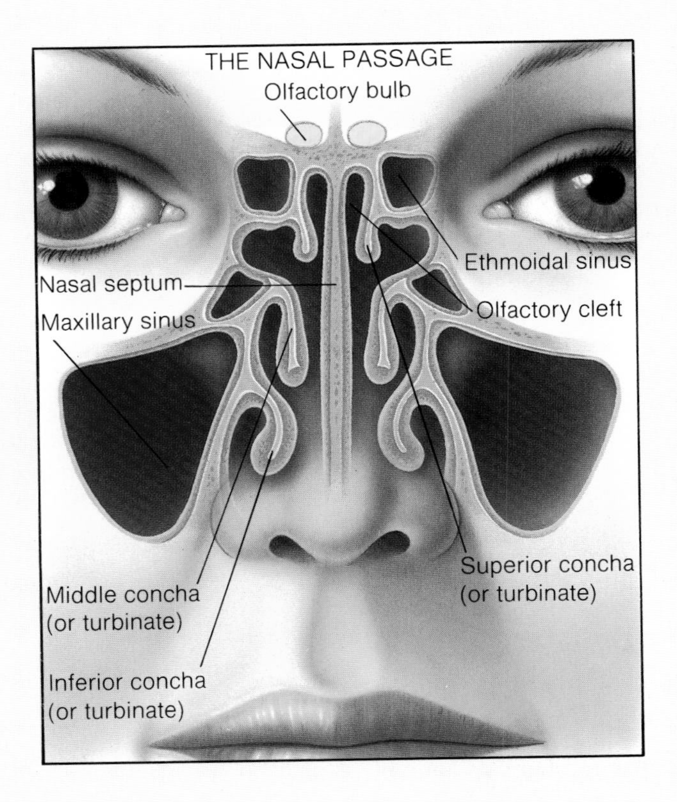

Convolutions of cartilage and bone form a labyrinth of cavities within the nasal area of the face. The superior conchae, or turbinate bones, and olfactory clefts are lined with olfactory membrane. Sensory hairs in the mucus-covered membrane send signals upward to the olfactory bulbs, located behind the bridge of the nose. Each bulb has a direct connection to the brain along the olfactory nerves.

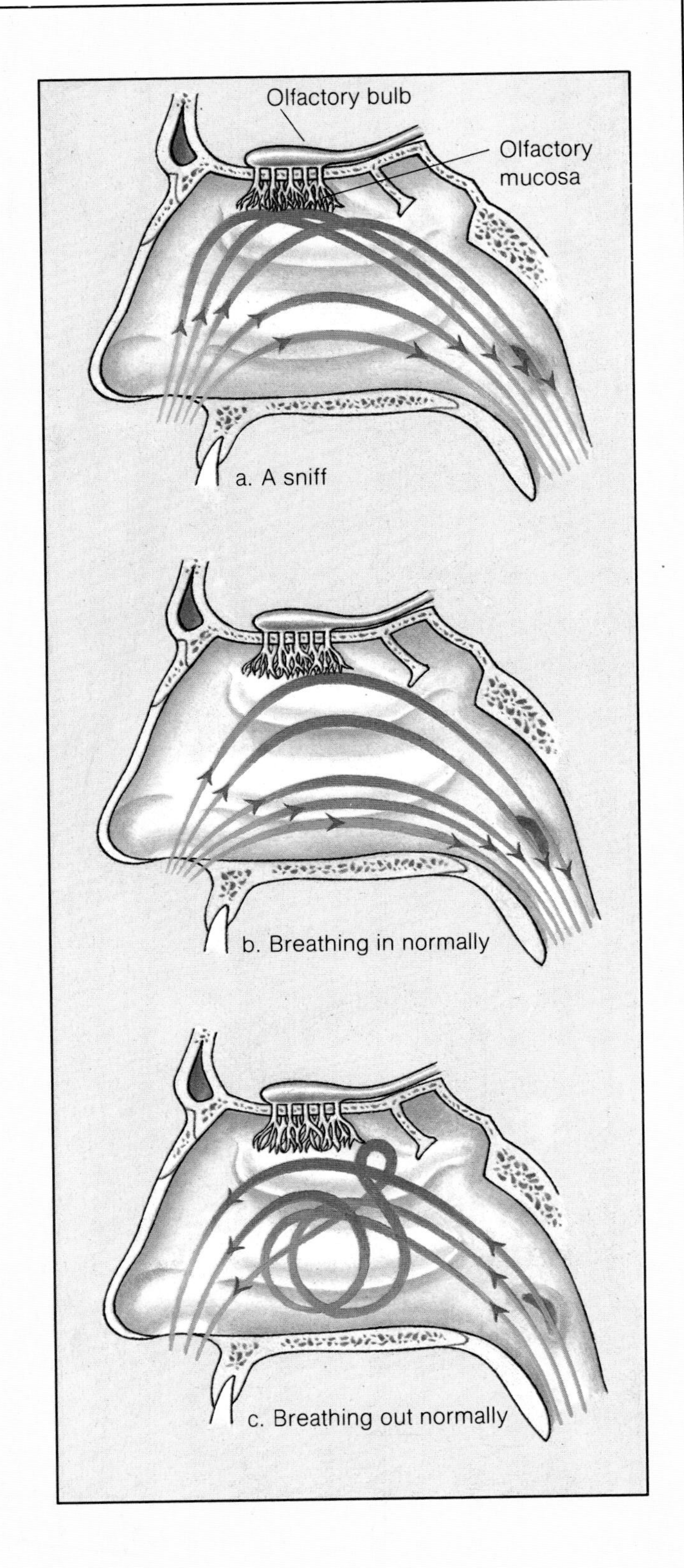

The flow of air through the nasal cavity *(right)* varies during sniffing, breathing in and breathing out. If you sniff hard to deliberately savor a smell (a), incoming air is forced up into the olfactory clefts to bombard the hairs of the olfactory membrane with odor molecules. During normal breathing in (b), little of the incoming air reaches the odor detectors. Breathing out (c) causes turbulent air flow, though odor perception is still maintained.

tongue, and arranged in a V-shaped formation directed toward the throat, are a set of seven to 12 bulging vallate papillae. Under high magnification these projections look rather like the round, domed keep of a castle surrounded by a deep moat and an outer steep-sided dike. Large numbers of taste buds are arranged in tiers on the outer walls of the "keep" and on both inner and outer banks of the "dike." And at the bottom of the moat there are secretory gland cells that produce a watery mucus that envelops the taste buds. Measurements show that there are around 250 taste buds on each vallate papilla during childhood and early adult life. But aging processes take their toll and by the age of 80 less than half of the buds are left.

There are rather more of the mushroom-shaped fungiform papillae, which are scattered over the surface of the tongue with particular concentrations at the tip and along the sides. In structure the fungiform papillae look rather similar to the central part of the vallate type; there is an almost spherical body with a flattened top and a narrow base. Finally the foliate papillae are clustered into two groups positioned on each side of the tongue just in front of the "V" of the vallate papillae. Again their name is descriptive, with each of these papillae having an elongated fold that looks like a leaf seen edge-on. Taste buds on these two types are scattered over the surface of the fungiform papillae, and on opposed walls of the foliate papillae.

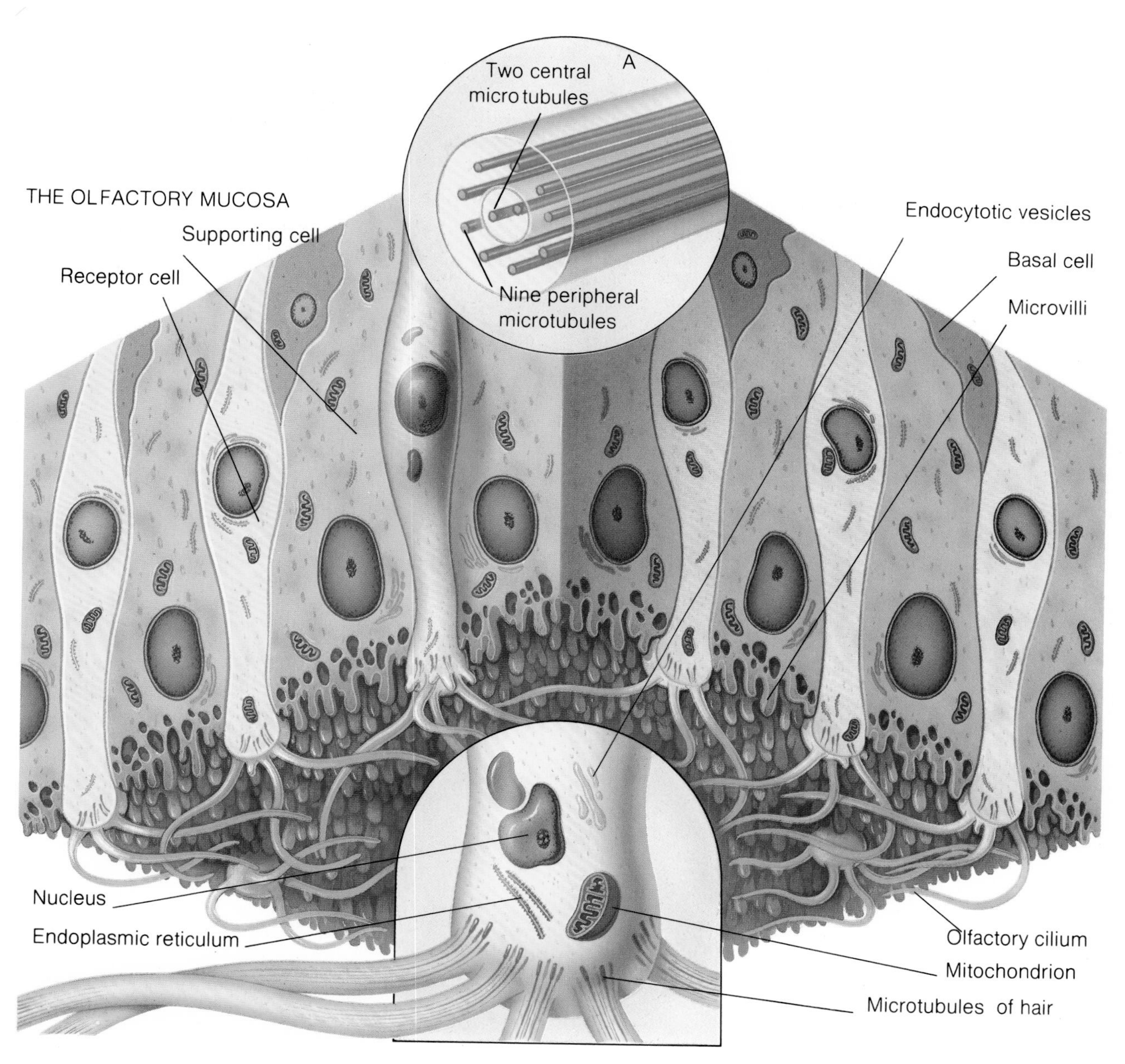

A close-up of the olfactory membrane shows the olfactory hairs, or cilia, spreading out from knobs (B) at the lower ends of their receptor cells. Each cell has a nucleus and the usual complement of cell contents, and is held in place by supporting cells surrounding it. The cilia have a fine structure composed of regularly aligned microtubules (A), and a number of receptor sites along their length. The arrival of an odor molecule at a receptor site triggers the receptor cell into generating a nerve impulse. Overexposure to a strong odor temporarily overwhelms the sense of smell.

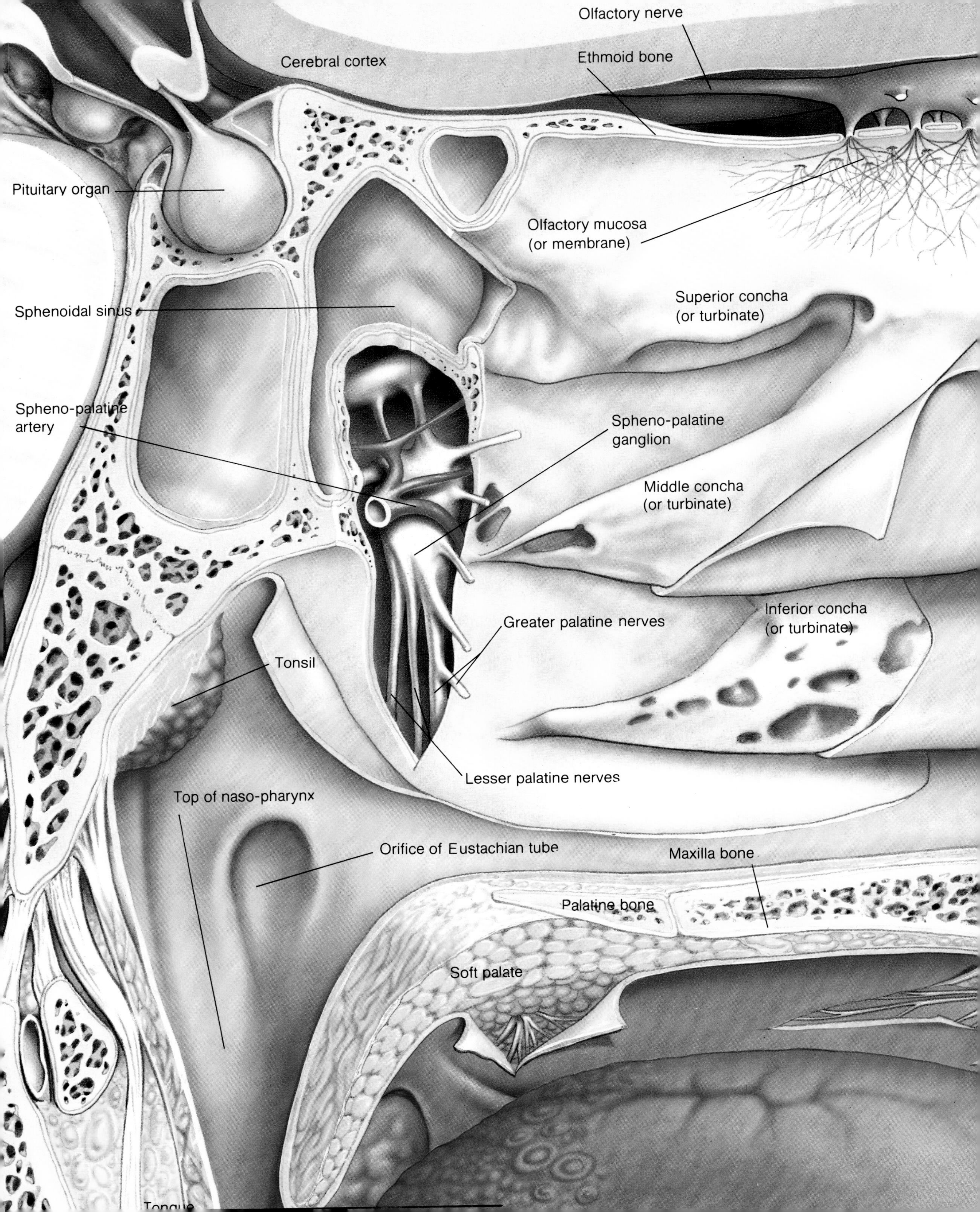
Olfactory nerve
Cerebral cortex
Ethmoid bone
Pituitary organ
Olfactory mucosa
(or membrane)
Sphenoidal sinus
Superior concha
(or turbinate)
Spheno-palatine
artery
Spheno-palatine
ganglion
Middle concha
(or turbinate)
Inferior concha
(or turbinate)
Greater palatine nerves
Tonsil
Lesser palatine nerves
Top of naso-pharynx
Orifice of Eustachian tube
Maxilla bone
Palatine bone
Soft palate

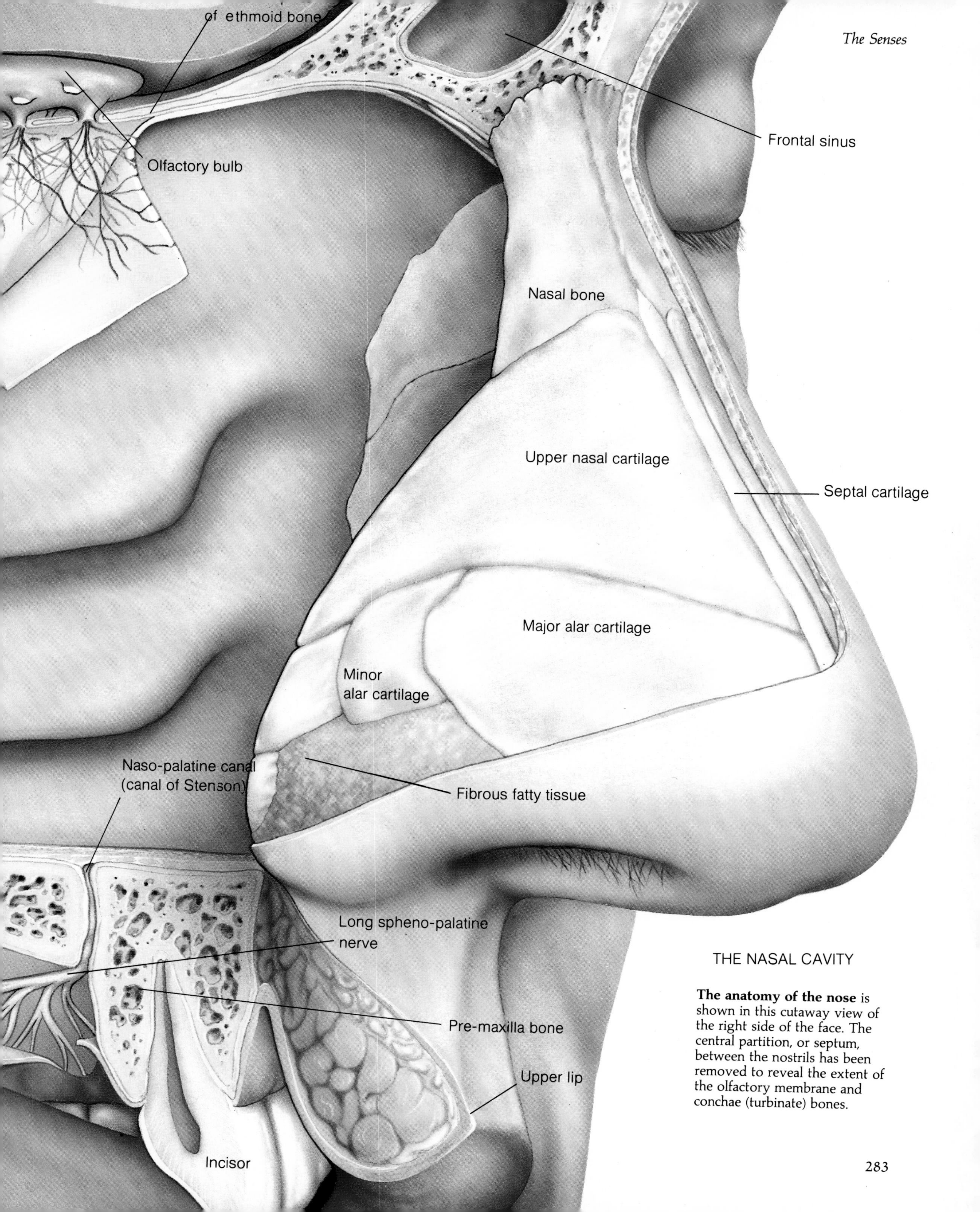

THE NASAL CAVITY

The anatomy of the nose is shown in this cutaway view of the right side of the face. The central partition, or septum, between the nostrils has been removed to reveal the extent of the olfactory membrane and conchae (turbinate) bones.

Receptors for taste

The actual taste buds are microscopic organs – no more than 20 to 40 millionths of an inch wide – containing 30 to 80 receptor cells, many of which are connected to the endings of separate nerve fibers. The buds are buried in the surface layers of the papillae and connect to the outside through a narrow duct known as the apical pore.

Detailed examination using an electron microscope shows that the taste buds have a complex structure. The body of the bud contains no less than three different cell types, known as Types 1, 2 and 3. All three types extend from the base of the bud upward to end in fingerlike projections in the apical pore.

The precise function of each cell type has not been firmly established, but it is believed that cells of Types 2 and 3 are the actual taste receptors, while Type 1 cells have a supporting function. All are renewed on a regular basis, surviving for only about ten days before being replaced. A fourth cell type, basal cells, are also found in the taste bud and these may be newly formed replacements that are about to develop into Type 1, 2 and 3 cells.

Taste receptor cells do not have their own nerve axons, instead they make contact with the ends of sensory nerve fibers running in the body of the tongue. Indeed in an embryo the taste buds form under the influence of the sensory nerve fibers. This close relationship is maintained in an adult; if the connection between nerve and bud is broken, the bud rapidly degenerates. But if the nerve regrows into the surface layer of the tongue, it stimulates the growth of new taste buds.

Types of taste

Everybody has favorite foods and favorite flavors, yet the taste buds themselves can recognize only four basic tastes: sweet, salt, sour and bitter. These characteristics are registered by different types of papillae on the tongue. Fungiform papillae respond to sweet and sour tastes; the vallate region responds best to sour and bitter sensations; and the foliate receptors also favor sourness.

Transferring these characteristics onto the surface of the tongue gives a taste map; it shows that the tip has peak sensitivity to sugars, the sides to acid tastes, and the back of the palatine section to bitterness. This arrangement explains why saccharine, and similar sweetish substances, leave a bitter aftertaste. Initially the tip of the tongue responds to the sweetness, but the bitter taste is then detected as the sensors at the back of the tongue come into action.

The actual sensation of taste depends on submicroscopic events taking place at the molecular level of body chemistry. Molecules of the chemical substances responsible for a taste interact with the membranes of receptor cells in the apical pore of a taste bud. This interaction generates an electrical disturbance in the cell, and in turn this triggers the response of an associated nerve cell which carries the message to the brain. In practice, however, it takes the impulse from several related sensor cells to produce a recognized taste response within the brain.

From tongue to brain

Nerve fibers for the taste buds on the tongue come from the seventh and ninth cranial nerves. Cranial nerve seven – the facial nerve – supplies the front regions of the tongue and so handles sensations that are mainly related to salt and sweet tastes from the fungiform papillae. Bitter and sour sensations come from vallate and foliate papillae on the rear part of the tongue, and are dealt with by the ninth cranial nerve. Most of the other taste buds scattered throughout the mouth cavity and throat are supplied by the tenth cranial nerve, the vagus.

Most children like sweet foods whereas many of their elders prefer savory dishes. This difference may be a primitive hang-over from the way in which taste sensations are processed in the brain. Signals from the taste receptors enter the central nervous system along two different routes, one to the subcortical regions of the brain and one to the cerebral cortex. Subcortical connections run to areas such as the medulla and cerebellum, which are some of the older parts of the brain in evolutionary terms. In particular the sensations route to the amygdala and hypothalamus regions, which appear to be responsible for nonconscious, or instinctive, reactions. A typical example is the way most people instinctively respond favorably to sweet tastes but reject bitter ones – unless they have consciously trained their tastes to more sophisticated responses, which favor bitter or savory sensations.

In contrast signals that are transmitted to the cerebral cortex pass through a taste center in the thalamus and on to at least two separate regions of the cortex. It is there that the direct taste sensations are experienced, where fine discriminations are made. And it is there that you deliberately educate your responses to different flavors. However the sensation described as taste also takes account of the smell of the food, its physical texture as you chew it and its temperature. All of these sensations fuse with the direct taste responses in a complex sensory experience.

Positive perceptions

The human response to individual tastes is at least partly colored by the foods needed in the diet, but deliberately superimposed preferences – such as the cravings some women experience during pregnancy – often have a much stronger effect. Even so, the inherent responses can still make you favor heavily salted foods when your body is short of salt, for example.

Taste perception can also be altered by disease, and there are several ways in which this can happen. One of them has little to do with specific taste receptors themselves, but involves the all-encompassing phenomenon of taste. Thus certain respiratory infections can disrupt the sense of smell, and when this occurs food seems to taste dull and insipid; the taste buds have not been affected, but the smell component is missing from the overall sensation. More direct is a process of taste aversion that acts unconsciously to make us actively dislike specific foods that have previously been associated with pain, nausea or illness.

Something in the air

Smell is often considered to be the least important of our senses, but it may well be one of the oldest and probably acts more directly on our subconscious than the other senses. There is little doubt that scents have important roles in human behavior. Indeed the body is provided with glands to produce specific odors. Many of these appear to be associated with sexual attraction and excitement – they generally start to operate during puberty – but there are others that have considerable significance. For example, the bond between baby and mother is thought to be enhanced by a form of scent imprinting. A baby suckling at the breast pushes his or her face into a bank of scent organs that surround the nipple. Equally, the smell of a baby is almost universally regarded as pleasant, providing an added

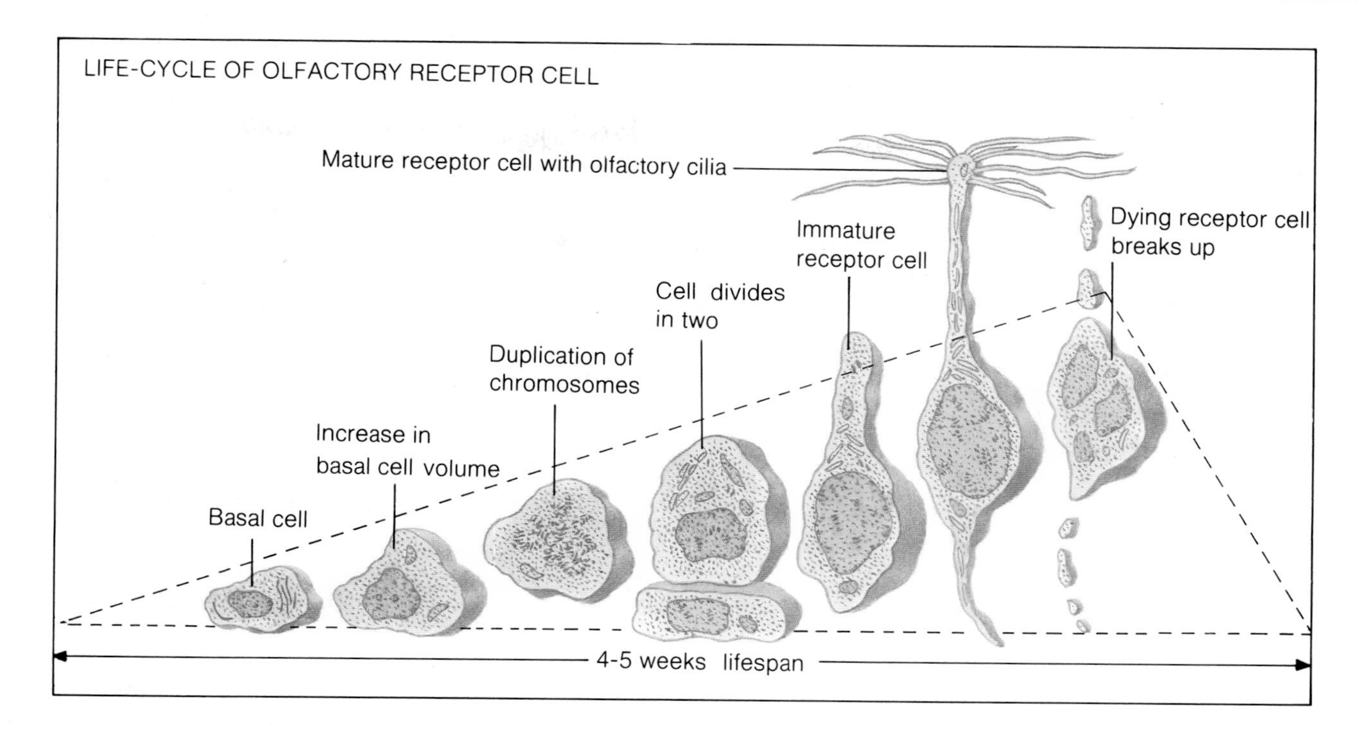

Olfactory receptor cells live for only about a month, and function as odor-detectors for only a small part of that time before they break up and die. They develop from basal cells, which are continually produced to maintain the supply.

source of pleasure for the mother when she cuddles her child.

More overtly a whole section of our culture is devoted to the carefully calculated stimulation of the sense of smell – and hopefully of other responses such as sexual attraction – with carefully blended perfumes and scents. And a further indication of the powerful potential of the sense of smell is the way it can become a major source of information in the absence of other more immediate senses, especially sight.

Nasal passages

Only a small part of the nose and nasal cavity (the internal space bounded by the floor of the cranium and the roof of the mouth) is taken up by the organs of smell – the rest of it is mainly concerned with processing the airflow on its way through to the lungs. The walls of the nasal cavity, and particularly the flaplike middle and inferior conchae, are coated with respiratory mucous membranes. These incorporate a vast number of tiny hairlike cells which act together to move sequential waves of mucus toward the throat. Dust, bacteria and chemical particles inhaled with the air are trapped by the mucus, carried back and swallowed; they are then dealt with by the gastric juices, which quickly nullify any potential harm.

The sense organs themselves consist of two yellowish-gray patches of tissue, the olfactory membranes, each about the size of a postage stamp. They lie in a pair of clefts located just under the bridge of the nose and at the top of the nasal cavity.

The yellowish hue of the olfactory membrane is due to the presence of two main kinds of pigment: carotenoids and free vitamin A, and phospholipids. The reasons for the coloration are not altogether clear, but the pigments seem to be necessary for the membrane to work. Certainly albino animals (ones which have no pigments at all) are known to have no sense of smell.

During normal breathing most of the air flows smoothly through the nose, with only a small fraction reaching the olfactory clefts – though this is enough to trigger a response to a new smell. Deliberate attempts to detect smells by sniffing make the air move through the nose much faster, increasing the flow that makes its way up to the olfactory clefts and so carrying more odor molecules to the sensors.

If you have a cold the production of respiratory mucus is several times greater than normal – so you have a runny nose. The mucus may obstruct the narrow openings to the olfactory tissue. This prevents odor-carrying air from reaching the sensory membranes and your sense of smell is restricted.

Inside the olfactory membrane

Viewed through an electron microscope the active surface of the nasal membrane looks rather like a plate of spaghetti in a sticky sauce. The spaghetti strands are olfactory cilia, or hairs, growing out from receptor cells, and the sauce is mucus produced by special glands (Bowman's glands) in the membrane. Each receptor cell – and there may be as many as 10 million of them – ends in a tiny swelling, or olfactory knob, which has about five olfactory hairs growing out of it. The hairs may be as long as one-hundredth of an inch but are normally much shorter. They lie tangled together in the mucus and may have slight swellings where they touch. Although the precise mechanism has not yet been fully established, the hairs appear to contain receptor sites that take up specific odor molecules and so trigger the receptor cell into producing a nerve impulse.

Just like the taste cells, smell receptor cells have only a short working life – they function for about a month and then start to break down and are removed by the mucous stream. Replacement cells are constantly being produced; they grow from cells in the base of the olfactory membrane, thrusting outward and developing olfactory hairs. One explanation for this short life is that the cells gradually become clogged up with absorbed odor molecules, but this concept is far from being universally accepted. Another possible reason is that the receptors are in a particularly exposed position for what are really sensitive nerve cells, and so are more likely to suffer damage. With senses such as vision and hearing the receptor cells are much better protected and do not normally deteriorate.

Why do smells smell?

Many theories have been proposed to explain just what gives a particular substance its specific odor. The most probable explanation is that the effect depends on the shape of the substance's molecules, and the way they lock onto (or into) the receptor sites on the olfactory hairs. Support for this theory is given by the way continued exposure to an odor rapidly results in odor fatigue so that the smell is no longer perceived. This fatigue may come about because the receptor sites are fully occupied, and a short break is needed before the sites are ready for further use and the specific smell sense can be reactivated.

The bulb that smells

The olfactory hairs are effectively a direct extension of the brain. The receptor cells they spring from run from the olfactory mucosa directly into a part of the brain called the olfactory bulb. The bulb is very small, about the size of a match head, and located slightly below the bridge of the nose, about four-tenths of an inch into the head. It is separated from the olfactory membrane by the cribriform plate, a wafer-thin section of bone at the front of the cranial cavity.

Nerve fibers from the receptor cells pass through openings in the cribriform plate and run to synapses (gaps between nerve connections) in the bulb. These are concentrated in structures known as glomeruli, with the 10 million or so receptor cells connecting to just 2,000 glomeruli. Two outputs run from each glomerulus: one set of 24 tufted cells connects to the olfactory bulb on the other side of the head, and another 24 cells – the so-called second olfactory neurons – run into deeper regions of the brain.

Each nerve cell in each group of 24 running from a glomerulus is like an electrical switch: it can be either on or off. Used like the binary code in a digital computer this arrangement allows a total of around 16 million permutations, giving plenty of scope for the identification of individual odors – if this really is the way that the odor signals are coded.

On line to the old brain

Nerve connections from the organs of smell reach into many parts of the brain. After leaving the glomerulus the second olfactory nerve runs along one of the lateral olfactory tracts and connects with the third olfactory neurons in the amygdala region. From there the nerves penetrate the basal and deeper areas of the brain, including the thalamus, entorhinal cortex (in the hippocampus region) and the lateral and basal nuclei of the amygdala itself. In addition an offshoot goes to the olfactory tubercle and on to the preoptic-hypothalmic region, which has further connections to the pituitary gland. These connections give the sense of smell direct access to many parts of the brain, in direct contrast to the senses of sight and hearing, which are much more localized. Signals from these sense organs are processed by relay centers in the cerebral hemispheres before they reach the rest of the brain.

A particular feature of the web of connections associated with smell is the way it is concentrated in the regions of the brain which, in evolutionary terms, are the most ancient. These parts are mainly associated with emotion and sexual behavior, and have little to do with the higher thought processes. But taking the brain as a whole only a minute part of it is devoted to the sense of smell – a reflection of the inferior standing it now has.

The scent or bouquet of a wine is as much a measure of its qualities as its taste. Only when the smell has been thoroughly appreciated will the wine taster put the glass to his lips. The "nose" of the wine taster is so highly developed that he can distinguish between many wines by their smell alone.

Sensitivity and discrimination
The ability to identify smells has two distinct sides to it – acuity and discrimination. Fairly high concentrations are needed for odors to register because the human sense of smell is not particularly acute – dogs can perceive many odors at concentrations that are several hundred million times weaker. But when it comes to discriminating between different smells, we can put up a much better performance. The ability to identify smells improves rapidly with practice and someone with a trained nose – say a perfumier – can distinguish between many thousands of odors, carrying a memory of the individual smells from one test or work session to the next.

Temporary loss of the sense of smell is a common phenomenon, often due to a heavy cold which causes swelling of the nasal membranes and produces large amounts of congesting mucus. Allergies and specific infections of the nasal membranes (rhinitis) have similar effects, all of which are associated with nasal congestion. Other conditions can alter the sensitivity to odors. In pregnancy, for example, smells previously thought of as pleasant may suddenly start to be repugnant.

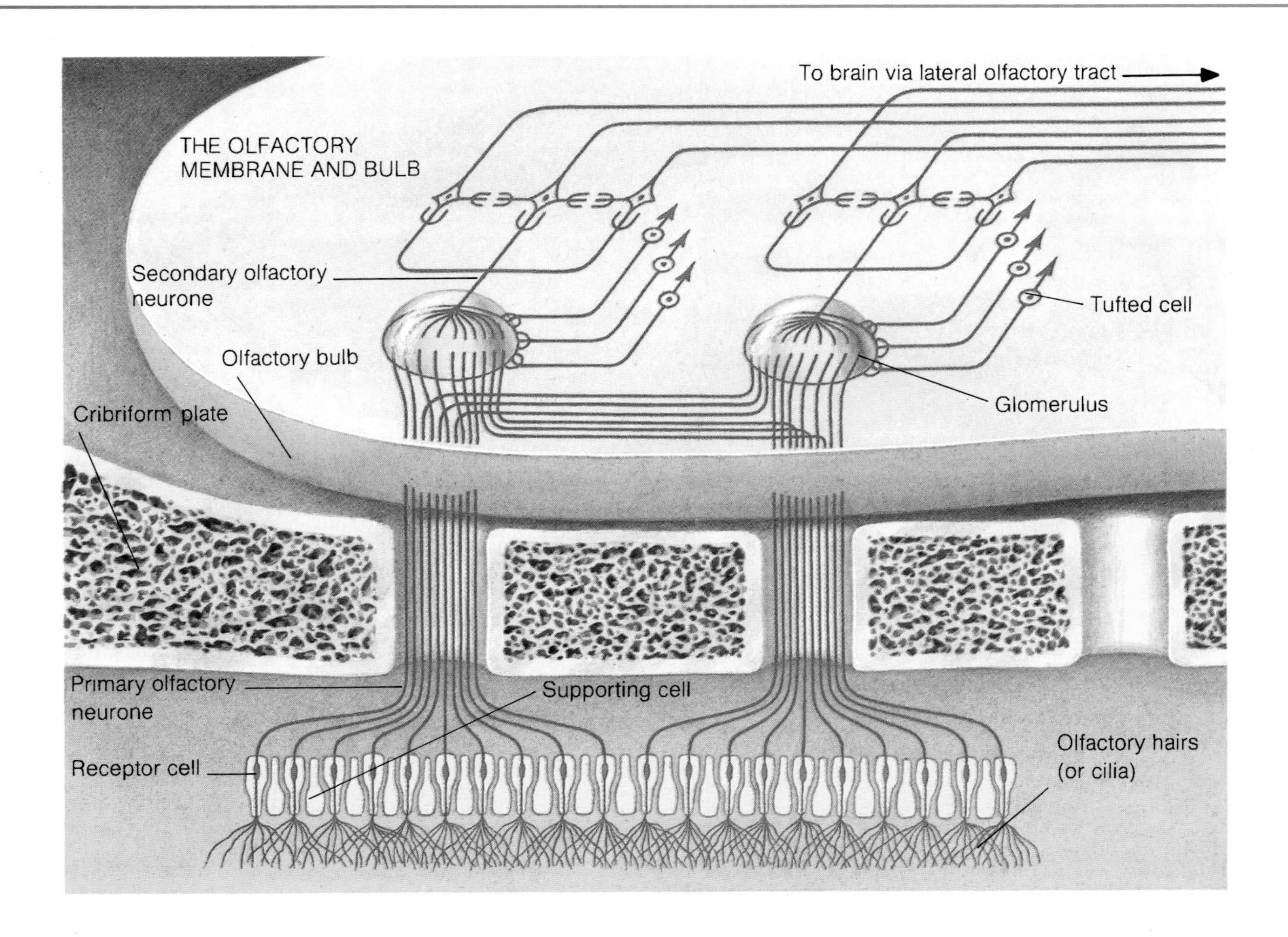

A series of holes in the cribriform plate bone of the skull allow primary neuron nerve fibers from the receptor cells to connect with glomeruli in the olfactory bulbs, before joining the olfactory nerve.

Much more serious are conditions such as brain tumors, which can destroy the sense of smell without directly affecting the olfactory organs, and epilepsy, which is associated with altered odor perception. In some cases the altered sensitivity occurs several days before the epileptic seizure. The reverse relationship also applies, with strong odors sometimes having the effect of stopping a seizure.

Changing preferences

Everybody has favorite scents or odors, but these odor preferences tend to change with age. Babies respond most positively to the odor characteristics of their mothers. Most young children are fond of sweet, fruit smells – such as that of strawberries – and are unimpressed by heavy odors such as musks. Producers of erasers and other novelties for children – as well as manufacturers of candies – are aware of these preferences and produce fruit "flavored" erasers and pencils. Perhaps the fondness for such odors reflects a more sensitive sense of smell. When exposed to a strong, but unfamiliar odor, children under ten years old tended to perceive it at slightly lower concentrations than adolescents did, and at much lower concentrations that did adults in the 21 to 39 years age range.

As young people enter adolescence, they begin to lose their fondness for fruity odors, and both boys and girls become attracted to heavy, oily odors such as those of musk and sandalwood. Dr. R. W. Moncrieff, in his classic research on human scent preferences, demonstrated that musk – a constituent of the sexual attractant odors of many species of animals – becomes attractive to adolescents at about the age of 15. But by about the age of 20, the attractiveness of musk has waned considerably, only to rise sharply thereafter.

Humans have a natural body odor that changes gradually throughout life. Baby smell is universally regarded as pleasant, and all mothers know the characteristically sweet smell of baby. This disappears during childhood, to be replaced by a more "adult" odor as a young person's apocrine and sebaceous glands begin working in earnest with the onset of sexual maturity. The odor is perfectly natural, although in modern society it is often deliberately removed by deodorants or modified or masked by perfumes.

The unpleasant smell commonly called "body odor" is not a product of scent glands but a consequence of poor personal hygiene. Bacteria get to work on stale sweat on unwashed bodies and convert it into substances that have, to many people, an extremely unpleasant smell. The cure is simple: a bath or shower and clean underclothes every day.

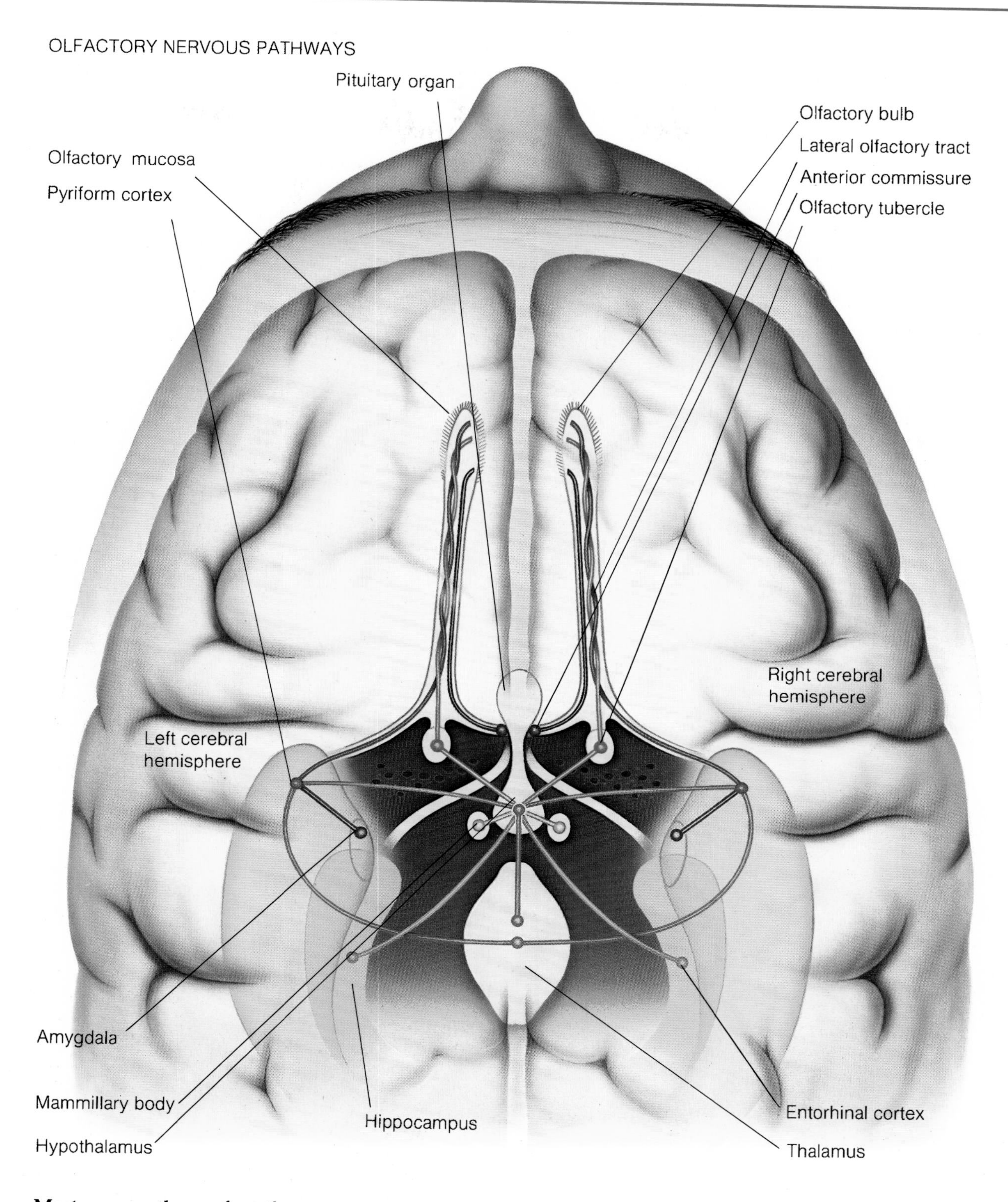

Most nerve pathways from the olfactory bulbs terminate in the "old" central area of the brain, which was responsible for directing primitive emotional and sexual behavior. Some connections are also made to the cerebral cortex, a "younger" part of the brain.

CHAPTER 15

Reproduction

The creation of human life is a wonder. Two minuscule cells – a male sperm and a female ovum – fuse together to form one new cell, so small that it is barely visible to the human eye. Within four days this cell implants inside a woman's womb where it rests, nourished by the mother. It grows and develops at a phenomenal rate until it emerges, nine months later, as a fully-formed, breathing, feeling new human being, a tiny replica of the adults who gave it form.

Since earliest times the wonder of conception and birth has fascinated thinkers, poets, dreamers and philosophers who have woven myth and fantasy about the event. But reproduction, though fantastic, is not haphazard. Instead it calls into play all the most sophisticated workings of the human body, relying for its achievement on timing, specialized anatomical structures, and the concerted efforts of glands and hormones.

Like yin and yang, reproduction involves the fusing into one whole of two halves – male and female. Even with today's advances in medical science that have enabled such developments as test-tube babies, the male and female contributions toward the creation of a new human being remain basically the same. The male body is designed to allow the manufacture, storage and transfer of male sex cells, or sperm. The woman's body is designed to manufacture, store and release eggs, or ova. Together the two bodies provide an anatomical structure that dependably unites sperm and eggs.

The male reproductive system

The sexual organs of a man are partly visible and partly hidden within the body. Visible parts are the penis and scrotum, a saclike pouch suspended below the penis which houses the two egg-shaped testicles, or testes. Hidden inside the body, and also part of the male reproductive system, are the prostate gland, seminal vesicles, and a number of tubes such as the vas deferens, or sperm duct, which links the whole system together.

The penis transfers sperm to the woman's body during sexual intercourse. Normally flaccid, but becoming erect during sexual excitement, the penis consists of spongy tissue loosely covered with skin. Erection results when extra blood is pumped into the spongy tissue during sexual arousal. The rounded head of the penis, or glans, is separated from the shaft by a rim of tissue called the corona, or crown. The foreskin, a retractable hood of skin, normally covers the glans but is often circumcised at birth, a procedure now believed to lessen the risk of penile cancer. A tube along the center of the penis, the ureter, serves to carry urine from the bladder. But at the climax of intercourse it functions as part of the sperm duct and carries semen.

The sperm industry

Sperm production takes place inside the testicles, at an astonishing rate. During a lifetime, from puberty onward, one man may produce as many as 12 trillion sperm. Each sperm takes about 72

A woman's changing contours during pregnancy can be recorded using the technique of stereophotography. Parallel lines, projected onto her figure from directly in front and behind, are molded by her curves to reveal the shape, just as contour lines on a map depict the changing topography of the landscape.

The male sexual apparatus is partly inside a man's body and partly outside. Sperm are manufactured in the testes, or testicles, which are kept at the correct temperature hanging in the external scrotum. From there they travel to be stored in the epididymes. Normally the penis is limp but during sexual arousal blood pumped into its spongy tissue causes it to become erect. As the moment of ejaculation approaches, Cowper's gland, the seminal vesicles and the prostate contribute fluids that mix with the sperm to lower its acidity and form semen, which muscular contractions squirt along the vas deferens and out of the urethra.

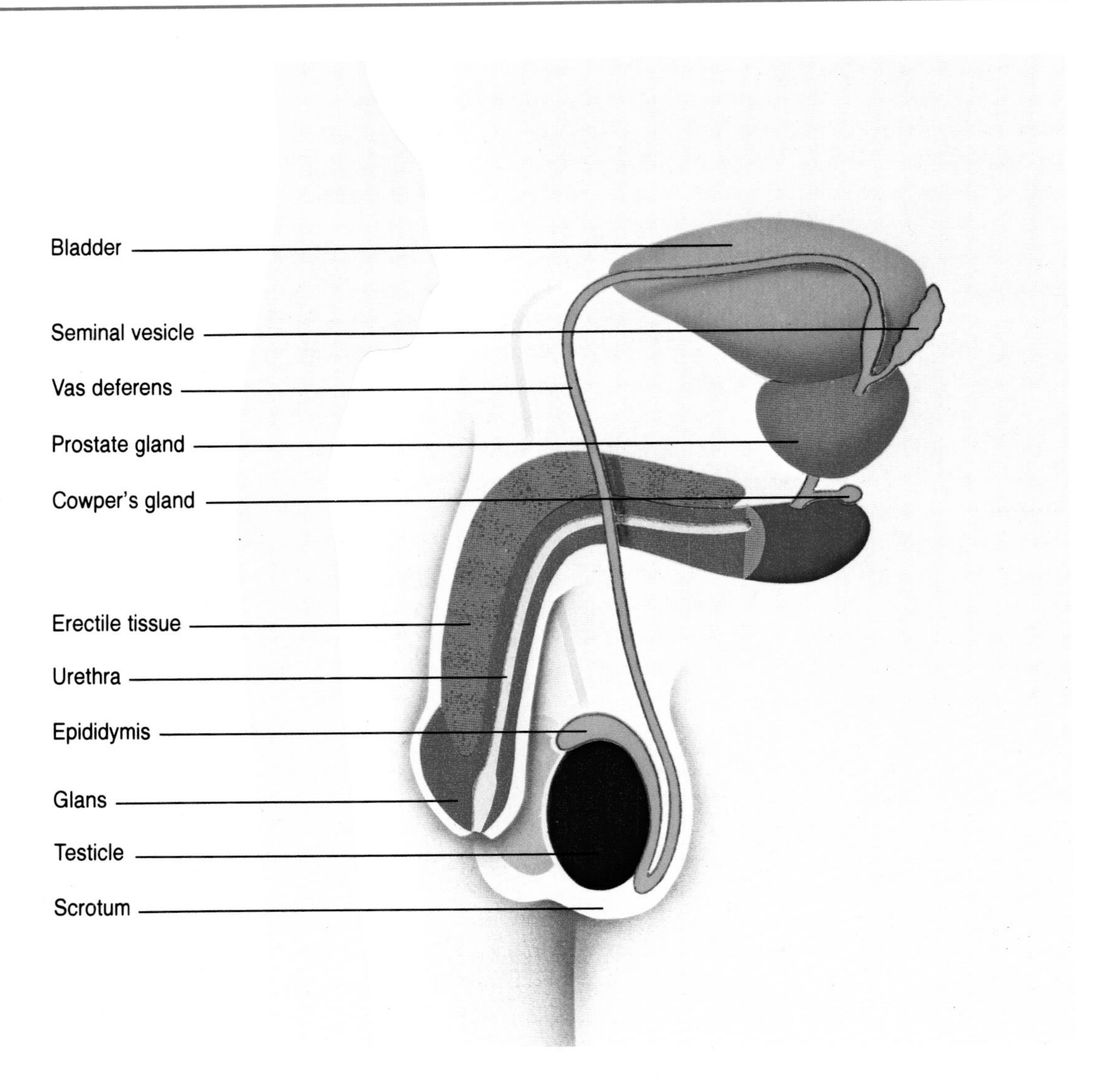

days to mature, and the entire operation is controlled by a complex interaction of hormones, and made possible by the thermostatlike quality of the scrotum, which keeps sperm at the correct temperature.

Sperm are manufactured from parent cells called spermatogonia lining the seminiferous tubules, which are coiled tubes within the testicles. When a spermatogonium divides, one half remains behind for future division and the other half migrates into a layer of specialized Sertoli cells, which secrete fluid and nourish the new cells. At this stage the young sperm cell, like others in the body, contains its full complement of 46 chromosomes. It divides further to produce four spermatids, each containing only 23 chromosomes.

Within the Sertoli cells the young sperm acquire the characteristic tadpole shape of mature sperm, developing an oval body housing the cell nucleus and forming a long whiplike tail. The maturing sperm then move into the epididymes, coiled tubes which lie on top of each testicle. The epididymis gives way to the vas deferens, a long tube connecting the testicle to a seminal vesicle behind the bladder. There are two seminal vesicles, in which the sperm cells are stored until fully mature; the vesicles also supply the fluid part of semen. Within the epididymis and vas deferens sperm remain fertile for some weeks.

Hormone control

A complex interplay of hormones triggers and controls sperm production. It begins at puberty as a boy matures into a man. The testes produce not only sperm but also testosterone, the male sex hormone responsible for the development of male characteristics such as hair on the face and body, the deepening of the voice, and the enlargement of the external sexual organs. The trigger for the teenage growth spurt in boys, testosterone also brings about its end, halting the growth of the long bones in the limbs at about the age of twenty.

Testosterone itself is secreted in response to chemical commands within the brain. Sparked off by these, the hypothalamus releases a hormone called gonadotropin-releasing factor. This stimulates the pituitary gland to release two hormones – luteinizing hormone (LH) and follicle-stimulating hormone (FSH). These enter the bloodstream and are carried to the testes, which begin to produce testosterone. Hormone levels are kept constant by an ingenious feedback system whereby rising testosterone levels suppress production of gonadotropin-releasing factor, which in turn reduces the release of LH by the pituitary. As testosterone levels gradually drop, so releasing factor and LH are stimulated, which again get to work on the testes and make them produce more sex hormone.

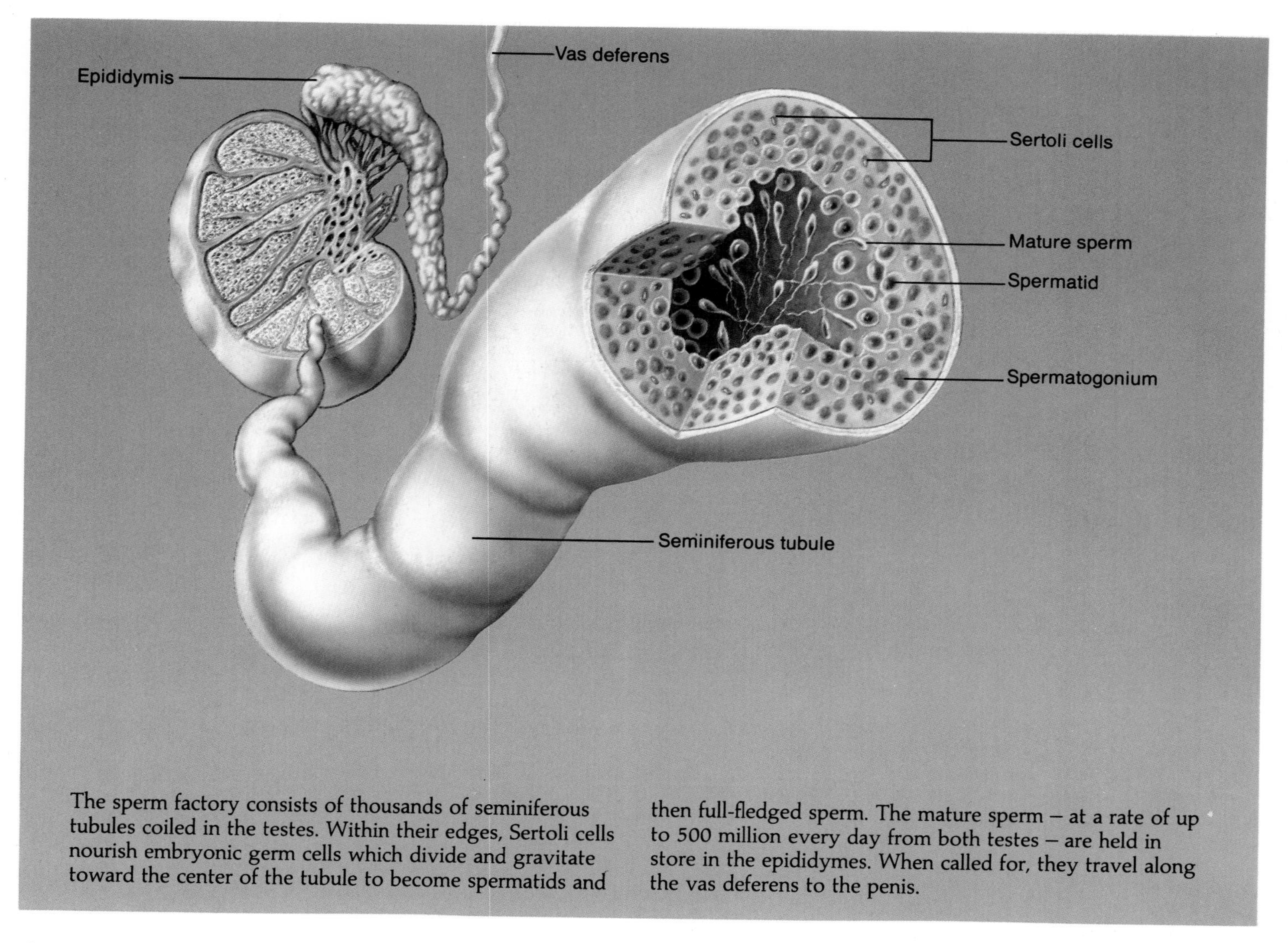

The sperm factory consists of thousands of seminiferous tubules coiled in the testes. Within their edges, Sertoli cells nourish embryonic germ cells which divide and gravitate toward the center of the tubule to become spermatids and then full-fledged sperm. The mature sperm – at a rate of up to 500 million every day from both testes – are held in store in the epididymes. When called for, they travel along the vas deferens to the penis.

The female reproduction system

The term "equal but different" so often applied to women and men is particularly apt when it comes to reproduction. For while both play an equal part in conception, it is the difference which enables successful reproduction. Whereas the man produces and transfers sperm, it is the woman's body which produces, stores and releases the vital other half, the ova. In addition it is the woman's body which receives sperm, shelters the developing fetus, and finally gives birth to the new infant.

Unlike those of a man, the woman's sexual and reproductive organs are almost entirely hidden, so that in a standing position the only visible sign is the pubic hair. A woman's sexual organs are known collectively as the vulva. They consist of a mound of fatty tissue, the mons pubis, which, from puberty, is covered with pubic hair. Extending from the mons are two folds of tissue, the outer lips, or labia majora, which protect the reproductive and urinary openings between them. Between them lie the labia minora, or inner lips, delicate, hairless and sensitive folds of skin. Below the mons they split into two folds to form a hood under which lies a small budlike organ, the clitoris. Descended from the same embryonic structure as the penis, the clitoris too swells during sexual arousal and is the most sensitive part of the woman's sexual organs.

Lying just below the clitoris is the exit of the urethra (which leads to the bladder) and the vaginal opening, the external entrance to the vagina. A muscular tube about 4–6 inches in length, the vagina is capable of great distension. It provides a port of entry for the penis and sperm as well as extending considerably during labor to allow a child to be born. A thin membrane, the hymen, seals the vagina in virginity, but is usually ruptured during the first sexual intercourse.

The upper end of the vagina leads into the cervix, the neck of the womb or uterus. A muscular pear-shaped organ which expands to a hundred times its normal size during pregnancy, the uterus lies in the lower abdomen. Extending outward and back from each side of the uterus are the Fallopian tubes, which extend their feathery ends towards the two ovaries.

The ovaries, like the male testes, have a dual function. They produce the female sex hormone, estrogen, and house the female sex cells, or ova. Unlike a man, however, a woman does not manufacture sex cells. Instead the newborn female is born with full complement. Known in their immature state as oocytes, they number about 600,000 at birth, and are contained in small saclike primordial follicles within the ovaries. In theory each one has the potential to ripen and be fertilized; in practice only about 400 ova ripen during a woman's lifetime.

The menstrual cycle

A woman's reproductive years begin with puberty and end with the menopause. Controlled by the hypothalamus in the brain, major changes occur from about the age of eight. The hypothalamus begins to secrete gonadotropin-releasing substance. This acts on the pituitary gland, causing it to release various hormones. Follicle-stimulating hormone (FSH) is the first to be released. This brings about the growth of the egg-containing follicles and makes them produce estrogen, which encourages the growth of breasts, widens the pelvis, and encourages the development of the external genitals.

The rising level of estrogen in the bloodstream has an effect on the hypothalamus known as "negative feedback," which causes a reduction of a second substance, luteinizing hormone (LH). This causes one of the follicles to burst and release an ovum for possible fertilization, the escape being known as ovulation. Ovulation frequently passes unnoticed but may be accompanied by a mild cramp or mitteldsmerz.

The remains of the follicle, now known as the corpus luteum, stay in the ovary secreting estrogen and a second hormone, progesterone, which prepares the uterus lining (the endometrium) to receive and nourish a fertilized egg. If the egg is not fertilized, levels of estrogen and progesterone drop and the uterus lining breaks down. It is then shed from the vagina together with mucus and the unfertilized ovum. The resulting bleeding constitutes the menarche, or first menstruation.

This same menstrual cycle then repeats itself roughly every 28 days throughout a woman's reproductive life. It begins around the age of 11 and ends with the menopause at about the age of 50, when the ovaries become exhausted and can no longer manufacture estrogen and progesterone. The menstrual cycle first becomes irregular and then ceases altogether, a change that is often accompanied by other symptoms such as hot flashes, depression, vaginal atrophy, and so on.

Conception

Walt Whitman has described sexual drive as "the procreant act of nature" and certainly sexual intercourse, while providing considerable sexual pleasure, is also nature's most effective way of transferring sperm from the male body to the female. Four stages of sexual arousal have been defined – excitement, plateau, arousal and orgasm – each of them accompanied by distinctive physical changes. In men the first stage begins with erection of the penis brought about when the spongy tissue becomes engorged with blood. In women arousal is also accompanied by changes in the external genitals and lubrication of the vagina. The penis penetrates the vagina which expands to accommodate it. Intensity builds and muscular contractions of the testicles, epididymes and vas deferens force sperm into the urethra. On their way sperm mix with seminal fluid secreted by the prostate gland and seminal vesicles to form semen which, at orgasm, is ejaculated into the woman's vagina.

In one ejaculation a man may send as many as 500 million sperm into a woman's vagina, although only one will fertilize an ovum. Once in the vagina, sperm swim upward by lashing their tails, aided by prostaglandins which dissolve the mucus plug at the entry to the uterus.

Many sperm fall by the wayside, but tens of thousands arrive in the uterus. There a glucose-rich environment helps them to move along the Fallopian tubes where, if the woman has ovulated within the previous 48 hours, one mature ovum will be

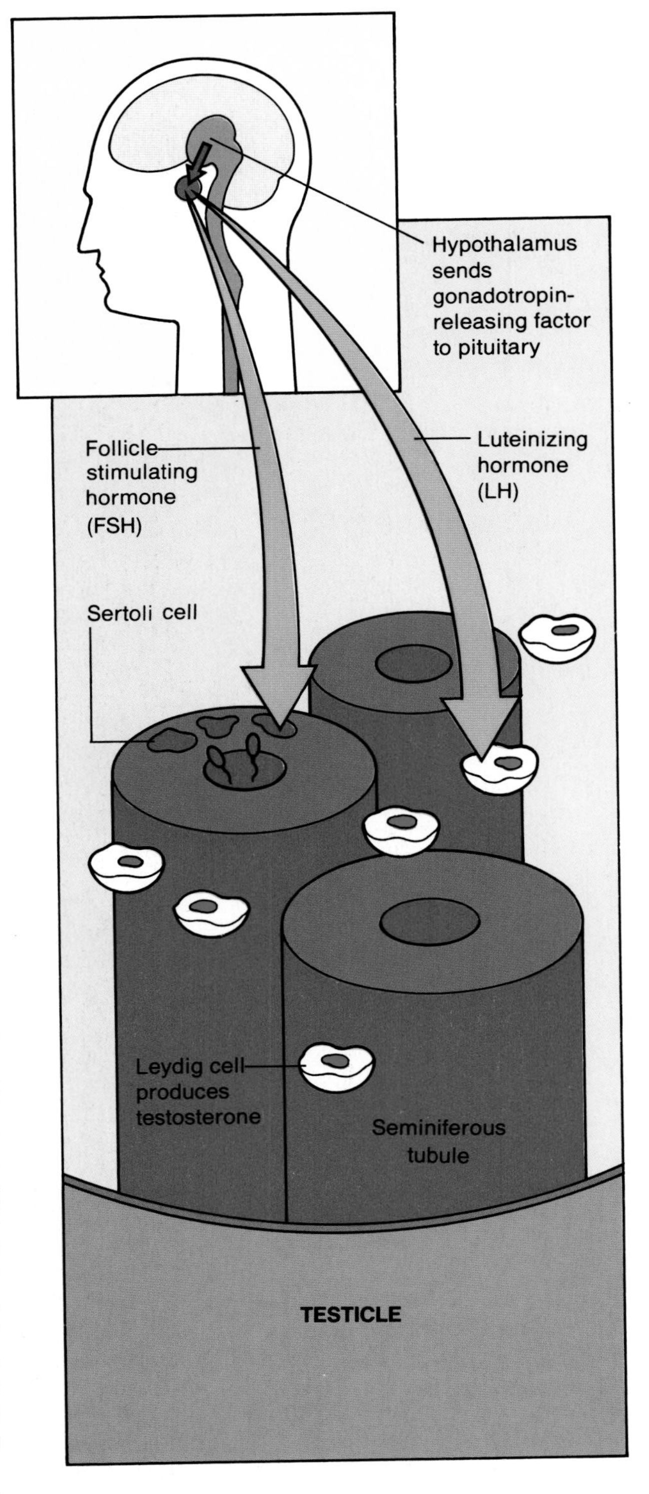

The testes *(above)* respond to and produce hormones. Sperm production by the testes is stimulated by hormones originating in and near the brain. The hypothalamus produces gonadotropin-releasing factor to make the nearby pituitary gland produce luteinizing hormone (LH) and follicle-stimulating hormone (FSH). These travel in the bloodstream to the testes and induce them to manufacture sperm and testosterone, which gives male characteristics. Sperm are seen *(right)* magnified more than 1,000 times.

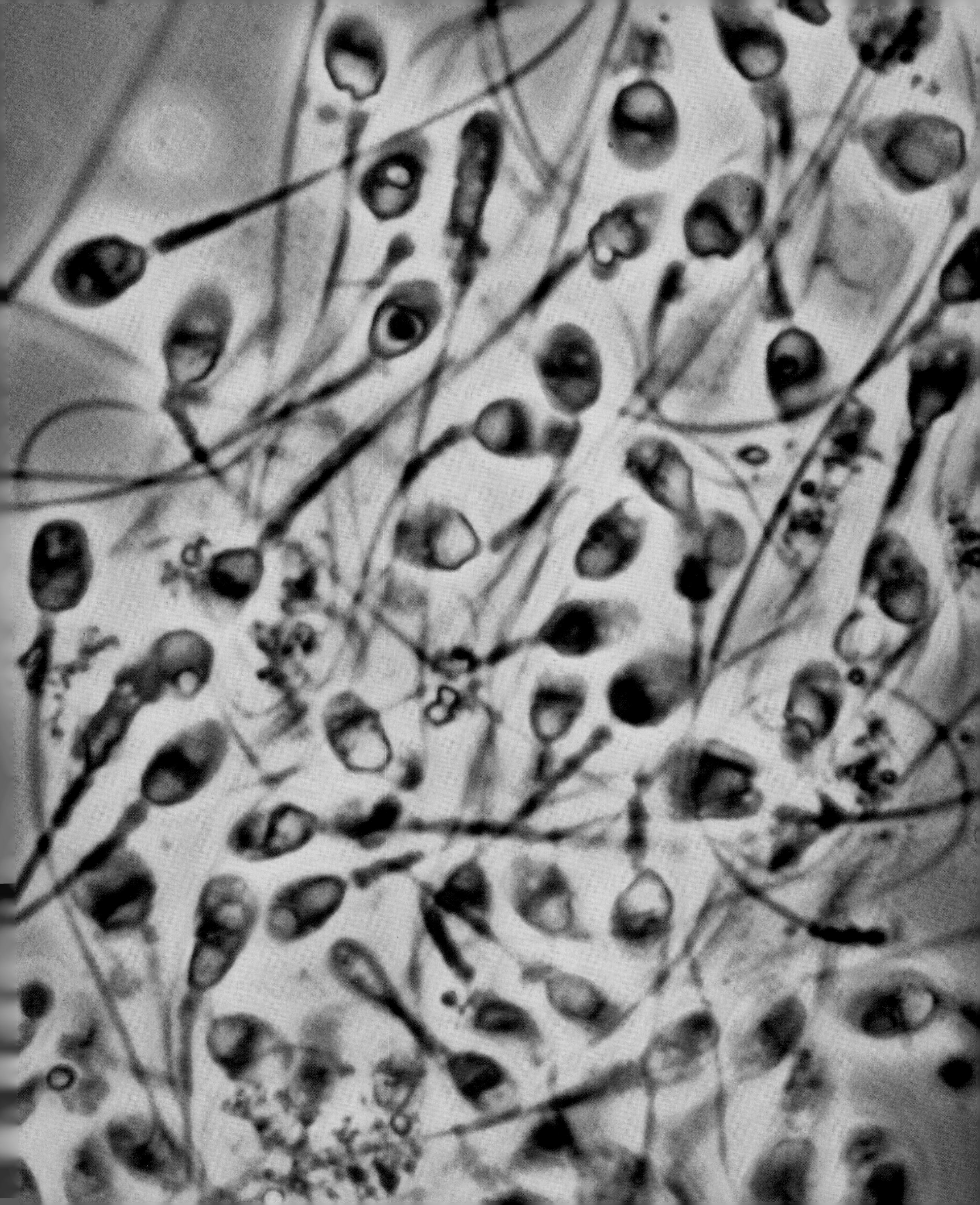

waiting. The first sperm to reach the egg penetrates its surface, releasing an enzyme called hyaluronidase. This slices through the chemical coating of the ovum and the two cells fuse. At the same time the ovum surface closes and enzymes are released which dislodge other sperm and create an impenetrable barrier. With this action the ovum is fertilized.

Twins occur either if the ovum, after fertilization, splits into two (in which case the babies will be identical twins), or if two ova have been released and fertilized. In this latter case each ovum is fertilized by a different sperm, and as a result the mother gives birth to non-identical or fraternal twins.

The sex of the future member of the human race is decided at the moment of conception. Both sperm and ovum contain only 23 DNA-carrying chromosomes, each half the number in all other body cells. The ovum's 23rd chromosome is either an X-chromosome or a Y-chromosome. If an X-sperm fuses with the ovum, the resulting embryo is female; if a Y-sperm fuses it is male. In this way, fertilization provides the new cell with its full complement of 46 pairs of chromosomes – all the genetic material needed to sustain life and determine the development of the future human being. It is at this point, too, that the genetic defects such as Turner's syndrome may be passed on.

Implantation

Within just a few hours the fertilized egg, now known as a zygote, begins to grow and develop, moving down from the Fallopian tube propelled by a fingerlike cilia. Cell division begins almost immediately as coils of deoxyribonucleic acid (DNA) carried on chromosomes reproduce themselves. The fertilized egg divides into two new cells, then four, then eight, and so on. By the fourth day the zygote, now comprising 16 cells, enters the uterus. As it continues to divide, a fluid-filled amniotic cavity develops within the structure, now known as a blastocyst. Some cells flatten out to form an outer mass, the trophoblast (from which the placenta will subsequently develop); the remainder cluster at one end to form the embryoblast (from which the embryo will develop).

Small projections, chorionic villi, form on the trophoblast and, about one week after fertilization, burrow into the uterine wall. The blood vessels in the villi tap into the maternal bloodstream across a band of tissue ultimately to become the umbilical cord. The chorionic villi also produce a third hormone, human chorionic gonadotropin (HCG); tests for the presence of this hormone in a woman's bloodstream give a reliable indication of pregnancy. HCG also maintains the corpus luteum in the ovary, so that it goes on producing estrogen and progesterone, which are responsible for the signs of early pregnancy such as breast tenderness, nausea, giddiness and the cessation of ovulation and menstruation.

Cell differentiation takes place with each cell division – a subtle but nevertheless crucial aspect of development can have a seemingly disproportionate repercussion such as a stunted limb – perhaps resulting from the loss of or damage to just one single cell. The embryoblast divides into two layers of cells, an outer endoderm and an inner ectoderm. Together they form the embryonic disk. During the third week this disk grows and lengthens, and a fissure develops along the length of the ectoderm. The ectoderm's cells slip toward this and spread sideways to form a third layer of cells, the mesoderm, sandwiched between the other two. The amniotic cavity also expands and encloses the entire embryonic disk and its connecting body stalk in a fluid-filled sac, within which the growing embryo has freedom of movement and is cushioned from the outside world.

The embryo

Over the next month the embryo undergoes dramatic changes. By the middle of the third week, it measures about one-eighth of an inch. Primordial, budlike forerunners of organs and tissues begin to appear and cells move through the mesoderm to create a rod which eventually becomes the spine. The nervous system also develops first as a neural rod, but then its upper end swells to form the brain.

Chunks of mesoderm called somites form in pairs on each side of the neural tube. From these begin to emerge muscle, skin and skeletal tissue. Heart and blood vessels also arise from the mesoderm, the heart beginning life as two tiny tubes which subsequently fuse into a single chamber.

By the end of the first month the primitive heart pumps blood around the embryo. During the second month sensory tissues, central and peripheral nervous systems develop from the ectoderm. Bone, cartilage, muscle and essential organs such as the liver, kidneys and spleen form from the mesoderm, while the endoderm gives rise to the lining of various tubes in the body: the digestive, excretory and respiratory tracts.

Arms and legs begin to appear by the end of the first month, first as primitive flippers, then developing hands, arms, shoulders and legs. Head growth is rapid – at this stage the embryo measures approximately half an inch, half of which is head. Within two more weeks the heart, now a series of chambers, is pumping blood, the brain continues to develop, and the respiratory system expands. At seven weeks, the embryonic skeleton begins to turn from cartilage into bone, and by the tenth week facial features – eyes, nose, ears and chin – have appeared.

Initially fishlike in appearance and life-maintained through the placenta, by the end of the second month the embryo has taken on distinctly human features.

The life-support system

The developing embryo is nourished through the placenta, a disk-shaped organ which develops during the first ten weeks of life at the point where the chorionic villi first burrowed into the endometrium. The placenta, which is about seven inches across and about a pound in weight, has an outer (maternal) surface divided into lobes of chorionic villi and tiny blood vessels. The flow of blood to and from these vessels is supplied from the mother's uterine artery and vein. The inner surface of the placenta is covered by a layer of amnion, and has a series of tiny vessels radiating out from the umbilical cord at its center. The umbilical cord, which links the embryo to the placenta, houses blood vessels which carry blood to and from the embryo.

The placenta acts as both pool and filter. Cells on the maternal surface fill with blood, and from this pool blood vessels on the embryonic surface draw not only oxygen but also, by diffusion, proteins, vitamins and other nutrients necessary for growth. Some, particularly proteins, are also manufactured by the placenta itself. At the same time waste products are drawn from the embryo's blood vessels. There are, however, a number of drugs and other harmful substances that, if taken by the mother, can cross the placenta and cause damage to the embryo.

The placenta also replaces the work of the corpus luteum, producing estrogen, progesterone and other hormones on which successful pregnancy depends.

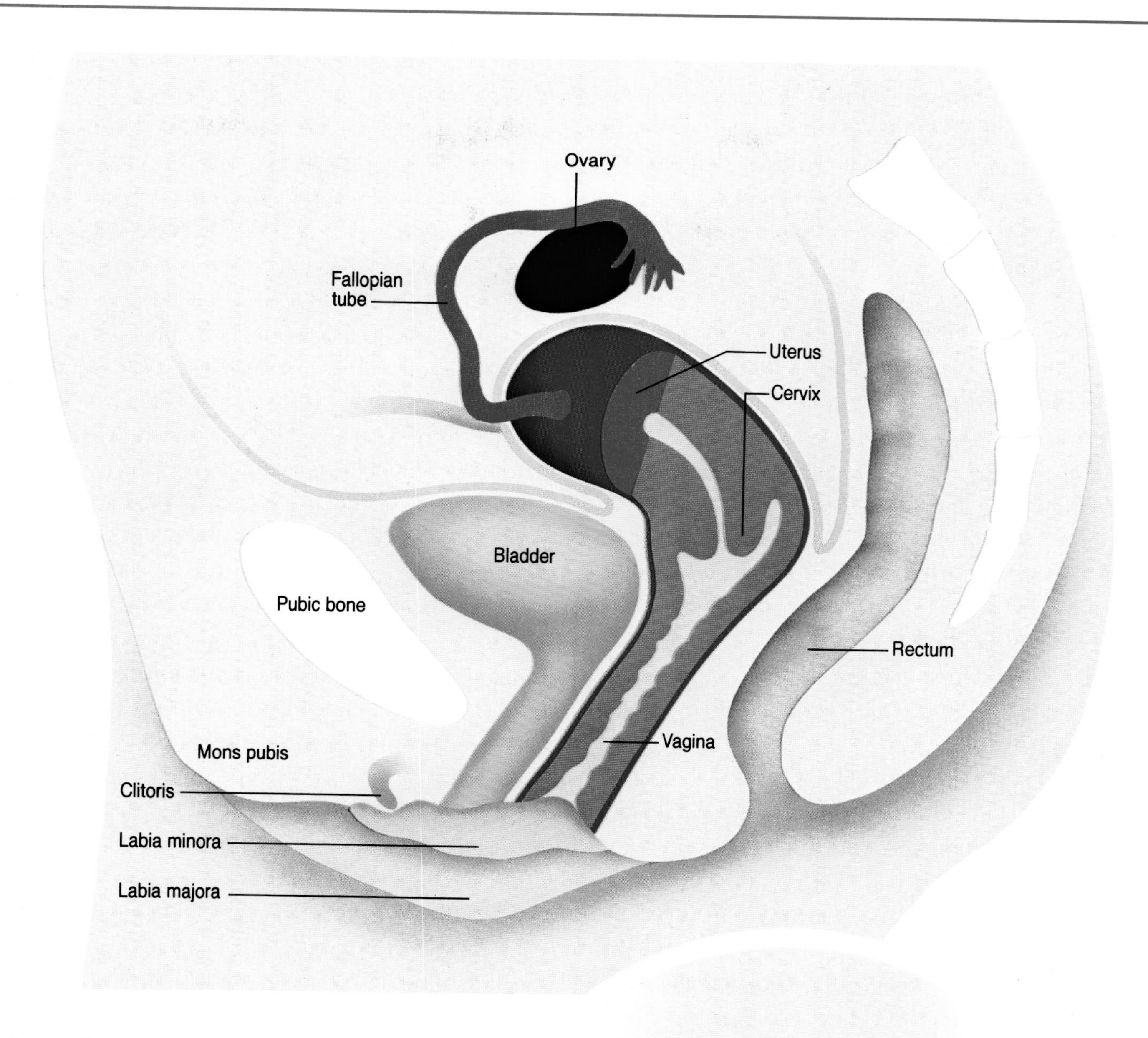

Female sexual organs are located entirely within a woman's body. They consist of two ovaries, situated near the feathery entrances of the Fallopian tubes, which lead to each side of the uterus, or womb. The cervix – the neck of the womb – leads in turn to the vagina, whose exterior opening is between the clefts of the liplike labia. The ovaries have two principal functions: to produce and periodically release ova, or eggs, and to manufacture the female sex hormones estrogen and progesterone. The vagina is the sheath that holds the man's penis during intercourse and receives ejaculated sperm.

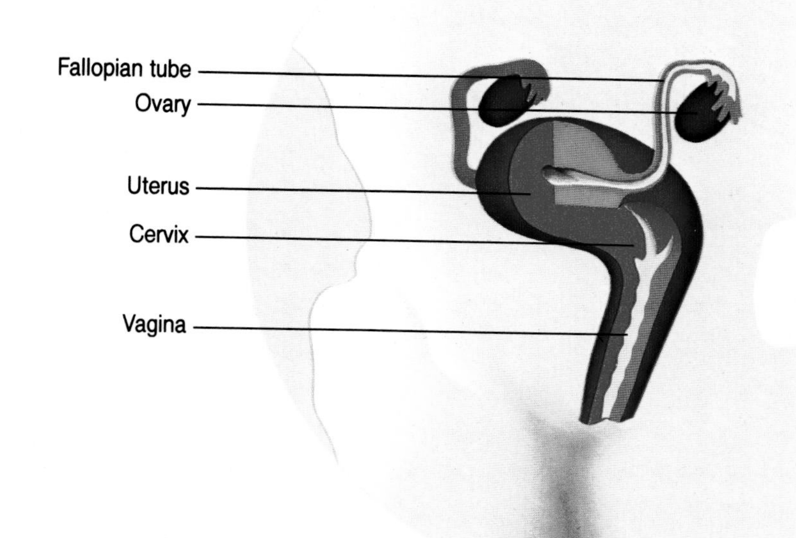

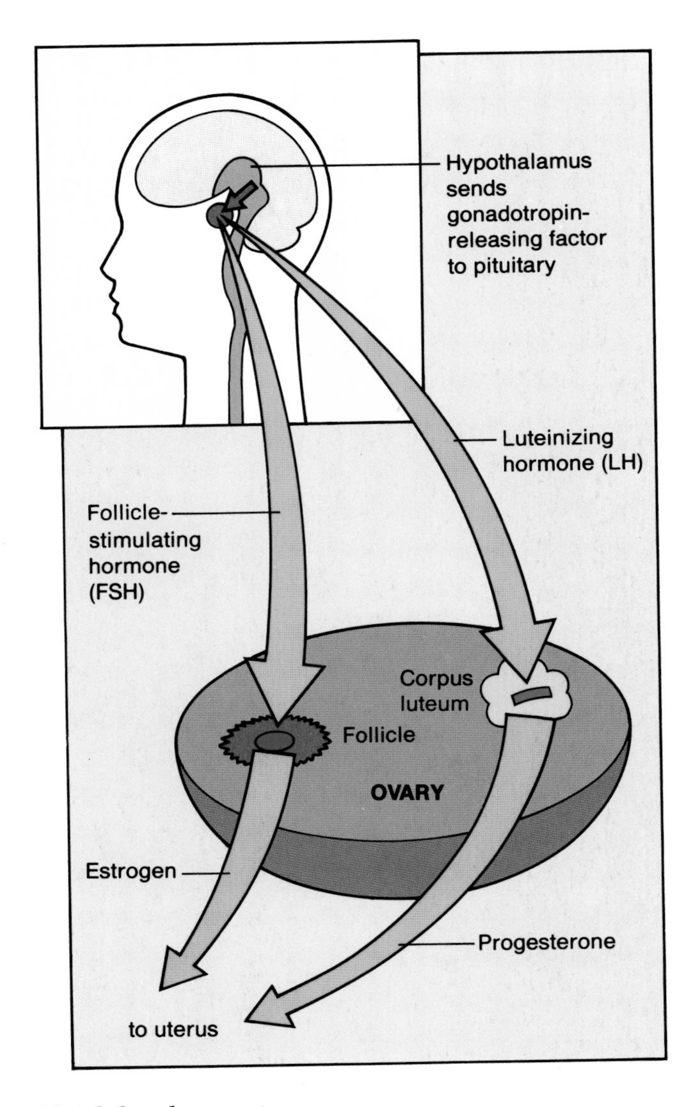

Hormonal stimuli in a woman control ovarian activity *(left)*. Gonadotropin-releasing factor from the hypothalamus triggers the pituitary to produce follicle-stimulating hormone (FSH), which commands the follicles in the ovary *(above)* to make estrogen. Next, luteinizing hormone (LH) acts on a follicle, which bursts to release the ovum, seen *(right)* magnified about 500 times. The empty follicle becomes a corpus luteum, making progesterone.

Fetal development

At eight weeks the embryo is a fully-formed minute infant, now called a fetus. Over the next 36 weeks growth takes precedence over development as the weight of the fetus increases 600 times from less than an ounce to around seven pounds, and it reaches its full length of 20 inches. Bodily proportions change, too, as limbs and trunk grow, reducing head proportions to one-fourth of the body length.

Hair, eyebrows and eyelashes are added during the 20th week, followed by fingernails and toenails. A fine, downy hair called lanugo covers the limbs and trunk.

Within its fluid-filled protection, the fetus also begins to move and swallow, taking in amniotic fluid. Thumb-sucking occurs during the fifth month, and one month later the fetus acquires the grasping reflex characteristic of a newborn infant. From 20 weeks the fetal heartbeat can be heard through a stethoscope. By 24 weeks the fetus can survive outside the womb if protected in an intensive care unit.

The changing body

While the fetus develops inside the mother, her body also undergoes the dramatic changes of pregnancy. For the first three months there are few outward signs of what is happening; it is not until the fetus begins its growth spurt that a woman's abdomen becomes increasingly swollen. But other changes have been taking place from the moment of conception, most of them controlled by the hormones estrogen and progesterone. The action of progesterone inhibits ovulation and so menstruation – a missed menstrual period is often the first sign of pregnancy. Other symptoms include nausea, giddiness, swollen or tender breasts, and frequent urination. As pregnancy progresses estrogen causes the softening and enlargement of the pubic joints, ligaments and tissues in preparation for labor. This action brings its own discomforts such as frequent urination, backache and stretch marks as connective tissue loosens.

Hormones also affect the blood circulation. Blood volume rises by nearly 50 percent, to compensate for loss during childbirth, but estrogen increases the risk of blood clotting or thrombosis. Varicose veins and hemorrhoids are other possible unpleasant side-effects.

The pregnant woman visits her obstetrician regularly throughout pregnancy, with the number of visits increasing during the eighth and ninth months. She receives a full physical examination and blood and urine tests. Blood is monitored for signs of anemia, quite common in pregnancy, and urine is checked for any signs of gestational diabetes, another condition

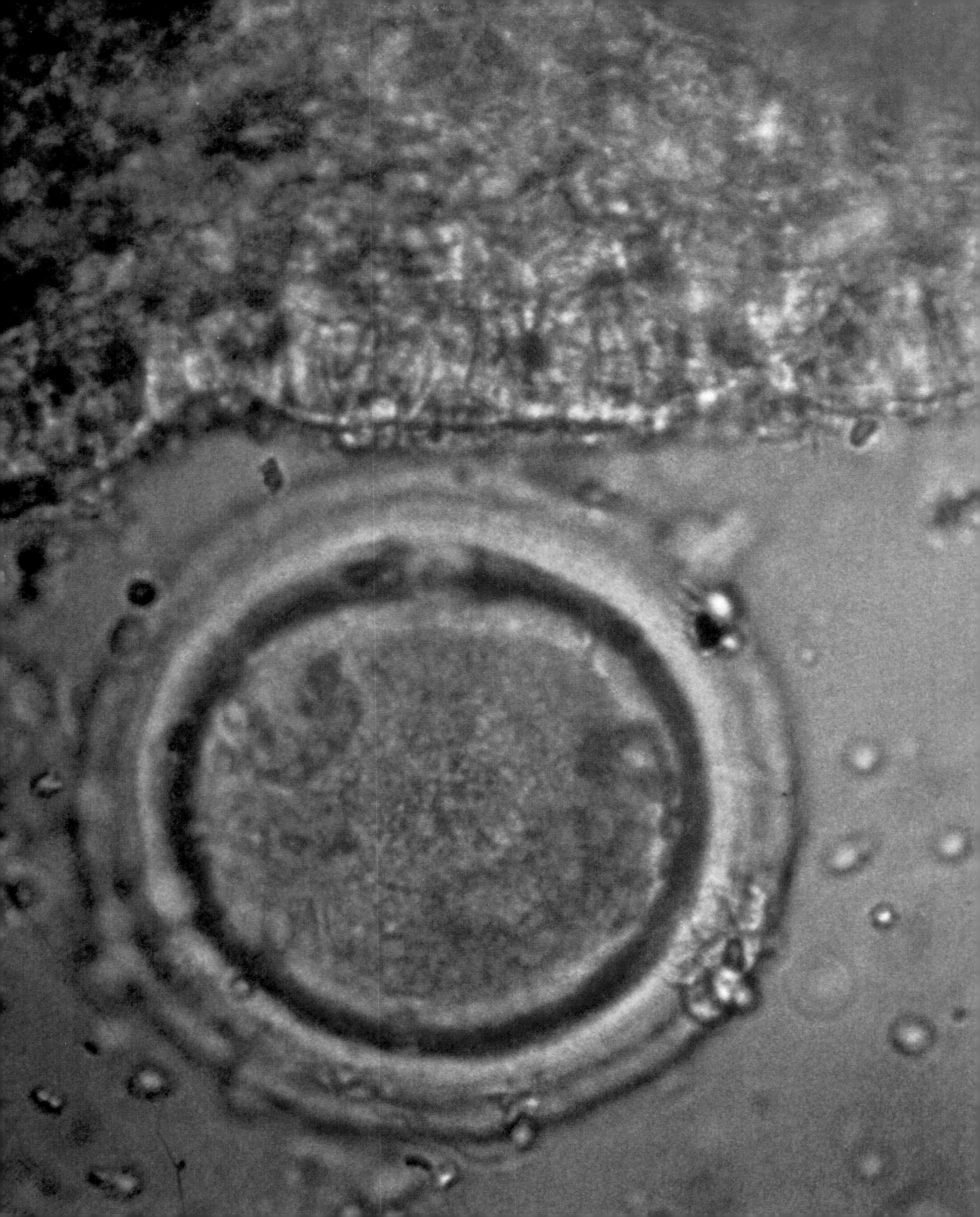

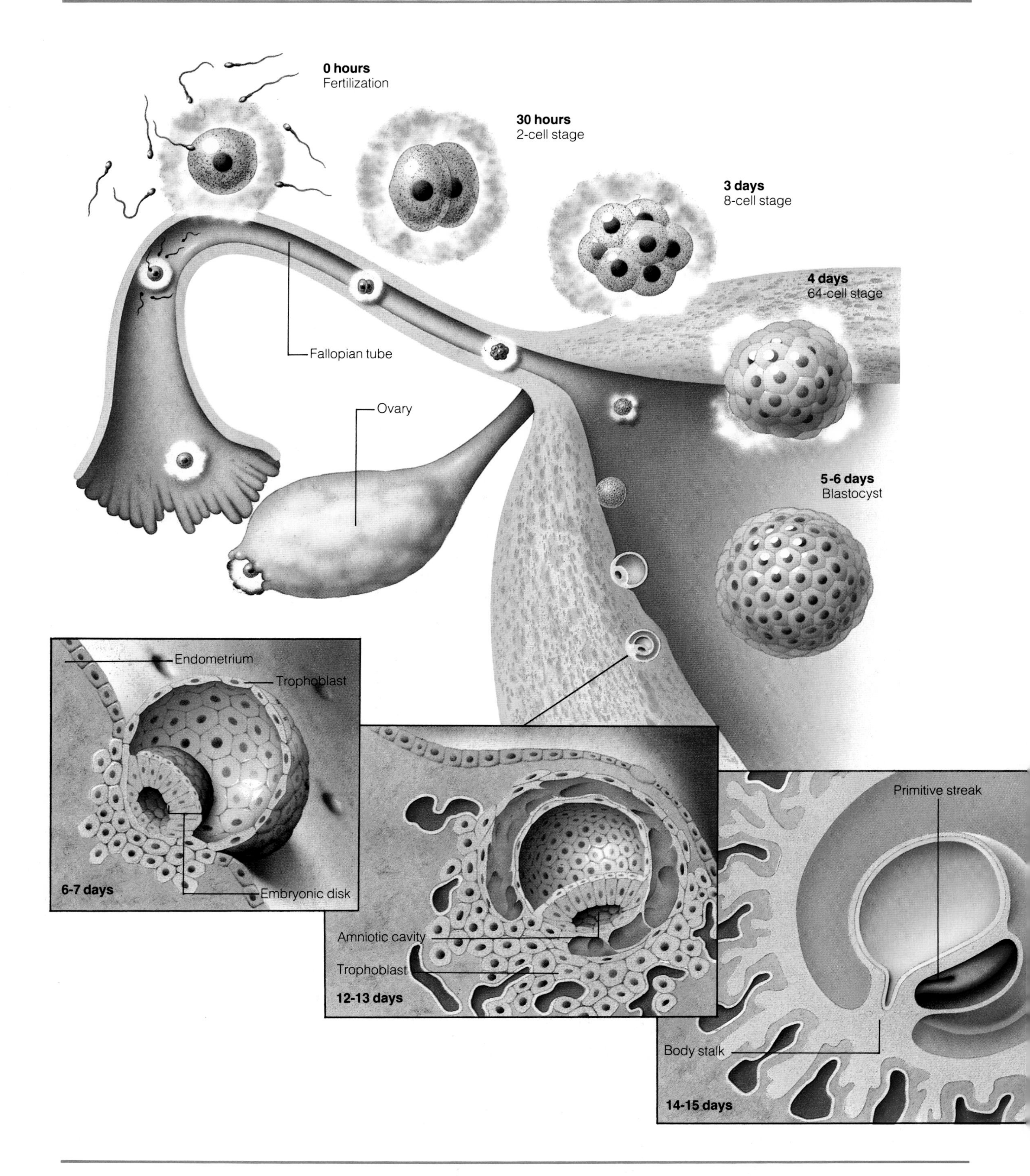
0 hours
Fertilization
30 hours
2-cell stage
3 days
8-cell stage
4 days
64-cell stage
5-6 days
Blastocyst
Fallopian tube
Ovary
Endometrium
Trophoblast
Embryonic disk
6-7 days
Amniotic cavity
Trophoblast
12-13 days
Primitive streak
Body stalk
14-15 days

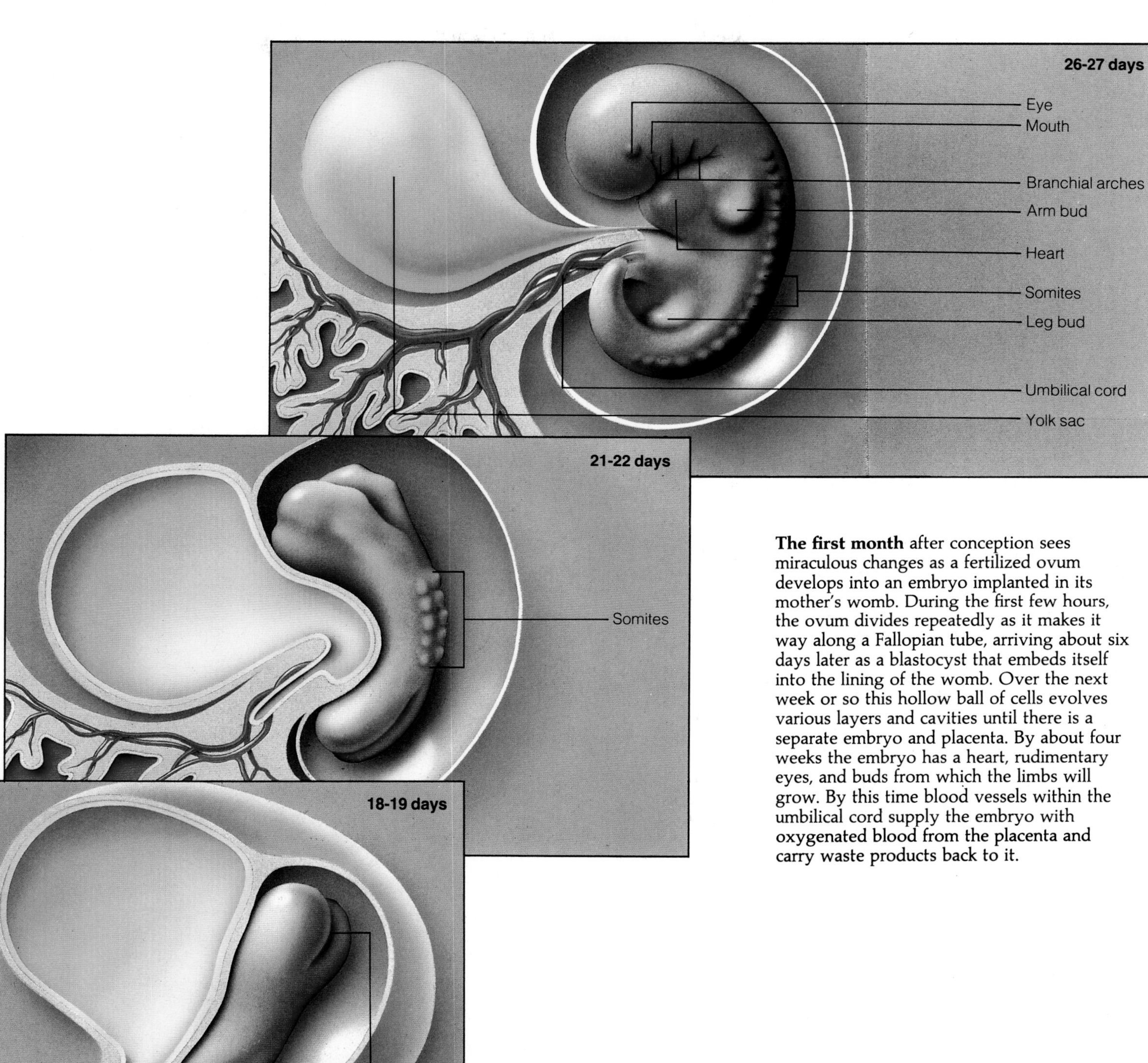

The first month after conception sees miraculous changes as a fertilized ovum develops into an embryo implanted in its mother's womb. During the first few hours, the ovum divides repeatedly as it makes it way along a Fallopian tube, arriving about six days later as a blastocyst that embeds itself into the lining of the womb. Over the next week or so this hollow ball of cells evolves various layers and cavities until there is a separate embryo and placenta. By about four weeks the embryo has a heart, rudimentary eyes, and buds from which the limbs will grow. By this time blood vessels within the umbilical cord supply the embryo with oxygenated blood from the placenta and carry waste products back to it.

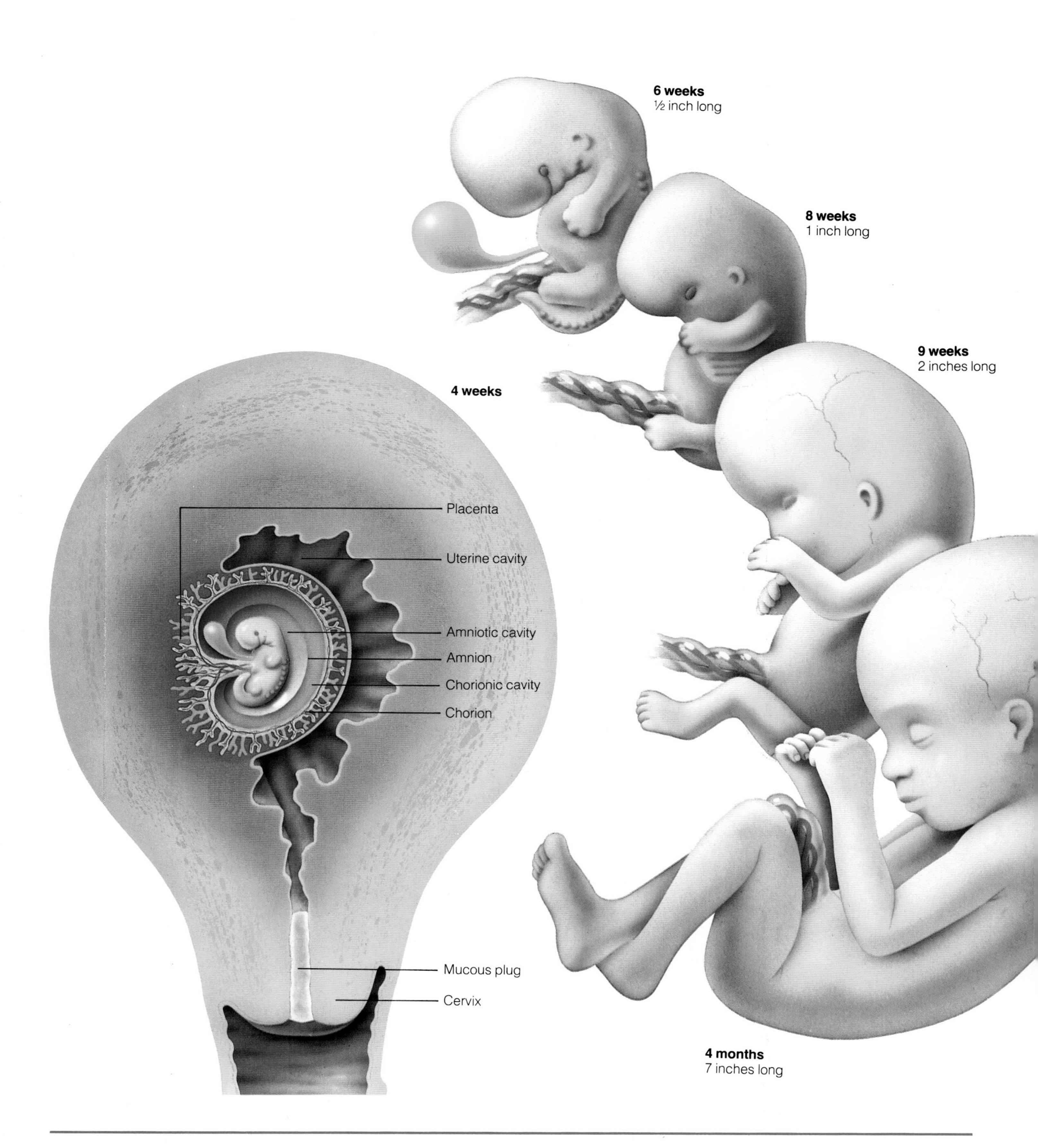
6 weeks
½ inch long
8 weeks
1 inch long
9 weeks
2 inches long
4 weeks
Placenta
Uterine cavity
Amniotic cavity
Amnion
Chorionic cavity
Chorion
Mucous plug
Cervix
4 months
7 inches long

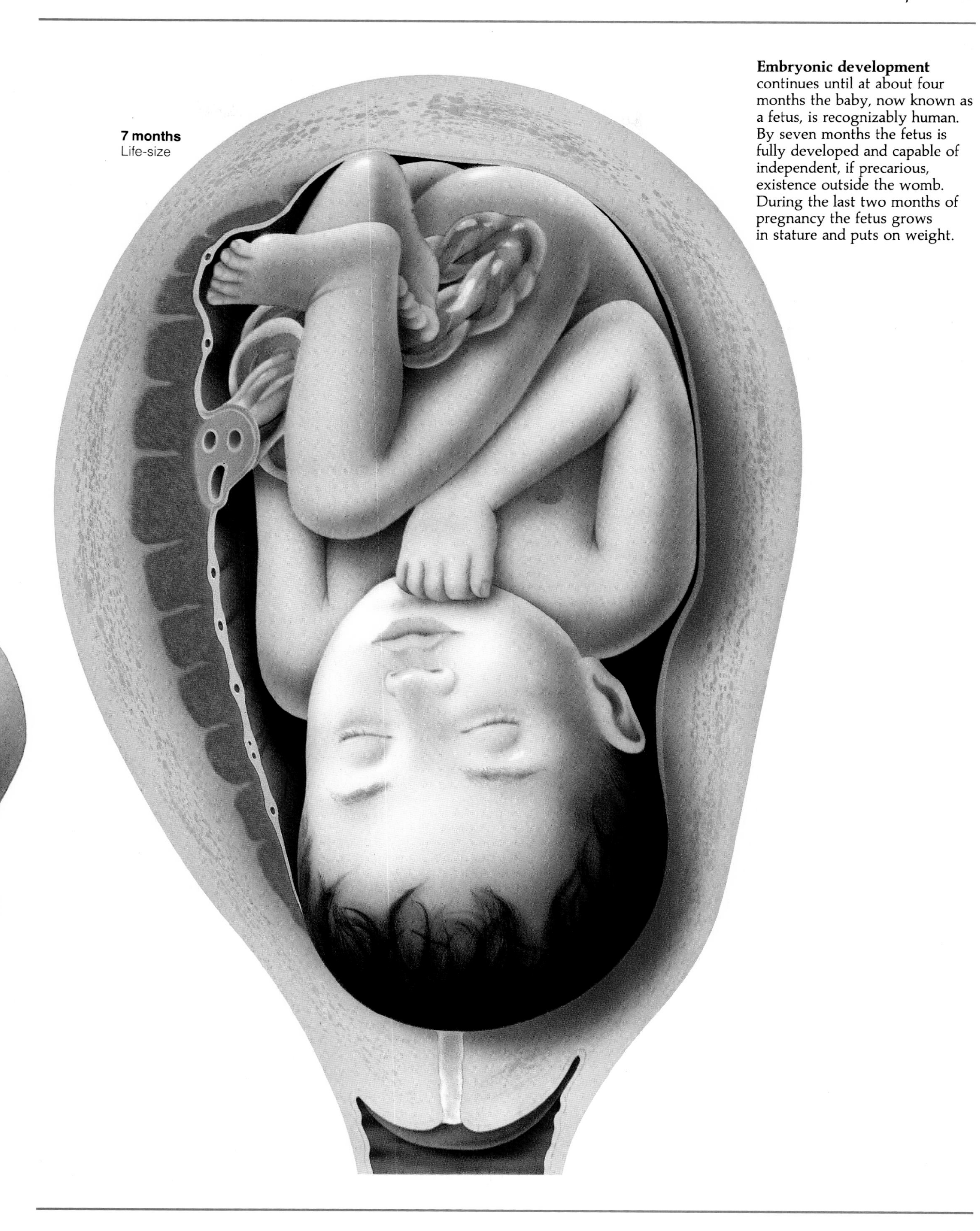

Embryonic development continues until at about four months the baby, now known as a fetus, is recognizably human. By seven months the fetus is fully developed and capable of independent, if precarious, existence outside the womb. During the last two months of pregnancy the fetus grows in stature and puts on weight.

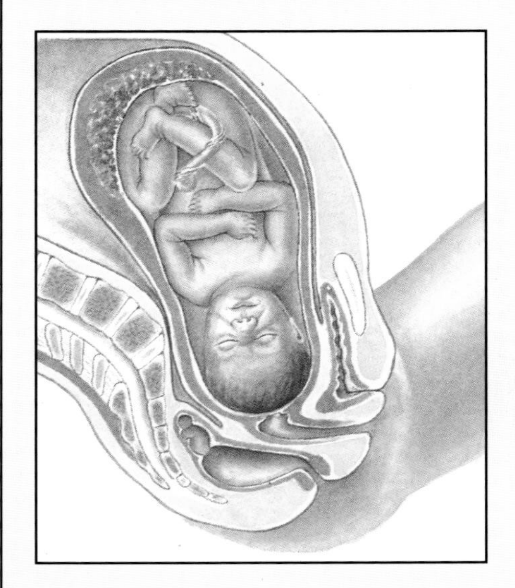
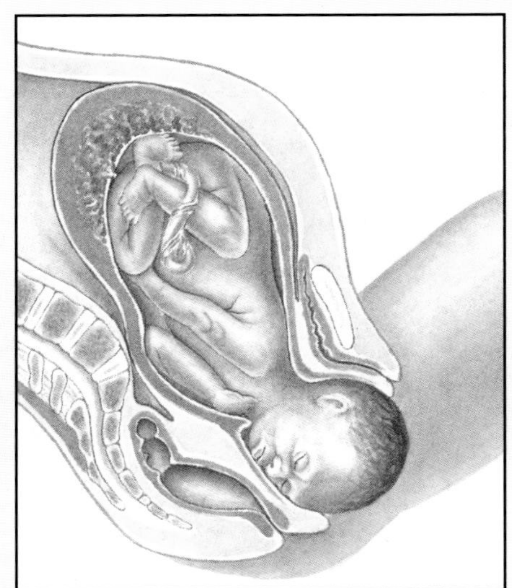
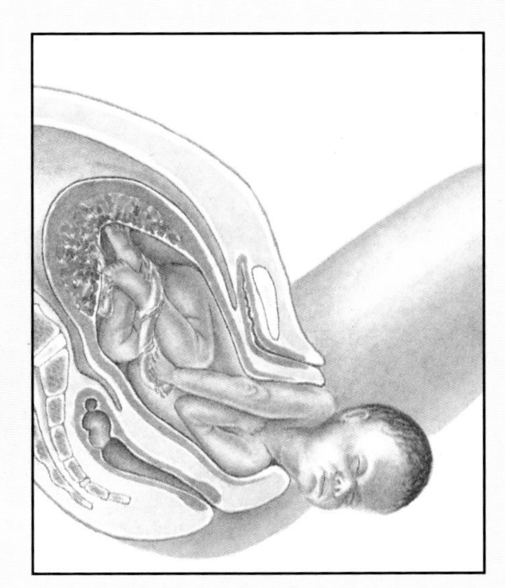

Childbirth, the short trip from the womb into the outside world, is a person's first and most important journey. For the parents, it is the culmination of nine months' patient waiting and a joyous experience. During the first stage of labor, contractions of the womb push the baby's head down as the cervix widens. During the second stage, the baby gently twists through the birth canal, its head emerges, and with one final push by the mother the miracle of birth is repeated once more. In the third and final stage the placenta is expelled.

peculiar to pregnancy. Later in pregnancy doctors watch carefully for signs of raised blood pressure that may indicate one of the most serious disorders – preeclampsia or eclampsia. Once a major cause of maternal mortality, eclampsia can now easily be detected and brought under control.

Fetal heartbeat is also carefully monitored, and the size of the fetus checked, most usually by ultrasound. By and large, however, the mother herself, by eating a balanced diet and following an exercise routine, can ensure a healthy pregnancy and a speedy return to her pre-pregnant shape. Later in pregnancy, too, most expectant mothers are encouraged to attend childbirth classes and learn the breathing routines so helpful in labor.

If a miscarriage occurs it is most likely to do so in the first three months of pregnancy. Causes are still uncertain but may include failure of the placenta to form, bacterial infections and hormonal dysfunction or chromosomal abnormality. Always a distressing experience, miscarriages are usually a sign that the fetus was defective, and after a first miscarriage the chances of a second, successful pregnancy are high.

Throughout pregnancy fetus and mother are closely interlocked, so that external dangers can be passed easily to the fetus, causing harm. Chief among these are viral infections, venereal diseases, and smoking, alcohol and drugs. The most hazardous viral disease is rubella (German measles), which, if contracted by the mother during the first month of pregnancy, can cause serious fetal deformity. Research now indicates clearly that smoking during pregnancy can cause miscarriage, fetal death, premature delivery, and respiratory problems. Alcohol too can cause low birth weight, and serious birth defects and deformities. The disabling effects of thalidomide during the early 1960s also highlighted the dangers of drugs which, if taken by the mother, can cross over the placenta with devastating effects.

Birth defects are probably a woman's greatest fear, but the advancement of medical technology does mean that a number of chromosomal or genetic abnormalities such as Down's Syndrome or spina bifida can be detected before birth. Amniocentesis, for instance, whereby a sample of amniotic fluid is removed and analyzed, can be carried out at 15 weeks. It is gradually being replaced by chorionic villus biopsy, which can be carried out between the 9th and 11th week. Chorionic villi contain a rich source of fetal DNA, so analysis of a small sample of chorionic tissue indicates possible genetic abnormalities.

Childbirth

The birth of a baby can be the most thrilling and most traumatic experience for both infant and mother. Fashion, cultures, historical time and even attitudes have shaped the method of childbirth in strange and wonderful ways. But as an event it is unique, and what seems evident today is that the more a woman is able to participate in the birth, the more joyous the experience.

Full-term pregnancy lasts nine months; it is followed by labor, during which the new infant is expelled from the mother's uterus. Throughout pregnancy the body prepares for labor with gentle contractions. During the last month these contractions gradually increase, culminating in labor. Quite what triggers labor is still uncertain. Some doctors believe that it starts once the uterus is fully extended, others that decreasing levels of progesterone initiate the event. Oxytocin, a hormone secreted by the pituitary, is also known to stimulate the uterus, while prostaglandins may also play a part. Recent work also suggests that the fetus itself may contribute to the onset of labor.

Whatever the actual cause, labor itself is heralded by various signs. In first pregnancies the fetus settles head down about two to three weeks before labor, the cervix softens, the lower part of the uterus expands and the cervix effaces – pulls up toward the body of the uterus – and dilates. As the cervix widens, a mucus plug, which has formed in the cervix to keep the uterus germ-free, slips out in what is known as the "show." It may be slightly tinged with blood. The amniotic sac slips down and ruptures, releasing the amniotic fluid, and labor begins.

Labor falls into three distinct stages. The first stage is the longest, an average of 14 hours for a first birth but less for second and subsequent births. It is characterized by persistent, regular contractions as the uterine muscles shorten and pull upward, pushing the fetus downward, and pulling the cervix upward. As the first stage progresses, contractions become increasingly intense, occurring every minute or two.

The second stage begins once the cervix is fully dilated, to about four inches. This stage may take anything from a few minutes to two hours. During it contractions slow down and the baby begins its journey along the birth canal, the mother working with the contractions to push the baby down. In a straightforward birth, the head emerges first, the shoulders turn, and further contractions push the baby out into the world. The baby's mouth and nasal passages are cleared of mucus, the baby breathes, takes its first cry and is given to the mother. Once the umbilical cord has finished pulsating, it is clamped and cut.

Ninety-five percent of all births are head-first in the so-called vertex position. But in just over three percent of births, the baby is in a breech position, with buttocks or feet presenting. In some limited instances it may be possible to turn the baby; otherwise Caesarian section – delivery of the fetus through an incision in the mother's abdomen – is the most usual procedure.

In the third and final stage of labor, the uterus continues to contract, leading to the painless expulsion of the placenta and umbilical cord. The mother is checked carefully to ensure that the entire placenta has been removed, and with this the process of pregnancy and birth is complete.

As soon as the baby is born, a nurse in the delivery room assesses its physical condition by assigning an Apgar score of zero to two for each of five tests. These include checking heartbeat, respiration, muscle tone, reflexes and complexion – any blueness could indicate a circulatory problem. The maximum score is ten, and babies with scores lower than six need some form of resuscitation.

Once the newborn is in the hospital nursery, it receives a more detailed physical. A nurse weighs the baby and measures the size of the fontanelles, the soft areas of the skull that have not yet hardened to bone. Nervous reflexes are also tested. Putting a finger at the side of the baby's mouth should make it turn and open its mouth. This response is the rooting reflex, which a baby uses to search out its mother's nipple to nurse. Touching the baby's palm elicits the grasping reflex – a firm grip by a tiny hand. And holding the infant upright with one foot touching the floor makes it take a tentative step, a reflex in the newborn but an ability that is soon lost and has to be relearned many months later as the nervous system matures.

All these tests testify to the perfection of the product of a nine-month miracle, the transition from a single cell to a brand-new member of the human race.

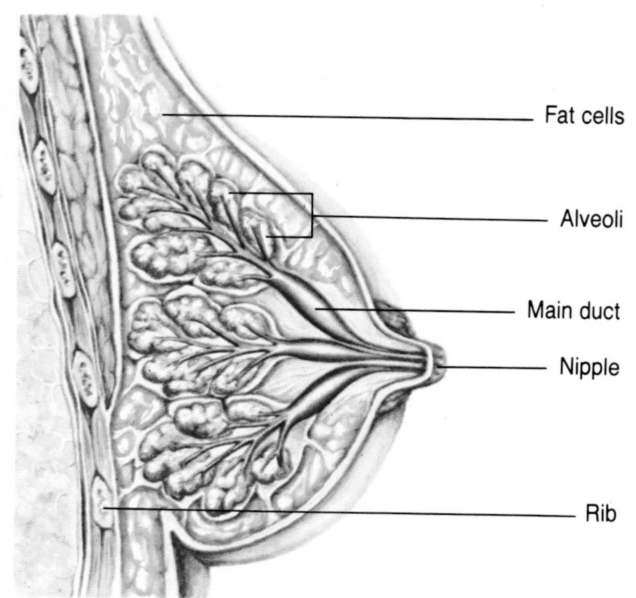

Mother's milk, it has been said, comes in convenient containers, at the right temperature, and is always available. But little milk is available until after birth, when the baby's suckling stimulates its production.

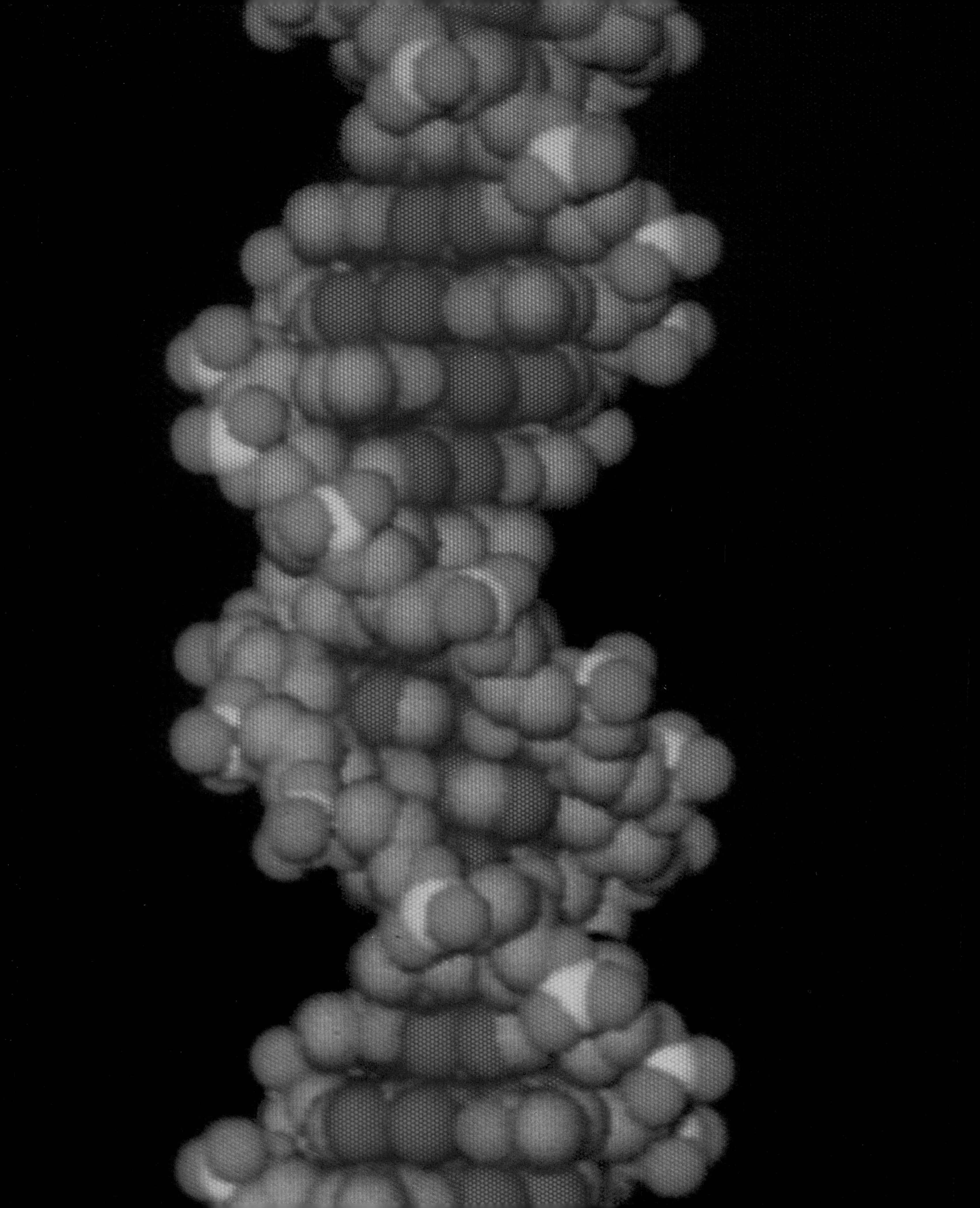

CHAPTER 16

Genetics and Heredity

In the whole complex and fascinating story of the human body, the reason why we are what we are is probably the most intriguing aspect of all. At conception, two sex cells – a sperm from a man and an ovum from a woman – fuse to produce a single fertilized cell. Within this one cell is contained all the information required to produce a new human being, and which determines every aspect of that human being from hair color and height to the essentials of artistic ability and intellect. This information is inherited from the previous parents, stored in the genes – thousands of which make up the threadlike chromosomes present in the cell – and carried in code on the double-helix molecule DNA (deoxyribonucleic acid) which in turn makes up the genes themselves.

Each human being is unique – and yet at the same time a person is the sum total of characteristics inherited from each parent and passed on from parent to offspring. The process by which this happens and the selection of particular characteristics is the result partly of precise programming and partly of random selection. Heredity is the mechanism by which characteristics are passed from one generation to the next. Inherited characteristics include everything from the obvious features such as hair color and physique to less obvious blood grouping, metabolism and enzyme grouping that determine body chemistry.

The detailed study or science of the mechanisms of heredity, and of variations that can occur both in animals and in plants, is called genetics. During the 20th century this science has blossomed, enabling us not only to understand why we inherit certain characteristics, but also to pinpoint and treat a variety of serious human disorders – and even, however controversially, to intervene in the creation of life itself.

Rules of inheritance

Heredity is controlled by the genes. Half of each person's genes come from the mother, half from the father. Like begets like and, as a result of genetic inheritance, offspring almost invariably resemble their parents. But there are also variations. Although the basic genetic components of heredity are passed equally, they combine themselves in an almost infinite number of ways so that there is very little likelihood of any two people – except identical twins – having exactly the same genetic make-up.

Random chance plays a large part in heredity, but there are also certain rules and patterns. The first of these were uncovered by the Moravian monk Gregor Mendel during the 19th century in a series of now famous experiments with garden peas. As a result of Mendel's work combined with subsequent knowledge of cell structure we know that inherited traits are passed on through the genes, and that each person has two copies of each gene, one copy inherited from the mother and one from the father. When the genes are passed from parent to offspring, either of the mother's copies or of the father's may be passed. By random mixing, a vast number of conformations is possible.

A convoluted coil of coded chemicals, a model of the DNA molecule spirals endlessly. Subtle variations in the order of atoms that make up the molecule determine the genetic message carried by DNA, which also has the property of being able faultlessly to replicate itself in order to pass on that message.

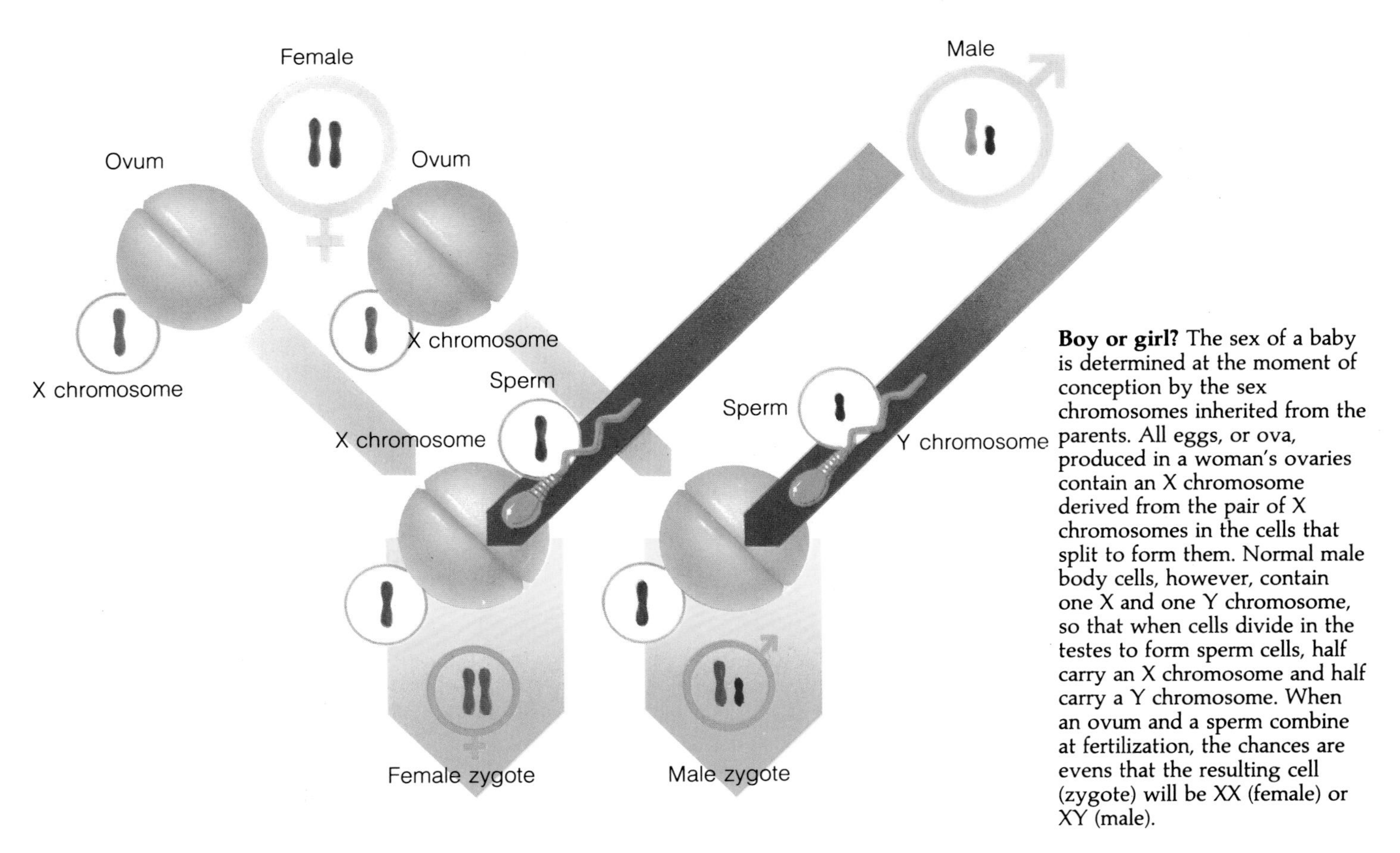

Boy or girl? The sex of a baby is determined at the moment of conception by the sex chromosomes inherited from the parents. All eggs, or ova, produced in a woman's ovaries contain an X chromosome derived from the pair of X chromosomes in the cells that split to form them. Normal male body cells, however, contain one X and one Y chromosome, so that when cells divide in the testes to form sperm cells, half carry an X chromosome and half carry a Y chromosome. When an ovum and a sperm combine at fertilization, the chances are evens that the resulting cell (zygote) will be XX (female) or XY (male).

The handover of genetic information from parents to child takes place at conception when the two cells, sperm from the father and ovum from the mother, each of which contains only half the number of necessary chromosomes, fuse together to form one cell carrying its full chromosomal complement.

Not all pairs of genes are, however, equal in effect. One may be what is known as dominant, the other may be recessive. When this happens the dominant gene suppresses the recessive one if both occur together. The recessive gene is apparent only if both genes are recessive. Red hair and very blond hair seem to be examples of recessive inheritance of "normal" features. Otherwise recessive inheritance of "normal" features is uncommon.

Genes may also be codominant. The best example of this is blood grouping. There are three possible forms of the major blood group gene – A, B and O. Any gene can have alternative forms – called alleles – but one person can have only two of them. Genes for group A and B are codominant, so that somebody who inherits both an A and a B gene has group AB blood. The group O gene is recessive and therefore can produce group O blood only in somebody who has two O genes.

Exceptionally, however, a person with group A blood conceiving with a partner with group B blood may produce an O blood group offspring if they both also have a "silent," or recessive, group O gene. It is this patterning that allows blood group genetics to help in some cases of disputed paternity.

Sex chromosomes and sex linkage

Genetic material is carried on the chromosomes – rodlike structures in the nucleus of every cell in the body. Each cell contains 46 chromosomes arranged in 23 pairs. The only exceptions are the sex cells, which contain only 23 unpaired chromosomes so that at conception, when sperm and egg fuse, they create one cell containing a full set of chromosomes. Of the 23 pairs produced, however, one pair – the sex chromosomes – differ from the others in that they are not always identical. In women the sex chromosome consists of two identical X chromosomes; in men the sex chromosome consists of one X chromosome and one, smaller, Y chromosome.

This difference in size is responsible for a number of sex-linked traits. Hemophilia is a blood-clotting disorder caused by the absence of a protein – factor VIII – usually carried on the X chromosome. A woman, with two X chromosomes, has two factor VIII genes, so that if one gene is abnormal, the other produces enough blood clotting factor to prevent excessive bleeding. A man, by contrast, with only one X chromosome, cannot produce enough clotting factor if the one gene is faulty because there is no second gene to replace it. For this reason, a woman carrying one abnormal gene does not develop hemophilia herself, but can pass on the trait to her male offspring.

Sex-linked traits can skip generations, notably in the case of color-blindness. The recessive gene for color-blindness is carried on the X chromosome. A color-blind man can pass the X chromosome with the abnormal, recessive gene only to his daughters so that all his children will have normal color vision because the daughters' other normal X chromosome will mask the recessive effect of the gene, while the sons will receive their X chromosome from the mother. But all the daughters are carriers of the defective gene – so their sons may inherit the X chromosome bearing it and so be affected. Grandfathers and grandsons, but not sons, may therefore be color-blind.

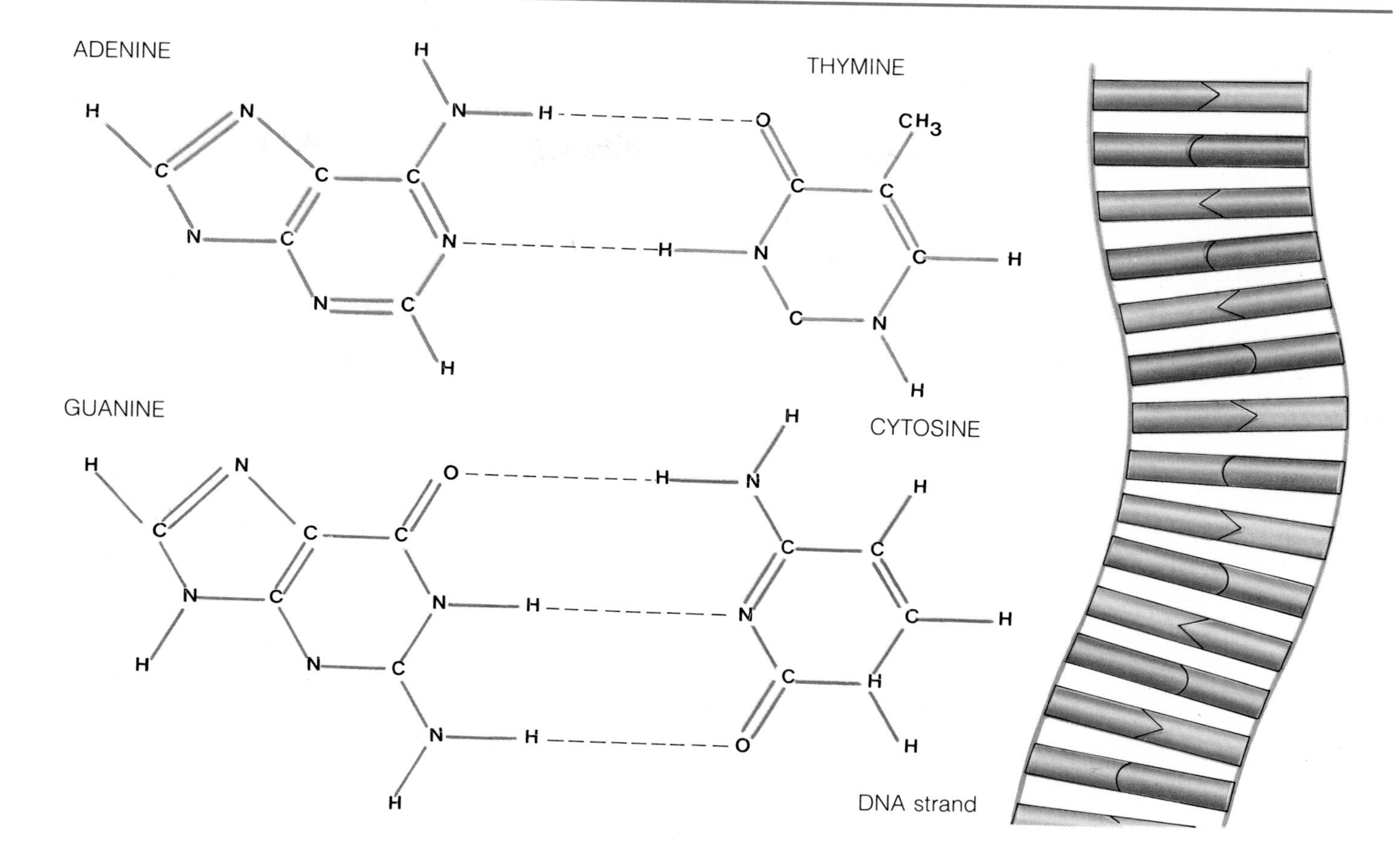

DNA's twisted ladder has sides made of alternate sugars and phosphate groups, and "rungs" consisting of pairs of nitrogen-containing bases held together by hydrogen bonds (indicated by dotted lines). The bases always pair with the same partner: adenine (drawn in blue) always joins with thymine (orange), and guanine (red) with cytosine (green).

Polygenic inheritance

Chance plays a large part in inheritance. There is no way of forecasting, for instance, whether an ovum will be fertilized by sperm containing an X chromosome to become a girl, or by a Y chromosome to become a boy. Either is possible. This occurs because chromosomes are sorted at random when packaged into eggs and sperms – a principle called "independent assortment." And it is believed that this principle applies to all genes.

Most human traits, such as height, are determined by groups of genes acting together – a pattern known as polygenic inheritance. But when this occurs, inheritance becomes complicated and virtually unpredictable. Perhaps only two genes – AA and BB – control height. But each gene may have two forms or alleles – A and a or B and b. (The capital notation is traditionally used to denote the dominant, the lower case the recessive.) A and B genes make a person taller; a and b genes make a person shorter.

An average sized person has one of each, producing a genetic make-up of AAaa BBbb. But when two average sized children grow up and themselves have children, the genes are sorted randomly. Each person's sex cells or gametes contain AABB, AAbb, aaBB or aabb genes. Each is equally likely, and when they fertilize each other the chances of any one egg meeting any one sperm are equal. So, statistically speaking, an AABB egg is fertilized by an AABB sperm once in 16 times; the same is true for the genes that make people shorter than average – one child in 16 inherits all four shortness alleles.

As a result, so-called average parents may have children with far from average height, although average parents produce average sized children just under 50 percent of the time.

Variation is continuous. Tall parents usually have tall children, but not as tall as their parents. Similarly, short parents often produce children who are taller than they are. This is known as "regression to the mean" and it also operates in many other human traits, such as intelligence. Although heredity is random, the laws of statistical probability are used by geneticists to investigate particular traits and in genetic counselling.

Changeable genes

All genes can change or mutate as a result of various environmental factors, the most talked about today probably being ionizing radiation. Most mutations are harmful, in that they usually result in the death of the mutated organism even before it is born, although some useful mutations have been used in selective crop and animal breeding to produce specific strains. Nevertheless, mutations provide an essential trigger for evolutionary change, even among human populations.

In human mutations and mutagens – the substances that cause mutation – are extremely significant. If mutations occur in the sex cells they are not apparent to the person carrying them but are likely to be inherited by the next generation. It has been argued, for instance, that Agent Orange, a chemical defoliant used during the Vietnam War, may have caused damage to the sex cells of people exposed to it. The claim that deformed offspring resulted is still under review. The same argument also rages over the descendants of those exposed to radioactive fallout after the bombing of Hiroshima and Nagasaki, and of those members of the military who were required to witness early nuclear test explosions.

Most mutations, however, occur in somatic or non-sex cells. These may not affect heredity, but can nevertheless cause dramatic changes. Body cells are constantly replacing and replicating themselves, so that if mutation takes place it is carried through to the new cells. As with other genes, mutant ones may be dominant or recessive. Dominant mutations probably affect any offspring, although recessive mutations may not become apparent until the carrier produces children with somebody carrying a similar recessive gene. The odds of this occurring are many thousands to one except in the case of close relatives – particularly first cousins, who have grandparents in common.

Cancer is now thought to be caused by genetic damage to cells. Instead of being programmed correctly, cancer cells are mutated ones with the capacity for unlimited growth and replication; they can also spread throughout the body. Agents that cause cancer – carcinogens – include chemicals, radioactivity, tobacco, asbestos and ultraviolet rays from the sun. The effects of radiation are the subject of considerable debate. As a result of recent studies it is now known that ionizing radiation does increase the risk of certain cancers, particularly leukemia (cancer of the blood cells), and that this effect can be produced by very low doses such as a medical X-ray of a pregnant woman. As a result there is much debate today as to whether there is, in fact, any "safe" dose of radiation.

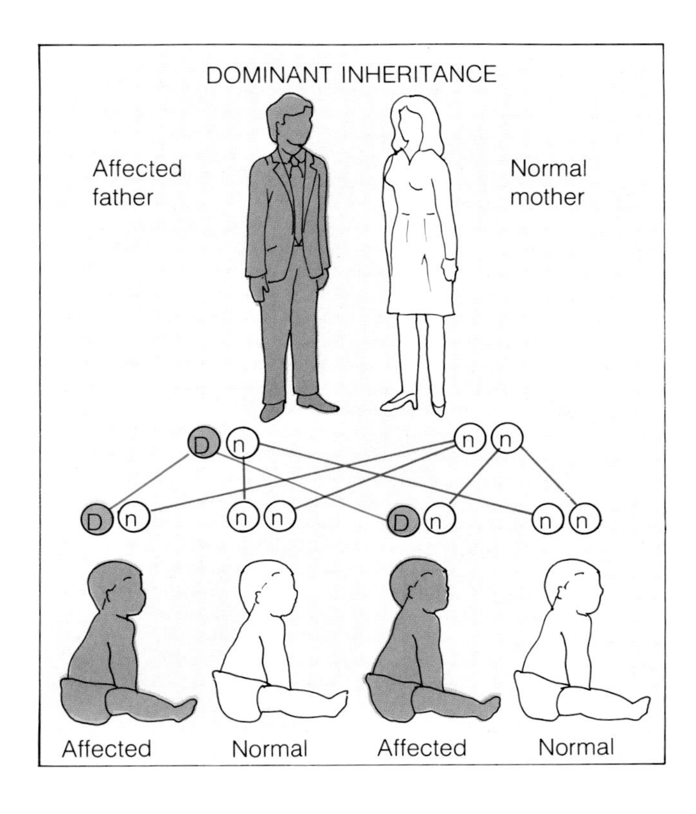

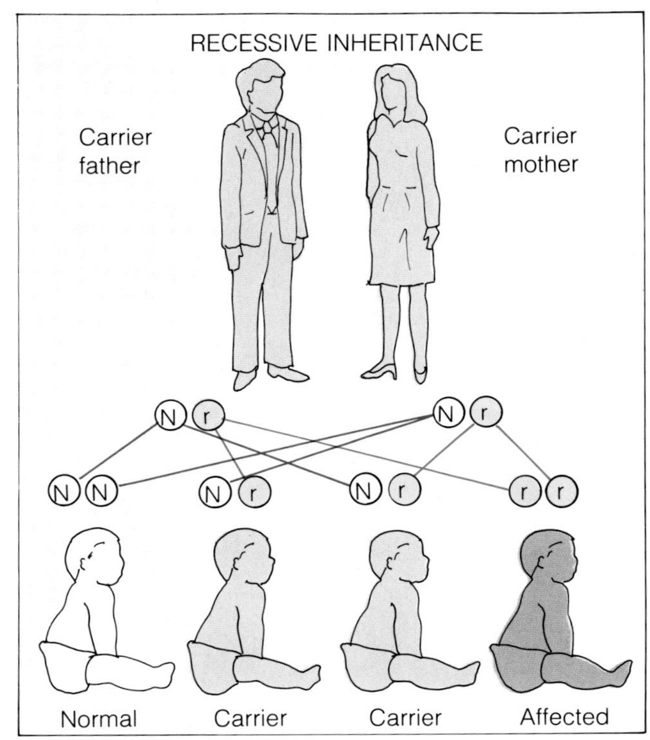

A gene reveals its presence, or fails to do so, depending on whether it is dominant or recessive, present alone or as a similar pair. A characteristic determined by a dominant gene *(above left)* manifests itself even if present only singly. A recessive gene, on the other hand, has to be part of a pair to reveal its presence. A person with a single recessive gene is a carrier, who does not display the characteristic but passes it on to half of his or her offspring. In the diagram *(above right)* both parents are carriers and so one child in four will inherit a pair of recessive genes and reveal the characteristic. The six possible combinations of the three genes for blood type *(see table, right)* result in the four common blood groups; I^A and I^B are dominant (producing the A, B and AB groups) over the recessive I^O.

INHERITANCE OF HUMAN BLOOD GROUPS

Possible Gene Combinations	Results in	Blood Group
I^AI^A or I^AI^O	Antigen A only	A
I^BI^B or I^BI^O	Antigen B only	B
I^AI^B	Antigens A and B	AB
I^OI^O	No antigens	O

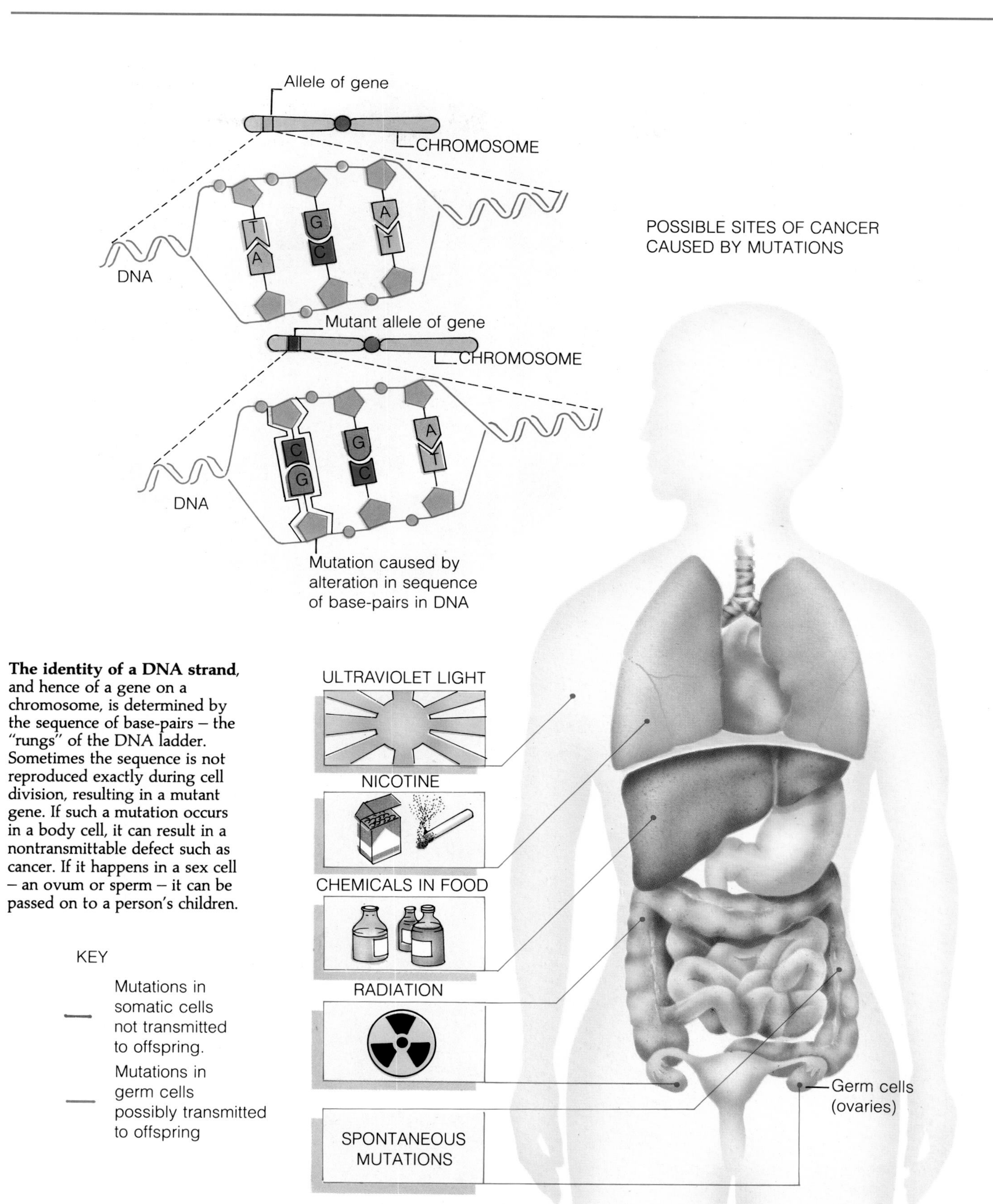

The identity of a DNA strand, and hence of a gene on a chromosome, is determined by the sequence of base-pairs – the "rungs" of the DNA ladder. Sometimes the sequence is not reproduced exactly during cell division, resulting in a mutant gene. If such a mutation occurs in a body cell, it can result in a nontransmittable defect such as cancer. If it happens in a sex cell – an ovum or sperm – it can be passed on to a person's children.

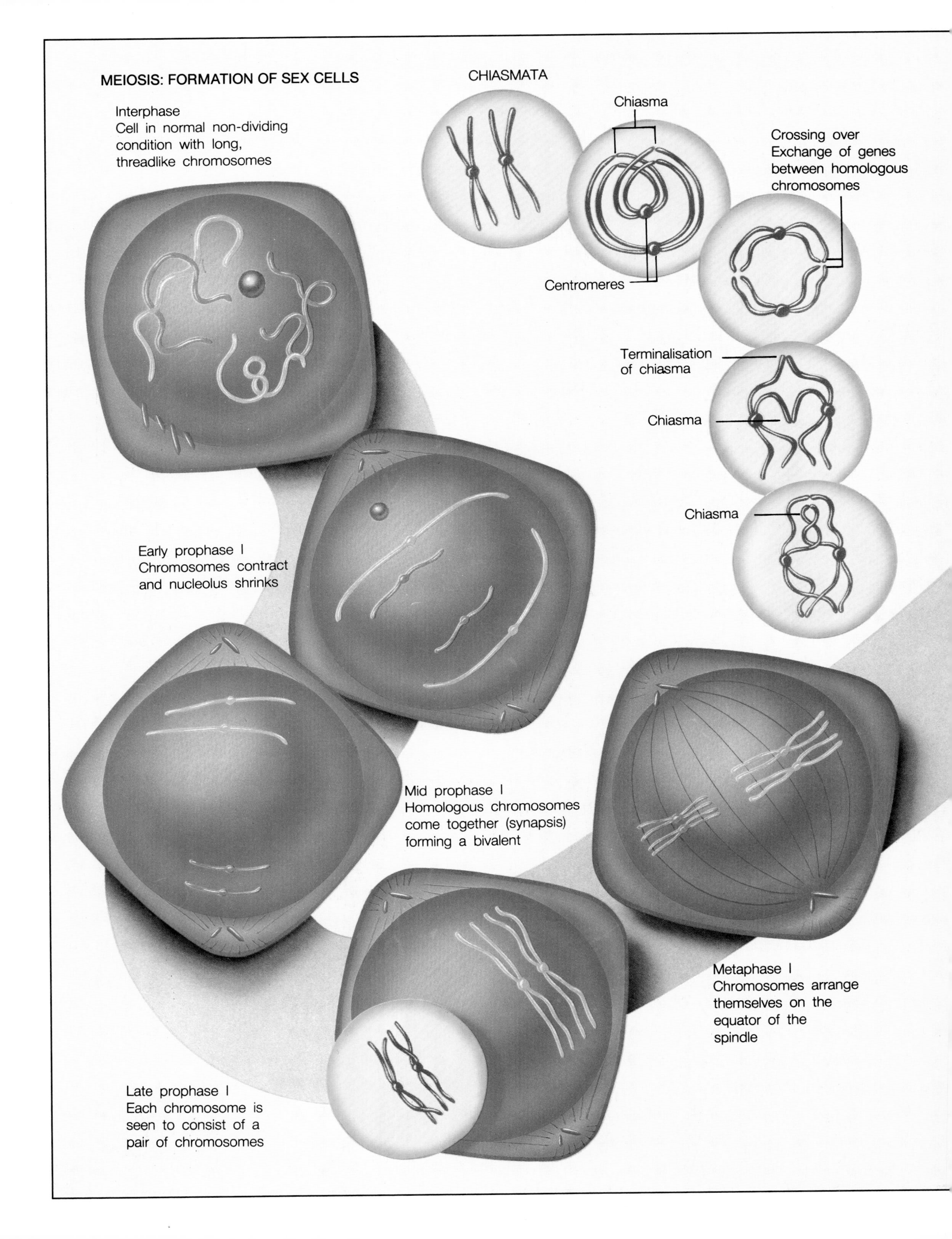
MEIOSIS: FORMATION OF SEX CELLS
Interphase
Cell in normal non-dividing condition with long, threadlike chromosomes
CHIASMATA
Chiasma
Crossing over
Exchange of genes between homologous chromosomes
Centromeres
Terminalisation of chiasma
Chiasma
Chiasma
Early prophase I
Chromosomes contract and nucleolus shrinks
Mid prophase I
Homologous chromosomes come together (synapsis) forming a bivalent
Metaphase I
Chromosomes arrange themselves on the equator of the spindle
Late prophase I
Each chromosome is seen to consist of a pair of chromosomes

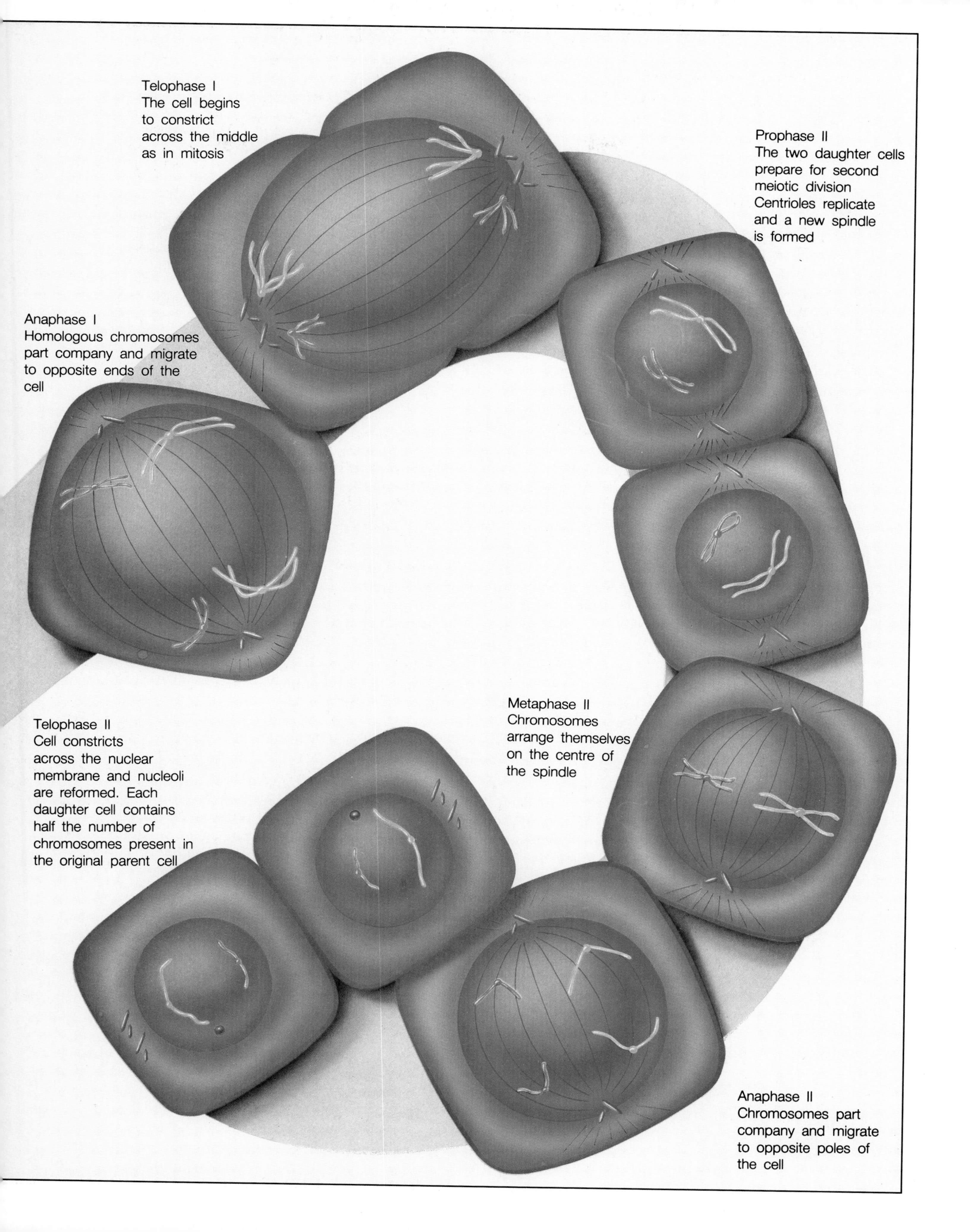
Telophase I
The cell begins to constrict across the middle as in mitosis
Prophase II
The two daughter cells prepare for second meiotic division
Centrioles replicate and a new spindle is formed
Anaphase I
Homologous chromosomes part company and migrate to opposite ends of the cell
Metaphase II
Chromosomes arrange themselves on the centre of the spindle
Telophase II
Cell constricts across the nuclear membrane and nucleoli are reformed. Each daughter cell contains half the number of chromosomes present in the original parent cell
Anaphase II
Chromosomes part company and migrate to opposite poles of the cell

DNA and the genetic code

Cell division is the basis of heredity. Each cell in the body contains within its nucleus the information that controls the minute-by-minute activity of each cell and of the whole body. It also contains all the genetic information passed from parent to offspring through the reproductive process. This information is encoded in the form of a substance called deoxyribonucleic acid, or DNA. This remarkable biological chemical has two essential characteristics: it can store information and make an exact copy of itself.

The DNA molecule is a giant in chemical terms but even so is much too small to be seen with even the most powerful optical microscope. Its shape resembles that of a twisted ladder – the famous double helix – with millions of rungs. The "sides" of the ladder are composed of alternating units of phosphate and a sugar (deoxyribose); the "rungs" consist of a linked pair of chemical compounds called nucleic acid bases. There are four bases, designated adenine (A), thymine (T), cytosine (C) and guanine (G). A can link only with T; C links with G.

At cell division DNA can be seen because it is located on the chromosomes which become visible at this time – and genes are short segments in the threads of chromosomes. During the normal process of mitotic cell division, by means of which cells multiply, chromosomes first duplicate themselves and are then pulled apart during the splitting of the "mother" cells so that each of the new "daughter" cells gets exactly one complete set.

But there is one exception to this normal process. This occurs when the male and female sex cells, or gametes, are formed. The purpose of division in this case is to produce a cell containing only half the normal complement of chromosomes. Known as meiosis, this type of cell division is accomplished by the chromosomes splitting twice, first to separate the duplicate chromosomes into separate cells, and secondly to separate the chromosome pairs into the daughter cells.

Because there are 23 pairs of chromosomes in humans, there are more than ten million different ways that the chromosomes can become sorted during meiosis. In a fertilized egg, each chromosomal pair is made up from halves from each parent, consequently the chances of children of the same parents getting exactly the same set of chromosomes are so infinitesimally small that it effectively never happens. Only identical twins – from a single fertilized egg – can have identical chromosomes.

Secret information

Normally hidden inside the opaque nucleus of a cell, chromosomes become apparent only when they separate from the nucleus immediately before cell division. At this point they can be chemically stained (the word "chromosome" means "colored body") and observed through a microscope. At high magnifications individual genes may sometimes show up as dark bands across the chromosomal threads. By a happy chance for scientists, the chromosomes in the saliva cells of *Drosophila*, the common fruit fly, are among the largest known. And the fruit fly has a conveniently short, 19-day life cycle.

Much of the pioneering work on genetics, during the first half of this century, was carried out by studying endless generations of fruit flies, counting the rates of natural mutations, such as white eyes as opposed to the normal red ones, and slowly drawing the first faint maps of linear-linked genes – the footpaths of heredity.

Glimpsing a faint track on a faded map does not mean that you could necessarily find and follow the way in pitch darkness. Blurred footprints caused by gross physical abnormalities showed conclusively that genes were the envelopes containing the basic instructions handed on from generation to generation; but how was the information written?

The medium is the message

Obviously an enormous amount of very precise information was being transferred, and it seemed equally obvious that the messages must be contained within the large, complex protein molecules in the gene substance. The strands of DNA that held them together were regarded as too simple, with only four basic subunits, to have any significance. For 30 years scientists pursued the protein trail. Then, in 1953, James Watson and Francis Crick published a short, 1,000-word paper which proposed the structure of deoxyribonucleic acid and predicted its role. The DNA revolution, which was to lead to the greatest ever advances in biological science, had begun.

The DNA molecule is very long; a minute and continuous spiral ladder that, in humans, is made up of about five billion pairs of nucleotide rungs. The four types of nucleic acid bases give the language of life an alphabet of just four different letters; but with 10 billion characters per molecule this represents about 20 billion bits of information. A single human DNA molecule equates to a library containing some 4,000 500-page books: enough to describe and define a person in perfect detail.

Simpler organisms, such as bacteria, require less genetic information to define them, and their chromosomes are often simple circles of DNA. The message is less complex, but the language is the same. Every life on Earth is written in DNA.

The language of life

All living things contain DNA. Between the simple circle of the DNA in a bacterium and the tightly coiled spirals of human chromosomes, DNA runs in an everlasting thread of life on earth. The model of DNA structure produced by Watson and Crick stemmed from research into work on the identification of bacteria begun at the time of the introduction of antibiotics to combat bacterial disease during World War II. Bacteria were ideally useful to geneticists because they reproduce so quickly – in minutes rather than days, months or years. Other laboratory experiments using fungi, which also grow and multiply rapidly, showed that individual genes controlled the production of specific amino acids, the building blocks of protein, inside a cell.

DNA works all the time. To use computer jargon: during reproduction DNA is arranged into disks and files (chromosomes and genes) outside the nucleus, and these transfer the master program (including the secret of reproduction) to the new cell. At all other times DNA acts as the cell's operating system, residing in so-far undetermined form inside the nucleus, and constantly managing the correct sequencing and assembly of amino acids in the protein factory of the cytoplasm.

Special delivery

Coded into the DNA is a series of instructions which determine the production of enzymes (themselves proteins), which ensure that each protein in a cell is assembled from the correct primary sequence of amino acids. Proteins are the cell's fundamental building blocks and direct operations within its walls. If assembled in the wrong order, an abnormal protein results which may disrupt some vital function inside the cell.

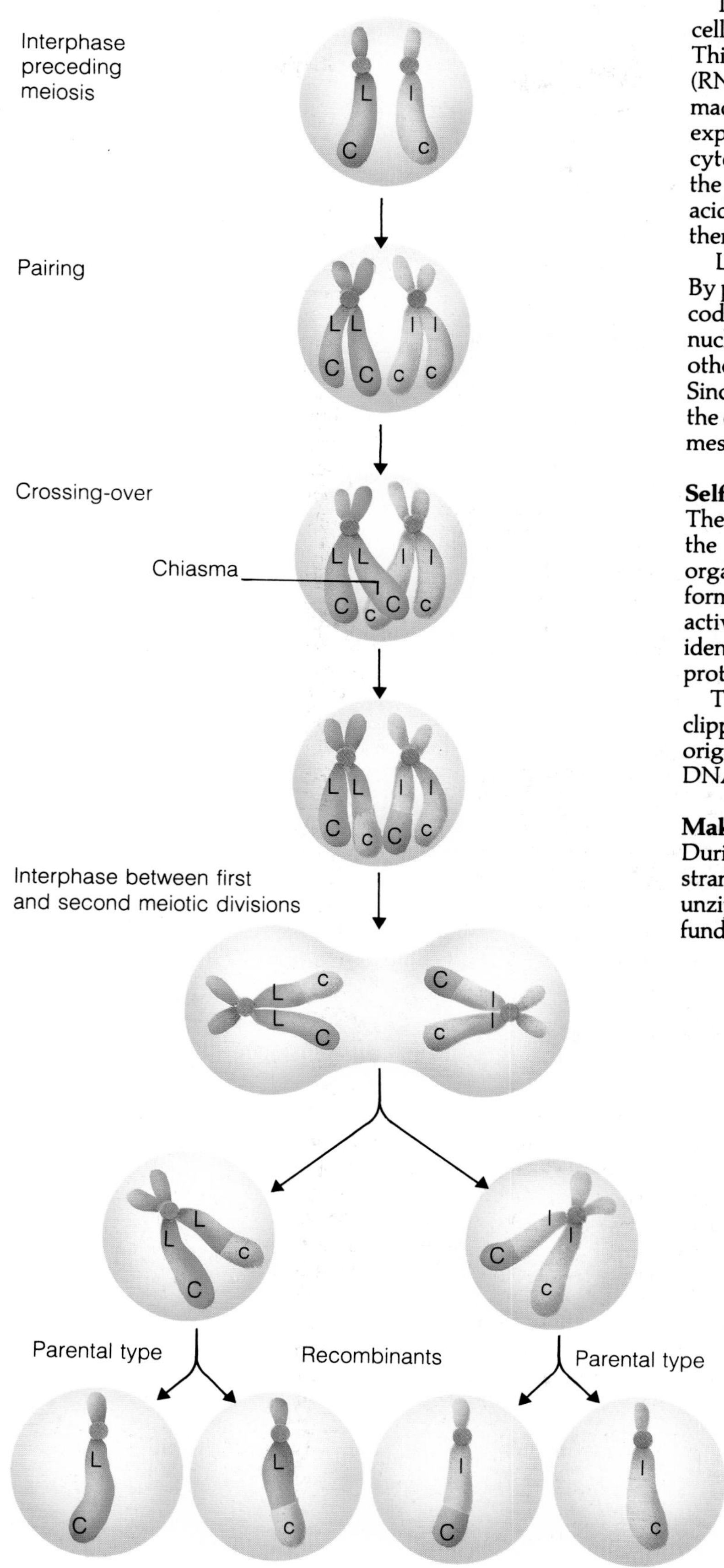

Information from the DNA in the nucleus is carried out to the cell cytoplasm to do its work by means of a chemical messenger. This resembles DNA in structure and is called ribonucleic acid (RNA). Three types of RNA are involved: ribosomal RNA is made in the nucleus. It is a direct copy of a gene which is exported along messenger RNA (mRNA) pathways through the cytoplasm. Transfer RNA (tRNA) acts like an adapter, allowing the information-coded ribosome to attach to successive amino acids and trigger the release of the appropriate enzymes to fix them onto a growing protein chain.

Like DNA, ribonucleic acid contains four bases or nucleotides. By patient experimentation, scientists have been able to crack the code with artificial mRNA made up of strings of a single nucleotide. Specific nucleotides produce specific amino acids, other codes switch the factory "off" when a protein is finished. Since the average protein contains about 400 amino acids, and the coded signal is known to be three bases in a row, each protein message must be about 1,200 nucleotides long.

Self-repairing mechanism

The library contained within the double helix of DNA includes the coded templates for all the proteins and their subsequent organization within the cells of the human body. It also holds the formulae for all the many enzymes which regulate the cell's activity: forming the ribosome RNA into a copy of a gene, identifying the sequence of amino acids, and stitching them into protein chains.

There is even a set of enzymes which police DNA itself, clipping out damaged nucleotides or filling in gaps so that the original sequence is restored. Each cell has at least 50 of these DNA maintenance enzymes on call at all times.

Making copies

During normal cell division and duplication by mitosis, the two strands of the DNA molecule are pulled apart, rather like being unzipped down the middle. In the copying process that is fundamental to life, and is the basis of all growth and reproduc-

Crossing-over of gene pairs during meiosis – a possibility for all the many thousands of genes on a chromosome – results in a huge variation of human genetic makeup and subtle differences between members of the same family. In this diagram, the letters L and l stand for alleles for long and short noses, and C and c stand for curly and straight hair. L and C are dominant; l and c are recessive. In this example, crossing-over has produced two chromosomes similar to the parental types (L/C and l/c), and two others (L/c and l/C) different from the parents.

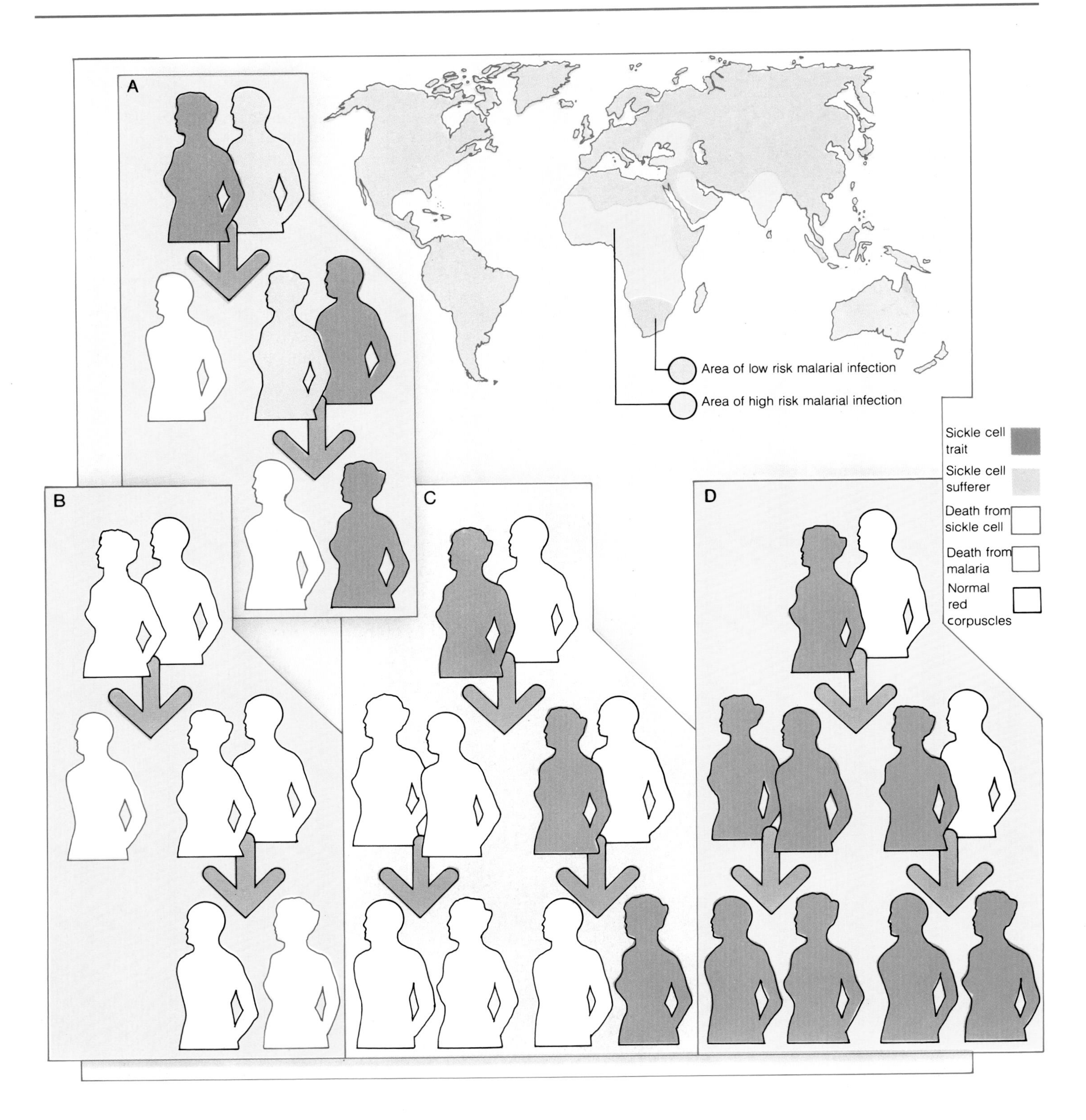

Immunity to malaria is an "advantage" conferred on sufferers from sickle-cell anemia. The disease first arose in tropical and semitropical areas of the world where malaria-carrying mosquitoes were prevalent. The blood of people with sickle-cell disease (A) has red corpuscles that have a collapsed, crescent shape. They clump together and easily clog the narrow capillary blood vessels, often proving fatal (figure with blue outline). People in the same area but without sickle-cell anemia (B) often succumb to malaria (red outline). If people with the sickle-cell trait move to an area where there is little risk of malaria (C), the sickle-cell gene eventually passes out of the population. Among people who are carriers of the disease (D), fewer than one percent of their red corpuscles are abnormal, and neither do they become infected by or die from malaria.

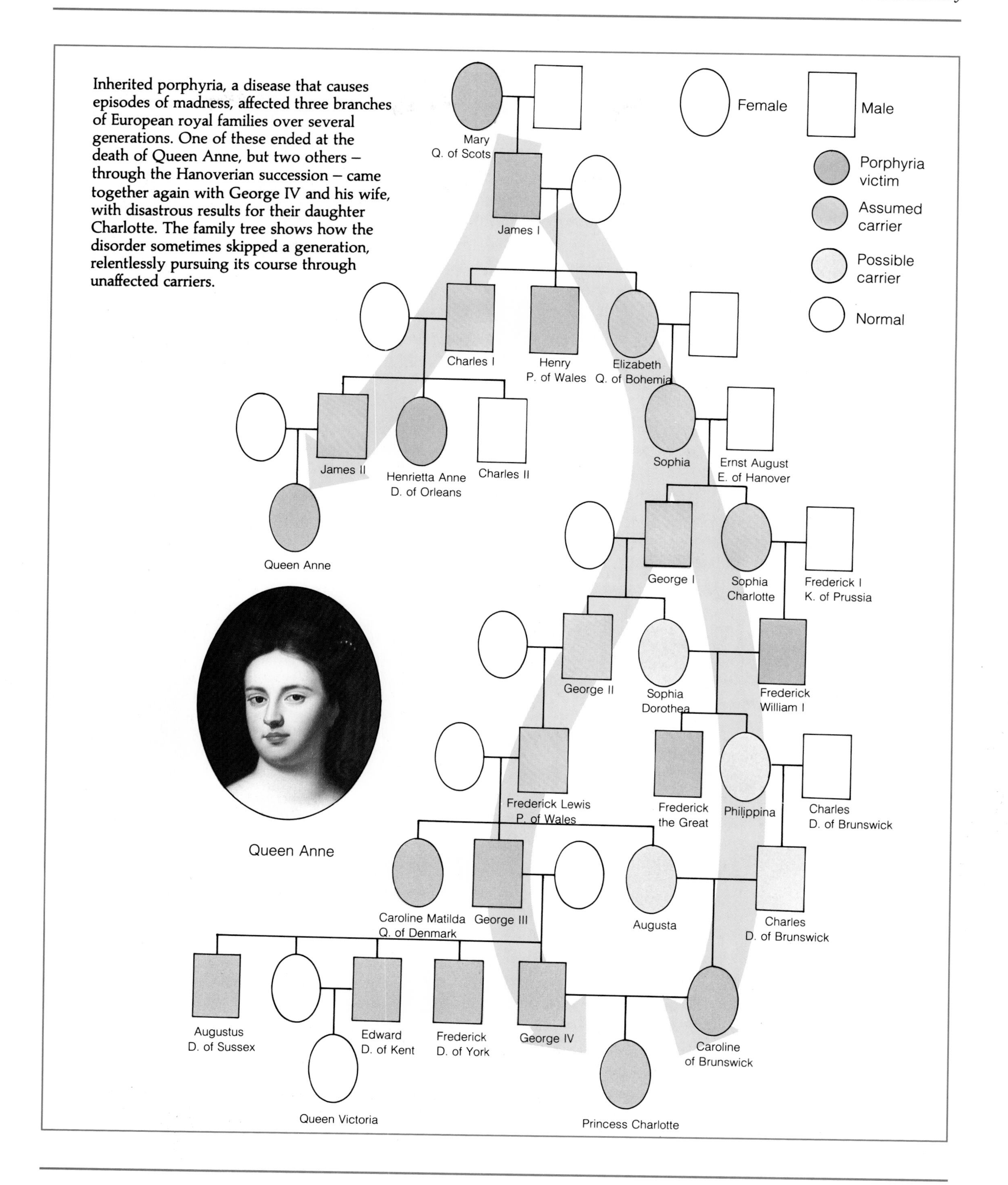

Inherited porphyria, a disease that causes episodes of madness, affected three branches of European royal families over several generations. One of these ended at the death of Queen Anne, but two others – through the Hanoverian succession – came together again with George IV and his wife, with disastrous results for their daughter Charlotte. The family tree shows how the disorder sometimes skipped a generation, relentlessly pursuing its course through unaffected carriers.

tion, a replica is assembled on each strand, corresponding exactly to the missing one. Thus each new DNA molecule contains one old strand and one new strand, the whole being identical to the original double strand.

Most of the time DNA is faithfully copied so that identical genes are formed. But a small mistake in the sequence of bases results in the wrong message being copied off by the RNA, the wrong instructions being sent out and the wrong protein being made in the factory. This alteration of the code is the basis of mutation. Radiation, whether atomic or from sunlight, can also scramble the genetic code, but the DNA maintenance enzymes are so efficient that of the five billion pairs of nucleotide bases in a human sex cell, only about 15 change in a year.

Alterations in the genetic code – mutations – occur all the time. When they occur in sex cells, the mutation can be passed on to offspring. If the mutation affects a dominant gene, then the mutation has a chance of establishing itself throughout the population and eventually becoming the norm. The controlling process, which decides which mutations become permanently incorporated into the code, is known as natural selection.

The natural choice

The concept of natural selection is at the core of Charles Darwin's theory of evolution, summed up in the popular phrase "survival of the fittest." When Darwin first put forward his ideas, this was often taken to mean that the strongest and most aggressive creatures would become dominant – "nature red in tooth and claw." But the evolutionary mechanism is far more subtle than that. Darwin's studies of animal species on isolated islands led him to the conclusion that those members of a species which are best suited to their environment will be the most successful.

For example, the reason giraffes have long necks is not through years of stretching to reach the high branches of trees; but because over the years, those animals with longer necks could reach more food, and therefore lived longer and had more offspring. Thus the proportion of long necks in the population would inevitably increase. We now understand that over millions of giraffe generations, the nucleotide code for long necks, which possibly started as a mutation in a single animal, became solidly established in giraffe DNA.

Evolution in action

Nature constantly improves itself; not through any great master plan, but through constant trial and error. If a mutation is beneficial to the species, it is likely to succeed; if it is harmful, then the creature dies and the altered DNA disappears without trace. Without constant, random mutation creating improvement and diversity life on earth would cease to evolve, and a billion-year process would stop.

Evolution is still a continuing process. In humans it is too slow to be measured except in the broadest of terms, but in organisms with much shorter lifespans, evolution can be seen to be at work. A salutary example can be seen in medicine. Before antibiotics were in widespread use, most bacteria were susceptible to their effects; the age of the miracle drug seemed to have dawned. However, spontaneous mutation produced the occasional bacterium that was resistant to the normally devastating drugs.

Previously there had been no advantage to a bacterium having a gene giving this resistance, and the trait would have been bred out in a few generations. The introduction of antibiotics now meant that the gene conferred two distinct advantages: survival in the face of antibiotic attack and, because all its non-resistant brothers had been killed, an environment free from other similar organisms competing for food.

Super bugs versus super drugs

Naturally, such resistant bacterial forms have proved very successful, especially in antibiotic-rich environments such as hospitals, and special drugs have had to be developed to knock out resistant bacteria. Overall, however, the original nonresistant strains are more prevalent, which indicates that they are better adapted to life without antibiotics than their drug-resistant cousins are to life with them. Evolution involves a very subtle interaction with the environment: a genetic mutation that better adapts life to one factor in the environment may prove a handicap in dealing with all the other factors.

The development of resistance is not confined to bacteria. Repeated use of DDT against malaria-carrying mosquitoes has led to the emergence of DDT resistant insects. These have arisen from those mosquitoes having genes that provided enough natural resistance to survive onslaught from DDT.

The two-edged sickle

Just as mosquitoes have adapted to human attempts to eradicate malaria, so humans have adapted in the face of the malarial onslaught, though with less completely successful results.

Full-blown sickle-cell anemia is a disease that affects about eight percent of the black population of the United States. People do not "catch" sickle-cell anemia, they inherit it from their parents. What they actually inherit from both mother and father are genes with a "mistake" in the DNA code that controls the production of a certain amino acid which is a component of the blood protein hemoglobin. Consequently, their red blood corpuscles have a distinctive sickle shape. These cells are inefficient at carrying oxygen around in the blood and have a tendency to clog the body's capillaries, which can prove fatal.

The rules of natural selection suggest that such a life-threatening mutation ought to have died out many years ago. But it did not, and the reason lies with the mosquito. Sickle-cell anemia developed in those regions of the world infested with malaria-carrying mosquitoes. Those people who inherit a sickle-cell gene from just one, not both of their parents develop mildly abnormal red blood corpuscles. This is not enough to make them ill, but it is enough to protect them from malaria.

Consequently, people with the sickle-cell trait who do not suffer the disease are more likely to survive, and have more children, than those without the sickle-cell gene. Away from a malarial environment, the abnormal gene gradually declines in frequency as the reproductive advantage disappears.

Royal flush

Most hereditary disorders have no such compensating advantage. They have purely negative effects on the reproductive power of the sufferer. Statistically they are rare because the line has a tendency toward dying out before the mutation can become established in the DNA code.

During the 18th and 19th centuries, the royal houses of Europe were afflicted by hemophilia, the inability of the blood to clot, which is transmitted through the female X chromosome but affects male children (although at the time doctors had no understanding of the hereditary mechanism involved).

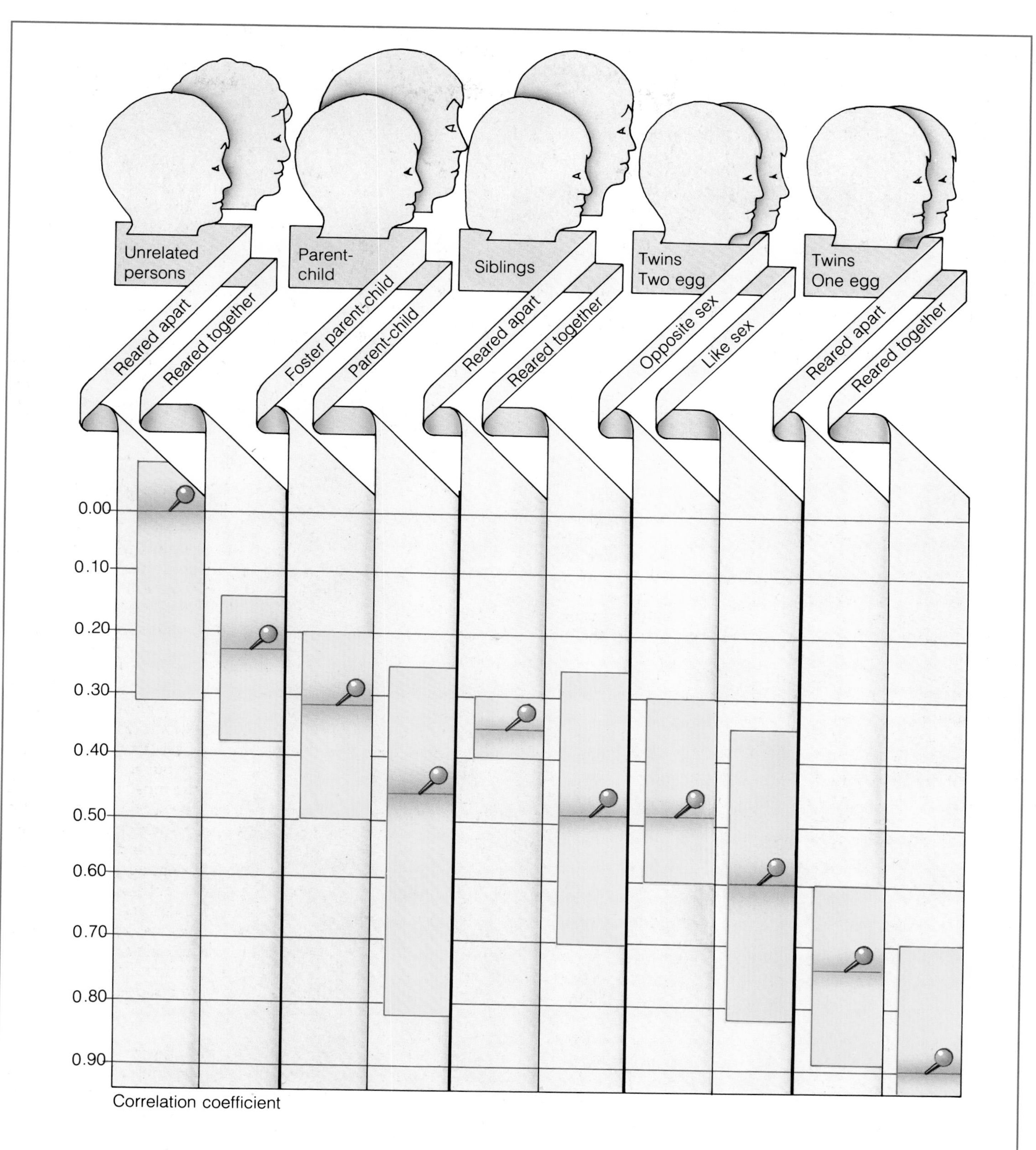

The relative importance of heredity – "nature" – and environment – "nurture" – can be assessed by comparing the correlation of intelligence test scores between pairs of individuals, from unrelated persons reared apart (zero correlation) to identical twins brought up together (correlation of 0.9). For two people with exactly the same score the correlation would be 1. As the chart indicates, the closer the relationship, the greater the correlation.

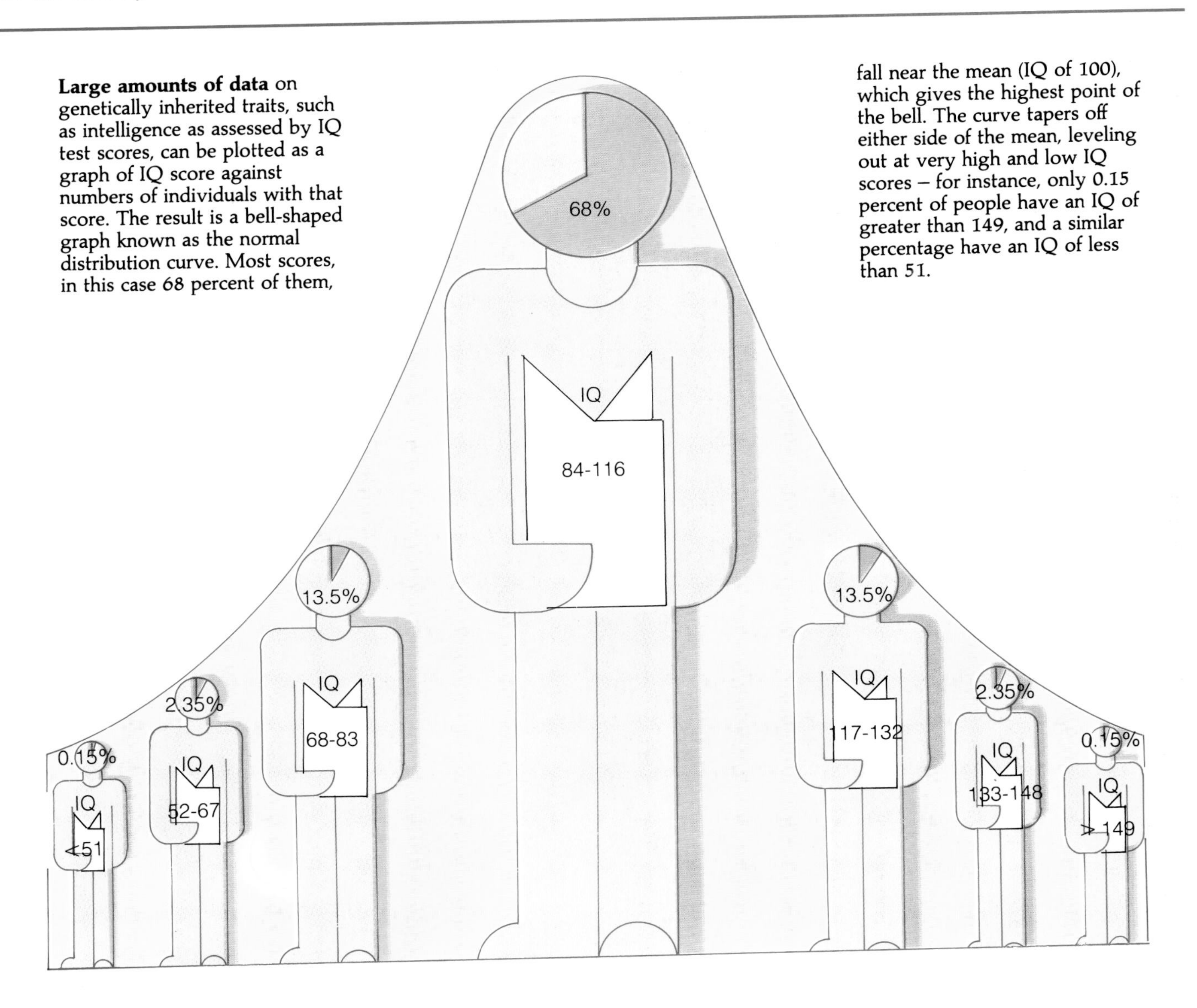

Large amounts of data on genetically inherited traits, such as intelligence as assessed by IQ test scores, can be plotted as a graph of IQ score against numbers of individuals with that score. The result is a bell-shaped graph known as the normal distribution curve. Most scores, in this case 68 percent of them, fall near the mean (IQ of 100), which gives the highest point of the bell. The curve tapers off either side of the mean, leveling out at very high and low IQ scores – for instance, only 0.15 percent of people have an IQ of greater than 149, and a similar percentage have an IQ of less than 51.

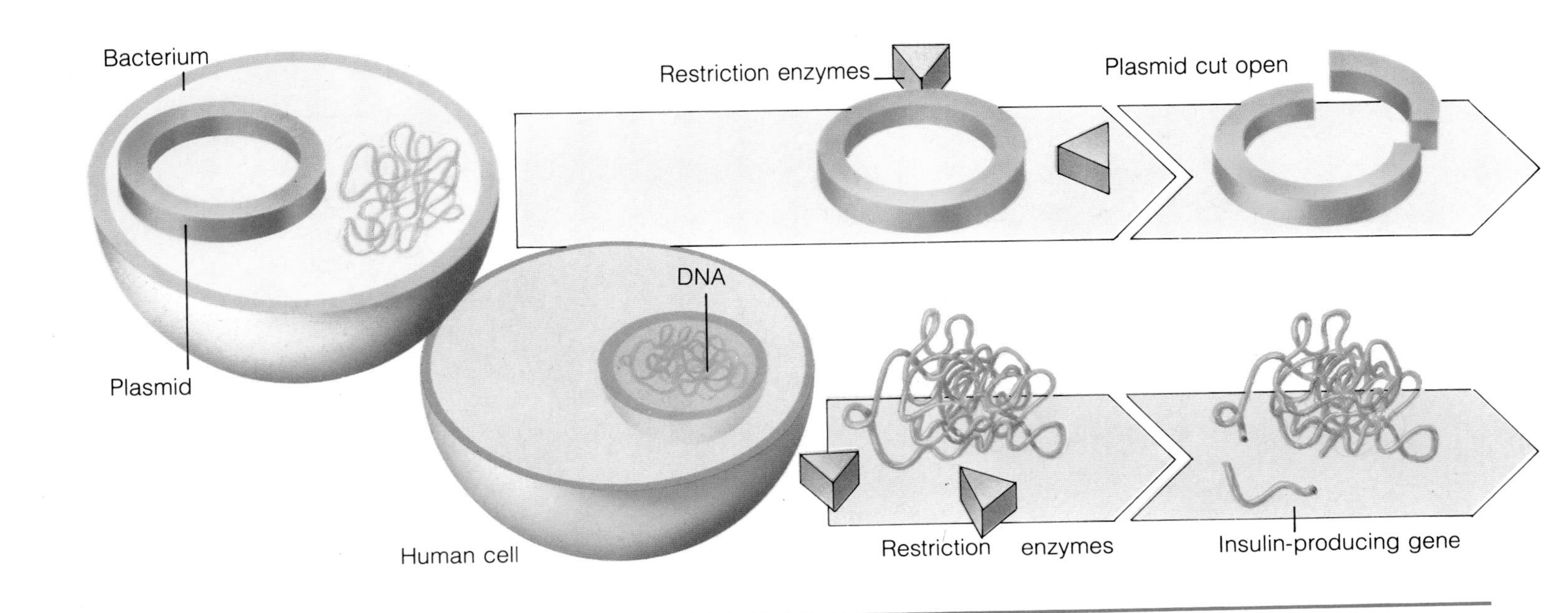

Bleeding was, however, only part of the inherited royal burden. From recent examination of direct descendants, medical experts have been able to show that King George III of England suffered from porphyria, a hereditary metabolic disorder.

Porphyrins are purple-red pigments contained in every cell in the human body. Normally, the body retains most of its porphyrins, but a rare genetic defect can cause them to become concentrated and excreted in large amounts. The accumulation of porphyrins causes severe pain, paralysis, and hallucinations, as well as giving the skin an extreme sensitivity to light. Little wonder that King George was thought mad, spending his time in a darkened room, or shaking hands with trees.

Through a piece of failed evolution – the mutated code sequence in one of his genes – King George was not fit in 1775 to deal with the crisis in his American colonies. The following year a new nation was born, and the history of the world was altered for ever.

Only by extreme coincidence does individual heredity have such a dramatic effect on the lives of so many others. Mostly, heredity is a personal thing. The random shuffling of our parents' genes is the hand in life we are dealt. It affects nobody but ourselves and our children. With our knowledge of genes we can now predict some of the ways they will affect offspring; other patterns of inheritance remain puzzling.

Identical hands

Abnormal conditions, disorders and diseases provide researchers with little markers along the trails of inheritance. Medicine's pathological approach, tracing everything back to its causes, has enabled scientists to separate the hereditary effects of a disorder from the effects of environment, diet, and so on.

When searching for the effects of heredity on normal human qualities, such as height, weight or intelligence, there are no such convenient signposts, and it is much more difficult to separate the effects of the environment, to differentiate between the inherited and the acquired.

For this reason, children born with the same hand of genetic cards, identical twins from the same egg, have a special significance for geneticists – they provide a natural benchmark. If differences exist between identical twins, they must be due to environment because the genetic material, the endless coiled sequence of coded DNA, is exactly the same in each twin.

Nature versus nurture

If you are tall, do you inherit tallness from your father, from your mother, or from a high protein diet? Do fat parents have fat children because of their genes, or because the children learn bad eating habits from their parents? In the nature versus nurture argument, "nature" refers to a person's genes; "nurture" represents the sum of the environmental factors that surround them, be they social, educational, or nutritional.

The study of twins is limited by the fairly small number of twin children born each year (about one birth in 80). Identical twins are even rarer, and only infrequently are these parted at birth and raised in different environments. There have therefore been only a few extensive twin studies, and these have not yet settled basic questions in favor of "nature" or "nurture."

Double vision

A twin study is conducted in one of two ways: either a particular quality, such as height, is measured; or the presence or absence of a particular feature is noted. Disease, itself an abnormal condition, is the feature most often looked for in these studies.

The crippling joint disease rheumatoid arthritis occurs in both of identical twins only about ten percent of the time. This is more often than the disease occurs in the rest of the population, but it

Insulin is a hormone that controls the way the body metabolizes sugars, and it is lacking in the blood of people with one form of diabetes. Many of them must have daily injections of insulin, and formerly this was extracted from the pancreases of dead cattle or hogs. But some diabetics are allergic to animal insulin, and to help them recombinant DNA techniques are now being used to make human insulin.

Plasmid DNA is removed from a bacterium, at the same time as DNA is removed from a human cell. Restriction enzymes cut open the plasmid and slice the insulin-producing gene from the human DNA. This gene is then spliced into the plasmid, which is put back into the bacterium. Several bacteria engineered in this way are cultured in a fermenter. The insulin they produce is purified for human use.

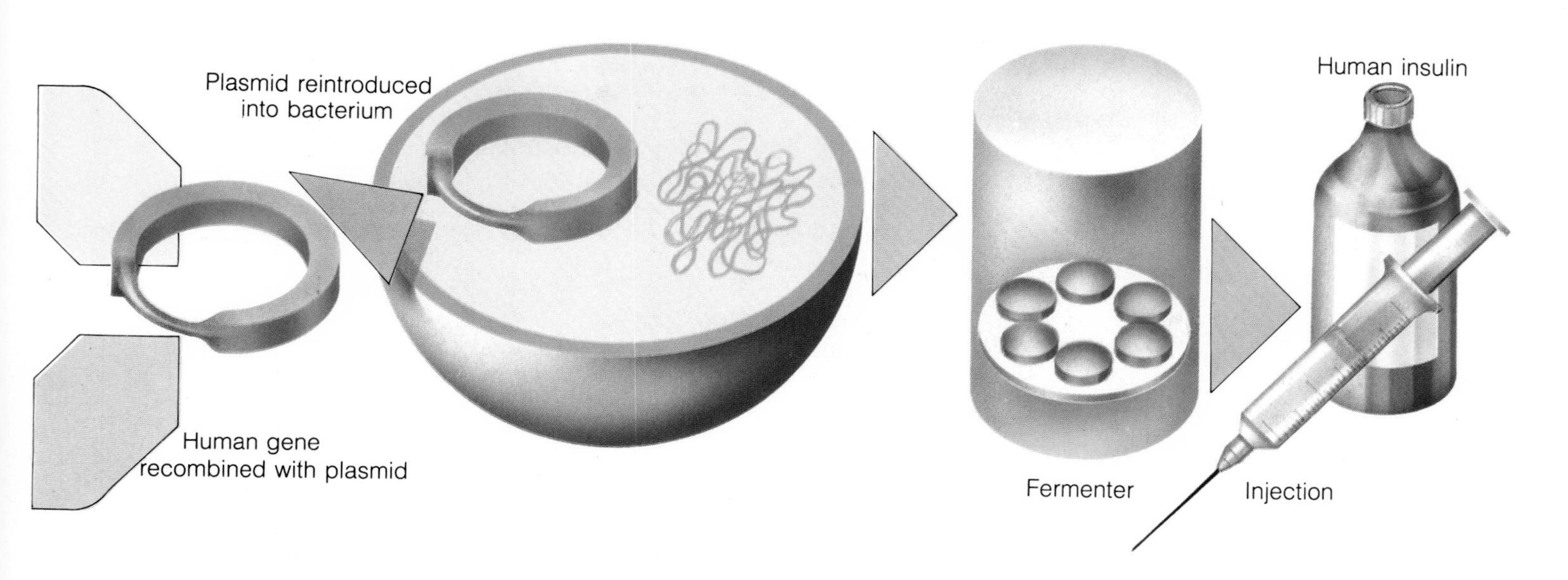

Down's syndrome, or trisomy 21, is a common genetic disorder, particularly among mothers over the age of 40 in whom it occurs in 1,000 out of every 100,000 births. The chart shows the incidences, per 100,000 births, of other trisomies and of various sex chromosomal abnormalities such as Klinefelter's syndrome and the very rare Turner's syndrome. It should not be overlooked that out of those same 100,000 births, on average 98,590 of them are perfectly normal.

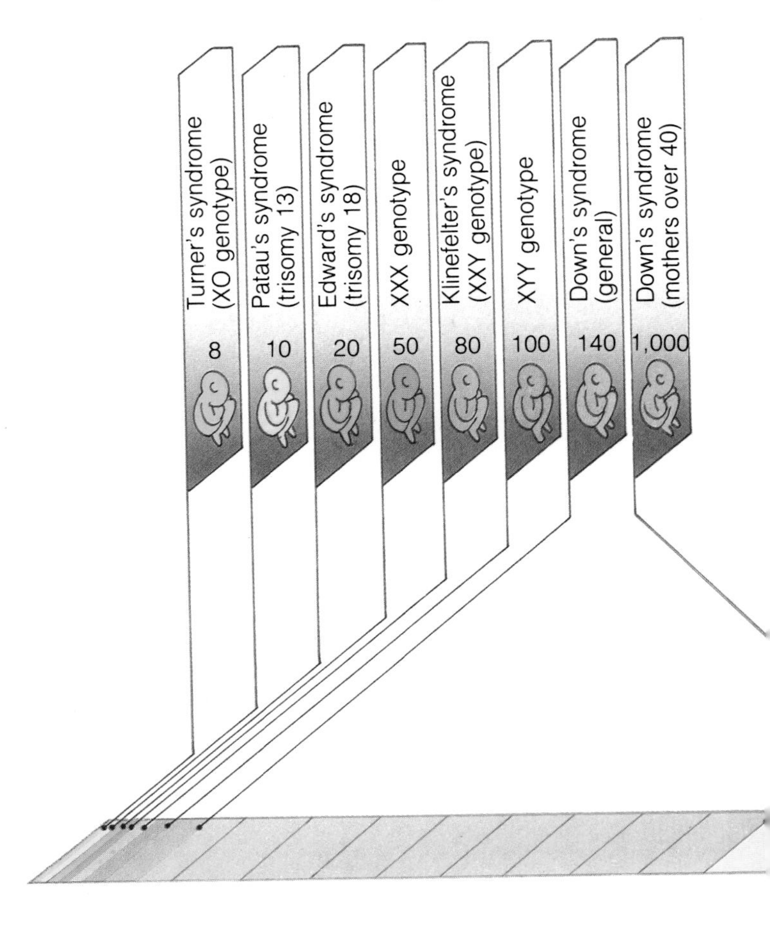

cannot be due to genetic factors alone. If it were, every identical twin would have the disease if the cotwin did. Other medical conditions, such as manic depression, demonstrate a much stronger genetic influence.

Height and weight can be easily measured and plotted against human norms. Height has been shown to be largely due to genes, weight much less so. However, this does not tally with the rapid increases in the average height of Japanese adults during this century, coinciding with the popularization of Western food in the Far East.

Intelligence was thought to be very much genetic, even despite the difficulties in measuring it. Correlation of IQ scores by identical twins raised apart is much closer than that predicted by the bell-curve of normal human distribution. This aspect of twin studies has, however, been dogged by controversy. Among the many criticisms raised has been the objection that such studies ignore the first nine months of life in the environment of the womb. It is virtually certain that intelligence is a factor affected both by nature and nurture.

The relationship between the nature and nurture may also be interactive. Changes in diet or climate could allow genetic changes already programmed into the DNA to manifest themselves as observable characteristics, such as height gain, in a whole population over a short span of time.

Through the discovery of new inherited traits scientists have recently been able to catch a glimpse of human evolution at work. These traits are central to certain critical interactions between a person and the modern world, and involve genes that govern the speed with which the body breaks down certain substances. These substances are nearly all drugs or chemical pollutants; the human species is beginning to evolve in response to the synthetic additives in the modern environment.

Human intervention

The brain power of humans has a huge role to play in the evolution of our species. Rather than allowing nature's incessant process of trial and error to direct humankind in response to environmental pressures, it is likely that people themselves will have taken a hand. By using techniques that are so new, and which promise such fundamental change, humans will be able to manipulate genes – and with them evolution – in a way that still seems to be in the realm of science fiction.

As soon as DNA was discovered, scientists began looking for ways to manipulate it. It is not the easiest substance to deal with – when extracted from a cell the long strings of DNA molecules have a tendency to form sticky, gluelike blobs. The strings get hopelessly entangled and, to make matters worse, the coded sequence of nucleotide bases has no obvious starting point. The breakthrough was made in 1970 at Johns Hopkins University, Baltimore, with the discovery of the first restriction enzyme. Usually isolated from bacteria, restriction enzymes have the ability to slice DNA up into regular, reproducible pieces of a manageable size.

Restriction enzymes have become the basic tool of a completely new profession, the genetic engineer. Hundreds of different restriction enzymes are now available to the scientist, each capable of cutting DNA in a slightly different way, producing a different fragment that can be analyzed. In this way,

the stuff of life has become visible and malleable.

These chemical scissors work by cutting DNA only when an exact recognition sequence of nucleotides occurs. When the effects of different restriction enzymes are compared, each restricted to different nucleotide sequences, it becomes possible to subdivide large pieces of DNA into several smaller fragments. Study of these small fragments has enabled scientists to get much closer to understanding how genes are constructed, and how their activity might be controlled. One very important subsequent development has been the ability to produce large amounts of exactly the same genetic material.

Changing the recipe

The fragment of bacterial DNA to be duplicated is first stitched into a plasmid, a type of extra gene structure on a bacterial chromosome. Plasmids normally exist as tiny circles of DNA; these can be snipped open with a restriction enzyme and the extra fragment inserted. The plasmid can then be reinserted into its normal bacterial host.

If the bacteria are then grown in large quantities, the plasmid can be recovered and the required DNA fragment purified out. This process is called gene cloning. A single gene is isolated, inserted into a carrier microorganism, grown, and billions of identical clones produced. The technique is also known as recombinant DNA research, because it involves recombining several pieces of DNA.

The first genes were cloned in California in 1973, and an explosion of genetic experimentation was ready to take place. The rewards seemed enormous, recombinant DNA techniques could produce endless supplies of vital substances such as human insulin which could be used to treat diabetes.

Facing the worst

A public outcry halted any immediate experiments with human DNA, or even with DNA from bacteria that lived in humans. In other areas of research, notably in genetic counseling, geneticists found a much greater degree of public acceptance.

The heartbreak and tragedy caused by birth defects is universal to all human societies. It is estimated that about four percent of the world's population suffers from some genetic, or partly genetic, abnormality and would therefore benefit from genetic counseling.

Most birth defects result from some aberration in the distribution of chromosomes, and at least one percent of all infants have a chromosomal anomaly. This most frequently occurs during the sorting process of meiosis when egg and sperm cells are formed. Numerically, the commonest chromosomal anomaly (about one in every 600 live births) produces what is known as Down's syndrome, sometimes called mongolism. Affected children are short, have an extra layer of tissue on their eyelids, and are mentally retarded.

Such children have an extra chromosome (47 instead of the normal 46). What happens is that during the formation of the egg, both number 21 chromosomes end up in the same egg cell. When the egg is fertilized by a sperm cell with its single chromosome 21, it produces a child with three number 21 chromosomes per cell.

Having three (instead of the normal two) of a particular

chromosome is known as a trisomy, and trisomy 21 produces Down's syndrome. Other trisomies which cause severe mental handicap are trisomy 18 (Edward's syndrome) and trisomy 13 (Patau's syndrome). In most pregnancies, chromosomal abnormalities occur as chance and random events, and parents of such children can take a little comfort from the fact that their own chromosomes are normal and that the situation is very unlikely to recur. Certain types of abnormality, however, have a recurrence risk that is significantly higher.

Non-fatal errors

During normal mitotic cell division, breaks may occur in the strands of DNA which are then repaired by the maintenance enzymes. Sometimes the pieces of gene material are attached to the wrong chromosomes, resulting in what is known as a translocation. These can be passed on to children. A translocation involving chromosome 22 exchanging a small segment with chromosome 9 is frequently associated with one type of cancer – chronic myeloid leukemia.

Failure of proper separation during mitosis can result in two cell lines forming, one with 45 chromosomes and one with 47. Individuals with tissue containing two or more chromosomally distinct cell types are known as mosaics; their children run a higher risk of birth defects than those of normal parents.

The other common chromosomal aberrations are those of the sex chromosomes. A male with an extra X chromosome (Klinefelter's syndrome, with the genotype XXY) is likely to be lanky and sterile, an extra Y chromosome produces a man of significantly increased stature, but probably very few of the criminal tendencies popularly attributed to this genotype. In females the triple X genotype is usually clinically normal, but may be deficient in secondary sexual development.

Statistical signposts

Genetic counseling tries to minimize human suffering by examining the statistical probabilities. A couple with a handicapped child want to know whether a future brother or sister will suffer the same disability, and what about the grandchildren? Counseling provides a scientific basis for the decision whether to continue the family or not.

The giving of genetic advice is fairly straightforward in typical conditions, especially if one of the three basic types of Mendelian inheritance is involved. These are: autosomal dominant (autosomal means any other chromosomes than the sex chromosomes); autosomal recessive (risks to carrier siblings are small unless they marry a near relative); and X-linked (generally recessive with risks to male children of carrier women).

With many diseases, the risk of recurrence is usually quite simple to calculate. For example, cystic fibrosis is an autosomal recessive disorder with a one in four risk for subsequent children if both parents are carriers.

Classic hemophilia is an X-linked recessive in which the odds that a carrier woman's son will be a sufferer, or that her daughter will be a carrier, are in both cases 50-50. Had this sort of information been available 200 years ago, the arrangement of royal marriages in Europe might well have been very different.

When a trait is inherited through more than one factor, the counseling is necessarily less accurate. The statistics are calculated on the basis of the empiric risk – that is, the likelihood of recurrence based on past experience of the way the trait has been inherited within other families. Down's syndrome is empirically

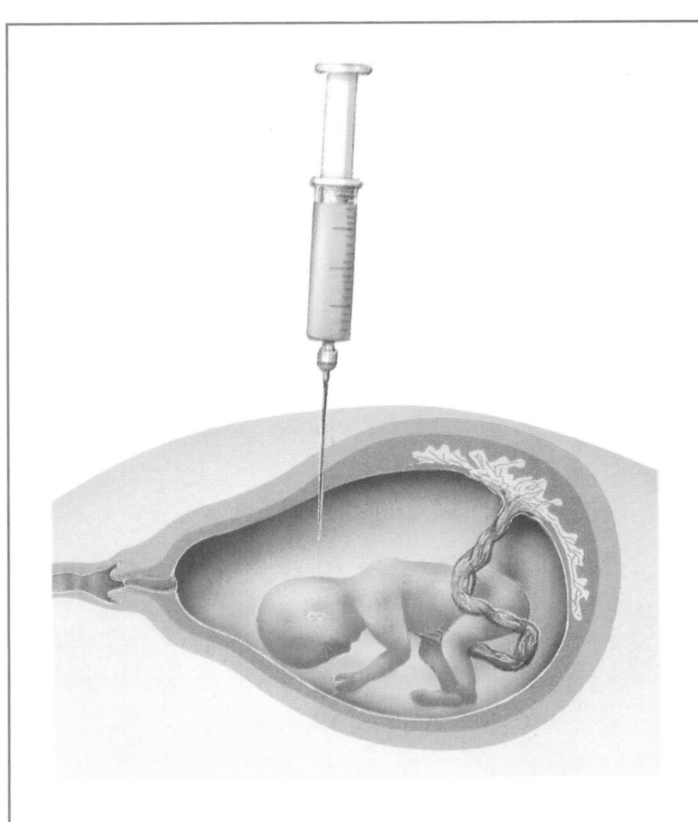

The genetic makeup of a fetus can be investigated by the technique of amniocentesis *(above)*. A syringe carefully inserted through the abdominal wall into the amniotic sac is used to draw off a small sample of amniotic fluid, which contains waste products and sloughed off skin cells from the fetus. The sample is split into fluid and cellular components, using a centrifuge, and part of it is cultured. Subsequent tests and analyses *(right)* provide information about the sex of the fetus and the nature of its chromosomes. Such screening techniques often form part of genetic counseling, which aims to provide parents at risk with information about their unborn children, and to detect any problems at an early stage of pregnancy.

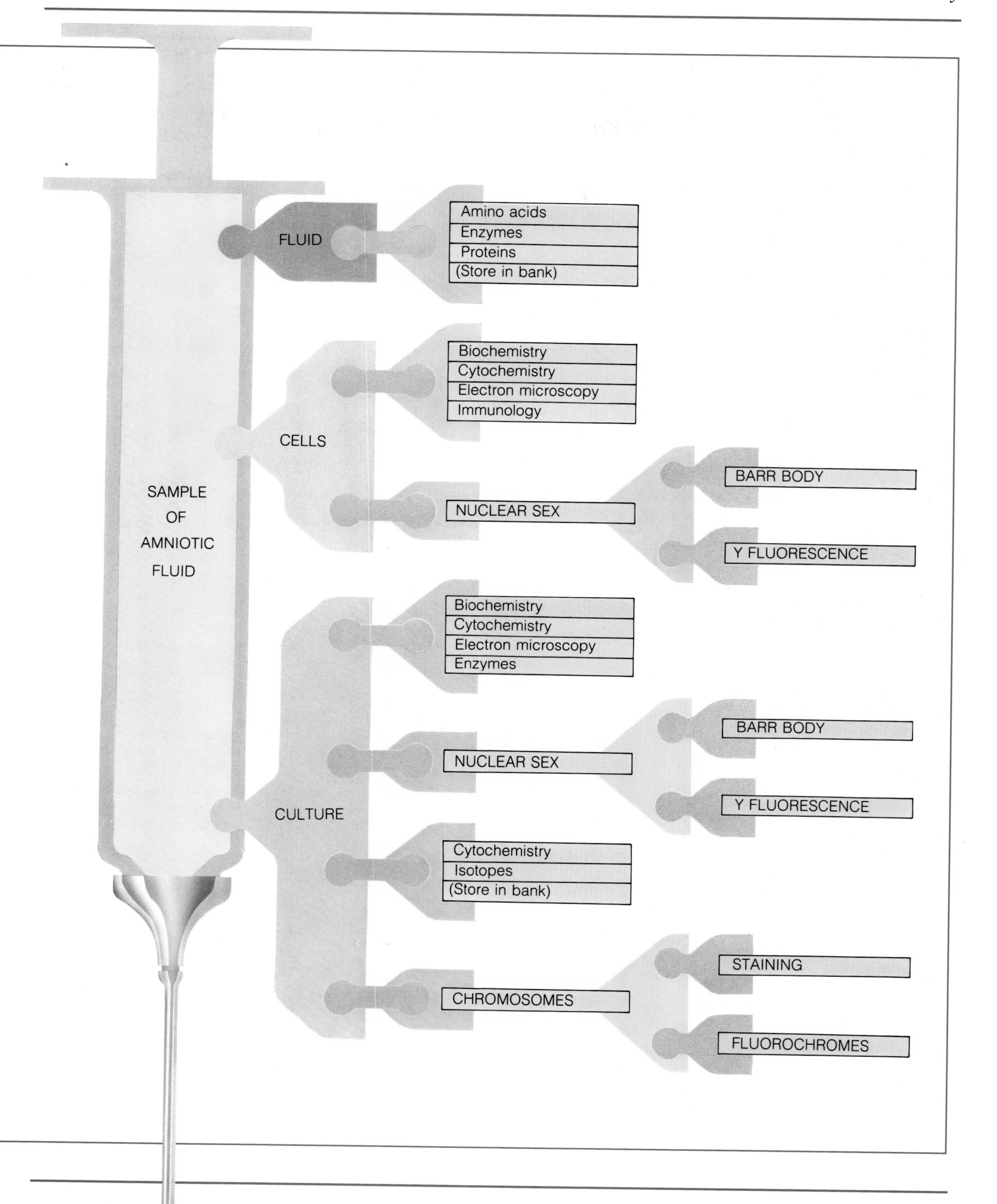
SAMPLE
OF
AMNIOTIC
FLUID
FLUID
Amino acids
Enzymes
Proteins
(Store in bank)
CELLS
Biochemistry
Cytochemistry
Electron microscopy
Immunology
NUCLEAR SEX
BARR BODY
Y FLUORESCENCE
CULTURE
Biochemistry
Cytochemistry
Electron microscopy
Enzymes
NUCLEAR SEX
BARR BODY
Y FLUORESCENCE
Cytochemistry
Isotopes
(Store in bank)
CHROMOSOMES
STAINING
FLUOROCHROMES

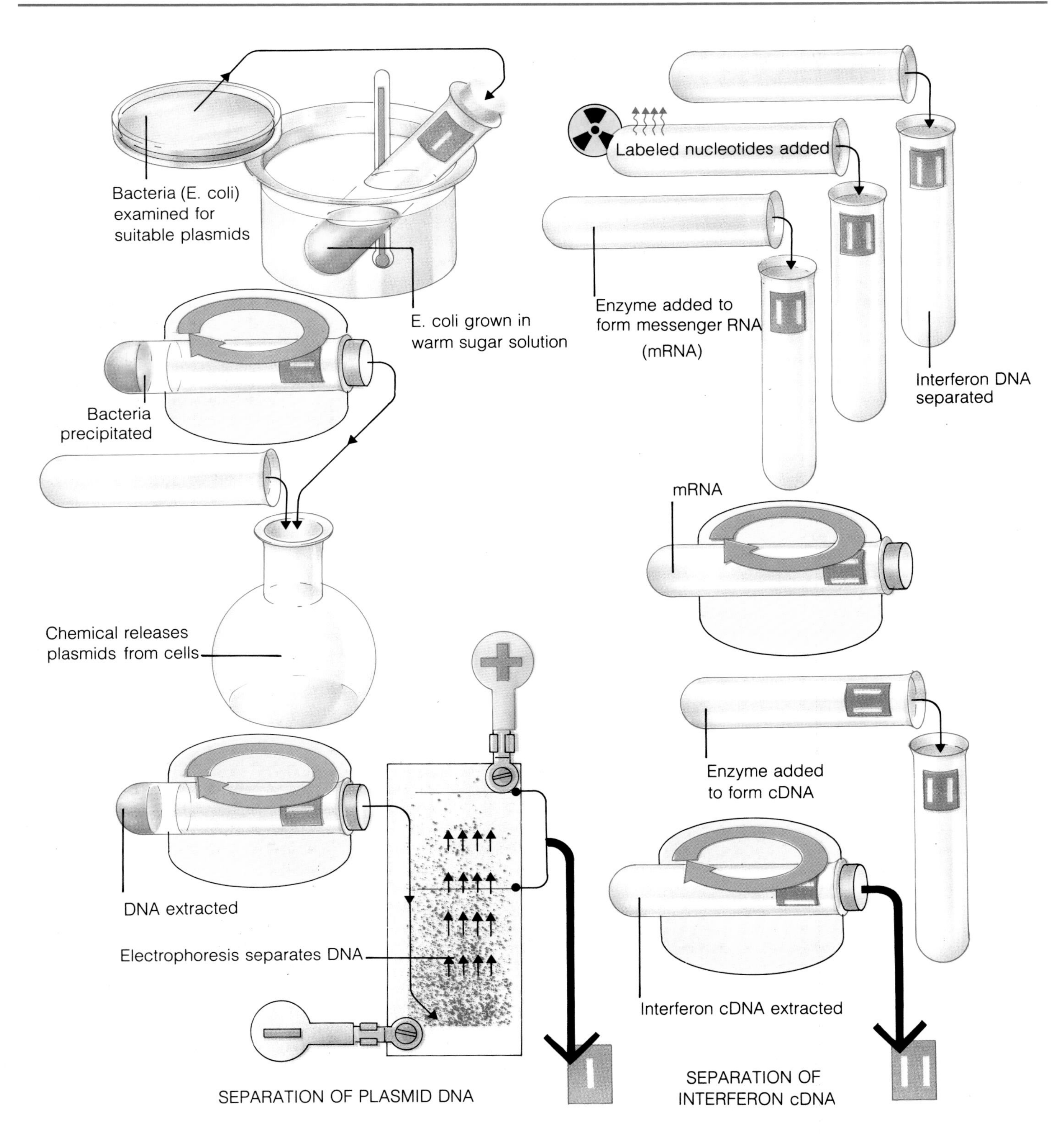

Interferon is a substance that may prove useful in the treatment of various virus diseases. It is a protein produced within cells under attack, but the production of the substance in useful amounts can be achieved only by genetic engineering techniques. Plasmid DNA from the bacterium *E. coli*, common in the human intestine, is broken open by suitable enzymes. The DNA is extracted using electrophoresis, in which an electric field separates out the required plasmid compound. This is further treated with enzymes and separated again to make it receptive to interferon cDNA. The interferon DNA and plasmid are combined, and if

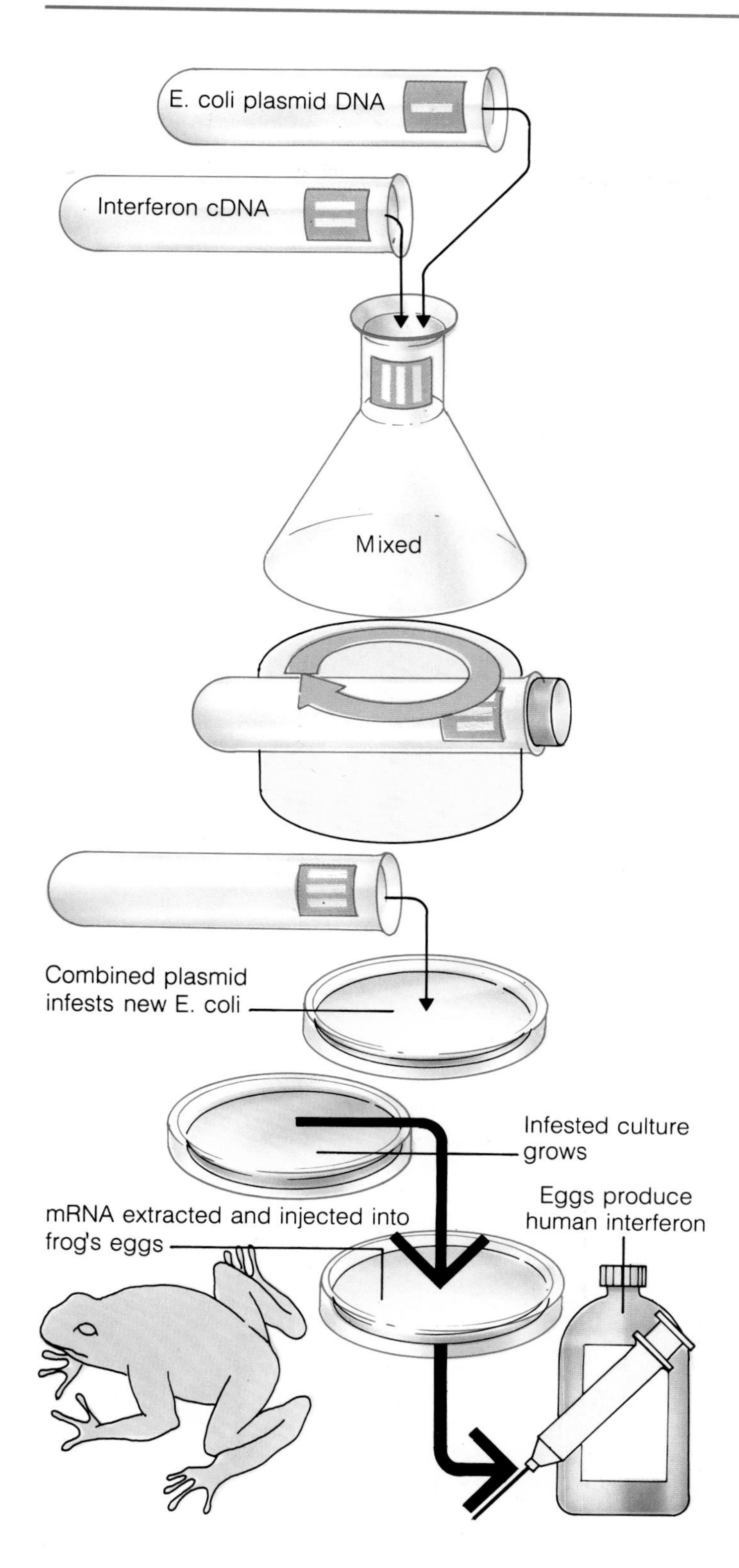

tests show that they will infest a new *E. coli* culture, their RNA is injected into frog's eggs. The eggs do not regard it as "foreign," and go on to manufacture human interferon.

associated with women over the age of 35; because all the eggs are formed when a girl baby is born, it is thought that over time there is some degeneration in the instructions for the maturing process of the egg. This process, which involves meiosis and the halving of the chromosome number, may be incomplete in such cases, and produce the extra chromosome typical of trisomy 21. However, post-35 pregnancies do not account for all the cases of Down's syndrome, so other factors must be at work.

Divining the waters

Amniocentesis, the taking of a sample of the amniotic fluid surrounding a developing fetus, is another powerful tool in the hands of a genetic counselor. Analysis of the fluid itself, and of skin cells which have sloughed off the fetus, can detect chromosomal, immunological and enzymatic disorders well before birth, and can even detect faults in specific genes. Together with other prenatal diagnostic techniques, including ultrasound scans and biopsy, amniocentesis seems to provide the opportunity to repair some of nature's mistakes before birth.

Unfortunately, even with the sort of advances promised by genetic engineering, the effective treatment of a complex condition like Down's syndrome lies a long way in the future, if ever. For the time being, and for some time to come, prenatal analysis will serve only two functions: to increase the sum of human knowledge through unexpected findings, and to provide the relevant weight of scientific probability to any discussion about the premature termination of the pregnancy.

When first discovered and put into practice, the procedures and necessary safeguards involved in recombinant DNA research brought about an outcry which nearly led to legislation banning all such experiments. As it was, the fear of some type of bio-disaster meant that for several years recombinant DNA research was permitted only if it conformed to very strict guidelines, involving crippled organisms that could not survive outside the test-tube.

Most of these restrictions have now been lifted, and one result has been valuable new insights into human cancer, but some voices still argue from principle that no genetically engineered organisms should be released into the environment. Their argument is that it is not possible to predict all the possible environmental consequences.

Arguing in favor are groups such as market gardeners who see that the introduction of a slight genetic modification to a bacterium that lives on the leaves of crops such as lettuce, would lower the temperature at which ice forms around it, and consequently on the leaves. Although the temperature difference is only two degrees, it could save a lot of money for commercial grows faced with a sudden cold snap.

High yield investment

Recombinant DNA technology offers agri-industry many other developments. Animals of the pig family do not normally eat grass. But the microflora that inhabit a hog's digestive tract could be modified to enable them to digest cellulose – a much more cost effective way of producing pork. Merino sheep could be made "self-shearing" by injecting them with a protein extracted from the salivary glands of male mice. Gene cloning would make the protein available at an economic price.

There is even the real possibility of giant cattle and other food animals; either by directly manipulating a breed's genes, or by low-cost, volume production of animal growth hormones grown

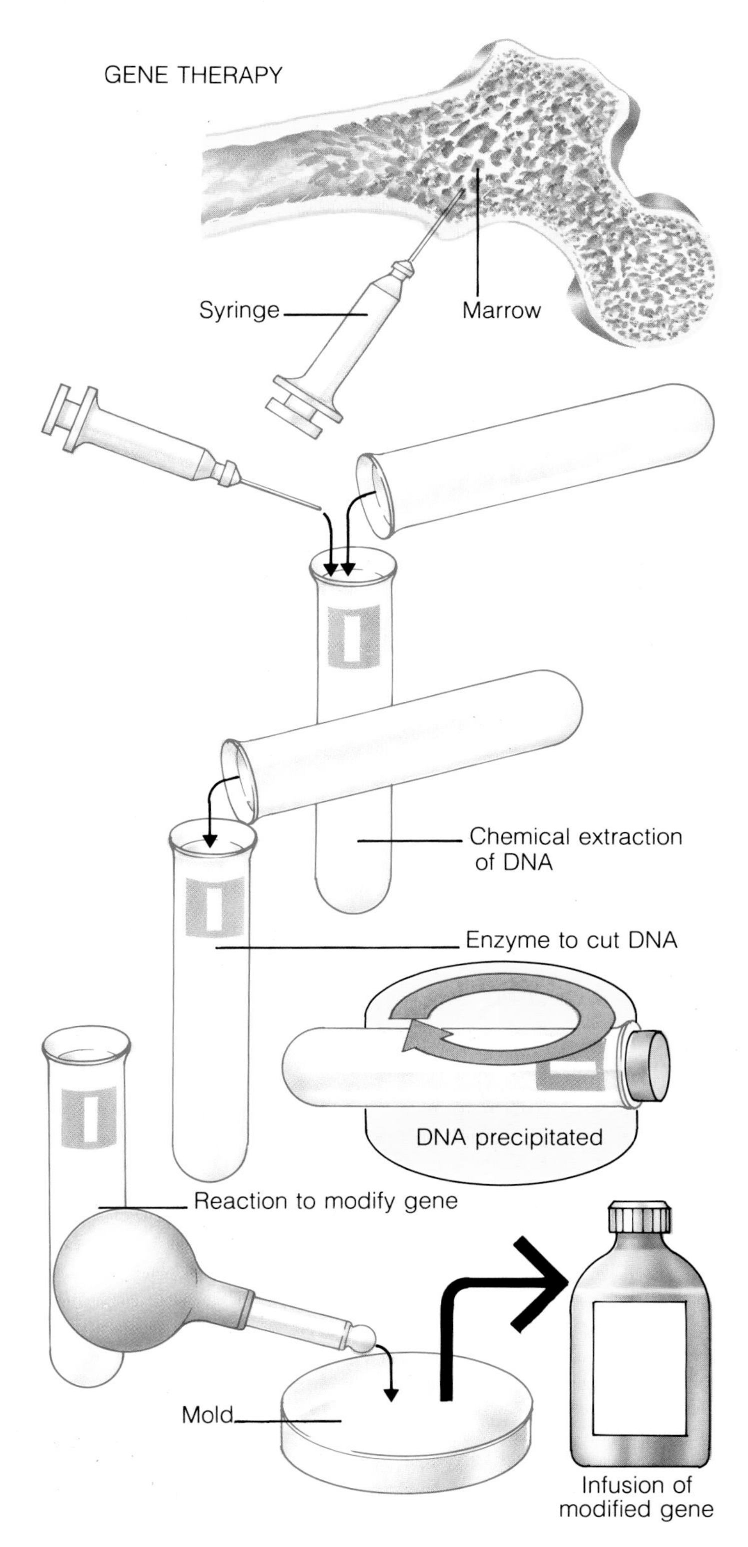

Gene therapy is a futuristic treatment undergoing clinical testing. In a treatment for cancer, DNA is extracted from a sample of the patient's bone marrow, the genes treated with DNA-cutting enzymes, and then injected back into the patient.

in vats of bacteria. However, public concern is likely to keep such engineered mutations safely in the realm of theory, certainly for the near future.

The most likely ground for the current crop of recombinant DNA technology lies in the field of medicine. Duchenne-type muscular dystrophy is a severe wasting disease that causes early death in boys. Carried on one of the mother's X chromosomes, there is a 50 percent chance that her sons will be affected. Using restriction enzymes to define a fragment of DNA associated with the defective gene, scientists can determine from a biopsy whether a male fetus is afflicted with this crippling and fatal disease. The untreatable disorder Huntington's chorea can also be genetically detected in an adult or a child.

Low cost miracle

A more dramatic example of DNA technology at work is provided by the interferon story. When a cell has been infected with one virus it seems to produce a substance which prevents adjacent cells from becoming similarly infected. Something interferes with the viral infection – hence the name interferon.

For many years the very existence of such a seemingly miraculous substance which could combat viral disease was a matter of fierce controversy. Eventually, using conventional techniques, a tiny amount of impure interferon was produced at enormous cost (a single grain had an estimated price tag of 100,000,000 dollars).

Then along came genetic engineering. Using complex recombinant DNA technology, bacteria and frog's eggs, several groups successfully cloned the interferon genes and used them to make previously unimaginable quantities. The price has now dropped ten-thousandfold, making it cheap enough for physicians to determine that, though useful, interferon is not the miracle drug it was once hoped.

Changing your genes

Using recombinant DNA under laboratory conditions to produce a useful substance has become acceptable practice. Meddling in the DNA affairs of living people poses ethical problems of many kinds. An unsuccessful attempt at such unauthorized intervention was made in 1980 by a professor of medicine at the University of California.

Two patients were suffering from a fatal genetic blood disease, in this case thalassemia, which is similar to sickle-cell anemia in that it also confers some immunity to malaria. Bone marrow cells were removed and the defective gene was replaced with a normal hemoglobin gene. When the cells were reintroduced into the patient's bone marrow, however, the gene transfer did not work, and the bone marrow continued to produce defective red blood corpuscles.

Other routes to gene therapy continue to hold hope. Some day soon it may be possible to treat certain enzyme deficiencies and cancers using purposely modified genes. The prospect of going beyond merely repairing genetic damage, and actually improving the human type by genetic surgery on a fertilized egg, is even more ethically ambiguous.

The possibility arises from the in vitro fertilization techniques that were developed to enable some infertile couples to have their own children, so-called "test-tube babies." Where the infertility arises from physical obstruction of the Fallopian tubes, eggs and sperm can be mixed together outside the mother and one of the resulting embryos implanted in her uterus. Inevitably

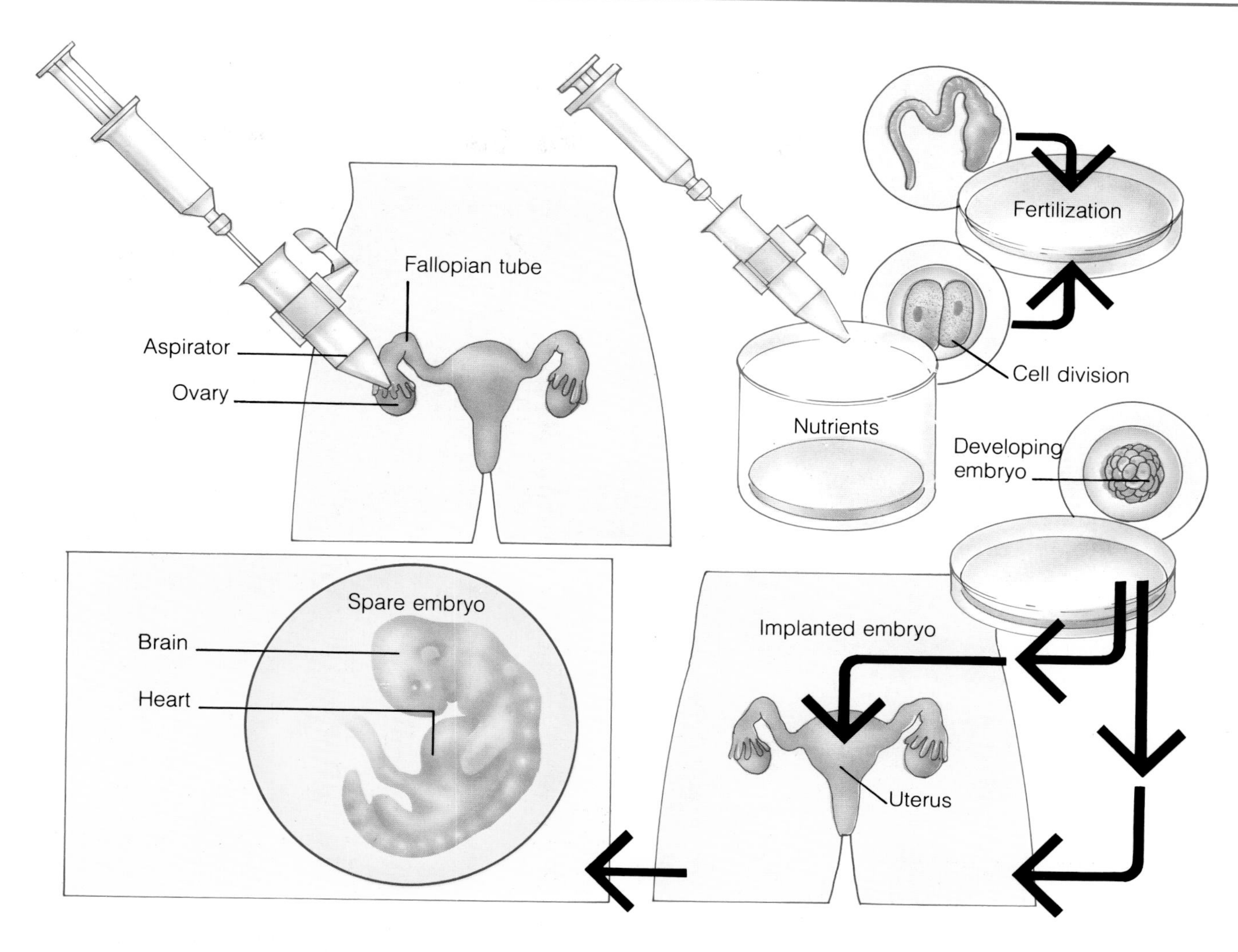

"Test-tube babies" result from in vitro fertilization of a woman's ovum outside her body. The young developing embryo is implanted into her uterus, where it continues to grow during the normal period of pregnancy. Tissue from spare embryos could, in theory, be grafted into adult organs to promote the growth of new cells to replace worn out or damaged ones.

this process leaves "spare" fertilized eggs which are not put back in the mother.

Some scientists argue that an extensive research program using cells and tissues from some of these spare embryos is the only way to make rapid progress toward solving the problems of genetic disorders.

A better copy

Cloning techniques complicate the matter even further. Genes are not the only things that can be cloned – already scientists can produce hundreds of genetically identical plants and, among animals, identical frogs. By using a very fine micromanipulator, the nucleus of a frog's egg can be removed and replaced with the nucleus of another cell. Admittedly, frog eggs are much larger than human eggs, and much more numerous; but the theoretical possibility does exist that an "improved" human embryo could be endlessly replicated.

A bacterium that normally lives harmlessly in the human gut is frequently used in connection with recombinant DNA research. This bacterium, scientific name *Escherichia coli*, could be modified to produce alcohol in industrial fermentation plants. If any bacteria escaped, they could theoretically reestablish themselves in human hosts and continue to produce alcohol. The resulting effect might seem amusing, but the real consequences would be alarming, and possibly fatal.

Brave new world

When Aldous Huxley described a future in which all the children were grown in vats with predetermined levels of intelligence, graded from alpha plus to epsilon minus, just enough bred into them to fit them only for their planned role in society, he was writing fiction to the limits of his imagination.

A little over 50 years later, the limits of human ingenuity have caught up with the limits of human imagination. Many wonders are now possible, we have the power to shape life by manipulating genes in living organisms. Yet from this new knowledge we inherit new responsibilities.

Glossary

A

accommodation the automatic adjustment of the curvature of the eye's lens by contraction or relaxation of muscles to bring images from varying distances into sharp focus on the retina.

acetylcholine a chemical transmitter released at nerve endings or nerve-muscle junctions to pass on a nerve impulse.

Achilles' tendon the strongest tendon in the body. It runs from the calf muscles to the heel bone.

adrenal gland one of two glands, each above a kidney, that secrete hormones, epinephrine (adrenaline) and other substances.

aerobic describes organisms or metabolic activities that require oxygen.

alveolus one of the thousands of tiny air sacs in the lungs where blood exchanges carbon dioxide for oxygen.

amino acid one of the chemicals from which a protein is built. Twenty different amino acids occur in body proteins in significant quantities.

amnion the membranous, fluid-filled sac that surrounds a fetus as it develops in the womb.

anaerobic describes organisms or metabolic activities that do not require oxygen.

androgen a general name for a hormone that produces masculine characteristics, such as testosterone.

antibody a substance produced in the blood in order to destroy or neutralize "foreign" antigens – substances such as toxins or certain bacteria.

antigen a substance capable of eliciting an immune response when introduced into the body.

aorta the principal artery in the body; which carries blood from the heart to other major arteries.

apocrine gland a gland that contributes part of its own cellular substance to its secretion, such as certain axillary and genital sweat glands.

aqueous humor the transparent fluid that fills the front chamber of the eye.

arachnoid membrane the middle of the three membranes that cover and protect the spinal column and brain.

areola the lightly pigmented ring that surrounds the nipple of the breast.

artery a blood vessel that carries blood away from the heart.

atrium one of two upper chambers of the heart. The right atrium receives deoxygenated blood returning to the heart from the body. The left atrium receives oxygenated blood from the lungs.

autonomic nervous system the sympathetic and parasympathetic part of the nervous system, which work "automatically," without conscious control.

axon the highly elongated extension of a nerve cell which conducts the nerve impulses.

B

basal ganglion a cluster of neurons at the base of the brain, which helps regulate body movements.

basophil a type of white blood cell.

bile a thick, brown-green fluid made by the liver, stored in the gallbladder and released into the intestine to aid digestion of fats.

blastocyst an early stage in the development of an embryo, resulting from repeated division.

blind spot a small portion of the retina that is insensitive to light; located at the point at which the optic nerve passes through the back of the eyeball.

blood cells various types of cells within the blood. Red cells (erythrocytes) are chiefly concerned with carrying oxygen; platelets, small clear cells, are important factors in blood clotting, and finally a variety of white cells, corpuscles, all of which help to combat infection.

brainstem the bottom section of the brain that joins onto the spinal cord. It controls many body functions automatically.

bronchus a tube that carries air into and out of the lungs.

bronchiole tiny air-conducting tubes branching from the bronchi and terminating in alveoli.

bursa a small sac containing synovial fluid that helps ease friction between moving skeletal parts.

C

capillaries the smallest blood vessels.

carbohydrate one of the three main types of substances that make up food. The others are protein and fat.

cardiac muscle the involuntary muscle of which the heart is composed.

cardiovascular system the general term for the circulatory system, the arteries and veins that carry blood to and from the heart.

carpus one of the bones at the wrist.

carotid artery a large blood vessel in the neck that supplies blood.

cecum the pouch that forms the first section of the large intestine, situated at the bottom of the ascending colon.

central nervous system the spinal cord and brain, from which other nerves derive.

cerebellum an oval-shaped portion of the brain, concerned with equilibrium and movement.

cerebral hemispheres the two halves of the cerebrum.

cerebrospinal fluid a clear liquid that cushions and protects the central nervous system.

cerebrum the two hemispheres of the forebrain.

cervix the neck of the uterus.

cholesterol a fatlike substance essential to the structure and metabolism of cells.

chromosome DNA-containing structure found in every cell. Chromosomes carry genes, which dictate hereditary characteristics.

cilia microscopical hairlike processes of many cell types.

clavicle the collar-bone; a slim, S-shaped bone at the base of the neck which, along with the scapula, forms the shoulder girdle.

clitoris a small oval of erectile tissue at the head of the vulva.

coccyx the "tailbone." A small, tapered bone located at the base of the spinal column, comprised of four fused vertebrae.

cochlea the part of the inner ear which contains the sensory organs for hearing.

collagen a tough, elastic protein found in bone, skin and all other connective tissue.

colon the part of the large intestine between the cecum and the rectum.

cones the light-sensitive retinal cells that respond best to bright light and via which colors are detected.

conjunctiva the mucous membrane that lines the eyelids and covers the surface of the sclera.

connective tissue the body tissue that surrounds, supports, separates and protects the various body organs and structures.

cornea the transparent structure of the eye's outer coat that covers the iris and pupil.

coronary arteries arteries ascending from the aorta, and curving down over the top of the heart. They transport blood to the heart muscle.

corpus callosum a structure composed of millions of nerve fibers that links the two cerebral hemispheres.

corpus luteum a mass of cells that forms in an ovary to produce the hormones estrogen and progesterone, and to help maintain pregnancy.

cortex the surface or outer layer of an organ, as opposed to the inner medulla.

cranium the vault of the skull, consisting of eight bones which lodge and protect the brain.

cuticle the outermost layer of a hair, or the area at the base of a nail.

cytoplasm the parts of a cell outside the nucleus.

D

dendrite a narrow projection of a nerve cell that can receive and conduct chemical impulses.

dentine the bonelike substance that forms the major part of a tooth.

dermis the layer of tissue beneath the epidermis containing blood vessels, nerves, hair follicles, sweat and sebaceous glands; also called the corium.

diaphragm the muscle that permits breathing by contracting and relaxing during respiration.

diastole the relaxation stage of each heartbeat between contractions, when the heart muscle fills with blood.

DNA (deoxyribonucleic acid) the basic genetic material that is passed from generation to generation in the genes. DNA is found in cell nuclei and controls protein manufacture in all cells.

duodenum the first part of the small intestine between the stomach and the jejunum.

dura mater the outermost of the three membranes that cover and protect the spinal column and brain.

E

embryo the developing baby from the time of fertilization up until the time that organs begin to form (about three months), after which it is termed a fetus.

endocrine glands glands that secrete substances directly into the bloodstream rather than through ducts.

endometrium the membrane forming the lining of the uterus.

endorphin a substance released by the brain thought to be of importance in pain control, and in the release of some hormones.

enzyme a protein that promotes a specific biochemical reaction, while itself remaining unchanged.

epidermis the outer layer of the skin.

epididymis the twisted duct lying behind, and leading from each testicle.

epiglottis a thin flap of cartilage that covers the entrance to the larynx during swallowing, and prevents food from entering the trachea.

epinephrine a hormone produced by the adrenal glands. Its main effects are to increase heart rate, relax air passages to the lungs and to redirect blood from the skin to the brain and muscles. Often called adrenaline.

epithelium a surface layer of cells on most internal and external systems of the body.

erythrocyte a red blood cell.

esophagus the gullet; a tube extending from the pharynx to the stomach.

estrogen one of a group of hormones which gives rise to female sexual characteristics and affects female reproductive activity.

exocrine gland any gland that secretes substances into the body or onto the surface of the skin through a duct.

extensor the opposite of flexor. It is any muscle that moves bones away from each other.

extracellular fluid fluid in the body not enclosed in cells.

F

Fallopian tube one of two tubes located in the female abdomen that leads from close to each ovary into the uterus.

fat a type of food that is a rich source of energy in the body.

fatty acid one of the basic units from which a lipid is constructed. It consists of long strings of carbon and hydrogen atoms.

femur the anatomical name for the thighbone.

fetus the baby developing in the uterus from the time that organs begin to form until birth.

fibrin the protein that forms the essential part of a blood clot.

fibrinogen the inactive form of a coagulation protein, converted to fibrin during clot formation.

fibula the smaller of the two bones in the lower leg, running parallel to, and articulating with, the tibia.

flexor a muscle that causes flexion (bending) of a joint.

follicle a small sac or cavity, such as the hair follicle from which a hair grows.

forebrain the front part of the brain that develops into the cerebrum.

G

gallbladder a sac located just below the liver. It stores bile, which is released to aid the digestion of fats.

ganglion a cluster of neurons, generally of those of the peripheral nervous system; however the term can refer to specific groups of nerve cells in the brain or spinal cord.

gene the smallest factor responsible for passing the inherited characteristics from parents to offspring.

globulin any of a group of proteins found in the blood.

glottis the area in the larynx where sound is produced.

glucose a simple sugar produced as the end product of starch and carbohydrate digestion. It is the principal source of energy in the body.

glycogen a carbohydrate stored in the liver and muscles that can be converted to glucose to supply energy.

growth hormone, or somatropin. A chemical produced by the pituitary gland which regulates the growth of bones and other tissue.

H

hair follicle a pitlike segment in the epidermis that produces a hair. It extends into the dermis, where the hair root is nourished through blood vessels.

hamstrings five tendons that course from the hip to the rear of the knee.

Haversian system the basic structural unit of compact bone, consisting of a Haversian canal and its surrounding concentric lamellae.

hemoglobin the iron-containing molecule of red blood cells responsible for oxygen and carbon dioxide transport.

hepatic of the liver.

hepatic portal vein vein through which blood from the capillaries of the spleen, pancreas, stomach and intestine are taken via the liver. The blood is filtered by the liver before it reenters the circulation.

hip bone one of two bones which form the pelvic girdle. Each hip bone is made up of three bones (ilium, ischium and pubis) which fuse together during early adulthood.

hippocampus a part of the limbic system of the brain.

hormones secretions of ductless (endocrine) glands within the body which may affect local or distant parts of the body, or the body as a whole.

humerus the long bone of the upper arm extending from the shoulder to the elbow.

hymen a membrane wholly or partially across the vaginal opening.

hypothalamus a vital area in the brain located beneath the thalamus, primarily concerned with hormone secretions, body temperature, and water balance.

I

ileum part of the small intestine between the jejunum and the cecum.

impulse the chemical message conducted along a nerve fiber.

incisors chisel-shaped teeth at the front of the mouth.

inner ear collective name for the semicircular canals and the cochlea that deal respectively with balance and hearing.

insulin a hormone produced by the pancreas that regulates sugar levels in the body.

intervertebral disks strong pads between the bones of the spine which helps absorb shock and prevents the vertebrae from grating against each other.

involuntary muscle any muscle of the body that cannot be consciously controlled.

ion an atom or molecule that carries an electric charge.

iris the pigmented diaphragm located behind the cornea and perforated by the pupil.

J

jejunum the part of the small intestine between the duodenum and the ileum.

joint the site at which two or more bones meet, usually permitting movement between them.

jugular vein one of four veins that conveys blood from the head and neck to the heart.

L

lacrimal gland a spongelike organ in the eye that secretes tears.

lactation the process by which milk is produced from the breasts of a mother to feed her baby. The sucking motion of the baby releases the hormone prolactin, which stimulates the milk ducts.

large intestine the continuation of the small intestine, where the final processes of digestion are completed. It consists of the cecum, the colon and the rectum.

larynx the voice box, located below the root of the tongue.

lens a transparent structure that is convex front and back. Located between the vitreous humor and the iris, and it focuses light onto the retina.

leukocyte a white blood cell.

ligament a band of fibrous tissue that connects bone or cartilage and supports joints.

limbic system a group of structures at the base of the forebrain involved in emotions and behavior.

lipid a fat or fatlike substance, generally insoluble in water.

lumen the space within, or the interior of, a hollow, tube shaped structure of the body.

lymph a transparent, watery liquid that circulates through the body in a system of tiny vessels, in a manner similar to blood.

lymphatic system a system of glands and vessels which drain fluid from the tissues of the body together with dead cells and bacteria. The system returns lymph to veins in the neck region, after it has passed through the lymph nodes, which act as cleansing filters.

lymphocyte a type of white blood cell that plays a key role in the body's defense system.

M

mandible the lower jawbone; the largest and strongest facial bone.

marrow a soft substance that fills the bone cavities. Red marrow, located in the ribs, vertebrae, pelvis and skull bones, produces blood cells; yellow marrow, found in the center of long bones, consists of fatty material.

mastoid a spongy air-filled bone located behind the ear.

maxilla one of the two main bones of the face that constitute the upper jaw.

medulla the central core, or inner layer of an organ.

medulla oblongata part of the brainstem linking the spinal cord below with the pons above; it controls respiration and blood circulation.

meiosis the type of cell division taking place during the formation of sex cells, in which the number of chromosomes in the cell is halved.

melanin the dark pigment present in skin, hair and the iris of the eye.

meninges the membranes which form an envelope around the brain and spinal cord.

menstrual cycle the near-monthly periodic discharge of bloody fluid from the uterus through the vagina, occurring between puberty and menopause.

metabolism the biochemical process, concerned with body chemistry that takes place within each living organism. Through metabolism tissue is built up or broken down.

metacarpals a group of five bones that extend from the wrist to form the palm of the hand.

metatarsals five bones that articulate with those of the ankle and toes to form the arch of the foot.

midbrain the uppermost part of the brain stem lying immediately below the cerebral hemispheres.

middle ear the part of the ear that contains the eardrum and the three ossicles; the malleus, incus and stapes.

mitosis the division of a cell nucleus into two identical nuclei.

monocytes mononuclear phagocytes formed in the bone marrow and transported to tissues where they develop into macrophages.

morula the early embryo, when it is little more than a ball of dividing cells.

motor end plate the end of a motor nerve fiber where it joins with a muscle fiber.

motor nerve a nerve carrying information from brain or spinal cord to a muscle.

mucous membrane a membrane containing mucus secreting glands which line body passages and organs that are open to external infection.

mucus a slimy secretion that lubricates body linings and helps protect them against infection.

myelin the fatty substance that makes up much of the sheath round many nerve fibers, increasing the speed of neural impulse transmissions. Nerves so covered appear white in color.

myofibril one of many slender fibrils, consisting mainly of protein, that fill a muscle fiber.

N

nerve fiber strand of nervous tissue along which are conducted the electrical impulses. These convey sensory and other information from the body to the central nervous system, and motor impulses from the central nervous system to the body.

neuroglia the numerous supporting cells of the nervous system in which the nerve cells are embedded.

neuron the basic conducting unit of the nervous system, consisting of a cell body and threadlike projections, dendrites, that conduct electrical impulses. Also known as a nerve cell.

neurotransmitter chemical substance that transfers impulses between neurons.

neutrophil the most common white blood cell, concerned with protecting the body against bacterial infections.

node of Ranvier one of the gaps in the insulating myelin sheath around a nerve.

nucleolus structure within the nucleus of the cell.

nucleus the part of the cell which contains the genetic information in the form of DNA.

O

olecranon the funny bone; the large pointed end of the ulna at the elbow.

olfactory concerning the sense of smell.

olfactory nerve the nerve that extends from the top of the nose to the center of the brain, via the olfactory bulb in the cranium.

omentum a two-layered fold in the peritoneum, which protects the intestines and limits the spread of abdominal infection.

optic chiasma the junction of the two optic nerves.

osmosis the diffusion of a solvent through a semipermeable (porous) membrane from a weak to a strong solution.

ossification hardening; the formation of bone from soft hyaline cartilage or fibrous membrane.

ovary one of a pair of glands in the female, on each side of the pelvic cavity, that produces ova and hormones.

ovulation the shedding of a mature ovum (egg) from the ovary.

ovum egg; the female reproductive cell.

oxyhemoglobin the molecule formed by the binding of oxygen to hemoglobin.

oxytocin the hormone produced by the pituitary gland that stimulates contractions of the uterus and release of milk at the breasts.

P

pacemaker a small cell cluster in the right atrium of the heart which emits electrical signals controlling contractions of the heart.

palate the roof of the mouth which separates it from the nasal cavity. There are two distinct parts, the hard palate at the front of the mouth, the soft palate at the back.

pancreas a large digestive gland which opens into the small intestine and contains tissue which produces the hormone insulin, important in the regulation of blood sugar.

papilla **1.** a tiny projection that presses from the dermis into the epidermis, nourishing a hair follicle. **2.** one of the four types of minute elevation that cover the tongue, three of which contain taste buds.

parathormone a hormone produced by the parathyroid glands, which raises levels of calcium in the blood and assists in controlling calcium metabolism.

parathyroid gland one of four endocrine glands in the neck whose hormones regulate the metabolism of calcium and phosphorus in the body.

parotid gland one of two saliva-producing glands, located in front of the ears.

pelvis bony structure consisting of the two hip bones, the sacrum and the coccyx, which supports the trunk and connects the legs to the rest of the body.

pepsin an enzyme produced in the stomach that assists in the digestion of protein.

peristalsis the synchronized contraction of muscles to cause the contents of a tube to move along it, such as the movement of digestive products along the alimentary canal.

peritoneum the lining membrane that covers the organs contained within the abdominal cavity.

phagocyte any cell whose function is to absorb and neutralize foreign bodies.

phalanges the bones in the toes and fingers.

pia mater the innermost of the three meninges, membranes that cover the spinal cord and the brain.

pinna the external part of the ear.

pituitary gland a small endocrine gland found at the base of the brain. It controls many body activities, including growth, metabolism and maturation.

placenta a spongy structure with many blood vessels that forms in the uterus during pregnancy. It supplies nutrients to the fetus and collects waste products from it via the umbilical cord.

plasma the clear fluid content of blood and lymph.

platelet a cell fragment found in the blood that plays a major role in clotting.

pleura the thin membranous covering of the lungs and the inside of the chest wall.

plexus a network of interwoven nerves, especially formed from the main branches of nerve trunks, such as those in the brachial, lumbar and sacral regions.

pons a part of the brain stem.

portal vein carries blood from the intestines to the liver.

premolar a chewing tooth located at the side of the mouth. The premolars are present in both milk and adult sets of teeth.

progesterone a hormone produced in the ovary by the corpus luteum and in the placenta, which prepares the endometrium for implantation and breasts for lactation, and which maintains the pregnancy.

prolactin a hormone that encourages growth of breast tissue and the release of milk.

proprio(re)ceptor a sense organ that responds to stimuli from within the body, such as those that monitor the state of muscular contraction.

prostate gland a gland surrounding the urethra at the neck of the bladder.

prostaglandins fatty acid derivatives; substances widespread in body tissues having many functions in reproduction.

protein organic compounds containing nitrogen; a major constituent of living cells.

puberty the stage when an individual matures sexually and so becomes capable of reproduction.

pubis the anterior (front' portion of each hip bone, joining at the front to form the pelvic arch.

pulmonary of the lungs.

pulse the evidence of a heartbeat as felt in some arteries, particularly those of the forearm and neck.

pylorus the muscular band at the base of the stomach that controls the flow of food into the duodenum.

R

radius the shorter of the two lower arm bones, located on the inside of the arm.

rapid eye movement (REM) the movement of the eyeballs during sleep that indicates dreaming.

recessive genes that are secondary in determining inherited characteristics. The opposite of dominant.

rectum the lower part of the alimentary canal, between the colon and the anus.

red blood cell a disk shaped blood cell containing hemoglobin. It transports oxygen in the blood. Also known as an erythrocyte.

reflex arc the path traveled by a reflex, where it is converted from an incoming sensory impulse to an outgoing motor impulse.

renin an enzyme secreted by the kidneys, which indirectly regulates salt metabolism and the retention of fluid.

respiration the process of breathing in and out. Oxygen is taken into the lungs and then carried round the bloodstream to tissues of the body. After providing energy for the cells it is released through the lungs as carbon dioxide.

respiratory center the regulator of breathing located in the brain.

retina the small patch of tissue at the back of the eyeball containing light-sensitive rods and cones.

ribosome small cell organelles, which may be free in the cytoplasm, or on the surface of the endoplasmic reticulum, involved in communication within the cell and in protein synthesis.

rib any of 12 pairs of curved bones extending from the thoracic vertebrae toward the sternum, protecting the chest cavity.

RNA ribonucleic acid; a long-molecule compound found in all cells. It carries instructions from the genes of DNA and regulates the assembly of amino acids into proteins.

rod a highly specialized, cylindrical, light-sensitive cell containing rhodopsin; it is most sensitive to dim light.

S

sacroiliac joint the point where the sacrum and the pelvis bones meet, at the base of the spine.

sacrum a triangular bone, consisting of five fused vertebrae, that forms a wedge between the two hip bones.

saliva a slightly alkaline mixture of watery mucus secreted by the salivary glands. Its functions are to moisten the mouth, soften the food, permit taste and commence digestion of carbohydrates.

scapula the shoulder blade. The scapula and clavicle together form the shoulder girdle.

scrotum the external muscular sac surrounding the testes.

sebaceous gland a skin gland associated with hair follicles. It produces sebum.

sebum an oily substance made up of fatty acids, cholesterol and debris from skin cells. Secreted by the sebaceous glands, it protects against bacteria and lubricates the hair and skin.

semen a sticky white fluid made by the male reproductive organs that carries sperm.

semicircular canal one of the three tubelike structures that constitute the organ of balance in the ear.

seminiferous tubule any of the small tubes of the testicles that produce sperm.

sensory nerve a nerve carrying information from a sense organ to the brain or spinal cord.

septum a wall or partition between two parts of an organ or cavity.

serum the clear component of blood which separates into liquid and solid elements.

sex hormone any of the chemical controllers of sexual characteristics and reproduction. Male sex hormones (androgens) are released by the testes, and the female (estrogens and progesterones) by the ovaries.

sinus a general term for a cavity, channel or depression within the body.

skeletal muscle, or striated muscle. The most prevalent type of muscle in the body, usually anchored to bone. Its function is to carry out voluntary movement.

skull a skeletal case enclosing and protecting the brain. The skull is comprised of the bones of the cranium and the face.

small intestine, or ileum. A tube in which most of the processes of digestion take place.

smooth muscle involuntary muscle that cannot be consciously controlled, such as the muscles of the intestines and blood vessels.

sperm the male reproductive cell.

sphincter a circular muscle that controls the opening and closing of a hollow organ, such as the heart.

spinal cord nerve tissue contained within the vertebral canal in the backbone.

spinal nerve one of the nerve trunks that emerges from the spinal cord.

spine the chain of vertebrae running from the cranium to the coccyx, within which is the spinal cord.

spleen an organ situated in the abdominal cavity near the liver; it is one site of the breakdown of the body's red blood cells, and is part of the immune system.

sputum principally, mucus that is coughed up from the lungs and windpipe.

squamous cell a flat cell of the epithelium, such as a cell lining the digestive tract. It often has a protective role.

sternum the breastbone; a daggerlike structure to which most of the ribs are attached.

subcutaneous just beneath the surface of the skin.

sympathetic nervous system part of the autonomic nervous system that prepares the body for emergency action by, among other things, increasing the heart rate and dilating the pupils.

synapse the "gap" or region where two nerve cells come into close contact, and across which an impulse can be transmitted.

synovial fluid a lubricating substance secreted by the membrane of a joint.

systole the contraction stage of the heart's beating cycle.

T

talus an ankle bone; the highest of the tarsal bones of the ankle.

tarsals seven short bones that make up the ankle.

tendon a fibrous cord of connective tissue that attaches muscle to bone.

testicles the two reproductive glands that produce male reproductive cells and testosterone.

testosterone one of the sex hormones produced in men; responsible for development of secondary sexual characteristics.

thalmus the relay center of the brain from where sensory impulses are directed to other brain areas.

thorax the chest; the bony cage formed by the ribs, protecting the heart and lungs.

thymus gland a ductless gland of the lower neck region which disappears by the time of adulthood.

thyroid gland a gland in the neck. Thyroxin, the hormone it produces, is important in controlling the use of energy by the body.

tibia the shinbone; the larger of the two bones of the lower leg.

tissue the name given to any group of cells of the same general type, which have their own specific function.

trachea the windpipe, leading from the larynx to the two bronchi.

tympanum the eardrum.

U

ulna the longer of the two bones of the forearm, located on the side opposite the thumb.

umbilical cord the connection in a pregnant woman between the placenta and the fetus.

urea an end product of the chemical breakdown of proteins in the body. It is excreted in the urine.

ureter a tube extending from the kidney along which urine passes to the bladder.

urethra the duct by which urine passes from bladder to the exterior; in the male it also conveys semen.

urinary tract both the passage and the organs connected with the production and discharge of urine.

urine a watery solution of waste products removed from the blood by the kidneys, stored in the bladder and finally expelled from the body via the urethra.

uterus the womb, the part of the female reproductive system in which the fetus develops.

uvula the visible projection suspended from the midpoint of the soft palate arch at the back of the mouth.

V

vagus nerve a nerve that controls intestinal movement and the secretion of digestive juices.

valve a flap or pocket of tissue that controls the directional flow of blood in tubular structures or cavities of the body, such as those of the heart.

vas deferens the excretory duct of each testicle that conveys sperm.

vasoconstriction the constriction of the smooth muscles in the walls of blood vessels.

vasodilation the widening of the walls of blood vessels to allow the blood to flow closer to the surface of the skin.

vein a vessel that carries deoxygenated blood toward the heart.

vena cava either of the two major veins that empty deoxygenated blood into the right atrium of the heart.

ventricles the two lower chambers of the heart.

vertebra one of the 33 bones that form the spinal column.

vesicle a small sac, such as the ones that hold chemicals at the tips of nerves.

villus a small protrusion from the surface of a membrane.

viscera the contents of the abdomen.

vitamin chemical substance required for the proper running of the body.

vitreous humor the gel-like substance within the eye.

vocal cord one of two bands of tissue in the larynx; the vocal cords produce sound when air is forced between them.

voluntary muscle another name for skeletal muscle.

W

white blood cell any of several types of blood cell that have a nuclei. Also known as a leucocyte.

wisdom teeth molars located at the back of the upper and lower jaws.

X, Y, Z

X chromosome a sex chromosome. Unlike the Y chromosome it carries major genes, which show sex-linkage. Females have two X chromosomes.

Y chromosome a sex chromosome. Males have one Y chromosome and one X chromosome.

zygote the fertilized ovum.

Index

Figures in italics refer to illustrations or captions to illustrations.

Acknowledgments

Photographic credits

6 Morton Beebe/The Image Bank; 8 Bull Publishing Consultants/Nick Birch; 16 Rex Features; 23 Mareshal/The Image Bank; 26 Photograph Agfa-Gevaert made on Agfa-contour film; gatefold Lennart Nilsson from his book *Behold Man*, published in the U.S. by Little, Brown & Co., Boston; 38 Rafael Beer 1980; 44 Drs John Heuser + Roger Cooke; 48 London Scientific Fotos; 56 J Stevenson/ Science Photo Library; 58 Biophoto Associates; 63 Dr A Leipins/Science Photo Library; 70 Science Photo Library; 75 Joe Baker/The Image Bank; 78 Dr Tony Brain/Science Photo Library; 79 Howard Sochurek/The John Hillelson Agency; 82 John Watney Photo Library; 87 American Cancer Society; 89 Zefa Picture Library; 91 Alex Hubrich/The Image Bank; 92 Manfred Kage/Peter Arnold; 102 Lennart Nilsson from his book *Behold Man*, published in the U.S. by Little, Brown & Co., Boston; 106 From *Tissues and Organs* by Richard G Kessel and Randy H Kardon, W H Freeman and Company; 108 Lennart Nilsson from his book *Behold Man*, published in the U.S. by Little, Brown & Co., Boston; 139 Arthur Seigleman/ Freelance Photographers Guild; 143 Dan McCoy/ Rainbow; 144 Clark/Goff/Science Photo Library; 160/161 Russ Kinne/Science Photo Library; 170 Biology Media/Science Photo Library; 176 Science Photo Library; 187 Hank Morgan/Science Photo Library; 188 Jan Hinsch/Science Photo Library; 206 Biophoto Associates; 213 Biophoto Associates; 214 Gene Cox/Science Photo Library; 230 Howard Sochurek/Woodfin Camp & Associates; 237 Manfred Kage/Science Photo Library; 242 Dan McCoy/Rainbow; 244 Manfred Kage/Peter Arnold; 252 Lennart Nilsson from his book *Behold Man*, published in the U.S. by Little, Brown & Co., Boston; 257 Kanaehara Shuppan Co. Ltd; 265 Dr Goran Bredberg/Science Photo Library; 286/7 Adam Woolfitt/Susan Griggs Agency; 290 A R Williams, Charing Cross Hospital/London Scientific Fotos; 295 John Walsh/Science Photo Library; 299 Petit Format/Nesile/Science Photo Library; 306 Dr A Lesk/Laboratory of Molecular Biology/Science Photo Library; 317 National Portrait Gallery, London; 323 Sally & Richard Greenhill.

Illustration credits

10 Mick Gillah; 11 Mick Saunders; 12 Mick Gillah, Norman Swift, Mick Saunders; 13 Mick Saunders; 14 Norman Swift; 15 Mick Saunders; 16 Norman Swift; 17 Norman Swift (top), Mick Saunders (bottom); 18, 19 Mick Gillah; 20, 21 Mick Saunders; 22, 23 Mick Gillah; 24, 25 Norman Swift; 28 Pat Kenny (left), Graziella Becker (right); 29 Jane Gordon; 30 Susan Sanford; 31 Frank Kennard; 32 Ed Musy; 33 Jane Gordon; Gatefold – Frank Kennard; 35 Robert J Demeresk; 36 Carol Donner (top), Scott Barrows (bottom); 37 Ray Srugis; 40–43 Frank Kennard; 44 Elsie Hennig; 46, 47 Mark Seidler; 50 Carol Donner; 52–55 Modern Artz; 58 (left), 59, 60 Mick Gillah; 61 Greensmith Associates; 62 Mick Gillah, Aziz Khan; 64–66 Mick Gillah; 67 Greensmith Associates; 68 Mick Saunders; 69 Mick Saunders (left), Aziz Khan (right); 72 Michael Courtney; 73 Les Smith; 74, 76, 77, 80, 81 Michael Courtney; 83 Les Smith; 84 Michael Courtney; 85 Les Smith; 86 Les Smith (top), Michael Courtney (center); 88 Michael Courtney; 89, 90 Les Smith; 94 Michael Courtney; 95, 96, 97 Frank Kennard; 98 Ian Bott; 99 Ken Goldammer (top), Michael Courtney (bottom); 100 Jennifer Arnold (top), Jack Lanza (bottom); 101 Michael Courtney; 103, 104 Lewis E Calver; 105 Virginia L Schoonover; 106 Lou Bory & Associates; 107 Virginia L Schoonover; 110 Les Smith; 111 Graziella Becker; 112 Ken Goldammer; 113 Masako Herman; 114 John Murphy; 115, 116 George Kelvin; 117 Frank Kennard; 118 Doug Cramer (top); 118–121 Frank Kennard; 123 Scott Barrows; 124, 125 Permut Rappaport & Associates; 126 Lou Barrow; 127 George Kelvin; 128, 129 David Gifford; 130, 131 Andrew Popkiewicz; 132 Michael Courtney; 133 Walter Hortens; 134 Les Smith; 135 Ian Bott; 136 Michael Courtney; 137 Manuel Bekier; 139 Masako Herman (bottom); 140, 141 Michael Courtney; 142 Lorelle Roboni (top), Jane Gordon (bottom); 146 David Gifford; 147 John Bavosi; 148, 149 Ivan Hissey; 150 Michael Courtney; 151 John Bavosi; 152, 153, 155 Frank Kennard; 156, 157 Michael Courtney; 158 Les Smith; 159 Michael Courtney; 161 Les Smith; 162, 163 John Bavosi; 165 Les Smith; 166, 167 John Bavosi; 168 Frank Kennard; 172 Mick Saunders; 174 Mick Gillah; 175 Mick Saunders; 176, 177 Norman Swift; 178, 179, 180 Mick Gillah; 181, 183, 184, 185, 186 Mick Saunders; 191 Aziz Khan; 192, 193, 194, 195 Mick Saunders, Shirley Willis; 198 Mick Gillah; 199 Aziz Khan; 200 Mick Gillah (left), Aziz Khan (bottom); 201 Aziz Khan; 202, 203 Mick Gillah; 204 Mick Saunders (top), Mick Gillah (bottom); 205 Mick Saunders; 208 Mick Gillah; 210 Mick Gillah, Aziz Khan; 212 Mick Saunders; 213 Mick Gillah; 215 Aziz Khan; 216, 217 AWKA Popkiewicz & Susan Smith; 218, 219 Mick Gillah; 221 Aziz Khan; 222 Mick Gillah; 223 Mick Saunders; 225 Aziz Khan; 226–227 Mick Gillah; 232–233 BB; 234 Louis Bory Associates; 236 Edward Allgor, Jim Ruttencutter; 237 Ian Bott; 238–239 Louis Bory Associates; 241 Frank Kennard; 242–243 Leonard Dank/Medical Illustrations Company; 245 Frank Kennard; 246 Edward Allgor; 247 Edward Allgor, Jim Ruttencutter; 248–249 Adolf Brotman; 250 Louis Bory Associates; 251 Edward Allgor; 254 Ray Srugis; 255 Frank Kennard; 256 Paul Richardson (top), Esperance Shatarah (bottom); 258 Another Color; 261 Chip Coblyn/Ed Musy; 263 Another Color; 264 Michael Woods; 266, 267 Frank Kennard; 268 Kangehara Shuppan Co Ltd, Tokyo; 269 Norman Barber; 270–271, 273 Michael Courtney; 274, 275 Frank Kennard; 276, 277 Michael Courtney; 278 Norman Barber; 279 Michael Courtney; 280 Michael Courtney (left), Norman Barber (right); 281, 282, 283 Michael Courtney; 285, 288 Norman Barber; 289 Michael Courtney; 292 Mark Seidler; 293 Frank Kennard; 294 Ian Bott; 297 Mark Seidler; 298 Ian Bott (left), Paul Richardson (right); 303 Les Smith; 304 Paul Richardson; 305 Joyce Hurwitz; 308 Mick Gillah; 309, 310, 311 Mick Saunders; 312, 313 Mick Gillah; 315–317, 320–329 Mick Saunders.